Bioremediation and Nanotechnology for Climate Change Mitigation

Arafat Abdel Hamed Abdel Latef
Ehab M. Zayed • Ahmad Alsayed Omar
Editors

Bioremediation and Nanotechnology for Climate Change Mitigation

Springer

Editors
Arafat Abdel Hamed Abdel Latef
Botany and Microbiology Department,
Faculty of Science
South Valley University
Qena, Egypt

Ehab M. Zayed
Field Crops Research Institute, Department
of Cell Study Research
Agricultural Research Center
Giza, Egypt

Ahmad Alsayed Omar
Biochemistry Department, Faculty of
Agriculture
Zagazig University
Zagazig, Egypt

Citrus Research and Education Center,
Department of Horticultural Sciences,
Institute of Food and Agricultural Sciences
University of Florida
Lake Alfred, FL, USA

ISBN 978-981-96-3071-4 ISBN 978-981-96-3069-1 (eBook)
https://doi.org/10.1007/978-981-96-3069-1

This Springer imprint is published by the registered company Springer Nature Singapore Pte Ltd.
The registered company address is: 152 Beach Road, #21-01/04 Gateway East, Singapore 189721, Singapore

Preface

In an era marked by escalating environmental challenges, the relationship between human activities and the health of our planet has never been more critical. Climate change, widespread pollution, and the degradation of ecosystems demand an urgent reassessment of our practices and a united effort to pursue sustainable solutions. This book, *Bioremediation and Nanotechnology for Climate Change Mitigation*, underscores the necessity of innovative approaches to environmental remediation and sustainable development.

The chapters provide an in-depth examination of contemporary environmental pollution and climate change issues. Insights from diverse experts highlight the complexity of these challenges. The opening chapter explores the state of global environmental governance, emphasizing the growing integration of ecological stewardship and climate action. Subsequent chapters address the impacts of population growth on pollution, sources of air and water contaminants, and sustainable waste management strategies.

Advanced technologies such as nanomaterials and artificial intelligence in bioremediation are presented, demonstrating their transformative potential in mitigating environmental degradation. The interplay between pollution and climate change is examined through agricultural and industrial practices, underscoring the need for sustainable methodologies. Further, the book discusses innovative bioremediation and phytoremediation strategies for restoring ecosystems, advocating the integration of nanotechnology and renewable energy to combat global warming.

This volume is a call to action for researchers, policymakers, and stakeholders, presenting challenges and actionable pathways toward environmental sustainability. It urges humanity to transcend economic and ideological divides, embracing global cooperation to confront environmental crises and achieve climate justice. Its pages inspire readers to reconsider their role in the ecosystem and champion collective stewardship for a healthier, more equitable planet.

Qena, Egypt — Arafat Abdel Hamed Abdel Latef
Giza, Egypt — Ehab M. Zayed
Zagazig, Egypt — Ahmad Alsayed Omar

Contents

About the Editors

Arafat Abdel Hamed Abdel Latef is a Professor at the Department of Botany and Microbiology, Faculty of Science, South Valley University, Egypt. He has published more than 90 research articles in peer-reviewed journals and 27 book chapters in international books. He served as editor of 4 published international books. He serves as an editorial member of several reputed journals. Furthermore, he is an expert reviewer for over 150 international journals. He was one of the top 2% of scientists in the world, according to Stanford University, in 2020, 2021, 2022, 2023, and 2024. He has received several awards, such as the Egypt State Award of Excellence in Agricultural Sciences 2021, the South Valley University Award of Scientific Excellence 2019, the Top Peer Reviewer Award 2019 Powered by Web of Science and Publons, and the South Valley University Award of Scientific Publication for the years from 2010 to 2024.

Ehab M. Zayed is a Professor of Biotechnology for 31 years and works in the fields of genetics, agricultural genetic engineering, and biotechnology. He graduated from ZU, Department of Agricultural Genetic Engineering, in 1993. In 1998, he obtained a Master's in Genetics in the Study of Genetic Variation in Soybean Plant. Besides, he has worked with many private sector companies in the field of biotechnology and the National Soybean Program. He presented a PhD thesis on "Improving Soybean Plants Using Genetic Engineering Techniques" in 2007 at GEBRI, MU, formerly, and SU currently. In addition to working at the scientific culture in Egypt through published and electronic newspapers, he has worked as Head of the Genetic Resources Department. He was born in Egypt in 1971 and has been working at the ARC since 1994. He now has 54 research papers in different areas of crops, most of which are available on Research Gate and Google Scholar.

Ahmad Alsayed Omar is a Professor of Plant Biochemistry and Molecular Biology at the Biochemistry Department, Faculty of Agriculture, Zagazig University, Egypt, and a Research Assistant Scientist at the University of Florida, USA. He has published over 80 peer-reviewed articles, including in *Nature Communications* and *Frontiers in Plant Science*. Dr. Omar serves as an Associate Editor for several journals, such as *Frontiers in Plant Science, International Journal of Forestry Research*, and *In Vitro Cellular & Developmental Biology – Plant*, and as a review editor for many others. He has received the Zagazig University Award for Scientific Publication annually from 2009 to 2023 and is an active member of the American Society for Horticultural Science and the Society for In Vitro Biology. Recently, he chaired the plant biotechnology section at the 2024 World Congress on In Vitro Biology and has been elected as the chair for the Society for In Vitro Biology's 2026 Meeting.

The Current Situation and Future of International Relations in Environmental Issues and Climate Change

1

Murad Muhammad, Shumaila Batool, Muqadas Batool, Sana Ullah, Abdul Basit, and Abdul Wahab

Abstract

Central policymakers, governments, climate researchers, and research funders emphasize the need to develop, formulate, and implement climate change policies. This chapter examines the present situation and prospects of global relations concerning environmental concerns, including climate change. It discusses the correlations between environmental change and armed conflict and the global policy obstacles associated with the interaction between resources and international relations. The significance of climate change and its consequences are underscored, highlighting the lack of claritty surrounding how it could escalate into a catalyst for global conflict. The advancements in Earth system research

M. Muhammad (✉)
Xinjiang Key Laboratory of Biodiversity Conservation and Application in Arid Lands, Xinjiang Institute of Ecology and Geography, Chinese Academy of Sciences, Urumqi, People's Republic of China

University of Chinese Academy of Sciences, Beijing, People's Republic of China

State Key Laboratory of Desert and Oasis Ecology, Key Laboratory of Ecological Safety and Sustainable Development in Arid Lands, Xinjiang Institute of Ecology and Geography, Chinese Academy of Sciences, Urumqi, People's Republic of China
e-mail: Muradbotany1@uop.edu.pk

S. Batool
Department of Zoology, Wildlife and Fisheries, University of Agriculture, Faisalabad, Pakistan

M. Batool
Department of Botany (Govt. Graduate College Layyah), Bahauddin Zakariya University, Multan, Pakistan

S. Ullah
Department of Agricultural Leadership, Education & Communication, College of Agricultural & Environmental Sciences, University of Georgia, Athens, GA, USA

A. A. H. Abdel Latef et al. (eds.), *Bioremediation and Nanotechnology for Climate Change Mitigation*, https://doi.org/10.1007/978-981-96-3069-1_1

and the recognition of the Anthropocene necessitate a reassessment of environmental security concerning mitigating the most severe risks. However, the focus on mitigating the most severe consequences of climate change in the future has not entirely replaced the previous energy security efforts. If the shift is not achieved, the climate will be more disrupted, potentially necessitating geoengineering techniques, which could lead to new conflicts. Despite their efforts, securing climate change and responding to it as an emergency has not been a primary focus for most countries or the United Nations. However, the Paris Agreement recognized the gravity of the problem. Certain viewpoints are more optimistic, proposing that environmental initiatives can be utilized to advance peacebuilding and positively influence the planet's future.

Keywords

Climate change · International relations · Global issues · Planet's future

1.1 Introduction

International relations (IR) is a broad discipline that focuses on studying the interactions and transactions between governments and non-state actors on a global scale (Stengel and Baumann 2017). International relations involve the intricate web of connections and interactions, including diplomacy, economics, politics, and social dimensions, that exist among states worldwide. Cooperation, conflict, alliances, negotiations, and the pursuit of national interests globally are all examples of these interconnections (Farrell and Newman 2016). Analyzing a nation's foreign policy and its interactions with other countries is vital to international relations. The discipline of international relations offers an understanding of the political, economic, and security efforts of the United States, a key participant in global events (Brooks and Wohlforth 2016). In recent years, the United States has transformed its commercial partnerships with foreign nations called "America First. This trade policy led to trade disputes and instilled a feeling of unpredictability in international markets. The diplomatic tensions and subsequent measures taken by other countries have substantially impacted the worldwide economy and international relations, such as ecosystem protection laws.

Participation in international alliances and organizations is crucial to the United Kingdom's foreign policy. The United Kingdom participates in various global institutions, such as the Commonwealth of Nations, the United Nations (UN), and the

A. Basit
Department of Horticultural Science, Kyungpook National University, Daegu, South Korea

A. Wahab
University of Chinese Academy of Sciences, Beijing, People's Republic of China

Shanghai Centre for Plant Stress Biology, CAS Centre for Excellence in Molecular Plant Sciences, Chinese Academy of Sciences, Shanghai, People's Republic of China

North Atlantic Treaty Organization (NATO) (Curran and Williams 2018). These alliances augment the United Kingdom's diplomatic scope and sway. The United Kingdom's participation in international climate change activities shows its commitment to global stability and safety. The United Kingdom has given £11.6 billion in climate finance from 2021 to 2025 to help poor nations decrease emissions and adapt to climate consequences, demonstrating its leadership and commitment to global environmental sustainability (UK Ministry of Defence 2020) (Goodall and Army 2019). The source of this information is a report published by the UK Ministry of Defence in 2020.

The field of international relations encompasses a wide array of exchanges and interactions that occur on a global level. Japan exemplifies its solid economic influence, security partnership with the United States, active engagement in international organizations (IOs), contributions to global governance, and commitment to addressing climate change (Wirth 2015). These advancements in Japan's foreign policy highlight its adherence to international conventions and agreements and its intricate role in redefining the global arena. Japan's solid economic influence, security partnership with the United States, active engagement in international organizations, and commitment to climate change, including its net-zero emissions pledge by 2050, underscore its adherence to international conventions and its role in global governance.

Organizations focused on safeguarding the environment at both national and international levels are confronted with substantial obstacles due to the rising number of recently identified environmental concerns in the past few decades. Examining the historical ecological challenges encountered by advanced societies during the 1950s and 1970s can provide valuable insights (Grove et al. 2018). Growing globalization drives economic, social, and technical progress. In both developed and developing nations, new social, political, and environmental movements advocate for innovative sustainable development strategies, as shown by the rise of innovative city initiatives and renewable energy investment, which reached $2.7 trillion globally in 2010 (Mowforth and Munt 2015).

Over the past 50 years, significant environmental challenges have been identified, highlighting the critical importance of taking preemptive measures (Tham 2016). Instances of ecological degradation include the reduction of the ozone layer in the stratosphere over Antarctica, the erosion of soil and conversion of land into desert, the discharge of industrial and municipal sewage wastewater in developed nations, the occurrence of harmful acid rain in lakes in Scandinavia, the problem of excessive population growth and urban development in most continents, the increasing difficulties in managing solid waste in major cities, the destruction of extensive areas of valuable tropical forests through deforestation, and the loss of biodiversity (Singh and Singh 2017). These instances were identified before their detrimental effects on human health and vulnerable ecosystems. Some specific scenarios necessitated costly remediation endeavors; in particular, implementing remedies proved exceedingly challenging.

Over the past few decades, scientific research and legislative actions have led to innovative solutions and substantial advancements in environmental protection.

Global solar capacity increased from 40 GW in 2010 to more than 710 GW in 2020, for example, due to scientific research and legislative initiatives that have improved environmental protection in the last several decades (Council, Earth et al. 2012). Nevertheless, the overall growth of the economy typically surpasses these developments. For example, the global vehicle count is rising in urban regions due to advancements in automobile fuel economy. Before severe harm or potentially dangerous conditions in natural resources and delicate ecosystems, specific issues about environmental pollution were discovered.

Nevertheless, there have been occurrences where the natural surroundings include substantial amounts of detrimental pollutants, especially in bodies of water such as rivers, lakes, and seas. These situations have been demonstrated to require significant financial resources for restoration. Several newly identified environmental problems caused by human activities have detrimental impacts on both living organisms and human health, often leading to illnesses and premature death (Kampa and Castanas 2008).

Environmental conservation organizations at the national and global levels prioritize ecological concerns, particularly those with the most significant negative impact on biodiversity, natural resources, sustainability, and human well-being. These organizations advocated for investigations and subsequent studies on environmental patterns, substantially increasing awareness about the dangers. Nanotechnology has great potential to combat global warming and climate change, especially in the changing international environmental relations scene. This chapter examines nanotechnologies' strategic role in reducing and adapting to climate change and the existing and future conditions of global collaboration and policy-making in this area. The goal is to study nanoscale technology's use in renewable energy generation, carbon capture and storage, water purification, and sustainable materials to prevent global warming. The chapter will also examine how international agreements, treaties, and frameworks affect these technologies and how global institutions regulate them. The chapter will analyze nanotechnology's strategic, ethical, and societal implications for climate change applications, including safety, environmental effects, and equitable access. This chapter incorporates empirical data, statistical studies, and predictive modeling to examine the future of nanotechnology, global warming, and international relations goals.

1.2 Current State of Environmental Issues

The current state of the ecosystem is complex and challenging, with many environmental problems affecting the world. Significant issues include climate change, deforestation, water and air pollution, biodiversity loss, and plastic trash (Singh and Singh 2017).

There has been a lack of uniformity and speed in the actions taken by authorities worldwide to address these environmental challenges. Several countries have passed laws to reduce emissions of greenhouse gases, protect natural areas, and encourage sustainable development. Following the Paris Agreement, several countries have committed to reducing their carbon emissions; meanwhile, many cities have implemented plans to reduce air pollution and promote the use of renewable energy (Haszeldine et al. 2018). However, environmental protections are lax in some countries, and economic interests sometimes precede ecological concerns. Failures in environmental protection can occur when governments lack the resources, political will, or public knowledge to address the issue adequately.

Although environmental conditions are concerning, many initiatives are underway to save the planet and ensure its viability. Though governments play a crucial role, people, businesses, and non-profits must work together to address environmental issues and see lasting change (Edwards 2010).

1.2.1 Global Environmental Challenges

The Earth and its inhabitants face significant danger due to global environmental issues, necessitating urgent and collaborative actions to mitigate their impacts. Global warming, biodiversity loss, pollution, forest loss, and resource depletion are only a few of the interconnected issues encompassed by these crises (Pörtner et al. 2023).

The environment is currently seeing the most significant effects from heat waves and other manifestations of climate change. Increasing temperatures exacerbate the frequency of storms, floods, heat waves, and droughts. Although natural factors may contribute to these consequences, the Industrial Revolution resulted in a significant rise in greenhouse gas emissions and the depletion of crucial natural resources due to human activities, overcrowding, resource exploitation, and excessive utilization of fossil fuels (Xu and Zhao 2023). Global warming and the increasing frequency and intensity of natural disasters indicate that anthropogenic climate change is underway. The scientific consensus acknowledges that climate change is a contributing cause to wide-ranging consequences, such as rising sea levels and the loss of numerous species within the Earth's ecosystems. The impacts of increasing temperatures and climate change are already experienced by the most susceptible populations in developing and impoverished third-world countries (Baker 2012).

Scientists have engaged in extensive deliberation over the necessity of Earth's diverse array of life for human survival. Biodiversity and healthy ecosystems are crucial in mitigating natural disasters, regulating climate, and providing humans access to food, fertile soil, and medicines (World Health Organization 2015). Species exhibit mutual interdependence and coexistence in all regions of the globe. Ecosystems are complex networks of interdependence in which all living organisms, including humans, contribute. Healthy ecosystems perform vital functions such as air filtration, soil preservation, temperature regulation, nutrient recycling, and provision of nutritious food. In addition, ecologically sustainable ecosystems provide a source of medicinal plants, seeds, and other material resources, making them a form of natural capital. In the last decade, approximately thousands of documented animal species have gone extinct, with an additional tens of thousands endangered due to habitat destruction, illegal hunting, and the introduction of non-native species, illustrating a significant loss of biodiversity (Sax et al. 2002). The United Nations has called for prompt and committed actions to safeguard endangered species and their natural habitats, such as tropical forests and other valuable natural resources.

In conjunction with anthropogenic deforestation and desertification, climate change has profoundly affected millions of individuals' lives and livelihoods and poses substantial challenges to achieving sustainable development. Forests are crucial for Earth's ongoing habitability and have an enormous impact on reducing climate change and its adverse effects. Forests cover around 31% of the Earth's land area. Forests comprise a total area of 4.06 billion hectares (ha). The Russian Federation, Brazil, Canada, the United States, and China collectively possess more than 50% of the world's forests (Rossato et al. 2018).

Since 1990, deforestation has led to the depletion of around 420 million hectares (ha) of forest worldwide, but the rate of forest loss has dramatically declined (Guppy 1984). The rate of deforestation has dropped to an anticipated 10 million hectares annually, compared to the preceding 5-year period (2015–2020) when it was 12 million hectares per year, and the rate of 12 million hectares per year from 2010 to 2015. Forests fulfill the fundamental survival requirements and serve as sources of employment, income, and job prospects for around 1.6 billion individuals, accounting for a quarter of the global population (Khan et al. 2020). For centuries, individuals and communities have depended on forests as crucial sources of support and financial stability throughout periods of turmoil.

Research indicates that states are failing to meet international targets to stop the decline and deterioration of forests globally by 2030 (Gasser et al. 2022). This assessment is the first to monitor progress since the targets were established during last year's United Nations Climate Change Conference of the Parties (COP26) in Glasgow, United Kingdom. As a component of a more comprehensive initiative to address climate change, safeguarding forest cover aids in the sequestration of atmospheric carbon dioxide (CO_2) and provides local cooling benefits (Cohen et al. 2022). Pollution in the environment is a global problem. However, it is terrible in developing nations due to poverty, a lack of funding for healthcare innovation, and lax environmental regulations.

Worldwide, environmental contamination has grown to alarming proportions due to developments in the past two centuries. Toxic, poisonous, and carcinogenic pollutants account for most ecological contaminants that harm people's health. We are using more energy (from fossil fuels) and releasing more harmful waste due to urbanization, industrialization, and economic progress. There has been a dramatic rise in the worldwide environmental contamination of urban air, water supplies, and agricultural soils (Khanna 2011). A primary global public health concern that transcends national boundaries is the accumulation of plastics and microplastics in the world's oceans and seas. Contemporary work environments and new ways of living have also influenced dangerous lifestyle behaviors with negative health implications.

According to experts, the most significant contributors to these environmental hazards are polluted river water and inadequate sanitation and hygiene practices in underdeveloped nations (Adelodun et al. 2021). Many diseases and premature deaths are caused by environmental pollutants and contaminants, which are known to have toxic adverse health effects after long-term exposure. These pollutants affect almost every human organ, including the lungs, liver, and kidneys, resulting in infant mortality, perinatal disorders, allergies, endothelial dysfunction, respiratory disorders, cardiovascular disorders, increased oxidative stress, and mental disorders.

Over the past few decades, the world's oceans and seas have transformed into enormous trash cans, collecting everything from industrial and municipal waste to soil runoff, fuel spills, and plastics (Dąbrowska et al. 2021). Ocean-related environmental crises also harm ecosystems from climate change, pollution dumps, wastewater, and fuel spills. The United Nations has issued several demands related to the oceans, including better management of protected areas, less garbage disposal, coastal cleanups, less overfishing, less harmful pollution, and less acidification due to global warming.

Plastics and microplastics are commonly found in water due to many sources, such as those originating from land and waste discarded by ships. Debris has been found in seafloor sediments, the uppermost layer of water, the water column of deep ocean regions, and coastal places that are distant from any pollution created by humans. This debris has been found worldwide, spanning from the equator to the poles, and it can become trapped in the ocean's ice. According to UN Environment, around 15% of marine debris is found on the water's surface, another 15% is suspended in the air, and the other 70% falls on the seafloor (Madricardo et al. 2020). Another study reveals that there are currently 5.25 million plastic items, with a combined weight of 268,940 tons, drifting in the Earth's oceans. Overfishing (Fig. 1.1) occurs when the number of fish harvested exceeds the population's ability to replenish itself through reproduction. The impact on the diversity of marine life is significant due to this. According to the World Wildlife Fund (WWF), aquatic species have experienced a drop of 39% during the past 40 years (McCauley et al. 2015). Ocean pollution, encompassing a diverse range of discarded goods and liquid spills, poses a significant hazard to marine biodiversity.

Fig. 1.1 "Overfishing" describes taking more fish from the ocean than the natural replenishment rate allows. According to a 2014 study, overfishing has wiped off 88 of 97 fish stocks in the Mediterranean Sea. According to the European Commission, overfishing has reached crisis proportions in the Mediterranean Sea, affecting 91% of fish species. This catastrophic scenario directly results from the European Union's (EU's) botched execution of the Common Fisheries Policy (CFP)

1.2.2 Human Activities and Climate Change

The present state of global warming may be attributed solely to human interference with the natural world. Every single person on this planet is affected by this environmental economics problem. Global warming is caused by human activity. Factories, cars, cutting down trees, and, most significantly, cattle all add to the problem of climate change. The Intergovernmental Panel on Climate Change (IPCC) highlighted in its Fifth Report (2013) that human activity is the primary cause of different climate trends (Bindoff et al. 2014). Increases in atmospheric and oceanic temperatures, shifts in the global hydrological cycle, decreased snow and ice cover, rising sea levels, and more intense and frequent weather events are all manifestations of these changes. Researchers found that human activity has been the main cause of global warming since the mid-twentieth century. Extreme weather events can potentially have a wide range of global repercussions, including but not limited to changes in sea levels, changes in precipitation patterns, and the spread of desert-like conditions (Warner 2009). Glaciers, permafrost, and sea ice are melting due to the warming, most noticeable in the Arctic. The Arctic permafrost layer has warmed from −10 to −5 °C within the last half-century. Half of the original size of the Arctic ice sheet was lost between 1970 and 2018, with a 25% reduction in surface area and a 1.3 m loss in thickness (where its concentration is 50–100 times greater than in the Earth's atmosphere) (Racoviteanu et al. 2022). Extreme weather events, including heatwaves, droughts, and heavy rainfall, are becoming more frequent due to global warming. Other effects include acidification of the oceans and the loss of species due to temperature changes. An inevitable future consequence of climate change could be the discharge of carbon dioxide from the oceans, where its concentration is 50–100 times greater than that of Earth's atmosphere (Chen and Drake 1986). Ecosystems can be upended and this occurrence can produce a greenhouse effect comparable to Venus's.

Listed below are several resources that address the environmental economics topic of global warming and its distribution. The concentration of freon in the Earth's atmosphere, as well as emissions of greenhouse gases, solid aerosol particles, deforestation, methane (CH_4), and nitrous oxide (N_2O), all play a role in the variations in global temperatures (Evseeva et al. 2021). All of it results from people doing things; therefore, people need to control it. For environmental economists to avoid unintended repercussions, the topic of human activity's impact on climate change must be carefully investigated.

1.3 International Cooperation and Agreements

Amid the growing trend of globalization, the importance of implementing internationally recognized environmental standards by renowned organizations such as the United Nations, G8/G20, European Union, and Environmental Protection Agency (EPA), becomes increasingly evident (Kirton 2018). Scientists and environmental groups concur that international collaboration is vital to tackling most ecological concerns. A thorough study has revealed various worldwide problems, such as the adverse effects on natural resources and human populations caused by certain activities and evaluations. Consequently, international environmental policy must address these concerns. Scientists have discovered numerous significant environmental issues in recent decades. Climate change, characterized by increasing global temperatures, is currently one of the most urgent challenges confronting the natural world (Hoegh-Guldberg et al. 2019). There is a widespread recognition of the importance of sustainable energy policy, biodiversity preservation, forest, sea, and soil conservation, and other initiatives to protect the environment. Other relevant subjects are desertification, sustainable waste management, and protecting consumers and staff against hazardous chemicals. To tackle these environmental issues, it is imperative to coordinate worldwide efforts to mitigate adverse consequences and halt natural resource depletion (Panel, Consumption et al. 2011).

1.3.1 Key International Treaties and Conventions

International treaties and agreements are crucial in tackling environmental challenges and mitigating climate change worldwide. These accords function as frameworks for nations to collaborate and strive toward shared objectives in safeguarding our planet. The Paris Agreement is a notable convention in which almost all countries committed to restricting global warming and implementing measures to decrease greenhouse gas emissions (Höhne et al. 2021). Another key accord is the Kyoto Protocol, which sought to diminish emissions from developed nations. The Montreal Protocol is crucial because it aims explicitly to eliminate compounds that cause the depletion of the ozone layer, thereby safeguarding the Earth's atmosphere (Velders et al. 2007). Furthermore, conventions such as the Convention on Biological Diversity and the Ramsar Convention specifically emphasize biodiversity conservation and the safeguarding of wetlands, respectively. These treaties and conventions provide a framework for collaboration, enabling nations to exchange information, resources, and obligations in protecting the environment for present and future generations (Haas et al. 1993).

1.3.1.1 Kyoto Protocol

Kyoto Protocol is a critical international convention that aims to combat climate change. The United Nations Framework Convention on Climate Change (UNFCCC) ratified the Kyoto Protocol in 1997, in Kyoto, Japan (Gupta 2016). The agreement's principal goal is to reduce the output of greenhouse gases, which are the primary causes of climate change and global warming. The Kyoto Protocol primarily targets developed countries, often known as Annex I nations, because of their long history of being at the forefront of greenhouse gas emissions. These countries have pledged to decrease emissions of six greenhouse gases, including carbon dioxide (CO_2), methane (CH_4), and nitrous oxide (N_2O) (Jones et al. 2023). The reductions will be achieved in precise amounts and within specified time frames. The Kyoto Protocol used unique strategies such as emissions trading, the clean development mechanism

(CDM), and joint implementation (JI) to achieve these carbon reductions. Nations can achieve their targets more flexibly through carbon trading, which allows them to buy and sell carbon credits. Developed countries can earn credits for their investments in projects that lower emissions in poor countries under the clean development mechanism (CDM) (Humphrey 2004).

Despite its significance, the Kyoto Protocol ran into problems, such as the withdrawal of certain powerful nations and doubts about its ability to address climate change adequately. Nevertheless, the Paris pact and other subsequent global climate accords and talks had their roots in this pact (Pavone 2018).

1.3.1.2 Paris Agreement

A globally binding treaty, the Paris Agreement addresses climate change. Over 266 Parties confirmed their backing on December 12, 2015, at the COP21 UN Climate Change Conference in Paris, France (Morano 2018). On November 4, 2016, the document became effective. The primary objectives of this program are to keep the average global temperature increase below 2 °C over pre-industrial levels and to keep the rise below 1.5 °C above those levels (Valone 2021). The urgency of keeping global warming below 1.5 °C by the century's end has been a theme among world leaders. Being the first legally enforceable agreement that brings all nations together to fight against and adapt to the effects of climate change, the Paris commitment marks a turning point in the fight against global warming. To implement the Paris Agreement, it is necessary to undergo a fundamental transformation in society and the economy based on the latest scientific discoveries. Under the Paris Agreement, countries gradually increase their efforts to combat climate change, known as "ratcheting up," every 5 years (Framework and Stocktake 2016). Nationally determined contributions (NDCs) refer to the strategies devised by countries to address the issue of climate change. These contributions were expected to be submitted by governments starting in 2020. With the introduction of each new nationally determined contribution (NDC), we hope to witness a more impressive demonstration of ambition than previous ones. The temperature goal of the Paris Agreement must be achieved by the conclusion of 2023 without considering individual country conditions. Considering this fact, the COP27 cover resolution requests that Parties reassess and improve the 2030 targets in their nationally determined contributions (NDCs) (Zarkik 2022).

The Paris Agreement now includes provisions that promote voluntary contributions from countries other than the original signatories. However, it emphasizes that developed nations should lead in providing financial assistance to less fortunate and more vulnerable countries. Climate finance is crucial for mitigation initiatives because of the high cost of reducing emissions. Climate funding plays a vital role in adapting to and mitigating the worst effects of climate change, as it requires substantial financial resources (Linnerooth-Bayer and Hochrainer-Stigler 2015).

The Paris Agreement aims to decrease greenhouse gas emissions and enhance the ability to withstand climate change by fully implementing technology advancements and facilitating their transfer (Segger 2016). The Technology Mechanism establishes a framework to facilitate the efficient functioning of technology. The mechanism expedites technology advancement and dissemination through its policy and implementation branches. Numerous challenges arising from climate change surpass the capacities of impoverished nations. Therefore, it is strongly recommended that all prosperous countries enhance their support for initiatives to increase less developed nations' skills and abilities (World Health Organization 2019). The Paris Agreement expressly emphasizes the need for developing nations to prioritize capacity-building efforts connected to climate change.

The implementation of a more resilient system of global supervision was established upon the signing of the Paris Agreement by states European Training Foundation (ETA). Starting in 2024, countries will be obligated to transparently report their initiatives to mitigate the consequences of climate change, adapt to its impacts, and participate in the exchange or provision of support through the ETF (Horster 2018). The report submission process is detailed, including international procedures. The Global Stocktake will utilize the data gathered through the ETF to assess the overall progress made toward the long-term climate goals. Consequently, nations will be motivated to set higher objectives for the next round (Hegglin et al. 2022).

Although the Paris pact's goals necessitate a significant increase in efforts to combat climate change, new markets and low-carbon solutions have emerged since the pact came into force. More and more nations, regions, towns, and businesses are establishing carbon neutrality goals. Zero-carbon solutions are rapidly gaining popularity in an economy where emissions constitute 25% of the total (Yuan et al. 2022). This advancement has created a multitude of fresh economic prospects for those who promptly adopted it, especially in the domains of power and transportation. By 2030, zero-carbon alternatives could be economically competitive in industries that account for over 70% of global emissions (Anika et al. 2022).

1.3.2 Successes and Challenges in Implementation

International cooperation and agreements have significantly increased global awareness of the pressing need to address climate change and environmental damage. The extensive involvement of nations in agreements such as the Paris Agreement indicates a collective dedication to implementing measures (Streck et al. 2016). This shared dedication has facilitated the mobilization of resources, encouraged cooperation, and nurtured a feeling of mutual accountability for safeguarding the Earth.

Furthermore, cooperation between nations has enabled exchanging knowledge, technology, and optimal methods for dealing with climate change and environmental concerns (Olabi and Abdelkareem 2022). This interaction has resulted in notable progress in the development of renewable energy, sustainable agriculture, and green technologies, which have substantially contributed to global efforts to reduce the impact of climate change and adapt to its effects. Using combined knowledge and resources, nations have successfully expedited advancements toward shared objectives and stimulated ingenuity in environmental sustainability (Labadi et al. 2021).

Furthermore, international accords establish a structure for nations to create and execute strategies to diminish greenhouse gas emissions, safeguard ecosystems, and foster sustainable development. These frameworks facilitate the synchronization of national priorities with global goals and promote the government's adoption of ambitious climate and environmental policies. International agreements set explicit objectives, specific timeframes, and effective systems for monitoring progress, so they offer a strategic plan for synchronized efforts and ensure that governments are held responsible for fulfilling their obligations.

Nevertheless, considerable obstacles still exist in executing global collaboration and treaties. Although countries have made obligations under these agreements, many face challenges in effectively implementing policies and procedures to achieve their aims (Castles 2019). Implementation gaps occur due to insufficient financial resources, inadequate technical capacity, and lack of political determination. Limited financial resources, insufficient physical structures, and conflicting interests frequently impede advancements in attaining climate and environmental goals (Eizenberg and Jabareen 2017).

Furthermore, global collaboration frequently encounters obstacles to equality and justice, specifically regarding allocating obligations and advantages among nations. Underdeveloped countries, which are commonly impacted to a greater extent by climate change, may face a shortage of resources and assistance required to engage in worldwide projects actively. To tackle these inequalities, it is necessary to foster more vital unity, provide financial aid, and enhance the ability of all nations to actively participate in and gain advantages from global collaboration on climatic and environmental matters (Jinping 2022).

Moreover, geopolitical rivalries and divergent interests among nations might impede efficient collaboration and advancement in tackling climate change and environmental concerns (Mbeva et al. 2023). Disputes on the distribution of responsibilities, trade-offs between promoting economic growth and protecting the environment, and geopolitical competitions could hinder collaborative efforts. To overcome these difficulties, engaging in diplomatic discourse, seeking compromise, and demonstrating leadership is necessary to establish a consensus and promote collaboration on shared priorities. Diplomatic discourse, compromise, and leadership are needed to reach consensus and collaborate on shared priorities. The Paris Agreement, signed by 197 nations, shows a global commitment to fighting climate change through cooperation and understanding.

1.4 Role of International Organizations

International organizations (IOs) have proven extremely useful in identifying environmental threats and encouraging research. They have advocated for countries to work together to address issues such as acid rain, water and air pollution, and global warming. Some organizations—the United Nations, specialized agencies, and nongovernmental organizations (NGOs)—work to safeguard the environment and take action against climate change through their official policies, programs, and initiatives (McCormick 2023). The UN has put together critical global meetings to discuss climate change, such as the United Nations Framework Convention on Climate

Change (UNFCCC), the Kyoto Protocol, and the Agreement on Climate Change in Paris (Matemilola et al. 2020). These conferences encourage people worldwide to collaborate and build solid global unity to cut greenhouse gas emissions and help countries affected by climate change.

1.4.1 United Nations Framework Convention on Climate Change (UNFCCC)

The United Nations Framework Convention on Climate Change is referred to as UNFCCC. It is an international agreement that the United Nations made in 1992 (Bodansky and Rajamani 2018). Its objective is to mobilize global cooperation in the fight against climate change and its aftermath. The UNFCCC guides nations on how to cooperate to maintain stable atmospheric concentrations of greenhouse gases at a level that prevents humans from dangerously interfering with the climate system (Goreau 2014).

The UNFCCC is meant to help poor countries get used to the destructive effects of climate change, which will happen anyway. It also encourages action on climate change, including action on adaptation. It also requires all Parties to make, execute, publish, and regularly update national and, if necessary, regional plans that include steps to help people adjust to climate change in the best way possible. As part of the UNFCCC, the secretariat helps member countries keep their promises under the Convention. It also organizes meetings, collects and shares statistics and other climate-related information, and helps carry out the duties of the Convention, its subsidiary bodies, and its ad hoc working groups (Pouffary et al. 2017).

1.4.2 Intergovernmental Panel on Climate Change (IPCC)

The UN has formed a supranational body called the Intergovernmental Panel on Climate Change (IPCC) (Beck and Mahony 2018). Improving our knowledge of climate change as a result of human activities is the primary goal of this undertaking. In 1988, the World Meteorological Organization (WMO) and the UN Environment Program (UNEP) collaborated to establish the IPCC, the Intergovernmental Panel on Climate Change. Later the same year, the IPCC was formally approved by the United Nations. Geneva, Switzerland, is home to the organization's secretariat, supported by the World Meteorological Organization (WMO) (Sokona et al. 2015). One hundred ninety-five member nations oversee the Intergovernmental Panel on Climate Change. Experts, government representatives, and scientists serve on the IPCC in a dual capacity. Renowned experts on the subject support the reports produced by the IPCC. All member states must reach a consensus to approve the reports. In addition to its scientific mission, the IPCC acts as a coalition of governments (Bhandari 2022). The goal here is to disseminate climate change information to policymakers. It also delves into possible solutions to the problem of climate change and examines its effects. The IPCC achieves this by reviewing peer-reviewed scientific articles.

The IPCC has not conducted primary research. It produces comprehensive assessments of our present knowledge about climate change. It compiles data on a variety of topics related to climate change. On top of that, it comes up with methods. Greenhouse gas emissions and sink absorption can be better estimated using these methods. Prior reports and scholarly publications form the basis of the assessments. Throughout six assessments, the publications provide mounting evidence of climate change. More than that, they show how human behaviors directly relate to this occurrence (McMichael 2003).

The *Principles Governing IPCC Work* publication is where the IPCC has laid out its process rules. We shall assess the following claims by consulting the IPCC.

1. The risk of climate change caused by human activities.
2. Its potential impacts and risk assessment.
3. Possible options for prevention.

The evaluations adhere to IPCC regulations and are characterized by their comprehensive, unbiased, and transparent nature (Swart et al. 2009). They address all the crucial aspects of our present scientific understanding of climate change. The data utilized in this study are derived from the disciplines of science, technology, and socioeconomics. The assessments conducted by the IPCC should avoid advocating any specific course of action. However, they must consider the essential objective factors while enacting legislation.

The IPCC has a "Gender Policy and Implementation Plan" to guarantee that it prioritizes attention to gender-related issues (Environment and Secretariat 2023). While performing its tasks, it endeavors to be comprehensive and considerate. The IPCC emphasizes the need to maintain a balanced participation in its work. All participants must be afforded an equitable opportunity (Beck and Forsyth 2015).

The Intergovernmental Panel on Climate Change (IPCC) has published six assessment reports, comprehensively presenting the latest discoveries in climate science, spanning 1990–2023. The United Nations operations, costing billions annually, rely heavily on contributions from member states, with the United States being the largest funder, accounting for approximately 22% of the organization's budget. Each country's share of funding is determined by its economic capacity and commitment to international cooperation, averaging around 0.001% of global GDP (Arias et al. 2023). Furthermore, the IPCC has released 14 specialized assessments on specific topics. Each evaluation report consists of four sections. The contributions from all three groups are incorporated, along with a comprehensive synthesis report. The synthesis report consolidates the group's work and includes any further reports produced within that assessment period.

1.5 Diplomacy and Climate Negotiations

Climate diplomacy is critical in international relations, involving negotiations, agreements, and diplomatic efforts to address climate change. Diplomats and negotiators engage in intimate talks to secure commitments, resolve disputes, and advance shared goals (Berridge 2022). These diplomatic efforts are an integral part of climate policy and shape the relations between nations.

1.5.1 Climate Diplomacy

Climate change, primarily caused by human activities such as fossil fuel combustion and clearing forests, has become one of the most pressing global challenges in the twenty-first century (Fakana 2020). Climate change diplomacy has become increasingly recognized as a crucial approach to addressing the intricate challenges of the worldwide climate issue. Climate change diplomacy involves a wide range of diplomatic initiatives that promote cooperation between countries in dealing with climate change and its related effects. Diplomatic engagement can be through discussions between two countries or many countries, as well as the creation of international agreements and the establishment of collaborative relationships that transcend national boundaries (Kaltofen and Acuto 2018). Climate change diplomacy is essential to successfully navigating the complexities of a warming globe. Diplomats can help alleviate the effects of climate change and make progress toward a more sustainable future by fostering international cooperation, resolving competing national interests, and obtaining promises of financial and technical support. Addressing the complex global difficulties climate change poses requires a meticulous and necessary process known as climate change diplomacy. Climate change diplomacy calls for strategic engagement and skilled diplomacy due to the complexity and difficulty of the task (Prantl and Goh 2022). By fostering collaboration, guaranteeing equity, and coordinating national goals, diplomats can pave the way for creating innovative and long-lasting climate change solutions. If we want to lessen the impact of climate change and bring the world together to tackle one of the most significant issues of this century, we need effective climate change diplomacy. By consistently pursuing diplomatic efforts, the global community may work together to build a future resilient to environmental hazards that can endure time (Biden 2021).

Moving money and technology from one place to another is a significant obstacle in climate change diplomacy. Developed countries need to help poor countries adapt to climate change and implement sustainable practices by providing financial aid and easing the transfer of technology (Jakob and Steckel 2014). This will help achieve the global climate goals. This calls for in-depth conversations to ensure transparency and accountability in resource allocation and the development of robust systems to monitor and evaluate support effectiveness.

1.5.2 COP Meetings and Negotiations

Delegates from almost every country get together in COP meetings—also called Conferences of the Parties—to discuss and negotiate climate change-related issues under the UN Framework Convention on Climate Change (UNFCCC) (Wallbott and Schapper 2017). Global actions to combat climate change are mainly promoted at these conferences. Annually, the COP gathers for a series of meetings, each given a unique number, such as COP26 and COP27. The first COP summit occurred in Berlin in 1995; since then, other cities worldwide have hosted the annual gathering (Ivanova 2016).

The UNFCCC, the Kyoto Protocol, and the Paris Climate Agreement are all products of major international conferences that the UN has convened to address climate change (Hermwille et al. 2017). To aid nations hit hard by climate change and reduce emissions of greenhouse gases, these conferences aim to forge more robust international engagement and build solid global cooperation.

The UN Environment Program (UNEP) and the UN Framework Convention on Climate Change (UNFCCC) are two of the UN's specialized agencies and programs that seek to combat climate change (Bergesen et al. 2018). Furthermore, as part of the deal, once a year, representatives from different governments get together at what is called the "Conference of the Parties" (COP) to discuss climate change policy and make decisions. Some groups are working to help countries adapt to climate change and lessen its impacts; others, like the Global Weather and Climate Forecasting Technology Initiative, are trying to improve nations' capacities and provide financial and technical aid.

1.5.2.1 Expectations for Future COPs

Given the United Nations' anticipated continued leadership in combating climate change, it is reasonable to assume that existing frameworks and mechanisms for international collaboration will undergo revision and improvement. Among them, we must prioritize bolstering international action to accomplish climate goals, encouraging innovation and technology sharing, and increasing funding for adaptation and mitigation. To combat climate change, it is crucial to strengthen partnerships and cooperation across nations, organizations, and other sectors, such as the commercial sector and civil society (Waddell 2017). Diplomatic tactics are crucial for fighting climate change, but they typically lack enforcement measures and fail to address urgency. Thus, these contacts can unintentionally worsen climate change. To effectively tackle climate change, governments, organizations, and sectors, including the commercial sector and civil society, must increase partnerships and cooperation to foster holistic and practical solutions.

1.6 Geostrategic Considerations

The latest report from the Intergovernmental Panel on Climate Change (2013) confirms that the Earth is experiencing and will continue to experience human-caused, rapidly increasing, and profoundly changing global warming (Duffy et al. 2019). Food, water, health, energy, the environment, and livelihood security are among the many areas in which it is becoming more intertwined with human life. As a result, climate change is becoming a national security and defense concern and a political hot potato. In light of this, international relations are facing strain from global warming as a new geostrategic concern that may impact current government systems. This is why climate change, as an increasingly complex issue, is rising to the top of international and national agendas (Gemenne et al. 2014).

1.6.1 Power Dynamics in Climate Diplomacy

Multiple occurrences worldwide exist where climate change intersects with different social, political, economic, and strategic elements, resulting in their convergence in unique and unexpected ways. There is a significant association between the destabilization of the planet's climate and the rapid and unpredictable shift in global power and conflicts. Climate change is a notable alteration on Earth, among other necessary geophysical and biological transformations like biodiversity loss, overusing land and freshwater resources, and ocean acidification (Upadhyay 2020). The rapid pace of change is causing significant disruptions to well-established norms and giving rise to new, more influential strategic elements. It is crucial to understand these factors to uphold international stability fully. According to Valentin (2013), these variables are changing the structure of states, societies, and communities and the entire system of international interactions (Garrett et al. 2020). They are sparking innovative tension that will drive forthcoming conflicts, revolutions, wars, and struggles.

The WBGU's 2008 study contended that the world community might unite on climate change by recognizing the perils of manufactured climate change and promptly implementing a dynamic and integrated climate strategy to avert detrimental consequences (Arnold 2018). If prompt and resolute measures are not taken, climate change could trigger numerous conflicts, both on a global and national scale, regarding matters such as the distribution of resources, management of migration, and financial reparations between nations that are most affected by the climate crisis and those suffering the most severe consequences. The paper states that climate change worsens the variables contributing to instability and violence, especially in developing countries with weak governance systems and those facing rapid population growth and resource scarcity. Neighboring nations might experience instability due to the impact of domestic or regional warfare through refugee movements, arms trading, and the withdrawal of fighters. Consequently, climate change can expand the geographical extent of conflict and crisis areas, even outside national borders, impacting society (Schleussner et al. 2016).

1.6.2 Environmental Security

In response to climate change's complex repercussions, states and non-state actors successfully employ environmental geostrategic techniques to address this issue. This novel approach aims to direct strategic analysis toward comprehending how the intersection of ecological, social, political, and economic pressures can give rise to fresh tensions and violent conflicts. For instance, there is a growing focus on the strategic ramifications of the Arctic's summer melting, which may facilitate the opening of the renowned "north-west passage" in the future and incite a fresh rivalry for resources in the area. The rationale for deploying distinct security measures in critical circumstances is predicated on this assertion.

Physical infrastructure security is intricately linked to climate security and the security of vital resources such as food, water, and energy (Burns 2019). Humanity's vast and complex challenges in the twenty-first century cannot be adequately addressed using traditional and coercive security measures. To effectively address and mitigate the impact of these growing concerns, it is critical to include other governmental departments and agencies in charge of environmental, development, science and technology, and economic and social policies. This inclusive approach should be incorporated into a complete security strategy.

1.6.3 Geopolitical Implications of Climate Change

Timber production, biofuel production, food production, and land allocation for wind or solar power are just a few of the land use choices that are becoming increasingly competitive as the population keeps climbing (Dale et al. 2011). When we think about how mountain water supplies could run dry, which could lead to more conflicts between neighboring countries, and how devastating floods could make coastal areas or island nations uninhabitable, forcing a mass migration of people in search of food and shelter, the geopolitical implications of climate change become apparent. One of the most critical factors in geopolitics has always been controlling land and maritime resources. In its belief that its continued dominance depends on sustained reliance on oil and gas, the United States has prioritized maintaining this position over solving climate change. For China's climate change policy to work, its economy must keep growing to reach its long-term goals. The United States, China, and Russia, all highly influential geopolitically and having an enormous demand for fossil fuels, are also investing heavily in their military might (Sivaram and Saha 2018).

Regarding controlling its power, China's policies have shown a lot of realism and restraint. But they have also demonstrated brutality in gaining control of vital resources. As the world's largest economic region, the European Union (EU) sees itself primarily as a long-term player committed to bolstering democracy and stability, with a focus on implementing effective climate change mitigation strategies. Significant action is required to reduce greenhouse gas emissions dramatically; this includes establishing a global carbon tax and investing heavily in R&D and the commercialization of low-carbon technologies (Ockwell et al. 2008). The present tendency could be turned around if these steps are taken. An appropriate European Union climate change plan should include a fair carbon price and ensure that all low-carbon technologies are competitive. The European Union has to step up its game to deal with the consequences of climate change and make the most of them. The European Union's (EU) dwindling real power, dependence on energy imports, aging population, and limited military forces make its capacity to impact global climate change policies questionable. China, the world's top greenhouse gas emitter, has exceeded its carbon intensity reduction and non-fossil fuel energy targets in renewable energy deployment. It leads worldwide renewable energy investment and aims to peak carbon emissions by 2030 and reach carbon neutrality by 2060.

Conversely, energy import dependence, demographic trends, and limited military capabilities may limit the EU's ability to influence global climate policies (Hill 2010).

1.7 The Future of International Relations in Environmental Issues

In the following years, nations will increasingly require collaborative efforts to address environmental challenges and combat climate change (Osofsky 2013). This entails collaborating to mitigate pollution, transitioning to renewable energy sources such as solar and wind power, and safeguarding ecosystems such as forests and oceans. International treaties such as the Paris Agreement will remain crucial in establishing objectives and ensuring that countries are held responsible (Khan et al. 2016). The prevalence of climate change mitigation technologies, such as electric vehicles and renewable energy sources, will increase. It is imperative to ensure that all individuals, particularly those most impacted by environmental issues, can participate in decision-making processes about their resolution. To effectively tackle environmental concerns in the future, it will be crucial to prioritize cooperation, innovation, and fairness.

1.7.1 Anticipated Developments

Several anticipated developments are expected to impact world affairs concerning climate change and environmental issues in the not-too-distant future. A greater emphasis on environmentally friendly recovery and sustainable growth is anticipated after the COVID-19 pandemic (Rume and Islam 2020). Infrastructure, conservation efforts, and renewable energy sources will receive national funding priorities. In addition, the development of renewable energy sources like wind and solar power is anticipated to accelerate, resulting in a global shift away from fossil fuels and toward low-carbon economies. As countries work to increase their resolve and achieve the goals of the Paris Agreement, international cooperation is expected to expand, particularly in the time leading up to significant climate summits like

COP26 (Meinshausen et al. 2022). It is believed that non-state actors, including cities, corporations, and civil society groups, will play a significant role in pushing for climate action and holding governments to account. These anticipated developments show that people worldwide are realizing the critical need to solve environmental problems and move toward a more sustainable and resilient future.

1.7.2 Emerging Trends

Several emerging tendencies influence international cooperation on climate change and environmental protection (Haas et al. 1993). There is growing consensus that climate change seriously threatens global security. As a result, nations are beginning to plan to protect their citizens against climate-related hazards, such as severe weather and food shortages. Another development is the increasing number of lawsuits filed against governments and corporations alleging that they are complicit in the causes of climate change and have failed to take sufficient action to combat the issue. People also utilize nature in their fight against climate change. Carbon dioxide is a greenhouse gas; planting trees can help mitigate its effects. Combating climate change is also increasingly dependent on financial resources. Clean energy and initiatives to assist communities in adapting to climate change are receiving increasing funding. Finally, computing power and satellite imagery advances enhance our environmental literacy and guide us toward more informed responses to climate change (Scholz and Binder 2011). All of these tendencies point to the fact that people and nations are getting creative in their fight against climate change and for environmental protection.

1.7.3 Prospects for Increased Cooperation

Cooperation among nations to address climate change and other environmental emergencies provides a promising outlook. A significant component of this is the growing recognition of the interconnectedness of various situations. Countries directly impacted by climate change are fostering an increasing need to cooperate on solutions that benefit all nations. Technological and scientific advancements have made it easier for countries to work together by exchanging information, collaborating on research, and developing new methods to adapt to and reduce the impacts of climate change (Thomalla et al. 2006).

Both citizens and the international community are increasingly exerting pressure on governments to tackle environmental concerns effectively. The public's heightened awareness of climate change and its accompanying action have generated significant momentum for governmental changes and international agreements. Due to the climate action grassroots movement, governments are collaborating more closely with NGOs and other non-profits to achieve common goals (Mingers and van Koppen 2019).

Economic incentives also motivate cooperation on environmental concerns. Many governments view transitioning to a low-carbon economy as an opportunity to generate employment and stimulate economic expansion. Countries may reduce their ecological footprint, boost their economies, and improve public health by investing in renewable energy, sustainable infrastructure, and green technologies.

In addition, the Paris Agreement and other international frameworks enable states to collaborate and synchronize their efforts to address climate change. Countries can cooperate to accomplish climate objectives and overcome common challenges through technology transfer and emissions trading (Karakosta et al. 2010). A concerted effort is needed to combat climate change, blending physical measures like renewable energy adoption with diplomatic collaboration via agreements like the Paris Agreement. Additionally, fostering sustainable business relations, including emissions trading and technology transfer, is crucial for achieving global climate objectives.

1.8 Conclusion

To classify a matter as a threat, one must first determine which political faction needs protection and then approve and construct the institutions that give that protection. Not every aspect has been resolved, nor have all other security logics—such as the Copenhagen School's defensive, crisis-oriented logic—been investigated. This arose because, in a particular environment, it provided the most effective defense against a specific type of threat. Despite this, they will continue to bargain when faced with new environments and threats. The hazards of climate change are unclear, complex to identify, and potentially catastrophic. This, according to Beck, is the basis for a risk society. This chapter investigated security practitioners' difficulties when interpreting climate change and environmental crises as security issues. It has also looked at how appeals to security concerns have drawn attention to the significance of preventative, non-violent actions and the role of non-state actors in ensuring public safety. An analysis of the rise to prominence and subsequent galvanization of social action in response to diverse environmental crises could be aided by a potentially non-essentialist, securitization-style technique focusing on the social process that detects dangers. The Copenhagen School, on the other hand, contends that realist logic determines a concept's "security." As a result of this problematic fixity, educational institutions have legitimized security approaches based on industry-specific and time-restricted security paradigms.

References

Adelodun B, Ajibade FO et al (2021) Assessment of socioeconomic inequality based on virus-contaminated water usage in developing countries: a review. Environ Res 192:110309

Anika OC, Nnabuife SG et al (2022) Prospects of low and zero-carbon renewable fuels in 1.5-degree net zero emission actualisation by 2050: a critical review. Carbon Capture Sci Technol 5:100072

Arias P, Bellouin N et al (2023) Intergovernmental Panel on Climate Change (IPCC). Technical summary climate change 2021: the physical science basis contribution of Working Group I to the Sixth Assessment Report of the Intergovernmental Panel on Climate Change. Cambridge University Press, pp 35–144

Arnold A (2018) Climate change and storytelling: narratives and cultural meaning in environmental communication. Springer

Baker JL (2012) Climate change, disaster risk, and the urban poor: cities building resilience for a changing world. World Bank Publications

Beck S, Forsyth T (2015) Co-production and democratizing global environmental expertise: the IPCC and adaptation to climate change science and democracy. Routledge, pp 113–132

Beck S, Mahony M (2018) The IPCC and the new map of science and politics. Wiley Interdiscip Rev Clim Chang 9(6):e547

Bergesen HO, Parmann G et al (2018) United Nations Framework Convention on Climate Change (UNFCCC) Year Book of International Co-operation on Environment and Development. Routledge:73–77

Berridge GR (2022) Diplomacy: theory and practice. Springer

Bhandari MP (2022) Getting the climate science facts right: the role of the IPCC. River Publishers

Biden JR (2021) Interim national security strategic guidance. The White House 8

Bindoff NL, Stott PA, et al. (2014) Detection and attribution of climate change: from global to regional. Climate change 2013: the physical science basis

Bodansky D and Rajamani L (2018) The evolution and governance architecture of the United Nations climate change regime

Brooks SG, Wohlforth WC (2016) America abroad: the United States' global role in the 21st century. Oxford University Press

Burns MG (2019) Managing energy security: an all hazards approach to critical infrastructure. Routledge

Castles S (2019) Why migration policies fail 1Celebrating 40 years of ethnic and racial studies. Routledge:300–320

Chen C-TA, Drake ET (1986) Carbon dioxide increase in the atmosphere and oceans and possible effects on climate. Annu Rev Earth Planet Sci 14(1):201–235

Cohen R, Eames PC et al (2022) Briefing: the 2021 Glasgow Climate Pact: steps on the transition pathway towards a low carbon world. Proc Inst Civ Eng-Energy 175(3):97–102

Council NR, Earth Do et al (2012) Science for environmental protection: the road ahead. National Academies Press

Curran D, Williams PD (2018) The United Kingdom and United Nations peace operations. European Approaches to United Nations Peacekeeping. Routledge, pp 34–55

Dąbrowska J, Sobota M et al (2021) Marine waste—sources, fate, risks, challenges and research needs. Int J Environ Res Public Health 18(2):433

Dale VH, Efroymson RA et al (2011) The land use–climate change–energy nexus. Landsc Ecol 26:755–773

Duffy PB, Field CB et al (2019) Strengthened scientific support for the endangerment finding for atmospheric greenhouse gases. Science 363(6427):eaat5982

Edwards M (2010) Small change: why business won't save the world. Berrett-Koehler Publishers

Eizenberg E, Jabareen Y (2017) Social sustainability: a new conceptual framework. Sustainability 9(1):68

Environment CCfICo and Secretariat D (2023) Global Climate Governance and China's role green recovery with resilience and high quality development: CCICED Annual Policy Report 2021. Springer, pp 3–57

Evseeva O, Evseeva S et al (2021) The impact of human activity on the global warming E3S web of conferences. EDP Sciences:11017

Fakana ST (2020) Causes of climate change. Glob J Sci Front Res 20:7–12

Farrell H, Newman A (2016) The new interdependence approach: theoretical development and empirical demonstration. Rev Int Polit Econ 23(5):713–736

Framework T, Stocktake G (2016) Empowering the Ratchet-up mechanism under the Paris Agreement
Garrett RD, Ryschawy J et al (2020) Drivers of decoupling and recoupling of crop and livestock systems at farm and territorial scales. Ecol Soc 25(1):24
Gasser T, Ciais P et al (2022) How the Glasgow declaration on forests can help keep alive the 1.5 C target, Proc Natl Acad Sci. 119(23):e2200519119
Gemenne F, Barnett J et al (2014) Climate and security: evidence, emerging risks, and a new agenda. Clim Chang 123:1–9
Goodall MT and Army B (2019) Friend or foe: Chinese influence in sub-Saharan Africa
Goreau TJ (2014) Global biogeochemical restoration to stabilize CO2 at safe levels in time to avoid severe climate change impacts to Earth's life support systems: implications for the United Nations Framework Convention on Climate Change. Geotherapy: Innovative Technologies for Soil Fertility Restoration, Carbon Sequestration, and Reversing Atmospheric CO2 Increase
Grove M, Ogden L et al (2018) The legacy effect: understanding how segregation and environmental injustice unfold over time in Baltimore. Ann Am Assoc Geogr 108(2):524–537
Guppy N (1984) Tropical deforestation: a global view. Foreign Aff 62(4):928–965
Gupta A (2016) Climate change and Kyoto protocol: an overview. Handbook of environmental and sustainable finance: 3–23. London: Academic Press
Haas PM, Keohane RO et al (1993) Institutions for the earth: sources of effective international environmental protection. MIT Press
Haszeldine RS, Flude S et al (2018) Negative emissions technologies and carbon capture and storage to achieve the Paris Agreement commitments. Philos Trans R Soc A Math Phys Eng Sci 376(2119):20160447
Hegglin MI, Bastos A et al (2022) Space-based earth observation in support of the UNFCCC Paris Agreement. Front Environ Sci 10
Hermwille L, Obergassel W et al (2017) UNFCCC before and after Paris–what's necessary for an effective climate regime? Clim Pol 17(2):150–170
Hill S (2010) Europe's promise: why the European way is the best hope in an insecure age. Univ of California Press
Hoegh-Guldberg O, Jacob D et al (2019) The human imperative of stabilizing global climate change at 1.5 C. Science 365(6459):eaaw6974
Höhne N, Kuramochi T et al (2021) The Paris agreement: resolving the inconsistency between global goals and national contributions climate policy after the 2015 Paris Climate Conference. Routledge:30–46
Horster M (2018) Climate change as a topic for impact investing. Positive impact investing: a sustainable bridge between strategy, innovation, change and learning, pp 97–111
Humphrey J (2004) The clean development mechanism: how to increase benefits for developing countries. IDS Bull 35(3):84–89
Ivanova M (2016) Good COP, bad COP: climate reality after Paris. Global Pol 7(3):411–419
Jakob M, Steckel JC (2014) How climate change mitigation could harm development in poor countries. Wiley Interdiscip Rev Clim Chang 5(2):161–168
Jinping X (2022) Hold high the great banner of socialism with Chinese characteristics and strive in unity to build a modern socialist country in all respects report at the 20th national congress of the communist party of China Qiu Shi
Jones MW, Peters GP et al (2023) National contributions to climate change due to historical emissions of carbon dioxide, methane, and nitrous oxide since 1850. Scient Data 10(1):155
Kaltofen C, Acuto M (2018) Rebalancing the encounter between science diplomacy and international relations theory. Global Pol 9:15–22
Kampa M, Castanas E (2008) Human health effects of air pollution. Environ Pollut 151(2):362–367
Karakosta C, Doukas H et al (2010) Technology transfer through climate change: setting a sustainable energy pattern. Renew Sust Energ Rev 14(6):1546–1557
Khan M, Sagar A et al (2016) Capacity building under the Paris Agreement. European Capacity Building Initiative, Oxford

Khan N, Fahad S, et al. (2020) Analysis of livelihood in the world and its impact on world economy. Available at SSRN 3717265
Khanna P (2011) Assessment of heavy metal contamination in different vegetables grown in and around urban areas. Res J Environ Toxicol 5(3):162
Kirton JJ (2018) Advancing global openness: G7 governance of globalization. The G7, anti-globalism and the governance of globalization. Routledge, pp 22–44
Labadi S, Giliberto F et al (2021) Heritage and the sustainable development goals: policy guidance for heritage and development actors. Int J Herit Stud
Linnerooth-Bayer J, Hochrainer-Stigler S (2015) Financial instruments for disaster risk management and climate change adaptation. Clim Chang 133:85–100
Madricardo F, Ghezzo M et al (2020) How to deal with seafloor marine litter: an overview of the state-of-the-art and future perspectives. Front Mar Sci 7:505134
Matemilola S, Fadeyi O et al (2020) Paris agreement. Encycl Sustain Manage 2020:1
Mbeva K, Makomere R et al (2023) The evolving geopolitics of climate change. Africa's right to development in a climate-constrained world. Springer, pp 85–126
McCauley DJ, Pinsky ML et al (2015) Marine defaunation: animal loss in the global ocean. Science 347(6219):1255641
McCormick J (2023) The role of environmental NGOs in international regimes. The global environment. Routledge, pp 52–71
McMichael AJ (2003) Climate change and human health: risks and responses. World Health Organization
Meinshausen M, Lewis J et al (2022) Realization of Paris Agreement pledges may limit warming just below 2 C. Nature 604(7905):304–309
Mingers N and van Koppen CK (2019) ENGO collaboration on climate change: stronger together?
Morano M (2018) The politically incorrect guide to climate change. Simon and Schuster
Mowforth M, Munt I (2015) Tourism and sustainability: development, globalisation and new tourism in the third world. Routledge
Ockwell DG, Watson J et al (2008) Key policy considerations for facilitating low carbon technology transfer to developing countries. Energy Policy 36(11):4104–4115
Olabi A, Abdelkareem MA (2022) Renewable energy and climate change. Renew Sust Energ Rev 158:112111
Osofsky HM (2013) The geography of solving global environmental problems: reflections on polycentric efforts to address climate change. NYL Sch L Rev 58:777
Panel UNEPIR, Consumption UNEPS et al (2011) Decoupling natural resource use and environmental impacts from economic growth. UNEP/Earthprint
Pavone IR (2018) The Paris Agreement and the Trump administration: road to nowhere? J Int Stud 11(1):34–49
Pörtner H-O, Scholes R et al (2023) Overcoming the coupled climate and biodiversity crises and their societal impacts. Science 380(6642):eabl4881
Pouffary S, Antonini A, et al (2017) United Nations framework: convention on climate change: twenty-third Conference of the Parties (COP23): 6 to 17 November 2017, Bonn, Germany. Guident negotiations (21)
Prantl J, Goh E (2022) Rethinking strategy and statecraft for the twenty-first century of complexity: a case for strategic diplomacy. Int Aff 98(2):443–469
Racoviteanu AE, Glasser NF et al (2022) Recent evolution of glaciers in the Manaslu region of Nepal from satellite imagery and UAV data (1970–2019). Front Earth Sci 9:767317
Rossato FGF, Susaeta A et al (2018) Comparison of revealed comparative advantage indexes with application to trade tendencies of cellulose production from planted forests in Brazil, Canada, China, Sweden, Finland and the United States. Forest Policy Econ 97:59–66
Rume T, Islam SD-U (2020) Environmental effects of COVID-19 pandemic and potential strategies of sustainability. Heliyon 6(9):e04965
Sax DF, Gaines SD et al (2002) Species invasions exceed extinctions on islands worldwide: a comparative study of plants and birds. Am Nat 160(6):766–783

Schleussner C-F, Donges JF et al (2016) Armed-conflict risks enhanced by climate-related disasters in ethnically fractionalized countries. Proc Natl Acad Sci 113(33):9216–9221
Scholz RW, Binder CR (2011) Environmental literacy in science and society: from knowledge to decisions
Segger M-CC (2016) Advancing the Paris Agreement on climate change for sustainable development. Camb Int Law J 5(2):202–237
Singh RL, Singh PK (2017) Global environmental problems. Principles and applications of environmental biotechnology for a sustainable future. Springer, Singapore, pp 13–41
Sivaram V and Saha S (2018) The geopolitical implications of a clean energy future from the perspective of the United States. Geopolit Renew: 125–162
Sokona Y, Calvo E et al (2015) IPCC expert meeting on potential studies of the IPCC process. World Meteorological Organization 28:29
Stengel FA, Baumann R (2017) Non-state actors and foreign policy. Oxford research encyclopedia of politics
Streck C, Keenlyside P et al (2016) The Paris Agreement: a new beginning. J Eur Environ Plann Law 13(1):3–29
Swart R, Bernstein L et al (2009) Agreeing to disagree: uncertainty management in assessing climate change, impacts and responses by the IPCC. Clim Chang 92:1–29
Tham KW (2016) Indoor air quality and its effects on humans—a review of challenges and developments in the last 30 years. Energ Buildings 130:637–650
Thomalla F, Downing T et al (2006) Reducing hazard vulnerability: towards a common approach between disaster risk reduction and climate adaptation. Disasters 30(1):39–48
Upadhyay RK (2020) Markers for global climate change and its impact on social, biological and ecological systems: a review. Am J Clim Chang 9(03):159
Valone TF (2021) Linear global temperature correlation to carbon dioxide level, sea level, and innovative solutions to a projected 6 C warming by 2100. J Geosci Environ Protect 9(03):84
Velders GJ, Andersen SO et al (2007) The importance of the montreal protocol in protecting climate. Proc Natl Acad Sci 104(12):4814–4819
Waddell S (2017) Societal learning and change: how governments, business and civil society are creating solutions to complex multi-stakeholder problems. Routledge
Wallbott L, Schapper A (2017) Negotiating by own standards? The use and validity of human rights norms in UN climate negotiations. Int Environ Agreem: Politics Law Econ 17:209–228
Warner TT (2009) Desert meteorology. Cambridge University Press
Wirth C (2015) Securing Japan in the ‘west’: the US-Japan alliance and identity politics in the Asian century. Geopolitics 20(2):287–307
World Health Organization (2015) Connecting global priorities: biodiversity and human health: a state of knowledge review
World Health Organization (2019) Global action plan on physical activity 2018–2030: more active people for a healthier world. World Health Organization
Xu Y, Zhao F (2023) Impact of energy depletion, human development, and income distribution on natural resource sustainability. Resour Policy 83:103531
Yuan X, Su C-W et al (2022) The race to zero emissions: can renewable energy be the path to carbon neutrality? J Environ Manag 308:114648
Zarkik A (2022) COP27: a brief account of contemporary climate adaptation and mitigation policies, a view from the south. Policy Center for the New South, Salé 2022

Increased Population Is a Source of Environmental Pollution and Climate Change

2

Sumaira Manzoor, Muhammad Mughees Ud Din, Mohammed Gamal, Muhammad Danish Toor, and Abdul Basit

Abstract

Global population increase is not a new issue. Rather, it has been around for a few decades, with roots in both prehistoric and modern periods. Recognized specialists long ago made up the term "overpopulation" and predicted terrible consequences if the world continues its current path. According to predictions, scientists developed the birth control pill and applied eugenics to regulate the population. Despite this, the population continued to rise, and diseases persisted. Migration has also contributed to population expansion and the considerable environmental dangers it entails. Urbanization destroys natural habitats and increases carbon dioxide emissions, which contribute to climate change and global warming. Humanity is at risk of losing the habitat it has created for itself,

S. Manzoor
College of Agronomy, Sichuan Agricultural University, Chengdu, People's Republic of China

M. M. U. Din
Institute of Soil and Environmental Sciences, University of Agriculture, Faisalabad, Pakistan

M. Gamal
Department of Soil Science and Plant Nutrition, Faculty of Agriculture, Ondokuz Mayıs University, Samsun, Türkiye

Department of Soil and Agricultural Chemistry, Faculty of Agriculture Saba Basha, Alexandria University, Alexandria, Egypt

M. D. Toor
Faculty of Science and Technology, Institute of Ecology and Earth Science, University of Tartu, Tartu, Estonia

A. Basit (✉)
Department of Horticulture, Graduate School, Kyungpook National University, Daegu, South Korea
e-mail: abdulbasit97_lily@knu.ac.kr

A. A. H. Abdel Latef et al. (eds.), *Bioremediation and Nanotechnology for Climate Change Mitigation*, https://doi.org/10.1007/978-981-96-3069-1_2

and species are becoming extinct. Global inequality causes a lack of food, water, and employment opportunities, as well as inadequate education. Even with the support of international organizations and agencies, unequal distribution of financial resources, natural resources, and individual rights causes poverty and defines global culture as greedy. National institutions can address the issue of overpopulation by implementing policies that are consistent with the recommendations provided by international organizations dedicated to serving the interests of the global community. People in this global network act in their own best interests, leaving others in abject poverty and shortages.

Keywords

Climate change · Environmental Pollution · Global population · Population increase

2.1 Introduction

The world has experienced unparalleled population expansion in recent times, along with growing apprehensions regarding environmental deterioration and the imminent threat of climate change. Researchers, legislators, and environmentalists all focus on this connection between population growth and environmental degradation. The consequences for the natural balance of our planet are growing more significant as the world's population rises to almost eight billion people and shows no indications of abating (Ehrlich et al. 1993). There are many facets and complexities in the interaction between population dynamics and environmental deterioration. An urgent concern arising from the swift expansion of the population is the increased demand placed on natural resources. Growing demands on land, water, energy, and food put a great deal of strain on ecosystems, resulting in habitat destruction, deforestation, and a decline in biodiversity. Moreover, the effects of population expansion go beyond regional environmental changes to include global phenomena like climate change. With far-reaching effects, the exponential rise in greenhouse gas emissions, mostly caused by human activity, has set off a chain reaction of climate disruptions. Anthropogenic climate change is causing existential challenges to ecosystems, economies, and human cultures globally (Myers 2014). Some of these expressions include melting ice caps, extreme weather events, rising temperatures, and altered precipitation patterns. It becomes essential to investigate the complex interactions between population dynamics, environmental degradation, and climate change considering these interconnected concerns. By comprehending the mechanisms underlying these occurrences and their interdependence, we may create strategies and policies that effectively reduce their negative consequences and promote sustainable development paths for coming generations.

Population mobility is a common strategic response that populations adopt to cope with stress, poverty, and environmental risk. There is currently a renewed interest in the topic due to several factors, including worries about how global environmental change (GEC) will affect human well-being and population mobility, the idea that migration is a practical adaptation strategy, the theory that environmental displacement could lead to problems with security and governance, and recent empirical findings (Laczko and Aghazaram 2009). The global context (Meyerson et al. 2007) in which fast urbanization and high population movement are changing population distribution and its ecological footprint (Tacoli 2009) is set against the backdrop of the resurgence of environmental migration in the research agenda.

It is critical to define the term "crisis" to comprehend the current state of our world. "We cannot make the world uninhabitable for other forms of life and have it habitable for ourselves," one renowned ecologist succinctly stated (Sorvall 1971). It is doubtful whether a man will survive if this happens, and he becomes the sole living thing on the planet. Vivian Sorvall is one of many academics who address the challenge of overpopulation and question why crises occur. The "obvious answers," such as population expansion, the scarcity of clean water, air pollution, and the crisis, are rooted in human evolution. She claims that "man and other forms of life have existed on this planet in relative harmony and balance for thousands of years." Why has this equilibrium been upset mostly by Westerners during the past 400 years? She also argues that one should understand the evolution of a man from three time periods: as a nomad, as a farmer, and finally as a technocrat. To understand this, the modern man distinguishes himself from other life forms by believing he can exploit nature. A man would associate himself with animals while he was a nomad. His ability to use tools was essential to his survival, and, thanks to these skills, he made the shift from a nomadic to an agricultural lifestyle. Man was seen as distinct from and superior to the rest of the natural order due to the influence of Greek and Judaic beliefs (Sorvall 1971). As a result, man began to procreate, quickly exceeding the earth's and civilization's capacity. The history of overcrowding from prehistoric times to the current decade is discussed in this study section.

2.1.1 Background History

In the 1960s and 1970s, Western scholars agreed that the rapid population growth in developing countries was a serious global catastrophe. Since then, the world has changed significantly. Rapid population expansion is one of the main factors contributing to a country's environmental degradation, as it negatively impacts the environment and natural resources. The problem of sustainable development is posed by population growth and environmental degradation. The process of socioeconomic development can be accelerated or hindered by the presence or absence of favorable natural resources. The growing population and the deteriorating environment pose the problem of sustainable development. The process of socioeconomic development can be accelerated or hindered by the presence or lack of advantageous natural resources (Ray and Ray 2011). Changes in the size, composition, and distribution of the population are caused by three basic demographic factors: birth (birth), deaths (mortality), and migration and immigration (the arrival of a population in a country leads to an increase in the population). These changes also raise some important questions about cause and effect. Global population and economic expansion are the leading causes of many severe environmental disasters. These include extreme land use, deforestation, habitat damage, and biodiversity loss. Energy needs are increasing due to changing consumption patterns. The consequences are water scarcity, air pollution, climate change, and global warming (Omer 2008).

2.1.2 Critical Implications for Environmental Sustainability and Climate Stability

According to Forsyth, "Frame assessments should consider alternative frames that may not be considered relevant in policy debates, rather than just those labeled as relevant at the time" (Forsyth 2003). Considering this in perspective, the notion of "climate stabilization" has gained prominence in political and strategic fields, with a focus on mitigation rather than adaptation, as well as quantification instead of dynamic decarbonization targets for energy systems. Climate stabilization has emerged as a central idea for climate change policy, entangling policy-making in unmanageable scientific ambiguities that give rise to many problematic hypotheses and recommendations that impact environmental economics and climate science (Forsyth 2003). It is well demonstrated how Levy et al. (2001) characterize the process of policy and problem development based on simple but inflexible and unscientific objectives, which are good for predicting policies but ultimately outlive their usefulness and are difficult to replace with more complex objectives due to the stable socialization of the former. As the narratives of reactionary interests against decarbonization have strengthened, scientists who want to act on climate change may be concerned about any possible destabilization of the dominant framework (Poumanyvong and Kaneko 2010a).

Many scientists, economists, and policymakers have adopted these ways of thinking because of the science-policy debate focused on stabilization, which has become an attractive approach when they are concerned with defining the reality and (approximate) scale of the problem. According to Tony Blair's comment at the beginning of the article (Gore 2008), stabilization is a concept that resonated with the natural intuition of economists and scientists. Stabilization's historical origins, however, are arguably best understood as a notion that resulted from an uncomfortable fusion of economics, politics, and physical science, which provided a particular method to frame the subject that was susceptible to these studies. In the 1980s, several crossing intellectual heritages, a dearth of alternatives, and model shortcomings made climate stabilization an intuitively sound heuristic. Although it has been beneficial in that it has drawn attention to reducing the negative human contributions to climate change, it has also encouraged policymakers to focus on long-term equilibrium targets, which have stymied short-term action. At the same time, they wait for the answers to all the questions regarding scientific, political, and economic uncertainty. This has led to comfortable situations for political and policy cycles (Frame et al. 2006). That target must soon be strengthened to 350 parts per million

(Gore 2008). However, during this meeting in Poznan, Poland, there were few short-term commitments and measures.

From an environmental perspective, what matters to people is not the specific stocks of natural capital under study but the capacity of capital in general to perform environmental activities essential to human well-being. It is, therefore, logical to characterize environmental sustainability as the maintenance of important environmental functions and, therefore, the maintenance of the capacity of the capital stock to perform these functions (Ekins et al. 2003). This is essentially in line with the definition of environmental sustainability given by English Nature (1994), which states that it is "the maintenance of the inherent qualities and characteristics of the environment and its capacity to perform its full function"—conservation of biodiversity. Certain pools of natural capital may not necessarily execute environmental services in a unique way. It is possible that, as was covered, different kinds of capital could provide flows that can fairly replace certain environmental services. It is also unnecessary to presume that every environmental function is vital to human welfare and should be preserved (Sober 1986).

2.1.3 Population Growth and Environmental Degradation

Although population growth is essential for progress when it exceeds the capacity of support systems, it also contributes significantly to environmental degradation. Even with their creative nature, development programs are unlikely to produce the desired effects until the relationship between population growth and the life support system is stabilized—population (2007). Population growth is linked to environmental stressors such as biodiversity loss, air and water pollution, and increased demand for arable land. It also affects the environment through the consumption of natural resources and waste production. Human population problems significantly impact our way of life and the future of this planet (Ray and Ray 2011). Environmental degradation is considered both a cause and a consequence of poverty. The complex relationship between poverty and the environment is circular. Because the poor are more dependent on natural resources than the rich and are less likely to have access to alternative resources, inequality can contribute to the depletion of natural resources faster than would otherwise be the case. Furthermore, environmental degradation can accelerate the process of impoverishment, as the poor are highly dependent on natural resources. An increasing number of poor households are migrating to cities due to environmental challenges and the lack of good economic options in rural areas (Deshingkar 2010).

Urban slums are growing, and megacities are emerging. Urban ecology has deteriorated due to rapid and unplanned urban expansion. This has led to a decline in cities' valuable environmental resource base, widening the gap between the supply and demand for infrastructure services such as electricity, housing, transport, communications, education, water supply, sanitation, and recreational facilities. As a result, there is a growing trend of air and water quality degradation, waste generation, slum expansion, and adverse land use changes, exacerbating urban poverty (Boadi et al. 2005).

2.2 Direct and Indirect Impacts of Population Increase on the Environment and Climate

Farming practices that cause soil erosion, salinization of the land, and nutrient loss directly influence the ecosystem. The overuse of water and land resources, as well as a multiplication in applying pesticides and fertilizers, has accompanied the development of the Green Revolution. Another significant factor contributing to land degradation has been shifting cultivation. One (Chitra and Priya 2020) of the main ways pesticides and fertilizers contaminate water bodies is by leaching. Salinization, alkalization, and water logging are caused by intensive farming and irrigation. Technological, institutional, and socioeconomic activity interact dynamically to cause environmental degradation. Numerous variables, such as population expansion, urbanization, intensification of agriculture, increased energy consumption, and transportation, can contribute to environmental changes. Poverty is still at the core of many environmental issues (Rai 2019).

Intense economic development is given more consideration in modern urbanization through economic urbanization. Both pollution emissions and environmental quality can be enhanced by this mode (Chen et al. 2018). Social urbanization emphasizes changing one's lifestyle and quality of life (Cui and Guo 2021). It has reinforced the idea of sustainable economic development and increased public readiness to participate in environmental conservation. Air pollution has unfavorable externalities. The local environmental quality can be improved with more public participation in regions that are geographically close by (Ge et al. 2021). The government has raised the cost of enterprise pollution discharge by strengthening the environmental supervision system (Syed et al. 2021); with widespread social supervision, the people have embraced a green lifestyle (Sun and Duman 2017). The environment is now less polluted thanks to all these actions. Through the promotion of green technologies, urbanization also directly impacts the reduction of environmental pollutants (Gu et al. 2018).

2.2.1 Resource Consumption

It is interesting to note that worries about a constantly expanding human population are nothing new. Malthus (1798), who saw that the population of humans was growing faster than the resources required for subsistence, predicted that this would become a problem in the future more than 200 years ago. However, Malthus downplayed the technological advancements that took place in the twentieth century, particularly in agriculture (the Green Revolution), which allowed people to generate an abundance of food and so allowed for a population explosion. The human population grew from 1 billion in 1830 to over 2.5 billion by 1950, which is currently very close to 8 billion. But feeding so many people has unavoidably had a significant negative influence on the environment, affecting all the planet's ecosystems, particularly food (Crist et al. 2021). According to Tilman et al. (2001), Meyer and Butaud (2009), Harter et al. (2015), and Urban et al. (2016), among others, the main causes of biodiversity loss today are: (1) changes in land and sea use, (2) unsustainable exploitation of organisms, (3) climate change, (4) pollution (e.g., eutrophication, pesticides, waste), and (5) invasive species. However, most of these variables are closely linked to food production by modern agriculture (Pingali 2012; Smith et al. 2013; Stefen et al. 2015), which is in turn responsible for at least 20% of global anthropogenic greenhouse gas emissions (Steinfeld et al. 2006; Schwarzer 2013). A known factor contributing to habitat degradation is the continuous conversion of land for agricultural and livestock production (Mora and Zapata 2013; Maxwell et al. 2016; Estrada et al. 2017).

2.2.2 Waste Production

Numerous researchers have examined the global issue of overpopulation in considerable detail, examining it from various angles. When discussing the problems brought about by global population expansion, garbage production is rarely brought up. However, the statistics from the 105 nations examined make evident what we already know: when all other things are equal, waste creation and population growth increase (Karak et al. 2012). Although the data do not match the GDP statistics exactly due to the wide variations in population sizes among countries, the regression model accounts for a significant portion of the waste generation in relation to the population. When (Brown and Wildenthal 1988) the populations of the United States, China, and American Samoa are compared in terms of how much waste each country produces in relation to its size, it becomes evident that waste generation is correlated with population size but also carries cultural significance in terms of how acceptable waste generation is perceived. By calculating the amount of garbage generated daily by everyone, the cultural perception of trash may be experimentally measured. The population of the United States is estimated to be 315 million. Although Americans produce the most volume and rank sixth in daily waste output per person, this number only accounts for slightly more than 4% of the global population (Rogoff et al. 2009).

The average American generates about 2 kg (4.4 lbs) of garbage each day, compared to 1.12 kg (2.64 lbs) for Samoans and 0.32 kg (0.7 lbs) for Chinese people, placing them 43rd and 88th, respectively. If the amount of waste produced was only dependent on population size, American Samoa, with its 57,000 residents, would rank in the hundreds rather than 43rd; China would rank first, not 88th, with its 1.35 billion people (or 19% of the world's population); and India, with its 1.25 billion people, would be the top two producers. The (Wiedinmyer et al. 2014) fact that the average American generates more than six times as much rubbish per person as the average Chinese person makes it evident how important cultural perspectives on waste are. Thus, the sheer number of people contributes to the amount of waste generated, but the acceptability of waste generation by society also plays a significant role (Hamano and Osman 2012).

2.2.3 Urbanization and Infrastructure Development

In addition to being a crucial phase of economic development, urbanization is an essential decision society takes in response to climate change. To achieve the goals of climate-resilient cities and sustainable human development, we can promote sustainable social ideas such as consumption moderation, the creation of green communities, the development of a low-carbon and innovation-driven economy, and raise environmental awareness to bridge the gap between urban and rural areas, and creative development and urban planning. We can also prioritize urban planning and encourage the integration of ecology and urbanism. China's approach to urban planning in the face of climate change involves the development of low-carbon urban planning frameworks methodically and hierarchically to study mitigation and adaptation technologies. Megacities (Yang et al. 2024) and urban agglomerations should achieve harmonious coexistence between nature and people and sustainable development of the regional economy through urban planning and construction, considering the increasingly apparent contradiction between climate change, human activities and the environment, ecology, and resource-carrying capacity. It is necessary to assess the impact of urban planning implementation on climate and environment, strengthen the construction of a green and liable environment with scientific evidence, regulate the relationship between urban and ecological environmental developments, and optimize social and economic benefits (Hao et al. 2016).

Second, we need to improve innovation capacity. The creation of zero-carbon energy and the increase in energy efficiency rely on technological innovation. Therefore, the problems of emissions and climate change can be solved through technological advances. Third, there is a need to focus on achieving balanced development between rural and urban areas and reducing income disparities between them. The government can implement preferential policies to support the growth of urban enterprises, encourage farmers to work in urban areas, and encourage the establishment of small towns. Finally, we need to make energy-saving policies more popular. The state should actively promote energy saving by increasing public awareness and education, integrating energy-saving knowledge into the national education and training system, popularizing scientific knowledge on energy saving, and raising public awareness of the urgency of environmental protection and energy saving, to encourage public opinion to prioritize resource conservation, slow down global warming, and attach great importance to energy conservation and emission reduction (Percival et al. 2021) (Fig. 2.1).

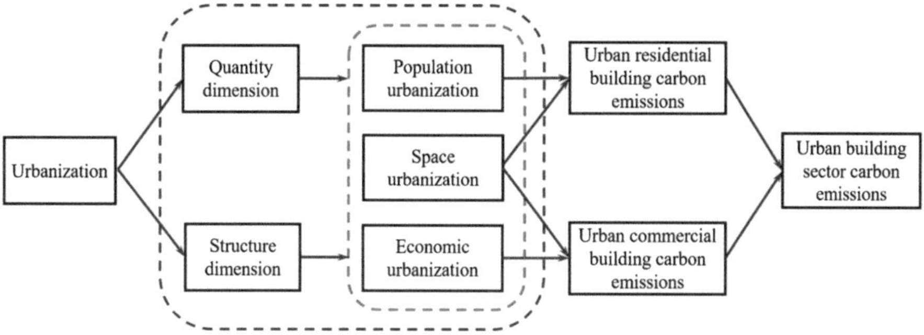

Fig. 2.1 The influencing mechanism of urbanization on carbon emissions in the urban building sector

2.2.4 Carbon Emission

The growth of carbon dioxide and the decrease in the ozone in the atmosphere are clear indicators of humanity's impact on the environment. Carbon dioxide is produced by burning coal, oil, and wood, as well as by volcanic eruptions and the respiration of plants, animals, and microbes. Carbon dioxide has been rising in the atmosphere exponentially for the last 200 years, increasing by 25% over this time. However (Baes et al. 1977), the natural process of releasing carbon dioxide into the atmosphere remained balanced until humans began burning fossil fuels and changing the rate of carbon dioxide production. The carbon dioxide that plants consume during their lives is stored in the soil and then converted into underground deposits of coal, gas, and oil. "Microscopic photosynthetic plants use carbon dioxide absorbed from the oceans to produce carbon-containing compounds that make up their cells" (Archer et al. 2009). After that, the carbon is buried at the bottom of the ocean. A natural balance of carbon dioxide emission and absorption is formed by plants absorbing this excess carbon dioxide through photosynthesis and storing it in the soil. Meanwhile, volcanoes have been highly active throughout history, releasing carbon dioxide into the atmosphere over time where extracts from natural sources produce carbon dioxide (Goeppert et al. 2012).

2.2.5 Transportation

In numerous European towns, vehicular traffic is the primary cause of localized air pollution, which can harm the ecosystem (Boston 2010). Internal combustion engines account for over 60% of all air pollutants in metropolitan areas (Stankovic 2012). One of the main issues in big cities these days is air pollution from motor vehicles' fuel combustion. Pollutants are released into the atmosphere depending on the fuel's composition, flammability, and evaporability. Furthermore, the amount of traffic, the accessibility of the roads, and the weather all impact the toxins that cars emit. State traffic bodies worldwide prioritize reducing the detrimental effects of transport on the environment as well as regular traffic jams or congestion in global metropolises (Selmic and Macura 2013). Several towns adopt tactics to promote public transportation and other non-motorized forms of transportation while simultaneously discouraging the use of cars in most metropolitan areas (Ivanović et al. 2013). The number of dangerous components that automobiles emit into the air has been significantly reduced by the European Union's enforcement of exhaust gas emission regulations

Regretfully, things are not the same in Serbia. Based on some evaluations, the fleet's average age is approximately 15 years, with only 10% of cars being older than 10 years. The fact that the percentage of newly registered automobiles barely makes up about 1.5% of all annually registered vehicles is very worrisome (Opsenica 2011). Large polluters, such as hazardous landfills in urban areas and industrial parks, are the focus of most studies on the inequality in environmental quality (Perlin et al. 1999). In contrast, the tiny polluters that the people of the City of Niš deal with daily are the main subject of this study. Specifically, the focus is on air pollution and vehicle frequency, two contaminants from road traffic. Automobile traffic is the main cause of pollution in the City of Niš's boundaries. The fact that most vehicles in use in Serbia produce enormous amounts of pollution and are often older models, many of which have been in service for more than 20 years, presents another issue. Due to capacity and condition constraints, there is currently no way to decrease the usage of passenger cars for commuting within towns and enhance the use of public transport. This study examined the impact of population expansion on carbon dioxide emissions in various regions. The effect of traffic on environmental quality was then examined. The technology employed in this study was the neuro-fuzzy technique, which is very resistant to errors in experimental data and capable of handling highly nonlinear data (Jang 1993).

2.2.6 Migration

Migration has caused and been a remedy for overcrowding ever since the development of modern humans. Mass migration began after Columbus discovered America, and in the first 4 years of the country's freedom, 250,000 people arrived. The Naturalization Act of 1790 mandated a two-year waiting period to obtain citizenship. Though migration was postponed after the Revolutionary, French, and Napoleonic Wars, immigrants were more welcomed than ever. "The bosom of America is open to accept not only the Opulent and respectable Stranger but also the afflicted and persecuted of all Nations and Religions," according to George Washington. Immigration began to rise once hostilities stopped; in the 1820s, "only 150,000 arrived, but the following decade it grew to 600,000." Twenty million people were already living there when 3 million immigrants arrived two decades later, creating one of the largest immigration waves in history. A large influx of European immigrants arrived between the end of the Civil War and the start of World War I. The first arrived from northern Europe, specifically Scandinavia, Germany, and Britain. Later, people from Italy and Imperial Russia, who were from southern

Europe, settled in America between 1890 and 1914. Besides, 300,000 Chinese immigrants arrived in 1848 to participate in the California Gold Rush, but only five million Italian immigrants entered during the same years (Lee 2011).

They were hired to build the transcontinental railway. Still, because white Americans felt threatened by their impending citizenship, the United States decided to impose a ten-year immigration ban on them in 1882 by passing the Chinese Exclusion Act. However, Asian immigration continued unabated, with hundreds of Japanese entering the country through San Francisco each month, increasing anti-immigrant sentiment among Americans. The Japanese emperor and President Theodore Roosevelt established the Gentleman's Agreement in 1907 to avoid granting laborers passports because Whites feared the Japanese would steal their employment. America prioritized patriotism following World War I, enacting the Johnson-Reed Act in 1924 to limit immigration to 150,000 per year and set nationality quotas based on percentages of the country's population. Fearing Nazism and, subsequently, Communist regimes, refugees fled west to the United States at the start and finish of World War II. Only 214,000 immigrants were permitted under the Refugee Relief Act of 1953, compared to 400,000 under the Displacement Persons Act of 1948. Following the 1956 revolution in Hungary, an additional 200,000 immigrants, including 30,000 more refugees, arrived. Meanwhile (Cardoso 1980), immigrants from South American nations were brought to the United States in the 1940s by the government to work in railway construction and agriculture. Most immigrants were from Puerto Rico, Jamaica, the Caribbean, and Mexico. International immigration to the United States increased significantly during the 1950s and 1980s. Under the pretext of the Civil Rights Movement, a lack of trained labor, and its capacity to serve as a shelter for refugees, the United States continued to admit immigrants to project an image of a free world in opposition to a Communist one. Around 800,000 Vietnamese immigrants arrived in the United States in 1975. The Refugee Act of 1980 was passed by Congress, which eliminated the parole authority, set a quota of 50,000, and expanded the criteria to encompass more people than just those fleeing Communism (Martin 1982). The 1980s' illegal immigration crisis, which has persisted unresolved to this day, was brought on by decades of continuous immigration. More than 25% of Miami, New York, and Los Angeles inhabitants were projected to be immigrants. Their presence in the United States had a detrimental impact on population growth. Using statistics from the Census Bureau, Leon Bouvier estimated that the United States population in 2000 would have been 232 million if immigration had not occurred after 1950, a difference of 43 million from the actual total of 275 million. The 1960s saw a 34% increase in immigrant women's childbearing rates compared to native-born American citizens, which led to a 61% increase in population because of immigration. Currently, 12% of the population of the United States consists of 34 million legal immigrants. According to Davis (2007), of them, 43% are Hispanic, 26% are White, 25% are Asian and Pacific Islanders, and 7% are Black chosen in the structural dimension to represent the process of spatial urbanization. The area of urban buildings (i.e., the sum of the floor area of urban residential buildings and the sum of the area of urban public buildings) is chosen to measure spatial urbanization in the quantitative dimension. The theoretical model shows how urbanization affects carbon emissions from urban buildings (Huo et al., 2020).

2.3 Indirect Impacts of Population Growth

2.3.1 Agricultural Sustainable Intensification

Farmers were compelled to embrace sustainable intensification due to the scarcity of land resources due to population pressure (Kassie et al., 2013). Food consumption rose because of overpopulation, and the rivalry for available land from other human endeavors like urbanization made it more difficult to use newly acquired land for farming (Ndiritu et al. 2014). As a result, excessive use of land resources is made to grow food to suit human requirements. A significant transition from sustainable agriculture to sustainable intensification is driven by the fall in productivity brought on by soil erosion, soil pollution, land degradation, and reduced soil carbon sequestration capacity (David 2019). In addition to influencing household-level farming decisions, population pressure may endanger the nation's food security. For instance, when a family's size grows, families with less land are more likely to implement sustainable intensification techniques on that land, such as conservation tillage, enhanced varieties, legume intercropping, and fertilization (Kassie et al., 2013). Furthermore, the susceptibility of small farmers is emphasized by their capacity to withstand hazards. Low crop yields brought on by climate change directly impact per capita income. Research conducted in Southern Mali by Falconnier et al. (2017) revealed that to raise the poverty line for all farmers, policies and interventions for sustainable intensification must be integrated with farmers' practices (Falconnier et al. 2018).

2.3.1.1 Greenhouse Gas Emissions

Numerous factors, including urban shape and spatial organization, can affect a city's emissions of greenhouse gases. The dense populations and economic activity found in urban areas and regions can result in "economies" of scale, closeness, and aggregation, which can be advantageous in influencing energy consumption and related emissions. However, the closeness of residences and commercial buildings can promote the use of public transport in place of private automobiles, as well as walking and cycling (Satterthwaite 2008). According to some research, for every doubling of the average density of a neighborhood, average car use drops by 20–40%. A proportional decrease in emissions accompanies this decrease. While Newman and Kenworthy (1989) argue that "by midcentury, the combination of green buildings and smart growth could produce the deepest reductions that many see as necessary to mitigate climate change," a negative correlation exists between per capita gasoline consumption and urban density. However, Brown and Southworth (2008) argue that "by midcentury, the combination of green buildings and smart growth could produce the deepest reductions many see necessary to mitigate climate change."

However, cities are often considered responsible for producing most of the world's greenhouse gas emissions and a disproportionate share of their contribution to climate change. Regarding climate change, the executive director of UN-Habitat has stated that cities are responsible for "80% of greenhouse gas emissions and 75% of global energy consumption"; the Clinton Foundation estimates that cities contribute "about 75% of all greenhouse gas emissions that trap heat in our atmosphere, even though they represent only 2% of the Earth's mass" (Brown and Southworth 2008). However, extensive studies of urban greenhouse gas emissions for specific cities show that, on average, city dwellers produce significantly fewer carbon emissions per person than people living in other regions of the same country (Satterthwaite 2008).

2.3.2 Industrialization

In 1997, the World Bank made the phasing out of lead in gasoline a top priority among its ten goals for improving health and the environment. Exposure to lead has been increasing in urban environments for decades; lead comes from industrial emissions, household paints, and lead gasoline. Several high-income countries, including the United States and Australia, have recently established new, less stringent environmental exposure standards to protect children. However, childhood lead poisoning, a particular risk to children's neurocognitive development, is a growing problem in many low-income countries, especially in urban settings (Naranjo et al., 2020). High blood lead concentrations have been observed in Bangkok, Jakarta, Taipei, Santiago, and Mexico City. In Dhaka, Bangladesh, the lead concentration in the air is one of the highest in the world, and the average blood lead concentration of 93 randomly selected rickshaw drivers was 53 mg/dL, 5 times the acceptable limit in high-income countries. The lead content of gasoline sold in Africa is the highest globally and is associated with high lead concentrations in the air, dust, and soil. Many other African exposures come from industrial, artisanal, and household sources (Tong et al. 2000).

According to recent surveys, over 90% of children in the Cape Province of South Africa have blood lead concentrations above 10 mg/dL. The best estimate of low-dose lead neurotoxicity during childhood comes from cohort studies in industrialized urban populations. These studies show that preschool children whose blood lead concentrations are in the highest and lowest quintiles, and therefore differ by about 10 mg/dL, have a consistent difference of 2–3% on intelligence measures. Therefore, lead-induced intellectual deficits in children are widespread in cities in developing countries with persistent environmental levels of lead exposure (Koller et al. 2004).

2.3.2.1 Degradation of the Environment

Population growth, which is expected to continue into the next century, threatens the natural environment, even though overall population growth rates are expected to decline. Care must be taken. A long-term perspective is important, as historians have shown. The epidemiological and agricultural transitions that contributed to higher survival rates are known to be reversible. However, Campbell (2007) challenges us when he notes that the recent literature in many disciplines is surprisingly silent on population growth. This inequality has already led to human suffering here. It has created forces that drive people to migrate within their own countries in search of food, water, and often safety. Transnational and trans-regional migrations are common. While the lure of better opportunities elsewhere is a major motivator, above-average rates of population growth in already resource-poor regions of the world are causing forces that are further pushing populations away from their countries of origin, putting pressure on regions where resources are more abundant and/or where Indigenous populations are declining. Developed countries are not immune to the consequences of their behavior on the natural environment. For at least half a century, the trend toward higher consumption has been a major factor in environmental degradation. The capacity to have more creates stress on the natural environment not only in the developed world but also in developing countries and regions, which are often important suppliers of specific natural resources (Venables 2016).

2.3.3 Energy Consumption

The energy demand is expected to double over the next few decades, with wealthy countries using twice as much energy as their population increases. When there is high per capita energy consumption, overall energy demand will rise on its own, negating the need for population expansion to be particularly large to justify excessive energy use. "The 75 million additional people expected to join the US population by 2050, for instance, will increase energy demand to approximately the current energy consumption of Africa and Latin America" (Brown and Wildenthal 1988). Since oil is the single fuel that dominates the world, price levels will be impacted by the prediction that global oil production will peak again by 2025, as in 1979. The most economically active nations, like Asia, will be the source of climate change elements shortly. Population growth will only be 50%, but consumption is predicted to increase by 361%. Additionally, increases in energy consumption are anticipated in Africa (by 326%) and Latin America (by 340%) (Hegab et al. 2017).

2.3.3.1 Global Warming

The nation's high population and rapidly rising energy consumption contribute significantly to global warming. Significant physical, environmental, and social effects of global warming can be beneficial and detrimental. These implications are difficult to estimate and fraught with uncertainty. Sea level rise, soil moisture, water availability, ocean circulation, and marine systems would all alter due to climate change (Miller 2010). These would affect forestry, agriculture, and natural ecologies such as fisheries and wetlands. The worldwide population would be more susceptible to health issues due to rising temperatures, which would also cause an increase in heat stress and a change in the patterns of vector-borne diseases. This would disrupt settlement patterns and lead to mass migrations, which would have significant socioeconomic consequences (Compendium of Environmental Statistics 2000).

2.4 Regional Case Studies

2.4.1 Comparative Analyses of Developing Countries

2.4.1.1 China

Although China has had a successful family planning policy, it was one of the most extreme approaches seen in human history. In 1978, Chinese scientist Jian Song urged the Chinese government to support a reduction in population growth by the year 2000. Otherwise, women would have three children, and China's population would reach four billion by 2080 (Baird and Fugelsang 2004).

"If China does not reduce its birth rate to 1.5 or even one child per woman, the depletion of resources would be catastrophic." Thus, there was strict government control, and women had to undergo monthly gynecological examinations. Women who chose abortion received 14 days of paid leave and "40 days if the abortion occurred in the second trimester and was quickly followed by sterilization" (Baird and Fugelsang 2004). In addition, parents with one child had priority in housing, better health care, and educational opportunities, while families with a second child had to pay for these benefits. "Those with more than two saw their wages cut by 10% over 14 years." They also risked losing their jobs. In rural areas, children needed to be born male. The one-child policy led parents to kill the new girl or to want the second child to be a boy, but before this somewhat "lenient" policy of 1983, a decade ago, women were forced to abort, regardless of the stage of pregnancy. China made 7.9 million abortions, 13.5 million intrauterine devices (IUDs), and about 7 million sterilizations. The situation worsened in 1983, when "more than 16 million women and more than 4 million men were sterilized, almost 18 million women were fitted with an IUD, and 14 million had an abortion." As medicine and technology allowed ultrasound, abortions could be performed on female fetuses, increasing the abortion rate among boys from 6% to 17% in 1995 (Sullivan 2006).

An interesting but strange fact is that Chinese citizens did not receive protection and support from international agencies. Even as reports of these abuses began to reach the West, family planning agencies such as the United Nations Population Fund (UNFPA) continued to fund the Chinese program with generous assistance from Japan. The International Planned Parenthood Foundation (IPPF) was aware of the abuses reported by the BBC. Still, it failed to take action to prevent the abortions of eight-month pregnant women and the suicides of the women. International institutions have shamefully failed to acknowledge the lack of respect for human rights. Instead, they have financially supported mass coercion while, ironically, the Catholic

Church and Protestant groups have called for morality and justice. "UNFPA, IPPF, and USAID have been targeted, with some success; aid grants have been cut, and department heads have lost their jobs." Figure 2.3 below shows China's population before and after the one-child policy (Baird and Fugelsang 2004).

2.4.1.2 India

In 2016, India's population was 1.34 billion, making it the second most populous country in the world after China (Indiaonlinepages.com) (Baus 2017). Despite India's improved agriculture and agro-industry, population growth has reduced India's chances of achieving an acceptable standard of living. During the Independence movement, Mahatma Gandhi drew attention to population growth and focused on abstinence as a solution. However, the government was the first in the world in the 1920s to implement a family planning program that included first birth control and then sterilization. By 1956, seven thousand people had been sterilized, and a strong advocacy campaign and incentives were aimed at encouraging couples to limit their number to two children per family (Baus 2017). When this policy proved ineffective, the government ordered the sterilization of women with more than three children but soon extended sterilization to the homeless and beggars. "Up to 8 million vasectomies were performed in India in 1976, most of them under dubious circumstances" (Aaseng 2010). The population strongly opposed the entire family planning movement, and when Indira Gandhi returned to power in 1980, her approach to population control was different from before, and she remained silent on controlled births. In addition to the lack of attention, the different religions in India have played a crucial role in the population growth. "About 85% of the country's population is Hindu and 10% is Muslim. Added to this is the small (14 million out of 820 million) but very open Sikh population" (Aaseng 2010). The Muslim religion demands reproduction at a high and rapid rate, so government policies seem to target only this segment of the population. Muslim men have been forced to leave their homes and have undergone vasectomies against their will. Another problem hindering population growth is poverty and illiteracy. Because of the high cost of contraceptives, couples would rather forgo contraception than be sterilized. Recognizing this fear, the government offered to pay for sterilization. Still, only the poor were attracted to the idea, as were illiterate women who did not understand the benefits of family planning groups. Rumors about contraceptives and family planning efforts caused misunderstandings and affected the population. "Nearly forty years after the efforts began, almost two-thirds of India's 120 million married women are not using any method of contraception." With a poor economy and low standard of living, the population is growing by "15 million Indians every year," almost allowing India to surpass China as the most populous country (Baus 2017).

2.4.1.3 United States

Aiming to improve agriculture and health in foreign countries, the Rockefeller Foundation failed to realize that its actions abroad contributed to a global problem. To prevent further growth, the National Academy of Sciences was asked to create a

Population Council with experts in agriculture, health, nutrition, and demography (Davis and Howard 2005). In 1946, the UN created a division focused on population and collected statistics. "In 1952, it projected a population of 3.6 billion by 1980, which seemed high. In fact, the real figure was 4.5 billion." After the First World Population Conference, the United Nations proposed solutions to control population, but "the combined opposition of Catholic and Communist countries rejected their advice, and the division of the population was reduced to branches." According to Catholic Leon (2004), religion has always been an obstacle to the effective implementation of birth control. Under President Eisenhower, the administration was called immoral because it wanted to invest money in the planning and implementation of birth control. The president was concerned about the issue but chose not to speak to the Church because his second term was approaching. During Kennedy's term, research into birth control began, and the Agency for International Development (AID) was tasked with working with foreign governments and private organizations on the issue of population growth. As part of its assistance to private organizations such as the Ford Foundation, Planned Parenthood International, the Population Council, and other universities, AID provided $2 million in grant; however, the money was never used for contraceptives. AID refused to buy condoms, which the Indian government requested, but it did buy jeeps so that local instructors could travel to rural villages to give lessons. During the famine in India, the United States sent grain to both governments, who agreed to start a birth control program. As a Roman Catholic, Kennedy did not show full support for contraceptives until his successor, Nixon, emphasized the importance of development aid programs and recognized population growth as a global problem that could no longer be ignored. By the late 1960s, "an informed public began to make a connection between population and environmental protection." In 1968, AID finally received $35 million in funding from Congress through the Foreign Assistance Act, which allowed it to work with governments, private organizations, United Nations agencies, and universities. During this decade, discussions about population control were intense, but culminated in the legalization of contraceptives. In the following decade, a second World Population Conference was organized by the United Nations in Bucharest, where the United States invited developing countries to decide on a common goal of sustainable growth. Although criticized for its desire to take control of other countries, the United States cut all funding for abortion and doubled investment in international population programs during President Carter's administration. At the Third Mexico Conference, held in 1984, President Reagan's administration declared that population growth was not a development problem and announced that it would no longer support any private agency or UN organization that supported abortion. A year later, the United States no longer wanted to fund abortions in China, so it withdrew $10 million from the United Nations Population Fund. Support for nativism came from anti-abortion groups such as Human Life International and the Pro-Life Action League. In 1986, Planned Parenthood International lost $20 million in government funding because the United States withdrew support for the program (Brenner 1985). During the administration of President George H. W. Bush, anti-abortion policies continued, preventing proper population control. In 1993, President

Bill Clinton legalized abortion, but by the turn of the century, the global focus had shifted to women's rights and needs. The International Conference on Population and Development (ICPD) was held in Cairo in 1994 and brought together 180 countries to discuss the health implications, gender relations, and social justice. The Cairo Program of Action emphasized the importance of non-governmental organizations and addressed the need to define population issues regarding reproductive health. The program managed to ignore the "neo-Malthusian" vision and instead called for "gender equality and affirmed women's rights to bodily integrity, informed consent, and free sexual intercourse." Neo-Malthusians, aware of population growth, criticized the program, arguing that there were insufficient resources to ignore the problem and that underdeveloped countries might lack a medical approach, as well as the idea that "reproductive health is a broad and nebulous concept." This was the last conference on world population. The Bush administration did not intend to support the United Nations, nor did the organization find support among its other members (Bush 1992).

2.4.1.4 Pakistan

Like most developing countries, Pakistan faces serious environmental problems. Rapid population growth and impressive GDP growth have put enormous pressure on the country's natural resource base and have significantly increased pollution levels. The rapid expansion of industrial production and urbanization has led to increasing levels of industrial waste, water pollution, solid waste, and vehicle emissions, which have caused serious health problems in many parts of the country (Brandon 1995)—background Paper on the Pakistan Report 2010.

An attempt has been made to estimate the environmental costs in Pakistan, and the environmental damage has been estimated at between $1 billion and $2 billion per year, or 2.6–5.0% of GDP in 1992. There is a close relationship between rapid population growth and poverty. If employment opportunities stagnate or deteriorate, continued population growth will increase relative and absolute poverty. The most recent data on poverty incidence show that poverty, which had declined in the 1970s and 1980s, increased in the 1990s, negatively affecting the demand of poor households for education, health, and housing conditions. The issue of poverty in Pakistan is important for sustainable development. Long-term development is not possible without protecting the rights of vulnerable groups and without the participation of the entire population in the development process. The incidence of poverty increased in the 1990s, mainly due to disappointing economic growth. Growth has always contributed to poverty reduction. The dominant effect of growth has been poverty reduction (Bhatti et al. 1999). Poverty can be combated if the economy achieves respectable economic growth of more than 6%. Various policy options can be used to reduce the incidence of poverty. Policies have been adopted to reduce concentrated control of assets and unequal access to education and income opportunities. Rapid economic and population growth are incompatible, so this growth must be reduced.

2.5 Mitigation Strategies and Managerial Policies

However, the criteria of importance, criticality, and sustainable use must be assessed differently. De Groot's four categories of environmental functions relate to very different aspects of the natural capital that provides them. Furthermore, each of the criteria must be interpreted to reflect the inherently dynamic nature of ecosystems for regulatory functions (e.g., maintaining ecosystem resilience, recycling waste, preventing erosion, maintaining air quality); criteria such as maximum carrying capacity, conservation of biodiversity, and the integrity of essential life-support processes are included; a geographical component (minimum size of the critical ecosystem) is included for habitat purposes (species conservation). For production functions (resource extraction), the maximum sustainable level of yield is a decisive criterion. The social sciences are more influential and are the source of criteria for information functions (perception of value landscapes, cultural and historical values, etc.). When every function's contribution to human welfare is fully understood, its significance can be assessed in these terms, and the functions that are thus determined to be highly significant are linked to the specific environmental capital stocks that are accountable for the human. It is, unfortunately, very difficult to measure the functions' contribution to human welfare since there is so much uncertainty regarding which are crucial and why, particularly when it comes to the habitat and regulating functions that are thought to support life processes. While monetary valuation methodologies can capture some environmental values, these methods and the figures they yield are controversial and rife with interpretive issues (Kumar 2012).

2.6 Technological Improvements

Instead of employing such methods, it would appear better to designate any environmental functions as "important" or critical (and thus necessary for environmental sustainability). There should only be three of these functions: (1) their loss would be permanent; (2) their loss would truly involve "immoderate losses"; and (3) they cannot be substituted in terms of welfare generation by any other function, whether environmental or not. It has long been recognized that a key component of environmental policy is the combination of uncertainty, irreversibility, and the possibility of high costs or excessive losses (Mallawaarachchi et al., 2020).

2.6.1 Renewable Energy Resources

Ciriacy-Wantrup (1952) developed the concept of "the safe minimum standard," which foreshadowed many of the contemporary problems around sustainability. For many resources, particularly renewable ones, Ciriacy-Wantrup (1952) first identifies the existence of "critical zones," which are defined as "a more or less clearly defined range of rates (of the flow of the resource) below which a decrease in flow cannot be reversed economically under currently foreseeable conditions." Such irreversibility is frequently technological as well as economic (Ciriacy-Wantrup, 1952). One may also add that it is biological in relation to extinct species. This suggests that the loss of environmental functions might be irreversible, according to the language used here.

2.6.2 Waste Management

As part of the human environment, wastes and byproducts of human activities—including domestic/household, agricultural, construction, manufacturing, commercial, institutional, and retail activities—have grown to be a major concern in many urban settlements, especially in West Africa (Akpan and Olukanni 2020). Noor et al. (2020) surveyed hazardous wastes, including hospital, electronic, and/or radioactive ones. According to Akpan and Olukanmi, as of 2020, Nigeria (2469 kt annually), Benin (428 kt annually), and Ghana (419 kt annually) are the top three countries in West Africa for the generation of hazardous waste. Additionally, it demonstrates that Benin had a greater average generation than other nations in the region (65 kg/person/year). The Republic of Benin's high level of cross-border trade is likely the reason why its citizens produce more hazardous waste (65 kg/person/year) than the typical person in the region (20 kg/person/year). In many of the region's nations, wastes are predicted to rise by 40% by 2050 because of the failure or lack of current waste management protocols and frameworks (Khan et al. 2022). Because less than half of the solid waste created is collected, Simelane and Mohee defined West African metropolitan settings as having inefficient garbage collection, management, disposal, and reuse that detracts from their appeal. There are not many properly managed landfills in the entire area, according to a review by Akpan and Olukanni. Urban municipal waste is another type of waste that has dominated the area. These wastes are typically defined by having more than 50% organic components and 30% recyclable or reusable materials (such as mixed wood, garden waste, leather, and rubber), as opposed to the average for European countries, which is 30% organic and 51% recyclable (primarily glass and metals) (Khan et al. 2022).

2.6.3 Sustainable Agricultural Practices

From the conceptual and historical perspective of sustainable agricultural intensification and ecological intensification, ecological intensification is more focused on exploiting ecological processes, ecosystem services, and resource use efficiency. On the other hand, since the emphasis is mainly on the optimal management of inputs and outputs in the production process, resource use efficiency is less expressed in terms of sustainable intensification. However, despite the different formulations, the definitions of ecological intensification and sustainable agricultural intensification overlap considerably in key elements such as increasing production and minimizing environmental impacts (Stakman et al. 1967). The general framework of sustainable intensification has much more conceptual connotations than ecological intensification. Ecological intensification emphasizes environmental concerns and rational production and consumption. However, ecological intensification pays more attention to ecological principles and environmental sustainability (Tittonell 2014). The conceptual overlap between sustainable agricultural and ecological intensification leads to confusion between the two. In many cases, sustainable intensification represents a relatively broad category that only addresses sustainability concerns to a certain extent, so most current agricultural practices fall into this category (Wezel et al., 2015). However, ecological intensification offers a clearer definition in general. For example, ecological intensification focuses on understanding and intensifying biological and ecological processes and functions in agroecosystems and extends its scope to landscape use and ecosystem services.

2.6.4 Policy Measures

2.6.4.1 Regulations on Gas Emission

The notion of a "critical area" is similar to the concept of "critical load" found in contemporary environmental policy (Ekins (2000), for example, in relation to the reduction of SO_2 emissions stipulated by the Second Sulphur Protocol. The possibility of "immoderate losses" due to environmental degradation was later identified by Ciriacy-Wantrup (1952), who stated that "an important objective of conservation decisions is to avoid possible inappropriate losses - even if only the probability of low * /accept the possibility of moderate losses * /even though the latter are more likely." The "minimax" criterion, which minimizes the maximum number of losses that can occur, is a decision rule that achieves this goal. The "Minimum Safety Standard" (MSS) was proposed by Ciriacy-Wantrup (Chap. 18) as an objective of conservation policy (or what we call environmental policy) when this criteria applies to sources presenting critical zones: where avoiding critical zones or physical conditions caused by human activity that would make it impossible to stop and reverse depletion, which is the minimum level of conservation certainty. The critical zone can be compared to a threshold in complex systems, beyond which an ecosystem can shift to a different stability field. To avoid exceeding the threshold, management must increase buffer capacity or resilience (Scheffer et al. 2001).

2.6.4.2 Zoning Laws

"Achieving sustainability requires considering policies that constrain the day-to-day operations of the economy in ways that improve the natural resources of future generations but bearing in mind the economic implications of implementing these policies," says Bishop et al. (1993), who sets the safe minimum standard (SMS) approach in the context of current environmental discourse. In this case, a standard of durability replaces the minimum safety criterion. Activities with the potential for irreversible consequences and excessive costs are now classified as environmentally unsustainable in previously defined categories. According to the SMS approach, regulations that constrain or direct these activities toward sustainability should be considered within a framework aimed at prevention. The importance of environmental functions for human well-being and the potential irreversibility and high costs associated with their loss support the argument that environmental sustainability should take precedence over other policy goals. However, as already mentioned, not all environmental functions can or should be maintained everywhere. Sustainability policies must focus on the most vital functions for the well-being and maintenance of human life, and some evaluation of these functions is necessary. The importance of the environmental function must determine the importance of its sustainable use. These types of important functions can be called "environmentally critical functions." When the capital actions needed to perform these tasks cannot be replaced by other environmental capitals or capitals that perform the same functions, they can be called "critical natural capitals" (CNC).

2.6.4.3 Family Planning

Some studies have shown that women's education reduces a country's fertility rate. More educated women are more likely to use effective methods of contraception (Stefoff 1993). So, the more educated women are, the fewer children they have. This is related to traditional customs and the opportunities available to women. In addition to education, the average number of children per woman depends on her professional status and gender equality. "Today, the fertility rates of European and North American women are the lowest in the world, with an average of 1.7 children per woman in Europe and 2.0 children per woman in the United States, compared to 3.5 in Latin America, 3.9 in Asia, and 6.1 in Latin America." In the past, Western societies had higher birth rates. Still, due to the implementation of laws giving equal rights to women and men, marriages began to occur later, and women had fewer children. In addition, scientists interested in population control believe that educational and work opportunities provide women with satisfaction, self-confidence, and financial security, and "they feel less pressure to 'prove' their worth to their husbands and families, or society at large." Compared to other Central American countries, Costa Rica has greatly developed its education system, which has put its fertility rate at 3.3 while other countries have struggled at 4.1.

2.7 Future Challenges and Recommendations

Distinguishing between vital and non-vital environmental functions is extremely difficult given the current unclear level of knowledge about ecosystems. Since it is unclear how natural systems function when they malfunction, all natural capital's regulatory "functions" are likely crucial. However, recent research indicates that environmental thresholds exist, and irreversible changes occur when resilience is lost (Holling 1994; Holling and Meffe 1996; Kates and Clark 1996; Scheffer et al. 2001). A given habitat may have an ecological surplus, even a significant surplus because not all species living there are essential to the functioning of the habitat. However, it is unclear ex-ante which species are or may be redundant. Therefore, science suggests great caution in categorizing environmental functions (by extent and elements).

The main question in solving the overcrowding problem is how to deal with the problem outside its boundaries. Environmental activists and population experts argue that global governance is key to enforcement measures. Still, the global network cannot enforce laws and exert social pressure equally across nations. To make global demographic changes visible, international institutions must set targets that national institutions must achieve within their borders. This seems impossible because some countries try reducing their carbon emissions while others use natural resources to obtain material benefits. However, if nations bind themselves together to ratify treaties for the benefit of the global community, then the world has already taken a step forward in controlling environmental degradation. Treaties have the power to impose new environmental efforts. They are likely to guide countries toward effective population control policies that focus on national shortcomings, such as the lack of education of young women.

2.8 Conclusion

This chapter has given us a better understanding of the growing problem of overpopulation. Although scientists cannot predict the future world, they argue that the devastating effects of population growth cannot be ignored. Disease, climate change, food and water shortages, unequal distribution, lack of family planning, and economic decline are warning signs that national and global institutions must act quickly. Since overpopulation is not the only global threat, it would not be wrong to conclude that the apocalypse is coming faster than ever. However, it is widely believed that technology may be the only factor that can save the planet from destruction. Technological innovations and devices produce a wealth of data that helps predict actions to maintain environmental sustainability and support humanity. However, these innovations have not yet been recognized by governments, so their use, apart from the scientific aspect, seems pointless. To find a solution to population growth, nations rely on each other to contribute to the well-being of the global community, while at the same time individuals within this circle act in their own self-interest. The real solution to overpopulation, or the apocalypse, is answering whether natural resources and financial means can be distributed equitably to maintain sustainability and implement measures for future fertility rates. If humanity capitulates in its fight against overpopulation, this problem could end the world.

References

Aaseng N (2010) Overpopulation: crisis or challenge? Franklin Watts, Inc., New York

Akpan VE, Olukanni DO (2020) Hazardous waste management: an African overview. Recycling 5:15

Archer D, Eby M, Brovkin V, Ridgwell A, Cao L, Mikolajewicz U et al (2009) Atmospheric lifetime of fossil fuel carbon dioxide. Annu Rev Earth Planet Sci 37:117–134

Baes CF, Goeller HE, Olson JS, Rotty RM (1977) Carbon dioxide and climate: the uncontrolled experiment: possibly severe consequences of growing CO2 release from fossil fuels require a

much better understanding of the carbon cycle, climate change, and the resulting impacts on the atmosphere. Am Sci 65(3):310–320
Baird AA, Fugelsang JA (2004) The emergence of consequential thought: evidence from neuroscience. Philos Trans R Soc Lond B Biol Sci 359:1797–1804. https://doi.org/10.1098/rstb.2004.1549
Baus, D. (2017). Overpopulation and the impact on the environment
Bhatti AM, Haque R, Javed T (1999) A sectoral analysis of poverty in Pakistan. Pak Dev Rev 38:859–872
Bishop KH, Lundström US, Giesler R (1993) Appl Geochem. Special Issue No. 2. 11
Boadi K, Kuitunen M, Raheem K, Hanninen K (2005) Urbanisation without development: environmental and health implications in African cities. Environ Dev Sustain 7:465–500
Boston MA (2010) HEI panel on the health effects of traffic-related air pollution, traffic-related air pollution: a critical review of the literature on emissions, exposure, and health effects. Special Report 17, Health Effects Institute
Brandon C (1995) Valuing environmental costs in Pakistan. The economy-wide impact of environmental degradation 1995. Background paper for Pakistan 2010 report. Washington, DC, World Bank, Asia Technical Department
Brenner MT (1985) International family planning. Hous J Int'l L 8:155
Brown MA, Southworth F (2008) Mitigating climate change through green buildings and smart growth. Environ Plan A 40(3):653–675
Brown BA, Wildenthal BH (1988) Status of the nuclear shell model. Ann Rev Nucl Part Sci 38:29–66. https://doi.org/10.1146/annurev.ns.38.120188.000333
Bush G (1992) The United Nations: forging a genuine global community. Address before the UN General Assembly, New York City, pp 721–724
Campbell JL (2007) Why would corporations behave in socially responsible ways? An institutional theory of corporate social responsibility. Acad Manag Rev 32(3):946–967
Cardoso LA (1980) Mexican emigration to the United States 1897–1931: socio-economic patterns. University of Arizona Press, Tucson
Chen M, Liu W, Lu D, Chen H, Ye C (2018) Progress of China's new-type urbanization construction since 2014: a preliminary assessment. Cities 78:180–193. https://doi.org/10.1016/j.cities.2018.02.012
Chitra K, Priya BS (2020) Agricultural pollution. A primer on earth pollution: pollution types and disposal. Bentham Science Publishers, Sharjah, pp 46–58
Ciriacy-Wantrup SV (1952) Resource conservation: economics and policies. University of California Press, Berkeley
Compendium of Environmental Statistics (2000) By Central Statistical Organisation (CSO), Ministry of Statistics and Programme Implementation, Govt. of India. Environmental Statistics – India – Reports
Crist E, Kopnina H, Cafaro P, Gray J, Ripple WJ, Safina C et al (2021) Protecting half the planet and transforming human systems are complementary goals. Front Conserv Sci 2:761292
Cui J, Guo G (2021) Impact of new-type urbanization on industrial land use efficiency based on spatial econometric model: evidences from industrial enterprises in Jiangsu Province. Res Environ Yangtze Basin 30:565–574. https://doi.org/10.11870/cjlyzyyhj202103005
David RM (2019) Farming revolution: let the soil change machine; M. Zhang Ganlin, Trans. Shanghai Science and Technology Press, Shanghai. (In Chinese)
Davis DE (2007) Containing the cost of infant formula to the WIC program. Issue Briefs, South Dakota State University. No. 2007491
Davis AK, Howard E (2005) Spring recolonization rate of monarch butterflies in eastern North America: new estimates from citizen science data. J Lepid Soc 59:1–5
Deshingkar P (2010) Migration, remote rural areas and chronic poverty in India, vol 323. ODI, London
Ehrlich PR, Ehrlich AH, Daily GC (1993) Food security, population and environment. Popul Dev Rev 19:1–32

Ekins P (2000) Economic growth and environmental sustainability: the prospects for green growth. Routledge, London/New York

Ekins P, Simon S, Deutsch L, Folke C, De Groot R (2003) A framework for the practical application of the concepts of critical natural capital and strong sustainability. Ecolog Econ 44(2–3): 165–185. Environ Impact Assess Rev 71:70–83

Estrada A, Garber PA, Rylands AB, Roos C, Fernandez-Duque E, Di Fiore A et al (2017) Impending extinction crisis of the world's primates: why primates matter. Sci Adv 3(1):e1600946

Falconnier GN, Descheemaeker K, Van Mourik TA, Adam M, Sogoba B, Giller KE (2017) Co-learning cycles to support the design of innovative farm systems in southern. Mali Eur J Agron 89:61–74

Falconnier GN, Descheemaeker K, Traore B (2018) Agricultural intensification and policy interventions: exploring plausible futures for smallholder farmers in Southern Mali. Land Use Policy 70:623–634

Forsyth, T., 2003. Critical political ecology: the politics of environmental science

Frame DJ, Stone DA, Stott PA, Allen MR (2006) Alternatives to stabilization. Geophys Res Lett 33:L14707

Ge T, Hao X, Li J. 2021. Effects of public participation on environmental governance

Goeppert A, Czaun M, Prakash GS, Olah GA (2012) Air as the renewable carbon source of the future: an overview of CO 2 capture from the atmosphere. Energy Environ Sci 5(7):7833–7853

Gore, A., 2008. Speech in Poznan. Transcript Available at http://blog.algore.com/

Gu S, Li H, Wu H (2018) Strategic thinking on promoting the development of green urbanization in new era. J Beijing Technol Bus Univ 33:107–116. https://doi.org/10.12085/j.issn.1009-6116.2018.04.011

Hamano A, Osman W (2012) The Bureau of Economic Analysis (BEA) releases estimates of gross domestic product, gross domestic product by industry, compensation by industry, and detailed consumer spending for American Samoa. Bureau of Economic Analysis, U.S. Department of Commerce, BEA 12–37:1–11

Hao Y, Chen H, Zhang Q (2016) Will income inequality affect environmental quality? Analysis based on China's provincial panel data. Ecol Indic 67:533–542

Harter S (2015) The construction of the self: developmental and sociocultural foundations. Guilford Publications, London

Hegab A, La Rocca A, Shayler P (2017) Towards keeping diesel fuel supply and demand in balance: dual-fuelling of diesel engines with natural gas. Renew Sust Energ Rev 70:666–697

Holling CS (1994) Simplifying the complex: the paradigms of ecological function and structure. Futures 26(6):598–609. https://doi.org/10.1016/0016-3287(94)90031-0

Holling CS, Meffe GK (1996) Command and Control and the Pathology of Natural Resource ManagementComando-y-control y la patología del manejo de los recursos naturales. Conserv Biol 10(2):328–337. https://doi.org/10.1046/j.1523-1739.1996.10020328.x

Huo T, Li X, Cai W, Zuo J, Jia F, Wei H (2020) Exploring the impact of urbanization on urban building carbon emissions in China: evidence from a provincial panel data model. Sustain Cities Soc 56:102068

Ivanović I, Grujičić D, Macura D, Jović J, Bojović N (2013) One approach for road transport project selection. Transp Policy 25:22–29

Jang J-SR (1993) ANFIS: adaptive-network-based fuzzy inference systems. IEEE Trans On Syst Man Cybern 23:665–685

Karak T, Bhagat RM, Bhattacharyya P (2012) Municipal solid waste generation, composition, and management: the world scenario. Crit Rev Environ Sci Technol 42(15):1509–1630

Kassie M, Jaleta M, Shiferaw B (2013) Adoption of interrelated sustainable agricultural practices in smallholder systems: Evidence from rural Tanzania. Technol Forecast Soc Chang 80:525–540. [CrossRef] Land 2019: 8, 157 27 of 27

Kates RW, Clark WC (1996) Environmental surprise: expecting the unexpected? Environ Sci Policy Sustain Dev 38(2):6–34

Khan I, Chowdhury S, Techato K (2022) Waste to energy in developing countries—a rapid review: opportunities, challenges, and policies in selected countries of sub-Saharan Africa and South Asia towards sustainability. Sustainability 14:3740

Koller K, Brown T, Spurgeon A, Levy L (2004) Recent developments in low-level lead exposure and intellectual impairment in children. Environ Health Perspect 112(9):987–994

Kumar P (2012) The economics of ecosystems and biodiversity: ecological and economic foundations. Routledge, London

Laczko F, Aghazarm C (2009) Migration, environment and climate change: assessing the evidence. International Organization for Migration (IOM)

Lee JH (2011) Chinese immigrants. Multicultural America [4 volumes]: an encyclopedia of the newest Americans [4 volumes], 277

Leon SM (2004) Beyond birth control: Catholic responses to the eugenics movement in the United States, 1900–1950. University of Minnesota

Levy MA, Cavendar-Bares J, Clark WC (2001) Goal and strategy formulation in the management of global environmental risks. In: Anon (ed) Learning to manage global environmental risks: vol. 2: a functional analysis of social responses to climate change, ozone depletion and acid rain. MIT Press, Cambridge, pp 87–113

Mallawaarachchi T, Auricht C, Loch A, Adamson D, Quiggin J (2020) Water allocation in Australia's Murray–Darling Basin: Managing change under heightened uncertainty. Econ Anal Policy 66:345–369

Martin DA (1982) Refugee Act of 1980: its past and future. Mich YBI Legal Stud 3:91

Maxwell SL, Fuller RA, Brooks TM, Watson JE (2016) Biodiversity: the ravages of guns, nets and bulldozers. Nature 536(7615):143–145

Meyer JY, Butaud JF (2009) The impacts of rats on the endangered native flora of French Polynesia (Pacific Islands): drivers of plant extinction or coup de grâce species? Biol Invasions 11:1569–1585

Meyerson FA, Merino L, Durand J (2007) Migration and environment in the context of globalization. Front Ecol Environ 5(4):182–190

Miller DA (2010) Conserving the environment. Greenhaven Press/Cengage Learning, Farmington Hills/Gale

Mora C, Zapata FA (2013) 17 anthropogenic footprints on biodiversity. In: The balance of nature and human impact. Cambridge University Press, Cambridge, pp 239–257

Myers TC (2014) Understanding climate change as an existential threat: confronting climate denial as a challenge to climate ethics. De Ethica 1(1):53–70

Naranjo VI, Hendricks M, Jones KS (2020) Lead toxicity in children: an unremitting public health problem. Pediatr Neurol 113:51–55

Newman P, Jeffrey RK (1989) Cities and automobile dependence: an international sourcebook, economics, geography. https://api.semanticscholar.org/CorpusID:127450584

Ndiritu SW, Kassie M, Shiferaw B (2014) Are there systematic gender differences in the adoption of sustainable agricultural intensification practices? Evidence from Kenya. Food Policy 49:117–127. [CrossRef]

Noor T, Javid A, Hussain A, Bukhari SM, Ali W, Akmal M, Hussain SM (2020) Types, sources and management of urban wastes. In: Urban ecology. Elsevier, Amsterdam, pp 239–263. [CrossRef]

Omer AM (2008) Energy, environment and sustainable development. Renew Sust Energ Rev 12(9):2265–2300

Opsenica M (2011) Impact of traffic air pollution on safety and health, First Scientific Conference POLITEHNIKA, Proceedings, pp 259–262. http://www.politehnika.edu.rs/politehnika2013/HQCD_Zbornik_radova_Politehnika_2011.pdf

Percival RV, Schroeder CH, Miller AS, Leape JP (2021) Environmental regulation: law, science, and policy [connected EBook with study center]. Aspen Publishing

Perlin SA, Sexton K, Wong DW (1999) An examination of race and poverty for populations living near industrial sources of air pollution. J Expo Anal Environ Epidemiol 9(1):29

Pingali PL (2012) Green revolution: impacts, limits, and the path ahead. Proc Natl Acad Sci USA 109:12302–12308

Poumanyvong P, Kaneko S (2010a) Does urbanization lead to less energy use and lower CO2 emissions? A cross-country analysis. Ecolog Econom 70(2):434–444
Rai J (2019) Understanding poverty-environment relationship from sustainable development perspectives. J Geogr Environ Earth Sci Int 19(1):1–19
Ray S, Ray IA (2011) Impact of population growth on environmental degradation: case of India. J Econom Sustain Develop 2(8):72–77
Rogoff M, Nicholls M, Keyser M, Lafaele P (2009) WTE in American Samoa: developing a plant. Waste Manage World:1–31
Satterthwaite D (2008) Cities' contribution to global warming: notes on the allocation of greenhouse gas emissions. Environ Urban 20(2):539–549
Scheffer M, Carpenter S, Foley JA, Folke C, Walker B (2001) Catastrophic shifts in ecosystems. Nature 413(6856):591–596
Schwarzer D (2013) The euro area crises, shifting power relations and institutional change in the European Union1. Glob. Policy 3:28–41. https://doi.org/10.1111/1758-5899.12013
Selmic M, Macura D (2013) Model for reducing traffic volume: case study of Belgrade, Serbia. J Transp Eng 140(2). https://doi.org/10.1061/(ASCE)TE.1943-5436.0000639
Smith A, Fressoli M, Thomas H (2013) Grassroots innovation movements: contrbutions and challenges. Journal of Cleaner Production (in press)
Sober E (1986) Philosophical problems for environmentalism, pp 173–195
Sorvall V (1971) Overpopulation: how many are too many? Pendulum Press, Inc., West Haven
Stakman E, Bradfield R, Mangelsdorf P (1967) Campaigns against hunger. Harvard University Press, Cambridge, MA
Stankovic S (2012) Model of sustainable air pollution management in urban areas, Doctoral Dissertation. University of Belgrade. Faculty of Organizational Sciences, Serbian
Steffen W, Richardson K, Rockström J, Cornell SE, Fetzer I, Bennett EM, Biggs R, Carpenter SR, Folke C, Gerten D, Heinke J, Mace GM, Persson LM, Ramanathan V, Reyers B, Sörlin S (2015) Planetary boundaries: guiding human development on a changing planet. Science https://doi.org/1259855
Stefoff R (1993) Overpopulation. Chelsea House Publishers, New York
Steinfeld H, Gerber P, Wassenaar TD, Castel V, De Haan C (2006) Livestock's long shadow: environmental issues and options. Food & Agriculture Org
Sullivan PA (2006) Commercial, political, and cultural uses (and abuses?) of diagnostic medical ultrasound: Unforeseen consequences, 1970–2005. University of Nebraska Medical Center. sumption, trade and urbanization in new EU member and candidate countries: a panel data analysis. Econ Model 44(44):97–103
Sun H, Duman Z (2017) The impact of heterogeneous environmental regulation. Sustain Energy Rev 70:666–697
Syed Abdul RK, Ponce P, Zhang Y (2021) Technological innovation and environmental taxes toward a carbon-free economy: an empirical study in the context of Cop-21. J Environ Manag 298:113418. https://doi.org/10.1016/j.jenvman.2021.113418
Tacoli C (2009) Crisis or adaptation? Migration and climate change in a context of high mobility. Environ Urban 21(2):513–525
Tilman D, Fargione J, Wolff B, D'antonio C, Dobson A, Howarth R et al (2001) Forecasting agriculturally driven global environmental change. Science 292(5515):281–284
Tittonell P (2014) Ecological intensification of agriculture-sustainable by nature. Curr Opin Environ Sustain 8:53–61. [CrossRef]
Tong S, Schirnding YEV, Prapamontol T (2000) Environmental lead exposure: a public health problem of global dimensions. Bull World Health Organ 78(9):1068–1077
Urban MC, Bocedi G, Hendry AP, Mihoub JB, Pe'er G, Singer A et al (2016) Improving the forecast for biodiversity under climate change. Science 353(6304):aad8466
Venables AJ (2016) Using natural resources for development: why has it proven so difficult? J Econ Perspect 30(1):161–184

Wezel A, Soboksa G, Mcclelland S (2015) The blurred boundaries of ecological, sustainable, and agroecological intensification: a review. Agron Sustain Dev 35:1283–1295. [CrossRef]

Wiedinmyer C, Yokelson R, Gullett B (2014) Global emissions of trace gases, particulate matter, and hazardous air pollutants from open burning of domestic waste. Environ Sci Technol 46:9523–9530

Yang A, Zhao J, Lin Y, Chen G (2024) Coupling and coordination relationship of economic–social–natural composite ecosystem in Central Yunnan Urban agglomeration. Sustain For 16(7):2758

Types of Fumes, Vapors, and Gases and a Vision to Overcome Them

3

Essam Zaki Mohammed

Abstract

Air pollution and environmental changes are the most important issues for human life. To deal with these issues, one should consider the effects of hazards of fumes, vapors, and gases and how to overcome them. Gases can be classified into several categories based on various characteristics such as their chemical composition, physical properties, application, and so on. Toxicity, asphyxiation, flammability, and health effects are some common hazards of gases. Source control, engineering controls, monitoring, and testing can reduce the hazard of gases. Vapors are gases typically in a gaseous state at room temperature and pressure but can also exist as liquids or solids under different conditions. Water, solvent, fuel, and volatile organic compound (VOC) vapors are examples of vapors. Health risks and environmental impact are common hazards of vapors. Source identification and control are important solutions to overcoming vapor hazards. Fumes generally refer to airborne particles or gases that are generated from various sources, and they often connote being harmful or irritating. Welding, VOC, and exhaust are some types of fumes. Health risks, chemical exposure, asphyxiation, and environmental impact are common hazards of fumes. To overcome these hazards, personal protective equipment (PPE), substitution, and engineering controls are taken into consideration. Overcoming gases, vapors, and fumes hazards contribute to reducing air pollution and improving human health life. Carbon footprint and artificial intelligence applications are the most modern techniques for monitoring and controlling the emissions of gases, vapors, and fumes. The Internet of Things is the most important way to analyze and control air pollution. PM2.5, or particulate matter with a diameter of 2.5 micrometers or

E. Z. Mohammed (✉)
Department of Computer Engineering, Modern Academy for Engineering and Technology, Cairo, Egypt
e-mail: ESSAM.ZAKY@eng.modern-academy.edu.eg

A. A. H. Abdel Latef et al. (eds.), *Bioremediation and Nanotechnology for Climate Change Mitigation*, https://doi.org/10.1007/978-981-96-3069-1_3

smaller, is a key component of air pollution. Fossil fuel emissions account for 65% of global CO_2 emissions and are also the primary cause of the majority of PM2.5-related deaths, highlighting the interconnected relationship between air quality and climate change.

Keywords

Gases · Vapors · Fumes · Carbon footprint · Internet of Things · Climate change

3.1 Introduction

Human health has been impacted by environmental climate change because of the elevated levels of carbon dioxide (CO_2) in the atmosphere. These modifications activate the human mind (Akimoto and Tanimoto 2023). Since human health is impacted by air pollution, environmental emissions and air quality monitoring must be taken into account (Padhy et al. 2024). The primary constituents of air pollution that contribute to climate change include gases, vapors, and fumes. There are numerous classifications used to classify gases, vapors, and fumes. To combat the risks of gases, vapors, and fumes, health risk studies must be taken into account (Khoshakhlagh et al. 2023). The most significant contributor to air pollution is CO_2. The atmospheric control studies are based on the measured CO_2 levels in the atmosphere (https://www.co2.earth/daily-co2).

Table 3.1 keeps track of the atmospheric CO_2 recorded highs and contrasts them with the most recent CO_2 recorded values.

Due to air pollution from vapors, gases, and fumes, airborne particulate matter poses a significant risk for a number of diseases (Jalasto et al. 2024). In order to establish the fact that vapors, gases, fumes, or mineral dusts alone have a higher risk of mortality for cardiovascular, neurological, and respiratory diseases, prior research on airborne exposure has primarily focused on extensive epidemiological studies of environmental exposure and less on occupational exposure. Studies of gases, fumes, and vapors must take into account the quality of the indoor environment, which is

Table 3.1 CO_2 highest daily readings for the last 10 years

Highest year reading			Previous highest year reading		
Year	Date	Reading	Date	Reading	% Changes
2024	26/4/2024	428.59 ppm	28/4/2023	425.01 ppm	0.84
2023	28/4/2023	425.01 ppm	26/4/2022	422.06 ppm	0.7
2022	26/4/2022	422.06 ppm	8/4/2021	421.36 ppm	0.17
2021	8/4/2021	421.36 ppm	1/6/2020	418.56 ppm	0.67
2020	1/6/2020	418.56 ppm	15/5/2019	415.87 ppm	0.65
2019	15/5/2019	415.87 ppm	14/5/2018	412.64 ppm	0.78
2018	14/5/2018	412.64 ppm	26/4/2017	412.87 ppm	−0.06
2017	26/4/2017	412.87 ppm	4/10/2016	409.55 ppm	0.81
2016	4/10/2016	409.55 ppm	14/6/2014	402.78 ppm	1.68
2015	14/6/2014	402.78 ppm	26/5/2013	400.88 ppm	0.47

Table 3.2 Air quality parameters statistics

Year	PM2.5 (μg/m³)	PM10 (μg/m³)	NO_2 (μg/m³)	SO_2 (μg/m³)	CO (ppm)	O_3 (μg/m³)
2020	68.5	109.2	56.7	9.8	2.2	43.6
2021	72.3	116.7	58.9	10.2	2.5	45.1
2022	65.8	105.4	54.3	9.4	2.2	42.7

defined by the temporal and spatial variability of pollutants (Fromme 2023). Wind and precipitation patterns may vary as a result of climate change. The removal and dispersion of PM2.5 from the atmosphere may then be impacted. PM2.5 frequently contains ammonium, nitrates, black carbon, and sulfates. A worldwide analysis of 2023 air quality data is given in the 2023 World Air Quality Report.

Data on PM2.5 air quality from 7812 cities across 134 nations, regions, and territories are included in the study (World Air Quality study, 2023). The average air quality parameters (PM2.5, PM10, NO_2, SO_2, CO, and O_3) observed between 2020 and 2022 are summarized in Table 3.2 (Deshpande 2024). High pollution levels are depicted in this table, with PM2.5 and PM10 continuously beyond advised thresholds.

3.2 Overview of Gases

Based on a number of attributes, including their chemical makeup, physical qualities, intended use, and more, gases can be divided into multiple groups. According to Pierre L. Fauchais et al. (2014), gas temperatures and velocities acquired using various thermal spray methods aid in improving gas risks. Emissions of hazardous air pollutants have a detrimental impact on human health and life (Sapuan et al. 2022).

3.2.1 Types of Gases

The following categories apply to gases.

3.2.1.1 Noble Gases

Helium (He), neon (Ne), argon (Ar), krypton (Kr), xenon (Xe), and radon (Rn) are examples of inert gases. These gases typically have low reactivity and are chemically inert. Despite being inert, noble gases are used in a variety of applications, such as medical imaging (xenon), welding (argon), and lighting (neon signs). Krypton is utilized in lighting and photography, while helium is essential for cryogenics and balloon inflation.

The Most Significant Noble Gases' Statistics

- Helium (He): It is the 71st most prevalent element and is found in the Earth's crust at a concentration of roughly 8 parts per billion. With 23% of the universe's normal matter made up of it, it is the second most plentiful element after hydrogen.
- Neon (Ne): The 80th most prevalent element, it is found in the Earth's crust at 70 parts per trillion (ppt). About 18 ppm (parts per million) of the atmosphere is made up of neon.
- Argon (Ar): With a concentration of 1.2 ppm, Ar is the 56th most prevalent element in the crust of the Earth. By volume, argon makes up around 0.93% of the atmosphere.
- Krypton (Kr): With a concentration of 10 ppt, it is the 81st most prevalent element in the crust of the Earth. The atmospheric concentration of krypton is 1 ppm.
- Xenon (Xe): The 83rd most prevalent element, it is found in the Earth's crust at 2 ppt. About 90 parts per billion of the atmosphere is made up of xenon.
- Radon (Rn): One of the least common elements, it is found in trace amounts (10–9 ppt) in the atmosphere and Earth's crust.

3.2.1.2 Gases in the Atmosphere

The atmosphere of Earth contains various gases. Nitrogen (N_2), oxygen (O_2), and trace amounts of other gases like carbon dioxide (CO_2), methane (CH_4), and noble gases make up the main constituents.

Examples of Atmospheric Gases

Examples of gases found in the atmosphere include:

- Nitrogen (N_2): Makes up around 78.084% of the atmosphere on Earth. Since nitrogen makes up most of the air we breathe, it is vital to life.
- Oxygen (O_2): Makes up about 20.947% of the atmosphere. Oxygen is vital for combustion and respiration.
- Argon (Ar): Present at about 0.934% of the atmosphere. Argon is inert and does not readily combine with other elements.

3.2.1.3 Industrial Gases

Gases used for a variety of functions in different sectors. Among the elements in this category are hydrogen (H_2), carbon dioxide (CO_2), oxygen (O_2), nitrogen (N_2), and argon (Ar). Among their many uses are in food processing, manufacturing, and healthcare. Demand growth, technology developments, and the need to comply with regulations are some of the variables that impact this sector's statistics and trends. This article examines the industrial gas industry, covering both the more general category of "others," as well as particular gases like argon, and the effects of regulatory compliance on the market. The market for industrial argon gases was valued at 117,330.4 million standard cubic feet (SCF) in 2022, which is projected to grow at a compound annual growth rate (CAGR) of 4.4% from 2023 to 2030.

3.2.1.4 Hydrocarbons

These are compounds composed of carbon and hydrogen. Butane (C_4H_{10}), propane (C_3H_8), ethane (C_2H_6), methane (CH_4), and several other alkanes, alkenes, and alkynes are examples. Fossil fuels, which are high in hydrocarbons, are the main source of carbon dioxide (CO_2) emissions, according to greenhouse gas (GHG) emissions. CO_2 accounts for about 65% of all greenhouse gas emissions. Rising temperatures, changed weather patterns, and an increase in the frequency and severity of natural disasters are all consequences of these emissions, which are a major contributor to climate change. The greenhouse effect, which traps heat and boosts global temperatures, is brought on by the increase in CO_2 concentrations in the atmosphere as a result of climate change and hydrocarbon combustion. Wide-ranging effects of this warming trend include melting polar ice caps, increasing sea levels, and altered precipitation patterns that have an impact on biodiversity and agriculture. Although CO_2 emissions are the main focus, it is crucial to remember that hydrocarbons can also indirectly contribute to ozone depletion. When released into the atmosphere, several hydrocarbon compounds have the ability to degrade stratospheric ozone, which exacerbates the problem of climate change.

3.2.1.5 Toxic Gases

Gases that, in certain amounts, are hazardous to people, animals, or the environment are toxic gases. Ammonia (NH_3), chlorine (Cl_2), hydrogen sulfide (H_2S), and carbon monoxide (CO) are a few examples. For instance, fine particulate air pollution alone is thought to be responsible for 100,000 premature deaths in the United States each year, as it contributes to lung cancer, strokes, respiratory infections, and cardiopulmonary disease (School of Public Health and Tropical Medicine, 2024).

3.2.1.6 Greenhouse Gases

Greenhouse gases (GHGs) are substances that trap heat in the Earth's atmosphere, causing the greenhouse effect and global warming. Nitrous oxide (N_2O), methane (CH_4), carbon dioxide (CO_2), and fluorinated gases are a few examples. The Global Warming Potential (GWP) of any greenhouse gas compares how well it retains heat in the atmosphere over a specific time frame, often 100 years.

The GWP makes it possible to compare the effects of various gases on global warming. Per unit of mass emitted, gases with a higher GWP cause more global warming. One major worry regarding climate change is the rising quantities of these greenhouse gases, which are mostly caused by human activity. Addressing global warming and its effects requires an understanding of these gases' atmospheric concentrations and GWPs.

3.2.1.7 Refrigerant Gases

Gases are used in refrigeration and air conditioning systems to help transfer heat. Common refrigerant gases include hydrofluorocarbons (HFCs), hydrochlorofluorocarbons (HCFCs), and chlorofluorocarbons (CFCs), as well as natural refrigerants like ammonia (NH_3) and carbon dioxide (CO_2).

Regulations are being implemented to limit the usage and emissions of these gases due to their significant environmental impact. With high greenhouse gas potentials (i.e., GWPs), refrigerant gases such as R404a, R410a, and R32 trap heat in the atmosphere far more efficiently than carbon dioxide (CO_2). R32 and R410a, for instance, have GWPs of 675 and 2080, respectively, suggesting that their release into the atmosphere can have a substantial impact on global warming. Refrigerant gas-dependent cooling systems account for a large amount of national energy consumption.

For example, in the UK, air conditioners use 20% of all electricity used annually. According to estimates, the energy required to run cooling systems will increase by 30 times globally by 2100, underscoring the growing significance of sustainable and effective cooling solutions. The primary source of the effects of air conditioning and refrigeration systems on stratospheric ozone is the release of ozone-depleting refrigerants.

Global warming is caused by both the release of refrigerants and the emission of greenhouse gases (GHGs) as a result of associated energy use (Calm 2002).

3.2.1.8 Medical Gases

Medical gases include carbon dioxide (CO_2) for anesthesia and medical operations, oxygen (O_2), nitrous oxide (N_2O), and medical air (a combination of oxygen and nitrogen).

Specialty Gases

High-purity gases with particular uses in the semiconductor manufacturing, electronics, and scientific research sectors. Ultra-high purity gases such as argon (Ar), nitrogen (N_2), and helium (He) are examples.

Flammable Gases

Gases that have the capacity to ignite and burn when they come into contact with oxygen or another source of ignition. Examples include methane (CH_4), hydrogen (H_2), propane (C_3H_8), and butane (C_4H_{10}). In 2018, 350,000 premature deaths occurred in the United States alone due to pollution from fossil fuels. This highlights the detrimental health effects of air pollution caused by flammable gases and other pollutants generated during the burning of fossil fuels, according to the Environmental and Energy Study Institute (EESI).

Certain communities are disproportionately impacted by air pollution, especially communities of color and those with lower incomes. For instance, Black and Hispanic Americans are exposed to 56% and 63% more particulate matter pollution, respectively, than their production. This disparity draws attention to the issues with social justice caused by fossil fuel pollution.

3.2.2 Gases and Their Hazards

Depending on their chemical makeup, concentrations, and exposure circumstances, gases can present a range of risks. The following are a few typical risks connected to gases.

3.2.2.1 Toxicity

Inhaling or absorbing a lot of gases might make them harmful. High quantities of toxic gases, including carbon monoxide (CO), hydrogen sulfide (H_2S), ammonia (NH_3), chlorine (Cl_2), and hydrogen cyanide (HCN), can result in poisoning, respiratory irritation, organ damage, or even death (Al-Sarraj et al. 2024).

3.2.2.2 Asphyxiation

In restricted areas, some gases, including helium (He), carbon dioxide (CO_2), and nitrogen (N_2), can displace oxygen, resulting in asphyxiation and an oxygen shortage. If oxygen deprivation persists, it can result in lightheadedness, disorientation, unconsciousness, and even death (2024).

3.2.2.3 Flammability

In the presence of an ignition source, flammable gases, like acetylene (C_2H_2), hydrogen (H_2), propane (C_3H_8), and methane (CH_4), can ignite and burn, resulting in fires or explosions. Under some circumstances, explosive gases, including hydrogen, methane, and other volatile organic compounds (VOCs), can combine with air to generate explosive mixtures.

3.2.2.4 Corrosivity

Materials, machinery, and exposed surfaces may be corroded by certain gases, including hydrogen fluoride (HF), sulfur dioxide (SO_2), and hydrogen chloride (HCl). Corrosive gasses can harm infrastructure, machinery, and metal buildings, posing a risk to public safety and contaminating the environment.

3.2.2.5 Reactivity

Reactive gases, such as fluorine (F_2), ozone (O_3), and chlorine (Cl_2), can combine chemically with other molecules to produce violent reactions or dangerous consequences. Toxic, corrosive, or flammable compounds can be created when reactive gases react with air, water, or other substances.

3.2.2.6 Health Impacts

Acute or long-term health impacts, such as lung damage, neurological diseases, cardiovascular issues, cancer, and respiratory irritation, can result from exposure to specific gases. The chance of acquiring chronic illnesses or other medical issues may also rise with prolonged exposure to low concentrations of dangerous gases.

3.2.2.7 Environmental Impact

A number of gases have the potential to cause ozone depletion, acid rain, air pollution, smog, and climate change. Methane (CH_4), nitrous oxide (N_2O), and carbon dioxide (CO_2) are examples of greenhouse gases that trap heat in the Earth's atmosphere, causing global warming and other climate-related effects.

3.2.2.8 Risk of Release

Unintentional gas leaks or releases from storage tanks, transportation networks, or industrial facilities can seriously endanger public health, safety, and the environment. Natural disasters, terrorist attacks, human error, and equipment failure can all cause gas leaks.

3.2.3 Overcoming the Hazards of Gases

It is crucial to recognize, evaluate, and manage the risks through appropriate handling, storage, transportation, and disposal procedures in order to deal with the hazards related to gases. To guarantee the safe management of gases in diverse contexts, this entails putting in place engineering controls, ventilation systems, personal protective equipment (PPE), emergency response plans, and regulatory compliance measures. Preventive measures, control tactics, and mitigation approaches are frequently used in conjunction to overcome gases, whether the goal is to manage gas emissions, minimize exposure to hazardous gases, or handle gas-related difficulties. The following are some broad strategies for getting around gases.

3.2.3.1 Source Control

To stop or reduce the discharge of gas emissions into the environment, identify and address their sources. This could entail adopting cleaner production methods, modernizing machinery to cut emissions, and putting pollution control systems into practice. Examples of source control include emergency alarms, continuous gas monitoring systems, suitable pipework, and auto shut-off valves. The likelihood of gas dangers can be considerably decreased by using less dangerous equipment, such as swapping out damaged tools or machinery for safer models. Emissions of hazardous gasses can be decreased, for instance, by replacing outdated boilers with more modern, efficient versions that burn cleaner fuel. Installing gas detection alarms can also notify employees when dangerous gasses are present, enabling prompt action and evacuation if required. In order to keep an eye on locations where gas risks are most likely to occur, these alarms should be positioned strategically.

3.2.3.2 Ventilation

To properly dilute and distribute gases, make sure enclosed spaces have enough ventilation. Appropriate ventilation systems can lower the concentration of dangerous gases and help remove contaminants from indoor spaces. General exhaust ventilation (GEV) and local exhaust ventilation (LEV) are the two main ventilation system types that are frequently used. Each has a distinct function and is essential to controlling air quality and reducing the dangers posed by dangerous gasses.

Local Exhaust Ventilation (LEV)

By capturing and eliminating pollutants at their source, LEV systems reduce the possibility of contamination spreading and exposing employees. This focused strategy works especially well in settings like welding, painting, chemical processing, and carpentry where dangerous gasses or particles are produced. Benefits of this kind include regulation and direct source capture.

General Exhaust Ventilation (GEV)

GEV systems, which work on the dilution principle, also contribute to better air quality. This implies that in order to reduce the concentration of pollutants, they mix tainted air with fresh air. GEV may not always be enough to protect worker health in workplaces with high concentrations of dangerous compounds, even though it is beneficial in some situations. The particular requirements of the workplace and the type of dangers present will determine whether to use an LEV or a GEV. LEV systems provide a more focused and efficient solution in circumstances where the production of dangerous gases cannot be completely stopped.

By capturing toxins as close to their source as feasible, they reduce the possibility of dispersion and operator exposure. LEV systems also improve operational efficiency by cutting energy consumption and operating expenses by eliminating the need for substantial air purification throughout the whole facility. GEV systems, on the other hand, work better in settings where the concentration of dangerous materials may be reduced to levels that are acceptable by mixing them with fresh air. However, because of their direct source capture capacity, LEV systems are recommended in situations where dangerous gas concentrations are significant.

3.2.3.3 Personal Protective Equipment (PPE)

To reduce exposure to dangerous gases at work, give employees the proper PPE, such as respirators and protective clothes. The particular risks present and the necessary level of protection should be taken into consideration when choosing PPE. For instance, self-contained breathing apparatuses (SCBAs) give firemen and other responders who handle hazardous materials a sealed supply of breathable air. These devices are made to guard against a variety of gases, such as sulfur dioxide, chlorine, and carbon monoxide. An SCBA's respirator component ensures that the wearer breathes clean air by filtering incoming air to remove pollutants.

3.2.3.4 Engineering Controls

Put engineering controls in place to stop the escape of gases into the environment by capturing or containing them at the source. Scrubbers, exhaust ventilation systems, and containment systems are a few examples (Wiegleb 2023).

3.2.3.5 Substitution

Whenever possible, substitute safer gases for dangerous ones. This could entail reducing the use of hazardous chemicals, converting to cleaner fuels, or implementing other technologies that emit less pollutants.

3.2.3.6 Monitoring and Testing

Keep a close eye on gas emissions and air quality to spot possible problems and evaluate how well control measures are working. Testing for air quality and analyzing gas can assist find areas for improvement and guarantee adherence to regulations.

3.2.3.7 Training and Education

Educate employees, community members, and stakeholders on the dangers of gases, safe handling practices, and emergency response procedures. Knowledge and awareness enable people to take preventative action against gas-related disasters.

3.2.3.8 Regulatory Compliance

Keep up with pertinent laws, rules, and policies pertaining to gas emissions and workplace safety and health. Minimizing environmental effects and safeguarding human health depend on adherence to legislative standards and industry best practices.

3.2.3.9 Emergency Preparedness

To handle gas-related catastrophes like leaks, spills, or releases, create and execute emergency response plans. Create procedures for communication, containment, and evacuation to lessen the possible effects of emergencies.

3.2.3.10 Cooperation and Stakeholder Engagement

Encourage cooperation among stakeholders, such as governmental organizations, business associates, neighborhood associations, and environmental organizations, in order to jointly address issues pertaining to gas. Involving stakeholders in the decision-making process encourages openness, responsibility, and a sense of collective ownership for overcoming challenges.

Individuals, groups, and communities can collaborate to efficiently manage gasses and establish safer, healthier surroundings for everybody by utilizing these tactics and methods.

3.3 Introduction to Vapors

Although they can exist as liquids or solids under some circumstances, vapors are gases that are normally in a gaseous state at ambient temperature and pressure. Vapors can be categorized according to their characteristics and uses, just as gases.

3.3.1 Vapor Types

The following are some typical vapor types.

3.3.1.1 Water Vapor

Water vapor is water in its gaseous state. It is the most prevalent greenhouse gas in the atmosphere and, via the water cycle, is essential to the Earth's climate system.

Mapping Anomalies: Understanding spatial differences in water vapor content is aided by visual representations, such as maps displaying water vapor anomalies. Shades of purple indicate regions with higher-than-normal levels of water vapor, while shades of green indicate regions with lower-than-normal levels.

Temporal Trends: Bar graphs represent deviations from the long-term average in the water vapor anomaly for the whole Arctic region. This graphic tool facilitates monitoring variations in the amount of water vapor over time.

3.3.1.2 Solvent Vapors

Ethanol, acetone, and methanol are examples of volatile liquids whose vapors are utilized as solvents. These vapors are frequently found in cleaning goods, labs, and industrial environments. Because of their volatility and potential for absorption by eating, inhalation, and skin contact, solvent vapors provide serious health and safety issues. The features of the particular solvent, such as its volatility, solubility, and the concentration of vapor in the workplace, all affect the risks connected with solvent vapors.

Here is a thorough analysis of the data and suggestions for controlling these risks. Solvents can cause dermatitis and other skin conditions by cleaning and defatting the skin. Certain solvents can harm organs like the liver, kidneys, heart, blood vessels, bone marrow, and the circulation by penetrating the skin and entering the bloodstream.

Solvent vapor inhalation can be harmful, impacting the respiratory system and perhaps resulting in long-term health problems. It is essential to perform a risk assessment for chemical vapor threats. The health risks of the chemical, exposure pathways, operating and ambient temperatures, flammability, vapor pressure, vapor density, and the number of chemicals being used are all important factors to take into account. Exposure to solvent vapors can be considerably decreased by putting into practice sensible precautions including closed processes, capping solvent reservoirs, gathering hazardous waste in special containers inside fume hoods, and using the right adapters to connect hand pumps to solvent drums.

3.3.1.3 Fuel Vapors

These are the vapors released by fuels such as kerosene, diesel, and gasoline. In addition to being crucial for combustion processes, these vapors are a major contributor to greenhouse gas emissions and air pollution. When gasoline evaporates, fumes are released, making it a highly flammable and poisonous liquid. Air pollution is caused by these vapors as well as by the compounds that are released when gasoline is burned, including particulate matter, nitrogen oxides (NOx), carbon monoxide, and unburned hydrocarbons. Carbon dioxide (CO_2) is a greenhouse gas that is produced when gasoline is burned.

The combustion of one gallon of gasoline without ethanol releases about 19 pounds of CO_2. About 22% of all energy-related CO_2 emissions in the United States in 2022 came from motor gasoline combustion and aircraft. An average passenger car releases roughly 4.6 metric tons of CO_2 per year, according to annual CO_2 emissions. This computation is predicated on an average yearly mileage of approximately 11,500 miles and a fuel economy of roughly 22.2 miles per gallon.

3.3.1.4 Volatile Organic Compounds (VOCs)

Organic chemical vapors that can evaporate at room temperature and pressure are known as vapors. Paints, solvents, cleaning products, and automobile exhaust are just a few of the sources that release volatile organic compounds (VOCs). They can have negative health impacts and add to air pollution. Large amounts of volatile organic compounds (VOCs) are released by industrial processes, such as the production of chemicals, polymers, and synthetic textiles. These emissions may originate from the actual production process or from the materials' transportation and storage.

VOCs are released during combustion in vehicles, especially those powered by gasoline or diesel. One of the main sources of VOCs in outdoor air is exhaust fumes from automobiles, trucks, and aircraft. Despite having a brief atmospheric lifetime, volatile organic compounds (VOCs) contribute significantly to global warming because of their high reactivity and capacity to destroy the ozone layer, which intensifies the greenhouse effect.

3.3.1.5 Mercury Vapor

It is the hazardous heavy metal mercury in gaseous form. Coal combustion, industrial operations, and some consumer goods can all emit mercury vapor. The U.S. Environmental Protection Agency (EPA) states that breathing in mercury vapor can cause major health issues.

Coal-Burning Power Plants: Similar to the United States, the majority of man-made mercury emissions—roughly 44% of all emissions—come from coal-burning power plants. Because coal and other fossil fuels naturally contain mercury, when these fuels are used for energy, the mercury is released into the atmosphere. Mercury has historically been found in a variety of consumer goods, including as thermometers, barometers, manometers, sphygmomanometers, dental amalgams, batteries, and certain cosmetics and medical devices. A further important source of mercury is electronic garbage, or "e-waste." Mercury is present in many electrical gadgets, such as computers, phones, switches, fluorescent lights, and batteries. Around 50 tons of mercury are thought to be discovered in unreported e-waste worldwide each year. These products may emit mercury into the environment when they are disassembled or burned for recycling.

3.3.1.6 Metal Vapors

These are the vapors of metals like cadmium, lead, and mercury. Metalworking, welding, and smelting processes can produce these fumes. Human health may suffer if metal fumes are inhaled. Usually composed of nickel, chromium, and other elements, e-cigarette heating coils have the potential to emit hazardous concentrations of metals into the vaping aerosols. These aerosols had a median lead content of almost 15 μg/kg, which was more than 25 times higher than the median amount in the refill dispensers. The Environmental Protection Agency's health-based guidelines for lead concentrations were exceeded in over half of the aerosol samples.

3.3.1.7 Aerosol Vapors

These are vapors that contain airborne aerosol particles. Aerosols can contain a wide range of things, such as germs, pollutants, and allergies. The consequences of aerosol fumes on the environment and human health can vary. Vapors are released into indoor spaces by aerosol chemicals like insecticides, disinfectants, and air fresheners. The formulation of the product and the environment in which it is used can have a significant impact on the concentration of these vapors. Aerosol vapors are released into the outdoors by activities including operating machinery, burning fossil fuels, and driving automobiles. These vapors contribute to air pollution and have a long range.

3.3.1.8 Cologne/Perfume Vapors

Fragrant substances used in colognes and fragrances are known as vapors. These vapors can have sensory and cultural importance and add to the scent of personal care products. Fragrances frequently use phthalates and their derivatives to improve the stability and duration of their scents. They have been connected to a number of health concerns, such as hormone imbalances and reproductive disorders.

3.3.1.9 Pharmaceuticals Vapors

Pharmaceuticals vapors are the volatile compounds that are utilized in medicine, including inhalants, anesthetics, and vaporized drugs. Inhalation is one way to provide these vapors for therapeutic purposes. Albuterol and corticosteroid-containing asthma inhalers, for example, produce fine aerosols that are deeply inhaled into the lungs. Depending on how the patient inhales and the settings of the device, different amounts of medication may be administered. For inhalable drugs to be safe and effective, the amounts and proportions of pharmaceutical vapors are meticulously regulated. In order to manage any adverse effects and maximize treatment outcomes, healthcare providers must have a thorough understanding of these characteristics. Inhalants, anesthetics, and vaporized drugs are examples of volatile chemicals used in medicine. It is possible to deliver these vapors therapeutically by breathing. Fine aerosols produced by asthma inhalers, such as those that include corticosteroids or albuterol, are deeply inhaled into the lungs. The patient's breathing technique and the device's settings can affect how much medication is administered. To guarantee both therapeutic efficacy and safety, the amounts and proportions of medicinal vapors in inhalable drugs are meticulously regulated. Healthcare professionals must comprehend these factors in order to manage possible side effects and maximize therapeutic results.

3.3.1.10 Essential Oil Vapors

Vapors of aromatic chemicals derived from plants are known as essential oil vapors. Aromatherapy, personal care products, and natural medicines frequently use essential oil vapors. When essential oils are utilized in aromatherapy or domestic items, they emit vapors into indoor spaces. The type of oil, how it is applied, and how long it is exposed can all affect how quickly these vapors are released. The environmental impact of essential oil manufacturing can be decreased by using sustainable agricultural methods and purchasing essential oils from certified organic farms.

These are but a handful of the several varieties of vapors that are known to exist, each with unique characteristics, origins, and impacts.

3.3.2 Vapors' Danger

Since both involve releasing compounds into the air in a gaseous form, the risks connected with vapors and fumes are comparable. However, under room temperature and pressure, vapors are usually defined as the gaseous form of substances that are normally liquid or solid. Here are a few risks associated with vapors.

3.3.2.1 Hazards to Health

Numerous health problems, such as headaches, nausea, dizziness, respiratory irritation, and, in extreme situations, organ damage or central nervous system impairment, can result from breathing in poisonous vapors. Workers may be exposed to high concentrations of specific vapors in the workplace, which can cause immediate health effects such headaches, dizziness, eye, nose, and throat irritation, and, in extreme situations, unconsciousness or death. More severe health consequences, including as neurological conditions, respiratory illnesses, and, in rare instances, cancer, might arise from prolonged or chronic exposure to certain vapors. Depending on the particular substance and the degree of exposure, the precise percentages of risk linked to these events can differ significantly.

3.3.2.2 Fire and Explosion

When combustible material vapors come into contact with an ignition source, they can catch fire or explode. This is especially risky in small areas where vapors can build up and become explosively concentrated. Waste management programs and public safety are seriously threatened by the potential for lithium-ion batteries used in vaping devices to explode and start fires in garbage trucks and waste management facilities. A sudden release of energy, typically accompanied by a shock wave, causes explosions. As the vapor concentration gets closer to the upper flammable limit—the highest concentration of a gas or vapor in air that will ignite and continue to burn—the likelihood of an explosion rises. The primary cause of explosion-related injuries and fatalities, including jet fires, pool fires, and BLEVEs (Boiling Liquid Expanding Vapor Explosion), is heat radiation.

3.3.2.3 Asphyxiation

Like fumes, some vapors have the ability to displace oxygen in the air, which can cause asphyxiation if inhaled in large quantities, particularly in confined or poorly ventilated areas. About 21% oxygen, 78% nitrogen, and 1% other gases make up the air we breathe. An oxygen-deficient environment, which poses a risk of asphyxiation, is defined as having oxygen levels below 21%. Workers who have oxygen levels between 11 and 18% suffer from mental and physical difficulties, such as impaired coordination and slower cognitive processes. Workers find it challenging to identify the impairment because there are no sensory cues of this alteration. The chance of losing consciousness rises sharply below 11% oxygen levels. The likelihood of surviving without irreversible brain damage is significantly decreased at levels below 8%, where unconsciousness is all but guaranteed.

3.3.2.4 Chemical Exposure

When vapors come into contact with the skin or eyes, they may include dangerous chemicals that might irritate the skin, burn the skin, or have other negative effects. Dermal dangers are categorized in the document according to the concentration of the material that comes into contact with the skin. For instance, Category 2 indicates moderate hazard at concentrations between 0 and 50 mg/kg, whereas Category 5 indicates severe hazard at concentrations between 2000 and 5000 mg/kg. The International Labour Organization (ILO) has a similar classification system for inhalation dangers, classifying gases according to their ppm (parts per million) concentrations. For example, a concentration of 100–500 ppm is classified as moderately hazardous (Category 2), whereas a concentration of 5000 ppm or above is classified as severely hazardous (Category 5).

3.3.2.5 Environmental Impact

When vapors are released into the atmosphere, they can damage ecosystems, human health, and air quality. Implementing suitable ventilation systems, using the right personal protective equipment (PPE), and adhering to safe handling practices are crucial for reducing the risks associated with vapors when working with or near substances that emit them. Employers should guarantee adherence to pertinent safety rules and regulations and offer training on the safe use of chemicals.

3.3.3 Overcoming the Vapor Hazard

It takes a combination of mitigation strategies, control tactics, and preventive measures to overcome vapor dangers, particularly those that are dangerous or unpleasant. The following are some broad strategies for defeating vapors.

3.3.3.1 Source Identification and Control

Determine the origins of vapor emissions and, if practical, take action to reduce or eliminate them. To capture or eliminate vapors at their source, this may entail the use of less volatile materials, the use of technical controls, or the enhancement of ventilation systems.

3.3.3.2 Ventilation

To efficiently dilute and disperse vapors, make sure indoor rooms have enough ventilation. To reduce exposure to dangerous vapors, ventilation systems should be built to remove contaminated air and offer sufficient air exchange rates.

3.3.3.3 Containment

To stop vapors from leaking into the surrounding area, put containment measures in place. To contain volatile materials and prevent vapor emissions, this may entail the use of sealed containers, coverings, or enclosures.

3.3.3.4 Personal Protective Equipment (PPE)

Give employees the right PPE to guard against exposure to dangerous fumes, such as respirators, gloves, and goggles. The particular risks present and the necessary level of protection should be taken into consideration when choosing PPE.

3.3.3.5 Replacement

Whenever possible, swap out dangerous fumes for safer ones. This could entail reducing the use of hazardous chemicals, converting to non-toxic or low-VOC products, or implementing different procedures that result in lower emissions.

3.3.3.6 Engineering Controls

Put engineering controls in place to stop or eliminate vapors at their source or on route. Scrubbers, vapor recovery units, and local exhaust ventilation systems are a few examples of devices that are used to collect and clean vapors prior to their discharge into the environment.

3.3.3.7 Testing and Monitoring

Keep a close eye on vapor emissions and air quality to spot possible contamination sources and evaluate how well control measures are working. Analyzing and sampling the air can help find areas for improvement and guarantee adherence to legal requirements.

3.3.3.8 Training and Education

Educate employees, stakeholders, and the general public on the dangers of vapors, safe handling practices, and emergency response procedures. Knowledge and awareness enable people to take preventative action against problems related to vapor.

3.3.3.9 Regulatory Compliance

Keep up with pertinent laws, rules, and policies pertaining to vapor emissions and workplace safety and health. Minimizing environmental effects and safeguarding human health depend on adherence to legislative standards and industry best practices.

3.3.3.10 Emergency Preparedness

Create and carry out emergency response plans to handle vapor-related situations such as spills, leaks, or releases. To lessen the possible effects of emergencies, set up procedures for evacuation, containment, and communication.

Communities, organizations, and individuals can collaborate to manage vapors and make everyone's surroundings safer and healthier by putting these tactics and methods into practice.

3.4 Introduction of Fumes

"Fumes" primarily refer to gases or airborne particles that are produced from a variety of sources and are frequently associated with negative connotations, such as being unpleasant or hazardous.

3.4.1 Various Fume Types

The following are some typical fume types.

3.4.1.1 Smoke

A mixture of gasses and solid and liquid particles released into the air during pyrolysis or combustion. Cigarettes, burning wood, industrial operations, and automobile exhaust are all sources of smoke. Approximately 12% of all smoke and hazardous fume inhalation injuries throughout the study period were caused by attempts to put out the fire. Smoke and toxic fume inhalation injuries were more common in men between the ages of 20 and 85, but more common in women under the age of 20 and over the age of 85. The 0–4 and 5–9 age groups experienced fewer inhalation injuries from smoke and poisonous fumes, and these were unrelated to attempts to put out the fire or use drugs or alcohol (Taylor and Fielding 2023).

3.4.1.2 Chemical Fumes

These fumes are gases or vapors that are emitted by chemicals as a result of heating, evaporation, or chemical reactions. Solvents, paints, insecticides, cleaning products, and industrial chemicals can all release chemical vapors. Depending on how hazardous the chemicals are, breathing in chemical vapors can be harmful to your health. One of the best strategies to lower chemical fume concentrations indoors is to make sure there is enough ventilation. This may entail using air purification devices, exhaust fans, or window openings. Unnecessary fume discharge can be avoided by handling chemicals in accordance with safety regulations, storing them appropriately, and maintaining tightly sealed containers.

3.4.1.3 Welding Fumes

These are gases and particulates generated during welding and associated procedures. Metal oxides, gasses (including ozone and nitrogen oxides), and other dangerous materials are present in welding fumes. Long-term welding fume exposure can cause respiratory disorders as well as other health problems. The initial particles produced by welding have a diameter of 5–40 nm and aggregate into 0.1–1 μm. Approximately 10% of welding is done with stainless steel, whereas 90% of welding is done with mild steel. Iron (Fe) and manganese (Mn) particles make up the majority of fumes from mild steel welding, while chromium (Cr) and nickel (Ni) are also present in fumes from stainless steel welding (Dauter et al. 2024).

3.4.1.4 Volatile Organic Compounds (VOCs)

Vapors released by volatile organic compounds (VOCs), which are substances that readily evaporate into the atmosphere at room temperature, are known as VOC fumes. Numerous home goods, building supplies, and industrial operations include volatile organic compounds (VOCs). Long-term exposure to volatile organic compounds (VOCs) can have negative health impacts and contribute to indoor air pollution. Individual sensitivity, exposure length, and concentration all affect the health impacts of volatile organic compounds (VOCs). Asthma symptoms may intensify, and headaches, nausea, dizziness, and irritation of the eyes, nose, and throat may result after brief exposures. Long-term exposure spanning years to decades can raise the risk of central nervous system damage, liver and kidney damage, and cancer.

3.4.1.5 Soldering Fumes

Solder, a filler metal, is utilized to fuse metal surfaces together utilizing fumes produced during the soldering process. Metal oxides, flux vapors, and other soldering process byproducts are found in soldering fumes. Soldering vapors may contain lead or other dangerous materials, and inhaling them can irritate the respiratory system. Soldering fumes are an example of occupational asthma, which is brought on by flux gasses and manifests as coughing, wheezing, chest pain, and shortness of breath.

3.4.1.6 Paint Fumes

Paint fumes are vapors released during the drying and curing process of wet paint. Solvents, volatile organic compounds, and other substances used in paint formulas are all present in paint fumes. When working with paints, proper ventilation is essential to reduce exposure to paint fumes, which can irritate the respiratory system and induce headaches and nausea. Because of higher indoor VOC concentrations and longer exposure periods, paint fume hazards are increased in production environments. Industry Safety and Hygiene News (ISHN) advises employers to utilize Heating, Ventilation, and Air Conditioning (HVAC) system filters with a minimum efficiency reporting value (MERV) of 13 or above, make sure there is adequate ventilation, and mandate that employees wear the right protective gear, such as safety goggles and N95 respirators.

3.4.1.7 Exhaust Fumes

Vehicle exhaust systems release gases and particulate debris known as exhaust fumes. Carbon monoxide, nitrogen oxides, particulate matter, and volatile organic compounds are all found in exhaust fumes. Exposure to exhaust fumes can lead to respiratory health issues and air pollution, particularly in cities with high traffic volumes. Because they contain a variety of heterocyclic aromatic amines and long-chain aldehydes, cooking oil fumes (COFs) are a significant indoor environmental pollutant and have strong mutagenic or carcinogenic health effects (Li et al. 2022). A simulation that compares statistics on the dangers of cooking emissions and their impact on human health has been taken into consideration.

3.4.1.8 Acid Fumes

Industrial operations, the burning of fossil fuels, and chemical reactions can all release gases that contain acidic substances, such as sulfur dioxide (SO_2) and nitrogen oxides (NOx). Surfaces and respiratory tissues can be adversely affected by acid fumes. For instance, the health effects of different concentrations of hydrochloric acid fumes can differ. At 50–100 parts per million (ppm), moderate symptoms might result in breathing difficulties, chest pain, and potentially fatal situations within 30–60 min. At 1000–1300 ppm, severe symptoms render breathing impossible and increase the risk of mortality within 30 min to an hour (sentryair.com).

These are but a handful of the several kinds of vapors that can be found in different settings, each with its own sources, makeup, and possible health risks. To reduce exposure to dangerous vapors, control mechanisms, personal protective equipment, and proper ventilation are frequently required.

3.4.2 Fume Hazards

The type of fumes and the environment they are present in determine their hazard. The following are some typical risks connected to fumes.

3.4.2.1 Health Risks

Exposure to harmful gases can cause a number of health issues, such as headaches, nausea, dizziness, respiratory irritation, and, in extreme situations, long-term harm to the nervous system, respiratory system, or other organs.

3.4.2.2 Fire and Explosion

Certain vapors have a significant potential for combustion or explosion. They can result in fires or explosions when they come into contact with ignition sources like sparks, flames, or hot surfaces.

3.4.2.3 Asphyxiation

If absorbed in high concentrations, particularly in tight areas, some fumes, like carbon monoxide, can displace oxygen in the air and cause asphyxiation.

3.4.2.4 Chemical Exposure

Chemical fumes may contain dangerous compounds that, when in contact with the skin, might result in chemical burns or irritation.

3.4.2.5 Environmental Impact

Air pollution from emitted fumes can have an adverse effect on ecosystems and human health.

3.4.3 Overcoming Fume Dangers

When working with or near materials that emit fumes, it is crucial to employ appropriate ventilation systems, personal protective equipment (PPE), and safe handling practices to reduce these risks. Furthermore, avoiding risks requires knowledge of the characteristics of the particular gasses involved as well as adherence to pertinent safety rules and recommendations. The following are some general strategies for getting rid of fumes.

3.4.3.1 Source Identification and Control

Determine the origins of fume emissions and, if practical, take action to reduce or eliminate them. This could entail reducing the use of hazardous materials, putting engineering controls in place, or enhancing ventilation systems to catch or eliminate emissions at their source.

3.4.3.2 Ventilation

To efficiently dilute and disperse pollutants, make sure indoor rooms have enough ventilation. To reduce exposure to dangerous pollutants, ventilation systems should be built to remove contaminated air and offer sufficient air exchange rates.

3.4.3.3 Engineering Controls

In order to catch or eliminate fumes at their source or along their course, it is recommended that engineering controls be used. Scrubbers made to collect and purify fumes prior to their discharge into the environment are examples, as are fume hoods and local exhaust ventilation systems.

3.4.3.4 Personal Protective Equipment (PPE)

To guard against exposure to dangerous vapors, give employees the proper PPE, such as respirators, gloves, and goggles. The particular risks present and the necessary level of protection should be taken into consideration when choosing PPE. For instance, firefighters handling dangerous materials have access to a vast supply of breathable air because of self-contained breathing apparatus (SCBA).

3.4.3.5 Replacement

Whenever possible, swap out dangerous products or procedures for safer ones. This could entail reducing the use of hazardous chemicals, converting to fuels with lower emissions, or implementing substitute technologies that emit less emissions.

3.4.3.6 Containment

Put containment procedures in place to stop fumes from leaking into the environment. This can entail employing enclosures, coverings, or sealed containers to keep volatile materials contained and prevent fume emissions.

3.4.3.7 Testing and Monitoring

Keep a close eye on fume emissions and air quality to spot possible contamination sources and evaluate how well control measures are working. Analyzing and sampling the air can help find areas for improvement and guarantee adherence to legal requirements.

3.4.3.8 Training and Education

Educate employees, stakeholders, and the public on the dangers of fumes, safe handling practices, and emergency response procedures. Knowledge and awareness enable people to take preventative action against fume-related occurrences.

3.4.3.9 Regulatory Compliance

Keep up with pertinent laws, rules, and policies pertaining to fume emissions and workplace safety and health. Minimizing environmental effects and safeguarding human health depend on adherence to legislative standards and industry best practices.

3.4.3.10 Emergency Preparedness

To handle fume-related incidents like leaks, spills, or releases, create and execute emergency action plans. Create procedures for communication, containment, and evacuation to lessen the possible effects of emergencies.

Individuals, groups, and communities can collaborate to efficiently manage fumes and establish safer, healthier surroundings for all by putting these tactics and methods into practice.

3.5 Conclusions and Future Strategies

A comprehensive strategy that incorporates monitoring, education, personal protective equipment, prevention tactics, and regulatory compliance is needed to overcome the difficulties presented by vapors, gases, and fumes. It is feasible to drastically lower the hazards connected to these airborne pollutants, safeguarding the environment and public health, by putting in place thorough safety procedures and keeping up with industry best practices. Hazards and dangers include effects on the environment and health. Prevention and control measures, personal protective equipment, monitoring and detection, education, and training are some techniques for overcoming obstacles. The carbon footprint must be taken into account when comparing contemporary engineering strategies to the previously outlined classical methodologies (Zaki et al. 2023). Analyzing carbon footprints helps to improve air quality.

Additionally, the Internet of Things (IOT) and other artificial intelligence techniques effectively address air pollution caused by gasses, vapors, and odors. It will be more efficient to regulate these pollutant dangers if database analysis is used to determine the concentration of these hazards and if the IOT paradigm is connected to specialized sensors that can measure various pollutants. Indoor air pollution can be monitored with IOT by creating an air quality database (Lambor et al. 2023).

References

Akimoto H, Tanimoto H (2023) Handbook of air quality and climate change. Springer, Cham. ISBN 978–981–15-2759-3 ISBN 978–981–15-2760-9 (eBook). https://doi.org/10.1007/978-981-15-2760-9

Al-Sarraj A, Al Soubaihi R, Saoud KM, Ahmad I, Ayesh, Bermak A (2024) Sub-ppm of toxic gases detection on Ag-doped WO3 nanosensor. Emerg Mater 7:2921. https://doi.org/10.1007/s42247-024-00766-2

Calm JM (2002) Emissions and environmental impacts from air-conditioning and refrigeration systems. Int J Refrig 25(3):293. https://www.sentryair.com/blog/industry-applications/chemical-solvents/hydrochloric-acid-vapor-health-risks/

Dauter UM, Gliga AR, Albin M, Broberg K (2024) Longitudinal changes in cardiovascular disease-related proteins in welders. Int Arch Occup Environ Health 97(7):803. https://doi.org/10.1007/s00420-024-02086-8

Deshpande A (2024) Ashish Tomar and Akash Pagare (2024) quantitative analysis of air pollution levels and its health implications in Delhi-NCR: a longitudinal study. Int J Sci Res Arch 11(02):759–768. https://doi.org/10.30574/ijsra.2024.11.2.0488

Pierre L. Fauchais, Joachim V.R. Heberlein and Maher I. Boulos (2014) Thermal spray fundamentals. ISBN 978-0-387-28319-7 ISBN 978-0-387-68991-3 (eBook) https://doi.org/10.1007/978-0-387-68991-3. Springer New York

Fromme H (2023) Indoor air quality occurrence and health effects of contaminants. Springer, Cham. ISBN 978–3–031-40077-3 ISBN 978–3–031-40078-0 (eBook). https://doi.org/10.1007/978-3-031-40078-0

Jalasto J, Luukkonen R, Lindqvist A, Langhammer A, Kankaanranta H, Backman H, Rönmark E, Sovijärvi A, Piirilä P, Kauppi P (2024) Occupational exposure to vapors, gasses, dusts, and fumes in relation to causes of death during 24 years in Helsinki, Finland. Int Arch Occup Environ Health 97:145–154. https://doi.org/10.1007/s00420-023-02031-1

Khoshakhlagh AH, Yazdanirad S, Saberi HR, Liao P-C (2023) Health risk assessment of exposure to various vapors and fumes in a factory of automobile manufacturing. Heliyon 9(8):e18583. www.cell.com/heliyon

Lambor S, Mahajan A, Bodhankar A, Prasad D, Mahajan S, Pujari A, Dhakalka R (2023) Indoor air pollution monitoring. ICISAT 2022, LNNS 624, pp 434–441. https://doi.org/10.1007/978-3-031-25344-7_40

Li S, Wang L, Guan S, Zhou S, Chen Y (2022) In vitro and in vivo low-dose exposure of simulated cooking oil fumes to assess adverse biological effects. Sci Rep 12:15691. https://doi.org/10.1038/s41598-022-19558-x

Padhy PK, Niyogi S, Patra PK, Hecker M (2024) Air quality and human health. WHO, Geneva. ISBN 978–981–97-1362-2 ISBN 978–981–97-1363-9 (eBook). https://doi.org/10.1007/978-981-97-1363-9

Sapuan SM, Ilyas RA, Asyraf MRM (2022) Safety and health in composite industry. ISSN 2662–1819 ISSN 2662–1827 (electronic) Composites Science and Technology ISBN 978–981–16-6135-8 ISBN 978–981–16-6136-5 (eBook). Springer, Cham. https://doi.org/10.1007/978-981-16-6136-5

Taylor M, Fielding J (2023) An analysis of domestic fire smoke/toxic fumes inhalation injuries. Fire Technol 59:2667–2681. https://doi.org/10.1007/s10694-023-01446-z

Wiegleb G (2023) Gas measurement technology in theory and practice. Springer, Cham. ISBN 978–3–658-37231-6, ISBN 978–3–658-37232-3 (eBook). https://doi.org/10.1007/978-3-658-37232-3

Zaki E, Metwally M, Magdy G, Baset AA (2023) carbon footprint study on renewable power plants: case study on Egypt's Benban Solar Park. Int J Appl Energy Syst 5(1):6

Types of Running Water, Groundwater, and Polluted Rain and a Vision to Overcome Them

4

Naseem Akhtar, Ataur Rehman, Muhammad Abdullah, and Aznan Fazli Ismail

Abstract

This chapter has been prepared specifically to assist human beings with a fascination for scientific and natural science in addition to fundamental concepts of water quality degradation that are crucial to sustainable management of running water and aquatic environments. Pollution is an essential issue that influences groundwater quality and the composition of other aquatic habitats in running water systems. Running water accounts for rainwater pollution associated with wet and dry precipitation and groundwater contamination by various sources of natural and anthropogenic phenomenon. However, limited attention has been given to performance knowledge of running water quality monitoring initiatives. This chapter aims to concentrate on groundwater contamination, rainfall pollution, and types of running water that have been addressed. Thus, the water quality enhancement project regarding running water significantly contributes to the effective oversight of initial groundwater and rainfall contamination, which will

N. Akhtar (✉)
Nuclear Science Program, Faculty of Science and Technology, Universiti Kebangsaan Malaysia, Bangi, Selangor, Malaysia

Division of Environmental Technology, School of Industrial Technology, Universiti Sains Malaysia, Gelugor, Pulau Pinang, Malaysia

A. Rehman
Department of Geosciences, College of Petroleum Engineering and Geosciences, King Fahd University of Petroleum and Minerals (KFUPM), Dhahran, Saudi Arabia

M. Abdullah
Department of Earth and Environmental Science, Indiana University - Purdue University Indianapolis (IUPUI), Indianapolis, IN, USA

A. F. Ismail
Nuclear Science Program, Faculty of Science and Technology, Universiti Kebangsaan Malaysia, Bangi, Selangor, Malaysia

A. A. H. Abdel Latef et al. (eds.), *Bioremediation and Nanotechnology for Climate Change Mitigation*, https://doi.org/10.1007/978-981-96-3069-1_4

improve water quality and function as a practical technical benchmark for managing sustainable water resources and ecological environments.

Keywords

Groundwater and rainwater contamination · Sustainable water resources development · Running water types

4.1 Introduction

Running water is an alternative acronym for freshwater or lotic communities (moving water), whereas lotic communities emerge through water moves from various sources, including rainfall, surface water, and groundwater (Malmqvist and Rundle 2002; Akhtar et al. 2020b; Hanh Nguyen et al. 2023). Although rainwater directly enters the running water community, surface water is also driven by previous rainwater that entered the running water, and groundwater that is absorbed into the soil subsequently moves to the water table, which can provide water for the running water community (Freeman et al. 2022). Additionally, there are numerous or less significant pathways through which water may infiltrate the river, such as through outlet pipe interference caused by humans or the transpiration of water by nearby plants (Bunn and Davies 2000; Sendzimir and Schmutz 2018). Even though a negligible percentage (0.006%) of the global freshwater supply is continuously sourced from rivers and streams, this statistic belies the critical importance of the lotic community for humanity and the natural environment (Malmqvist and Rundle 2002).

Running water provides a diverse range of benefits to humanity, such as water sources for residential, agricultural, and industrial applications, facilitating the production of electricity and waste disposal, supplying pathways for navigation, and distributing spots suitable for outdoor activities (Akhtar et al. 2020a; Yang et al. 2023). These are significant benefits that running water supplies to human beings and the natural environment. Running water is mainly associated with rainfall in a specific area, which indicates a flow state (Kurwadkar 2019). However, running water is polluted by natural and anthropogenic sources worldwide (Akhtar et al. 2021c). Volcanic actions, forest decomposition, and ionic reactions are all examples of natural sources (Müller et al. 2020). Further, by changing the geological condition through the mechanisms of weathering and the deposition of rocks and minerals, the movement of water influences the natural environment (Tong et al. 2022).

Anthropogenic sources include vehicular emissions, industrial emissions, biomass combustion, urban activities, and agriculture practices (Qadir et al. 2020; Li et al. 2021). Numerous recent studies have examined this phenomenon globally, including in the United States, Asia, China, India, Europe, and Africa (Bunn and Davies 2000; Akhtar et al. 2021b; Das et al. 2021; Hanh Nguyen et al. 2023). Polluted water has emerged as an increasingly critical environmental concern, endangering communities' usability and obstructing their sustainable development (Allan et al. 2021). The study's most important emphasis has been the issues associated with rainwater and groundwater pollution. Moreover, understanding an area's atmospheric composition is crucial for mitigating pollution from the atmosphere. From this perspective, studying rainwater pollution acts as an attempt to quantify and categorize atmospheric processes to avoid harm to natural ecosystems via acidification of soils and water bodies, disruption of the nutrient cycle, and health concerns (Kim et al. 2005).

Initial rainfall is the primary cause of rainwater pollution. Large amounts of gaseous pollutants are dissolved in rainwater and released into rivers during the earliest phase of rainfall, leading to severe contamination of river water (Sanmi 2018). In addition, rainwater pollution originates from the leaching and scouring of contaminants released on the surface and subsurface. Furthermore, several studies have demonstrated how domestic effluent contaminates initial rainfall (Exall and Vassos 2012). Moreover, the pipeline-based immediate release of rainfall into collecting water can rapidly decrease water quality. The main cause of surface runoff pollution is initial rainwater, which has been reported by the United States Environmental Protection Agency (USEPA). In addition, the level of surface runoff pollution in rivers and lakes has increased from 9 to 18% (Nemčić-Jurec et al. 2022). Further, rainfall significantly balances hydrological systems by interacting with various chemical and physical processes, including movement and removal, in addition to atmosphere interactions. However, ions diluted by rainwater may contribute to the increased acidity due to predominantly nitrate and sulfate substances through chemical processes (Exall and Vassos 2012; Li et al. 2021).

Although rainfall may affect groundwater quality characteristics through dilution mechanisms within aquifers, differences in precipitation events, climate change, and local geology complicate the resultant fluctuations in physical and chemical parameters (Akhtar et al. 2023). In groundwater research, chemical contamination has become an increasing problem over the last 30 years (Müller et al. 2020). A substantial amount of the impurities present in groundwater originate from geological processes, specifically the dissolution of naturally occurring mineral deposits located within the Earth's crust (Martins et al. 2019). However, because of the exponential growth of the population, urbanization, and industrialization, agricultural output and economy are confronted with the formidable obstacle of adverse effects caused by anthropogenic contaminants in groundwater (Akhtar et al. 2021c). Emerging contaminants, including but not limited to toxic metals, hydrocarbons, trace organic contaminants, pesticides, nanoparticles, and microplastics, pose a significant risk to sustainable socioeconomic development, ecological services, and human health (Akhtar et al. 2020a, 2021a). Even though groundwater contamination constitutes a significant threat to human populations, it also provides an excellent opportunity for scientists to gain a deeper understanding of the evolutionary processes of subsurface aquifers and for policymakers to comprehend the mechanism of safeguarding the quality and quantity of groundwater (Li et al. 2021). On the other hand, groundwater assumes a critical role as a resource in arid and semiarid regions characterized by a scarcity of surface water and rainfall.

The mechanisms of water pollution due to human and anthropogenic activities have been studied for over 50 years. Although a comprehensive overview of contributing pollution factors associated with running water sources remains in progress, it

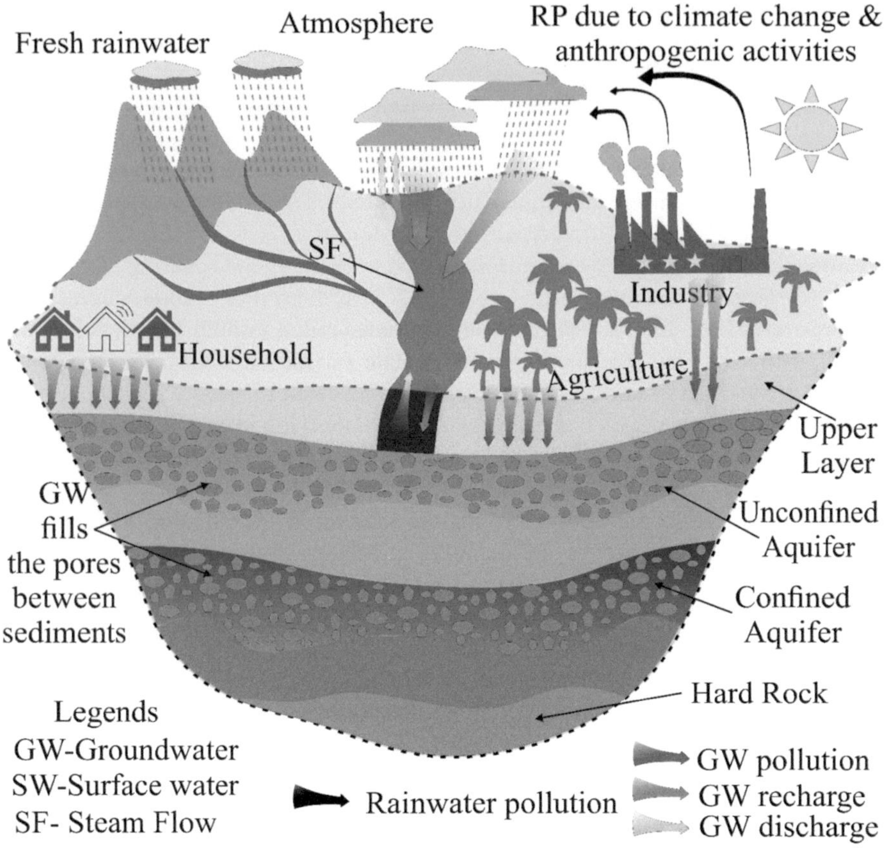

Fig. 4.1 Graphic depicting the pollution of aquatic environments caused by human interference

has not been extensively evaluated in the scientific literature. In this context, the current status of awareness of the numerous causes that contribute to water pollution from rainfall and groundwater leads to a connection to this gap. Thus, this chapter has addressed several types of running water sources, rainfall pollution, and groundwater contamination, as well as provides better knowledge to understand the pollution processes, types, pathways, and a vision to overcome their consequences on water resources (Fig. 4.1).

4.2 Types of Running Water

Running water is a dynamic aspect that constantly shapes the Earth's surface, sculpting landscapes and sustaining life all over the planet. Its movement, defined by speed, volume, and characteristics, leads to the classification of several varieties of moving water, each having a distinct purpose in the natural world.

(a) **Rivers**

Rivers, the lifeblood of landscapes, move water from higher to lower elevations and vary significantly in size, velocity, and structure. Rivers are classified based on flow patterns, channel forms, and discharge volumes. They can be classed as meandering, braided, or straight channels.

(b) **Meandering Rivers**

These rivers have winding channels with gentle bends and slow-moving water. They generate distinct twists and complicated patterns as they move across landscapes.

(c) **Braided Rivers**

In contrast to meandering rivers, braided rivers are made up of several channels intertwined between gravel bars and islands. The dynamic environment is formed by periodic channel shifts caused by a large sediment load.

(d) **Straight Rivers**

Some rivers follow a relatively straight path with little meandering. These straight rivers frequently have more consistent flow patterns and are governed by geological features such as fault lines or structural restrictions. A complex network of minor watercourses traverses the landscape, including streams, creeks, brooks, and rivulets. These water bodies are classified based on their size, discharge, and ecological relevance.

(e) **Streams and Creeks**

Streams and creeks are typically used interchangeably to represent tiny, narrow watercourses originating from springs or rainfall and flowing into bigger rivers. They play an important role in maintaining local ecosystems and providing habitat for aquatic creatures.

(f) **Brooks and Rivulets**

Brooks and rivulets are tiny bodies of water in mountainous areas with shallow depths and rapid movement. Torrents and rapids are fast-flowing, turbulent portions generated by barriers that draw explorers to activities such as rafting.

(g) **Rapids**

Rapids are parts of rivers where water runs quickly across uneven ground, causing turbulent and tumultuous circumstances. These places, typically created by rocks or impediments, attract adventurers who participate in kayaking and whitewater rafting.

(h) **Torrents**

Torrents are river portions with rapid, churning flows caused by steep grades or abrupt changes in channel shape. These powerful water currents play an important role in erosion and sediment transfer.

(i) **Waterfalls**

Waterfalls are spectacular phenomena formed when water flows over a steep drop, which occurs when rivers experience elevation shifts. Beyond their aesthetic value, waterfalls play an important role in creating landscapes through erosive processes.

(j) **Estuaries and Tidal Channels**

Estuaries and tidal channels are unique settings where freshwater meets saltwater, providing essential habitat for marine life. The ecological value of estuaries and tidal channels emphasizes their role as transitional zones between freshwater and marine habitats (Dyer and Huntley 1999).

(k) **Delta Channels**

Delta channels are the complicated network of distributaries generated at the mouths of rivers that shape landscapes. Deltas are wetlands that occur when rivers discharge their water and debris into another body of water. Understanding differences among kinds of water is essential for enjoying nature's beauty and acknowledging its ecological significance. They help sustain different ecosystems, make mobility easier, provide habitat, and provide recreational possibilities. Studying their traits contributes to long-term conservation and management.

4.2.1 Structure and Function of Running Water Systems

Running water, often known as flowing water, is essential to ecosystems and the Earth's hydrological cycle. It can take on many different shapes, including creeks, brooks, rivers, and streams. Studying freshwater ecosystems and managing water resources require understanding the composition and operation of running water. Studying the "Structure and Functioning of Running Water" topic is a fascinating and important way to learn about the complex world of freshwater ecosystems and how important they are to the environment (Kipnusu et al. 2023). This section attempts to give a summary of the subject, emphasizing its importance and the main points covered. The present status of our knowledge of stream ecosystem structure and function is based on a number of generalizations that have been tested primarily in woodland streams of the temperate zone (Hayes et al. 2019).

The flowing water structure comprises a few elements, including the streambed, banks, riparian zone, and the channel itself. A water body's features are shaped by various factors, ranging from the channel's size and flow dynamics to the stability and biological diversity of the surrounding land and banks. The overall characteristics of water courses are shaped by the interaction of multiple components within its complex and dynamic structure. For example, meanders introduce sinuous patterns to river channels through erosional and depositional processes, affecting sediment transport and the landscape's appearance (Ielpi et al. 2022). Within the channel, riffles are shallow, fast-flowing sections that often provide oxygenation and serve as ideal spawning areas for certain fish species, while pools are deeper, slower-moving zones that offer refuge and habitat diversity (Archdeacon and Reale 2020). The water body's total biodiversity is enhanced by the niche opportunities that these differences in flow velocity and depth provide for aquatic creatures. Point bars, usually located inside meander bends, are created when sediment is deposited during reduced flow. The ensuing vegetated patches can serve as significant habitats for species.

Meander loops may occasionally get disconnected from the main channel, giving rise to oxbow lakes—unique bodies of water with unusual ecological traits. In addition to supporting a variety of riparian flora, floodplains located next to the channel are essential for managing floods and replenishing groundwater. Deltas in river mouth situations show how freshwater and marine environments interact dynamically. Larger bodies of water, such as seas or oceans, receive silt from rivers, which causes the formation of deltas—complex, dynamic landforms. Because they frequently serve as hubs for transportation, agriculture, and fisheries, deltas are crucial not only for biological reasons but also for human populations. The flowing water of rivers, streams, and other bodies of water plays a vital and diverse role in our natural environment. It is a dynamic force that actively shapes the environment, influences geological processes, and gives life on Earth its means. This article examines the various functions that flowing water does, from its essential support of wildlife and human civilization to its critical part in the hydrological cycle.

The hydrological cycle of the Earth, a continuous process that controls the amount of water dispersed globally, essentially depends on the flow of water. Precipitation, which can be either rain or snow, initiates this cycle. A portion of this moisture percolates into the ground, replenishing groundwater supplies and providing the necessary moisture for terrestrial flora to thrive (Hess et al. 2021). Even yet, a sizable amount of this precipitation makes its way over the Earth's surface and eventually collects in rivers and streams. These flowing waterways act as vital conduits, facilitating water movement into larger bodies of water like lakes and seas. This intricate dance of water keeps the balance of freshwater availability, a crucial aspect of Earth's hydrology.

Apart from its function in the hydrological cycle, flowing water has an amazing ability to erode. It can shape the Earth's surface across geological time scales, forming canyons, valleys, and a wide variety of other landforms. Rocks and sediments gradually break down into smaller pieces during this erosional process. Running water becomes a carrier as it moves downstream, carrying these sediments and transferring vital nutrients and minerals along the way. In addition to influencing the physical environment, this dual function of artist and courier enhances the productivity of floodplains and delta regions by accumulating silt. Rivers and streams and other running water environments are extremely important because they offer a wide variety of aquatic organisms with vital habitats. These ecosystems are remarkably adept at forming a variety of microhabitats inside their boundaries. A variety of aquatic species, including fish, insects, amphibians, and microbes, can find shelter and breeding grounds in riffles, which are defined by shallow, swiftly flowing waters, and pools, which include deeper, more leisurely moving parts (Stock et al. 2021). The richness of life present in these environments serves as the cornerstone of complex food webs, maintaining the delicate balance of life in aquatic environments. Furthermore, flowing water is essential to cycling nutrients, especially essential elements like phosphorus and nitrogen.

There are several ways that these vital nutrients get up in water bodies, including wastewater and runoff from agricultural regions. These nutrients are actively used by aquatic organisms, such as algae and aquatic plants, to support their growth. Because these nutrients eventually find their way to land through various mechanisms, the cycling of nutrients within rivers and streams not only forms the foundation of the aquatic food web but also offers vital support to terrestrial ecosystems. Running water also has a major role in controlling temperature. During times of high outside temperatures, these pools of water act as cooling agents, assisting in regulating local temperatures (Das et al. 2003). They frequently create colder microclimates around them, providing a haven for both land animals and aquatic life. On the other hand, moving water can be somewhat warmer than the surrounding air in colder weather, which makes it an essential source of warmth and stability for the surrounding area. Flooding water is extremely significant to human cultures besides its ecological purposes. Streams and rivers have always been important routes for transportation, enabling the flow of people and products (Wondzell 2011). These bodies of water still significantly impact transportation today, fostering connectedness and economy. Running water also enhances the well-being and standard of living for communities that live near watercourses by providing a variety of leisure possibilities, such as swimming, boating, kayaking, and fishing. Finally, a powerful renewable energy source is flowing water. As a clean and sustainable energy source, hydropower plants use the kinetic energy of flowing water to create electricity.

The worldwide endeavors to decrease dependence on fossil fuels and alleviate the effects of climate change are greatly aided by this contribution to sustainable energy generation. In summary, flowing water serves various purposes, including biological roles, geological processes, and significant advantages to human cultures (Anderson et al. 2019). Understanding their critical role emphasizes how urgent it is to protect and manage these priceless natural resources. Preserving the benefits of flowing water is still vital for the health of the environment and the next generation of humans as we navigate an ever-changing planet.

4.2.2 Factors Affecting Running Water Ecosystems

Numerous natural aspects of the environment impact the ecological condition of running water environments. Multiple simulations at broad scales demonstrate important factors that impact the basics of the natural world of running water bodies, such as trophic energy supply, organic matter transportation, nutrient cycling, floodplain dynamics, and hydraulic patterns. Regional and local-scale research has highlighted the significance of water chemical science, hydrology, microhabitat structure, and biological interactions in running habitats. Human activities can change the natural balance of elements that affect the flow of rivers, such as by modifying the types of materials on the streambed or modifying the average yearly temperature patterns. In addition, these interactions can contribute to alterations that exceed the typical conditions predicted for a certain running environment.

Several instances can include an unexpected rise of harmful metals into a river or a complete disappearance of a certain microhabitat, resulting in the final consequence of damaging or eradicating the natural environment, either partially or completely. Four basic categories of ultimate driving factors lead to changes in flowing waters: ecosystem destruction, physical habitat alteration, water chemistry modification, and actual species additions or removals. Numerous proximate causes affect these variables (Table 4.1).

4.3 Groundwater Pollution

Groundwater, an essential natural reserve, refers to the subterranean water found in saturated soils and geological formations below the water table. Groundwater is a crucial source of freshwater for people worldwide and is utilized for domestic, agricultural, and industrial uses (Elsawah and Guillaume 2016). A third or so of the world's population gets their drinking water from groundwater. However, poor management, population expansion, urbanization, and industrial and agricultural activities have made this essential resource more vulnerable to contamination (Constant et al. 2016). Groundwater contamination is a serious environmental issue that is becoming more urgent and has detrimental effects on ecosystems, the economy, and public health. A comprehensive strategy with strict rules, ethical industrial and agricultural practices, proactive monitoring, and early pollution identification is required to address this issue. While chemical pollutants and hazardous microorganisms from urban development, and household and industrial wastes are the main causes of groundwater contamination in urban areas, fertilizer application, pesticide use, and careless groundwater pumping for irrigation are the main causes of groundwater contamination in rural areas of developed countries.

Groundwater contamination can influence socioeconomic development, environmental quality, and human health. Many studies, for instance, have demonstrated that communities of humans are at risk for health problems due to excessive concentrations of metals, persistent organic pollutants, fluoride, and nitrate (Wu et al. 2014). Compared to adults, babies and children are more susceptible to the effects

Table 4.1 The major changes triggered by human factors driving modifications to running water environments

Ultimate forcing factors	Subfactors	Biotic implications	Abiotic modification	Proximate causes
Water chemistry	Nutrient mixing	Increased primary production and algal blooms	Increased N and P	Agriculture, NO_3 emissions into the atmosphere, deforestation, landfills, industry, and sewage systems
	Toxic elements	Direct physiological toxic effects	Many trace metals (e.g., cu, hg, Zn, Al, cd, Pb)	Mining, sewage works, landfills, and industrial wastewater
	Organic metals	Reduced habitat availability	O_2 solids decreased and suspended solids increased	Agriculture, sewage works, and urbanization
	Organochloride toxins	Toxic effects through biomagnification	Polychlorinated biphenyls (PCBs) and organochlorine pesticides (e.g., Dichlorodiphenyl-trichloroethane (DDT), dieldrin)	Industrial effluents, garbage incineration, landfills, sewage systems, and agriculture
	Acidification	Impacts on the body that are direct and indirect (food chain)	Reduced pH and increased Al^{3+}	Industrial emissions (SO_2) and exhaust waste (NO_3)
	Endocrine disruptors	Interference with naturally produced hormones	Pharmaceuticals (estrogens), pesticides (DDT, dieldrine), and organohalogens (dioxines, furanes, PCBs)	Agriculture, waste incineration, and industrial waste
Habitat modification	Siltation	Altered habitat conditions	Reduced substratum complexity	Development of agriculture, urbanization, and deforestation
	Hydrology	Altered habitat conditions and reduced dispersal	Reduced likelihood of drought, disruption of upstream–downstream connections, and loss of normal flow periodicity	Channelization, deforestation, water transfer scheme, damming, and water abstraction
	Riparian corridor changes	Modified habitat conditions and trophic dynamics	Modified light and organic matter energy inputs as well as in-stream marginal habitat	Agriculture, urbanization, and channelization

(continued)

Table 4.1 (continued)

Ultimate forcing factors	Subfactors	Biotic implications	Abiotic modification	Proximate causes
Ecosystem destruction	–	Species and population extinction	Complete ecosystem loss	Agriculture, urbanization expansion, and water abstraction
Species removal and addition	–	Heightened or diminished competition, changed ecosystem dynamics, and riparian energy inputs	Invasive species	Fisheries, sport fishing, aquaculture trade, and horticulture plants

of these contaminants, thus this is especially important for them (He et al. 2020; Karunanidhi et al. 2021). For example, "blue baby syndrome," also known as infant methemoglobinemia, is caused by excessive nitrate concentrations in the drinking water used to make baby formulas. Groundwater pollution can also have an impact on human health by having an impact on the food production chain. Human health hazards can arise from the buildup of toxic components in grains and vegetables due to irrigation with groundwater polluted by heavy metals and wastewater carrying persistent pollutants (Jenifer and Jha 2018; Yuan et al. 2019).

Land and forest quality can also be adversely impacted by groundwater pollution. Soil pollution and deterioration of land quality can result from contaminated groundwater. The quality of surface water can deteriorate due to the transportation of toxins from groundwater through interactions with surface water (Teng et al. 2018). This chapter explores the origins, effects, and possible remedies for the growing issue of groundwater pollution by utilizing data from scientific studies and professional opinions.

4.3.1 Sources of Pollution in Groundwater

Various human activities and natural sources can pollute groundwater (anthropogenic sources). Since the early Holocene, some 10,000 years ago, when humans started to become sedentary and establish towns and farms, groundwater contamination has been an issue brought on by or exacerbated by human activity (Postigo 2018). Most sources of pollution include the removal of impurities produced by the usage of water. It is simpler to identify sources of surface water contamination. On the other hand, the origins of subterranean water pollution are hard to pinpoint and persist for many years. Groundwater contaminants are often odorless and colorless. Furthermore, polluted groundwater has long-lasting and elusive harmful effects on human health (Chakraborti et al. 2015). Due to the lengthy residence durations and underlying geological layers in which groundwater is found, cleanup is difficult and expensive once contaminated (Su et al. 2020; Wang et al. 2020). Even when the source of contamination is isolated, the natural cleansing procedures for polluted groundwater can take decades or even hundreds of years to complete (Tatti et al. 2019). Regarding groundwater pollution, agencies and scientists have frequently adhered to their own classification. Subclassifications have also been created, and other source classification techniques have been established. Table 4.2 lists the several classification schemes for sources of contamination.

The categorization of sources based on their origin has been often employed in recent decades. Currently, there is a wider range of categories based on origin, and it is typical to have subclassifications based on both origin and/or kind of chemical constituent. For example, municipal garbage may be a significant contributor to the pollution of groundwater. This section provides a concise documentation of the prevalent causes of groundwater pollution, encompassing both natural and human origins and sources like air deposition and surface drainage.

4.3.1.1 Natural and Anthropogenic Pollution Sources

Groundwater pollution often arises from anthropogenic activity. These operations have the potential to pollute groundwater when materials or trash are encountered. Indeed, while it is accurate to state that most instances of groundwater contamination stem from human activities, it is important to note that natural factors, such as

Table 4.2 Classification of groundwater contamination sources

Parameters	Classification	Examples
Origin	Natural or human induced (anthropogenic)	Natural: Sea water intrusion, rock erosion, and atmospheric deposition Anthropogenic: Urban runoff, industrial effluents, and waste discharge
Character/source	Point, linear, and diffused sources	Point: Landfills, waste dumps, septic tanks, and underground tanks Linear: Roads Diffused: Excess fertilizers, and herbicides and insecticides from agricultural lands and residential areas
Release	Discharge and transport sources	Discharge sources: Industrial effluents, oil spills, municipal sewage outfalls, agricultural runoff, and urban runoff Transport sources: Vehicle emission and industrial stack emission
Location	Above and below groundwater surface	Above groundwater: Industrial discharge, agricultural runoff, urban runoff, acid rain, and atmospheric deposition. Below groundwater: Leakage from underground storage tanks, landfills and waste disposal sites, and irrigation practices
Chemical type	Heavy metals, hydrocarbons, and pesticides	Iron, manganese, chloride, sulfate, etc.

cyclical processes or natural events, can also contribute to groundwater pollution. Freshwater sources can be contaminated by several natural processes, including climate change, natural catastrophes, soil matrix, rock–water interactions, and geological issues (Akhtar et al. 2021c). Contaminated natural chemicals frequently infiltrate groundwater through several pathways. An important factor to consider is the disintegration of minerals that contain noxious compounds. Various naturally occurring elements, such as iron, manganese, arsenic, fluoride, chloride, sulfate, or radionuclide, can dissolve in groundwater and cause natural contamination of this essential resource. Additionally, there is the interaction between plant detritus, peat, and humus as decayed organic matter that flows through groundwater as particles. Depending on the surrounding environment, groundwater may include one or more of these chemicals. Excessive ingestion of some chemicals can have detrimental effects on your health, while other substances can alter the flavor, color, or scent of your body.

Anthropogenic pollutants are compounds that are generated directly from human activities, mostly stemming from land-use patterns. The predominant causes of pollution are those caused by human activities. This area often includes the management of wastewater and solid waste, the treatment and disposal of industrial wastewater, the application of fertilizers, herbicides, and insecticides, the handling of by-products and waste generated by mining activities, and the management of nuclear energy waste. Anthropogenic sources can arise from various activities, including excessive pumping, unrestricted use of fertilizers, mining operations, waste disposal sites, extensive urban expansion, alteration of microclimate due to climate change, improper use of chemicals, disposal of organic and inorganic substances, accumulation of waste and sewage, disruption of river networks, processing of radioactive minerals, and presence of cemeteries. Human-induced alterations significantly impact the water cycle by altering the intensity of existing circumstances and introducing new factors in several aspects, such as leakages, irrigation, extraction, and wastewater (Khatri and Tyagi 2015). Anthropogenic activities result in the concentration of pollutants, including trace elements, in discharge material or water. These pollutants can be very hazardous and potentially fatal to people.

4.3.1.2 Atmospheric Deposition

Groundwater, an essential drinking water resource for millions of individuals globally, faces continuous peril from contamination. Although industrial discharges and agricultural runoff are well-documented causes of groundwater pollution, air deposition is a lesser recognized yet substantial component. This section examines the phenomenon of atmospheric deposition as a contributor to groundwater contamination. It delves into the many types of pollutants involved, their origins and causes, and the resulting environmental consequences. Atmospheric deposition refers to the process by which atmospheric pollutants are transferred into the terrestrial and aquatic surfaces (Gunawardena et al. 2013; Han et al. 2014). It comes in two main forms:

(a) Dry deposition refers to the direct settling of airborne particles and gasses onto the ground without precipitation. Arid locations and drought times are particularly prone to experiencing high levels of dry deposition.
(b) Wet deposition is how precipitation, including rain or snow, carries contaminants to the Earth. Particulate and dissolved pollutants may be present.

As a result, there are wet and dry components to air deposition (Fig. 4.2), with the former immediately causing pollution from surface runoff and the latter possibly causing pollution from materials and chemicals washed off catchment surfaces (Marsalek 2014). In urban areas, where there are many sources of air pollution and a high probability of pollutants washing off into stormwater runoff due to the existence of impermeable surfaces and runoff conveyance networks, atmospheric deposition plays a significant role in runoff pollution (Hobbie et al. 2017).

4.3.2 Consequences of Groundwater Contamination

Groundwater contamination arises when pollutants infiltrate subterranean aquifers, compromising the quality of this essential ecological asset. Groundwater

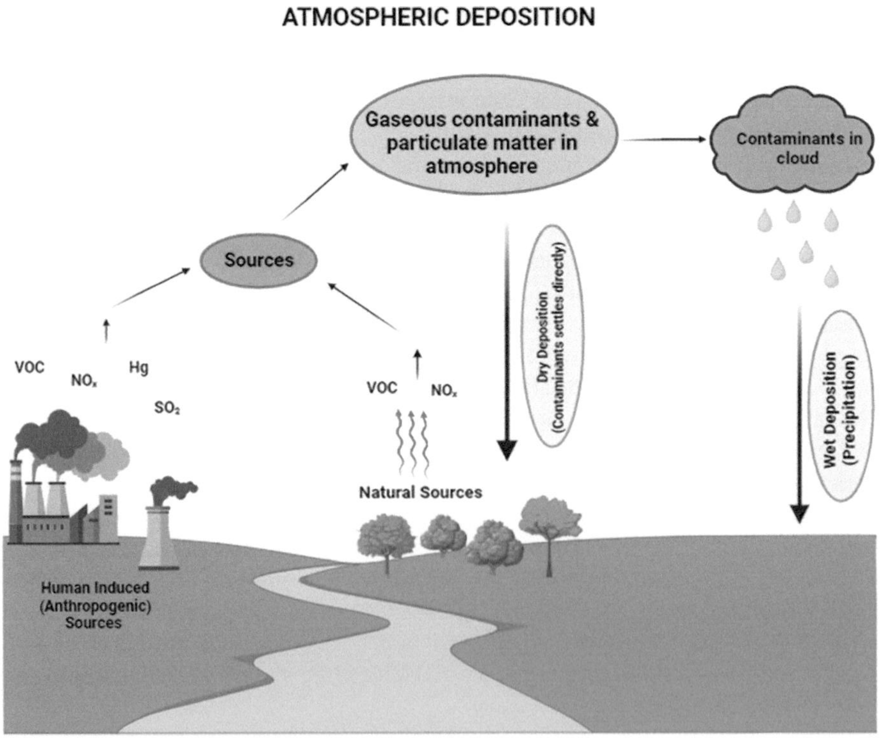

Fig. 4.2 Atmospheric deposition process as wet and dry deposition

contamination has many repercussions that can profoundly affect the environment and human health, resulting in substantial economic and social ramifications. Groundwater contamination can lead to diminished drinking water quality, reduced water availability, impaired surface water systems, expensive cleanup efforts, increased prices for alternative water sources, and significant health hazards. Multiple studies have demonstrated that elevated concentrations of fluoride, nitrate, metals, and persistent organic pollutants pose a significant health hazard to human populations. This is particularly crucial for newborns and children since they are more vulnerable to these pollutants' impacts than adults (He et al. 2020; Karunanidhi et al. 2021). One instance is the condition called "blue baby syndrome," scientifically known as newborn methemoglobinemia, which arises due to high levels of nitrates in the drinking water used for preparing baby formulae.

The pollution of groundwater can also impact human health by affecting the food production chain. Utilizing groundwater polluted with heavy metals and wastewater containing persistent toxins for irrigation can lead to the buildup of harmful elements in grains and vegetables, posing health hazards to humans (Jenifer and Jha 2018; Njuguna et al. 2019; Yuan et al. 2019). Groundwater contamination may disrupt the ecosystem by impacting the food chain. For instance, if pollution causes the death of all fish in a river, crocodiles that rely on fish as their food source may suffer from starvation or be forced to migrate. Groundwater pollution can have adverse effects on the quality of land and forests. Groundwater that is polluted can result in soil pollution and the deterioration of land quality. In several agricultural districts located in dry locations, elevated levels of salt in groundwater are a significant contributing component to the process of soil salinization (Wu et al. 2014). Groundwater contamination not only endangers water quality but also exacerbates the shortage of this essential resource.

A significant proportion of the world's population depends on groundwater for their daily requirements, encompassing drinking, agricultural, and industrial applications. Nevertheless, in the event of extensive contamination, communities are confronted with the harsh truth of having to cease using polluted water, which compels them to seek alternate water sources. Finding alternate sources is frequently plagued with difficulties, particularly when serving a sizable population. The repercussions of polluted groundwater or deteriorated surface water extend beyond obvious health risks and environmental effects. Estuaries that have been impacted by elevated levels of nitrogen resulting from groundwater contamination have experienced significant declines in oyster habitats. The disturbance to fragile ecosystems underscores the complex interdependence between groundwater pollution and biodiversity decline. Severe groundwater pollution might compel communities to forsake groundwater as a feasible source of potable water entirely. Nevertheless, in instances where pollution levels are not excessively high, remedial endeavors may be pursued.

Remediating polluted groundwater requires significant monetary allocations from municipalities or relevant governing bodies. Even after implementing cleanup operations, it is crucial to continuously monitor water quality for an extended period to verify the effectiveness of remediation measures and the restoration of water safety standards. Due to the often-sluggish movement of groundwater, pollution frequently goes unnoticed for extended durations. Treating polluted groundwater to ensure its suitability for drinking purposes incurs significant expenses. Moreover, contaminated water supplies can lead to a decline in property values in impacted regions and escalate healthcare costs because of waterborne illnesses, so imposing economic hardships on communities and governments. Achieving sustainable economic growth necessitates maintaining an equilibrium between the pace at which natural resources are replenished and the level of human consumption.

Freshwater is arguably the most precious of all natural resources. Nevertheless, persistent groundwater pollution can diminish freshwater accessibility, disrupting the equilibrium between water supply and demand and potentially resulting in socioeconomic crises and even armed conflicts. In the future, water scarcity resulting from pollution might potentially contribute to individual disputes (Schillinger et al. 2020), potentially impeding the progress of a nation's socioeconomic advancement. Groundwater pollution goes beyond environmental issues and has become a complicated societal issue that requires collaboration between natural and social scientists. The far-reaching ramifications of groundwater contamination include environmental disturbances, public health risks, financial costs, and societal difficulties.

Addressing these negative effects requires collaborative multidisciplinary actions to minimize them. The pollution causes ecosystem disturbances, putting biodiversity and ecological equilibrium at risk. Simultaneously, contaminants pose health hazards, affecting susceptible populations and burdening healthcare systems. The economic consequences stem from the expenditures associated with cleaning up, the decrease in property value, and the expenses related to healthcare. On the other hand, the social issues include the lack of water and the disruptions within the community. Collaborating with environmental scientists, sociologists, economists, and politicians is essential for understanding the complexities and developing comprehensive policies. Adopting sustainable water management techniques, such as strict legislation, environmentally friendly methods, technological innovations, and public awareness campaigns, is crucial to protect this vital resource and guarantee a healthier and more prosperous future for communities worldwide.

4.3.3 Groundwater Risk Thresholds

The escalating demand for water resources, driven by population expansion, industry, and agriculture, poses a significant danger to water safety. Freshwater availability is declining in many regions globally, accompanied by a deterioration in quality. Additionally, significant freshwater resources in the form of groundwater reserves are depleted owing to excessive extraction rates surpassing the normal recharge rate of aquifers (Niu et al. 2019). Additional stress on groundwater resources is caused by inadequate sanitation, inadequate infrastructure, and pollution from natural sources. Addressing the development of management techniques to safeguard groundwater and evaluating their potential risks for specific purposes are major global problems, as they enhance public availability of potable water.

Groundwater risk thresholds are defined as the specific amounts or boundaries of pollutants in groundwater that, when exceeded, negatively affect human health, ecosystems, or the environment (Bulut et al. 2020). Setting these limits is essential for evaluating and controlling the hazards linked to groundwater pollution. The idea entails the identification of pollutant concentrations or exposure levels that may pose risks to human health or result in ecological damage. The assessment of groundwater danger thresholds necessitates a multidisciplinary approach, integrating knowledge from hydrogeology, environmental science, toxicology, and public health. The objective is to establish scientifically substantiated thresholds for diverse pollutants found in groundwater, including heavy metals, nitrates, pesticides, industrial chemicals, and pathogens. These thresholds function as benchmarks for regulatory requirements, assisting in safeguarding water quality and public health. Groundwater risk thresholds are often established by using risk assessment frameworks, statistical analysis, hydrogeological modeling, and toxicological research. Scientists engage in comprehensive studies to comprehend the behavior of pollutants, their movement and distribution in underground water sources, potential routes of exposure, and the impacts on human populations and ecosystems.

This research aids in establishing the permissible thresholds of pollutants that can exist in groundwater without inducing any detrimental effects. Exceeding these risk thresholds can lead to ecological consequences such as disturbances in aquatic ecosystems, loss of biodiversity, degradation of habitats, and changes in ecosystem processes. Potential human health hazards linked to pollutants in groundwater above threshold levels may encompass acute or chronic ailments, developmental impairments, neurological disorders, or heightened susceptibility to cancer. Nevertheless, setting groundwater risk levels presents certain difficulties. Establishing universal thresholds is hard due to variations in geographical circumstances, pollutant properties, and varying sensitivity among groups (Hinsby et al. 2008). Furthermore, the issue of establishing exact thresholds is compounded by the scarcity of data, uncertainty in risk assessment, and the continuous evolution of scientific knowledge. It is crucial to progress by enhancing monitoring approaches, developing risk assessment procedures, and performing multidisciplinary research. These endeavors aim to enhance the comprehension of contaminant behavior, their impact on ecosystems and human health, and the setting of more precise and all-encompassing criteria for groundwater danger. In summary, groundwater risk thresholds are crucial for directing laws, legislation, and management techniques that attempt to protect groundwater resources. It functions as a crucial instrument in guaranteeing the sustainable utilization of this irreplaceable natural asset while safeguarding human health and ecosystems from the harmful consequences of groundwater pollution.

4.4 Rainwater Pollution

Rainwater pollution arises from the interaction between raindrops, air pollutants, and toxins, giving rise to substantial environmental and health risks. Rainwater, which initially forms from the condensation of water vapor, can become contaminated when it meets different airborne contaminants, industrial discharges, dust, and suspended particulate matter in the atmosphere. Rainwater pollution is caused by acidic compounds emitted by industries and chemicals, pesticides, and particles that are washed off from urban surfaces and agricultural areas (Deng 2021). This pollution hurts ecosystems, leading to the acidity of water bodies, deterioration of soil, and damage to aquatic life and flora. Moreover, using polluted rainwater for drinking might jeopardize human well-being due to hazardous heavy metals, chemicals, or diseases. The corrosive effects of acid rain, caused by the combination of pollutants, result in the deterioration of infrastructure, which creates significant economic consequences (Bakhshipour et al. 2016). Addressing rainwater pollution requires a comprehensive approach that includes implementing measures to control emissions, adopting sustainable agricultural practices, enhancing industrial processes, and implementing monitoring initiatives. These actions aim to maintain water quality, protect ecosystems, and safeguard human health from the negative effects of polluted rainwater. Rainwater pollution arises from various sources, encompassing natural and anthropogenic factors that introduce contaminants into rainwater. Some primary sources of rainwater pollution include:

(a) **Airborne Pollutants**

Human activities are a major source of atmospheric emissions that greatly contribute to the contamination of rainfall. Emissions of sulfur dioxide (SO_2), nitrogen oxides (NOx), carbon monoxide (CO), and particulate matter occur because of industrial activities, automobile exhaust, power generation, and other combustion processes. These contaminants react with atmospheric moisture, resulting in the creation of acidic rainwater.

(b) **Industrial Emissions**

Factories and manufacturing facilities release various pollutants into the air, such as heavy metals, chemicals, and particulate matter. These emissions can be carried by wind and then blended with precipitation, resulting in pollution.

(c) **Agricultural Runoff**

Precipitation can effectively remove chemicals, insecticides, fertilizers, and animal excrement from agricultural areas. Dissolved compounds and phosphates, which are pollutants, dissolve in rainwater and contribute to water contamination, especially in regions with extensive agricultural operations.

(d) **Urban Runoff**

Within urban areas, rainwater meets a range of surfaces, including roads, pavements, roofs, and parking lots. The rainwater accumulates contaminants such as oil, grease, heavy metals, trash, and litter during this interaction. The runoff transports these pollutants into bodies of water or drainage systems, contributing to precipitation contamination.

(e) **Natural Sources**

Contaminants from natural sources can also contaminate rainwater. Volcanic eruptions, wildfires, and dust storms emit particulate matter and chemicals into the atmosphere. Pollen, spores, and naturally occurring mineral particles can also cause rainwater contamination.

(f) **Atmospheric Dust and Particles**

Aerosol particles, including dust, soot, and pollen, can combine with rainwater during precipitation, introducing different pollutants and impurities.

(g) **Transported Substances**

Pollutants emitted in a specific spot can be carried across vast distances by wind and then deposited with rains in a different region, resulting in contamination in regions distant from the initial pollution source.

Comprehending and dealing with these sources of rainwater contamination are essential in executing efficient measures to safeguard water quality, conserve ecosystems, and guarantee the health and welfare of humans. Efforts focused on mitigating emissions, adopting sustainable practices, and controlling runoff can effectively reduce the pollution of rainfall and mitigate its negative effects on the ecosystem.

4.5 Vision to Overcome Water Resources

A vision for water refers to a systematic innovation and research program created by stakeholders in the global freshwater industry. The vision aims to promote the development of sustainable running water resources and enhance the provision of sustainable water-related services worldwide. A vision for water is a collaborative expression outlining the desired path for the water sector. However, it cannot prohibit everyone involved from surpassing the targets set within the specified timeframe of the vision (IUCN 2000). A vision for water ought to be viewed as a first step toward fostering collaboration among companies, public sector stakeholders, colleges and universities, research centers, and other organizations engaged in the creative continuum for sustainable water infrastructure. Nevertheless, water is an indispensable resource for different sectors. Increasing water demands will place significant strain on the country's water resources because of population expansion and economic progress. While formulating its National Water Resources Policy, the government has presumably referred to European authorities and has articulated the aim of achieving sustainable and efficient water management through establishing a vision (Water Europe 2017):

- Widespread possession of secure, sufficient, and affordable potable water, cleanliness, and sanitation.
- Adequate water supply is essential in the areas where it is required to maintain food security, sustain the lives of individuals, and support business operations.
- Ensuring the safeguarding of lives and economic sustenance for everyone by mitigating the risks of droughts and floods.
- An uncontaminated aquatic habitat that sustains thriving fisheries and aquatic ecological systems.

The water industry provides an ideal atmosphere for conducting investigations and fostering innovation through effective stakeholder collaboration. Revolutionary global solutions for the water sector will have an important presence in the international marketplace and support providing safe drinking water, balanced water resource management, and effective wastewater treatment that minimizes the environmental impact (Hellström 2014). All individuals, including municipalities and business organizations, have unrestricted access to a sufficient supply of excellent drinking water that is safe for prolonged consumption. Various solutions and amenities have been developed to respond to the needs of a variable environment and provide more secure analysis in water treatment facilities. Metropolitan areas have implemented climate-resilient water sources that effectively harness precipitation as a valuable resource (Sivanappan 2006). These networks prioritize the purification of groundwater and rainfall while minimizing any negative effects on the environment. Wastewater networks can decrease the release of harmful substances, even in the face of a fluctuating climate, to maintain healthy ecological and chemical conditions across groundwater and sources of water such as lakes and rivers (UNW-DPAC 2015). In addition, drainage systems enhance resource efficiency by extracting biogas, nutrients, and freshwater.

The socioeconomic importance of water becomes apparent among citizens and the authorities. It has resulted in improved consciousness, enhanced utilization of resources, and greater awareness of the environment. To achieve the objective of water sustainability, the stakeholders in the water industry have set specific targets that indicate the projected status in 2020 and 2050 in terms of ensuring a secure source of water worldwide, implementing climate and environmentally friendly water resource solutions, and adopting climate-adapted and resource-efficient water treatment techniques (Hellström 2014).

4.6 Conclusion

This chapter has discussed the running water system in the context of groundwater and rainfall pollution. Running waters such as groundwater and rainfall are distinguished by significant dynamism, being susceptible to significant natural phenomena including storms, droughts, and ice formation. Human exploitation normally alters flowing water by decreasing groundwater and organic material, disrupting flow patterns, fragmenting ecological systems, and reducing biodiversity.

Modernization, which is associated with a swiftly growing human population, will have the most significant influence on running water environments. Upcoming deterioration will be focused on regions around the world where water resources for management have become most restricted. Also, understanding of running water environments is the lowest; most of the harm is expected in lowland streams, which are additionally not well studied. Improved methods for management and a greater public understanding have strengthened running water habitats in developed nations and have the potential to support initiatives to conserve these in underdeveloped countries because of the appropriate implementation of oversight safeguards.

An expanded civilization will have numerous and wide-ranging effects on running water systems. As humanity's population grows, the immediate reasons for changes caused by humans will get worse, consisting of the growing number of towns and industries, growth in farming operations, and an elevated diversion or damming of water. Eliminating or decreasing one factor affecting rivers may not provide significant improvements due to several changes caused by human alterations causing ecological alterations. Enhancing the understanding of running water systems worldwide is crucial, and it is equally important to enhance public awareness of the challenges related to water resource management. Various hazards to running water ecosystems vary significantly. There is not a single approach for addressing the various issues that endanger rivers and streams worldwide. Everyone proposes increasing the focus on educational and scientific study in emerging nations. Researchers look for further information about the species, their biological requirements and difficulties, their impact on running water ecosystems, and the physical and chemical reactions inside these networks. Essential data need to be acquired, together with a practical study on the effects of human actions on running waters.

References

Akhtar N, Mislan MS, Syakir MI, et al (2021a) Characterization of aquifer system using electrical resistivity tomography (ERT) and induced polarisation (IP) techniques. In: IOP conference series: earth and environmental science. Malaysia, pp 1–14

Akhtar N, Syakir MI, Ahmad MI et al (2021b) Modification of the Water Quality Index (WQI) process for simple calculation using the multi-criteria decision-making. Water 13:1–34. https://doi.org/10.3390/w13070905

Akhtar N, Syakir MI, Anees MT ct al (2020a) Characteristics and assessment of groundwater. In: Kalantar B (ed) Groundwater management and resources, 1st edn. Intech Open, London, pp 1–20

Akhtar N, Syakir MI, Bhawani SA, Umar K (2021c) Various natural and anthropogenic factors responsible for water quality degradation: a review. Water 13:1–35. https://doi.org/10.3390/w13192660

Akhtar N, Syakir MI, Rai SP et al (2020b) Multivariate investigation of heavy metals in the groundwater for irrigation and drinking in Garautha Tehsil, Jhansi District, India. Anal Lett 53:774–794. https://doi.org/10.1080/00032719.2019.1676766

Akhtar N, Syakir MI, Tweib SA et al (2023) Particle size distribution and composition of soil sample analysis in a single pumping well using a scanning electron microscope coupled with an energy dispersive X-ray. Water 15:1–20

Allan JD, Castillo MM, Capps KA (2021) Rivers in the anthropocene. In: Allan JD, Castillo MM, Capps KA (eds) Stream ecology, 3rd edn. Springer, Cham, pp 131–161

Anderson EP, Jackson S, Tharme RE et al (2019) Understanding rivers and their social relations: a critical step to advance environmental water management. Wiley Interdiscip Rev Water 6:1–21. https://doi.org/10.1002/WAT2.1381

Archdeacon TP, Reale JK (2020) No quarter: lack of refuge during flow intermittency results in catastrophic mortality of an imperiled minnow. Freshw Biol 65:2108–2123. https://doi.org/10.1111/fwb.13607

Bakhshipour Z, Asadi A, Huat BBK et al (2016) Effect of acid rain on geotechnical properties of residual soils. Soils Found 56:1008–1020. https://doi.org/10.1016/j.sandf.2016.11.006

Bulut OF, Duru B, Çakmak Ö et al (2020) Determination of groundwater threshold values: a methodological approach. J Clean Prod 253:120001. https://doi.org/10.1016/j.jclepro.2020.120001

Bunn SE, Davies PM (2000) Biological processes in running waters and their implications for the assessment of ecological integrity. Hydrobiologia 422–423:61–70. https://doi.org/10.1007/978-94-011-4164-2_5

Chakraborti D, Rahman MM, Mukherjee A et al (2015) Groundwater arsenic contamination in Bangladesh—21 years of research. J Trace Elem Med Biol 31:237–248. https://doi.org/10.1016/j.jtemb.2015.01.003

Constant T, Charrière S, Lioeddine A, Emsellem Y (2016) Use of modeling to protect, plan, and manage water resources in catchment areas. Environ Sci Pollut Res 23:15841–15851. https://doi.org/10.1007/s11356-015-5459-6

Das A, Joardar M, De A et al (2021) Pollution index and health risk assessment of arsenic through different groundwater sources and its load on soil-paddy-rice system in a part of Murshidabad district of West Bengal, India. Groundw Sustain Dev 15:1–16. https://doi.org/10.1016/j.gsd.2021.100652

Das SK, Putra N, Roetzel W (2003) Pool boiling characteristics of nano-fluids. Int J Heat Mass Transf 46:851–862. https://doi.org/10.1016/S0017-9310(02)00348-4

Deng Y (2021) Pollution in rainwater harvesting: a challenge for sustainability and resilience of urban agriculture. J Hazard Mater Lett 2:100037. https://doi.org/10.1016/j.hazl.2021.100037

Elsawah S, Guillaume JHA (2016) Integrated groundwater management. Cham, Springer. Springer Link.com

Exall K, Vassos TD (2012) Integrated urban water management: water use and reuse. In: Metropolitan sustainability: understanding and improving the urban environment, 1st edn. Woodhead Publishing Limited, Canada, pp 319–349

Freeman MC, Bestgen KR, Carlisle D et al (2022) Toward improved understanding of streamflow effects on freshwater fishes. Fisheries 47:290–298. https://doi.org/10.1002/fsh.10731

Gunawardena J, Egodawatta P, Ayoko GA, Goonetilleke A (2013) Atmospheric deposition as a source of heavy metals in urban stormwater. Atmos Environ 68:235–242. https://doi.org/10.1016/j.atmosenv.2012.11.062

Han Y-J, Kim J-E, Kim P-R et al (2014) General trends of atmospheric mercury concentrations in urban and rural areas in Korea and characteristics of high-concentration events. Atmos Environ 94:754–764. https://doi.org/10.1016/j.atmosenv.2014.06.002

Hanh Nguyen H, Venohr M, Gericke A et al (2023) Dynamics in impervious urban and non-urban areas and their effects on run-off, nutrient emissions, and macroinvertebrate communities. Landsc Urban Plan 231:104639. https://doi.org/10.1016/j.landurbplan.2022.104639

Hayes JW, Goodwin EO, Shearer KA, Hicks DM (2019) Relationship between background invertebrate drift concentration and flow over natural flow recession and prediction with a drift transport model. Can J Fish Aquat Sci 76:871–885

He X, Li P, Ji Y et al (2020) Groundwater arsenic and fluoride and associated arsenicosis and fluorosis in China: occurrence, distribution and management. Expo Heal 12:355–368. https://doi.org/10.1007/s12403-020-00347-8

Hellström D (2014) A vision for water—research and innovation agenda for the water sector in Sweden

Hess A, Wadzuk B, Welker A (2021) Evapotranspiration estimation in rain gardens using soil moisture sensors. Vadose Zo J 20:1–12. https://doi.org/10.1002/vzj2.20100

Hinsby K, Condesso de Melo MT, Dahl M (2008) European case studies supporting the derivation of natural background levels and groundwater threshold values for the protection of dependent ecosystems and human health. Sci Total Environ 401:1–20. https://doi.org/10.1016/j.scitotenv.2008.03.018

Hobbie SE, Finlay JC, Janke BD et al (2017) Contrasting nitrogen and phosphorus budgets in urban watersheds and implications for managing urban water pollution. Proc Natl Acad Sci 114:4177–4182. https://doi.org/10.1073/pnas.1618536114

Ielpi A, Lapôtre MGA, Gibling MR, Boyce CK (2022) The impact of vegetation on meandering rivers. Nat Rev Earth Environ 3:165–178. https://doi.org/10.1038/s43017-021-00249-6

IUCN (2000) Visions for water and nature, a world strategy for conservation and sustainable management of water resources in the 21st century

Jenifer MA, Jha MK (2018) Comprehensive risk assessment of groundwater contamination in a weathered hard-rock aquifer system of India. J Clean Prod 201:853–868. https://doi.org/10.1016/j.jclepro.2018.08.005

Karunanidhi D, Aravinthasamy P, Deepali M et al (2021) Groundwater quality evolution based on geochemical modeling and aptness testing for ingestion using entropy water quality and total hazard indexes in an urban-industrial area (Tiruppur) of southern India. Environ Sci Pollut Res 28:18523–18538. https://doi.org/10.1007/s11356-020-10724-0

Khatri N, Tyagi S (2015) Influences of natural and anthropogenic factors on surface and groundwater quality in rural and urban areas. Front Life Sci 8:23–39. https://doi.org/10.1080/21553769.2014.933716

Kim RH, Lee S, Kim YM et al (2005) Pollutants in rainwater runoff in Korea: their impacts on rainwater utilization. Environ Technol 26:411–420. https://doi.org/10.1080/09593332608618546

Kipnusu EK, Erimba CMM, Mbaka JG (2023) Impact of weirs on decomposition rates of Dombeya goetzenii (K. chum) leaf litter in the River Njoro Kenya. African J Pure Appl Sci 4:1–11. https://doi.org/10.33886/ajpas.v4i3.424

Kurwadkar S (2019) Occurrence and distribution of organic and inorganic pollutants in groundwater. Water Environ Res 91:1001–1008. https://doi.org/10.1002/wer.1166

Li P, Karunanidhi D, Subramani T, Srinivasamoorthy K (2021) Sources and consequences of groundwater contamination. Arch Environ Contam Toxicol 80:1–10. https://doi.org/10.1007/s00244-020-00805-z

Malmqvist B, Rundle S (2002) Threats to the running water ecosystems of the world. Environ Conserv 29:134–153. https://doi.org/10.1017/S0376892902000097

Marsalek J (2014) Urban water cycle processes and interactions. CRC Press, Washington, DC

Martins EH, Nogarotto DC, Mortatti J, Pozza SA (2019) Chemical composition of rainwater in an urban area of the southeast of Brazil. Atmos Pollut Res 10:520–530. https://doi.org/10.1016/j.apr.2018.10.003

Müller A, Österlund H, Marsalek J, Viklander M (2020) The pollution conveyed by urban runoff: a review of sources. Sci Total Environ 709:136125. https://doi.org/10.1016/j.scitotenv.2019.136125

Nemčić-Jurec J, Ruk D, Oreščanin V et al (2022) Groundwater contamination in public water supply wells: risk assessment, evaluation of trends and impact of rainfall on groundwater quality. Appl Water Sci 12:1–13. https://doi.org/10.1007/s13201-022-01697-1

Niu J, Zhu X-G, Parry MAJ et al (2019) Environmental burdens of groundwater extraction for irrigation over an inland river basin in Northwest China. J Clean Prod 222:182–192. https://doi.org/10.1016/j.jclepro.2019.03.075

Njuguna SM, Makokha VA, Yan X et al (2019) Health risk assessment by consumption of vegetables irrigated with reclaimed waste water: a case study in Thika (Kenya). J Environ Manag 231:576–581. https://doi.org/10.1016/j.jenvman.2018.10.088

Postigo C (2018) Reference module in earth systems and environmental sciences. Elsevier, New Washington, DC, p 489

Qadir A, Tahir M, Ahmad Abir I et al (2020) Land change detection of Delhi area, India: a GIS approach. Pjaee 17:7769–7780
Sanmi AM (2018) Effect of air pollution on rain water: a case study of Ado-Ekiti, Nigeria. Int J Adv Eng Res Sci 5:19–24. https://doi.org/10.22161/ijaers.5.8.3
Schillinger J, Özerol G, Güven-Griemert Ş, Heldeweg M (2020) Water in war: understanding the impacts of armed conflict on water resources and their management. WIREs Water 7:1–12. https://doi.org/10.1002/wat2.1480
Sendzimir J, Schmutz S (2018) Challenges in riverine ecosystem management. In: Huisman J (ed) River ecosystem management: science for governing towards a sustainable future, 1st edn. Springer International Publishing, Switzerland, pp 213–213
Sivanappan RK (2006) Rain water harvesting, conservation and management strategies for urban and rural sectors. In: National Seminar on Rainwater harvesting and water management, pp 1–5
Stock W, Callens M, Houwenhuyse S et al (2021) Human impact on symbioses between aquatic organisms and microbes. Aquat Microb Ecol 87:113–138. https://doi.org/10.3354/AME01973
Su Z, Wu J, He X, Elumalai V (2020) Temporal changes of groundwater quality within the groundwater depression cone and prediction of confined groundwater salinity using Grey Markov Model in Yinchuan area of Northwest China. Expo Heal 12:447–468. https://doi.org/10.1007/s12403-020-00355-8
Tatti F, Petrangeli Papini M, Torretta V et al (2019) Experimental and numerical evaluation of groundwater circulation wells as a remediation technology for persistent, low permeability contaminant source zones. J Contam Hydrol 222:89–100. https://doi.org/10.1016/j.jconhyd.2019.03.001
Teng Y, Hu B, Zheng J et al (2018) Water quality responses to the interaction between surface water and groundwater along the Songhua River, NE China. Hydrogeol J 26:1591–1607. https://doi.org/10.1007/s10040-018-1738-x
Tong X, Mohapatra S, Zhang J et al (2022) Source, fate, transport and modelling of selected emerging contaminants in the aquatic environment: current status and future perspectives. Water Res 217:118418. https://doi.org/10.1016/j.watres.2022.118418
UNW-DPAC (2015) Water and sustainable development from vision to action: means and tools for implementation and the role of different actors. Zaragoza, Spain
Wang D, Wu J, Wang Y, Ji Y (2020) Finding high-quality groundwater resources to reduce the Hydatidosis incidence in the Shiqu County of Sichuan Province, China: analysis, assessment, and management. Expo Heal 12:307–322. https://doi.org/10.1007/s12403-019-00314-y
Water Europe (2017) Water Europe Water Vision 2030—the value of water, Brussels
Wondzell SM (2011) The role of the hyporheic zone across stream networks. Hydrol Process 25:3525–3532. https://doi.org/10.1002/hyp.8119
Wu J, Li P, Qian H, Fang Y (2014) Assessment of soil salinization based on a low-cost method and its influencing factors in a semi-arid agricultural area, Northwest China. Environ Earth Sci 71:3465–3475. https://doi.org/10.1007/s12665-013-2736-x
Yang L, Wang Y, Wang Y et al (2023) Water quality improvement project for initial rainwater pollution and its performance evaluation. Environ Res 237:116987. https://doi.org/10.1016/j.envres.2023.116987
Yuan Y, Xiang M, Liu C, Theng BKG (2019) Chronic impact of an accidental wastewater spill from a smelter, China: a study of health risk of heavy metal(loid)s via vegetable intake. Ecotoxicol Environ Saf 182:109401. https://doi.org/10.1016/j.ecoenv.2019.109401

Recent Advances in Sustainable Waste Management Practices

5

Habib Ur Rahman, Mahiuddin Khan, and Allah Ditta

Abstract

Recently, the increasing quantity of municipal solid trash has been causing major environmental problems, necessitating a better strategy for dealing with the waste that is generated. An effective waste management strategy is critical in reducing negative environmental, social, and economic repercussions. Sustainable waste management is introduced to preserve the balance between the environment, social, and economic factors in a variety of methods, including the application of legislation and waste management practices. As a result, identifying the present waste management system used by the industry is critical for making adjustments and improvements in the transition to sustainable waste management. This chapter discusses the current waste management system including different types of waste, their management strategies, and the problems of incorporating the concept of sustainability into waste management by evaluating previous similar research and recent developments in sustainable waste management around the globe. After going through the literature, it has been found that the current waste management systems and factors that hinder the concept of sustainability in waste management need to be considered seri-

H. U. Rahman
Department of Biotechnology, Shaheed Benazir Bhutto University Sheringal, Dir Upper, Khyber Pakhtunkhwa, Pakistan

M. Khan
Department of Environmental Sciences, Shaheed Benazir Bhutto University Sheringal, Dir Upper, Khyber Pakhtunkhwa, Pakistan

A. Ditta (✉)
Department of Environmental Sciences, Shaheed Benazir Bhutto University Sheringal, Dir Upper, Khyber Pakhtunkhwa, Pakistan

School of Biological Sciences, The University of Western Australia, Perth, WA, Australia
e-mail: allah.ditta@sbbu.edu.pk

A. A. H. Abdel Latef et al. (eds.), *Bioremediation and Nanotechnology for Climate Change Mitigation*, https://doi.org/10.1007/978-981-96-3069-1_5

ously using an actively community-based approach. It allows a major shift in waste management by improvising current waste management technology more sustainably.

Keywords

Sustainable waste management · Climate change · Waste management strategy · Recycling

5.1 Introduction

Waste management is conserving resources and protecting human and environmental resources (Cucchiella et al. 2017). Any item left over from industrial and human activity that is worthless is referred to as waste (Sukholthaman and Shirahada 2015). The tremendous rise in trash creation that has resulted from the years of urbanization and population expansion presents a serious concern. For correct disposal, it's also critical to have an effective waste monitoring and management system (Shreya et al. 2023). Globally, 1.3 billion tons of solid trash were produced in 2012; by 2025, that amount is predicted to have doubled (Sadef et al. 2016). Most industrialized nations, including several in Europe and Japan, have adopted cutting-edge waste management technologies that aim to reduce garbage's eventual disposal while creating heat or power and preserving energy resources elsewhere (Dijkgraaf and Vollebergh 2004). The majority of the time, trash incineration technology is used for energy recovery; if this method is unsuccessful, the garbage is finally dumped into landfills. Landfilling carries some concerns, including the need for huge tracts of land, increased potential for leaks into the air, water, and soil, and a lower energy recovery rate than burning. On the other hand, burning results in lower emissions, especially when using waste-to-energy (WTE) recovery methods (Miranda and Hale 1997).

The Global Waste Management Outlook, a collaborative effort between the United Nations Environment Program and the International Waste Management Association, is a groundbreaking scientific assessment of waste management worldwide and a call to action for all countries. As a follow-up to the Rio + 20 Summit, the document was produced in response to the United Nations Environment Program (UNEP) Governing Council decision GC 27/12. It outlines the rationale behind a holistic approach to waste management as well as how resource and waste management can be recognized as a key factor in sustainable development and mitigating climate change (Wilson et al. 2015). After materials are recovered and recycled, all garbage is regarded as a sustainable resource. Furthermore, renewable waste materials from agriculture, homes, and industries may be converted into biogas, bioalcohols, and hydrogen, among other usable energy forms, by using the right WTE technology (Mubeen and Buekens 2019). Infrastructure development and systems planning will be greatly impacted by rising waste quantity trends (Kumar and Goel 2009). Careless handling and management of rubbish significantly negatively influence its ecological imprint. In underdeveloped countries, waste is either burned or dumped outdoors since land is readily available and the economy is viable. Municipal solid waste (MSW) landfills are the third-largest source of anthropogenic methane emissions, according to the United Nations (UN) Climate Summit (2014). Groundwater, surface water, and soil are contaminated by landfill leachate due to its toxic materials and metal content (Cheela et al. 2021). Solid waste creation in industrialized countries has resulted in severe environmental strain due to population increase and changing consumer habits. Proper trash disposal has become essential for maintaining the environment, human health, and sustainable means of subsistence (Al-Dailami et al. 2022).

WTE technology has shown to be a crucial tactic in addressing this problem and is likely to play a bigger role in the future. With a recycling rate of 51%, high-income countries (HICs) have made great strides in the recycling of garbage. On the other hand, just 16% of garbage gets recycled in low-income countries (LICs), with the remaining 93% being disposed of in open landfills. This implies that solid waste management (SWM) practices are inadequate in LICs, which can seriously jeopardize livelihoods, the climate, and human well-being (Shovon et al. 2024). Effective waste management has emerged as a significant concern in the twenty-first century's quickly urbanizing environments (Thompson 2023). An incredible 2 billion tons of solid waste are produced worldwide each year. This amount is expected to rise to about 3.5 billion tons by 2025. It is alarming to read that in developed countries, just 2% of waste is properly disposed of, but in poor or low-income countries, that number may be as high as 93% (Yatoo et al. 2024).

5.2 Prospects of Sustainable Waste Management

All societies should aim to ensure sustainable waste management. Approximately 80% of garbage produced worldwide is now disposed of in landfills, many of which are improperly designed or contained. The growing focus on how human activity affects the environment and the rising need for resources and energy has led to a new understanding of waste streams (Mulya et al. 2022). Utilizing waste streams to recover minerals and energy is becoming increasingly common, particularly in industrialized nations like Japan, the United States, and Europe. Even if these initiatives now have little effect on trash disposal, as society grows more aware of its alternatives, the utilization of waste streams to extract value may expand (Castaldi 2014).

5.3 Global Waste Generation and Projection Comparisons

A major consequence of rising production and consumption is an excessive quantity of garbage being produced. An enormous quantity of garbage is being disposed of into the environment due to the need for more products and services to fulfill human demands. Waste production often rises with societal urbanization and economic growth (Ahmed and Ali 2004; Beolchini et al. 2012). Global production of MSW is now anticipated to be 1.3 billion tons per year; by 2025, this figure is expected to increase to 2.2 billion tons. According to conservative estimates, metropolitan areas in Asia are anticipated to generate 1.8 million tons of MSW daily by 2025 (Hoornweg and Bhada-Tata 2012). A major issue is the growing amount of MSW generated, especially in emerging country towns with limited capacity and diminishing landfill places (Sukholthaman and Shirahada 2015). Garbage management is approaching a breaking point, particularly in many emerging Asian nations. According to the U.S. Environmental Protection Agency, the creation of MSW has increased by 2.6 times between 1960 and the present (UN-Habitat 2010).

5.4 Criteria for Classification of Wastes

The source or origin of trash formation is a crucial factor in garbage classification, and its significance is supported by the fact that the location and time of waste generation significantly influence the waste's characteristics. Waste may be categorized as production, special, commercial, institutional, street, and residential waste based on where it comes from. Any material, chemicals, products, damaged items, etc. created at the household level due to individual or household activities are referred to as domestic trash (Torkayesh et al. 2022). Similarly, waste from establishments, companies, and roadways is considered. Production waste is defined as any garbage generated because of related technological processes or because of extracting and processing natural resources. Waste that falls into the special waste category includes waste from electric and electronic equipment, bulky waste, construction and demolition waste, waste generated within the health system, and other waste that, due to its unique nature, require special collection, recycling, and treatment installations (Vlachokostas et al. 2021). Special waste is typically collected from the generation source through separate collection. Because certain waste kinds are more hazardous or harmful than others due to their nature and structure, the degree of waste toxicity needs to be taken into account when classifying waste (Bolan et al. 2023). The greater risk to the environment and public health is directly correlated with the level of toxicity of the materials found in the waste that is produced and gathered. Consequently, three categories of waste may be considered potentially harmful: semi-hazardous, inert, and hazardous (Bilgilioglu et al. 2022). Garbage designated as hazardous is defined as garbage that significantly harms environmental components and increases pollution. Combustible, corrosive, reactive, or poisonous waste falls into this category and poses a severe risk of contamination in the area (Khoaele et al. 2023). The opposing pole is occupied by inert waste, which does not pose a concern to the environment or public health due to its physical and chemical characteristics since it does not undergo structural or morphological changes due to outside influences (Hoang and Fogarassy 2020). Glass, shattered bottles, ceramic debris, and similar materials are common examples of inert waste. Waste that presents an average environmental risk is classified as semi-hazardous waste. This category could include garbage that, although in theory less dangerous, could become more dangerous if mismanaged. Semi-hazardous trash is waste that does not fit the hazardous or inert waste categories. Examples of this type of waste include biodegradable waste and garbage that can burn (Burcea 2015).

5.5 Categories of Wastes

5.5.1 Solid Waste and Liquid Waste

Wastes that are generated in a solid state are referred to as solid wastes. Global waste creation is increasing daily. Urban populations produced 2.01 billion tons of solid trash in 2016, with an average daily contribution of 0.74 kg per person (Khan et al. 2016). Because of growing urbanization, the amount of rubbish produced annually is expected to increase by 70% between 2016 and 2050. It is projected that by 2050, the amount of waste produced will have risen from 2.01 billion tons in 2016 (Akacem et al. 2020).

Wastewater from stores, fuel depots, mines, quarries, ships, industries, offices, and residences is referred to as liquid waste. Wastewater comes from three primary sources: residential, commercial, and agricultural (Noor et al. 2020). The growing shortage of water poses an inevitable threat to human existence. On the other hand, water contamination has grown to be a serious issue, particularly in emerging nations (Zhang et al. 2016). Wastewater generation is rising dramatically because of urbanization and population growth. Nearly 90% of wastewater in underdeveloped nations is untreated in lakes, rivers, and the ocean (Ahmad et al. 2019). Every year, illnesses carried by wastewater claim the lives of six to eight million individuals, of whom 1.8 million are children under the age of five (Khan et al. 2016). Furthermore, it is predicted that the amounts of gases produced from wastewater, such as nitrous oxide and methane (CH4), will rise by 25% and 50%, respectively, and will have a detrimental effect on the climate. The largest wastewater stream comes from the food, paper, petrochemical, and agricultural industries. It contains severely polluted inorganic or organic salts and oil (Salgot and Folch 2018). The industrial revolution, coupled with population increase, caused a rise in the demand for freshwater supplies. Although several strategies exist to address it, such as chemical, biological, and physical ones, freshwater is becoming scarce in the ecosystem (Tetteh et al. 2019).

5.5.2 Municipal Solid Waste

The majority of garbage produced by homes and businesses that is not naturally gaseous or liquid is categorized as municipal solid waste. MSW is sometimes referred to as "garbage" or "trash," and it includes wastes from stores, businesses, merchants, and offices, except hazardous, industrial, and construction wastes (Yang et al. 2018). Municipal solid wastes, sometimes called garbage or rubbish, are commonplace items like batteries, paints, furniture, appliances, newspapers, food scraps, bottles, clothing, and packaging that we discard after using them (Khan et al. 2016). Paper, glass, metals, textiles, wood, plastics, and food waste make up MSW, a kind of biomass waste. The two most popular MSW management techniques are landfilling and open dumping. Both approaches have drawbacks, including contaminated environments, the production of methane gas that contributes to global warming, and labor disputes (Murtaza et al. 2017).

5.5.3 Organic Waste

Quick-decomposing garbage is referred to as organic waste. Organic wastes can come from various sources, such as agricultural waste, market waste, kitchen waste, and municipal solid garbage. Improper handling of this garbage might negatively affect the ecosystem (Gonawala and Jardosh 2018).

5.5.4 Inorganic Recyclable Waste

Plastic, tin glass, aluminum, and other metals are among the recyclable inorganic waste materials (Narayanamoorthy et al. 2024).

5.5.5 Inorganic Non-recyclable Waste

Non-biological plastic bags, random plastics, random glass, sanitary wastes, disposable diapers, and polystyrene are examples of non-recyclable garbage (Ojeda-Benitez et al. 2003).

5.5.6 Hazardous Waste

Hazardous wastes have many qualities, such as corrosivity, reactivity, toxicity, and ignitability. Hazardous waste is always a risk to the environment and public health. Because of the qualities mentioned above, breathing in or ingesting dangerous wastes can be detrimental to one's health (Chandra and Yadav 2015). According to the 2016 hazardous waste regulations, each industry handles, treats, and gets rid of hazardous waste. The number of hazardous industrial wastes has significantly increased along with the expansion of industrialization (Lieberman et al. 2018). Hazardous waste is mainly produced by companies that make distilleries, tiles, textiles, electronics, paints and pigments, chemicals, pharmaceuticals, and engineering products. Electroplating is a primary source of hazardous waste (Kwikiriza et al. 2019).

- Hazardous waste is defined as waste having one or more of the following qualities: it is oxidizing, explosive, highly flammable, flammable, irritant, harmful, toxic, carcinogenic, corrosive, infectious, toxic for reproduction, mutagenic; it releases toxic gas when it comes into contact with water, air, or acid; it is ecotoxic; it is sensitizing; or it has the potential to yield another substance after disposal that has any of the aforementioned properties (Shin et al. 2013).

- "Non-hazardous waste" is the phrase used to describe products or materials that a waste producer no longer needs but that are approved for processing at a waste processing facility or disposal at a landfill. Materials from both residential and non-residential sources are included in non-hazardous garbage (Wang et al. 2016).

5.5.7 Industrial Waste

Industrial regions generate solid waste that pollutes the air, water, and land, necessitating sound management. Waste residues from chemical processing, smelting, ore dressing, and mining waste are examples of industrial solid wastes created during industrial activities (Cai et al. 2018). The detrimental impacts of pollution start interfering with natural security and human health as society and economic standing improve. As industrialization speeds up, so does the amount of industrial waste produced, endangering human health (Prabakar et al. 2018).

5.5.7.1 Wastes from Agro-industry

Farm waste is the debris created during the production of agricultural products, their harvesting or processing, the rearing of animals, and their storage. It also includes dust, wastewater released into the environment, and leftovers from industrial waste (rice straw, coconut fiber) (Girelli et al. 2020; Mo JiaHao et al. 2018). Agriculture wastes, also known as agro wastes, are the residues from processing raw agricultural goods, such as cereal and grain crops, fruits and vegetable crops, animal waste, meat, poultry, and dairy products (Obi et al. 2016). Approximately one-third of all agricultural production is produced worldwide yearly in the form of 1.3 billion tons of agricultural trash. Produced each year from fruits, vegetables, roots, and tubers, 520–650 million tons of waste are generated, which make up 40–50% of the global quantitative loss (Ravindran et al. 2018). Agro-industrial processing in the European Union generates over 367 million tons of trash yearly, with Germany, the United Kingdom, Italy, France, and Spain leading the way in waste output (Correddu et al. 2020).

5.5.7.2 Wastes from the Food Industry

Various agricultural streams yield a diverse range of foods, necessitating the growth of processing businesses such as those focused on fruits and vegetables, cereals and pulses, edible oils, dairy, meat and poultry, and marine food. Each of these sectors generates a sizable quantity of garbage. Any stage of the process, including manufacturing, harvesting, handling, processing, packing, storing, and distribution, can result in waste (Poonam Sharma et al. 2020). Over six million tons of solid waste, consisting of leaves, skin, peels, stalks, and stems, are produced annually by the frozen food processing, fruit, and vegetable canning businesses (Sagar et al. 2018).

The poultry and meat processing businesses release large numbers of animals, such as 49% of cattle, 47% of sheep and lambs, 44% of pigs, and 37% of chicken, and discard them as waste (Adhikari et al. 2018). Every year, the food processing sectors within the European Union discharge around 88 million tons of food waste. Because of its chemical makeup, this garbage has the potential to aggravate issues like eutrophication, acidification, and global warming. As a result, it is known to have enormous detrimental consequences on the environment and the global economy (Salihoglu et al. 2018).

5.5.7.3 Wastes from the Paper and Pulp Industry

Making and recycling paper involves accumulating a wide range of dangerous chemical substances, including mineral oils, phenols, parabens, and phthalates (Sebastião et al. 2016). Large-scale trash generated during the papermaking and pulping processes is known as solid-state waste from the paper and pulp-processing sector, including paper sludge. De-inked paper wastes are also an organic by-product of paper recycling (Zhang and Sun 2018).

5.5.8 Agricultural Wastes

The residual materials produced during the growth and production of crops, fruits, vegetables, dairy products, meat, and poultry are known as agricultural wastes. During the development and processing of agricultural commodities, other waste materials may be produced, but their worth is lower, and their profit margin is smaller than the cost of production (Acharya et al. 2018). Agricultural wastes, or agro wastes, include wastes from food processing, animal waste (animal carcasses, manure), crop waste (pruning, vegetable and fruit droppings, sugarcane bagasse, and maize stalks), and hazardous waste (herbicides, insecticides, and pesticides). The quantity that agricultural waste contributes is rarely calculated, although, in industrialized locations, it accounts for a sizable portion of the trash (Obi et al. 2016). Crop leftovers, animal dung, and agricultural by-products all directly contribute to increased agricultural output. An estimated 998 million tons of garbage are produced by agriculture worldwide (Acharya et al. 2018). Almost 80% of this total solid waste is made up of organic trash. The manure output is 5.27 kg/day/1000 kg living weight on a wet weight basis (Obi et al. 2016).

5.5.8.1 Crop Residues

The term "crop residue" describes plant pieces left in the field after harvest (Lal 2005). Approximately 3.5–4 × 109 Mg of plant residues are generated yearly, with grains accounting for 75% of this total (Bhattacharyya and Barman 2018). The removed leftovers have a detrimental impact on soil fertility, agronomic productivity, and environmental quality since they are typically consumed for food and fiber (building materials, domestic fire, animal feed, and bedding, biofuel production, paper manufacturing, and mushroom cultivation) (Maw et al. 2019).

5.5.8.2 Farmyard Waste

Garden garbage is one of the municipal solid waste's main constituents (MSW) constituents. It is produced by natural parks, plant nurseries, gardens, and road sweepings, among other places. Grass clippings, fresh and dried leaves, tiny plant roots and shoots, and woody debris are the main components of yard waste. Large amounts of yard waste handled incorrectly might cause smell issues at the disposal locations (Bary et al. 2005; López et al. 2010). Typically, grass clippings make up the majority of garden trash (about 75%), most likely due to the clippings being produced all year. Yard trash contains nutrients that microbes may use to break down MSW biologically (Akinbile and Yusoff 2012).

5.5.8.3 Pesticides Waste

Modern science and technology have led to significant advancements in agricultural productivity and quality, but they have also resulted in increasing pesticide waste. The packing of pesticide residues is referred to as pesticide trash. Pesticide waste will negatively affect human health, the environment, and the land and water (Li and Huang 2018).

5.5.8.4 Fertilizer Waste

Fertilizers containing phosphate are widely used worldwide to provide crops with the necessary phosphorus. To produce phosphoric acid, which is the first stage in the manufacturing of commercial phosphate fertilizer, previously concentrated phosphate ores are usually treated with sulfuric acid through a wet digestion process. However, throughout this industrial process, a variety of by-products is produced. The main undesirable by-product is phosphogypsum (PG), a term used to refer to a mixture of big solid and small liquid waste components (Cánovas et al. 2017). Although unreacted phosphate rock, gangue mineral particles, and other minor solid phases that are reaction products of the acid-wet process, including alkali fluorosilicates and fluorides (e.g., quartz and feldspars), can also be included, gypsum makes up the majority of PG. In addition to industrial process fluids, composed of residual acids that have become trapped in the voids between mineral particles, PG also consists of solid components. During PG disposal, spent process waters are routinely injected into the PG stacks. Once these acidic fluids pass through the PG stacks, they usually collect in peripheral channels. Yet, edge outflows might appear from the bottom of the stacks if the draining mechanism is ineffective.

Phosphoric acid production is waste producing; for every ton of phosphoric acid generated, about five tons of PG are created. The massive output of these undesirable by-products—more than 100–280 million ton of PG produced annually—worldwide has prompted researchers to look for novel recycling options (Parreira et al. 2003).

5.5.9 Medical Waste

The United States Medical Waste Tracking Act of 1988 defines medical waste as "any solid waste that is generated in the diagnosis, treatment, or immunization of human beings or animals, in research pertaining thereto, or in the production or testing of biological" (United States Congress 1988). According to estimates from the World Health Organization (WHO), over 20% of these medical wastes are hazardous substances that might be radioactive, toxic, or infectious (Birchard 2002). Radioactive materials, leftover chemicals, and pharmaceutical packaging are among the hazardous wastes produced by hospitals. If infectious trash is not disposed of appropriately, there is a risk of a disease epidemic (Noor et al. 2020).

5.5.10 Biomedical Wastes

Biomedical wastes are thought to be hazardous and toxic substances. Any waste product obtained from the vaccination or treatment of live organisms is referred to as biomedical waste, and it can be either liquid or solid. Wastes from biology, pathological, and clinical laboratories and receptacles such as bowls, syringes, and ampullas are included in this (Kaur et al. 2019). Biomedical waste in urban areas is a health problem in developed nations. Infectious diseases like AIDS and hepatitis are more likely to spread when biological waste from medical laboratories is carelessly and randomly disposed of (Kwikiriza et al. 2019). Greater biomedical waste production is correlated with more extensive healthcare facilities in urban areas of developed countries. In India, the daily output of hospital waste is estimated to be 0.33 million tons, or 0.5–2 kg/person. The volume of hazardous trash is anticipated to range from 10% to 25% (Noor et al. 2020).

5.5.11 Universal Waste

Universal waste contains trace amounts of dangerous substances that are bad for the environment and human health, including arsenic, antimony, beryllium, cadmium, copper, lead, mercury, nickel, PCBs (polychlorinated biphenyl), and zinc. Even though these substances might not be found in great concentrations, they have the potential to accumulate to hazardous levels and damage the environment over time. Specific regulations are in place in several nations for the handling and disposal of universal waste (Ali et al. 2023a, b). Certain items are often classified as universal waste: fluorescent lamps and bulbs; paint; automobile and equipment fluids; cell phones; batteries (all types); small appliances; mercury-containing devices and equipment; non-empty aerosol cans (such as propane, butane, and pesticides); photovoltaic modules and solar panels; printer cartridges; antifreeze; cathode ray tubes (CRTs); ballasts; oil-based finishes; and hazardous waste pharmaceuticals (Adebambo 2017).

5.5.12 Construction Site Waste

Building and tearing down waste is a combination of extra materials left over when buildings and other structures—like roads and bridges—are built, renovated, or demolished (Kofoworola and Gheewala 2009). The huge volumes of building waste created due to the world's rapid economic development have many severe environmental repercussions, including land degradation, energy waste, carbon and greenhouse gas emissions, and the loss of natural resources (Wang et al. 2014). The primary waste from buildings is glass, plastics, wood, and steel, as well as excess mortar, concrete, broken bricks, green waste (grass, bushes), and excavated soil. Growing urbanization has various detrimental effects on the environment, from waste from construction to destruction (Ulubeyli et al. 2017).

5.5.13 Radioactive Waste

Radioactive wastes are by-products of several nuclear reactions in medicine and research, such as nuclear fission and nuclear power generation (Lennemann 1978). Radioactive materials that are bad for the environment and human health are among these wastes (Noor et al. 2020). These trash kinds are frequently retained to limit radioactive exposure to humans since their disintegration durations are longer than those of other types of waste. High-level radioactive waste should be stored for 50 years before being disposed of, unlike low-level radioactive waste, which requires less storage and is disposed of immediately (Lee et al. 2019). Fuel wastes are usually submerged for 45 years before being dried out and stored. Deep geographic areas appear to be the best places to dispose of radioactive waste (Noor et al. 2020).

The terms "low-level," "medium-, or intermediate-level," and "high-level" refer to different amounts of radionuclides or radioactivity in radioactive waste, and they can be used interchangeably. These terms, first used in the 1950s for operational purposes, usually indicated how the waste was being handled or processed and what was being done at a specific location under local operating conditions rather than accurately describing radioactive concentrations or contamination (Hemidat et al. 2022). This is still the case today, as the precise and/or numerical splits between the three groups vary from nation to nation and even within institutions within the same country. Although high radiation levels are undoubtedly a characteristic of high-level waste, its primary characteristic is likely that it needs specific handling and attention due to the radioactive decay heat load. Examples of these unique handling and considerations include substantial biological shielding and designed cooling systems. When the term "high-level waste" is used, it usually refers to the liquid effluent, or raffinate, from the first cycle of fuel reprocessing operations that recovers the plutonium and unburned uranium. Actinides, or transuranium elements that create alpha particles, are included in this category if their concentration of fission products is high enough to require cooling unless they are kept apart from the waste (Natarajan et al. 2020).

Liquid low- and medium-level radioactive waste (LLLRWs and LMLRWs) originates from several sources, including nuclear reactors, isotope laboratories, and nuclear medical institutions. Many of these wastes need to be treated to reduce their total levels of radioactive and nonradioactive contaminants to be disposed of in compliance with national and international requirements (Kumar et al. 2024). To securely deposit the concentrate in a radioactive "cemetery" following fossilization, the treatment methods for these wastes entail eliminating radioactive chemicals from the effluent and lowering its volume. This is due to the wastes' relatively huge origin volume and low specific activity content (Chmielewski et al. 2001).

5.5.14 Mining Waste

Significant amounts of residues are produced throughout the mining process, which need to be carefully handled to balance the need for environmental sustainability with economic efficiency (Daraz et al. 2023). Considerations include the need for energy, the environmental and public health hazards, the pressure on water supplies, and the necessary technologies. The waste produced by mining can be solid, slurry, or tailings; tailings, waste rock, slag, and tail ends are the most frequent types. Under some conditions, however, vegetation and overburden may also be deemed trash (Aznar-Sánchez et al. 2018).

5.6 Impact of Waste Incorporation on Soil Health

Mining, military operations, intensive industrial activity, poor urban waste management, and accidents introduce excessive levels of local soil and water pollutants. Soils could only handle these pollutants to a limited extent by filtering or transforming them. When this capacity is surpassed, problems including contaminated water, contaminated soil contact with humans, contaminated plants, and landfill gas hazards become more serious (Europe Environmental Assessment Agency 2007). Depending on the contaminants' inclination, they may start in soil-held water or seep into subterranean water. Hazardous waste may change the chemistry of the soil and affect plants and animals that rely on the soil for nutrition (Table 5.1). These waste materials include Cd, Cu, Ni, Pb, and Zn (Shayler et al. 2009). Human activities such as mining, smelting, manufacturing, farming, and burning fossil fuels introduce heavy metals (such as arsenic, cadmium, and mercury) into our soils (Murtaza et al. 2021a, b; Raza et al. 2023; Sabir et al. 2020; Sarfraz et al. 2019). Our disposal of heavy metal-containing products, including paint, technological waste, and sewage, increases the amount of heavy metal pollution even more (De Silva et al. 2023). The health of millions of people might be impacted by soil contamination. Malignancies brought on by arsenic, asbestos, and dioxins; brain damage and intelligence quotient (IQ) decrease from lead and arsenic; renal sickness from lead, mercury, and cadmium; and skeletal and bone disorders from lead, fluoride, and cadmium were among the grave health concerns (Taddese 2019).

Table 5.1 Application of garbage to land and how it affects the physicochemical characteristics of the soil

Soil property	Waste type	Outcome
Soil moisture/water movement and retention	Fly ash	Increase bulk density, increase porosity, increase hydraulic conductivity
	Dairy effluents, meat processing effluents	Change/reduce hydraulic conductivity
	Coal fly ash	Reduce hydraulic conductivity
Soil pH	Bauxite, cement, and lime kiln, fly ash, sugar beet, lime and wood ash, dairy waste, steel slag and paper pulp	Liming agent to reduce acidity
Redox potential	Black coal fly ash, tannery waste	Changes in redox affecting solubility and mobility of Contaminants
Organic matter	Fly ash, tannery sludge, dairy effluents	Mostly increased organic matter decreased in some cases
Soil salinity and sodicity	Tannery waste, dairy wastes, pulp and paper mill effluent, fly ash	Increase soil salinity and/or sodicity

The soil's chemical makeup is the main factor influencing changes in vegetation. Many variables, including soil fertility, acidity/alkalinity, oxidative/reductive conditions, and the presence of organic matter, affect the accumulation of chemical elements in plants. Sometimes, higher-intensity disruptions might risk certain species becoming extinct and leading to low richness (Ali et al. 2014).

5.7 Waste Collection Techniques

There are several types of garbage collection systems, and the influence of each kind on waste recovery varies. Comprehending the diverse rubbish collection methods is vital for salvaging resources from waste in an environmentally sound way. Depending on the situation, several trash-collecting systems have been put in place in developed economies. In the quest for resource recovery from trash, industrialized economies must recognize the importance of each collecting system (Mwanza et al. 2018). Numerous factors have affected waste recovery, and the effects of each aspect vary. Even though the effects of these factors may differ in how waste from households and other sectors is managed or recovered, it is important to remember that waste collection systems are essential to the management and recovery of end-of-life products. Several garbage collection techniques are available for handling residential solid waste, and each system has a different impact on waste recovery (Dahlén and Lagerkvist 2010). There are two types of household garbage collection systems: drop-off sites and property-close collection systems. However, the design of residential waste collection systems varies worldwide (Mbande 2003). Two forms of property-close collection systems include curbside and door-to-door collection systems. Under drop-off collection systems, residents must bring recyclables to the drop-off sites. A buyback collection system is a waste management strategy that relies on recycling and trash recovery operations. Recyclers who return their recyclables to the facilities (Zhang and Wen 2014). A deposit-reimbursement system combines a tax on product use with a refund when a product is returned in its original packaging or for appropriate disposal or recycling (Walls 2011).

5.7.1 Kerbside Collection System

Recyclable waste collection programs at the kerbside greatly affect residential waste management. In the kerbside collection system, trash receptacles such as bins, bags, or sacks are assigned to specific homes. Each family is in charge of setting the container at the curb on collection days and returning it to its storage location after emptying it (Gonzalez-Torre et al. 2003).

5.7.2 Drop-off Collection System

Homeowners must drop off garbage and recyclables at certain places under drop-off collection systems. Drop-off sites and drop-off centers are the two different categories of drop-off locations. While households keep separate trash streams in nearby containers at drop-off sites, they transport different separated waste streams to recycling center containers at drop-off facilities (Xevgenos et al. 2015).

5.7.3 Buyback Collection System

These are locations where consumers may deposit reusable goods or recyclables in exchange for cash. Buyback facilities often purchase secondary materials from the general population and resell them to brokers or manufacturers. Nevertheless, these facilities may or may not process the recyclables. The fact that these facilities may not handle recyclable materials does not mean their influence on resource recovery is negligible (Consult 2010).

5.7.4 Deposit Refund System

One type of deposit refund mechanism is the legislation about returnable containers. It combines a product fee with an incentive for recycling or appropriate disposal. Deposit return systems have been widely implemented as a tool for economics to increase and capture old packaging for recycling (Astrup and Hedh 2011).

5.8 Criteria for Selection of Appropriate Technology for Waste Treatment

Over the past few decades, waste management issues have grown increasingly complicated. Environmental managers are primarily driving the attainment of a sustainable waste management system through the rising volumes of garbage produced and societal environmental consciousness. Nevertheless, in reality, various variables and influences sometimes contradict standards for determining answers in practical applications (Achillas et al. 2013). Choosing a solid waste disposal technique is a difficult decision that involves many different qualitative and quantitative considerations. The main reason is that besides being a social issue, managing solid waste disposal is a multi-criteria decision problem that considers political, sociocultural, technical, socioeconomic, and environmental factors (Arıkan et al. 2017). To create appropriate solutions for solid wastes, various sectors, including politics, geography, economics, public health, sociology, engineering, and materials science, must work together in complex ways (Muşdal 2007).

5.9 Techniques for Solid Waste Management

5.9.1 3R Principle

The current state of the environment results from ongoing greenhouse gas emissions. Energy facilities that use fossil fuels, such as coal, oil, and natural gas, as well as the construction and industrial sectors, traffic, transportation, and agriculture, emit these gases (Dijkers 2019). Thus, with the support of business and legal regulations, strategies to reduce CO2 emissions into the atmosphere are being developed concurrently with electromobility's growth and energy generation from renewable sources. Numerous technologies are still releasing carbon dioxide, which adds to global warming (Ali and Yusof 2018). Global energy consumption has increased significantly, and the supply and demand curves show how anxious people are about increasing urbanization and limited resources. Demand is growing every day because of the heightened competition for limited resources (Dijkers 2019). Many factors contribute to this resource demand. The primary cause of issues is the world's rapidly growing population, which is expected to reach eight billion by 2030 and nine billion by 2050. The closed-loop ecosystem that supports the circular economy allows for efficient use of resources. This is related to the closed-loop ecosystem's linear resource utilization (Nelson et al. 2017). The 3R system (reduce, reuse, and recycle) is implied by the model (Fig. 5.1).

- Reduce: reducing the use of nonrenewable resources.
- Reuse: reusing items that still have potential for use after they have been used.
- Recycling: a process for sorting items after a client has used them.

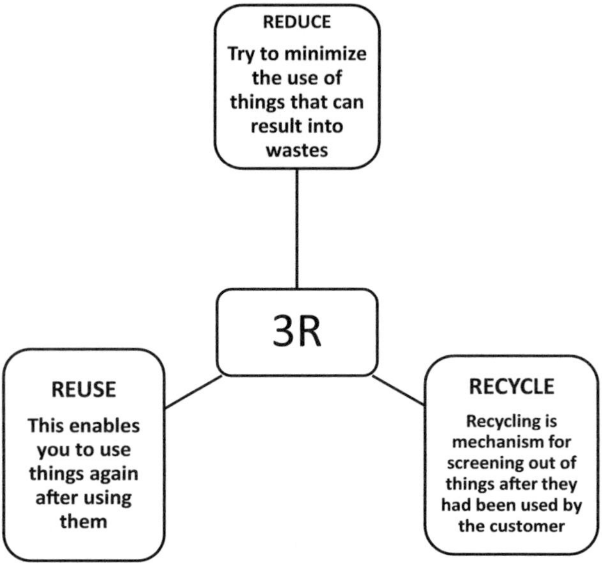

Fig. 5.1 3R—reduce, reuse, and recycle

5.9.2 Incineration

One technique for treating garbage is incineration, which involves burning the organic elements in the waste to provide thermal treatment. The products of incineration include ash, gas, and heat. Particulate matter or solid lumps scattered by flue gas may be present in the ash. Before being distributed, this incinerated flue gas needs to be free of any particle or gaseous contaminants (Ghasemi and Yusuff 2016). One of the main benefits of incineration is that it produces heat that can be utilized by electric power plants, which lowers the demand for fossil fuels. The ideal temperature range for incineration is between 900 and 1200C. Because incineration needs a high sample size and might cause non-target destruction such as disease outbreaks, it requires specialized knowledge (Khan et al. 2017; Khan et al. 2016). A grate system is used in the incineration process to burn the organic components in the crude waste. Hydraulic rams assist the boilers in these systems, pushing wastes into the ignition chamber (Makarichi et al. 2018). The grate systems assess the condition, while the waste is transported to the burning chamber. Drying the wastes and burning them further to produce ash on an abrade enhanced the system's overall accuracy. The gases produced by the grate system are released as steam at 850 °C, and the heat produced powers the grate system itself (Yasin et al. 2017).

5.9.3 Inertization

The increasing generation of toxic wastes is contributing to the ongoing degradation of the ecosystem. Considering that a large amount of hazardous waste is disposed of in landfills, it is essential to employ various technologies to lower the environmental risk related to waste disposal. The use of solidification/inertization (S/I) technology is one way to lessen the environmental risk associated with waste disposal. The phrase describes methods that may be applied to transform hazardous compounds into less soluble, transportable, or dangerous forms, improving their physical state and handling capabilities. The waste's physicochemical structure should considerably reduce the lixiviation of hazardous compounds after the solidification/inertization (S/I) process is applied. This is because of the decreased mobility, which makes it possible to regulate the release of hazardous materials into the environment.

Solidification/inertization (S/I) procedures are primarily influenced by the following factors: (a) waste properties; (b) binder utilized in the agglomeration technique (such as cement, lime, sand); (c) additives typically found in proprietary or patent-protected goods; d) conditions for processing. Hazardous waste management and the application of solidification/inertization (S/I) technologies are related. Essentially, these techniques are useful when certain materials' lixiviation is restricted by legislation (Andrés et al. 1991).

5.9.4 Microwaving

The microwave is a new method for handling biohazardous waste, such as material from medical institutions. Conventional microwave technology is frequently used to inactivate germs, which may be appropriate in some situations but may also give rise to the misconception that microwave systems cannot be utilized to inactivate solid "dry" waste. Nevertheless, traditional microwave ovens cannot regulate the inactivation process, particularly about moisture content. However, few highly developed microwave technologies with the right measurements enable the verified inactivation of biohazardous compounds. Some of these technologies are commercially accessible, and they are an effective inactivation technique. It is imperative to take into account that biohazardous waste needs to be carried exclusively in closed systems or that the waste be deactivated directly at the site of generation. Furthermore, compared to the more popular autoclaves, microwave technology has the potential to reduce energy expenses (Zimmermann 2017). The ability to supply energy directly to materials that absorb microwaves is the primary advantage of microwave energy since it enables the volumetric heating of samples. Long heating times, temperature gradients, and environmental energy loss are among the problems that can be reduced (Bélanger et al. 2007). Researchers are currently focusing on various uses for microwave technology, which is not limited to home cooking. Microwave radiation could be useful for microbial inactivation, even in a household setting. Microwave radiation was applied after wastewater was experimentally contaminated kitchen sponges, cleaning pads, and syringes. The overall bacterial count was decreased by more than 99% at a 100% power level in just 1–2 min, and longer exposure times were necessary to achieve full inactivation, depending on the organism (Park et al. 2006). As one of the newest methods to treat bio hazardous waste, including material from healthcare institutions, microwaves also have a lot of promise. In underdeveloped nations, microwave technology may be beneficial in addressing some waste-related problems (de Titto et al. 2012).

5.9.5 Land Disposal/Deep Burial

Nuclear energy is now seen as a clean energy source essential to human society's long-term viability. However, despite the rapid advancement of nuclear power, some challenges have also emerged, chief among them being the safe management of high-level radioactive waste (HLW). Strong toxicity, a long half-life, and high radioactivity characterize HLW. HLW disposal is difficult due to scientific, technical, and engineering concerns. As a result, every country that has a nuclear sector has focused heavily on the safe disposal of HLW (Wang et al. 2006). There have been many suggestions for alternatives, such as disposal in deep ocean depths, granite melting, ice sheets, and space. Currently, geological disposal at deep burial depths is seen to be a practical way to handle HLW securely. However, the challenging construction circumstances, strict safety requirements, and long operating life (tens of thousands of years) make producing such a geological record challenging. As a result, selecting a site, constructing it, and assessing its safety are complex tasks. Moreover, this unique disposal technique may be done without requiring any prior engineering expertise from finished projects (Latessa et al. 2023). Because of this, some countries have stated unequivocally that underground research facilities ought to be constructed before the disposal program is finished. These facilities are meant to be utilized for field research, verification, and feature assessment. They are also meant to provide the necessary research base and practical training. These facilities are known as underground research labs (URLs) and are utilized to dispose of heavy metal waste (HLW) (Zhang et al. 2020).

5.9.6 Waste Immobilization

The industrial sector in the United States and other affluent nations produces significant garbage from byproducts. It is imperative to dispose of these by-products to prevent any potential harm to the environment and public health (Adekola et al. 2021). The most current technology to be offered to stop pollutants from freely moving through garbage and into the surrounding media is the waste immobilization approach. Immobilization can be nearly permanent in the case of vitrification or transient, such as in confinement. Assessing some fundamental waste features is necessary before choosing an efficient immobilization method. The kind, quantity, and conditions of the trash and the place all influence the choice of a suitable waste immobilization approach (Ma et al. 2024). Chemical treatments can efficiently stabilize and solidify contaminated soils; however, the products may not necessarily be suitable for reuse as consumables or construction materials. Using a cutting-edge process called vitrification, polluted and toxic waste may be turned into a chemically stable product resembling glass. Both in-situ and ex-situ vitrification are possible (Ojovan 2024). Usually, containment tactics are employed to "buy time" in urgent or transient situations. Chemical treatments can efficiently stabilize and solidify contaminated soil; however, the products may not be suitable for reuse as consumables or construction materials. The Environmental Protection Agency has concluded vitrification is the "best demonstrated available technology" for handling heavy metals and high-level radioactive waste (Meegoda et al. 2003).

5.9.7 Source Land Filling

A landfill is an area designated for the disposal of waste. The goal is to keep the waste out of the surrounding environment—especially the groundwater.

Landfills may be divided into three categories:

(i) Open dumps, also known as open landfills, are the most common landfills in developing countries. They occur when debris is simply tossed into open places that are low-lying.
(ii) Waste is dumped, compacted, and frequently covered with topsoil to keep animals out in semi-controlled or managed landfills. Whether it is municipal, industrial, or clinical/hospital trash, none is disposed of separately. This type of landfill is not intended to release leachate or emit landfill gas.
(iii) Developed nations employ sanitary landfills, which are equipped with a network of ponds for leachate treatment and interception. Additionally, this kind of landfill has controls in place to manage the emissions from the breakdown of trash (Narayana 2009).

5.9.8 Composting

Composting is converting organic waste (such as leftover food and yard trash) into humus, similar to soil (Ullah et al. 2020; Ditta et al. 2018). Composting is thus acceptable as an additional recycling method. With the aid of microorganisms that have access to adequate oxygen, organic waste components begin a process that produces heat and gas. Organic wastes may be organically recycled into fresh soil for various uses, including landscaping and vegetable and flower gardens. This process is known as composting (Farooq et al. 2018). The organic component of solid waste is retained and converted into humus-rich material during the composting process, making it a valuable soil conditioner for both gardening and farming (Ijaz et al. 2020; Javeed et al. 2021; Majeed et al. 2022). Both aerobic and anaerobic microbial decomposition can be used. Kitchen wastes and similar wastes are utilized for composting; solid wastes are broken down by labor at the stations using fresh vegetables (Zeb et al. 2018). This extremely microbially rich chemical is added to soil to boost its fertility and phosphate content. Fresh vegetables that are left over from grocery and food waste are the primary sources of organic elements in city life. Chipped grass and the shrubs left behind after pruning also significantly contribute. With the aid of microbes with access to adequate oxygen, organic components in the wastes begin to react, producing heat and gas. Homeowners have been urged to utilize this technique as a direct means of recycling, and farmers interested in sustainable agriculture are already using it as a valuable tool to boost yields (Naveed et al. 2021). Professional farmers are now learning that soil enhanced with compost may help suppress illnesses and keep pests away.

Growers may minimize their usage of pesticides, save money, and preserve natural resources by using compost for these advantageous purposes. Composting has also emerged as an affordable means of managing mortality in the chicken sector. Pathogenic microbes are eliminated, and a nutrient-rich product that may be utilized or sold is produced (Liu et al. 2023). On a larger scale, it now encompasses green waste from gardens and parks, sewage sludge, and municipal solid rubbish. Compost technology is now used in bioremediation to control storm water runoff, lessen smells, restore contaminated soils, and decompose volatile organic compounds.

The compost product may be used in parks, street flowerbeds, outdoor seating areas, and other urban green spaces. This will mainly benefit a closed recycling system inside the city by reducing transportation costs and the requirement for chemical pesticides and fertilizers (Cameron 2023). Composting solid waste requires both product preparation for sale and breakdown. Solid waste preparation includes receiving, sorting, separating, shredding, removing plastic, rubber, leather, and similar materials, and adding moisture and nutrients.

The breakdown of the waste is the first step in the composting process; however, this breakdown needs to be managed differently from the deposition at the landfill. This is essentially the action of aerobic microorganisms. The windrow or static pile technique of making compost can take anywhere from 6 months to a year, depending on the environmental conditions and composition of the organic part; an in-vessel device can produce compost in as little as 2 weeks (Hashim et al. 2022). The simplest method is windrow composting, which entails building long heaps that are 2.0–2.5 m wide and 1.5–2.0 m high. The piles are often moved to ensure enough aeration. The completed compost takes 2–3 weeks to prepare.

By employing mechanical digesters with forced aeration, seeding, moisture and nutrient management, and other specialist procedures, compost production may be sped up to take less than a week. Landfill gas is created when the organic portion of wastes, such as sewage sludge and household garbage, biodegrades in a landfill site (Duan et al. 2021). The main component of landfill gas is methane, which may be collected in a controlled and planned manner and used appropriately to form energy. Landfill gas also lowers the amount of organic waste produced (Akkucuk 2014).

5.9.9 Pyrolysis

Pyrolysis is breaking down fuel molecules using heat in an inert atmosphere (Fytili and Zabaniotou 2008). Pyrolysis is a process that turns sewage sludge into fixed carbon, ash, bio-oils, combustible gases, and water vapor at temperatures ranging from 300 to 900 °C (Fytili and Zabaniotou 2008). Pyrolysis is a more environmentally friendly way of eliminating fuel pollutants than the widely utilized combustion and incineration processes (Menéndez et al. 2002). The pyrolysis of organic materials, such as lignocellulosic waste, plastics, and municipal solid wastes, has generally drawn growing attention due to the possibility of converting these wastes into valuable chemicals or useable energy (Kalinci et al. 2009; Luo et al. 2010; Menéndez et al. 2004), but sludge pyrolysis has seen fewer published research studies. The pyrolysis of sewage sludge has attracted a lot of attention lately as a financially and environmentally sound approach to using sewage sludge for beneficial applications (Aslam et al. 2024; Awad et al. 2021; Kalinci et al. 2009; Luo et al. 2010; Menéndez et al. 2004). The product of the pyrolysis process may be grouped using the following fractions: The following steps take place when sewage sludge is pyrolysis:

(a) Stable (non-condensable) gases, mainly made up of hydrogen, carbon monoxide, carbon dioxide, methane, and, in trace amounts, low molecular weight hydrocarbons;
(b) Liquids (tar and/or oil), particularly hydrocarbons, organic acids, and carbonyl compounds of high molecular weight phenols, aromatic compounds, aliphatic alcohols, acetic acid, and water; and.
(c) Solids, mostly solid carbon and ash (with a notable quantity of heavy metals) (Fytili and Zabaniotou 2008).

The vaporization of volatile components is the initial stage of pyrolysis. The major breakdown of non-volatile components comes next, producing char, tar, and fumes. Ultimately, hydrocarbons and aromatic compounds are produced in the volatile phase by subsequent pyrolysis of the char at a higher temperature (Khiari et al. 2004). A schematic of the pyrolysis process for sewage sludge is shown in Fig. 5.2 (Khiari et al. 2004).

5.10 Techniques for Liquid Waste Management

5.10.1 Mechanical Methods

Larger oils and fats, larger floating and dragging solids, and granular particles larger than 0.1 mm are eliminated by mechanical or preliminary treatment systems. To do this, various mechanical equipment that can collect and separate solid materials from the bulk wastewater is utilized, including blowers, pumps, screens, gates, and tanks (Yenkie et al. 2019). As will be covered below, screening and flotation activities are part of the mechanical treatment process (Afolalu et al. 2021).

Screening is among the oldest methods of wastewater treatment. This mechanical procedure removes big non-biodegradable solid pollutants from wastewater, such as clothing, hair, fiber, paper, wood, and fecal particles, to prevent damage or blockage to downstream facilities, machinery, and pipes. It also prevents offensive suspended contaminants from getting into primary settling tanks and interfering with plant functions. Several variables influence this demand for resources. The primary cause of issues is the world's rapidly growing population, which is expected to reach eight billion by 2030 and nine billion by 2050. The closed-loop ecosystem

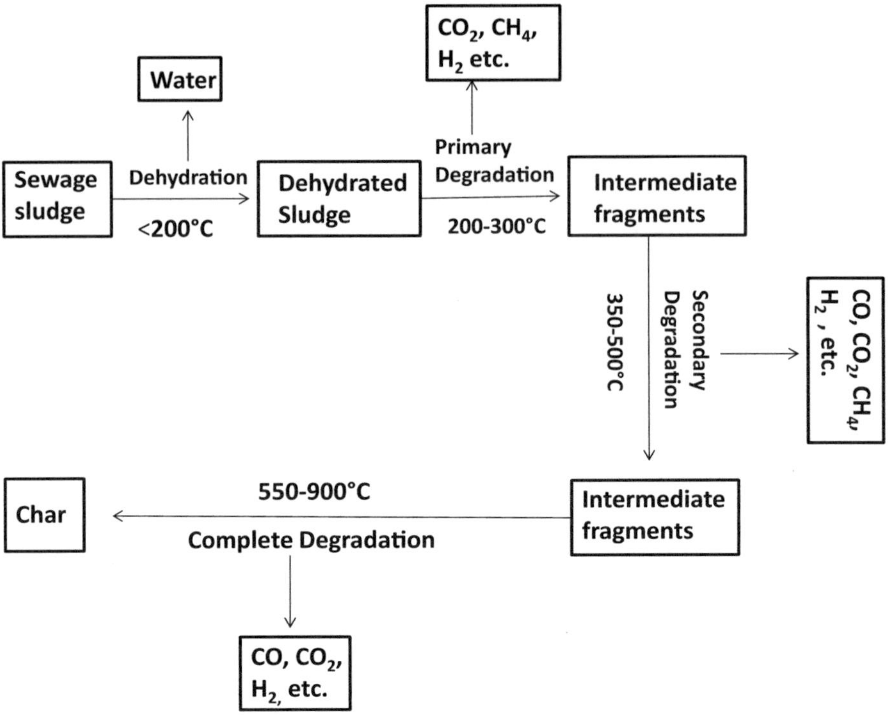

Fig. 5.2 Diagrammatic representation of the pyrolysis process. (Shao et al. 2008)

that supports the circular economy allows for efficient use of resources. This relates to the closed-loop ecosystem's linear resource utilization (Ikumapayi and Akinlabi 2018; Tabassum et al. 2015).

Flotation is a unit operation that involves injecting a very thin gas—usually air bubbles—into a fluid to remove finely suspended solids, fats, oils, and grease. This process encourages agglomeration and raises the materials to the surface. The materials and bubbles can then be removed with a cleaning instrument (Ikumapayi and Akinlabi 2018).

5.10.2 Chemical Methods

A crucial part of treating extremely toxic water is using chemical treatment techniques, often known as tertiary treatment. These techniques, which include coagulation, neutralization, and disinfection, employ a variety of chemical reactions to enhance the quality of the water (Afolalu et al. 2022).

The coliform guidelines, which specify that *Escherichia coli* should not be found in arbitrary water samples, are followed while disinfecting water (Sangeetha et al. 2020). Therefore, disinfection aims to lessen or immobilize pathogens, including bacteria, viruses, and parasites (Adam and Krüger 2018). As part of the disinfection treatment process, wastewater is treated with a strong oxidizing agent (such as alcohols, phenols, hydrogen peroxide, ozone, and chlorine compounds) to prevent the growth of bacteria (Saleh 2016) and lessen the rate at which the water deteriorates, eliminating any taste or odor problems (Pudasainee et al. 2020). However, various factors affect how successful disinfection is: such as the kind of disinfectant to be used, how much of it to use, how long to use it for, the quality of the water to be treated, and other environmental factors.

To meet the needs of different processing units within the wastewater treatment system, neutralization is employed to control the pH of wastewater (Afolalu et al. 2022). Acid or alkali is added as needed; acid effluent is sometimes neutralized with basic lime. Neutralization is crucial to protecting downstream equipment from corrosion attacks (Pal 2017).

One of the most significant physiochemical processes in wastewater treatment is the coagulation process (Ikumapayi and Akinlabi 2018). This method is used to speed up the precipitation of ions (heavy metals) and colloidal-dispersed organic and inorganic particles in wastewater held mainly by electric charges (Afolalu et al. 2022; Batool et al. 2023; Mehmood et al. 2023a, b; Murtaza et al. 2024a, b, c; Pal 2017). The addition of ions with opposite charges destabilizes the forces holding colloids apart. Alum ($Al_2(SO_4)_3.14H_2O$) and liquid alum ($Al_2(SO_4)_3.xH_2O$), two chemical substances referred to as coagulants, can also cause coagulation. The coagulation process results in large volumes of sludge that must be disposed of. The particles then gradually coarsen and agglomerate, decreasing in quantity within the wastewater volume (Pal 2017; Zahrim et al. 2019).

5.10.3 Biological Methods

Biological treatment methods, commonly called secondary treatment, convert the dissolved organic content in wastewater into flocculent or stable organic and inorganic particles (Afolalu et al. 2021). This approach's basic tenet is that biological activity eliminates pollutants from the environment. Anaerobic and aerobic biological treatments are the two varieties available (Saleh 2020).

The wastewater that settles is placed in a specially designed bioreactor, where, in the presence of oxygen, microorganisms such as bacteria, algae, and fungus break down biodegradable organic components chemically (Samer 2015). Aerobic treatment techniques include activated sludge, traditional trickling filters, and rotating biological contractors. This treatment method effectively removes phosphates, volatile, suspended, and dissolved organic matter, biochemical oxygen demand (BOD), and chemical oxygen demand (COD) from low-strength wastewater (Ikumapayi and Akinlabi 2018). After aerobic treatment, significant quantities of bio-solids may build up and require further treatment. Oxidation ponds and aeration lagoons are two instances of aerobic treatment techniques (Yildiz 2012).

Rich, complex organic matter is converted into simple organic molecules by microorganisms in the anaerobic treatment system without needing oxygen (Mehmood et al. 2018, 2021, 2022; Ikumapayi and Akinlabi 2018). Organic pollutants undergo anaerobic digestion to produce biogas (Zahrim et al. 2019), which mainly consists of 60–75% methane (CH_4), 25–40% carbon dioxide (CO_2), and trace quantities of hydrogen (H_2), nitrogen (N_2), and hydrogen sulfide (H_2S) (Sangeetha et al. 2020). When compared to aerobic treatment, this method of treatment takes longer, but it has several benefits: it can remove organic materials more effectively, can handle large amounts of organic matter, produces small quantities of typically very stable sludge, uses less energy, and produces biogas (CH_4) as a by-product (Samer 2015; Tetteh et al. 2019).

5.11 Advanced Technologies Used for Waste Treatment and Waste Disposal

Here are a few cutting-edge and creative waste treatment and disposal technologies (Mukherjee et al. 2021).

- Microbial fuel cells (MFC)
- Smart waste bin approach
- Mr. Trash Wheel
- Plasma gasification
- Incineration
- Vitrification
- Phosphate ceramics
- Ion-exchange
- Synroc
- Geologic disposal and transmutation
- Biological reprocessing
- Bio-refineries
- Swedish model for pollution-free earth

5.12 Limitations and Future Perspectives

Any type of waste treatment and disposal project requires significant financial outlays, including mass-burn municipal solid waste incineration, hazardous waste treatment facilities, and waste landfills. The substantial financial resources required, which might need tens of millions of pounds in investments, are often unavailable to local governments. These projects are increasingly being created as partnerships between the public and commercial sectors. Bank loans backed by ongoing firm earnings and assets would be necessary for financing from the private sector, especially for the bigger projects. The borrowing corporation bears the risks associated with such a loan. The following are a few restrictions on environmentally friendly trash management.

(i) Emphasizes gathering and evaluating immaterial data;
(ii) Use irreversible treatments in place of procedures to address emergent negative effects that can be changed;
(iii) Finding quick fixes rather than adopting a longer-term sustainable perspective;
(iv) Misreads delays in response to an intervention as a lack of reaction, which leads to the use of greater treatments and overcorrection that needs to be moderated;
(v) Ignores or undervalues the intervention's adverse effects;
(vi) Focuses more on individual issues than the sustainability of the waste management system; and.
(vii) Depends on linear forecasts of recent ephemeral occurrences.

Wastes produced by commercial or public operations either directly or indirectly endanger public health. Uncollected or unmanaged wastes have the potential to cause respiratory infections. The larger towns dispose of their refuse in an unsanitary manner by just dumping it, and in the rare cases where burial is done, it is done ineffectively. The selection of sites is typically predicated on a rudimentary comprehension of the possible impacts on surface and groundwater resources. Certain types of dumping impact domestic animal and human health as well as the quality of the water. A sizable rural populace still uses surface streams for drinking water. Because scavengers often search for food or other valuable resources, uncontrolled dumping endangers peri-urban, rural, and urban areas. The potential presence of chemicals, outdated medications, or spoilt food presents an increased danger. One reason for poor health is the absence of these services. The natural ecosystem has suffered as a result of many metropolitan areas' inadequate upkeep, regulation, and oversight. There are clear issues with pollution, habitat loss, and visual amenity. Therefore, the availability of these services needs to be ensured for controlling the above-mentioned environmental issues.

5.13 Conclusions

Sustainable waste management aims to minimize environmental harm and maximize resource recovery. There is a growing need for efficient waste management solutions because of the volume of rubbish produced worldwide. Wastes are classified according to their chemical and physical condition (liquid or solid). There are several types of waste: universal waste, hazardous waste, industrial trash, agricultural waste, construction site waste, organic waste, recyclable and non-recyclable inorganic garbage, and radioactive waste. The fertility, quality, and health of an ecosystem are all impacted by proper waste management. Waste-derived soil additives can improve or worsen soil characteristics. Curbside collection, recycling facilities, and drop-off locations are examples of effective garbage-collecting techniques. Technology decisions are made based on waste kind, volume, cost, and environmental effects. Waste is reduced by following the 3Rs (Reduce, Reuse, and Recycle). The methods include land disposal, waste immobilization, incineration, inertization, microwaving, source land filling, composting, and pyrolysis. Liquid waste is treated by biological (constructed wetlands, bioremediation), mechanical (filter, sedimentation), and chemical (coagulation, flocculation) techniques. Anaerobic digestion, plasma gasification, and waste-to-energy techniques are examples of innovations. Infrastructure development, public awareness, and regulatory compliance are among the difficulties. Sustainable practices and the concepts of the circular economy should be the main topics of future work. Diverse tactics are used in sustainable waste management to reduce environmental impact and increase resource efficiency.

References

Acharya J, Kumar U, Rafi PM (2018) Removal of heavy metal ions from wastewater by chemically modified agricultural waste material as potential adsorbent-a review. Int J Curr Eng Technol 8(3):526–530

Achillas C, Moussiopoulos N, Karagiannidis A, Banias G, Perkoulidis G (2013) The use of multi-criteria decision analysis to tackle waste management problems: a literature review. Waste Manag Res 31(2):115–129

Adam C, Krüger O (2018) Wastewater as a resource: from rare earth metals to phosphorus. Phosphorus: polluter and resource, p 241

Adebambo O (2017) Rochelle Owen, Executive Director, Office of Sustainability, Dalhousie University

Adekola PO, Iyalomhe FO, Paczoski A, Abebe ST, Pawłowska B, Bąk M, Cirella GT (2021) Public perception and awareness of waste management from Benin City. Sci Rep 11(1):306

Adhikari BB, Chae M, Bressler DC (2018) Utilization of slaughterhouse waste in value-added applications: recent advances in the development of wood adhesives. Polymers 10(2):176

Afolalu SA, Ikumapayi OM, Abioye AA, Yusuf OO, Emetere ME (2021) Evaluation of starch-biopolymer synthesized from chaffs of common beans (Phaseolus Vulgaris). Int J Des Nat Ecodyn 16(5):551–556

Afolalu SA, Ikumapayi OM, Ogedengbe TS, Kazeem RA, Ogundipe AT (2022) Waste pollution, wastewater, and effluent treatment methods–an overview. Mater Today: Proc 62:3282–3288

Ahmad K, Wajid K, Khan ZI, Ugulu I, Memoona H, Sana M, Nawaz K, Malik IS, Bashir H, Sher M (2019) Evaluation of potential toxic metals accumulation in wheat irrigated with wastewater. Bull Environ Contam Toxicol 102:822–828

Ahmed SA, Ali M (2004) Partnerships for solid waste management in developing countries: linking theories to realities. Habitat Int 28(3):467–479

Akacem M, Miller DD, Faulkner JL, Akacem M, Miller DD, Faulkner JL (2020) The population threat. In: Oil, institutions, and sustainability in MENA: a radical approach through the empowerment of citizens, pp 131–168

Akinbile CO, Yusoff MS (2012) Solid waste generation and decomposition using compost bin technique in Pulau Pinang, Malaysia. Waste Manag Res 30(5):498–505

Akkucuk U (2014) Handbook of research on developing sustainable value in economics, finance, and marketing. IGI Global

Al-Dailami A, Ahmad I, Kamyab H, Abdullah N, Koji I, Ashokkumar V, Zabara B (2022) Sustainable solid waste management in Yemen: environmental, social aspects, and challenges. Biomass Convers Biorefinery:1–27

Ali MSM, Yusof RNR (2018) Intention to practice reduce, reuse, and recycle (3R) among expatriates working in Malaysia. Int J Acad Res Bus Soc Sci 8(3):276–295

Ali SM, Pervaiz A, Afzal B, Hamid N, Yasmin A (2014) Open dumping of municipal solid waste and its hazardous impacts on soil and vegetation diversity at waste dumping sites of Islamabad city. J King Saud Univ-Sci 26(1):59–65

Ali M, Javeed HMR, Tariq M, Khan AA, Qamar R, Nawaz F, Masood N, Ditta A, Abbas T, Zamir MSI, Sabagh AE, Shahzad M, Naeem R, Mubeen M (2023a) Use of biochar for biological carbon sequestration. In: Jatoi WN, Mubeen M, Hashmi MZ, Ali S, Fahad S, Mahmood K (eds) Climate change impacts on agriculture. Springer, Cham, pp 393–411

Ali A, Malik SA, Shafiullah M, Malik MZ, Zahir MH (2023b) Policies and regulations for solar photovoltaic end-of-life waste management: insights from China and the USA. Chemosphere 340:139840

Andrés A, Viguri J, Irabien A (1991) Inertization of toxic wastes. Toxicol Environ Chem 31(1):593–601

Arıkan E, Şimşit-Kalender ZT, Vayvay Ö (2017) Solid waste disposal methodology selection using multi-criteria decision-making methods and an application in Turkey. J Clean Prod 142:403–412

Aslam A, Batool F, Noreen S, Abdelrahman EA, Mustaqeem M, Albalawi BFA, Ditta A (2024) Metal oxide impregnated biochar for azo dyes remediation as revealed through kinetics, thermodynamics, and response surface methodology. ACS Omega 9(4):4300–4316

Astrup N, Hedh A (2011) European refunding scheme for drinks containers. Joint Parliamentary Committee, European Economic Area

Awad M, El-Sayed MM, Li X, Liu Z, Mustafa SK, Ditta A, Hessini K (2021) Diminishing heavy metal hazards of contaminated soil via biochar supplementation. Sustain For 13(22):12742

Aznar-Sánchez JA, García-Gómez JJ, Velasco-Muñoz JF, Carretero-Gómez A (2018) Mining waste and its sustainable management: advances in worldwide research. Fortschr Mineral 8(7):284

Bary A, Cogger C, Sullivan D, Myhre E (2005) Characterization of fresh yard trimmings for agricultural use. Bioresour Technol 96(13):1499–1504

Batool F, Qadir R, Adeeb F, Kanwal S, Abdelrahman EA, Noreen S, Albalawi BFA, Mustaqeem M, Imtiaz M, Ditta A, Gondal HY (2023) Biosorption potential of Arachis hypogaea derived biochar for Cd and Ni as evidenced through kinetic, isothermal, and thermodynamics modeling. ACS Omega 8(43):40128–40139

Bélanger JM, Paré JJ, Poon O, Fairbridge C, Ng S, Mutyala S, Hawkins R (2007) Remarks on various applications of microwave energy. J Microw Power Electromagn Energy 42(4):24–44

Beolchini F, Fonti V, Dell'Anno A, Rocchetti L, Vegliò F (2012) Assessment of biotechnological strategies for the valorization of metal-bearing wastes. Waste Manag 32(5):949–956

Bhattacharyya P, Barman D (2018) Crop residue management and greenhouse gas emissions in tropical rice lands. In: Soil management and climate change. Elsevier, pp 323–335

Bilgilioglu SS, Gezgin C, Orhan O, Karakus P (2022) A GIS-based multi-criteria decision-making method for the selection of potential municipal solid waste disposal sites in Mersin, Turkey. Environ Sci Pollut Res:1–17

Birchard K (2002) Out of sight, out of mind… the medical waste problem. Lancet 359(9300):56

Bolan S, Padhye LP, Kumar M, Antoniadis V, Sridharan S, Tang Y, Singh N, Hewawasam C, Vithanage M, Singh L, Rinklebe J (2023) Review on distribution, fate, and management of potentially toxic elements in incinerated medical wastes. Environ Pollut 321:121080

Burcea ŞG (2015) A new waste classifying model: how waste classification can become more objective? Annals Univ Oradea 1:1060

Cai W, Liu C, Zhang C, Ma M, Rao W, Li W, He K, Gao M (2018) Developing the ecological compensation criterion of industrial solid waste based on energy for sustainable development. Energy 157:940–948

Cameron R (2023) "Do we need to see gardens in a new light?" Recommendations for policy and practice to improve the ecosystem services derived from domestic gardens. Urban For Urban Green 80:127820
Cánovas C, Pérez-López R, Macías F, Chapron S, Nieto J, Pellet-Rostaing S (2017) Exploration of fertilizer industry wastes as a potential source of critical raw materials. J Clean Prod 143:497–505
Castaldi MJ (2014) Perspectives on sustainable waste management. Ann Rev Chem Biomol Eng 5:547–562
Chandra R, Yadav S (2015) The use of PMDE with sugar industries pressmud for composting: a green technology for safe disposal in the environment. Environmental Waste Management. CRC Press Taylor and Francis Group, Boca Raton, pp 1–32
Cheela VS, Ranjan VP, Goel S, John M, Dubey B (2021) Pathways to sustainable waste management in Indian smart cities. J Urb Manag 10(4):419–429
Chmielewski AG, Harasimowicz M, Tyminski B, Zakrzewska-Trznadel G (2001) Concentration of low-and medium-level radioactive wastes with three-stage reverse osmosis pilot plant. Sep Sci Technol 36(5–6):1117–1127
Consult L (2010) Integrated waste management plan report on recycling version V2. Lephalale Municipality
Correddu F, Lunesu MF, Buffa G, Atzori AS, Nudda A, Battacone G, Pulina G (2020) Can agro-industrial by-products rich in polyphenols be advantageously used in the feeding and nutrition of dairy small ruminants? Animals 10(1):131
Cucchiella F, D'Adamo I, Gastaldi M (2017) Sustainable waste management: waste to energy plant as an alternative to landfill. Energy Convers Manag 131:18–31
Dahlén L, Lagerkvist A (2010) Pay as you throw: strengths and weaknesses of weight-based billing in household waste collection systems in Sweden. Waste Manag 30(1):23–31
Daraz U, Li Y, Ahmad I, Iqbal R, Ditta A (2023) Remediation technologies for acid mine drainage: recent trends and future perspectives. Chemosphere 311:137089
De Silva S, Carson P, Indrapala DV, Warwick B, Reichman SM (2023) Land application of industrial wastes: impacts on soil quality, biota, and human health. Environ Sci Pollut Res 30(26):67974–67996
de Titto E, Savino AA, Townend WK (2012) Healthcare waste management: the current issues in developing countries, vol 30. SAGE Publications Sage UK, London, pp 559–561
Dijkers M (2019) Reduce, reuse, recycle: good stewardship of research data. Spinal Cord 57(3):165–166
Dijkgraaf E, Vollebergh HR (2004) Burn or bury? A social cost comparison of final waste disposal methods. Ecol Econ 50(3–4):233–247
Ditta A, Muhammad J, Imtiaz M, Mehmood S, Qian Z, Tu S (2018) Application of rock phosphate enriched composts increases nodulation, growth, and yield of chickpeas. Int J Recycl Org Waste Agric 7(1):33–40
Duan Z, Scheutz C, Kjeldsen P (2021) Trace gas emissions from municipal solid waste landfills: a review. Waste Manag 119:39–62
Europe Environmental Assessment Agency (2007) Progress in management of contaminated sites (CSI 015). Europe Environmental Assessment Agency, Kongan, Denmark [Internet]
Farooq N, Kanwal S, Ditta A, Hussain A, Naveed M, Jamshaid MU, Iqbal M (2018) Comparative efficacy of KCl blended composts vs. sole application of KCl or K2SO4 in improving K nutrition, photosynthetic capacity, and growth of maize. Soil Environ 37(1):68–74
Fytili D, Zabaniotou A (2008) Utilization of sewage sludge in EU application of old and new methods—a review. Renew Sust Energ Rev 12(1):116–140
Ghasemi MK, Yusuff RBM (2016) Advantages and disadvantages of healthcare waste treatment and disposal alternatives: Malaysian scenario. Pol J Environ Stud 25(1):17
Girelli AM, Astolfi ML, Scuto FR (2020) Agro-industrial wastes as potential carriers for enzyme immobilization: a review. Chemosphere 244:125368
Gonawala SS, Jardosh H (2018) Organic waste in composting: a brief review. Int J Curr Eng Technol 8(1):36–38

Gonzalez-Torre PL, Adenso-Dıaz B, Ruiz-Torres A (2003) Some comparative factors regarding recycling collection systems in regions of the USA and Europe. J Environ Manag 69(2):129–138
Hashim S, Waqas M, Rudra RP, Akhtar Khan A, Mirani AA, Sultan T, Ehsan F, Abid M, Saifullah M (2022) On-farm composting of agricultural waste materials for sustainable agriculture in Pakistan. Scientifica 2022(1):5831832
Hemidat S, Achouri O, El Fels L, Elagroudy S, Hafidi M, Chaouki B, Ahmed M, Hodgkinson I, Guo J (2022) Solid waste management in the context of a circular economy in the MENA region. Sustain For 14(1):480
Hoang NH, Fogarassy C (2020) Sustainability evaluation of municipal solid waste management system for Hanoi (Vietnam)—why to choose the 'waste-to-energy' concept. Sustain For 12(3):1085
Hoornweg D, Bhada-Tata P (2012) What a waste: a global review of solid waste management
Ijaz M, Rizwan MS, Sarfraz M, Ul-Allah S, Sher A, Sattar A, Ali L, Ditta A, Yousaf B (2020) Biochar reduced cadmium uptake and enhanced wheat productivity in alkaline contaminated soil. Int J Agric Biol 24:1633–1640
Ikumapayi OM, Akinlabi ET (2018) Composition, characteristics and socioeconomic benefits of palm kernel shell exploitation-an overview. J Environ Sci Technol 11(6):1–13
Javeed HMR, Ali M, Ahmed I, Wang X, Al-ashkar I, Qamar R, Ibrahim A, Habib-Ur-Rahman M, Ditta A, Sabagh AE (2021) Biochar enriched with buffalo slurry improved soil nitrogen and carbon dynamics, nutrient uptake, and growth attributes of wheat by reducing leaching losses of nutrients. Land 10(12):1392
Kalinci Y, Hepbasli A, Dincer I (2009) Biomass-based hydrogen production: a review and analysis. Int J Hydrog Energy 34(21):8799–8817
Kaur H, Siddique R, Rajor A (2019) Influence of incinerated biomedical waste ash on the properties of concrete. Constr Build Mater 226:428–441
Khan MN, Luna IZ, Islam MM, Sharmeen S, Salem KS, Rashid TU, Zaman A, Haque P, Rahman MM (2016) Cellulase in waste management applications. In: New and future developments in microbial biotechnology and bioengineering. Elsevier, pp 237–256
Khan M, Nizami A, Rehan M, Ouda O, Sultana S, Ismail I, Shahzad K (2017) Microbial electrolysis cells for hydrogen production and urban wastewater treatment: a case study of Saudi Arabia. Appl Energy 185:410–420
Khiari B, Marias F, Zagrouba F, Vaxelaire J (2004) Analytical study of the pyrolysis process in a wastewater treatment pilot station. Desalination 167:39–47
Khoaele KK, Gbadeyan OJ, Chunilall V, Sithole B (2023) The devastation of waste plastic on the environment and remediation processes: a critical review. Sustain For 15(6):5233
Kofoworola OF, Gheewala SH (2009) Estimation of construction waste generation and management in Thailand. Waste Manag 29(2):731–738
Kumar KN, Goel S (2009) Characterization of municipal solid waste (MSW) and a proposed management plan for Kharagpur, West Bengal, India. Resour Conserv Recycl 53(3):166–174
Kumar P, Yadav S, Yadav A (2024) Radioactive waste minimization and management. In: Green chemistry approaches to environmental sustainability. Elsevier, pp 165–184
Kwikiriza S, Stewart AG, Mutahunga B, Dobson AE, Wilkinson E (2019) A whole systems approach to hospital waste management in rural Uganda. Front Public Health 7:454509
Lal R (2005) World crop residue production and implications of its use as a biofuel. Environ Int 31(4):575–584
Latessa SH, Hanley L, Tao W (2023) Characteristics and practical treatment technologies of winery wastewater: a review for wastewater management at small wineries. J Environ Manag 342:118343
Lee S, Hrma P, Pokorny R, Klouzek J, Eaton WC, Kruger AA (2019) Glass production rate in electric furnaces for radioactive waste vitrification. J Am Ceram Soc 102(10):5828–5842
Lennemann WL (1978) The management of high-level radioactive wastes. IAEA Bull 21(4):1–16
Li Z, Huang J (2018) How to effectively improve pesticide waste governance: a perspective of reverse logistics. Sustain For 10(10):3622

Lieberman RN, Knop Y, Izquierdo M, Palmerola NM, de la Rosa J, Cohen H, Muñoz-Quirós C, Cordoba P, Querol X (2018) Potential of hazardous waste encapsulation in concrete with coal fly ash and bivalve shells. J Clean Prod 185:870–881

Liu Q, He X, Li D (2023) Unearthing the mechanisms underlying calcium carbonate therapies for eliminating pathogens during composting. Chem Eng J 451:139087

López M, Soliva M, Martínez-Farré FX, Bonmatí A, Huerta-Pujol O (2010) An assessment of the characteristics of yard trimmings and recirculated yard trimmings used in biowaste composting. Bioresour Technol 101(4):1399–1405

Luo S, Xiao B, Hu Z, Liu S (2010) Effect of particle size on pyrolysis of single-component municipal solid waste in fixed bed reactor. Int J Hydrog Energy 35(1):93–97

Ma S, Fu S, Yang H, He P, Sun Z, Duan X, Jia D, Colombo P, Zhou Y (2024) Exploiting bifunctional 3D-printed geopolymers for efficient cesium removal and immobilization: an approach for hazardous waste management. J Clean Prod 437:140599

Majeed A, Muhmood A, Niaz A, Ditta A, Rajpar MN (2022) Comparative efficacy of different biochars and traditional manures in the attenuation of cadmium toxicity in rice (Oryza sativa L.). Arab J Geosci 15:209

Makarichi L, Jutidamrongphan W, Techato K-A (2018) The evolution of waste-to-energy incineration: a review. Renew Sust Energ Rev 91:812–821

Maw MJ, Goyne KW, Fritschi FB (2019) Soil carbon changes following conversion to annual biofuel feedstocks on marginal lands. Agron J 111(1):4–13

Mbande C (2003) Developing areas. J South Afr Inst Civ Eng 45:3

Meegoda JN, Ezeldin A, Fang H-Y, Inyang HI (2003) Waste immobilization technologies. Pract Period Haz Toxic Radioact Waste Manag 7(1):46–58

Mehmood S, Rizwan M, Bashir S, Ditta A, Aziz O, Yong LZ, Dai Z, Akmal M, Ahmed W, Adeel M, Imtiaz M, Tu S (2018) Comparative effects of biochar, slag, and ferrous–Mn Ore on Lead and cadmium immobilization in soil. Bull Environ Contam Toxicol 100(2):286–292

Mehmood S, Ahmed W, Rizwan M, Imtiaz M, Elnahal ASMA, Ditta A, Irshad S, Ikram M, Li W (2021) Comparative efficacy of raw and HNO3-modified biochar derived from rice straw on vanadium transformation and its uptake by rice (Oryza sativa L.): insights from photosynthesis, antioxidative response, and gene-expression profile. Environ Pollut 289:117916

Mehmood S, Ahmed W, Alatalo JM, Mahmood M, Imtiaz M, Ditta A, Ali EF, Abdelrahman H, Slaný M, Antoniadis V, Rinklebe J, Shaheen SM, Li W (2022) Herbal plants- and rice straw-derived biochars reduced metal mobilization in fishpond sediments and improved their potential as fertilizers. Sci Total Environ 826:154043

Mehmood S, Ahmed W, Alatalo JM, Mahmood M, Asghar RMA, Imtiaz M, Ullah N, Li W-D, Ditta A (2023a) A systematic review on the bioremediation of metal contaminated soils using biochar and slag: current status and future outlook. Environ Monit Assess 195:961

Mehmood S, Ahmed W, Rizwan M, Ditta A, Irshad S, Chen DY, Bashir S, Mahmood M, Li W, Imtiaz M (2023b) Biochar, slag, and ferrous manganese ore affect Pb, Cd, and antioxidant enzymes in water spinach (Ipomoea aquatica) grown in multi-metal contaminated soil. Crop Pasture Sci 74(1–2):132–146

Menéndez J, Inguanzo M, Pis J (2002) Microwave-induced pyrolysis of sewage sludge. Water Res 36(13):3261–3264

Menéndez J, Domınguez A, Inguanzo M, Pis J (2004) Microwave pyrolysis of sewage sludge: analysis of the gas fraction. J Anal Appl Pyrolysis 71(2):657–667

Miranda ML, Hale B (1997) Waste not, want not: the private and social costs of waste-to-energy production. Energy Policy 25(6):587–600

Mo Jia Hao MJ, Yang Qi YQ, Zhang Na ZN, Zhang Wen Xiang ZW, Zheng Yi ZY, Zhang Zhien ZZ (2018) A review on agro-industrial waste (AIW) derived adsorbents for water and wastewater treatment

Mubeen I, Buekens A (2019) Energy from waste: prospects toward sustainable development. In: Current developments in biotechnology and bioengineering. Elsevier, pp 283–305

Mukherjee AG, Wanjari UR, Chakraborty R, Renu K, Vellingiri B, George A, CR SR, Gopalakrishnan AV (2021) A review of modern and smart technologies for efficient waste disposal and management. J Environ Manag 297:113347
Mulya KS, Zhou J, Phuang ZX, Laner D, Woon KS (2022) A systematic review of life cycle assessment of solid waste management: methodological trends and prospects. Sci Total Environ 831:154903
Murtaza G, Habib R, Shan A, Sardar K, Rasool F, Javeed T (2017) Municipal solid waste and its relation with groundwater contamination in Multan, Pakistan. IJAR 3(4):434–441
Murtaza G, Ditta A, Ullah N, Usman M, Ahmed Z (2021a) Biochar for the management of nutrient impoverished and metal contaminated soils: Preparation, applications, and prospects. J Soil Sci Plant Nut 21:2191–2213
Murtaza G, Ahmed Z, Usman M, Tariq W, Ullah Z, Shareef M, Iqbal H, Waqas M, Tariq A, Wu Y, Zhang Z, Ditta A (2021b) Biochar-induced modifications in soil properties and their impacts on crop growth and production. J Plant Nut 44(11):1677–1691
Murtaza G, Ahmed Z, Usman M, Iqbal R, Zulfiqar F, Tariq A, Ditta A (2024a) Physicochemical properties and performance of non-woody derived biochars for the sustainable removal of aquatic pollutants: a systematic review. Chemosphere 359:142368
Murtaza G, Ahmed Z, Valipour M, Ali I, Usman M, Iqbal R, Zulfiqar U, Rizwan M, Mahmood S, Ullah A, Arslan M, Rehman MHU, Ditta A, Tariq A (2024b) Recent trends and economic significance of modified/functionalized biochars for remediation of environmental pollutants. Scient Rep 14:217
Murtaza G, Usman M, Eldin SM, Valipour M, Rizwan M, Iqbal R, Zulfiqar U, Ali I, Ahmed I, Majeed A, Munir A, Alwahibi MS, Elshikh MS, Ditta A (2024c) Effectiveness of corn stalk biochar in amending the contaminated soil attributes and enhancing sustainable grass growth. Glob NEST J 26(1):5429
Muşdal H (2007) Tıbbi atıkları işleme ve bertaraf etme teknolojisi seçme problemine bulanık analitik hiyerarşi prosesi ve bulanık analitik ağ prosesi yaklaşımı. YT Ü, İstanbul
Mwanza BG, Mbohwa C, Telukdarie A (2018) The influence of waste collection systems on resource recovery: a review. Procedia Manufact 21:846–853
Narayana T (2009) Municipal solid waste management in India: From waste disposal to recovery of resources? Waste Management 29(3):1163–1166
Narayanamoorthy S, Anuja A, Pragathi S, Sandra M, Ferrara M, Ahmadian A, Kang D (2024) Assessment of inorganic solid waste management techniques using full consistency and extended MABAC method. Environ Sci Pollut Res 31(7):9981–9991
Natarajan V, Karunanidhi M, Raja B (2020) A critical review of radioactive waste management through biological techniques. Environ Sci Pollut Res 27:29812–29823
Naveed M, Tanvir B, Xiukang W, Brtnicky M, Ditta A, Kucerik J, Subhani Z, Nazir MZ, Radziemska M, Saeed Q, Mustafa A (2021) Co-composted biochar enhances the growth, physiological, and phytostabilization efficiency of Brassica napus and reduces associated health risks under Cr stress. Frontiers in Plant Science 12:775785
Nelson PM, Van Norman ER, VanDerHeyden A (2017) Reduce, reuse, recycle: the longitudinal value of local cut scores using state test data. J Psychoeduc Assess 35(7):683–694
Noor T, Javid A, Hussain A, Bukhari SM, Ali W, Akmal M, Hussain SM (2020) Types, sources, and management of urban wastes. In: Urban ecology. Elsevier, pp 239–263
Obi F, Ugwuishiwu B, Nwakaire J (2016) Agricultural waste concept, generation, utilization, and management. Niger J Technol 35(4):957–964
Ojeda-Benitez S, de Vega CA, Ramırez-Barreto ME (2003) Characterization and quantification of household solid wastes in a Mexican city. Resour Conserv Recycl 39(3):211–222
Ojovan MI (2024) Vitrification as a key solution for immobilisation within nuclear waste management. Arab J Sci Eng:1–9
Pal P (2017) Chapter 2-Chemical treatment technology. In: Industrial water treatment process technology. Butterworth-Heinemann, pp 21–63
Park D-K, Bitton G, Melker R (2006) Microbial inactivation by microwave radiation in the home environment. J Environ Health 69(5):17

Parreira AB, Kobayashi ARK, Silvestre O Jr (2003) Influence of Portland cement type on unconfined compressive strength and linear expansion of cement-stabilized phosphogypsum. J Environ Eng 129(10):956–960

Poonam Sharma PS, Gaur V, Kim SangHyoun KS, Ashok Pandey AP (2020) Microbial strategies for bio-transforming food waste into resources

Prabakar D, Manimudi VT, Sampath S, Mahapatra DM, Rajendran K, Pugazhendhi A (2018) Advanced biohydrogen production using pretreated industrial waste: outlook and prospects. Renew Sustain Energy Rev 96:306–324

Pudasainee D, Kurian V, Gupta R (2020) Coal: past, present, and future sustainable use. Future Energy:21–48

Ravindran R, Hassan SS, Williams GA, Jaiswal AK (2018) A review on bioconversion of agro-industrial wastes to industrially important enzymes. Bioengineering 5(4):93

Raza MAS, Ibrahim MA, Ditta A, Iqbal R, Aslam MU, Muhammad F, Ali S, Çiğ F, Ali B, Ikram RM, Muzamil MN, Rahman MH, Alwahibi MS, Elshikh MS (2023) Exploring the recuperative potential of brassinosteroids and nano-biochar on growth, physiology, and yield of wheat under drought stress. Scientific Rep 13:15015

Sabir A, Naveed M, Bashir MA, Hussain A, Mustafa A, Zahir ZA, Kamran M, Ditta A, Núñez-Delgado A, Saeed Q, Qadeer A (2020) Cadmium mediated phytotoxic impacts in Brassica napus: managing growth, physiological and oxidative disturbances through the combined use of biochar and Enterobacter sp. MN17. J Environ Manag 265:110522

Sadef Y, Nizami A, Batool S, Chaudary M, Ouda O, Asam Z-u-Z, Habib K, Rehan M, Demirbas A (2016) Waste-to-energy and recycling value for developing integrated solid waste management plan in Lahore. Energy Sour Part B: Econ Plan Policy 11(7):569–579

Sagar NA, Pareek S, Sharma S, Yahia EM, Lobo MG (2018) Fruit and vegetable waste: bioactive compounds, their extraction, and possible utilization. Compr Rev Food Sci Food Saf 17(3):512–531

Saleh TA (2016) Nanomaterials for pharmaceuticals determination. Bioenergetics 5(1):1000226

Saleh TA (2020) Nanomaterials: classification, properties, and environmental toxicities. Environ Technol Innov 20:101067

Salgot M, Folch M (2018) Wastewater treatment and water reuse. Curr Opin Environ Sci Health 2:64–74

Salihoglu G, Salihoglu NK, Ucaroglu S, Banar M (2018) Food loss and waste management in Turkey. Biores Technol 248:88–99

Samer M (2015) Biological and chemical wastewater treatment processes. Wastewat Treat Eng 150(10.5772):61250

Sangeetha T, Rajneesh CP, Yan W-M (2020) Integration of microbial electrolysis cells with anaerobic digestion to treat beer industry wastewater. In: Integrated microbial fuel cells for wastewater treatment. Elsevier, pp 313–346

Sarfraz R, Hussain A, Sabir A, Fekih IB, Ditta A, Xing S (2019) Role of Biochar and plant growth-promoting rhizobacteria to enhance soil carbon sequestration–a review. Environ Monit Assess 191:251

Sebastião D, Gonçalves MS, Marques S, Fonseca C, Gírio F, Oliveira AC, Matos CT (2016) Life cycle assessment of advanced bioethanol production from pulp and paper sludge. Biores Technol 208:100–109

Shao J, Yan R, Chen H, Wang B, Lee DH, Liang DT (2008) Pyrolysis characteristics and kinetics of sewage sludge by thermogravimetry Fourier transform infrared analysis. Energy Fuels 22(1):38–45

Shayler, H., McBride, M., and Harrison, E. (2009). Sources and impacts of contaminants in soils

Shin SK, Kim W-I, Jeon T-W, Kang Y-Y, Jeong S-K, Yeon J-M, Somasundaram S (2013) Hazardous waste characterization among various thermal processes in South Korea: a comparative analysis. J Haz Mater 260:157–166

Shovon SM, Akash FA, Rahman MA, Rahman W, Chakraborty P, Monir MU, Karim KMR, Habib MA, Khan MFH (2024) The potential for sustainable waste management and energy recovery in Bangladesh: a review. Sustain Energy Technol Assess 64:103705

Shreya M, Nimal Yughan V, Katyal J, Ramesh R (2023) Technical solutions for waste classification and management: a mini-review. Waste Manag Res 41(4):801–815. https://doi.org/10.1177/0734242x221135262
Sukholthaman P, Shirahada K (2015) Technological challenges for effective development towards sustainable waste management in developing countries: A case study of Bangkok, Thailand. Technol Soc 43:231–239
Tabassum S, Zhang Y, Zhang Z (2015) An integrated method for palm oil mill effluent (POME) treatment for achieving zero liquid discharge–a pilot study. J Clean Prod 95:148–155
Taddese S (2019) Municipal waste disposal on soil quality. A review. Acta Sci Agri 3(12):09–15
Tetteh EK, Rathilal S, Chetty M, Armah EK, Asante-Sackey D (2019) Treatment of water and wastewater for reuse and energy generation-emerging technologies. Water Wastewat Treat:53–80
Thompson PG (2023) Armed groups: the twenty-first-century threat. Rowman and Littlefield
Torkayesh AE, Rajaeifar MA, Rostom M, Malmir B, Yazdani M, Suh S, Heidrich O (2022) Integrating life cycle assessment and multi-criteria decision making for sustainable waste management: key issues and recommendations for future studies. Renew Sustain Energy Rev 168:112819
Ullah N, Ditta A, Khalid A, Mehmood S, Rizwan MS, Mubeen F, Imtiaz M (2020) Integrated effect of algal biochar and plant growth promoting rhizobacteria on physiology and growth of maize under deficit irrigations. J Soil Sci Plant Nut 20:346–356
Ulubeyli S, Kazaz A, Arslan V (2017) Construction and demolition waste recycling plants revisited: management issues. Procedia Eng 172:1190–1197
Un-Habitat (2010) Solid waste management in the world's cities: water and sanitation in the world's cities 2010. Routledge
Vlachokostas C, Michailidou AV, Achillas C (2021) Multi-criteria decision analysis towards promoting waste-to-energy management strategies: a critical review. Renew Sustain Energy Rev 138:110563
Walls M (2011) Deposit-refund systems in practice and theory. Resources for the future discussion paper (11–47)
Wang J, Su R, Chen W, Guo Y, Jin Y, Wen Z, Liu Y (2006) Deep geological disposal of high-level radioactive wastes in China. Chinese Journal of Rock Mechanics and Engineering 25(4):649–658
Wang J, Li Z, Tam VW (2014) Critical factors in effective construction waste minimization at the design stage: a Shenzhen case study, China. Resour Conserv Recyc 82:1–7
Wang Y, Ng KTW, Asha AZ (2016) Non-hazardous waste generation characteristics and recycling practices in Saskatchewan and Manitoba, Canada. J Mater Cycl Waste Manag 18:715–724
Wilson DC, Rodic L, Modak P, Soos R, Carpintero A, Velis K, Iyer M, Simonett O (2015) Global waste management outlook. UNEP
Xevgenos D, Papadaskalopoulou C, Panaretou V, Moustakas K, Malamis D (2015) Success stories for recycling of MSW at municipal level: a review. Waste Biomass Valoriz 6:657–684
Yang Y, Heaven S, Venetsaneas N, Banks C, Bridgwater A (2018) Slow pyrolysis of organic fraction of municipal solid waste (OFMSW): characterisation of products and screening of the aqueous liquid product for anaerobic digestion. Appl Energy 213:158–168
Yasin H, Usman M, Rashid H, Nasir A, Randhawa IA (2017) Alternative approaches for solid waste management: a case study in Faisalabad Pakistan. Earth Sci Pakistan 1(2):7–9
Yatoo AM, Hamid B, Sheikh TA, Ali S, Bhat SA, Ramola S, Ali MN, Baba ZA, Kumar S (2024) Global perspective of municipal solid waste and landfill leachate: generation, composition, eco-toxicity, and sustainable management strategies. Environ Sci Pollut Res 31(16):23363–23392
Yenkie KM, Burnham S, Dailey J, Cabezas H, Friedler F (2019) Generating Efficient Wastewater Treatment Networks: an integrated approach comprising of contaminant properties, technology suitability, plant design, and process optimization. In: Computer Aided Chemical Engineering, vol 46. Elsevier, pp 1603–1608
Yildiz BS (2012) Water and wastewater treatment: biological processes. In: Metropolitan sustainability. Elsevier, pp 406–428

Zahrim A, Azreen I, Jie S, Yoiying C, Felijia J, Hasmilah H, Gloriana C, Khairunis I (2019) Nanoparticles enhanced coagulation of biologically digested leachate. In: Nanotechnology in water and wastewater treatment. Elsevier, pp 205–241

Zeb H, Hussain A, Naveed M, Ditta A, Ahmad S, Jamshaid MU, Ahmad HT, Hussain B, Aziz R, Haider MS (2018) Compost enriched with ZnO and Zn-solubilizing bacteria improves yield and Zn-fortification in flooded rice. Ital J Agron 13(4):310–316

Zhang L, Sun X (2018) Influence of sugar beet pulp and paper waste as bulking agents on physical, chemical, and microbial properties during green waste composting. Biores Technol 267:182–191

Zhang H, Wen Z-G (2014) The consumption and recycling collection system of PET bottles: a case study of Beijing, China. Waste Manag 34(6):987–998

Zhang M, Xie L, Yin Z, Khanal SK, Zhou Q (2016) Biorefinery approach for cassava-based industrial wastes: current status and opportunities. Biores Technol 215:50–62

Zhang Q, Liu C, Duan K, Zhang Z, Xiang W (2020) True three-dimensional geomechanical model tests for stability analysis of surrounding rock during the excavation of a deep underground laboratory. Rock Mech Rock Eng 53:517–537

Zimmermann K (2017) Microwave as an emerging technology for the treatment of biohazardous waste: a mini-review. Waste Manag Res 35(5):471–479

Agricultural Pollution and Its Relation to Climate Change

6

Muhammad Shaheryar, Momna Hayat,
Muhammad Nadeem Shah, Hafiz Saqib Hayat,
Haseeb Ur Rehman, Ali Sheryar,
Muhammad Saleem Kashif, and Muhammad Tanveer Altaf

Abstract

Agriculture contributes to climate change through greenhouse gas emissions, deforestation, soil deterioration, and water pollution. These factors threaten global food security and environmental sustainability. Sustainable agriculture techniques must be implemented immediately to reduce these effects. This mitigation technique relies on emission reduction, agricultural waste management, and pest management. Sustainable agriculture and ecosystems require soil and water quality preservation. Environmental issues must be addressed through eco-friendly agriculture. International cooperation is needed to address the complex relationship between agriculture and climate change. Inaction may worsen environmental issues, making activity urgent. Sustainable agriculture reduces greenhouse gas emissions and ensures food security. Maintaining agricultural output and environmental resilience requires promoting soil and water conservation. Policy interventions and incentives are needed to promote sustainable agriculture worldwide. Education and awareness efforts can assist farmers and consumers practice environmental stewardship. Developing sustainable agricultural tech-

M. Shaheryar · H. S. Hayat · H. U. Rehman
Institute of Agronomy, Bahauddin Zakariya University, Multan, Punjab, Pakistan
e-mail: haseeburrehman@bzu.edu.pk

M. Hayat
Department of Horticulture, Faculty of Agricultural Sciences and Technology, Bahauddin Zakariya University, Multan, Punjab, Pakistan

M. N. Shah (✉) · M. S. Kashif · A. Sheryar
Department of Agriculture, Government College University Lahore, Lahore, Punjab, Pakistan
e-mail: nadeemshah@gcu.edu.pk; saleem.kashif@gcu.edu.pk

M. T. Altaf
Department of Field Crops, Faculty of Agriculture, Recep Tayyip Erdoğan University, Rize/Pazar, Türkiye

A. A. H. Abdel Latef et al. (eds.), *Bioremediation and Nanotechnology for Climate Change Mitigation*, https://doi.org/10.1007/978-981-96-3069-1_6

nologies and practices requires research and innovation. Effective mitigation initiatives require collaboration between government, Non-governmental organizations (NGOs), and business sectors. Agriculture's environmental impact must be addressed holistically, considering social, economic, and environmental concerns. Sustainable land management can reduce agricultural biodiversity and ecosystem services in loss. Reforestation and agroforestry can sequester carbon and conserve biodiversity. Sustainable water management is vital for saving freshwater and reducing agricultural runoff pollution. Integrating pest management can reduce chemical pesticide use and increase natural pest control. Organic farming boosts nutrient cycling and reduces erosion. Promote agro-ecological concepts to improve climate change resistance and biodiversity and ecosystem services. Finally, agriculture's environmental impact must be addressed collaboratively and creatively to achieve long-term sustainability and resilience.

Keywords

Agricultural pollution · Climate change · Greenhouse gas emissions · Sustainable practices · Mitigation strategies · Food security

6.1 Introduction

Agricultural practices are a significant contributor to the planet's warming since they result in the emission of gases such as methane and nitrous oxide. If we can cut down on these emissions from farming, it will greatly assist in reducing the rate of climate change. However, we must have a clear understanding of the important distinction that exists between carbon dioxide (CO_2) and methane (CH_4). As time passes, carbon dioxide (CO_2) remains in the atmosphere for an extended time, accumulating like a stockpile. On the other hand, methane behaves more like a flow; it is released and decomposes relatively quickly. When we talk about emissions using a measurement called CO_2-equivalent, it can be misleading because it does not reflect how these gases affect the temperature in the same manner as it does when we talk about emissions using other measurements. Therefore, when we are trying to figure out how various businesses contribute to global warming and who needs to do what to minimize emissions, simply looking at numbers similar to CO_2 does not give us the complete picture. The individual gases that are being emitted need to be taken into consideration, as well as the many ways in which they influence the climate (Lynch et al. 2021). Agriculture plays a significant role in our economy and in providing employment opportunities for individuals; but there are instances when agriculture can be detrimental to both the environment and our health. In agriculture, the term "agricultural pollution" refers to the release of dangerous elements into the environment, whether in the air, water, or soil. Contemplate it in this manner: While farmers are working, they are putting pressure on the environment. The status of the environment and the resources it contains can be altered due to this stress. It is then that society responds to these shifts in various ways. In order to address this issue, we require accurate information regarding our environment, the various pollutants produced by farming, and how these pollutants impact both humans and the ecosystem. The more we know about all of this, the better we will be able to find out how to lessen the pollution that comes from agriculture (Abbasi et al. 2014).

As a result of the increased ambition of global climate policy, which is outlined in the objective of the Paris Agreement to "limit the rise in the global average temperature to well below 2 °C above preindustrial levels and to pursue efforts to limit the temperature increase to 1.5 °C above preindustrial levels" (UNFCCC 2015), there has been an increase in the scrutiny that is being placed on the potential contributions that all sectors could make towards mitigating climate change. Recent articles, such as those written by Poore and Nemecek (2018) and Springmann et al. (2018), have brought attention to the fact that this includes a particular emphasis on agriculture. In spite of this emphasis, there are a few critical factors about the unique consequences of methane (CH_4) and nitrous oxide (N_2O), the principal greenhouse gases emitted from agricultural operations, that appear to be disregarded or misinterpreted in discussions concerning the role that agriculture plays in climate change. Not only is it vital to understand these distinctions to develop mitigation methods effectively, but it is also essential to evaluate the efficiency of those policies.

6.2 Agriculture as a Culprit and Victim of Climate Change

Agriculture, which is both a problem and a victim of climate change, is like a coin with two sides since it is equally affected by both. Agricultural practices have an impact on climate, but they also contribute to the phenomenon of climate change. There are positive and negative aspects to the interaction that exists between agriculture and climate. It is therefore possible to assert that agriculture is both a contributor to and a victim of climate change. One of the primary contributors to the magnitude of the climate change is agriculture. For the most part, human activity is responsible for climate change. The agricultural sector is of critical significance to the economies of a great number of emerging nations. On the other hand, it is extremely sensitive to changes in the weather, such as extremely hot summers or particularly heavy rains. Climate change is a significant issue on a global scale. While it is warming the earth, it is also causing a great deal of trouble for ecosystems, the lives of people, and even the continued existence of certain species (Fereja 2016).

Agriculture contributes to the worsening of this problem by generating greenhouse gases, which trap heat in the atmosphere. There are a variety of ways in which the consequences of climate change can be observed, including temperatures that are higher throughout the whole year, shifts in the timing and amount of rainfall, longer summers, and an increase in the frequency and severity of extreme weather events such as floods, droughts, and heat waves. Additionally, sea levels are rising, and we are witnessing an increase in the frequency and severity of extreme weather occurrences. On the other hand, there are activities that we engage in that contribute to the worsening of climate change. These activities include the depletion of soil and the cutting down of trees. In addition to contributing to the issues of climate change, these actions are harmful to the environment. Greenhouse gas emissions, deforestation, changes in land use, and agricultural techniques such as cattle and rice production, which increase methane emissions, as well as waste management and pollution, are all examples of human activities that contribute to climate change. Other human activities include pollution and waste management. Carbon dioxide, methane, and nitrogen are examples of greenhouse gases that contribute significantly to the phenomenon of global warming, which is referred to as the greenhouse effect (Ruane and Rosenzweig 2022; McCoy 2020). They are responsible for the greenhouse effect. The French mathematician Jean Baptiste Fourier (1768–1830) is credited with the invention of the phrase "greenhouse effect." Fourier observed that certain gases can trap heat in the atmosphere. According to Leggett (1999), this effect is critically important for preserving a warm layer close to the surface of the Earth, which is necessary for the existence of life. On the other hand, the problem emerges when the equilibrium of heat in the Earth's energy budget becomes unstable due to international warming. It is believed that human activities are contributing to an increase in greenhouse gases, which is breaking this balance and leading to climate change, which in turn affects agriculture. However, this phenomenon is not without its consequences.

6.3 Factors Causing Agriculture Pollution

When we talk about pollution, we refer to the negative effects of three essential components of the natural world: the air, the water, and the soil. Regarding farms and the products they generate, every region of nature approaches the issue in a distinctly different manner. Therefore, it is essential to have a thorough understanding of how farm products can impact the air, water, and soil (StudySmarter Web Link).

6.3.1 Air Pollution Caused by Agriculture

Greenhouse gases are essential because they help maintain our world's temperature by retaining heat. But an excessive amount of these can be problematic, and the combustion of fossil fuels by humans results in an excessive amount of these gases being produced. When plants and soil breathe and decompose, they naturally emit greenhouse gases such as carbon dioxide through the process of respiration. This is a regular occurrence that contributes to equilibrium. On the other hand, certain farming practices result in the emission of additional greenhouse gases, which contributes to air pollution. The release of heat-trapping gases into the atmosphere stems from the carbon that has been stored in plants and soil, which is then released into the atmosphere. CO_2, CH_4, and N_2O are the primary pollutants that are produced by agricultural practices and are responsible for the contamination of the air.

Farms use a variety of methods to produce each of these gasses. Let us take a more in-depth look at the process that is used to create each one (Study Smarter Web Link).

6.3.2 Water Pollution Caused by Agriculture

Water pollution from agriculture occurs when contaminants such as chemicals from pesticides and fertilizers, animal feces, and soil particles are carried into rivers, lakes, and groundwater by runoff from irrigation systems or rains. In addition to causing environmental disruptions, these contaminants can harm aquatic life by poisoning it. The overuse of fertilizers can result in eutrophication, a condition in which the growth of algae reduces the amount of oxygen present in the water, which in turn impacts fish and other creatures. Through the introduction of dangerous compounds and diseases into water sources, animal waste poses a threat to the health of humans as well as to the ecosystems of aquatic environments (StudySmarter Web Link).

6.3.3 Soil Pollution Caused by Agriculture

The accumulation of toxic substances such as salts and heavy metals, which are frequently introduced into the soil through the use of chemicals and irrigation methods, is a threat to the soil's health caused by agriculture. This problem is made even more complicated by soil erosion, which is made worse by agricultural activities like tilling, destabilizes the soil structure and makes it easier for particles to run off. Therefore, the loss of soil through erosion not only depletes necessary nutrients but also adds to water pollution. This is because sediment accumulates in water bodies, which blocks sunlight and hinders the growth of aquatic plants, so disturbing ecosystems (StudySmarter Web Link).

6.4 Impact of Climate Change on Agriculture

In order to gain an understanding of the effects that climate change has on agriculture, it is necessary to take into account the interrelated systems that affect food security and agricultural commerce. These systems include biological, social, and political variables (Ruane and Rosenzweig 2022). Despite the fact that rapid socioeconomic growth around the world may create the idea that climate change is not damaging, it poses extra obstacles, particularly in industrialized countries where it impedes development progress. Agricultural output in the future is driven not only by the direct biological impacts of changing climate conditions on farms but also by economic incentives and policy decisions that include a variety of stakeholders at both the local and global levels (Ruane and Rosenzweig 2022). These effects are particularly constant not only in Malawi but also in other places. Specifically, the Environmental Affairs Department of Malawi emphasizes that the negative consequences observed in the country are comparable to those observed in adjacent countries and even further afield. As a result of climate change, numerous regions across Africa and the rest of the world are going through instances of flooding and the damages that are associated with it, as well as acute limitations in runoff that are leading to cutbacks in the amount of hydroelectric output. In addition, climate change is responsible for extreme weather events, changes in the patterns of rainfall, and variations in the quantities of carbon dioxide in the atmosphere, all of which contribute to the detrimental effects climate change has on agriculture.

6.4.1 Extreme Weather Conditions

Climate changes have caused extreme weather conditions during the past three decades, which have resulted in a global drop in agricultural productivity of approximately 1–5% per decade. This is in comparison to what would have been accomplished if these changes had not occurred with regard to agricultural production. According to Thornton et al. (2018), tropical cereal crops such as maize and rice have been significantly impacted. There is a high probability that agricultural productivity will decrease worldwide, with tropical regions facing particularly unfavorable impacts, even if the temperature rises by only 2 °C. At a temperature increase of 1.5 °C, it is anticipated that the repercussions on human and natural systems will be severe, ultimately affecting all agricultural systems. According to Thornton et al. (2018), due to temperature variations, major cash crops like coffee and cocoa in some tropical regions are anticipated to see changes in distribution and productivity. The consequences of heat waves, which manifest in climate change, are devastating to crops and livestock. The reduction in photosynthetic activity that occurs in plants as a result of extremely high temperatures can result in the loss of leaves and the potential failure of crops. Heat waves can render pollen infertile during the flowering process, also known as anthesis, which can lead to reproductive failure and a decrease in grain quantities (Ruane and Rosenzweig 2022). Livestock can experience heat stress during heat waves, which can result in stunted growth as well as decreased output of dairy products and meat (Fereja 2016). Because heat waves cause crops to fail and animals to suffer from heat stress, which results in the loss of foliage and fodder, this dual impact on crops and animals further contributes to the lack of food that is available. Furthermore, the loss of animals that produce methane due to heat stress reduces the production of methane, which is generally beneficial to the climate. Heat waves represent huge economic dangers in the Indo-Pak Region, with over 4.5% of GDP at risk by the end of the decade due to lost productivity and health repercussions. Heat waves are a major contributor to the region's economic woes. A significant amount of India's landmass is located in areas prone to intense heat, which results in decreased production and the capacity to work outside. Not only may extreme heat have a detrimental influence on the quality of life for millions of people, but it could also cut GDP by 2.8% by the year 2050. In addition, heat waves put a strain on power systems because of the high energy needs for air conditioning and groundwater pumping, which further exacerbates environmental problems (according to a report by McKinsey and Company 2020).

6.4.2 Changes in Rainfall Patterns

Alterations in the patterns of rainfall have enormous effects on agriculture, causing unstable planting seasons, droughts, and floods, which in turn affect food security, particularly among vulnerable communities. Crops and cattle are both negatively impacted by droughts, which are a consequence of changing rainfall patterns. Droughts have a negative influence on agricultural productivity. The production of livestock is especially susceptible to the effects of climate change since natural pastures are deteriorating in quality and there is less food available. In addition, the survival of cattle is further threatened by unstable water supplies, which are drying up as a result of high temperatures and decreased rainfall conditions. The loss of livestock can be attributed to various issues, including but not limited to severe heat, water scarcity, a lack of feed, and infections, all of which are made worse by droughts induced by climate change (Fereja 2016). An important study that was conducted by Karki and Gurung (2012) investigated the effects that climate change might have on agriculture in Nepal, a nation that is highly dependent on agriculture for its entire economy. According to the study's findings, agriculture is still very vulnerable to the effects of climate change, despite the fact that it significantly contributes to Nepal's exports and GDP. Rice, wheat, maize, millet, barley, and buckwheat are the cereal crops that form the backbone of Nepalese agriculture; yet, the country is facing greater vulnerability due to the weather unpredictability related to rising temperatures and changing precipitation patterns. Rice, in particular, is in danger due to changes in the dependability of stream flows, enhanced and irregular monsoon rains, and the effects of flooding. In addition, shifting climatic conditions contribute to the emergence of new pests and diseases, which further impacts the agricultural sector (Karki and Gurung 2012). This is because a significant portion of Nepal's cultivated areas depends on monsoon rainfall. Changes in the timing and duration of the monsoon could significantly impact agricultural production, particularly rice yields, which could lead to food insecurity in the country.

6.4.3 Pests and Diseases

Stable climatic conditions are extremely important for maintaining control over diseases that affect crops and animals. Various mechanisms, including shifts in the distribution and quantity of disease vectors and reservoirs, survival of infections, and farming techniques, are among the ways in which climate change can impact the health of cattle and crops. Temperature and precipitation changes can cause the introduction of illnesses and parasites into new areas or raise the prevalence of diseases that are already there. This can lead to a drop in productivity and increased mortality rates in crops and animals (Rust and Rust 2013). In many cases, the presence of disease vectors and climate change interacts with socio-economic aspects and human-induced climate change drivers. These drivers include the loss of habitats and changes in land use, which can have an impact on the migration of people, livestock, and crops. When there is an overlap between reservoirs, vectors, and humans, disease transmission can occur. This is especially true for diseases that are transmitted by ticks that are also zoonotic. Gray et al. (2009) and Randolph (2008) found that changes in climatic circumstances can alter these parameters and the interactions between them. According to the findings of Patz et al. (2005), climate change is anticipated to have an effect on heat-related health problems as well as the prevalence of climate-sensitive infectious diseases. Climate change is one factor contributing to the emergence of new diseases. As an illustration, the avian influenza virus has become a worldwide issue due to its ability to spread through the international movement of animals and products derived from animals. There is a possibility that this disease could spread from poultry to human hosts, which could result in a pandemic that extends over the entire world. Both direct and indirect losses are incurred due to avian influenza and other animal diseases (Fereja 2016). Direct losses include the mortality of animals and a reduction in fertility, while indirect losses include increased costs associated with treatment and loss of market access. The production of large ruminants, such as cattle, is essential in both developed and developing nations. These animals provide various functions, including the production of commercial goods such as meat and milk, as well as the provision of food, income, transportation, and fertilizer in developing regions. However, developing regions are particularly susceptible to the effects of climate change since they have a limited capability to successfully manage large ruminants, which further exacerbates their vulnerability (Fereja 2016).

6.5 Agriculture's Impact on Climate Change

It is important to note that agriculture plays a key part in contributing to climate change. The generation of greenhouse gases and the expansion of agricultural land into new places are two of the consequences that climate change will have on the agricultural sector. The agricultural sector is not only susceptible to the effects of climate change and weather-related dangers, but it is also a significant contributor to the production of greenhouse gases and the alteration of land use, both of which are factors that contribute to the phenomenon of climate change (Ruane and Rosenzweig 2022). Throughout human history, the growing demand for additional land for crop production and animal grazing pastures has been a primary impetus for deforestation. Interactions between carbon and nitrogen stocks in soils and fertilizers, as well as methane emissions from activities such as paddy rice cultivation and cattle digestion, are some of how agricultural systems that contribute to the release of greenhouse gases. According to Ruane and Rosenzweig (2022), the agriculture sector is accountable for approximately one quarter of the overall emissions of greenhouse gases. These greenhouse gases contribute to global warming, which in turn leads to climate changes that impact food security. This is especially true in vulnerable places, such as developing countries, which lack the economic capacity to deal with such changes. In addition, agricultural practices such as soil depletion and degradation are other factors that contribute to increased levels of climate change. These activities not only reduce the quality of the soil but they also release more carbon into the atmosphere, which exacerbates the effects of greenhouse gas emissions on climate change.

6.5.1 Agricultural GHG's Emission

Some of the most significant contributions to global warming are carbon dioxide (CO_2), methane (CH_4), and nitrous oxide (N_2O), according to Myhre et al. (2013). A variety of climatic pollutants primarily causes anthropogenic climate change. Even though agriculture and food production are linked to all three of these gases, direct emissions from agriculture are distinct in that CH_4 and N_2O mostly dominate them. It is frequently claimed that the global food system is responsible for around 21–37% of annual emissions. This information is derived from the 100-year Global Warming Potential (Mbow et al. in press). However, the composition of gases emitted by the food system differs from the total global emissions balance. Agricultural activity is responsible for approximately 50% of all anthropogenic methane emissions and approximately 75% of anthropogenic nitrogen dioxide emissions (Mbow et al. in press). When it comes to quantifying carbon dioxide emissions from the food system, the task is made more complicated by the existence of various processes and the problems associated with using consistent accounting methodologies or sectoral borders. Even while agricultural production is responsible for a negligible amount of carbon dioxide emissions, such as those that result from lime and urea, these sources only account for a very small fraction of the total carbon dioxide emissions. These CO_2 emissions are considered to be a part of the food system emissions; however, they are highly uncertain and are typically accounted for as energy or transport emissions within the framework of the Intergovernmental Panel on Climate Change (IPCC) (Vermeulen et al. 2012). CO_2 emissions in agricultural operations or embedded in inputs (e.g., fertilizer manufacture and transport) are also considered to be part of the food system emissions. Reducing these emissions is anticipated to be accomplished mostly through the decarbonization of energy generation as a whole, rather than through the implementation of specific agricultural adjustments. In addition, the food system is a large contributor to the change in land use that results in CO_2 emissions, primarily as a result of the clearing of land for the production of crops or grazing.

6.5.2 Soil Depletion and Degradation

Depletion and degradation of soil are terms that describe the worsening of soil quality as a result of incorrect usage of the soil. This phenomenon typically occurs in regions impacted by agriculture, industry, or urban expansion. This ecological problem is serious and pervasive, and it has the potential to become even more severe as a result of changes in weather patterns. In addition to having an effect on the climate as a whole, it poses a significant risk to ecosystems, reducing the productivity of certain ecosystems. The destruction of vegetation, which provides animals with both food and a place to live, results from soil degradation. It is important to note that the services that soil ecosystems provide are essential for the upkeep of the carbon and water cycles, which highlights the strong connection between climate change and soil health. Further, certain agricultural techniques have the potential to reduce the fertility of the soil. Continuous cropping or grazing can deplete the nutrients in the soil, which can disturb the natural cycle that refills the soil with mineral-rich organic matter from dead plants. Heavy machinery can also cause soil to become compacted, which can cause structural damage, decrease the amount of water that can penetrate the soil, and worsen erosion. Erosion of the soil is a significant problem all over the world, but it is most prevalent in regions where huge tracts of land are removed for agricultural purposes in order to satisfy the requirements of an expanding population. This imbalance in ecosystems makes it possible for greenhouse gases to be released into the sky. These gases trap heat in the atmosphere, contributing to changes in the global climate patterns. In light of this, addressing the degradation of soil is absolutely necessary in order to reduce the negative effects of climate change.

6.5.3 Deforestation

Agriculture has become especially susceptible to the effects of climate change as a result of the prevalent use of antiquated farming practices. A substantial danger to the forests of regions such as Central and South America, Southeast Asia, Russia, China, and Africa is posed by illegal logging, mostly done to satisfy the international demand for flooring and furniture market. These cleared lands are frequently converted into farmland that is not sustainable through the practice of slash-and-burn agriculture, which is predominantly driven by monocultures such as palm oil plantations and livestock farming (McCoy 2020). The palm oil plantations that are prevalent in Latin America and Southeast Asia have destroyed around 25% of the rainforests that are located in Indonesia. As a result of this process, enormous quantities of carbon dioxide are released into the atmosphere, which not only greatly contribute to the overall emissions but also put endangered species, such as orangutans, in danger by destroying their habitat. The role that forests play as carbon sinks is diminished due to deforestation, which further contributes to the acceleration of climate change (McCoy 2020). As a result of the demand for grazing, animal husbandry is another factor that contributes to the destruction of tropical forests. In addition, the production of livestock is a significant contributor to the emission of greenhouse gases. Cattle and sheep, through their digestive processes, discharge methane and nitrous oxide into the atmosphere. The warming potential of these gases is greater than that of carbon dioxide, and they are responsible for roughly 10% of the carbon emissions that occur in the United States (McCoy 2020). Although greater levels of carbon dioxide can have some beneficial impacts on agriculture, such as enhancing the efficiency with which plants use water and increasing the amount of photosynthesis that occurs, same levels also have certain negative effects. When there is a larger concentration of carbon dioxide in the atmosphere, the nutritional value of crops can decrease, resulting in increased production costs because supplements are required to keep the nutritional balance intact. According to Ruane and Rosenzweig (2022), this can intensify the problem of food insecurity and impede growth, particularly among extremely vulnerable communities. Overall, the negative implications of climate change outweigh the positive impacts, particularly with regard to sustainability and food security. This is the case despite the fact that certain aspects of climate change may be beneficial to agriculture.

6.5.4 Agricultural Burning

According to Jenkins et al. (1996), agricultural burning is a prevalent farming technique involving the combustion of waste material such as crop leftovers. This practice is used for a variety of goals, including the clearance of land, the control of pests, and the enhancement of soil nutrients for improved crop quality. On the other hand, this process is responsible for the emission of hazardous by-products into the atmosphere, such as chemicals, smoke, and particulate matter, which poses a threat to human health (Jenkins et al. 1996). According to Venkataraman et al. (2006), these emissions consist of carbon, carbon dioxide, carbon monoxide, and sulfur dioxide, all of which have the potential to negatively impact air quality and agricultural production. In addition, agricultural burning, which is frequently carried out at low temperatures, is a contributor to air pollution, particularly when residues from rice and wheat are burned (Werther et al. 2000). While farmers may turn to agricultural burning as a waste management method, it is essential to set rules to reduce the negative effects that this practice has on the environment.

6.5.5 Use of Fertilizers

According to Savci (2012), fertilizers are necessary for increasing the fertility of the soil and the levels of nutrients to increase crop output. Nitrogen, phosphorus, and potassium are the three most important components of fertilizers. However, the application of chemical fertilizers in excessive amounts might result in air pollution by generating nitrogen oxides such as nitrogen dioxide, nitrogen oxide, and nitrogen dioxide (Savci 2012). An excessive amount of fertilizer is still being used in impoverished regions, even though industrialized countries have decreased their use of fertilizer due to environmental concerns (Kongshaug 1998). According to Kongshaug (1998), the application of fertilizer is responsible for the emission of around 1.2% of greenhouse gases. Ammonium-based fertilizers are responsible for the emission of ammonia gas, which plays a role in the development of acidic rain, which harms crops. In addition, the processes of nitrification and denitrification that occur in soil result in the production of nitrous oxide, which contributes to the release of greenhouse gases (Kongshaug 1998).

6.5.6 Water Pollution and Acidification

As a result of the excessive use of chemical plant protection products such as fertilizers and herbicides, agriculture is a substantial contributor to the pollution of coastal water and the acidification of seas and lakes. The agricultural sector is responsible for 70% of all water abstractions worldwide, significantly contributing to water pollution. According to Mateo-Sagasta et al. (2017), farms discharge significant quantities of agrochemicals, organic debris, drug residues, sediments, and salty drainage into bodies of water, which threaten the health of humans and endangered aquatic ecosystems. Not only does this pollution result in the loss of biodiversity but it also causes disruptions in ecosystems, contributing to climate change. Agricultural pollution has superseded contamination from urban areas and businesses as the principal cause of water degradation in inland and coastal waters in many nations with high incomes and developing economies (Mateo-Sagasta et al. 2017). This is the case in many countries. The most common chemical contamination found in groundwater worldwide is nitrate, which is mostly derived from agricultural practices. Agricultural pollution is further exacerbated by increased sediment runoff and groundwater salinization, which affects the Earth's energy balance and contributes to climate change (Ruane and Rosenzweig 2022). As a result of population expansion and dietary shifts, there has been an increase in the amount of pressure placed on water quality by agricultural, livestock, and aquaculture systems (Mateo-Sagasta et al. 2017). This is in order to meet the increasingly high demand for food. Both the pollution of water and the acidity of water lead to the emission of greenhouse gases, which further upsets the equilibrium of the Earth's energy supply and exacerbates the effects of climate change. In developing countries, the pursuit of industrialization for economic growth presents a dilemma because there is a high probability that water pollution and acidification will continue. Furthermore, wealthy countries have not been able to stop the trend of agricultural industrialization, which is a key contributor to climate change. Since emerging nations are prioritizing industrialization, the gains that industrialized countries have made in carbon markets may be undercut as a consequence (Mateo-Sagasta et al. 2017). To reduce the effects of climate change and ensure that water management methods are sustainable on a global scale, addressing agricultural pollution is essential.

6.6 Mitigation Strategies to Control Agricultural Pollution

Population growth surpasses the Earth's capacity, resulting in environmental degradation and food insecurity. Given the significant increase in population since Malthus' time, it is imperative to prioritize agricultural sustainability in order to tackle the pressing concerns of food security and ecological issues. Nevertheless, the pursuit of immediate profits can jeopardize the long-term sustainability of agriculture. Expertise in long-term planning is crucial for ensuring sustainability. Key components include measures like soil conservation and crop rotation with nitrogen-fixing plants such as pulses and Sesbania rostrata (Kesavan and Swaminathan 2008). Regarding agricultural practices for sustainability, it is crucial to prioritize ecological stability and biodiversity conservation over short-term gains (Kesavan and Swaminathan 2008). According to Pretty (2007), sustainable practices should prioritize being friendly to the ecosystem, accessible to farmers, and supportive of food production. We need new and innovative herbicides to address the issue of weed resistance (Duke 2012). It is crucial to prioritize sustainable food production and organic farming in order to effectively tackle food security challenges (Paoletti et al. 2010). Achieving global emission reductions is essential to effectively address climate change and its related consequences (Lynch et al. 2021). Efforts to decrease emissions primarily revolve around reducing the release of greenhouse gases into the atmosphere, specifically carbon dioxide (CO_2), methane (CH_4), and nitrous oxide (N_2O).

6.6.1 Global Emission Reductions

Addressing agricultural greenhouse gas emissions is of utmost importance. It is essential to eliminate net food system CO_2 emissions and all other CO_2 emissions. Additionally, we must focus on reducing agricultural methane and N_2O emissions, as they have distinct climate benefits and should be actively promoted. The atmospheric concentrations of methane and N2O closely resemble their "worst-case" representative concentration pathways (RCPs), as indicated by studies conducted by Nisbet et al. (2019). In order to meet the climate goals of the Paris Agreement, every sector must take immediate and significant actions to reduce their emissions of all gases. Failure to sufficiently reduce agricultural emissions will jeopardize our capacity to restrict global warming to 1.5 °C (Leahy et al. 2020). The current projections for emissions from the food system alone pose a significant threat to achieving this target (Clark et al. 2020). Despite this context, there are still numerous uncertainties surrounding the establishment of targets for various greenhouse gases. Experts have proposed that the Paris Agreement should be understood as aiming to achieve net-zero greenhouse gas emissions on a global scale, as measured by the GWP100, even though it is not explicitly stated. Some experts have pointed out that there are varying opinions on how to balance different gases (Fuglestvedt et al. 2018). They have also suggested that the agreement could benefit from a more specific emphasis on achieving net-zero CO_2 instead of requiring net-zero emissions across all gases for the temperature targets (Tanaka and O'Neill 2018). These points can be debated because, as shown above, targets based on the GWP100 do not have a clear connection to temperature outcomes. Using policy accounting tools instead of focusing on the temperature goal itself carries certain risks. Not all gases are equal, so simply shifting efforts from one gas to another won't yield the same results. Experts have pointed out that prioritizing the reduction in methane emissions over CO_2 can have negative long-term consequences. This approach may provide some immediate climate benefits, but it ultimately leads to warmer temperatures in the years to come. It is important to note that reducing methane emissions alone can only significantly impact limiting peak warming if we are also making progress towards achieving net-zero CO_2 emissions. The temporal details are not revealed by a GWP100 accounting-based framework (Lynch et al. 2020a). When it comes to agriculture, there is a potential trade-off between supporting certain products or production methods and the lifespan of greenhouse gases. However, an even more significant concern is that addressing agricultural emissions might divert attention away from the crucial goal of de-carbonization. Suppose significant efforts are made to decrease agricultural emissions but prove to be costly in terms of monetary expenses, political capital, public support, or individual effort and divert attention from the goal of eliminating fossil CO_2 emissions. In that case, our climate situation will worsen.

6.6.2 Agri-waste Management

Expertise in agricultural waste management encompasses a range of factors, including waste treatment, agricultural practices, and rural development. In addition to farmers, other stakeholders also have a vital role in managing agricultural waste. Experts in the field of agriculture understand the potential negative impacts of non-natural wastes such as pesticides, fertilizers, and machinery waste on the environment and human health. Proper treatment and recycling of these wastes are essential to address agricultural pollution. Using reusable containers is an effective way to reduce waste. A study comparing waste management strategies in Europe and five other countries highlights various approaches (Sakai et al. 2011). Research conducted in Indonesia indicates that cassava waste pulp can be transformed into a highly efficient superabsorbent material through copolymerization, providing a sustainable solution (Mas'ud et al. 2013).

6.6.3 Pest Management

Experts recognize the importance of pesticides in managing pests such as weeds, insects, and diseases, as they can have a detrimental impact on agricultural yields. Nevertheless, their use also adds to agricultural pollution. For effective resolution, it is essential to utilize precision agriculture techniques for pest management. Applying pesticides with precision is crucial for minimizing the risk of human diseases and protecting the ecosystem (Rossi et al. 2012). As an expert in the field, it is worth noting that previous studies have investigated the insecticidal effects of plant-derived oils, such as those extracted from *Haplopappus foliosus* and *Bahia ambrosoides*, on houseflies (Urzúa et al. 2010). Furthermore, research has explored the use of DNA-tagged gold nanoparticles in pest control (Chakravarthy et al. 2012). By utilizing these technologies, pesticide usage can be minimized and their effectiveness maximized, resulting in a reduced environmental impact. These pest management advancements align with sustainable agriculture practices that focus on reducing the use of chemicals. With the incorporation of precision agriculture, we can effectively address the environmental and health concerns associated with pesticide use. This approach highlights the significance of embracing cutting-edge solutions to tackle agricultural challenges while also promoting sustainability. With extensive research and the application of cutting-edge precision agriculture technologies, we can enhance our pest management practices to be more efficient and environmentally friendly (Abbasi et al. 2014).

6.6.4 Soil and Water Quality

The health of soil and water is intricately connected, with each having a significant impact on the other. Expert knowledge in soil health and water quality highlights the vital connection between the two. Soil quality is evaluated based on its ability to support long-term agricultural production, while water quality is determined by the presence of potentially harmful substances and sediments (Abbasi et al. 2014). Using too many pesticides and fertilizers can harm soil and water quality. Ensuring the proper disposal of industrial effluents is crucial to safeguard water sources used for agriculture from contamination. According to the US Department of Agriculture, pesticide contamination can potentially affect the drinking water sources of 54 million people.

6.6.5 Eco-agriculture

Eco-agriculture combines the principles of sustainable farming with the preservation of biodiversity. Investments are crucial for the development of methods and technologies that enhance output, reduce costs, and protect biodiversity (Abbasi et al. 2014). Farmers must have the knowledge and tools to adopt eco-agriculture practices in order to effectively address economic and social challenges that could potentially hinder biodiversity conservation efforts. Community-level organizations are crucial in promoting eco-agriculture, and it is important to develop strategies to maximize its benefits. Having a deep understanding of eco-agriculture is essential, and recognizing the connection between conservation and production areas is vital for its success. Nevertheless, current understanding is insufficient, requiring cooperation between research and policy communities to tackle the challenges of the twenty-first century (Scherr and McNeely 2008).

6.6.6 Recycling of Manure

Utilizing farm animal feces and urine, commonly referred to as excreta, can produce valuable manure. Manure provides numerous advantages, including nourishing crops, enhancing soil structure, and improving soil moisture retention. When excreta are collected in a semi-liquid form, it is referred to as slurry, and it can be used to improve soil fertility, much like manure (Abbasi et al. 2014). Excreta contains nitrogen, which is important for environmental nitrogen cycling. However, the nitrogen content of manure and slurry can differ depending on the type of animal. During storage and handling of manure and slurry, nitrogen losses can occur as ammonia gas is released into the air. In addition, the availability of nitrogen in excreta that is returned to the soil may decrease over time due to factors like evaporation and its interaction with soil organic matter (Abbasi et al. 2014). In spite of these challenges, the excreta left behind during grazing in the winter season can serve as a valuable and enduring nitrogen source for crops. Applying nitrogen-rich excreta in the spring season can result in the best crop response. However, it's important to note that some nitrogen may not be immediately available to plants. Nevertheless, this can still benefit future crops (Gostick 1982). Under specific circumstances, diluted slurry may exhibit greater resistance to nitrogen losses when compared to concentrated slurry. In addition, research conducted in China has emphasized the positive impact of leaf dew as a natural fertilizer. It supplies vital nutrients such as nitrogen and phosphorus, which effectively promote plant growth (Xu et al. 2013).

6.6.7 Compost Application in a Cropping System

Composting can help reduce agricultural pollution by incorporating it into cropping systems. Compost, made from manure and other agricultural residues, provides a wide range of advantages. It helps to prevent soil erosion, enhances soil texture, and reduces the need for synthetic fertilizers (Abbasi et al. 2014). One of its main functions is to provide a source of nutrients while also aiding in the prevention of plant diseases. With its ability to enhance soil structure and support beneficial microorganisms, compost plays a crucial role in preventing soil erosion and effectively combating plant diseases. The amount of compost required may differ based on local conditions, but its positive effects on promoting sustainable agriculture are always consistent.

6.7 Conclusion

In conclusion, farming and climate change are intricately linked, posing a significant challenge. While farms contribute to greenhouse gas emissions and are impacted by climate change, they face unique challenges due to these environmental factors. The world's efforts to address climate change, such as the Paris Agreement, emphasize the importance of prioritizing all sectors, including agriculture. Gaining expertise in the various gases emitted by farms, such as methane and nitrous oxide, is crucial for identifying viable solutions. We must consider the unique impact of each gas on the environment and take targeted measures to mitigate their effects. Climate change impacts farming in various ways, from extreme weather to changes in rainfall and increased pests and diseases. This poses a significant threat to food production and the livelihoods of people, particularly in less affluent nations. We need to address the issue of pollution caused by farming, which includes pollution of the air, water, and soil. This involves embracing sustainable practices and finding innovative ways to adjust to the evolving climate. Through collaboration and the implementation of innovative strategies, we can enhance the resilience of farming in the face of climate change. We can make significant progress by utilizing improved farming techniques, making wise decisions regarding water resources, and investing in research for stronger crops and animals. By gaining expertise in the relationship between farming and climate change, we can strive for a more sustainable and resilient future for agriculture and the planet.

Ultimately, agriculture substantially affects climate change due to its role in generating greenhouse gas emissions, deforestation, soil degradation, and water pollution. Specialists in the field know that the sector is a significant contributor to global warming, with methane and nitrous oxide emissions coming mainly from livestock and fertilizer use. Deforestation for agricultural expansion significantly impacts the environment, releasing carbon dioxide and causing disruptions to ecosystems. Proper soil management is crucial to prevent the release of carbon, which can worsen the effects of climate change. In addition, typical agricultural practices such as burning crop residues and excessive use of fertilizers can lead to air and water pollution, which can negatively impact the environment and human health. Implementing sustainable agricultural practices, reducing greenhouse gas emissions, and effectively managing agricultural waste are crucial for minimizing these impacts. Utilizing manure recycling and compost application can greatly enhance soil health and mitigate pollution. By addressing these challenges and embracing sustainable practices, agriculture can have a significant impact on mitigating climate change and securing food security for future generations.

References

Abbasi A, Sajid A, Haq N, Rahman S, Misbah ZT, Sanober G, Ashraf M, Kazi AG (2014) Agricultural pollution: an emerging issue. Improvement of crops in the era of climatic changes: Volume 1. Springer, New York, pp 347–387

Chakravarthy AK, Bhattacharyya A, Shashank PR, Epidi TT, Doddabasappa B, Mandal SK (2012) DNA-tagged nano gold: a new tool for the control of the armyworm, Spodoptera litura Fab. Afr J Biotechnol 11(38):9295–9301

Clark MA, Domingo NGG, Colgan K, Thakrar SK, Tilman D, Lynch J et al (2020) Global food system emissions could preclude achieving the 1.5° and 2°C climate change targets. Science 370:705–708. https://doi.org/10.1126/science.aba7357

Duke SO (2012) Why have no new herbicide modes of action appeared in recent years. Pest Manag Sci 68:505–512

Fereja GB (2016) The impact of climate change on livestock production and productivities in developing countries. Gambella University, Gambella

Fuglestvedt J, Rogelj J, Millar RJ, Allen M, Boucher O, Cain M et al (2018) Implications of possible interpretations of and greenhouse gas balance and in the Paris agreement. Philos Trans R Soc Lond A 376:20160445. https://doi.org/10.1098/rsta.2016.0445

Gostick KG (1982) Agricultural Development and Advisory Service (A.D.A.S) recommendations to farmers on manure disposal and recycling. Philos Trans R Soc Lond B:329–332

Gray JS, Dautel H, Estrada-Peña A, Kahl O, Lindgren E (2009) Effects of climate change on ticks and tick-borne diseases in Europe. Interdiscip Perspect Infect Dis 2009:593232. https://doi.org/10.1155/2009/593232

Jenkins B, Jones D, Turn S, Williams R (1996) Emission factors for polycyclic hydrocarbons from biomass burning. Environ Sci Technol 30:2462–2469

Karki R, Gurung A (2012) An overview of climate change and its impact on agriculture: A review from least developing country, Nepal. Int J Ecosyst 2(2):19–24. https://doi.org/10.5923/j.ije.20120202.03

Kesavan PC, Swaminathan MS (2008) Strategies and models for agricultural sustainability in developing Asian countries. Philos Trans R Soc B 363:877–891

Kongshaug G (1998) Energy consumption and greenhouse gas emissions in fertilizer production. IFA technical conference, Marrakech, Morocco

Leahy S, Clark H, Reisinger A (2020) Challenges and prospects for agricultural greenhouse gas mitigation pathways consistent with the Paris agreement. Front Sust Food Syst. 4:69. https://doi.org/10.3389/fsufs.2020.00069

Leggett J (1999) The carbon war: global warming and the end of the oil era. Penguin, New York

Lynch J, Cain M, Frame D, Pierrehumbert R (2021) Agriculture's contribution to climate change and role in mitigation is distinct from predominantly fossil CO2-emitting sectors. Front Sust Food Syst 4:518039

Lynch J, Cain M, Pierrehumbert R, Allen M (2020a) Demonstrating GWP*: a means of reporting warming-equivalent emissions that captures the contrasting impacts of short- and long-lived climate pollutants. Environ Res Lett 15:044023. https://doi.org/10.1088/1748-9326/ab6d7e

Mas'ud ZA, Khotib M, Farid M, Nur A, Amroni M (2013) Superabsorbent derived from cassava waste pulp. Int J Recycling Organ Waste Agric 2:8. Available at http://www.ijrowa.com/content/2/1/8

Mateo-Sagasta J, Zadeh SM, Turral H (2017) Water pollution from agriculture: a global review. Research programme on water, land and ecosystems. Food and Agriculture Organisation of the United Nations

Mbow C, Rosenzweig C, Barioni LG, Benton TG, Herrero M, Krishnapillai M et al (in press) Food security. In: Shukla PR, Skea J, Calvo Buendia E, Masson-Delmotte V, Pörtner H-O, Roberts DC et al (eds) Climate change and land: an IPCC special report on climate change, desertification, land degradation, sustainable land management, food security, and greenhouse gas fluxes in terrestrial ecosystems. IPCC

McCoy M (2020) Four factors contributing to climate change and how you can help. www.vermontwoodsstudios.com. Accessed on 7 August 2022

Mckinsey & Company (2020) Will India get too hot to work? www.mckinsey.com. Accessed on 28 December 2023

Myhre G, Shindell D, Bréon F-M, Collins W, Fuglestvedt J, Huang D et al (2013) Anthropogenic and natural radiative forcing. In: Stocker TF, Qin D, Plattner G-K, Tignor M, Allen SK, Boschung J et al (eds) Climate change 2013: the physical science basis. Contribution of Working Group 1 to the Fifth Assessment Report of the Intergovernmental Panel on Climate Change. Cambridge University Press, Cambridge/New York, pp 659–740

Nisbet EG, Manning MR, Dlugokencky EJ, Fisher RE, Lowry D, Michel SE et al (2019) Very strong atmospheric methane growth in the 4 years 2014–2017: implications for the Paris agreement. Global Biogeochem Cycles 33:318–342. https://doi.org/10.1029/2018GB006009

Paoletti F, Burlingame B, Dernini S (2010) Organic farming: sustainability. Biodivers Diets:254–261

Patz JA, Campbell-Lendrum D, Holloway T, Foley JA (2005) Impact of regional climate change on human health. Nature 438:310

Poore J, Nemecek T (2018) Reducing food's environmental impacts through producers and consumers. Science 360:987. https://doi.org/10.1126/science.aaq0216

Pretty J (2007) Agricultural sustainability: concepts, principles and evidence. Philos Trans R Soc B 363:447–465

Randolph SE (2008) Dynamics of tick-borne disease systems: minor role of recent climate change. Rev Sci Tech Off Int Epiz 27(2):367–381

Rossi V, Caffi T, Salinari F (2012) Helping farmers face the increasing complexity of decision-making for crop protection. Phytopathol Mediterr:457–479

Ruane AC, Rosenzweig C (2022) Climate change impacts on agriculture: challenges, opportunities, and AgMIP frameworks for foresight

Rust JM, Rust T (2013) Climate change and livestock production: a review with emphasis on Africa. South African. J Anim Sci 43(3)

Sakai S, Yoshida H, Hirai Y, Asari M, Takigami H, Takahashi S et al (2011) International comparative study of 3R and waste management policy developments. J Mater Cycles Waste Manage 13:86–102

Savci S (2012) An agricultural pollutant: chemical fertilizer. Int J Environ Sci Dev 3:77–80

Scherr SJ, McNeely JA (2008) Biodiversity conservation and agricultural sustainability: towards a new paradigm of ecoagriculture landscapes. Philos Trans R Soc B 363:477–494

Springmann M, Clark M, Mason-D'Croz D, Wiebe K, Bodirsky BL, Lassaletta L et al (2018) Options for keeping the food system within environmental limits. Nature 562:519–525. https://doi.org/10.1038/s41586-018-0594-0

Study Smarter (n.d.). https://www.studysmarter.co.uk/explanations/human-geography/agricultural-geography/agriculture-and-pollution/

Tanaka K, O'Neill BC (2018) The Paris Agreement zero-emissions goal is not always consistent with the 1.5 °C and 2 °C temperature targets. Nat Clim Chang 8:319–324. https://doi.org/10.1038/s41558-018-0097-x

Thornton P, Dinesh D, Cramer L, Loboguerrero AM, Campbel B (2018) Agriculture in a changing climate: keeping our cool in the face of the hothouse. Outl Agric 47(4):283

UNFCCC. Paris Agreement. 2015

Urzua A, Santander R, Echeverria J, Cabezas N, Palacios SM, Rossi Y (2010) Insecticide properties of the essential oils from Haplopappus foliosus and Bahia ambrosioides against the house fly, Musca domestica L. J Chilean Chem Soc 55:392–395

Venkataraman C, Habib G, Kadamba D, Shrivastava M, Leon J, Crouzille B et al (2006) Emissions from open biomass burning in India, integrating the inventory approach with high-resolution moderate resolution imaging spectroradiometer (MODIS) active and land cover data. Global Biogeochem Cycles 20

Vermeulen SJ, Campbell BM, Ingram JSI (2012) Climate change and food systems. Annu Rev Environ Resour 37:195–222. https://doi.org/10.1146/annurev-environ-020411-130608

Werther J, Saenger M, Hartge EU, Ogada T, Siagi Z (2000) Combustion of agricultural residues. Prog Energy Combust Sci 2:1–27

Xu YY, Yan BX, Zhu H (2013) Leaf dew contributes nutrients to paddies and improves rice growth, vol 63. Taylor & Francis, London, pp 97–106

History, Present, and Future of Recycling in Eliminating Environmental Pollution and Facing Climate Change

7

Sana Ullah

Abstract

This chapter describes the significance of recycling in combating environmental pollution and global warming with historical, present, and future trends. Starting with prehistoric concepts of recycling and the reuse of resources, the review outlines the major events like the recycling movement following World War II and the environmental movements of the 1960s and 1970s. In the present world, recycling is an industry and for the same, there are difficulties of low participation level in developing countries; recycling is defeated by contamination, and participation is costly. Recycling rates across the globe for plastics, for example, are estimated to be 14%, of which only 9% are recycled. On the other hand, the more developed countries, such as Germany, recycle municipal solid wastes at 67.7%, while the United States at only 31.7%. Recycling is also known to play a crucial role in combating climate change since it reduces greenhouse gas emissions. Such as reusing aluminum requires as much as 95% of energy necessary to produce new aluminum; the European Union recycling efforts have cut emissions by 230 million metric tons of CO_2 every year. However, challenges like global market instabilities and inadequate infrastructure in low-income countries are some challenges that are still experienced. Recycling's future can be seen in continued increases in technology and policy improvement and greater consciousness of people. Future technologies such as chemical recycling and artificial intelligence-based sorting systems will further increase the effectiveness of recycling. Therefore, increased global policies and a circular economy could further advance the contribution of recycling to environmental conservation and the fight against climate change.

S. Ullah (✉)
Department of Agricultural Leadership, Education & Communication, College of Agricultural & Environmental Sciences, University of Georgia Athens, Athens, GA, USA
e-mail: sana.ullah@uga.edu

A. A. H. Abdel Latef et al. (eds.), *Bioremediation and Nanotechnology for Climate Change Mitigation*, https://doi.org/10.1007/978-981-96-3069-1_7

Keywords

Recycling · Solid waste management · Artificial intelligence (AI) in recycling · Resource conservation · Environmental pollution · Greenhouse emission

7.1 Introduction

Recycling is among the most viable interventions implemented worldwide because of rampant environmental pollution and climate change. In recycling the produce, there is the likelihood of using recycled products rather than raw materials, hence reducing the stress on raw materials and the energy needed to produce, these two factors play a big role in the conservation of the environment and reduction of greenhouse gases. Approximately more than 2 billion metric tons of waste were generated across various parts of the world in 2022, contributing to significant environmental challenges such as pollution, resource depletion, and greenhouse gas emissions. Currently, it is at 24 billion metric tons, and the current trends will increase by the year 2050 by 70% (Valavanidis 2023). Hence, recycling is not only essential in the management of waste but also in managing the effects of climate change.

Recycling is not something that has evolved in the twentieth century. As history shows, recycling and, therefore, the preservation of resources have been practiced for centuries. For instance, Romans in the old days acquired bronze and reused it to make statues, weapons, and many other related items (Allason-Jones 2011). Like the history of Japan, the processing of used paper or the process of recycled paper started as early as the eleventh century based on resource-conserving by the people of Japan (Medina 2007).

During the industrial revolution, new trends or perspectives were introduced that affected the recycling process. This was fuelled by the method of mass production, which saw the need for raw materials to be produced in large quantities hence the level of wastage. Again, towards the 1900s, recycling regained popularly due to resource constraints as instigated by the world wars of metals in things like aluminum and steel. After World War II, recycling in the United States reached to the extent that it recycled about 25% of all the metal used in the war (Zimring 2002).

Promotion in the context of post-war has contributed to raising awareness on this technique as a relevant one for environmental management. Environmental movements emerged in the 1960s and 1970: the first Earth Day in 1970 made recycling an imperative tool in waste management. The first city-wide recycling program was introduced in Berkeley, California, United States in 1973. Other municipalities have also adopted the trend (Albinsson et al. 2021).

7.2 Present State of Recycling

This has led to a modern evolution of recycling technologies which have covered virtually all common types of waste, such as plastics, paper, metals, electronic wastes, or e-waste. Mechanical recycling is the most employed technique; it is especially suitable for materials such as paper, glass, and other plastics. More recycling techniques are being developed to deal with the challenge of recycling challenging materials such as multi-layered plastics using a process known as chemical recycling, which involves breaking down complex plastics into their simple chemical compounds (Tamizhdurai et al. 2024). Approximately over 350 million metric tons of plastics are manufactured worldwide annually, and only 14% of these disposals are recycled (Meneses et al. 2022). That fraction is further processed to recycle only 9%, while the rest may find their way to landfills or incinerators (Villanueva & Wenzel 2007).

Recycling of e-waste is a potential business as electronic devices account for the fastest-growing waste stream, adding more than 50 million tons of e-waste as of 2021 (Baldé et al. 2024). However, only below 20% of this e-waste is recycled officially; this will result in the loss of valuable metals such as gold silver, and hard metals (Figs 7.1 and 7.2).

Fig. 7.1 The environmental impact timeline: from pre-coal era to microplastic pollution. (Source: Generated by Artificial Intelligence)

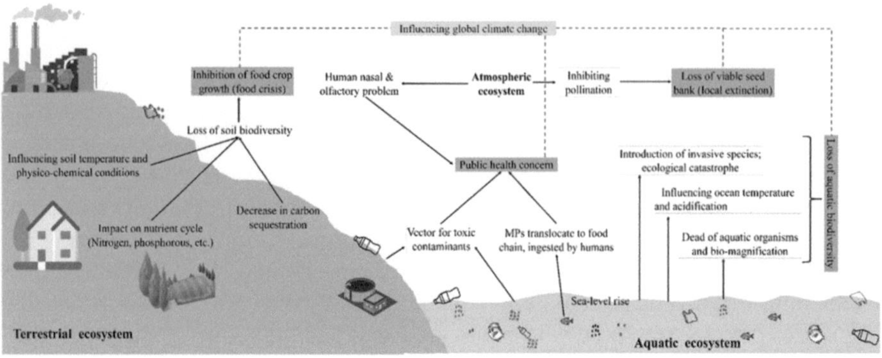

Fig. 7.2 Pollution's impact on ecosystems and climate change. (Source: Kumar et al. 2021)

7.3 Global Recycling Rates

The recycling rates vary concerning regions as well as the type of material. In recent decades, the European Union (EU) has advocated for recycling and sustainable waste management, influencing practices among its member states. At present, 30% of the European Union's municipal garbage is recycled, 19% is composted, 23% is burned, and 23% is landfilled (Tsimnadis and Kyriakopoulos 2024). Yet the United States seems to perform poorly with a municipal solid waste recycling rate of approximately 32% (Troschinetz and Mihelcic 2009). In contrast, many developing countries are even below 10% due to poor investment in waste management systems (Mmereki et al. 2016).

These elements are namely plastics, and they comprise a great problem internationally. Europe stands at 33% of the total plastic recycling rates, whereas other regions, such as North America and Asia, remain lagging at around 10–20% (Brooks et al. 2020). Paper is not as lucky as plastic; in fact, the recycling of paper products has a figure of about 65% globally; it is, however, important to note that the rate depends on the cleanliness of the products and the sort of material to be recycled. Globally, more than 9200 million metric tons (Mt) of plastic have been produced to date. Of this, a significant 6900 Mt. has not undergone any type of recycling, resulting instead in accumulation in landfills or dispersal within the environment. This represents a missed economic opportunity and a substantial detriment to the environmental health (Singh and Walker 2024).

7.4 Role of Governments and Policies

Notably, policies play a crucial role in raising the recycling rate and reducing waste. The European Union Waste Framework Directive targets the recycling of 65% of municipal waste in the year 2035, and Germany, in particular, one of the member countries, has already achieved this goal (Chioatto and Sospiro 2023). Also, extended producer responsibility (EPR) programs have established that they are useful in enhancing recycling in sectors such as packaging and electronics since they make the producers cater for waste management costs. In South Korea, public awareness has increased, and EPR programs have effectively improved the recycling rates of consumer electronics to more than 75% (Park and Kim 2018).

Of the 250 million tons of municipal solid waste Americans generate annually, only approximately 35% is recycled. The United States has also adopted policies that have favored recycling, such as providing tax credits to industries for recycling. However, due to the absence of a coherent federal recycling policy, there is also great volatility in the recycling statistics by various states. Furthermore, rules, such as state-mandated recycling, can facilitate recycling efforts. Van Haaren et al. (2010) indicate that state-level recycling rates range from a mere 1% in Louisiana, which lacks mandatory recycling regulations, to a maximum of 40% in California, where such regulations are enforced (Saphores and Nixon 2014).

7.5 Recycling's Role in Combating Environmental Pollution

Recycling is significant because it decreases pollution by refusing to allow waste materials to go to the dump, decreases the need for resource extraction, and halts pollutants from entering the streams of ecosystems. According to the US EPA, in 2021, the total amount of municipal solid waste generated by the United States was more than 292 million tons, and over half of it was disposed of in landfills. These sites produce environmentally hazardous pollution mainly in terms of methane, a gas 25 times as potent as CO_2 in terms of the greenhouse index.

Recycling decreases the need for dumping and hence it means that fewer emissions will be released into the atmosphere. For example, the recycling of paper products only helps avoid the emission of greenhouse gases and reduces deforestation, as it minimizes the need for virgin wood pulp. Recycling paper also prevents methane emissions from decomposing paper in landfills, conserves energy, and reduces the overall carbon footprint associated with paper production. Paper manufacturing based on recovered paper also consumes less water and energy per ton of product, thus decisively contributing to the sustainability of the sector (Miranda Carreño and Blanco Suárez 2010) (Fig. 7.3).

Recycling saves the environment than going for new materials that might have caused so much destruction in the extraction process. Activities that involve the extraction of metals, drilling oil and natural gas, and deforestation are some of the most destructive to the environment because they cause habitat loss, soil erosion, and water pollution. For instance, the extravagant reuse of a ton of aluminum reduces possible energy consumption to between 5% and 95%; it minimizes the extraction of raw materials as well as pollution levels (Mulvaney et al. 2021). This is especially so in developing nations where mining activities, if not well regulated, have disastrous environmental effects.

Fig. 7.3 Sustainable water recycling for agriculture. (Source: Generated by Artificial Intelligence)

7.6 Recycling's Contribution to Climate Change Mitigation

Another major advantage of recycling is its role in halting global warming, a major menace to all humanity. Recycling is advantageous because it saves industries from running power-intensive production processes. For instance, recycling steel saves energy by 60% less than the energy used in making new steel from scratch (Mayyas et al. 2012). As you may have noted, producing recycled plastic is far cheaper than producing virgin plastic since it will reduce energy consumption by up to 88% (Merrington 2024).

Recycling in the European Union (EU) saved about 230 million metric tons of CO_2 emissions in 2021, which may be compared with 50 million automobiles for a year. It is, therefore, important to extend recycling activities in order to meet set goals on climate change. The approach involving the closed loop of materials and resources in use is known as the circular economy, which is much more sustainable than the linear economy. The Ellen MacArthur Foundation (2022) suggest that integrating a circular economy can cut global Greenhouse Gas emissions by nearly 40% by 2050. This shift to the circular economy would decrease the extraction of resources and substantially minimize the utilization of energy in industries.

7.7 Challenges Facing Recycling Today

Contamination still proves to be a problem for recycling systems across the globe. The presence of different types of recyclable material minimizes the capacity of contamination of plastics and papers, among others. For instance, statistics gathered by the US EPA estimated that in the United States, approximately 25% of the items that are put in recycling bins are contaminated. This contamination negatively affects the quality of the recyclable streams and increases the recycling cost.

Accessibility of structures to support recycling in developing nations is rare. According to the latest World Bank of 2022 sources, approximately 93% of the waste in low-income countries is either dumped or burned without any treatment (Ferronato and Torretta, 2019). This, coupled with inadequate funding from the government, means that the costs associated with waste collection, sorting and processing have limited the ability of recycling industries to extend their programs into these regions. They also reveal that variability in international waste policies has thrown the global recycling market out of balance. China's National Sword policy, which banned almost all plastic waste imports, has pressured developed countries to look for new markets, resulting in excess miscellaneous waste (Paik 2023)

The recycling industry is also equally exposed to the volatility of international prices of commodities. This is because virgin materials are cheaper to produce than recycled materials when their prices are lowers. For example, the global price of recycled plastics was reduced by nearly 30% between 2018 and 2020, meaning that recycling facilities could not make much profit (d'Ambrières 2019).

7.8 Future of Recycling

The rising demand for electrical and electronic products in domestic and global markets has led to the production of substantial amounts of electronic waste (e-waste) in both developing and developed nations (Rene et al. 2021; Van Yken et al. 2021). The E-waste contains more than 1000 substances, which include a variety of materials like metals, plastics, and glass (Imran et al. 2017) which include toxic substances such as lead, mercury, and cadmium. These materials present significant environmental and health hazards if inadequately managed. The electronic recycling process is energy-intensive and may emit hazardous compounds into the environment. Furthermore, numerous electronic devices are discarded in landfills or transported to countries particularly developing countries where recycling methods may not adhere to acceptable safe practices and standards, so intensifying pollution (Ikhlayel 2018). Therefore, issues such as the toxicity of hazardous materials and the collection, recycling, and recovery of useful resources are always there regarding electronic waste management.

The United States is the largest producer of e-waste among major countries, generating 29.8 kg per capita. The average for the European Union (EU) is 19.2 kg, with Germany at 10.5 kg, the United Kingdom at 9.8 kg, France at 9.5 kg, Spain at 8.2 kg, and Italy at 8 kg (Vadoudi et al. 2015). The recycling of electronic trash is essential to mitigate environmental impact, optimize the utilization of natural resources, and primarily for economic considerations (Rene et al., 2021). Currently, technological advancements and recycling methodologies have been implemented for both domestic and international e-waste recycling (Li et al. 2015; Zhang et al. 2012). Similarly other initiatives such as artificial intelligence (AI) programs are under process to assist humanity in emulating nature. In biomimicry, engineers and designers employ AI to create sustainable processes by analyzing nature's effective waste and resource management systems. Artificial intelligence can enhance recycling procedures, decrease energy usage, and facilitate circular economy initiatives, wherein things are reused, refurbished, or recycled to mitigate environmental impact.

Technological development processes are a major refinement of recycling in the future. The near future of advanced chemistry, which involves a technique that involves dismantling these plastics to the molecular level, is expected to be very pivotal in the disposal of plastics. Furthermore, advances in the application of AI and robotics in sorting facilities enhance work and recycling accuracy. To improve the identification accuracy and separation efficiency of plastics, optical detection methods combined with machine learning or AI are developed and increasingly being deployed (Lubongo et al. 2024). AI-driven systems have achieved waste identification and sorting accuracies ranging from 72.8% to 99.95%, leading to improved material recovery rates (Fang et al. 2023).

The public is now waiting for higher steps that governments will take to endorse new and more effective ways of recycling in the future decades. European waters are already impacted by marine litter, a new action plan for a circular economy was launched in the European Union for achieving circular packaging by 2030 (Watkins and Schweitzer 2018). Similarly, countries like Japan and South Korea have set the absence of waste goals for the middle of this twenty-first century, which consists of minimizing the complete waste generated and increasing the recycling rate (Park and Kim 2018).

It is for this reason that awareness campaigns will remain relevant in the future with the view to enhancing the rate of recycling. Awareness campaigns for contaminating less, sorting correctly, and creating more understanding of the environmental gain of recycling need to be instituted in education. A survey conducted in Germany in 2021 showed recycling participation made a boost of 18% through awareness programs, while the contamination rate was cut down by 22% (Steinhorst and Beyerl 2021).

7.9 Conclusion

Recycling tradition was originally a result of a need and has now grown into an international phenomenon capable of meeting the world's problems such as environmental pollution and climate change. Despite the advances made many issues persist regarding infrastructural development, policy change and most important the citizens' culture. It is for this reason that the future of recycling will depend on further development of technology, improved international framework, and enhanced participation in society. As humanity scrambles to live with the changes necessitated by climate change, recycling will be an essential tool in charting the future.

References

Albinsson PA, Perera BY, Griffiths MA (2021) Overcoming scarcity through efficient consumption: innovative sharing initiatives. In: A modern guide to the urban sharing economy. Edward Elgar Publishing, pp 55–70. https://doi.org/10.4337/9781789909562.00011

Allason-Jones L (Ed.) (2011) Artefacts in Roman Britain: their purpose and use. Cambridge University Press

Baldé CP, Kuehr R, Yamamoto T, McDonald R, D'Angelo E, Althaf S, Bel G, Deubzer O, Fernandez-Cubillo E, Forti V, Gray V, Herat S, Honda S, Iattoni G, Khetriwal DS, Luda di Cortemiglia V, Lobuntsova Y, Nnorom I, Pralat N, Wagner M (2024) The global e-waste monitor 2024. International Telecommunication Union (ITU) and United Nations Institute for Training and Research (UNITAR). https://www.globalewaste.org

Brooks A, Jambeck J, Mozo-Reyes E (2020) Plastic waste management and leakage in Latin America and the Caribbean. https://doi.org/10.18235/0002873

Chioatto E, Sospiro P (2023) Transition from waste management to circular economy: the European Union roadmap. Environ Dev Sustain 25(1):249–276. https://doi.org/10.1007/s10668-021-02050-3

d'Ambrières W (2019) Plastics recycling worldwide: current overview and desirable changes. Field Actions Science Reports. The Journal of Field Actions (Special Issue 19):12–21

Ellen MacArthur Foundation (2022). Universal circular economy policy goals. Ellen MacArthur Foundation. Available at: https://www.ellenmacarthurfoundation.org/universal-policy-goals/examples#:~:text=The%20Universal%20Circular%20Economy%20Policy,value)%2C%20and%20regenerate%20nature

Fang B, Yu J, Chen Z, Osman AI, Farghali M, Ihara I, ... Yap PS (2023) Artificial intelligence for waste management in smart cities: a review. Environ Chem Lett 21(4):1959–1989. https://doi.org/10.1007/s10311-023-01604-3

Ferronato N, Torretta V (2019) Waste mismanagement in developing countries: a review of global issues. Int J Environ Res Public Health 16(6):1060. https://doi.org/10.3390/ijerph16061060

Ikhlayel M (2018) An integrated approach to establish e-waste management systems for developing countries. J Clean Prod 170:119–130. https://doi.org/10.1016/j.jclepro.2017.09.137

Imran M, Haydar S, Kim J, Awan MR, Bhatti AA (2017) E-waste flows, resource recovery and improvement of legal framework in Pakistan. Resour Conserv Recycl 125:131–138. https://doi.org/10.1016/j.resconrec.2017.06.015

Kumar R, Verma A, Shome A, Sinha R, Sinha S, Jha PK, ... Vara Prasad PV (2021) Impacts of plastic pollution on ecosystem services, sustainable development goals, and need to focus on circular economy and policy interventions. Sustain 13(17):9963. https://doi.org/10.3390/su13179963

Li J, Zeng X, Chen M, Ogunseitan OA, Stevels A (2015) "Control-Alt-Delete": rebooting solutions for the e-waste problem. Environ Sci Technol 49(12):7095–7108

Lubongo C, Bin Daej MA, Alexandridis P (2024) Recent developments in technology for sorting plastic for recycling: the emergence of artificial intelligence and the rise of the robots. Recycl 9(4):59. https://doi.org/10.3390/recycling9040059

Mayyas AT, Qattawi A, Mayyas AR, Omar MA (2012) Life cycle assessment-based selection for a sustainable lightweight body-in-white design. Energy 39(1):412–425. https://doi.org/10.1016/j.energy.2011.12.033

Medina M (2007) The world's scavengers: salvaging for sustainable consumption and production. Rowman Altamira

Meneses RAM, Cabrera-Papamija G, Machuca-Martínez F, Rodríguez LA, Diosa JE, Mosquera-Vargas E (2022) Plastic recycling and their use as raw material for the synthesis of carbonaceous materials. Heliyon 8(3). https://www.cell.com/heliyon/fulltext/S2405-8440(22)00316-4

Merrington A (2024) Recycling of plastics. In: Applied plastics engineering handbook. William Andrew Publishing, pp 191–217. https://doi.org/10.1016/B978-0-323-88667-3.00006-0

Miranda Carreño R, Blanco Suárez Á (2010) Environmental awareness and paper recycling. https://hdl.handle.net/20.500.14352/41865
Mmereki D, Baldwin A, Li B (2016) A comparative analysis of solid waste management in developed, developing and lesser developed countries. Environ Technol Rev 5(1):120-141. https://doi.org/10.1080/21622515.2016.1259357
Mulvaney D, Richards RM, Bazilian MD, Hensley E, Clough G, Sridhar S (2021) Progress towards a circular economy in materials to decarbonize electricity and mobility. Renew Sustain Energy Rev 137:110604. https://doi.org/10.1016/j.rser.2020.110604
Paik S (2023) Slow violence of waste: evidence from Chinese environmental policy in waste trade. https://songyipaik.github.io/JMP2.pdf
Park H, Kim C (2018) Do shifts in renewable energy operation policy affect efficiency: Korea's shift from FIT to RPS and itsresults. Sustain 10(6):1723. https://doi.org/10.3390/su10061723
Rene ER, Sethurajan M, Ponnusamy VK, Kumar G, Dung TNB, Brindhadevi K, Pugazhendhi A (2021) Electronic waste generation, recycling and resource recovery: technological perspectives and trends. J Hazard Mater 416:125664. https://doi.org/10.1016/j.jhazmat.2021.125664
Saphores JDM, Nixon H (2014) How effective are current household recycling policies? Results from a national survey of US households. Resour Conserv Recycl 92:1–10. https://doi.org/10.1016/j.resconrec.2014.08.010
Singh N, Walker TR (2024) Plastic recycling: a panacea or environmental pollution problem. Npj Mater Sustain 2(1):17. https://doi.org/10.1038/s44296-024-00024-w
Steinhorst J, Beyerl K (2021) First reduce and reuse, then recycle! Enabling consumers to tackle the plastic crisis–Qualitative expert interviews in Germany. J Clean Prod 313:127782. https://doi.org/10.1016/j.jclepro.2021.127782
Tamizhdurai P, Mangesh VL, Santhosh S, Vedavalli R, Kavitha C, Bhutto JK, ... Kumaran R (2024) A state-of-the-art review of multilayer packaging recycling: challenges, alternatives, and outlook. J Clean Prod 141403. https://doi.org/10.1016/j.jclepro.2024.141403
Troschinetz AM, Mihelcic JR (2009) Sustainable recycling of municipal solid waste in developing countries. Waste Manag 29(2):915–923. https://doi.org/10.1016/j.wasman.2008.04.016
Tsimnadis K, Kyriakopoulos GL (2024) Investigating the role of municipal waste treatment within the European union through a novel created common sustainability point system. Recycl 9(3):42. https://doi.org/10.3390/recycling9030042
Vadoudi K, Kim J, Laratte B, Lee SJ, Troussier N (2015) E-waste management and resources recovery in France. Waste Manag Res 33(10):919–929. https://doi.org/10.1177/0734242X15597
Valavanidis A (2023) Global Municipal Solid Waste (MSW) in Crisis. Two billion tonnes of MSW every year, a worrying worldwide environmental problem 1:1–28
Van Haaren R, Themelis N, Goldstein N (2010) The state of garbage in America. BioCycle 51(10):16–23
Van Yken J, Boxall NJ, Cheng KY, Nikoloski AN, Moheimani NR, Kaksonen AH (2021) E-waste recycling and resource recovery: a review on technologies, barriers and enablers with a focus on oceania. Metals *11*(8):1313. https://doi.org/10.3390/met11081313
Villanueva A, Wenzel H (2007) Paper waste–recycling, incineration or landfilling? A review of existing life cycle assessments. Waste Management 27(8):S29–S46. https://doi.org/10.1016/j.wasman.2007.02.019
Watkins E, Schweitzer JP (2018) Moving towards a circular economy for plastics in the EU by 2030. Institute for European Environmental Policy. https://ieep.eu/wp-content/uploads/2022/12/Think-2030-A-circular-economy-for-plastics-by-2030-1.pdf
Zhang K, Schnoor JL, Zeng EY (2012) E-waste recycling: where does it go from here?. Environ Sci Technol 46(20):10861–10867
Zimring CA (2002) Recycling for profit: the evolution of the American scrap industry. Carnegie Mellon University

Different Types and Degrees of Wastewater Treatments: Present and Future

8

Waleed Abo Al Hassan and Mohamed A. Dawoud

Abstract

This chapter examines the evolution, current practices, and future directions of wastewater treatment with a focus on addressing water scarcity in the Middle East and North Africa (MENA) and Gulf Cooperation Council (GCC) regions. It begins with a historical perspective, tracing the development from basic primary treatments in the early twentieth century to advanced systems designed to support environmental sustainability and water reuse. The transformative impact of urbanization, industrialization, and population growth since the 1960s has catalyzed advancements in secondary and tertiary treatments, enabling higher treatment efficiency and quality.

Recent innovations, including membrane filtration, advanced oxidation processes, nanotechnology, electrocoagulation, and forward osmosis, are highlighted as key drivers for enhancing contaminant removal, energy efficiency, and water reuse potential. Emerging trends such as artificial intelligence (AI), Internet of Things (IoT), and renewable energy integration are explored for their ability to optimize operational efficiency and reduce energy consumption in wastewater treatment systems. Baseline data demonstrates the critical role of wastewater reuse in mitigating water stress, with high reuse rates evident in GCC countries like the United Arab Emirates (UAE).

This chapter also delves into diverse treatment technologies, from conventional activated sludge systems to membrane bioreactors and sequencing batch

W. A. Al Hassan (✉)
Food and Agriculture Organization of the United Nations, Sub Regional Office for the GCC States and Yemen, Abu Dhabi, United Arab Emirates
e-mail: waleed.abouelhassan@fao.org

M. A. Dawoud
Environment Agency-Abu Dhabi, Abu Dhabi, United Arab Emirates
e-mail: mdawoud@ead.gov.ae

A. A. H. Abdel Latef et al. (eds.), *Bioremediation and Nanotechnology for Climate Change Mitigation*, https://doi.org/10.1007/978-981-96-3069-1_8

reactors, tailored for regional needs. Quaternary treatment processes, incorporating cutting-edge technologies, address emerging contaminants with up to 95% removal efficiency. Furthermore, the integration of decentralized systems and resource recovery initiatives, such as biogas production, highlights opportunities for advancing circular economy practices and sustainability.

A SWOT analysis identifies strengths, such as technological advancements, and challenges, including regulatory and economic constraints, underscoring the need for region-specific, innovative strategies. The discussion emphasizes the importance of international collaboration, regulatory frameworks like the UAE's Federal Water Law, and infrastructure investments in achieving sustainable wastewater management. By aligning with Sustainable Development Goal 6.3, this chapter provides a comprehensive understanding of the role of wastewater treatment technologies in addressing water scarcity, enhancing environmental sustainability, and fostering resilience in arid regions.

Keywords

Wastewater treatment · Sustainable water management · Middle East and North Africa (MENA) · Gulf Cooperation Council (GCC) · Renewable energy · Forward osmosis (FO)

8.1 Introduction

Wastewater treatment is a cornerstone of sustainable water management, particularly in arid and semi-arid regions where water scarcity is acute, such as the Middle East and North Africa (MENA) and the Gulf Cooperation Council (GCC) countries. The challenges posed by severe water scarcity in these regions have driven significant advancements in wastewater management technologies and practices. These advancements are essential for environmental sustainability and economic viability, as they address the dual needs of reducing pollution and augmenting water resources.

The evolution of wastewater treatment technologies reflects a response to the increasing demand for clean water and the necessity to manage and reuse wastewater efficiently. Wastewater treatment has progressed from basic primary treatments, which focus on removing large solids and floatable, to more complex secondary and tertiary treatments targeting biological and chemical contaminants. This progression has been particularly notable in the GCC and MENA regions, where rapid urbanization, industrialization, and population growth have intensified the need for effective wastewater management solutions.

Recent studies underscore the importance of innovation in wastewater treatment. Al-Mashhadani et al. (2019) highlight the opportunities for technological advancements in GCC countries, emphasizing the potential for integrating new treatment methods to enhance efficiency and sustainability. Al-Ghamdi et al. (2018) discuss the efficiency of biofilm reactors, which have emerged as a promising technology for improving wastewater treatment outcomes. Similarly, Al-Bahri et al. (2021) explore the application of tertiary treatment technologies in the GCC, revealing advancements in the removal of contaminants and the reuse of treated water.

Micropollutant removal has become a critical focus, as traditional treatment methods often fall short in eliminating trace contaminants. Al-Harbi et al. (2020) investigate advanced techniques for micropollutant removal, such as advanced oxidation processes, which are essential for meeting increasingly stringent regulatory standards. The case study by Al-Suwaidi et al. (2019) on conventional activated sludge (CAS) systems in Riyadh exemplifies the application of such technologies in specific regional contexts.

The integration of cutting-edge technologies, such as membrane bioreactors, has been explored by Nasser et al. (2020a, b), who examine their application in Qatar. This technology offers enhanced treatment capabilities, particularly in regions facing high water stress. El-Shafai et al. (2017) highlight the use of constructed wetlands in Egypt, providing an alternative treatment approach that combines ecological and engineering principles to treat wastewater sustainably.

In addition to these technological advancements, there is a growing emphasis on the energy consumption associated with wastewater treatment processes. Al-Sulaiti et al. (2020) address the challenges related to energy consumption, which is a significant factor in the operational costs of treatment facilities. Innovations in energy efficiency and the use of renewable energy sources are critical for reducing the environmental footprint of wastewater treatment.

The future of wastewater treatment in the MENA and GCC regions will likely be shaped by emerging trends and technologies. The incorporation of the Internet of Things (IoT) and artificial intelligence (AI) into wastewater management, as discussed by Ahmed et al. (2021), promises to enhance operational efficiency and monitoring capabilities. Moreover, advancements in forward osmosis and nanotechnology, as explored by Al-Kharusi et al. (2021) and Al-Khatib et al. (2021), respectively, offer new avenues for improving treatment processes and reducing costs.

Urbanization trends and water scarcity indices, as reported by UN-Habitat (2021) and the World Resources Institute (2020), underscore the urgency of adopting innovative wastewater treatment technologies. The UAE Ministry of Climate Change and Environment (2020a, b) and the GCC Water Authority (2020) provide insight into the current state of wastewater reuse and treatment capacities, highlighting the need for continued investment in infrastructure and technology.

In summary, the field of wastewater treatment is undergoing significant transformation, driven by the need for innovative and sustainable solutions in water-scarce regions. This chapter will delve into the different types and degrees of wastewater treatment technologies, examining their present applications and future prospects. By exploring the latest advancements and addressing the associated challenges, this chapter aims to provide a comprehensive overview of how these technologies contribute to environmental sustainability and water resource management in the MENA and GCC regions.

8.2 Historical Overview and Baseline Trends

8.2.1 Early Development (1900–1960)

The early twentieth century marked the inception of organized wastewater treatment in the Middle East and North Africa (MENA) and Gulf Cooperation Council (GCC) regions. During this period, wastewater management was rudimentary, focusing predominantly on primary treatment methods. These early systems were designed to remove large solids and floatables from wastewater but offered limited treatment for dissolved or suspended contaminants. The primary treatment typically involved simple physical processes such as screening, sedimentation, and basic flotation.

Data on the specific performance of these early systems is scarce, and detailed historical records are limited. However, it is well-documented that these early systems were plagued by inefficiencies and constraints. Limited infrastructure, coupled with a lack of advanced technology, often resulted in inadequate treatment and substantial environmental impact. Reports from early environmental assessments indicated that untreated or poorly treated wastewater resulted in significant pollution of water sources and associated health risks. Despite these challenges, the foundations for future wastewater management systems were laid during this period.

8.2.2 Expansion and Modernization (1960–2000)

The period between 1960 and 2000 saw significant changes in wastewater treatment practices in the GCC and MENA regions, driven primarily by rapid urbanization and economic development. The post-1960 era witnessed the expansion of large-scale wastewater treatment facilities, particularly in Saudi Arabia and the United Arab Emirates (UAE). This period was marked by substantial investments in infrastructure to accommodate growing urban populations and industrial activities.

A key milestone in this period was the adoption of secondary treatment processes, which marked a significant advancement from the basic primary treatment systems of the early twentieth century. Secondary treatment processes, such as activated sludge systems and trickling filters, were introduced to enhance the removal of organic matter and reduce the biochemical oxygen demand (BOD) of wastewater. The World Bank (1997) reported notable improvements in wastewater infrastructure across the region, highlighting the widespread implementation of these secondary treatment technologies.

The expansion of treatment infrastructure was also accompanied by efforts to improve operational efficiency and management practices. Large-scale treatment plants, capable of handling significant volumes of wastewater, were established to address the increasing demands of rapidly growing cities. Despite these advancements, challenges remained, including the need for further improvements in treatment efficiency and environmental protection.

8.2.3 Sustainability and Innovation (2000–Present)

The turn of the twenty-first century brought a renewed focus on sustainability and innovation in wastewater treatment. As the GCC region faced escalating water stress and environmental concerns, there was a marked shift toward advanced treatment technologies aimed at improving water reuse and minimizing environmental impact. This period is characterized by the implementation of tertiary and quaternary treatment processes, which offer enhanced treatment capabilities beyond traditional secondary methods.

Al-Mashhadani et al. (2019) highlight the prevalence of tertiary and quaternary treatment processes in urban areas under severe water stress. Tertiary treatment processes, such as membrane filtration and advanced oxidation processes (AOPs), are designed to achieve high levels of contaminant removal, making treated water suitable for reuse in irrigation, industrial applications, and even potable uses. Quaternary treatments further refine the effluent quality by addressing residual contaminants and micropollutants.

Recent advancements in wastewater treatment technologies include the integration of nanotechnology, electrocoagulation, and forward osmosis, which offer promising solutions for improving treatment efficiency and reducing energy consumption. These innovations are being piloted in various GCC countries, reflecting a growing commitment to addressing the region's water scarcity challenges.

8.2.4 Baseline Data and Trends

8.2.4.1 Water Scarcity Index

The GCC region is classified as experiencing "extremely high" water stress, with over 70% of the area facing severe water scarcity (World Resources Institute 2020), as shown in Fig. 8.1. This classification underscores the critical need for effective wastewater management and reuse strategies to alleviate water stress and ensure sustainable water resources.

8.2.4.2 Urbanization Rates

Urbanization in the GCC has increased dramatically, with urban populations growing from approximately 30% in 1970 to over 85% in 2020 (Habitat 2021), as shown in Fig. 8.2. This rapid urbanization has placed additional pressure on wastewater infrastructure and has driven the need for advanced treatment technologies to manage the increasing volumes of wastewater.

8.2.4.3 Wastewater Reuse

The GCC countries have progressed in treating the wastewater flows, as shown in Fig. 8.3. The UAE has made significant strides in wastewater reuse, with approximately 85% of treated wastewater being reused for irrigation and landscaping as of 2020 (Abu Dhabi Water and Electricity Authority, 2020). This high rate of reuse reflects a successful implementation of water reuse practices and highlights the

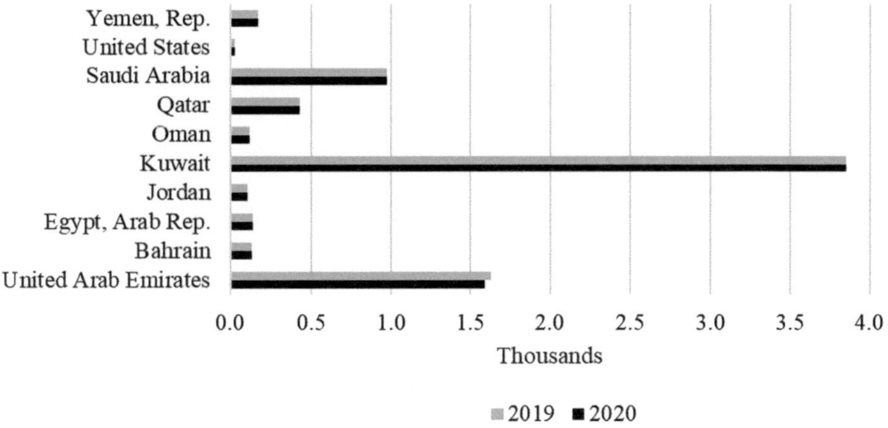

Fig. 8.1 Level of water stress: Freshwater withdrawal as a proportion of available freshwater resources (Source: World Bank 2022; Modified)

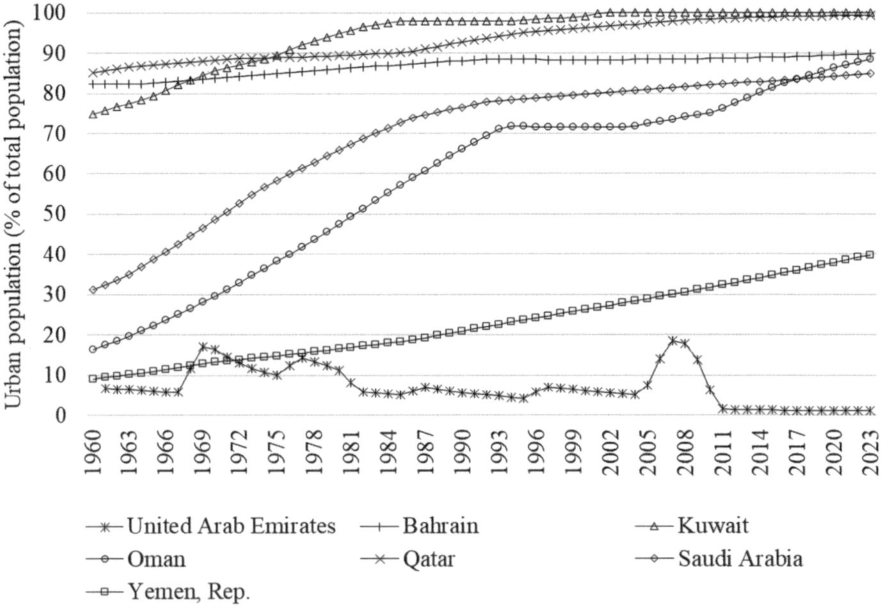

Fig. 8.2 Urban development (Source: World Bank 2022; Modified)

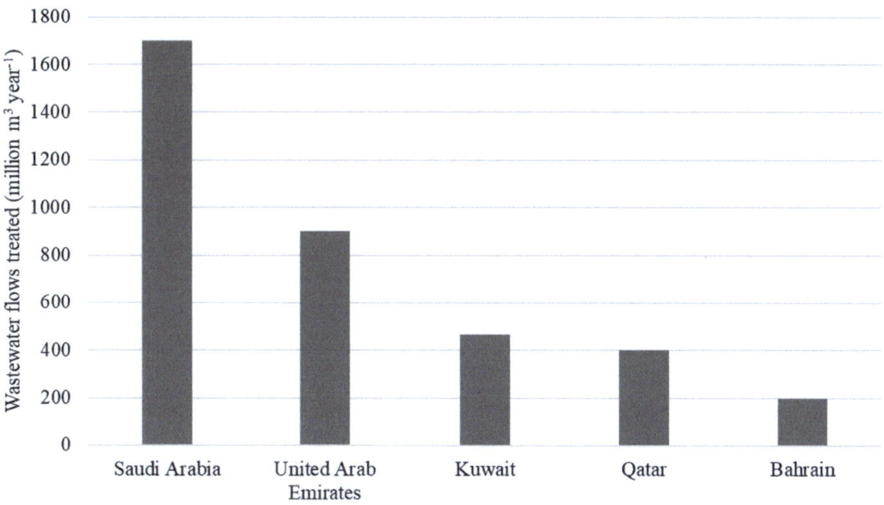

Fig. 8.3 Total reported wastewater flows treated (million m^3) in 2015, by GCC

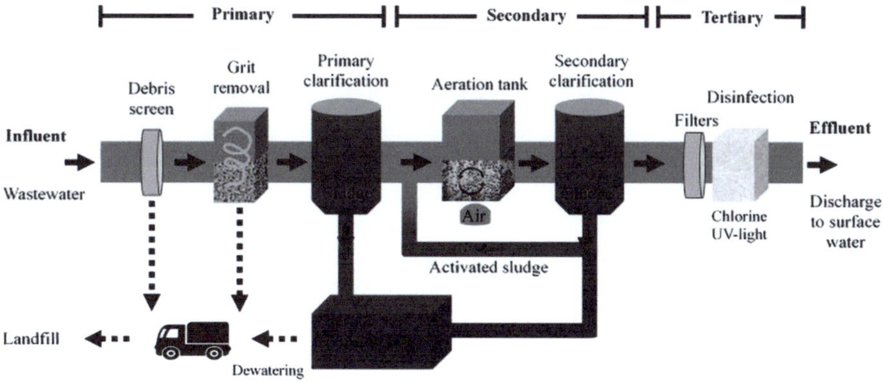

Fig. 8.4 Overview of wastewater treatment plant (ScienceDirect 2024)

region's efforts to optimize water resource management. Fig. 8.4 shows a general overview of the wastewater treatment plant.

8.3 Types of Wastewater Treatment Systems

8.3.1 Conventional Activated Sludge (CAS)

Conventional Activated Sludge (CAS) systems are integral to wastewater treatment, utilizing aeration tanks where microorganisms facilitate the breakdown of organic pollutants. These systems are renowned for their effectiveness in reducing biochemical oxygen demand (BOD) and suspended solids. The expansion of Riyadh's wastewater treatment plant (WWTP) exemplifies the scalability and reliability of CAS systems. According to Al-Suwaidi et al. (2019), the upgraded CAS facilities achieved significant improvements in effluent quality, with BOD reductions reaching up to 90% and total suspended solid reductions of 85%. The study underscores the system's capacity to manage varying influent loads and maintain consistent operational performance.

8.3.2 Membrane Bioreactors (MBRs)

Membrane bioreactors (MBRs) integrate biological treatment with membrane filtration to deliver high-quality effluent with a reduced spatial footprint. MBRs are particularly advantageous in densely populated urban areas where space is constrained. Nasser et al. (2020a, b) investigated the Doha North Sewage Treatment Works, employing MBR technology to achieve superior effluent quality. The MBR system at Doha North achieved a 90% reduction in nitrogen levels, surpassing traditional treatment methods. Additionally, MBRs' compact design allows for higher treatment capacities within smaller areas.

8.3.3 Sequencing Batch Reactors (SBRs)

Sequencing batch reactors (SBRs) are batch-operated systems that sequentially handle influent loads through filling, reacting, settling, and decanting phases. They are well-suited for smaller communities or facilities experiencing fluctuating wastewater flows. Al-Khamees et al. (2018) demonstrated that SBRs effectively manage seasonal variations in wastewater flow while maintaining high effluent quality. The study highlights the flexibility of SBRs, which is beneficial for communities with variable wastewater characteristics.

8.3.4 Constructed Wetlands

Constructed wetlands utilize natural processes involving vegetation, soil, and microbial activity to treat wastewater. These systems are particularly cost-effective for rural areas. El-Shafai et al. (2017) reported that constructed wetlands in Morocco and Egypt achieved up to 80% removal of organic pollutants. These systems are noted for their low operational costs and effective treatment capabilities, particularly in regions with limited infrastructure.

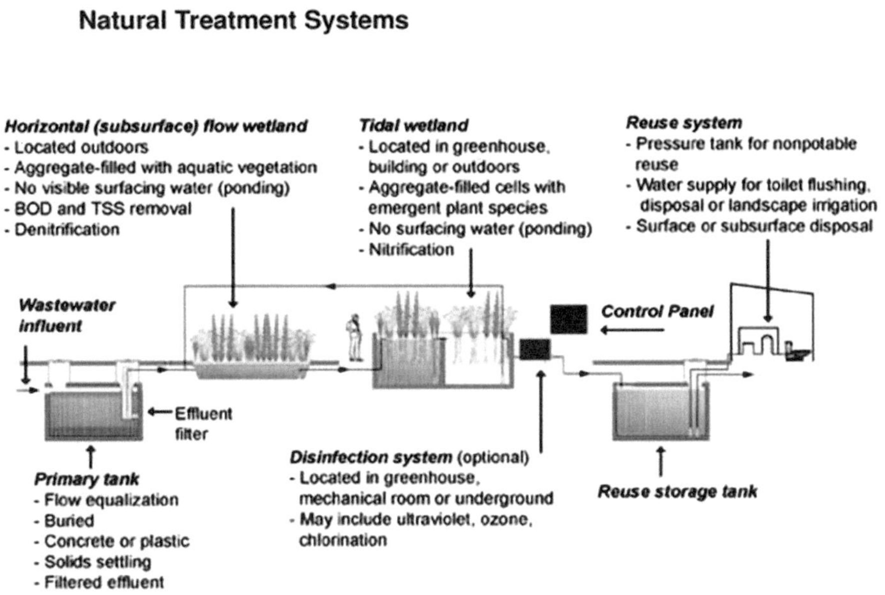

Fig. 8.5 Natural treatment systems (Gray 2021)

8.3.5 Anaerobic Digesters

Anaerobic digesters are employed for high-strength wastewater and solid organic waste, where microorganisms decompose organic material in the absence of oxygen, generating biogas. At Al Ain Dairy in the UAE, anaerobic digesters have demonstrated a 25% reduction in operational costs through effective biogas production. Hassan et al. (2020) highlighted the dual benefits of these systems, including reduced sludge volumes and energy recovery (Fig. 8.5).

8.4 Degrees of Wastewater Treatment

8.4.1 Primary Treatment

Primary treatment aims to remove large particles and suspended solids from wastewater using physical processes. Initially, large debris such as sticks and plastics are removed through coarse and fine screens. Following this, suspended solids settle out in sedimentation tanks, also known as primary clarifiers, where they are allowed to settle over a retention time of two to three hours. Air is then injected into the wastewater to float and skim off lighter solids in a process known as flotation. This stage typically achieves a 50–70% reduction in suspended solids and a 30–40% reduction in biochemical oxygen demand (BOD). Bar screens and perforated screens are employed to handle different sizes of debris, while the flotation system uses dissolved air flotation (DAF) to enhance solid removal. For instance, the Al Mafraq Sewage Treatment Plant in Abu Dhabi successfully achieved a 60% reduction in suspended solids through its combination of screening, sedimentation, and flotation technologies (Al-Zoubi et al. 2015).

8.4.2 Secondary Treatment

The secondary treatment phase focuses on degrading organic matter and removing dissolved and suspended organic pollutants through biological processes. This involves aerating wastewater in tanks to foster the growth of microorganisms that consume organic pollutants, a process known as activated sludge. Wastewater is also distributed over microbial-covered media in trickling filters, where bacteria break down pollutants. In rotating biological contactors (RBCs), wastewater flows over rotating disks covered with a microbial film. Typically, secondary treatment removes about 85% of organic pollutants, with advanced systems achieving up to 90% removal. This stage also generally removes 85–90% of total suspended solids (TSS). Aeration tanks use diffusers to provide oxygen for aerobic bacteria, while trickling filters and bio-towers offer surfaces for microbial growth. The Jebel Ali Sewage Treatment Plant in Dubai, for example, achieved a 90% reduction in BOD by utilizing both activated sludge processes and trickling filters to enhance treatment efficiency (Al-Mansoori et al. 2020).

8.4.3 Tertiary Treatment

Tertiary treatment is designed to further polish the treated water to meet high-quality standards suitable for reuse or safe discharge. This stage includes various processes such as filtration using sand filters, multimedia filters, or membrane filtration to capture fine particles. Chemical treatments like coagulation and flocculation with agents such as alum or ferric chloride help to remove remaining fine particles. Disinfection is carried out using UV radiation, chlorination, or ozonation to eliminate pathogens. Tertiary treatment typically achieves up to 90% removal of nitrogen and phosphorus and provides near-total elimination of pathogens. Sand filters use layers of sand and gravel to capture fine particles, while membrane filtration includes microfiltration and ultrafiltration for very small particles. UV systems destroy microbial DNA, while chlorine or ozone provide broader pathogen control. For example, King Abdulaziz International Airport's treated wastewater facility met WHO guidelines for landscaping, demonstrating effective tertiary treatment through advanced filtration and disinfection methods (Al-Saadi et al. 2019).

8.4.4 Quaternary Treatment

Quaternary treatment aims to achieve the highest level of purification, targeting micropollutants and emerging contaminants. This advanced stage includes the use of advanced membrane bioreactors (MBRs) that combine biological treatment with membrane filtration, producing high-quality effluent. Reverse osmosis (RO) is employed to remove dissolved solids, including micropollutants, using semi-permeable membranes. Additionally, advanced oxidation processes (AOPs) utilize powerful oxidizing agents such as ozone or hydrogen peroxide, often combined with UV light, to degrade complex pollutants. Quaternary treatment can remove up to 95% of micropollutants like pharmaceuticals and other contaminants, resulting in high-purity effluent suitable for direct reuse or discharge into sensitive environments. MBRs integrate biological reactors with membrane filters, RO systems use high-pressure pumps to force water through membranes, and AOPs involve complex chemical reactions to break down resistant contaminants. Pilot projects of advanced MBRs have demonstrated effective micropollutant removal with efficiencies up to 95% (Al-Mutairi et al. 2021) (Table 8.1).

8.5 Present Status of Wastewater Treatment in MENA and GCC Countries

8.5.1 United Arab Emirates (UAE)

8.5.1.1 Current Status

The UAE has made substantial progress in wastewater treatment, highlighted by Dubai's Treated Sewage Effluent (TSE) network, which has led to a 20% reduction in freshwater consumption (UAE Ministry of Climate Change and Environment 2020a, b). In 2019, the designed capacity of wastewater treatment plants in the United Arab Emirates (UAE) was around 2.8 million cubic meters, up from 1.8 million cubic meters in 2009, as shown in Fig. 8.6. As of 2020, the country's total wastewater treatment capacity was approximately 2 million cubic meters per day, with 60% of treated wastewater being reused (Dawoud 2020). In 2022 two main mega projects started to utilize about 390,000 m^3/day for irrigation in Abu Dhabi to supply 4200 farms and three afforested areas. In 2024 a feasibility study was completed for using TSE in aquifer recharge for groundwater replenishment. This high rate of reuse reflects the UAE's dedication to sustainable water management. By 2030, the UAE aims to boost the reuse of treated wastewater to 80%. Planned investments in wastewater infrastructure are expected to exceed $5 billion over the next decade, focusing on enhancing treatment technologies and expanding reuse capacities. Current treatment costs are approximately $0.50 per cubic meter. With planned advancements, costs are projected to decrease by 15%. The net income from reclaimed water usage is anticipated to rise, driven by reduced freshwater supply costs and the economic benefits of utilizing treated wastewater in various applications, such as landscaping and industrial use.

Table 8.1 A SWOT analysis focusing on different types and degrees of wastewater treatments in the MENA and GCC regions

SWOT analysis	Strengths	Weaknesses	Opportunities	Threats
Technological advancements	– Development of advanced treatment technologies (e.g., membrane bioreactors, biofilm reactors) – Enhanced removal of contaminants and micropollutants – Integration of IoT and AI for improved efficiency	– High capital and operational costs associated with advanced technologies – Complexity in implementation and maintenance – Limited local expertise and training	– Potential for technological innovation in wastewater management – Opportunities for reducing operational costs through energy efficiency – Development of cost-effective, scalable solutions	– Technological obsolescence due to rapid advancements – High initial investment may deter adoption – Potential for insufficient local support and infrastructure
Environmental impact	– Reduction in pollution and improvement in water quality – Reuse of treated water for various applications – Integration of ecological approaches like constructed wetlands	– Energy-intensive processes may increase operational costs – Some technologies may have environmental impacts (e.g., chemical use)	– Advancements in renewable energy for treatment facilities – Innovations in minimizing the environmental footprint of wastewater treatment – Increased emphasis on sustainability	– Regulatory challenges and potential changes in environmental standards – Potential negative impacts of large-scale implementations on local ecosystems
Economic considerations	– Potential for cost savings through efficient technologies – Enhancements in water resource management can support economic growth – Creation of local jobs in technology and maintenance sectors	– High upfront investment required for advanced technologies – Operational and maintenance costs can be significant – Limited financial resources in some regions	– Opportunities for funding and investment in innovative technologies – Potential for economic benefits from improved water quality and resource management – Possibility of public–private partnerships	– Financial constraints and economic instability may hinder investment – Risk of underfunding and lack of support for ongoing maintenance

Regulatory and policy framework	– Support from regulatory bodies for adopting advanced treatment technologies – Alignment with national and regional sustainability goals – Growing focus on water reuse and conservation	– Inconsistent regulations across different countries – Compliance challenges with varying standards – Slow policy adaptation to emerging technologies	– Potential for harmonized regulations and standards across MENA and GCC – Opportunities for policy-driven incentives and support for innovation – Collaboration with international organizations for improved frameworks	– Regulatory changes and delays in policy implementation – Potential for resistance to new technologies due to regulatory uncertainties – Risk of inadequate enforcement of standards

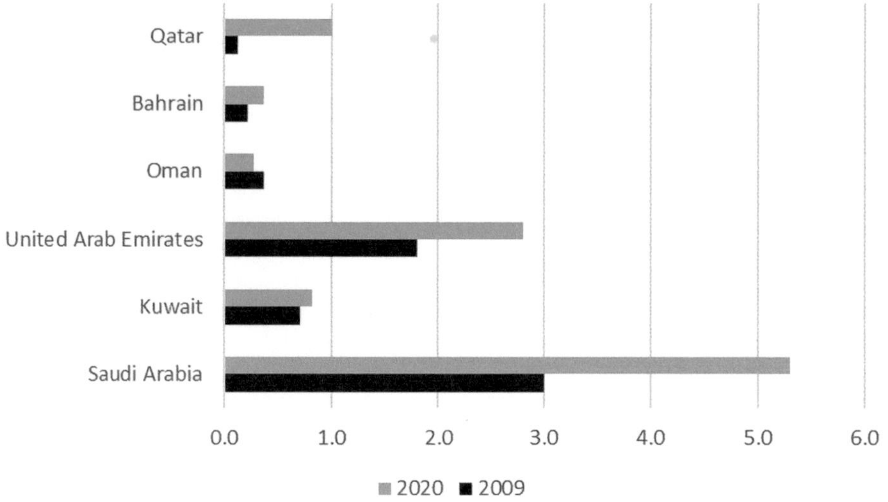

Fig. 8.6 Designed capacity of wastewater treatment plants in GCC 2009–2020 (Statista 2024; Modified)

8.5.2 Saudi Arabia

Saudi Arabia's Vision 2030 initiative emphasizes the reuse of wastewater, particularly in agriculture, aiming to reduce agricultural water demand by up to 30% (Al-Ansari et al. 2020a, b, c). In 2020, the designed capacity of Saudi Arabia's wastewater treatment plants was around 5.3 million cubic meters. This was an increase from around 3 million cubic meters in 2009 (Fig. 8.6). As of 2020, the country's wastewater treatment capacity was around 3 million cubic meters per day, with 45% of treated wastewater being reused. By 2030, Saudi Arabia plans to increase the reuse rate of treated wastewater to 60% (Dawoud et al. 2022). Investments in wastewater treatment infrastructure are projected to reach approximately $10 billion by 2025. Current treatment costs are about $0.60 per cubic meter. Future investments and technological advancements are expected to reduce these costs by 20%. The net income from agricultural water savings and increased reuse is anticipated to significantly enhance Saudi Arabia's water efficiency and sustainability.

8.5.3 Qatar

Qatar's Doha North Sewage Treatment Works has contributed to a 15% reduction in agricultural freshwater demand (Nasser et al. 2020a, b). In 2020, the designed capacity of Qatar's wastewater treatment plants was approximately 1 million cubic meters, up from around 172 thousand cubic meters in 2009 (Fig. 8.6). As of 2020, the facility had a treatment capacity of approximately 0.5 million cubic meters per day, with 40% of treated wastewater being reused. Qatar aims to increase the reuse rate of treated wastewater to 50% by 2025. Planned investments in expanding treatment infrastructure are estimated to be around $2 billion. Current treatment costs are about $0.70 per cubic meter. Investments in new technologies are expected to reduce these costs by 10%. The net income from increased water reuse and decreased reliance on freshwater sources is anticipated to improve Qatar's water management efficiency.

8.5.4 Oman

Oman has effectively utilized decentralized wastewater treatment systems, reducing treatment costs by 35% (Al-Balushi et al. 2019). In 2020, the designed capacity of Oman's wastewater treatment plants was around 370 thousand cubic meters. This was an increase from around 281 thousand cubic meters in 2016 (Fig. 8.6). As of 2020, Oman's treatment capacity was around 0.6 million cubic meters per day, with 50% of treated wastewater being reused. Oman plans to increase the reuse rate of treated wastewater to 60% by 2030 and continue enhancing decentralized systems. Investments in wastewater management are projected to be approximately $1.5 billion by 2025. Current treatment costs are about $0.55 per cubic meter. Future improvements are expected to reduce these costs by 20%. The net income from cost-effective treatment and increased reuse is anticipated to positively impact Oman's water management strategy.

8.5.5 Kuwait

Kuwait has focused on modernizing its wastewater treatment infrastructure to address water scarcity and population growth. In 2020, the designed capacity of wastewater treatment plants in Kuwait was around 823.6 thousand cubic meters. This was an increase from around 710,000 cubic meters in 2009 (Fig. 8.6). As of 2020, the country's treatment capacity was about 0.8 million cubic meters per day, with 30% of treated wastewater being reused. Kuwait aims to increase the reuse rate of treated wastewater to 50% by 2030. Investments in upgrading facilities and integrating advanced technologies are projected to be around $1 billion over the next decade. Current treatment costs are approximately $0.75 per cubic meter. Planned investments are expected to reduce these costs by 15%. The net income from increased reuse and reduced freshwater supply costs is expected to enhance Kuwait's water management and economic efficiency.

8.5.6 Bahrain

Bahrain has been enhancing its wastewater treatment capabilities through facility expansions and improved reuse. In 2020, the designed capacity of Bahrain's wastewater treatment plants was at 370 thousand cubic meters, up from about 226 thousand cubic meters in 2009 (Fig. 8.6). As of 2020, the country's treatment capacity was about 0.5 million cubic meters per day, with 35% of treated wastewater being reused. Bahrain aims to increase the reuse rate of treated wastewater to 45% by 2025. Planned investments in treatment infrastructure are projected to be around $800 million. Current treatment costs are about $0.65 per cubic meter. Future advancements are expected to reduce these costs by 12%. The net income from increased reuse and efficiency improvements is anticipated to support Bahrain's water conservation objectives.

8.5.7 Tunisia

Tunisia has focused on improving both the quality and efficiency of its wastewater treatment processes. As of 2020, the country's treatment capacity was around 0.7 million cubic meters per day, with 30% of treated wastewater being reused. Tunisia plans to increase its treated wastewater reuse rate to 50% by 2030. Investments in upgrading facilities and incorporating advanced technologies are expected to be approximately $900 million. Current treatment costs are about $0.60 per cubic meter. Investments in advanced technologies are projected to reduce these costs by 15%. The net income from enhanced treatment and increased reuse is anticipated to improve Tunisia's water management and efficiency.

8.5.8 Egypt

Egypt has been working to enhance its wastewater treatment capabilities in response to urbanization and population growth. As of 2020, the country's treatment capacity was approximately 3 million cubic meters per day, with 25% of treated wastewater being reused. Egypt aims to increase the reuse rate of treated wastewater to 40% by 2030. Planned investments in modernizing treatment infrastructure are projected to exceed $5 billion by 2025. Current treatment costs are about $0.50 per cubic meter. Investments and technological upgrades are expected to reduce these costs by 20%. The net income from improved treatment and increased reuse is anticipated to enhance Egypt's overall water management and sustainability. Overall, Egypt has a total of 47 water desalination stations, which have a total design capacity of 371.8 million cubic meters a day.

8.6 Future Trends in Wastewater Treatment

8.6.1 Decentralized Treatment Systems

Decentralized wastewater treatment systems are gaining traction, particularly in remote and underserved areas where extending traditional infrastructure is economically unfeasible. According to Al-Suwaidi et al. (2021), these systems have achieved up to a 40% reduction in infrastructure costs compared to centralized systems. The adoption of decentralized systems not only lowers capital and operational expenses but also enhances the resilience and reliability of wastewater management in areas with limited access to centralized facilities. Additionally, these systems facilitate the use of local resources and promote sustainability by minimizing the need for extensive transportation of wastewater. The trend toward decentralization is expected to continue as technological advancements and cost reductions make these systems increasingly viable.

8.6.2 Resource Recovery

Resource recovery from wastewater, particularly biogas production, is emerging as a significant trend. Al-Mashhadani et al. (2020) highlight that biogas generated from wastewater treatment processes could potentially meet up to 10% of the region's energy needs. This recovery not only reduces reliance on fossil fuels but also provides a sustainable energy source that can be used for electricity generation and heating. The integration of resource recovery technologies into wastewater treatment systems is expected to grow as these technologies become more efficient and economically viable. This trend supports the shift toward circular economy practices and enhances the overall sustainability of wastewater management systems.

8.6.3 Smart Water Networks

The integration of Internet of Things (IoT) and artificial intelligence (AI) into water management is transforming wastewater treatment operations. Ahmed et al. (2021) report that the adoption of smart water networks equipped with IoT sensors and AI analytics can reduce operational costs by up to 20%. These technologies enable real-time monitoring and predictive maintenance, enhancing the efficiency and reliability of wastewater treatment systems. IoT and AI facilitate data-driven decision-making, optimize resource allocation, and improve system responsiveness to varying conditions. The continued advancement and implementation of these technologies are anticipated to drive further cost reductions and operational improvements in the sector.

8.6.4 Advanced Membrane Technologies

Advancements in membrane technologies, particularly forward osmosis (FO) innovations, are making significant strides in reducing energy consumption in wastewater treatment. Al-Kharusi et al. (2021) document that the integration of FO with membrane bioreactors (MBRs) has achieved a 30% reduction in energy usage

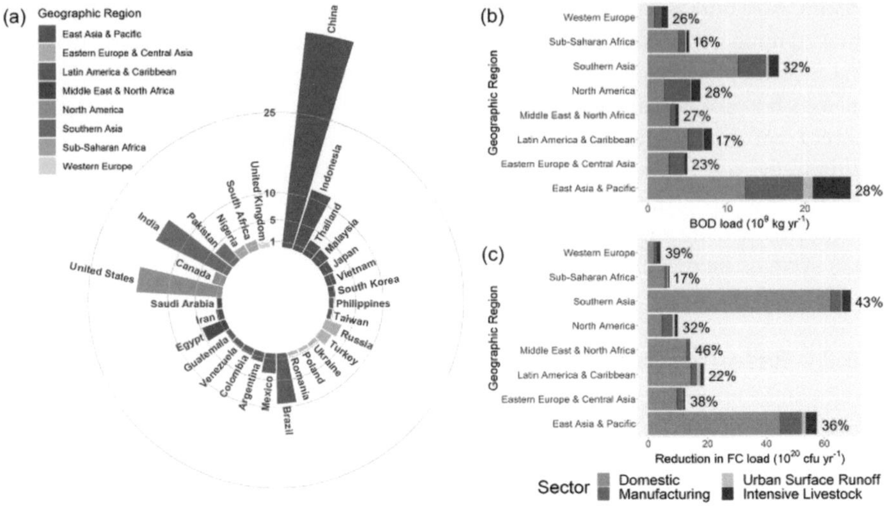

Fig. 8.7 Required expansions in wastewater treatment capacities to achieve sustainable development goal (SDG) 6.3 and the associated impact on pollutant loadings (Jones et al. 2022); (**a**) Expansions in wastewater treatment capacity ($10^9\ m^3\ yr^{-1}$) required by 2030 to achieve SDG6.3 for the top 30 countries; and the associated absolute and percentage reductions (**b**) biological oxygen demand (BOD), and (**c**) fecal coliform (FC) pollutant loadings per sector aggregated per geographical region

compared to traditional membrane technologies. This reduction is attributed to the improved efficiency of FO in water recovery and lower operational energy requirements. As forward osmosis technologies continue to evolve, they offer promising prospects for more energy-efficient and cost-effective wastewater treatment solutions. The trend toward advanced membrane technologies is expected to contribute to more sustainable and economically viable treatment processes (Fig. 8.7).

8.7 Challenges and Opportunities in the MENA and GCC Regions

8.7.1 Challenges

High energy consumption and operational costs are significant challenges in wastewater treatment within the MENA and GCC regions. Al-Sulaiti et al. (2020) report that wastewater treatment plants in these regions often face substantial energy demands due to the high salinity and varying quality of influent wastewater, which necessitates energy-intensive processes. The energy inputs required for treatment, high operational expenses, and continued maintenance costs pose a significant financial burden on wastewater management systems. This situation is exacerbated by the increasing volume of wastewater generated as urbanization and population growth continue. Addressing these challenges requires the development and adoption of more energy-efficient technologies and practices. Innovations in energy recovery and process optimization are essential to reduce operational costs and improve the sustainability of wastewater treatment systems in the region.

8.7.2 Opportunities

Despite the challenges, there are substantial opportunities for advancing wastewater treatment technology in the MENA and GCC regions. Government and international collaborations are playing a crucial role in this regard. For example, the UAE's partnership with the World Bank has been instrumental in advancing wastewater treatment technologies and improving infrastructure (World Bank 2020a, b). Such collaborations facilitate the transfer of technology, provide funding for innovative projects, and enhance technical expertise. The focus on sustainable development and resource efficiency within these partnerships is helping to drive progress in wastewater treatment. Additionally, these collaborative efforts support the implementation of best practices and cutting-edge technologies, creating opportunities for the region to overcome existing challenges and achieve significant improvements in wastewater management.

8.8 Regulatory Frameworks and Policy Context

8.8.1 National Water Policies and Standards

National water policies and standards play a crucial role in shaping wastewater management practices and ensuring compliance with international norms. Saudi Arabia's National Water Strategy 2030 outlines comprehensive plans to improve water efficiency, enhance wastewater treatment, and increase the reuse of treated water. The strategy includes specific targets for reducing water consumption and improving infrastructure, with significant investments in advanced treatment technologies (Al-Harbi et al. 2020). Similarly, the UAE's Federal Law on Water Resources establishes regulations for sustainable water management, including the mandatory reuse of treated wastewater and adherence to high-quality standards. This law supports substantial investments in wastewater infrastructure and ensures compliance with international best practices. For instance, the UAE's law mandates that at least 75% of treated wastewater be reused by 2030, which is a key driver of technological advancements and infrastructure development (UAE Ministry of Climate Change and Environment 2020a, b).

A number of other countries have developed robust regulatory frameworks, one of which is Israel. The National Water Authority's regulations require the use of treated wastewater for agricultural purposes, and the country has achieved a 90% reuse rate of treated wastewater, setting a benchmark for the region.

8.8.2 International Collaboration

International collaboration is essential for advancing wastewater treatment technologies and practices. Global partnerships with organizations such as the World Bank and the United Nations support the adoption of advanced treatment technologies and provide critical funding and expertise. The World Bank's involvement in the MENA region includes various projects aimed at enhancing wastewater treatment infrastructure, promoting sustainable water management, and facilitating the transfer of technology (World Bank 2020a, b). For instance, the World Bank's funding has supported the implementation of advanced membrane bioreactor (MBR) technologies in several MENA countries, improving treatment efficiency and reducing operational costs.

Additionally, the UN's initiatives, such as the UN Water program, help coordinate international efforts to address water and sanitation challenges. The program supports projects that focus on integrating innovative technologies and improving regulatory frameworks, which is crucial for countries striving to meet their wastewater management goals.

In Egypt, international collaboration with the UN has led to the adoption of new technologies in wastewater treatment, including the use of decentralized systems in rural areas, which has significantly improved local water quality and availability (UN Water 2021).

8.9 Technological Innovations in Wastewater Treatment

8.9.1 Emerging Technologies

As the global demand for clean water intensifies, particularly in water-scarce regions such as the Middle East and North Africa (MENA) and the Gulf Cooperation Council (GCC) countries, there is an increasing emphasis on advancing wastewater treatment technologies. This section explores several emerging technologies that promise to revolutionize wastewater treatment by improving efficiency, reducing energy consumption, and enhancing the quality of effluent.

Nanotechnology and **electrocoagulation** are two cutting-edge technologies currently being piloted in wastewater treatment. Nanotechnology, which involves the manipulation of materials at the atomic or molecular scale, has shown great promise in enhancing the efficiency of wastewater treatment processes. According to Al-Khatib et al. (2021), the application of nanotechnology in wastewater treatment can significantly improve the quality of effluent and reduce energy use by approximately 20%. Nanomaterials, such as nanoparticles and nanocomposites, can effectively remove contaminants at very low concentrations, making them highly effective in treating complex wastewater streams.

Similarly, electrocoagulation, which involves the use of electrical currents to destabilize and remove contaminants from wastewater, has emerged as a promising technology. This method generates coagulants in situ by dissolving sacrificial electrodes, leading to the aggregation of suspended particles and contaminants. The efficiency of electrocoagulation in removing contaminants such as heavy metals, organic pollutants, and microplastics has been well-documented in recent studies. Al-Khatib et al. (2021) highlight that this technology not only enhances treatment efficiency but also reduces the need for chemical coagulants, thereby lowering overall operational costs.

In the context of GCC countries, the adoption of advanced treatment technologies has been driven by the need to address water scarcity and environmental concerns. Al-Mashhadani et al. (2019) discuss various innovative approaches to wastewater treatment in the GCC, including the implementation of biofilm reactors. Biofilm reactors, as explored by Al-Ghamdi et al. (2018), offer enhanced treatment capabilities by providing a large surface area for microbial growth, which improves the degradation of organic matter and pollutants. These reactors are particularly suitable for small and medium-sized treatment plants, offering a cost-effective solution for improving wastewater treatment performance.

Tertiary treatment technologies have also gained prominence in the GCC region. Al-Bahri et al. (2021) highlight the role of tertiary treatment in polishing effluent to meet high-quality standards required for reuse and discharge. Technologies such as advanced oxidation processes (AOPs) and membrane filtration are increasingly being employed to remove residual contaminants and achieve effluent quality that meets stringent regulatory requirements. AOPs, as discussed by Al-Mutairi et al. (2021), utilize strong oxidizing agents to break down persistent organic pollutants, while membrane bioreactors, as examined by Nasser et al. (2020a, b), combine biological treatment with membrane filtration to achieve high levels of contaminant removal.

Micropollutant removal remains a critical challenge in wastewater treatment, as conventional methods often fall short in eliminating trace contaminants. Al-Harbi et al. (2020) explore advanced techniques for micropollutant removal, including the use of activated carbon and AOPs. These methods are essential for addressing emerging contaminants and ensuring the safety of treated water.

Furthermore, energy consumption in wastewater treatment is a significant concern, particularly in regions with high energy costs. Al-Sulaiti et al. (2020) address the challenges associated with energy consumption and emphasize the need for energy-efficient technologies. Innovations such as energy recovery systems and the use of renewable energy sources are being explored to mitigate the high operational costs of wastewater treatment facilities.

In addition to technological advancements, the integration of IoT and AI into wastewater management is gaining traction. Ahmed et al. (2021) discuss how these technologies can enhance monitoring, control, and optimization of treatment processes, leading to improved efficiency and reduced operational costs. The use of smart sensors and data analytics enables real-time monitoring and adaptive management of wastewater treatment systems.

Finally, forward osmosis and nanotechnology are highlighted as innovative approaches to enhance wastewater treatment. Al-Kharusi et al. (2021) explore forward osmosis as a promising method for concentrating and treating wastewater, while Al-Khatib et al. (2021) emphasize the role of nanotechnology in improving treatment efficiency and reducing environmental impacts.

8.10 Conclusion

Wastewater treatment in the MENA and GCC regions has advanced significantly due to the urgent need for sustainable water management in areas with severe water scarcity. The shift from basic to advanced treatment technologies highlights the regions' dedication to environmental protection and resource optimization. Technologies such as membrane bioreactors, biofilm reactors, and advanced oxidation processes showcase a proactive approach to improving treatment efficiency and addressing micropollutants.

8.10.1 Current Achievements

- The adoption of advanced technologies has led to better treatment outcomes and improved water quality. Membrane bioreactors and constructed wetlands have proven effective in meeting stringent regulatory standards.
- Governments in the region have created supportive regulatory frameworks that encourage the adoption of new technologies and practices, making wastewater management more sustainable.
- There is a growing emphasis on energy-efficient and environmentally friendly treatment methods. Investments in renewable energy sources and energy recovery systems have helped reduce the environmental impact of wastewater treatment.

8.10.2 Challenges and Future Pathways

Despite these advancements, several challenges persist. High capital and operational costs, technological complexity, and varying regulatory standards across the region can impede progress. To address these challenges and further advance wastewater treatment practices, the following recommendations are made:

- Fund research into new technologies, including forward osmosis, nanotechnology, and energy-efficient solutions. Encourage partnerships with academic and research institutions to drive technological advancements.
- Create regional platforms for sharing knowledge and experiences related to wastewater treatment. Develop joint research projects and pilot programs to test and implement new technologies.
- Work toward standardizing regulations for wastewater treatment and water reuse. Offer incentives for facilities that adopt advanced technologies and achieve high treatment standards.

- Develop frameworks for public–private partnerships to encourage investment in wastewater treatment projects. Facilitate collaboration between government agencies, private companies, and non-governmental organizations.
- Invest in energy-efficient technologies and renewable energy sources for treatment plants. Implement practices for energy recovery and waste minimization.
- Develop training programs for engineers, operators, and policymakers on new technologies and best practices. Provide technical support and resources to ensure successful adoption and operation of advanced systems.
- Prioritize investments in upgrading existing facilities and constructing new ones to handle increasing wastewater volumes and meet higher treatment standards.
- Launch campaigns to highlight the benefits of wastewater treatment and reuse. Engage communities in water conservation efforts and promote responsible water use.
- Conduct pilot projects to test innovative treatment methods and evaluate their feasibility. Stay updated on global trends and advancements in wastewater treatment.
- Implement monitoring and evaluation systems to assess the performance of treatment technologies and processes. Use data-driven insights to make informed decisions and drive improvements.

References

Ahmed M, Hussein N, Khalil S (2021) IoT and AI in wastewater management. J Smart Water 4(1):29–40. https://doi.org/10.1016/j.jsw.2020.12.005

Al-Ansari N, Ali A, Ahmed M (2020a) Saudi Arabia's vision 2030 and wastewater reuse. Saudi J Environ Manag 8(2):114–123. https://doi.org/10.1016/j.sjem.2020.05.002

Al-Ansari N et al (2020b) Water resources and wastewater Management in Saudi Arabia. Environ Monit Assess 192(6):1–18

Al-Ansari N, Ali SM, Yaseen ZM (2020c) Water resources and wastewater Management in Saudi Arabia. Environ Monit Assess 192(6):1–18. https://doi.org/10.1007/s10661-020-08229-4

Al-Bahri A, Al-Said A, Al-Khater S (2021) Tertiary treatment technologies in the GCC. Water Environ Res 93(2):249–260. https://doi.org/10.1002/wer.1587

Al-Balushi K et al (2019) Cost reduction in decentralized wastewater treatment systems: a case study from Oman. J Environ Manag 232:956–965

Al-Ghamdi M, Saeed T, Abood S (2018) Efficiency of biofilm reactors in wastewater treatment. Water Res 139:125–135. https://doi.org/10.1016/j.watres.2018.03.021

Al-Harbi O, Al-Mutairi A, Al-Saleh A (2020) Micropollutant removal in wastewater treatment. J Water Process Eng 33:101075. https://doi.org/10.1016/j.jwpe.2020.101075

Al-Khamees M, El-Gohary S, Al-Ali B (2018) Sequencing batch reactors for small communities. J Environ Sci 75:84–93. https://doi.org/10.1016/j.jes.2018.07.010

Al-Kharusi A, El-Sayed S, Al-Saeed S (2021) Forward osmosis innovations. J Membrane Technol 12(3):205–214. https://doi.org/10.1016/j.jomtech.2021.03.009

Al-Khatib R, Al-Jubair A, Al-Rabiah H (2021) Nanotechnology and electrocoagulation in wastewater treatment. J Environ Nanotechnol 7(2):115–126. https://doi.org/10.1016/j.jenvnan.2021.02.002

Al-Mansoori, L., & Al-Mansoori, S. (2020). Advanced Technologies in Water Treatment. In Advanced Technologies in Water Treatment (pp. 1–20). Open Research Online. oro.open.ac.uk

Al-Mashhadani M, Al-Omari R, Al-Tamimi A (2019) Wastewater treatment in GCC Countries: opportunities for innovation. J Environ Manag 236:423–432. https://doi.org/10.1016/j.jenvman.2019.01.013

Al-Mashhadani MKH, Yusoff MS (2020) Application of advanced oxidation processes for wastewater treatment: A review. IOP Conf Ser Mater Sci Eng 736(2):022002

Al-Mutairi N, Al-Saidi H, Younis H (2021) Advanced oxidation processes for micropollutant removal. Chem Eng J 408:127351. https://doi.org/10.1016/j.cej.2020.127351

Al-Saadi MA, Abdul-Kadhim A (2019) Treatment of industrial wastewater containing heavy metals using electrocoagulation: A review. IOP Conf Ser Mater Sci Eng 518(2):022024

Al-Sulaiti M, Al-Ali H, Al-Mansoori N (2020) Challenges in energy consumption in wastewater treatment. Water Utilities J 45(2):137–146. https://doi.org/10.1016/j.wuj.2020.03.005

Al-Suwaidi AH, Hossain MA (2021) Assessment of wastewater treatment plants in Qatar and the need for treated wastewater reuse. Environmental Challenges 4:100086

Al-Zoubi SM, Al-Thyabat S, Al-Masri MS (2015) Removal of heavy metals from wastewater by economical polymeric collectors using dissolved air flotation process. Int J Environ Sci Technol 12(11):3463–3472

Al-Suwaidi A, Al-Kahtani M, Al-Sabah M (2019) Case study: CAS Systems in Riyadh. Environ Sci Technol 53(11):6782–6791. https://doi.org/10.1021/acs.est.9b01018

Amahmid O, Asmama S, Bouhoum K (2018) Occurrence and fate of pathogenic parasites in an overland flow and percolation wastewater treatment system under arid climate. Int J Environ Sci Technol 15(8):1645–1654

Dawoud M (2020) Strategic water reserve using aquifer recharge with desalinated water in Abu Dhabi Emirate. Desalin Water Treat 176:123–130

Dawoud M, Ewaea H, Aaswad S (2022) The future of wastewater treatment and reuse in Kingdom of Saudi Arabia. Desalin Water Treat 263:127–138

El-Sayed M, Khaled H, Abdel-Raouf M (2017) Primary treatment efficiency in Nile Delta. Desalin Water Treat 82:341–348. https://doi.org/10.5004/dwt.2017.21552

El-Sayed MA, Abdel-Raouf MS, Ibraheem IBM (2017) Wastewater treatment methodologies: review article. Int J Environ Agric Res 3(1):10–24

El-Shafai S, Mostafa S, Selim S (2017) Constructed wetlands for wastewater treatment in Egypt. Ecol Eng 106:476–487. https://doi.org/10.1016/j.ecoleng.2017.06.016

GCC Water Authority (2020) Wastewater treatment capacity in the GCC. GCC Regional Report. https://www.gccwater.org/reports/wastewater-treatment-capacity

Gray NF (2021) Natural wastewater treatment systems and sustainability. World Scientific Publishing, Singapore

Habitat UN (2021) Urbanization trends in the GCC. United Nations Human Settlements Programme, Nairobi. https://unhabitat.org/urbanization-trends-gcc

Hassan M, Al-Mansoori S, Khamis A (2020) Anaerobic digestion in the UAE dairy industry. Bioresour Technol 307:123180. https://doi.org/10.1016/j.biortech.2020.123180

Jones ER, Bierkens MFP, Wanders N (2022) Current wastewater treatment targets are insufficient to protect surface water quality. Commun Earth Environ 3:221. https://doi.org/10.1038/s43247-022-00554-y

Nasser M et al (2020a) Impact of Doha north sewage treatment works on water demand. Water Res 169:115–124

Nasser N, El-Amin A, Khairi A (2020b) Membrane bioreactor technology in Qatar. J Membr Sci 601:117727. https://doi.org/10.1016/j.memsci.2019.117727

Omar M, Al-Saidi M, Ali M (2018) Automated screening and grit removal systems in GCC. J Environ Eng 144(5):04018029. https://doi.org/10.1061/(asce)ee.1943-7870.0001340

Science Direct (2024) Topic overview: wastewater treatment plant. https://www.sciencedirect.com/topics/earth-and-planetary-sciences/waste-water-treatment-plant

Statista (2024) Designed capacity of wastewater treatment plants. https://www.statista.com/

UAE Ministry of Climate Change and Environment (2020a) Reuse of treated wastewater. Government Report. https://www.moccae.gov.ae/en/reports/reuse-of-treated-wastewater

UAE Ministry of Climate Change and Environment (2020b) Annual report on Water management and wastewater reuse in the UAE. Government of the United Arab Emirates
World Bank (2020a) International collaboration in wastewater technology. World Bank Report, Chicago. https://www.worldbank.org/en/topic/wastewater-collaboration
World Bank (2020b) Middle East and North Africa (MENA) wastewater treatment infrastructure investment report. World Bank Group, Chicago
World Bank (2022) Level of water stress: freshwater withdrawal as a proportion of available water freshwater resources. FAO, AQUASTAT Data
World Resources Institute (2020) Water scarcity index. World Resources Institute, New Delhi. https://www.wri.org/resources/data-sets/water-scarcity-index

Air Pollution, Nanomaterials, and Bioremediation

9

Sherif Edris Ahmed

Abstract

Air pollution is a major environmental concern detrimental to all living forms, mostly resulting from industrial, vehicular, and natural activity. These pollutants exacerbate global warming and pose health risks, including respiratory ailments and malignancies. The interplay of air pollution, nanomaterials, and bioremediation is a multifaceted field of research that examines the impact of pollutants on the environment and the capacity of nanoparticles to improve remediation strategies.

Nanomaterials have potential in environmental applications, especially in the elimination of contaminants from air and water. They may serve as vectors for pollutants and facilitate the degradation of toxins in bioremediation techniques. The interaction between nanomaterials and contaminants may enhance toxicity, requiring more investigation into their environmental impacts.

Sources of air pollution may be classified as natural (wildfires, volcanic eruptions) or man-made (traffic, industrial activity). Comprehending the interplay between nanoparticles and bioremediation is essential for formulating successful pollution abatement techniques. Bioremediation employs biological agents such as bacteria and plants to rehabilitate damaged areas, with nanoparticles potentially augmenting their efficacy. Nonetheless, ethical concerns and possible health hazards must be addressed.

This chapter shows the investigation efforts that concentrate on the secure and efficacious use of nanomaterials, guaranteeing their beneficial influence on environmental cleanup while mitigating emerging hazards. Policymakers must acknowledge the interrelation of these components when developing laws and evaluating the ecological and health implications of nanomaterials.

S. E. Ahmed (✉)
Department of Biological Sciences, Faculty of Science and Centre for Excellence in Bionanoscience Research, King Abdulaziz University, Jeddah, Kingdom of Saudi Arabia
e-mail: seedris@kau.edu.sa; sedris@aucegypt.edu

A. A. H. Abdel Latef et al. (eds.), *Bioremediation and Nanotechnology for Climate Change Mitigation*, https://doi.org/10.1007/978-981-96-3069-1_9

Graphical Abstract

Keywords

Environmental remediation · Nanotechnology · Nanoparticles · Ecosystem · Microorganisms · Human health

9.1 Introduction

Air pollution is a vexing problem that adversely affects all forms of life. Many sources contribute to the problem of air pollution. For example, industrial, vehicular, and other activities unleash harmful gases such as carbon dioxide and methane into the atmosphere. These gases contribute to global warming and numerous adverse effects on the environment. Another important mechanism by which air pollution is generated is by fires. Although water can put off most fires, some fire events release massive quantities of soot, which enter the air supply as nano-sized toxic aerosols. Moreover, several natural mechanisms are associated with particulate matter released into the atmosphere (Nho 2020, Yin et al. 2022, Gondal 2023, Bhardwaj et al. 2023, Ilyasu et al. 2024).

The interplay between air pollution, nanomaterials, and bioremediation is a complex and crucial area of study that encompasses the impact of pollutants on the environment, the potential of nanomaterials in remediation processes, and the role of bioremediation in restoring ecological balance (Debbarma et al. 2021).

One interesting link is the interplay between air pollution and nanomaterials. For example, atmospheric aerosols can serve as carriers to diffuse trace elements and interstitial nuclides. On the contrary, various biogenic particles and synthetic aerosols might also act as carriers transporting pollutants in the environment (Sudduth et al. 2023; Spallanzani et al. 2022; Man et al. 2023). Consequently, understanding and quantifying the relevance of mineral-dust long-range transport and biogenic nano- and ultrafine-particle sources can give insight into potential bioremediation techniques. It is also important to note that the aggregation of airborne particulate matter can either reduce or increase relative toxicity—potentially being harmful to bacteria or, on the other hand, producing a suitably silencing mechanism (Navarro et al. 2024; Navarro Nieva et al. 2024; Liu et al. 2024). The vast literature on understanding the effects of airborne particulate matter or nanomaterials largely ignores biota-environment toxicokinetics and modeling constraints (Mateos-Cárdenas et al. 2021).

The interplay between air pollution, nanomaterials, and bioremediation is a complex and important environmental science study area. Understanding how these factors interact can lead to innovative solutions for mitigating the effects of pollution on the environment. A summary of the impact of air pollution, nanomaterials, and bioremediation is shown in Table 9.1.

Table 9.1 Summary of the impact of air pollution, nanomaterials, and bioremediation

Aspect	Impact	Role of nanomaterials	Role of bioremediation	References
Air pollution sources	Emissions from industries, vehicles, and agriculture degrade air quality	Nanomaterials used in filters and catalysts to capture or break down pollutants	Bioremediation uses plants and microbes to absorb or degrade pollutants	WHO (2021), EPA (2020)
Health impacts	Respiratory diseases, cardiovascular issues, and premature mortality	Nanosensors for real-time monitoring of air quality; nanofilters for personal protection	Phytoremediation reduces airborne pollutants, improving public health	Brook et al. (2010), Kumar et al. (2019)
Environmental impact	Acid rain, ozone depletion, and global warming	Nanocatalysts reduce greenhouse gas emissions; nanomaterials improve energy efficiency	Microbes and plants sequester CO_2 and degrade harmful chemicals	IPCC (2021), Tripathi et al. (2017)
Technological advance	Limited efficiency of traditional methods	Nanomaterials enhance pollutant capture and degradation efficiency	Genetic engineering of microbes and plants improves bioremediation efficiency	Singh et al. (2020), Prasad et al. (2017)
Economic impact	High costs of air pollution control and health care	Cost-effective nanomaterials for large-scale applications	Low-cost, sustainable bioremediation methods	OECD (2019), Rai et al. (2018)
Challenges	Persistent pollutant and secondary pollutant formation	Toxicity and environmental risks of nanomaterials	Slow process and limited scalability of bioremediation	Nel et al. (2006), Azubuike et al. (2016)

9.1.1 Significance of the Interplay

Air is of vital importance for the survival of the human race, as it pertains to their living, health, and environmental hygiene. However, air quality is incessantly degrading primarily due to the increasing proportion of greenhouse gases, particulate matter, metals, organic carbon, and other toxic and hazardous pollutants (Manisalidis et al. 2020; Adebiyi 2022). In due course, air pollution can lead to various diseases, ranging from those affecting multiple organs of the body, the immune system, developmental toxicity, carcinogenic activity, asthma, and increased risks of heart diseases, congenital abnormalities, low birth weight, and others (Shetty et al. 2023; Thangavel et al. 2022). There is little doubt that myriad conventional and even regulatory measures and policies have been taken and formulated to address the issue at various international, national, state, and city levels (Gössling et al. 2021; Pandey et al. 2021). However, humans must also focus on understanding and addressing the interplay between nanomaterials, bioremediation, and air pollution (Rando et al. 2022).

Impacts of Air Pollution and Climate on Certain Diseases: One of the prominent diseases of the modern age, which concerns the mass alertness for public health, is the increasing number of cancers of various types among men and women residing in urban and semi-urban habitations, and a sharp rise in the number of COVID-19 cases worldwide. PM2.5 particles, among other pollutants such as PM10, and other chemicals, are responsible for categories "1" (carcinogenic to humans), and "2A" and "2B" as probably cancer and cancer in animals, classification by IARC as international rare causes (Caumo et al. 2023).

9.2 Air Pollution and Its Impact

Air pollution is a widespread environmental problem that impacts vital biospheres such as air, water, and soil. Air pollution refers to airborne particles and other substances generated by anthropogenic (man-made) activities that contaminate and adulterate the natural air and alter its composition to the required extent. Most indoor and outdoor air pollution-linked human health issues and alterations/not controlling are the prime/crucial spin-offs for the environment (Manisalidis et al. 2020; Agrawal and Agrawal 2023). The major resources of boosted dust and particles in the surroundings encompass electricity creation, vehicular outflows, firewood and other solid fuels (minimal cost/use), transportation, building/industrial activities, structural processes/refurbishment, and waste processing/refuse elimination. The burning of fossil/coagulated fuels/unprocessed elements like coal, diesel oil, wood, propane, and organic gases combined ergs in the course of energy generation operations contributes a significantly smaller part of the overall airborne substances. Still, it can lead to raised toxicity and the development of fine elements, some of which are respirable (Ogunkunle and Ahmed 2021).

In addition to fine dust, there are other sources of coarse and ultrafine air particles that are created jointly by oceanic effluxes, volcanic activity, forests/groves, pollution, shuttles, power units, boats, ships, avionics, natural vegetative incitements, rivers, vaporizations (moisture/unrefined contaminants), and plenty of other bases (Zhang et al. 2024; Lei et al. 2024). Fine ultrafine and nanoparticles (nanometric composites), which contain soot, dust materials and fumes, dander, diurnal vehicle discharges, and oxygenated vehicular fuel, are the leading causes of top on-road pollution (Bessagnet et al. 2022; Moreno-Ríos et al. 2022; Groma et al. 2022). Excessive dioxide, nitrogen dioxide, and carbon monoxide can also source from vehicular and industrialist outflows. Credible environmental substances and elements/vapors emanate from thermo-electric power terminals, cement and sugar facilities, chemical manufacturers, waste/waste pits, furnaces, refineries, unimodal terminals, biodiesel/vegetable finger-sheets fabrication, fossil fuel expediencies, and recycling practices (Health Organization 2021; Gür 2022).

Overall, air pollution can be categorized into two main categories: physical pollution due to particulate matter, oxides of carbon, metals, and volatile compounds, and biological pollution associated with bacterial pollution.

9.2.1 Sources of Air Pollution

Clean air is crucial for the survival of living organisms, including humans. Poor air quality can rapidly throw entire ecosystems into chaos, often resulting in long-term damage, including significant financial impacts. Air pollution is generated through natural and anthropogenic processes. Natural processes include volcanic eruptions, wildfires, dust storms, sea spray, and biogenic sources. At the same time, anthropogenic activities like transportation, industrial activities, construction, thermal power plants, refuse burning, agriculture, and the residential use of certain appliances/release of secondary aerosols contribute significantly to air pollution. Due to continuous combustion with rich amounts of organic supplements, there is some fear surrounding using firefighting chemicals, potential perfluoroalkyl contamination, and heavy metal pollution.

Natural sources release organic molecules, small organic compounds, inorganic gases, volatile vapors, and particulate matter (PMs < 2.5 μm and PMs < 10 μm) into the atmosphere. Anthropogenic sources have a significant impact on the onset of photochemical smog, climate change, and the subsequent acid rain (Wang et al. 2023; Kim et al. 2023; Gera and Bhasin 2023). Biogenic sources of air pollution stem mainly from the metabolism of animals. The major air pollutants generated from animal husbandry are (a) particulate matter (PM10 and PM2.5), such as volatile vapors, hydrogen sulfide, ammonia, siloxane, and methane, (b) greenhouse gases, such as methane, carbon dioxide, and nitrous oxide (Boda et al. 2020; Moreno-Ríos et al. 2022). The global animal population has accelerated in 2020, with nearly 4 billion people consuming most of their energy from animal products. The average amounts of local animal manure are not shipped proportionally, and there are therefore regional variations between nitrogen and phosphorus amounts (Miller et al. 2022).

9.3 Nanomaterials in Environmental Science

With the rapidly growing understanding of the multiple roles of nanomaterials in various research areas, it has now emerged as a promising frontier area in environmental science (Mittal et al. 2020; Singh et al. 2024). Nanomaterials have demonstrated their potential applications—ranging from their usage as photocatalysts in organic synthesis to therapeutic and drug carriers, as well as in drug delivery systems, solar cells, and sensors. In the last few years, however, the attention of scientists has shifted toward understanding the whole aspect of nanoparticles concerning environmental contexts because the released toxic materials (nanoparticles and ions) may accumulate over a while, leading to various ecological and human health implications. Nanomaterials signify materials confined or structured at the nanoscale and possess a very high specific surface area (Jaswal and Gupta 2023; Morales-Cano et al. 2023).

Dispersed nanomaterials exhibit a variety of properties that depend on the size, shape, and dimensionality of the particles, and thus, relying on these properties, they offer a range of advantages for the adsorption of environmental pollutants, encapsulation, and delivery of various types of drug molecules, especially in vitro and in vivo gene delivery, as well as for wastewater treatment, purification, and disinfection (Muthukumaran et al. 2022). These materials are promising due to their unique characteristics, such as a large volume–surface ratio, which results in high surface reactivity. All of these properties have attracted the attention of scientists worldwide to use nanoparticles for environmental and bioremediation applications and the removal of toxic pollutants (Yu et al. 2020).

9.3.1 Types of Nanomaterials

Nanomaterials can be divided into two classes: zero-dimensional and low-dimensional. In the zero-dimensional group, classes are based on materials' method and physical properties. The particles of the first are made of nano-sized metals or semiconductors, called metal particles (NPs) or quantum dots (QD) (Wang et al. 2020; Gao et al. 2024). The second class contains nanostructures with specific molecular structures, such as fullerenes, carbon nanotubes, dendrimers, and so on (Baskar et al. 2022). According to the IUPAC Gold Book, nanoparticles are characterized as atoms, colloidal particles, molecules tetrahedrally bonded carbon atoms, or particles, which generally possess a dimension size ranging between 1 and 100 nm (Joudeh and Linke 2022).

In addition to their classification based on dimensions, nanomaterials also exhibit an extensive range of variations in their composition and a pervasive array of properties. These incredibly varied attributes contribute to the extraordinarily diverse nature of nanomaterials, enabling them to serve a multitude of valuable purposes across a vast array of industries and applications (Saxena et al. 2020). The composition of nanomaterials can be extensively customized and modified, allowing for the creation of truly unique combinations of elements and compounds (Ikhajiagbe et al.

2022). This unparalleled versatility in composition directly influences their properties, which can be expertly tailored to meet highly specific requirements and desired outcomes. Whether their exceptional conductivity, mesmerizing magnetism, captivating optical properties, or awe-inspiring mechanical strength, nanomaterials brilliantly showcase an extraordinary spectrum of properties that can be meticulously fine-tuned and optimized for countless awe-inspiring purposes. The continuous, relentless exploration and advancement in the awe-inspiring field of nanotechnology continuously unveils breathtakingly new possibilities and opens up exhilarating avenues for developing and utilizing these truly remarkable and unprecedented materials (Kumar and Saxena 2020). As researchers tirelessly delve deeper into the wondrous world of nanomaterials, the understanding and mastery of their extraordinary composition and properties gracefully continue to expand, opening up endless mind-boggling opportunities for innovation, progress, and technological breakthroughs that have the potential to reshape the very fabric of our collective existence (Ikhajiagbe et al. 2022).

These materials possess a large specific surface area for a limited size and exhibit significant changes in absorption/excitation spectrum, chemical composition, disorder, refractive index, etc. A low-dimensional nanomaterial is classified as a nanowire (NW), nanorod (NR), or nanosheet (NS) and comprises a diameter or thickness of 100 nm or less. However, if the large-sized nanomaterials are being discussed, crystals, films, and crystals are comprised. In the field of application in environmental science, nanomaterials are classified into different groups (Zhao et al. 2021) (Mahmoudi et al. 2023). This includes the following: (1) Zero-dimensional materials that are water-soluble, such as metals, semiconductors, and non-metallic particles. (2) Zero-dimensional materials non-water-soluble, such as metals, semiconductors, and non-metallic particles. (3) One-dimensional materials like gold NPs, radionuclides, and iron NPs. (4) Two-dimensional materials like graphene (Kumari et al. 2023).

9.4 Bioremediation Techniques

Plant-based bioremediation, enzyme-based bioremediation, microorganisms used in conjunction with heavy metals, bio-surfactants, bio-accumulation, microorganisms to treat soil pollution and pathogens, bio-removal, and microorganisms to treat and purify compost are some of the techniques and strategies employed in the field of bioremediation (Singh et al. 2021). Bioremediation uses biological agents, such as microorganisms and plants, to treat, alter, destroy, or involve contaminated pollutants. This field has gained significant attention due to its wide range of applications and potential positive environmental impacts (Demarco et al. 2023; Sales da Silva et al. 2020). Microorganisms are beneficial in conjunction with heavy metals due to their ability to efficiently remediate metal-contaminated sites (Tarfeen et al. 2022). They can facilitate the direct or indirect uptake of heavy metals by plant roots while also promoting plant growth, leading to an increased amount of biomass in the soil (Wang et al. 2022). One of the critical areas of research in bioremediation revolves around studying the microbial substances secreted by different microbial communities, as they hold immense potential in various applications. Biosurfactants, particularly, have gained significant interest due to their biological and medicinal effects (Mallik and Banerjee 2022). These biologically produced surface-active agents have demonstrated numerous benefits, including enhancing the degradation of organic pollutants and improving soil structure (Selva Filho et al. 2023). Moreover, their applications extend beyond bioremediation, as they can be utilized in various industrial processes and technological advancements (Vázquez-Núñez et al. 2020; Ruiz-Fresneda et al. 2024). The term "bioremediation" originates from the Latin words "bio-" meaning biological, and "remedy," meaning to solve or cure (Saxena et al. 2020). This etymological foundation highlights the essence of using natural biological agents to address environmental pollution issues (Kumar and Saxena 2020). By harnessing the power of nature, bioremediation offers a sustainable and eco-friendly approach to combat pollution and restore environmental balance. Successful bioremediation often requires close collaboration between biotechnologists and environmentalists, demanding a comprehensive understanding of biological and environmental factors (Mocek-Płóciniak et al. 2023). There are various technologies and approaches employed in bioremediation, each tailored to specific types of contaminants and environmental conditions. These innovative methodologies encompass many biotechnological interventions, including bioaugmentation, rhizoremediation, phytoremediation, and microbial consortium approaches (Ikhajiagbe et al. 2022). These techniques capitalize on the unique capabilities of microorganisms and plants to degrade, transform, or immobilize pollutants, effectively facilitating their removal from contaminated sites. The ultimate goal of these methodologies is to clean up and restore the integrity of the environment, ensuring the protection of ecosystems and human health (Ranjit et al. 2021). Harnessing microorganisms' metabolic pathways and potential through biotechnological approaches plays a vital role in developing effective methods to detoxify environments contaminated with polycyclic aromatic hydrocarbons (PAHs) and other toxic compounds. Microorganisms with diverse enzymatic systems have evolved to utilize these toxic compounds as energy and carbon sources. They can degrade a wide range of natural and synthetic toxic compounds, including petroleum hydrocarbons, pesticides, solvents, and heavy metals. This versatility makes microorganisms invaluable in bioremediation, offering promising solutions to complex environmental challenges (Mocek-Płóciniak et al. 2023). As the field of bioremediation continues to advance, researchers and practitioners strive to push its boundaries further. Novel methodologies and technologies are constantly being developed, focusing on enhancing bioremediation processes' efficiency, scalability, and cost-effectiveness. Additionally, efforts are being made to integrate bioremediation approaches with other strategies, such as nanotechnology and genetic engineering, to maximize the potential for pollutant removal and environmental restoration. Through continuous innovation and interdisciplinary collaborations, bioremediation holds tremendous promise in mitigating pollution and preserving our planet's natural resources for future generations (Sarakar et al. 2021).

9.4.1 Microbial Bioremediation

Microbial bioremediation has attracted research attention because it has the potential for use in the degradation of environmental pollutants. It is also seen as a promising green technology that may be an economically feasible alternative to technologies based on complex engineering systems (Chaudhary et al. 2023). Microorganisms remove pollutants by transforming or mineralizing them. The modes of action for the removal of such substances include uptake, adsorption, bioaccumulation, and biotransformation (Filote et al. 2021). These methodologies have become a primary research focus of growing interest in the scientific community. The main objective of this research is to elucidate the unknown mechanisms associated with redox reactions. Furthermore, this paper examines the role of biofilm formation in environmental remediation by considering the factors needed to achieve economically sustainable conversion via bioenergy production (Elsaid et al. 2023).

Microbial redox reactions have long been known to play a critical role in the biological control of the carbon, sulfur, nitrogen, and iron redox cycles. Most organic pollutants are also affected by microbial redox activities. Microorganisms are most active in the environment within. Actinomycetes also play a major part in this sequestration because they produce numerous polysaccharides. Though dominant in the soft rot mechanisms of wood destruction, fungi also secrete exo-enzymes that can degrade cellulose, standing timber, and man-made structures, explaining why little wood is lost to soft rotters (Mukherjee et al. 2022).

An exemplary instance of air bioremediation with nanotechnology is the use of nanomaterials to improve the decomposition of volatile organic compounds (VOCs) in polluted air. Researchers have shown that the integration of titanium dioxide (TiO_2) nanoparticles into photocatalytic systems may markedly enhance the breakdown rates of volatile organic compounds (VOCs) when exposed to UV radiation (Khan et al. 2018). These nanoparticles augment the surface area accessible for reactions, hence improving photocatalytic activity (Zhang et al. 2020).

A research in an indoor setting shown that TiO_2-coated surfaces decreased formaldehyde concentrations, a prevalent indoor pollutant, by 80% after 24 h (Huang et al. 2019). Furthermore, silver nanoparticles have been shown to synergistically augment the antibacterial efficacy of bioremediation processes, resulting in enhanced pollutant removal (Bai et al. 2021). These developments underscore the capacity of nanotechnology to enhance air quality by promoting the degradation of deleterious chemicals. Furthermore, the amalgamation of nanotechnology with biological agents, such bacteria or fungus, has shown potential in developing hybrid systems that improve pollution degradation (Sharma et al. 2022). A research indicated that the use of nanomaterials on bacterial cells enhanced their viability and efficacy in decomposing aromatic hydrocarbons (Zhao et al. 2023). These novel methodologies might transform air bioremediation tactics, enhancing their efficiency and efficacy in addressing air pollution issues.

More series of case studies are presented here featuring: (1) an adaptive management approach for the release of nano-photocatalytic paints in a historic industrial site; (2) an innovative application for the recovery of ancient stone surfaces in historical buildings; (3) air bioremediation in a green roof experiment; and (4) the design and implementation of a large-scale urban installation in a traffic intersection area (Chia et al. 2021; Shahid et al. 2020; Morais et al. 2021; Kumar et al. 2020; Shahi Khalaf Ansar et al. 2023; Goswami et al. 2022; Zainith et al. 2021).

Policymakers must acknowledge the interrelation of these components when developing regulations and evaluating the environmental and health consequences of nanomaterials.

9.5 Applications of Nanomaterials in Bioremediation

Nanomaterials have great potential in practical applications, such as the utilization of nano-Fe0 and carbonaceous materials, for in situ treatment of contaminated groundwater, soil, and sediment by various environmental contaminants, including organics, inorganics, and radionuclides (Roy et al. 2021). These applications have been extensively reviewed in the scientific literature. However, it should be noted that the use of nanomaterials in bioremediation tasks is not universal, as their interactions with environmental contaminants and the potential emission of nanowastes have not been fully studied in certain specific cases (Gonçalves and da Silva Delabona 2022). Thus, further research is necessary to determine the suitability of nanomaterials for different bioremediation scenarios. Moreover, if nanomaterials are employed for bioremediation through biotic processes at the molecular or atomic scale, their regulation should be based on the principles of biotic science (Hidangmayum et al. 2023). Therefore, in addition to physical- or chemical-science-based investigations, studies on molecular environmental toxicology, cellular biophysics, biochemistry, molecular biology, and genetic engineering or molecular biotechnology are indispensable for the proper regulation of nanomaterials in bioremediation within the field of environmental and ecological science (Vázquez-Núñez et al. 2020). Furthermore, using nanomaterials or nano-bioremediation shows the potential in mitigating anthropogenic air pollution and improve the global environment. They can effectively address substances with low bioavailability, such as gas-phase or semi-volatility subsolid or solid particles. The photocatalytic degradation of air pollutants can be achieved indoors, outdoors, and in automobiles using the energy of visible light through the utilization of UV-energized nano-photocatalysts (Qamar et al. 2022; Hidangmayum et al. 2023). Simultaneously, nanomaterials demonstrate promising capabilities as adsorbents or coadsorbents for the persistent removal of toxic microorganisms, volatile organic compounds (VOC), heavy metals, and odorous gases from gas-phase streams passing through fluid flush biofilters or packed bioreactors employed in biofiltration and bioventing processes. Nanomaterials have exhibited substantial applicability and efficiency when used in bioventing systems to eliminate indoor air contaminants and facilitate the in situ bioremediation of subsurface soil and groundwater with contaminated air (Hussain et al. 2022). They have the potential to contribute to the remediation of polluted environments significantly. Furthermore, the development and utilization of nanomaterials for bioremediation purposes are continually evolving through ongoing research and technological advancements. Various innovative strategies and techniques are being explored and implemented to enhance the efficacy and efficiency of nanomaterial-based bioremediation approaches (Ahmed et al. 2022; Shahi et al. 2021). One area of focus is the modification and functionalization of nanomaterials to boost their adsorption and catalytic properties. For instance, the surface of nanomaterials can be chemically engineered to increase their affinity toward target contaminants, allowing for enhanced capture and removal from environmental matrices. Additionally, the incorporation of specific functional groups onto nanomaterial surfaces can facilitate targeted interactions with contaminants, leading to improved remediation efficiency (Sahoo and Prelot 2020; Khan et al. 2021; Jawed et al. 2020; Tahoon et al. 2020). Moreover, efforts are being made to engineer nanomaterials with controlled release capabilities. This involves the development of smart nanomaterials that can release encapsulated remediation agents in a controlled manner, ensuring sustained treatment over extended periods (Su and Kang 2020). This approach minimizes the need for repeated application of nanomaterials and reduces the potential for environmental accumulation. Furthermore, nanomaterial synthesis techniques are also being refined to enable the production of materials with tailored properties and characteristics. These advancements allow for the design of highly selective nanomaterials toward specific contaminants, ensuring minimal interference with non-targeted substances. Additionally, the scale-up of nanomaterial synthesis processes is being explored to meet the demands of large-scale bioremediation applications (Harish et al. 2023). In addition to their direct role in contaminant removal, nanomaterials are also being investigated for their potential in monitoring and sense environmental pollutants (Roy et al. 2021). By incorporating nanosensors into remediation systems, real-time monitoring of contaminant levels can be achieved, enabling prompt and targeted intervention. This integration of nanotechnology and sensing capabilities holds immense promise for developing efficient and responsive bioremediation strategies (Sharma et al. 2021). Overall, the applications of nanomaterials in bioremediation are expanding and diversifying, offering immense potential for addressing environmental contamination challenges. Continued research and development efforts are crucial for maximizing the benefits and ensuring nanomaterial-based remediation technologies' safe and responsible deployment (Ahmed et al. 2022). Through interdisciplinary collaborations and comprehensive investigations, the full potential of nanomaterials in bioremediation can be realized, contributing to the preservation and restoration of our ecosystems (Xu et al. 2020).

9.5.1 Nanomaterials for Air Pollution Remediation

Nanomaterials have been extensively investigated and researched for their potential applications in environmental remediation, particularly in addressing water and air pollution issues (Roy et al. 2021). These remarkable materials have shown great potential in capturing harmful exhaust emissions from various industries and mitigating the impact of traffic-related pollution (Fussell et al. 2022). Furthermore, they have proven to be effective in developing filtration agents for large-scale facilities like cement industries and factories, as well as in creating mobile particulate mask filters and child mask filters. Beyond their usage as filtration agents, nanomaterials have also shown promise in diverse applications, such as developing clothes with integrated filtration properties. Additionally, they can be utilized as hydrophobic particulate collection agents and road fillers (Zeng et al. 2021). Nanomaterials' versatility and wide-ranging possibilities make them an invaluable resource for enhancing sustainable living practices. Although nanomaterials have demonstrated effectiveness in removing hazardous particles from the air, several studies have revealed some limitations regarding filtration efficiency and pressure performance. Thus, it becomes crucial to develop alternative materials that offer improved filtration efficiency while maintaining low pressure levels to effectively remove hazardous fumes and pollutants (Bhatt et al. 2022; Khan et al. 2021). In this context, nanomembranes and nanofibers have emerged as viable options for separating particles from the air, which can be further integrated into the design of filtration masks. In fact, when incorporated into biodegradable forms of nanocomposites, these nanomaterials have been found to enhance the filtration performance significantly. This is achieved through the attenuation of crystallite size and the amplification of surface area, both of which contribute to the improved performance of the filtration materials. Notably, a filter cut-off efficiency of 94.2% has been achieved for face mask filters, even when subjected to a high particulate matter (PM) 10 loading (Xiong et al. 2020). In conclusion, the results obtained from various studies have effectively demonstrated the tremendous potential of these newly developed nanomaterials. Their applications in the field of filtration have proven to be highly effective in removing a significant amount of airborne dust particles. Moreover, these filters have also been shown to possess the capability of degrading microorganisms, with degradation efficiency values of 10%, 50%, and 60% observed after 10 days, 20 days, and 40 days of incubation in the presence of *Pseudomonas* sp. (He et al. 2020). This highlights the added benefits and versatility of these nanomaterial filters, which can contribute to improving air quality. Additionally, it is worth noting that the prolonged exposure of the masks to *Pseudomonas* sp. for 40 days of incubation resulted in a reduced peak intensity of the crystallite lattice, as observed through XRD analysis, further supporting the effectiveness of these filtration masks. Nanomaterials have thus emerged as a powerful tool in the fight against pollution and promoting sustainable living practices (Ahmed et al. 2022). Their potential is immense, and further research and development in this field will undoubtedly lead to even more impressive advancements in the future. Nanomaterials hold immense promise in environmental remediation, particularly in addressing water and air pollution (Roy et al. 2021). Through extensive investigation and research, these exceptional materials have demonstrated their ability to capture harmful exhaust emissions from various industries and their potential to mitigate the impact of pollution caused by traffic-related activities. Moreover, these nanomaterials display efficacy in developing filtration agents for large-scale facilities, including cement industries and factories. Additionally, they contribute to the creation of mobile particulate mask filters and child mask filters, thereby catering to the need for diverse filtration solutions. Expanding on their multifunctionality, nanomaterials have exhibited promise in diverse applications, such as developing clothes with integrated filtration properties (Operti et al. 2021). Furthermore, they can be utilized as hydrophobic particulate collection agents and road fillers, showcasing their versatility and wide-ranging possibilities. These characteristics solidify the status of nanomaterials as an invaluable resource for enhancing sustainable living practices. Although nanomaterials have proven their efficacy in removing hazardous particles from the air, several studies have brought to light limitations regarding filtration efficiency and pressure performance (Bhatt et al. 2022; Khan et al. 2021). Consequently, it is imperative to develop alternative materials that offer improved filtration efficiency while maintaining low-pressure levels, ensuring the effective removal of hazardous fumes and pollutants (Deng et al. 2021). In this regard, nanomembranes and nanofibers have emerged as viable options for the separation of particles from the air, and they can be seamlessly integrated into the design of filtration masks. Remarkably, when incorporated into biodegradable forms of nanocomposites, these nanomaterials have significantly enhanced filtration performance (Ji et al. 2023). By attenuating the crystallite size and amplifying the surface area, these materials contribute to the overall improved performance of filtration materials. This commendable progress is emphasized by attaining a filter cut-off efficiency of 94.2% for face mask filters, even under high particulate matter (PM) 10 loading conditions. All in all, the results obtained from various studies effectively demonstrate the tremendous potential of these recently developed nanomaterials (Mallakpour et al. 2022). Additionally, it is worth mentioning that prolonged mask exposure to *Pseudomonas* sp. for 40 days of incubation led to a reduced peak intensity of the crystallite lattice, as observed using XRD analysis. This reinforces the effectiveness of these filtration masks. Ultimately, nanomaterials have emerged as a potent tool in the ongoing battle against pollution, catalyzing the widespread adoption of sustainable living practices (Wibowo et al. 2024). The potential of nanomaterials is immeasurable, and with further research and development in this domain, there is no doubt that even more impressive advancements will be achieved.

9.6 Challenges and Future Directions

Numerous challenges and needs require immediate consideration and attention. Due to the interconnectedness between aerosols and atmospheric components, a wide range of potential proposals exist for the development of intelligent stratified catalytic processes (Arfin et al. 2023). However, these proposals are rarely put into practice on an industrial scale. In these instances, technical challenges must be addressed, along with the establishment of collaborations with companies and political-smart communities, in order to advance environmental improvements truly. As a more significant percentage of the population migrates to urban areas, it becomes even more crucial to thoroughly analyze the significant and profound impact of diverse air pollutants, both those of natural origin as well as vehicular emissions, on the adjuvant properties and intricate dynamics of the earth's atmosphere. These meticulous and comprehensive investigations should encompass a wide array of methodologies, encompassing both meticulously conducted in situ observations and meticulously executed in vitro experiments. From a technological standpoint, it is imperative and indispensable to extensively deliberate and suggest the highly desirable and practically beneficial decentralization of heavily polluted areas. By employing cutting-edge and sophisticated atomistic simulations, synergistically combined with the European Union's imposingly advanced and contemporary technology of the Materials Genome, scientists and researchers can play an exceptionally crucial and pivotal role in generating a plethora of innovative and revolutionary ideas for controlling and mitigating the adverse impacts and potentially deleterious effects of volatile organic compounds (VOCs) and efficiently managing toxic and flammable waste within enclosed spaces (Jiao et al. 2023). Moreover, it is of utmost importance to emphasize the significance of implementing and integrating sustainable practices and renewable technologies that have a profound and tangible impact on citizens' lives. By facilitating and encouraging the widespread adoption and utilization of these groundbreaking and transformative technologies through means such as community engagement programs that emphasize their benefits and practicality, the dream of a cleaner and healthier environment can swiftly transform into a tangible reality that benefits all members of society and enhances their well-being. Furthermore, an ongoing challenge lies in safeguarding process workers against the potential risks associated with nanomaterials, exposure to crumb rubber, and the residues of nuclear bomb detonations. Automatic sampling techniques and introducing "personal protection wellness" strategies are of particular interest in this regard. Looking toward the future, it is imperative to conduct environmental impact statements and engage in research that focuses on identifying potentially valuable biological assets. These efforts align with companies' aspirations for sustainability and a green policy, aiming to foster innovation, improve competitiveness, and enhance visibility, all to create a safer socio-industrial landscape for future generations.

9.6.1 Ethical Considerations in Nanomaterial Use

Primarily, our purpose in this section is not to dissuade potential NPD (Nanotechnology Products Database), in favor of conventional application, but to foster critical reflection on the ethical dimensions of these issues. What risks does NPD bear in regard to low to mild conformity to ethical and/or policy implications? What are the ethical implications of using nanomaterials when less ethically contentious technologies are available? Conversely, are there actual moral obligations to retool the regulatory policy to include pathways incorporating nanomaterials or, on the part, featuring justice-centered discussions? In scenarios where the use of nanomaterials appears comparably safe, is it ethical and/or efficacious to promote NPD in favor of biotic or infrastructure technologies?

Much literature has recently been directed to discussing the potential trade-offs (benefits and risks) of using nanomaterials in environmental and human health contexts. Furthermore, a significant portion of the literature has expressed doubt and concern over the plausibility of using nanomaterials to solve these problems (Falinski et al. 2020; Venkatramanan et al. 2021; Malsch et al. 2024; Kim et al. 2024). Primarily focusing on the implications of using nanomaterials for environmental remediation has numerous ethical implications. The unorthodox use of nanomaterials to remediate as opposed to compounded biotic or infrastructure systems begs further discussion for multiple reasons. First, personal exposure to nanomaterials could foster unknown health and environmental consequences. Additionally, introducing nanomaterials into environments could lead to intergenerational exposures with unknown effects.

Likewise, the distribution of nanomaterial-based environmental remediations could exacerbate social injustices. In the context of NPD, it is equally important to consider unresolved conflicts or ongoing regulatory and environmental injustices, such as the continued struggle for clean air, clean water, housing, or remediation by marginalized communities. Given this, a critical examination of whether to support nanomaterial introduction into environmental contexts would map onto different advocacy frames, taking into account capacity building and more localized targeted campaigns against specific applications, using a justice frame, evidence generation, and a focus on the precautionary principle, invoking environmental health and environmental justice; and more general resistance to environmental manipulation, using a panel framing.

9.7 Conclusion and Recommendations

The conjoining of concepts such as pollution of the atmosphere, proper nanomaterials, and bioremediation potential and impedance effects has been approached in this essay. The short overview provides concise information on atmospheric pollution and the effect of nanoparticles, as well as commentary. In the bioremediation part, the information regarding the advantages and disadvantages of bioremediation is described. Additionally, the general terms of air pollution, as well as its intraversing, are defined. Afterward, the classification of air pollutants, as well as their effects on the global level, is studied. The potential and the pitfalls due to nanomaterials of three different sizes are discussed. The types of bioremediation and the bioremediation method are classified in the general part. Both of the bioremediation categories are briefly discussed, and one example is described in detail. In the end, bioremediation potential and pitfalls are observed and mostly written about. Furthermore, some approaches to combining nanomaterials and bioremediation will be given, which could prove problematic due to the potential of nanomaterials causing harm. Consequently, the goal of scaling down classical bioremediation methods is considered unsound, but there are new approaches that do not support this argument. We propose that the best way to utilize nanomaterials in environmental improvement is to integrate bioremediation and nanomaterial engineering results. Integrating scientific results from these two fields should not be seen as a synthesis, marina, or an effacing of the two approaches; rather, the integration should be based on new results. Policy implications of the potential interplay between ambient air pollution and nanomaterials concerning bioremediation are identified.

Based on the above, we recommend the following: (a) Using nanomaterials to enhance pollution must be assessed before implementation to determine the possible related risks. (b) Because the use of bioremediation methods can be risky, they are best utilized to remediate environments that have already been contaminated with various agents. (c) Efforts must be advanced to ensure that research and industrial applications of nanomaterials go hand-in-hand with studying the likely toxic effects on living organisms and alteration of ecosystems. (d) Comprehensive basic and applied investigations of ENM toxic impact on the special microorganisms and traditional bioremediation for their potential future removal are required. (e) Complete assessments of ENM impact on human health.

9.7.1 Policy Implications

In an overview, we have proposed: (1) the interconnection of air pollution, nanomaterials, and bioremediation as a triad that may have relevance on health and ecosystem quality; (2) two mechanisms that may be generally critical; (3) how, on the one hand, policies to limit air pollution, of course, should be relatively stricter; (4) the expressed uncertainty regarding the potentiality of inhalation exposure of NMs contained in the aerosol, a pollution type whose regulation has been discussed; (5) some other aspects of the regulation of NMs under the viewpoint of environmental and human toxicology. Moreover, the policy implications of the mutual connection between air pollution, nanomaterials, and bioremediation have been integrated with each other. An aspect that has not been examined so far here and that deserves consideration is the interconnectedness between these particular types of declining environmental quality (i.e., air pollution), technological products that may add to their degradation (i.e., nanomaterials), and processes that are relevant to counteract such degradation or to recover those portions of the environment that are affected by the decline.

References

Adebiyi FM (2022) Air quality and management in petroleum refining industry: a review. Environ Chem Ecotoxicol 4:89–96

Agrawal M, Agrawal SB (2023) Effects of air pollution on plant diversity. in environmental pollution and plant responses. Routledge, London, pp 137–152

Ahmed SF, Mofijur M, Rafa N, Chowdhury AT, Chowdhury S, Nahrin M, Ong HC et al (2022) Green approaches in synthesising nanomaterials for environmental nanobioremediation: technological advancements, applications, benefits and challenges. Environ Res 204:111967

Arfin T, Pillai AM, Mathew N, Tirpude A, Bang R, Mondal P (2023) An overview of atmospheric aerosol and their effects on human health. Environ Sci Pollut Res 30(60):125347–125369

Azubuike CC et al (2016) Challenges of bioremediation. Biotechnol Adv
Bai Y et al (2021) Synergistic effect of silver nanoparticles in bioremediation processes. J Hazard Mater 403:123757
Baskar AV, Benzigar MR, Talapaneni SN, Singh G, Karakoti AS, Yi J, Vinu A (2022) Self-assembled fullerene nanostructures: synthesis and applications. Adv Funct Mater 32(6):2106924
Bessagnet B, Allemand N, Putaud JP, Couvidat F, André JM, Simpson D, Thunis P (2022) Emissions of carbonaceous particulate matter and ultrafine particles from vehicles—a scientific review in a cross-cutting context of air pollution and climate change. Appl Sci 12(7):3623
Bhardwaj LK, Rath P, Choudhury M (2023) A comprehensive review on the classification, uses, sources of nanoparticles (NPs) and their toxicity on health. Aerosol Sci Eng 7:69–86
Bhatt P, Pandey SC, Joshi S, Chaudhary P, Pathak VM, Huang Y, Chen S (2022) Nanobioremediation: a sustainable approach for the removal of toxic pollutants from the environment. J Hazard Mater 427:128033
Boda E, Rigamonti AE, Bollati V (2020) Understanding the effects of air pollution on neurogenesis and gliogenesis in the growing and adult brain. Curr Opin Pharmacol 50:61–66
Brook RD et al (2010) Particulate matter air pollution and cardiovascular disease. Circulation
Caumo S, Yera AB, Alves C, Rienda IC, Kováts N, Hubai K, de Castro Vasconcellos P (2023) Assessing the chemical composition, potential toxicity and cancer risk of airborne fine particulate matter (PM2. 5) near a petrochemical industrial area. Environ Toxicol Pharmacol 101:104170. [HTML]
Chaudhary P, Ahamad L, Chaudhary A, Kumar G, Chen WJ, Chen S (2023) Nanoparticle-mediated bioremediation as a powerful weapon in the removal of environmental pollutants. J Environ Chem Eng 11(2):109591
Chia SR, Chew KW, Leong HY, Ho SH, Munawaroh HSH, Show PL (2021) CO_2 mitigation and phycoremediation of industrial flue gas and wastewater via microalgae-bacteria consortium: possibilities and challenges. Chem Eng J 425:131436
Debbarma P, Joshi D, Maithani D, Dasila H, Suyal DC, Kumar S, Soni R (2021) Sustainable bioremediation strategies to manage environmental pollutants. In: Removal of refractory pollutants from wastewater treatment plants. CRC Press, Boca Raton, pp 267–288
Demarco CF, Quadro MS, Selau Carlos F, Pieniz S, Morselli LBGA, Andreazza R (2023) Bioremediation of aquatic environments contaminated with heavy metals: a review of mechanisms, solutions and perspectives. Sustain For 15(2):1411
Deng Y, Lu T, Cui J, Samal SK, Xiong R, Huang C (2021) Bio-based electrospun nanofiber as building blocks for a novel eco-friendly air filtration membrane: a review. Sep Purif Technol 277:119623
Elsaid K, Olabi AG, Abdel-Wahab A, Elkamel A, Alami AH, Inayat A, Abdelkareem MA (2023) Membrane processes for environmental remediation of nanomaterials: potentials and challenges. Sci Total Environ 879:162569
EPA (2020) Air quality trends. Environmental Protection Agency
Falinski MM, Turley RS, Kidd J, Lounsbury AW, Lanzarini-Lopes M, Backhaus A, Zimmerman JB et al (2020) Doing nano-enabled water treatment right: sustainability considerations from design and research through development and implementation. Environmental science. Nano 7(11):3255–3278
Filote C, Roşca M, Hlihor RM, Cozma P, Simion IM, Apostol M, Gavrilescu M (2021) Sustainable application of biosorption and bioaccumulation of persistent pollutants in wastewater treatment: current practice. PRO 9(10):1696
Fussell JC, Franklin M, Green DC, Gustafsson M, Harrison RM, Hicks W, Zhu Y (2022) A review of road traffic-derived non-exhaust particles: emissions, physicochemical characteristics, health risks, and mitigation measures. Environ Sci Technol 56(11):6813–6835
Gao T, Qiang Y, Liu Y (2024) Applications of zero-dimensional carbon–based nanomaterials in bioimaging. Zero-dimensional carbon nanomaterials. Woodhead Publishing, London
Gera H, Bhasin M (2023) Photochemical smog-a review. J Sci Res Teaching 2(7):210–228

Gonçalves CR, da Silva Delabona P (2022) Strategies for bioremediation of pesticides: challenges and perspectives of the Brazilian scenario for global application–a review. Environ Adv 8:100220
Gondal AH (2023) Nanotechnology advancement in the elimination of chemical toxins from air spectrums. Int J Environ Sci Technol 20(11):12775–12792
Gössling S, Meyer-Habighorst C, Humpe A (2021) A global review of marine air pollution policies, their scope and effectiveness. Ocean Coastal Manag 212:105824
Goswami RK, Mehariya S, Karthikeyan OP, Gupta VK, Verma P (2022) Multifaceted application of microalgal biomass integrated with carbon dioxide reduction and wastewater remediation: a flexible concept for sustainable environment. J Clean Prod 339:130654
Groma V, Alföldy B, Börcsök E, Czömpöly O, Füri P, Kéri AH, Osán J (2022) Sources and health effects of fine and ultrafine aerosol particles in an urban environment. Atmospheric. Pollut Res 13(2):101302
Gür TM (2022) Carbon dioxide emissions, capture, storage and utilization: review of materials, processes and technologies. Prog Energy Combust Sci 89:100965
Harish V, Ansari MM, Tewari D, Yadav AB, Sharma N, Bawarig S, Barhoum A (2023) Cutting-edge advances in tailoring size, shape, and functionality of nanoparticles and nanostructures: a review. J Taiwan Inst Chem Eng 149:105010
He W, Guo Y, Gao H, Liu J, Yue Y, Wang J (2020) Evaluation of regeneration processes for filtering facepiece respirators in terms of the bacteria inactivation efficiency and influences on filtration performance. ACS Nano 14:13161
Health Organization, W (2021) WHO global air quality guidelines: particulate matter (PM2. 5 and PM10), ozone, nitrogen dioxide, sulfur dioxide and carbon monoxide. Health Organization, W
Hidangmayum A, Debnath A, Guru A, Singh BN, Upadhyay SK, Dwivedi P (2023) Mechanistic and recent updates in nano-bioremediation for developing green technology to alleviate agricultural contaminants. Int J Environ Sci Technol 20(10):11693–11718
Huang Y et al (2019) Indoor air quality improvement through TiO_2 photocatalysis: a case study. Build Environ 148:427–435
Hussain A, Rehman F, Rafeeq H, Waqas M, Asghar A, Afsheen N, Iqbal HM (2022) In-situ, ex-situ, and nano-remediation strategies to treat polluted soil, water, and air–a review. Chemosphere 289:133252
Ikhajiagbe B, Anoliefo GO, Igiebor FA, Musa SI (2022) Enhancement of bioprocesses for the reclamation of impacted environment. In: Bioenergy and environmental biotechnology for sustainable development. CRC Press, pp 57–80
Ilyasu US, Aworunse OS, Olusanya CS, Isibor PO, Gana M, Oyewole OA (2024) Nanoparticles in air and their impact on air quality. In: Environmental nanotoxicology: combatting the minute contaminants. Cham, Springer, pp 183–202
IPCC (2021) Climate change 2021: the physical science basis. Intergovernmental Panel on Climate Change
Jaswal T, Gupta J (2023) A review on the toxicity of silver nanoparticles on human health. Mater Today: Proc 81:859
Jawed A, Saxena V, Pandey LM (2020) Engineered nanomaterials and their surface functionalization for the removal of heavy metals: a review. J Water Proc Eng 33:101009
Ji K, Liu C, He H, Mao X, Wei L, Wang H, Zhou F (2023) Research progress of water treatment technology based on nanofiber membranes. Polymers 15(3):741
Jiao Y, Kramshøj M, Davie-Martin CL, Albers CN, Rinnan R (2023) Soil uptake of VOCs exceeds production when VOCs are readily available. Soil Biol Biochem 185:109153
Joudeh N, Linke D (2022) Nanoparticle classification, physicochemical properties, characterization, and applications: a comprehensive review for biologists. J Nanobiotechnol 20:262
Khan Y et al (2018) Titanium dioxide nanoparticles for photocatalytic degradation of volatile organic compounds. Environ Sci Technol 52(6):3485–3493
Khan S, Naushad M, Al-Gheethi A, Iqbal J (2021) Engineered nanoparticles for removal of pollutants from wastewater: current status and future prospects of nanotechnology for remediation strategies. Journal of environmental. Chem Eng 9(5):106160. [HTML]

Kim Y, Kim J, Wiebenga MH, Oh SH, Kim DH (2023) Abatement of photochemical smog precursors through complete hydrocarbon oxidation over commercial Pd catalysts under fuel-lean conditions with NO promoting. Environ Pollut 338:122721

Kim M, Goerzen D, Jena PV, Zeng E, Pasquali M, Meidl RA, Heller DA (2024) Human and environmental safety of carbon nanotubes across their life cycle. Nat Rev Mater 9(1):63–81

Kumar S et al (2019) Nanomaterials for air pollution mitigation. J Clean Prod

Kumar V, Saxena G (2020) Microbe-assisted phytoremediation of environmental pollutants: recent advances and challenges. AK Peters/CRC Press

Kumar R, Ghosh AK, Pal P (2020) Synergy of biofuel production with waste remediation along with value-added co-products recovery through microalgae cultivation: a review of membrane-integrated. Sci Total Environ 698:134169

Kumari S, Raturi S, Kulshrestha S, Chauhan K, Dhingra S, András K, Singh T (2023) A comprehensive review on various techniques used for synthesizing nanoparticles. J Mater Res Technol 27:1739

Lei T, Xiang W, Zhao B, Hou C, Ge M, Wang W (2024) Advances in analysis of atmospheric ultrafine particles and application in air quality, climate, and health research. Sci Total Environ 2024:175045

Liu H, Liu T, Chen S, Liu X, Li N, Huang T, Zhang H (2024) Biogeochemical cycles of iron: processes, mechanisms, and environmental implications. Sci Total Environ 2024:175722

Mahmoudi M, Landry MP, Moore A, Coreas R (2023) The protein corona from nanomedicine to environmental science. Nat Rev Mater 8(7):422–438

Mallakpour S, Azadi E, Hussain CM (2022) Fabrication of air filters with advanced filtration performance for removal of viral aerosols and control the spread of COVID-19. Adv Colloid Interf Sci 303:102653

Mallik T, Banerjee D (2022) Biosurfactants: the potential green surfactants in the 21st century. J Adv Scie Res 13:97

Malsch I, Isigonis P, Bouman E, Afantitis A, Melagraki G, Antunes D, Dusinska M (2024) Exploring ethical impacts of nanomaterials for wastewater remediation. In: Water management in developing countries and sustainable development. Springer, Singapore, pp 275–288

Man F, Tang J, Swedrowska M, Forbes B, de Rosales RT (2023) Imaging drug delivery to the lungs: methods and applications in oncology. Adv Drug Deliv Rev 192:114641

Manisalidis I, Stavropoulou E, Stavropoulos A, Bezirtzoglou E (2020) Environmental and health impacts of air pollution: a review. Front Public Health 8:14. frontiersin.org

Mateos-Cárdenas A, Jansen AR, O'Halloran J, van Pelt FNAM, Jansen MA (2021) Impacts of microplastics in the Irish freshwater environment. Environ Protec Agency Ireland Res Rep 377:61

Miller V, Reedy J, Cudhea F, Zhang J, Shi P, Erndt-Marino J, Lopez-Jaramillo P (2022) Global, regional, and national consumption of animal-source foods between 1990 and 2018: findings from the global dietary database. The lancet planetary. Health 6(3):e243–e256

Mittal D, Kaur G, Singh P, Yadav K, Ali SA (2020) Nanoparticle-based sustainable agriculture and food science: recent advances and future outlook. Front Nanotechnol 2:579954

Mocek-Płóciniak A, Mencel J, Zakrzewski W, Roszkowski S (2023) Phytoremediation as an effective remedy for removing trace elements from ecosystems. Plan Theory 12(8):1653

Morais EG, Cristofoli NL, Maia IB, Magina T, Cerqueira PR, Teixeira MR, Gouveia L (2021) Microalgal systems for wastewater treatment: technological trends and challenges towards waste recovery. Energies 14(23):8112

Morales-Cano KL, Hermida-Castellanos L, Adame-Adame CM, Peláez LAP, Peña-Montes C (2023) Micro (Nano) Plastics as carriers of toxic agents and their impact on human health. In: Advances and challenges in microplastics. IntechOpen, London

Moreno-Ríos AL, Tejeda-Benítez LP, Bustillo-Lecompte CF (2022) Sources, characteristics, toxicity, and control of ultrafine particles: an overview. Geosci Front 13(1):101147

Mukherjee S, Sarkar B, Aralappanavar VK, Mukhopadhyay R, Basak BB, Srivastava P, Bolan N (2022) Biochar-microorganism interactions for organic pollutant remediation: challenges and perspectives. Environ Pollut 308:119609

Muthukumaran P, Babu PS, Shyamalagowri S, Aravind J, Kamaraj M, Govarthanan M (2022) Polymeric biomolecules based nanomaterials: production strategies and pollutant mitigation as an emerging tool for environmental application. Chemosphere 307:136008

Navarro Nieva A, Moral García AID, De Pablos I, Delgado Calvo-Flores R, Párraga Martínez JF, Martín García JM, Martínez-Checa Barrero FJ (2024) Microorganisms isolated from Saharan dust intrusions in the Canary Islands and processes of mineral atmospherogenesis. Appl Sci 14:1862

Navarro A, del Moral A, de Pablos I, Delgado R, Párraga J, Martín-García JM, Martínez-Checa F (2024) Microorganisms isolated from Saharan dust intrusions in the Canary Islands and processes of mineral atmospherogenesis. Appl Sci 14(5):1862

Nel A et al (2006) Toxic potential of nanomaterials. Science

Nho R (2020) Pathological effects of nano-sized particles on the respiratory system. Nanomed Nanotechnol 29:102242

OECD (2019) The economic consequences of air pollution. OECD Publishing

Ogunkunle O, Ahmed NA (2021) Overview of biodiesel combustion in mitigating the adverse impacts of engine emissions on the sustainable human–environment scenario. Sustainability 13:5465

Operti MC, Bernhardt A, Grimm S, Engel A, Figdor CG, Tagit O (2021) PLGA-based nanomedicines manufacturing: technologies overview and challenges in industrial scale-up. Int J Pharm 605:120807

Pandey A, Brauer M, Cropper ML, Balakrishnan K, Mathur P, Dey S, Dandona L (2021) Health and economic impact of air pollution in the states of India: the global burden of disease study 2019. The lancet planetary. Health 5(1):e25–e38

Prasad R et al (2017) Bioremediation: a sustainable approach. Springer

Qamar SA, Bhatt P, Ghotekar S, Bilal M (2022) Nanomaterials for bioremediation of air pollution. In: Nano-bioremediation: fundamentals and applications. Elsevier, Amsterdam, pp 243–261

Rando G, Sfameni S, Galletta M, Drommi D, Cappello S, Plutino MR (2022) Functional nanohybrids and nanocomposites development for the removal of environmental pollutants and bioremediation. Molecules 27(15):4856

Ranjit P, Varma CASL, Maddela NR, Reddy KV (2021) Biotechnology of twenty-first century. In: Innovations in biotechnology for a sustainable future. Springer, Cham, pp 17–42

Rai PK et al (2018) Phytoremediation of air pollutants. Environ Sci Pollut Res

Roy A, Sharma A, Yadav S, Jule LT, Krishnaraj R (2021) Nanomaterials for remediation of environmental pollutants. Bioinorg Chem Appl 2021(1):1764647

Ruiz-Fresneda MA, González-Morales E, Gila-Vilchez C, Leon-Cecilla A, Merroun ML, Medina-Castillo AL, Lopez-Lopez MT (2024) Clay–polymer hybrid hydrogels in the vanguard of technological innovations for bioremediation, metal biorecovery, and diverse applications. Mater Horizons 11:5533

Sahoo TR, Prelot B (2020) Adsorption processes for the removal of contaminants from wastewater: the perspective role of nanomaterials and nanotechnology. In: Nanomaterials for the detection and removal of wastewater pollutants. Elsevier, Amsterdam, pp 161–222

Sales da Silva IG, Gomes de Almeida FC, Padilha da Rocha e Silva NM, Casazza AA, Converti A, Asfora Sarubbo L (2020) Soil bioremediation: overview of technologies and trends. Energies 13(18):4664

Sarakar S, Enamala MK, Chavali M, Sarma GVSS, Mannam KM (2021) Nanophytoremediation: an overview of novel and sustainable biological advancement. In: Importance & applications of nanotechnology. MedDocs Publishers, pp 47–61

Saxena G, Kumar V, Shah MP (2020) Bioremediation for environmental sustainability: toxicity, mechanisms of contaminants degradation, detoxification and challenges. Elsevier, Amsterdam

Selva Filho AAP, Converti A, Soares da Silva RDCF, Sarubbo LA (2023) Biosurfactants as multifunctional remediation agents of environmental pollutants generated by the petroleum industry. Energies 16(3):1209

Shahi Khalaf Ansar B, Kavusi E, Dehghanian Z, Pandey J, Asgari Lajayer B, Price GW, Astatkie T (2023) Removal of organic and inorganic contaminants from the air, soil, and water by algae. Environ Sci Pollut Res 30(55):116538–116566

Shahi MP, Kumari P, Mahobiya D, Shahi SK (2021) Nano-bioremediation of environmental contaminants: applications, challenges, and future prospects. In: Bioremediation for environmental sustainability. Elsevier, Amsterdam, pp 83–98. researchgate.net

Shahid A, Malik S, Zhu H, Xu J, Nawaz MZ, Nawaz S, Mehmood MA (2020) Cultivating microalgae in wastewater for biomass production, pollutant removal, and atmospheric carbon mitigation; a review. Sci Total Environ 704:135303. marineagronomy.org

Sharma P, Pandey V, Sharma MMM, Patra A, Singh B, Mehta S, Husen A (2021) A review on biosensors and nanosensors application in agroecosystems. Nanoscale Res Lett 16:1–24. springer.com

Sharma A et al (2022) Hybrid nanomaterial Systems for Enhanced Bioremediation of air pollutants. Environ Technol Innov 24:102032

Shetty SS, Deepthi D, Harshitha S, Sonkusare S, Naik PB, Madhyastha H (2023) Environmental pollutants and their effects on human health. Heliyon 9(9):e19496

Singh A et al (2020) Nanotechnology for air pollution control. Nano Today

Singh S, Singh P, Saifi A, Pratap SG, Singh PK (2021) Plant-based Enzyme-mediated Biodegradation of Azo dyes: a review. Int J Plant Environ 7(03):187–201. ijplantenviro.com

Singh A, Rajput VD, Varshney A, Sharma R, Ghazaryan K, Minkina T et al (2024) Revolutionizing crop production: nanoscale wonders-current applications, advances, and future frontiers. Egypt J Soil Sci 64(1):221. ekb.eg

Spallanzani R, Koga KT, Cichy SB, Wiedenbeck M, Schmidt BC, Oelze M, Wilke M (2022) Lithium and boron diffusivity and isotopic fractionation in hydrated rhyolitic melts. Contrib Mineral Petrol 177(8):74. springer.com

Su S, Kang PM (2020) Systemic review of biodegradable nanomaterials in nanomedicine. Nanomaterials 10:656. mdpi.com

Sudduth ER, Trautmann-Rodriguez M, Gill N, Bomb K, Fromen CA (2023) Aerosol pulmonary immune engineering. Adv Drug Deliv Rev 199:114831. sciencedirect.com

Tahoon MA, Siddeeg SM, Salem Alsaiari N, Mnif W, Ben Rebah F (2020) Effective heavy metals removal from water using nanomaterials: a review. PRO 8(6):645. mdpi.com

Tarfeen N, Nisa KU, Hamid B, Bashir Z, Yatoo AM, Dar MA, Sayyed RZ (2022) Microbial remediation: a promising tool for reclamation of contaminated sites with special emphasis on heavy metal and pesticide pollution: a review. PRO 10(7):1358. mdpi.com

Thangavel P, Park D, Lee YC (2022) Recent insights into particulate matter (PM2. 5)-mediated toxicity in humans: an overview. Int J Environ Res Public Health 19(12):7511. mdpi.com

Tripathi S et al (2017) Bioremediation of air pollutants. Environ Sci Pollut Res

Vázquez-Núñez E, Molina-Guerrero CE, Peña-Castro JM, Fernández-Luqueño F, de la Rosa-Álvarez MG (2020) Use of nanotechnology for the bioremediation of contaminants: a review. PRO 8(7):826. mdpi.com

Venkatramanan V, Shah S, Prasad R (2021) Exploring synergies and trade-offs between climate change and the sustainable development goals. Springer, Cham

Wang Z, Hu T, Liang R, Wei M (2020) Application of zero-dimensional nanomaterials in biosensing. Front Chem 8:320. frontiersin.org

Wang Y, Narayanan M, Shi X, Chen X, Li Z, Natarajan D, Ma Y (2022) Plant growth-promoting bacteria in metal-contaminated soil: current perspectives on remediation mechanisms. Front Microbiol 13:966226. frontiersin.org

Wang Y, Liu T, Gong D, Wang H, Guo H, Liao M, Wang B (2023) Anthropogenic pollutants induce changes in peroxyacetyl nitrate formation intensity and pathways in a mountainous background atmosphere in southern China. Environ Sci Technol 57(15):6253–6262

Wibowo YG, Ramadan BS, Taher T, Khairurrijal K (2024) Advancements of nanotechnology and nanomaterials in environmental and human protection for combatting the covid-19 during and post-pandemic era: a comprehensive scientific review. Biomed Mater Devices 2(1):34–57. springer.com

WHO (2021) Air pollution and health. World Health Organization

Xiong SW, Fu PG, Zou Q, Chen LY, Jiang MY, Zhang P, Gai JG (2020) Heat conduction and antibacterial hexagonal boron nitride/polypropylene nanocomposite fibrous membranes for face masks with long-time wearing performance. ACS Appl Mater Interfaces 13(1):196–206

Xu L, Xu M, Wang R, Yin Y, Lynch I, Liu S (2020) The crucial role of environmental coronas in determining the biological effects of engineered nanomaterials. Small 16:e2003691

Yin XH, Xu YM, Lau ATY (2022) Nanoparticles: excellent materials yet dangerous when they become airborne. Toxics 10:50. mdpi.com

Yu Z, Li Q, Wang J, Yu Y, Wang Y, Zhou Q, Li P (2020) Reactive oxygen species-related nanoparticle toxicity in the biomedical field. Nanoscale Res Lett 15(1):115

Zainith S, Saxena G, Kishor R, Bharagava RN (2021) Application of microalgae in industrial effluent treatment, contaminants removal, and biodiesel production: opportunities, challenges, and future prospects. In: Bioremediation for environmental sustainability. Elsevier, Amsterdam, pp 481–517

Zeng M, Chen M, Huang D, Lei S, Zhang X, Wang L, Cheng Z (2021) Engineered two-dimensional nanomaterials: an emerging paradigm for water purification and monitoring. Mater Horizons 8(3):758–802

Zhang Y et al (2020) Enhanced photocatalytic activity of TiO_2 nanoparticles for VOCs degradation. Appl Catal B Environ 264:118384

Zhang J, Chen Z, Shan D, Wu Y, Zhao Y, Li C et al (2024) Adverse effects of exposure to fine particles and ultrafine particles in the environment on different organs of organisms. J Environ Sci 135:449–473

Zhao J, Lin M, Wang Z, Cao X, Xing B (2021) Engineered nanomaterials in the environment: are they safe? Crit Rev Environ Sci Technol 51(14):1443–1478. tandfonline.com

Zhao J et al (2023) Nanomaterial-enhanced bacterial degradation of aromatic hydrocarbons. Biodegradation 34(1):45–58

Water Pollution, Nanomaterials, and Bioremediation

10

Maha S. A. Harridy, Nahid A. A. Morsi, Hussein M. Ahmed, and Mohamed A. El-Khateeb

Abstract

Lack of water resources and efficient wastewater treatment are two important requirements for emerging nations. Developing and applying cutting-edge, highly effective wastewater treatment technology is essential. Scientists have been interested in recent advanced methods in nanomaterial sciences, among other therapies. The use of nanoparticles in the treatment of wastewater is auspicious. Its unique ability to have a large surface area makes it helpful in filtering out harmful metal ions, bacteria that cause illness, and organic and inorganic solutes from water. This chapter describes the different nanomaterials used for the remediation of wastewater.

Keywords

Nanomaterials · Wastewater · Treatment · Reuse · Preparation

M. S. A. Harridy (✉)
Central Lab of Organic Agriculture, Agricultural Research Center (ARC), Giza, Egypt

N. A. A. Morsi
Cell Research Department, Field Crops Research Institute, ARC, Giza, Egypt

H. M. Ahmed
Housing and Building National Research Center, Cairo, Egypt

M. A. El-Khateeb
Water Pollution Research Department, National Research Centre, Dokki, Cairo, Egypt

A. A. H. Abdel Latef et al. (eds.), *Bioremediation and Nanotechnology for Climate Change Mitigation*, https://doi.org/10.1007/978-981-96-3069-1_10

10.1 Introduction

Supplying enough water to satisfy present and future demands is one of Egypt's most significant problems today. More water is needed due to the welfare lifestyle, limited freshwater supplies, climate change, agricultural usage, and rapid population increase. These concerns push the Egyptian government's desire for more water resources. The primary sources of water pollution include industrial effluents, nuclear power plants, fossil fuel facilities, sewage and other waste, agricultural runoff, and industrial wastes from chemical industries (El-Khateeb et al. 2022; Ahmed et al. 2022; El-Khateeb 2009). This exacerbates water pollution, which makes the aquatic environment unsuitable for aquatic life, agriculture, drinking, or other uses (El-Khateeb et al. 2022). An enormous amount of the environment's damage and contamination was caused by human activity, particularly industrialization and agricultural methods, which negatively impact the ocean and rivers essential to life. Consequently, various manufactured and natural worldwide occurrences and activities, including urbanization and industrial growth, have contributed to the accumulation of toxins in the environment, endangering both human health and the natural world (Sarker et al. 2021).

Over one billion people use contaminated drinking water sources, and over 2.6 billion people—40% of the world's population—do not have access to enough basic sanitation facilities. Thus, many suffer and get weaker from illness, and hundreds of children die every day from diarrhea and other diseases linked to water, sanitation, and hygiene (Nandi et al. 2017). The devastation and contamination of the environment have grown dramatically along with global industrial activity. River and seawater pollution has been especially hazardous because of industrial effluents from metallurgy, mining, tannery, and electroplating. These effluents contain a variety of heavy metals, such as chromium, lead, cadmium, zinc, copper, mercury, and nickel. Many removal techniques, including ion exchange, solvent extraction, electrochemical treatment, ultrafiltration, and chemical precipitation, have been employed, most of these are energy-intensive, unsuitable for usage in aqueous media, and not very cost-effective. Conversely, while desorption may restore part of the adsorbents, adsorption is a low-cost, effective process that can be reversed (González 2020; EL-Bady et al. 2023).

Metals are among the contaminants that are constantly and widely prevalent in the air, water, and soil. Given that water is an essential requirement for both plants and humans, the presence of large amounts of metal in water causes considerable worry. Metals enter and accumulate in plants through irrigation, and then they are consumed by people. There is a need for sustainable and feasible solutions to tackle the issue of metal contamination (Abdel-Shafy and El-Khateeb 2021). So, the treated wastewater, which has been utilized for irrigation with certain constraints, is one of the significant water resources (El-Khateeb et al. 2009, 2021). But if the treated wastewater's characteristics differed from the permitted limits, using it to irrigate crops could harm people's health (Sarker et al. 2021).

Egypt's wastewater treatment facilities can handle roughly 7.6 billion m^3/year of wastewater (Hellal et al. 2021). For many nations, pollution and water scarcity are major problems. The unique physicochemical features, high economic value, and high removal efficiency of nanomaterials make them attractive new methods for managing water quality. Water treatment involves the use of four distinct types of nanomaterials: reducing, adsorption, nanofiltration membranes, and photocatalytic. Nanoparticles are used to monitor water quality, particularly those utilized to identify pathogens and trace contaminants. Carbon nanotubes, magnetic nanoparticles, noble metal nanomaterials, and quantum dots are some examples of these nanomaterials (Xue et al. 2017).

Nanotechnology has entered numerous industries, including cosmetics, sensors, cleanup, and medicine. An efficient, practical method of purifying water tainted with metal and other substances is nanoremediation. Nanomaterials are incredibly versatile because of their variations in size, shape, surface chemistry, and chemical composition (González 2020).

Two specific categories of the remediation variants—nanoremediation and nano-bioremediation—have drawn particular attention recently. Because of their enormous surface area and reactivity, nanoparticles provide certain advantages over existing approaches in water treatment. However, biological nanoparticles still have lower removal efficiencies compared to inorganic nanoparticles. On the other hand, nano-bioremediation is associated with environmentally friendly materials, which is a significant advantage over remediation that uses metallic nanoparticle oxide (González 2020).

This chapter provides information on several nanoparticles (NPs) utilized in water treatment. The successful use of nanoparticles (NPs) in laboratory studies supports the potential use of these materials in small- to large-scale water remediation projects.

10.2 Nanoscience Definition

The field of nanoscience and technology is vast and multidisciplinary, and it has experienced rapid global growth in recent years. The foundation of both nanotechnology and nanoscience is nanomaterials. The main areas of research and development in these decades have been water technology, energy, the environment, and pharmaceuticals. Nanomaterials are widely used in medicinal delivery, water purification, fuel cells, defense, and navy detection of threats, among other applications. The industrial revolution has improved living conditions. Without the application of nanoscience and nanotechnology, modern high-speed personal computers and mobile communications would not have been feasible (Palit and Hussain 2018). Advanced research in nanotechnology enables the creation of a broad class of materials, including particulate matter with at least one dimension smaller than 100 nanometers (nm). Chemical compounds or materials that are produced and used on a very small scale are known as nanomaterials, or NMs (Palit and Hussain 2018).

Particles that are larger than 100 nm in all dimensions are considered materials. Different physical properties of NMs can rely on their size and shape, unlike bulk materials whose physical qualities are independent of size. When scientists discovered that a substance's size affects its physicochemical properties—its chemical, electrical, mechanical, and optical characteristics—they came to understand the significance of NMs. Due to these special qualities, nanoparticulate materials have generated much interest (Gleiter 2000). Water treatment facilities, oil refineries, petrochemical industries, industrial processes, catalytic processes, buildings and building materials, diagnostics, and medication delivery are just a few of the applications for which nanoparticles (NPs) show promise. Condensation, attrition, chemical precipitation, ion implantation, pyrolysis, and hydrothermal synthesis are some processes used to produce NPs (Tian and Li 2018).

10.3 Classification of Nanomaterials

Several groups of NMs can be created using different criteria. NMs are often classified by chemical composition, shape, state, and dimensionality (Gleiter 2000). Their size, which varies from 1 to 100 nm in at least one dimension, is another factor that determines their classification. These materials' general form and dimensions allow NMs to be further classified into four types. All of the dimensions of zero-dimensional nanomaterials are at the nanoscale, or smaller than 100 nm. Includes quantum dots (QDs), hollow spheres, metal, cubes, nano-rods, polygon, spherical, and core–shell nanometer particles (NMs). Materials such as metallic, polymeric, ceramic, nanotube, nano-rod filament or fiber, nanowires, and nanofibers are examples of one-dimensional nanomaterials; these materials have one dimension that is not on the nanoscale, and the other two dimensions are. Two-dimensional nanomaterials have two dimensions, but only one is present at the nanoscale. Single- and multilayered, crystalline or amorphous thin films, nanoplates, and nanocoatings are examples of 2D materials. Above 100 nm, three-dimensional (3D) materials come in various sizes (Palit and Hussain 2018; Aversa et al. 2018). Multiple nanocrystals are combined in different directions by 3D NMs. Examples of similar materials include foam, fibers, carbon nanobuds, nanotubes, fullerenes, pillars, polycrystals, honeycombs, and layer skeletons (El-Abeid et al. 2024; Shiau et al. 2018).

Aspect ratio, sphericity, and flatness are all characteristics of NP morphology. NPs can be categorized as scattered and agglomeration or isometric and inhomogeneous based on uniformity. This agglomeration depends on the electromagnetic properties of the NPs, specifically their magnetism. Depending on their chemistry and electromagnetic properties, magnetic NPs can exist in dispersed forms, suspensions, colloids, or an agglomerated state. Magnetic NPs prefer to cluster in an agglomerate state unless their surfaces are functionalized. NMs can be categorized into a number of groups based on their chemical makeup, including single constituent NPs and nanocomposites, which include more subtypes. Carbon, graphenes, CNTs, and fullerenes comprise most carbonaceous nanoparticles (NMs). Metals like silver, copper, iron, alumina, zinc, titania, and silica are used to prepare metallic NMs. Another type of NM with branchlike structures of nanoscale dimensions is branched dendrimers. In order to enhance characteristics like mechanical strength, toughness, and surface charge, materials known as nanocomposites add nanoparticles (NPs) to a matrix of conventional materials. Moreover, the shape and functionalization of NPs determine their ability to aggregate in a liquid, producing either hydrophobicity or hydrophilicity. Nanorods, nanozigzags, nanohooks, nanostars, nanocubes, nanohelices, and nanoplates are examples of NPs with various morphologies and electrical or thermal conductivity (Ullah and Lim 2022). Because of quantum physics, QDs—tiny semiconductor particles that are only a few nanometers in size—have optical and electrical characteristics that differ from those of bigger particles. Because QDs reemit light at a specified wavelength after absorbing white or ultraviolet light, they exhibit unique optical and electrical features. In this case, excitons, valence band holes, and conduction band electrons are limited to three spatial dimensions (Alkaç et al. 2021).

10.3.1 Metal Nanomaterials

Studies on metal nanoparticles (NPs), including gold, iron, silver, and other metallic NPs, have revealed that these materials differ from bulk materials in their chemical, optical, and electrical characteristics (El-Abeid et al. 2024). For instance, a sono-electrochemical process coupled with ultrasonic vibration can be used to synthesize and produce Au NPs (Shiau et al. 2018). Certain metallic nanoparticles (NPs) have varying sizes depending on how they are built. Once particle size is reduced to the nanoscale, the electronic structure bandgap changes, resulting in separate electronic levels with many atoms visible on the surface. These atoms on the surface materials will become increasingly active as their nanoscale size decreases because of the growing gap between their atomic coordinates and the unsaturated sites. The active surface area of metal nanoparticles (NPs) is an essential characteristic for several applications, including catalysis and adsorption procedures. In general, metal nanoparticles (NPs) can absorb light by intra-band transitions like Al, Ag, Au, and Cu, and inter-band transitions like Pt, Pd, Ni, and Ru. According to earlier research, light irradiation can increase the catalytic activity of metal nanoparticles (Kim and Lee 2018). Among these categories of nanomaterials are:

10.3.1.1 Silver Nanoparticles (Ag NPs)

Silver is the most widely used substance, which has been demonstrated to have antibacterial qualities due to its low toxicity and capacity to eradicate microorganisms in water. Ag NPs, which consist of silver salts like silver nitrate and silver chloride and are employed in various applications, are biocidal (Zhang et al. 2016).

Ag NPs can be made in several ways, such as physical synthesis, photochemical reduction (microwave radiation, pulse radiolysis, photo-reduction, X-ray radiolysis), biological synthesis (bacteria, plants, and fungi) (Zhang et al. 2016), chemical reduction (thermal, the oxidation of glucose, polyol process, Tollen, micro-emulsion), and electrochemical (polyol process) (Bachhav et al. 2020). Pb, Cr, and Cd ions were successfully removed from petroleum effluent by 0.75 g of walnut husk extract-silver nanoparticles (WHE-Ag NPs) over 5 h at 25 °C, with corresponding efficiencies of 72.6, 81.3, and 88.1%. The metal ion physical-sorption of

WHE-AgNPs on CN, NH, CO, CC, CO, and OH sites enabled this. According to a Freundlich isotherm analysis, AgNPs made from *Moringa stenopetala* leaf extract effectively eliminated 97% of Cr (VI) from the solution (Rajput et al. 2023).

10.3.1.2 Gold Nanoparticles (Au NPs)

Gold nanoparticles (Au-NPs) range in diameter from 1 nm to 8 μm. They can be spherical, octahedral, tetrahedral, nanotriangular, nanorod, and nanoprismatic, among other configurations. Au-NPs can be coated with different biomolecules and polymers for different applications. Au-NPs are mostly used to extract mercury, producing interesting transformations into AuHg, $AuHg_3$, and Au_3Hg. The behavior of citrate-based Au-NPs for removing mercuric ions was investigated after they were immobilized with Al_2O_3. Au NPs were additionally reinforced with sodium borohydride and other adsorbents to increase their capacity to extract mercuric ions (Lisha et al. 2009). Au-NPs supported on aluminum could extract Hg(0) with a removal capacity of up to 4.065 mg/g, much greater than the removal capacity of other adsorbents. This was achieved using sodium borohydride to convert Hg(II) to Hg(0). Au NPs remove more Hg from the body than any other adsorptive substance. This kind of Au NP supported on aluminum could be used to clean wastewater in a, valuable, affordable way since Au particles are easily recovered (Lo et al. 2012).

10.3.1.3 Iron Zero Valent

Zero-valent iron nanoparticles (nZVI) are the most commonly utilized metallic nanoparticles due to their low cost and high effectiveness against heavy metals, insecticides, and chlorinated compounds. The immobilization of lead, zinc, and cadmium in aqueous media is commonly used for iron nanoparticles. During the nanoremediation process, iron hydroxides and oxides are produced, and nZVI go through an oxidation phase (from Fe^0 to Fe^{+2} or Fe^{+3}). The pollutants can be removed by reduction, adsorption, or co-precipitation with species generated from Fe^0 (Pasinszki and Krebsz 2020). Before co-precipitation, chemisorption takes place. For example, nZVI is widely used to extract Cd+2 and create the nZVI-Cd complex in the process. It is important to remember that after oxidation, the majority of nZVI aggregate becomes less reactive; however, a very small fraction endures and can be hazardous to the environment. Collateral damage must be avoided by examining the residual concentration, and techniques like spectroscopy, colorimetry, and microscopy can now be used to do this (Vidmar et al. 2018).

10.3.2 Carbon Nanoparticles

Since carbon is necessary for life, it has long been a well-studied element, but recently, its properties at the nanoscale have drawn special attention. In recent years, several carbon-based nanoparticles have been created; the most often employed ones for cleanup include fullerenes, nanotubes, graphene, and quantum dots. Carbon nanoparticles have exceptional electrical properties, pore structure, and chemical and physical endurance, which enable them to filter out a range of pollutants from water efficiently (Kotia et al. 2020).

Carbon atoms are arranged in a hexagonal pattern and form graphene, a monolayer that is separated from graphite. Graphene, graphene oxide, and reduced graphene oxide have all been widely utilized in nanoremediation; the latter are mostly used for adsorption. However, carbon nanotubes have strong adsorption powers because of their one-dimensional structure and can draw out heavy metals and organic dyes from water, among other pollutants (Chai et al. 2021).

The following are examples of nanoporous carbon-based materials with suitable physicochemical properties for use in water treatment processes to remove pollutants such as fluorides, heavy metals, textile dyes, and pharmaceutical products: activated carbons, carbon nanotubes (CNTs), including single- and multi-walled CNTs, graphene, and its oxide. MWCNTs' ability to adsorb hexavalent chromium in contaminated environments, for instance, was evaluated in a study. Every year, some 700,000 tons of dyes are manufactured, of which 2% end up in wastewater. It is well established that dyes have the potential to cause mutagenesis, carcinogenesis, and teratogenicity, among other health issues in humans. The water's refractive index is altered by dyes, which lowers light penetration and affects photosynthesis, ultimately leading to an eutrophication issue (Berradi et al. 2019). Numerous techniques, including adsorption, oxidation, sedimentation, coagulation, micro- and ultrafiltration, and reverse osmosis, have been demonstrated for the management of dyes. However, the adsorption technique has drawn much attention recently due to its simple design, affordability, and ease of use; it is also effective against soluble and insoluble organic, inorganic, and biological pollutants. Compounds known as metal-organic frameworks (MOFs) are produced when metal ions interact with multidentate organics that exhibit high porosity, pore size and shape tenability, and a large surface area (Berradi et al. 2019).

The general properties of carbon-base nanoparticles can be enhanced by mixing them with these structures. One example of a metal-organic framework is nickel-benzene dicarboxylate (Ni-BDC), which has been coated over carbon nanotubes (CNT) and graphene oxide to be used as an absorbent of methylene blue, a dye commonly found in wastewaters. The adsorption capacity is 39.27% when utilized unadorned; it increases to 71.25% when decorated over carbon nanotubes and can reach 96.78% when decorated over graphene oxide (Ahsan et al. 2020).

Due to their special qualities, carbon-based NMs are important in several multidisciplinary domains. Amorphous carbon, diamond, and graphite are just a few of the solid-state allotropes of carbon (Li et al. 2019). These hybridized sp2 carbon atoms, which have been generated in multiple dimensions, make up these carbon-based nanomedicines (NMs) (Li et al. 2019; Nehra et al. 2019; Xie et al. 2019). Different chemical and physical features, such as conductivity, mechanical qualities, chemical stability, and thermal properties, are displayed by carbon-based NMs of nanoscale dimensions. Because of their many applications, carbon-based NMs have garnered a lot of attention. The following categories of materials can be applied based on the form of carbon-based NMs:

1. The carbon (C) allotrope fullerene (0D) has 60 C atoms organized in a bucky ball configuration. Because of their special qualities, fullerene derivatives can neutralize reactive species like oxygen and nitrogen (Sumi and Chitra 2019).
2. Nanoscale in one dimension

   Surface coatings or thin films are materials with one dimension in the nanoscale range. For many years, thin films have been produced and employed in various applications, such as fiber-optic systems, chemical and biological sensors, information storage systems, electronics, and magneto-optic and optical devices. Different techniques can be used to deposit thin films, which can also be developed in an atomic-level manner (monolayer). 2D materials are those nanomaterials (Battaglia et al. 2020);
3. Nanoscale in two dimensions

   Nanomaterials have two dimensions on the nanometer scale. These include, among others, fibers, fibrils, nanotubes, and nanowires (sometimes referred to as 1D material). One-dimensional nanomaterials also include free particles with huge aspect ratios and diameters in the nanoscale range. Less is known about the characteristics of 1D materials, and their manufacturing technology is less sophisticated (Sudha et al. 2018; Yan et al. 2016).
4. Nanoscale in three dimensions

   The 0D nanomaterials are generally characterized as nanoscale materials in all three dimensions. These consist of particles, fullerenes, precipitates, colloids, and quantum dots, or nanocrystals. While some 0D systems, like carbon black, titanium oxide (TiO_2), natural nanomaterials and combustion products, metallic oxides, and zinc oxide (ZnO), are well known, others, like fullerenes, dendrimers, and quantum dots, pose the biggest production and property-understanding challenges (Silva et al. 2018).

10.3.3 Silica Nanomaterials

Numerous procedures provide spherical and non-agglomerated silicon nanoparticles (Si NPs), the typical precursors of which are sodium silicate solution and tetraethyl orthosilicate (TEOS). Mesoporous silica materials are useful for environmental cleanup procedures due to their high surface modification ease, capacitance, and huge pore volume. Numerous studies have demonstrated the potential of these materials to remove contaminants from the gas phase due to their remarkable adsorption characteristics. Furthermore, a number of investigations have discovered that materials made of mesoporous silica display noticeable surface changes. Silica nanoparticles (NPs) made from the leaves of *Saccharum officinarum* (SOL), *S. ravannae* (SRL), and *Oryza sativa* (OSL) were utilized in a batch adsorption experiment to remove 60.1% (200 ppm) and 70.3% (30 ppm) of nickel. The NPs also removed Pb(II) and Cu(II). For SRL (5 mg), Pb(II) and Cu(II) had the highest adsorption capabilities of 98.8 and 78.8 mg/g, 99.5 and 97.52 mg/g for SOL (5–10 mg), and 99.75 and 97.99 mg/g for OSL SNPs (5–6 mg) (Sopi and Hassan 2024).

10.3.4 Nanoparticles Plasmonics

Because heterogeneous photocatalysis can remove or reduce a variety of pollutants, including dyes, pesticides, and Cr(IV), it has been the subject of intensive research for many years. This environmentally friendly method is regarded as green because it uses only sunlight to create harmful disinfection byproducts. However, as with any new technology, there are some major disadvantages, such as photoabsorption being limited to the ultraviolet region of the electromagnetic spectrum and that charge carrier recombination occurs quickly (Gellé and Moores 2019). To tackle this problem, noble metal nanoparticles that span a larger portion of the electromagnetic spectrum have been incorporated into semiconductors. The process by which conduction electrons on noble metal nanoparticles oscillate collectively when excited is known as surface plasmon resonance. This technique extends the semiconductors' exciton (electron–hole) period, increasing photocatalytic activity by injecting hot electrons into the semiconductors (Gellé and Moores 2019; Krishchenko et al. 2020).

Due to its many benefits, TiO_2 is the primary semiconductor used in photocatalysis. This oxide is nontoxic, has chemical stability, and has strong photocatalytic activity. Recently, the focus of research has been on visible light photocatalysis with Au nanoparticles. The Au/TiO_2 photocatalyst has been found to include oxygen molecules, OH radicals, and superoxide anion radicals (•O^{2-}) under bright light. The presence of this species facilitates the breakdown of pollutants. An interesting combination that has also been used with chitosan is Au/TiO_2. When this polymer's nanoparticles come into touch with Au and TiO_2, they demonstrate improved adsorption and photodegradation capabilities in addition to mechanical and chemical stability (El-Khateeb 2009; Wang et al. 2022).

A plasmonic fiber that can degrade a variety of pollutants, including metronidazole, methylene blue, and carbofuran, can be produced using this technique (González 2020). About 5% of sun radiation is necessary to activate TiO_2. A further workaround to overcome this problem and increase the exciton's lifetime is to decorate the oxide with plasmonic nanoparticles. The electrons in these nanoparticles act as redox centers, promoting the reactions required for photocatalysis. Since Au nanoparticles' surface plasmon resonance is visible in the 400–600 nm electromagnetic spectrum (depending on the conditions of synthesis), Au electrons are transferred to TiO_2 via plasmonic effects when visible light is shone on Au/TiO_2 complexes (Wang et al. 2022).

However, the presence of Au is also beneficial when the structures are subjected to UV light because the nanoparticles act as a trap for the photoexcited electrons. Bismuth titanate oxide is another catalyst that has been used a lot lately. It has a multilayer shape and exceptional optical and electrical properties. The wide bandgap of 3.1 eV and low charger-carrier mobility of this photocatalyst are the reasons for its subpar performance. As a result of its limited stability and propensity to convert Ag^+ ions to Ag^0, which results in further pollution, silver phosphate is a photocatalyst that has been thoroughly researched. Heterojunction structures seem promising to address problems with both photocatalysts; Ag_3PO_4/Ti_3C_2 hybrid material simultaneously increases Ag^+ ion stability and improves the low harvesting capacity associated with titanium composites. Plasmonic nanoparticles can be added to the heterojunction to further increase its efficiency. For example, $Bi_4Ti_3O_{12}$/Ag/Ag_3PO_4 can effectively remove pollutants such as rhodamine B and tetracycline hydrochloride, which the photocatalysts are unable to remove on their own. Certain “new contaminants” cannot be addressed using conventional methods; these include contaminants that come from the personal care and pharmaceutical sectors (Barrocas et al. 2016).

While plasmonic properties can sometimes be further exploited, plasmonic nanoparticles are a promising solution for treating different sorts of pollutants. It is possible to combine many nanoforms; specific morphologies allow for the acquisition of multiple plasmon resonances. Au nanorods were recently added to TiO_2 in an environmental cleanup study. The findings showed that a higher amount of visible and even near-infrared light could be absorbed than Au in nanosphere form. Another study examined the impact of adding Au to nanospheres, nanorods, and nanostars in SiO_2/Au/TiO_2 systems. The system, including the Au nanostars, had the highest photocatalytic activity due to its multiple plasmon resonance (González 2020).

10.3.5 Metal Oxide Nanomaterials

Numerous types of metal oxides, including ZnO, Fe_2O_3, Al_2O_3, TiO_2, and SiO_2, have been produced by hydrothermal or sol–gel processes. Metal oxide offers a major advantage in several applications, including catalysts, chemical sensors, and semiconductors, because of the change in its surface characteristics, which affects the bandgap energy of materials (Saleh and Fadillah 2019). The biocompatibility of these materials in offering a highly active surface area is a more significant feature. A polymer chain attachment, coupling agent addition, or metal ion doping are just a few of the simple reactions that can alter the surface of NPs (Das et al. 2020; Qi et al. 2019). Furthermore, using environmentally friendly methods like Al_2O_3, fatty acids can be employed to alter the surface of NPs (Liu et al. 2018). The properties of the materials can be changed by surface modification of NPs. For example, when Al_2O_3 was modified with a different kind of fatty acid, the materials precipitated in a water solution because the modifier was hydrophobic. A surface that has been changed by adding a functional organic group can also have special chemical and physical characteristics, like high density and targeted analyte binding. Various organic compounds modify the surface of nanoparticles (NPs). These chemicals include amines (Gaur and Banerjee 2019), anionic compounds (Chu et al. 2017), thiols (Zeng et al. 2019), and epoxies (Hur et al. 2019).

10.3.5.1 Aluminum Nanoparticles' (Al_2O_3 NPs') Function

Aluminum is one of the elements that are most common on Earth. Al NPs have a range of useful uses due to their favorable physical and chemical characteristics and their reactivity changes with size. The three stages of Al_2O_3 NP production are solid (mechanical ball milling and mechano-chemical), liquid, and gas. The demonstrates of Zn-doped Al_2O_3 NPs exhibit performance to be used in different applications as adsorebents. In a different investigation, the Freundlich isotherm demonstrated that 97% of Pb(II) and 87% of Cd(II) were effectively eliminated, with adsorption capacities of 47.08 and 17.22 mg/g on γ-Al_2O_3 NPs. The use of aluminum hydroxide nanoparticles (NPs) resulted in removal efficiencies of 97% (pH 5) and 95% (pH 5.5) for Cr(VI) and Pb(II), respectively. The concentration of the metal ions was 10 mg/L, while the adsorbent was 1 g/L (González 2020).

There are many different ways to treat water, ranging from chemical processes like flocculation and coagulation-reduction to physical processes like distillation and reverse osmosis. However, most of these strategies are insufficient and must be combined with additional approaches to increase effectiveness, which drives up the expense. Absorption is one of the most promising technologies because of its large-scale reproduction capabilities, low cost, and minimal environmental impact. Four characteristics are necessary for an absorbent to be considered good: (1) high removal capacity, (2) affordable, (3) environmentally friendly, and (4) strong structure to avoid disintegration. Graphene is a material that precisely fits all four characteristics due to its large surface area, low cost of manufacture, and simplicity of structural change (Xue et al. 2017).

The removal effectiveness and selectivity of graphene derivatives can be enhanced by adding thiols, amines, or other functional groups, thanks to their unique structure, which has been extensively researched in the field of environmental remediation. Many methods have been developed to synthesize graphene, such as heat treatment, chemical vapor deposition, and mechanics; but, due to their high cost, most of these are unsuitable for large-scale manufacturing. However, the harmful and environmentally dangerous compounds utilized in chemical reduction result in "impure" graphene. Despite this, chemical reduction seems to be a workable, affordable way to create graphene. Therefore, research has focused on developing safe and reasonably priced reducing medicines (Silva et al. 2018). L-cysteine is a prime example of this reducing agent; it has been used in the functionalization and production of graphene oxide. The end product removed mercury from water, with an absorption capacity of up to 85.3% (Nhlane et al. 2020).

10.3.5.2 Titanium Dioxide Nanoparticles (TiO_2-NPs)

The adsorption potentials of the TiO_2-NPs for Au and Ag were 22.63 and 14.06 mg/g, respectively. According to Weiping Qin, Daisheng Zhang et al. (2018), metal-oxide nanoparticles are used to reduce the accumulation of dangerous metals such As, Cd, and Cu (Qin et al. 2018). Yaqoob et al. (2018) reported that TiO_2-NPs at 25 mg/L considerably decreased the negative impacts of wastewater on maize development metrics ($p < 0.05$). TiO_2, the naturally occurring titanium oxide, is included in the crystalline polymorphs rutile, anatase, and brookite that are frequently seen. Except for rutile's lighter weight and stronger corrosion resistance, the three polymorphic forms of anatase are virtually identical. The instability of brookite, which turns into rutile at very high temperatures, contributes to its scarcity. Hydrothermal, sol–gel, chemical vapor deposition, microemulsion, and chemical precipitation techniques can all be used to make it. Moreover, it can be organically produced by plants, bacteria, fungi, and biologically derived compounds (Yaqoob et al. 2020).

10.3.5.3 Zinc Oxide Nanoparticles (ZnO NPs)

Physical (pulsed laser deposition, thermionic vacuum arc, thermal evaporation), biological (plant extracts and microorganisms), and chemical (precipitation, sol–gel, hydro/solvo-thermal, micro-emulsion, sono-chemical, chemical vapor deposition (CVD), microwave-assisted method, and electrochemical deposition) methods can all be used to produce ZnO NPs. The ZnO NPs are low in toxicity, have a rigid, stiff structure, and are biodegradable. The adsorption capabilities of silica-based zinc oxide nanocomposites (nano-SZO) for Cu(II), Ni(II), and Cd(II) were 32.53, 32.10, and 30.98 mg/g, respectively. The maximal adsorption capabilities of casein-based ZnO NPs for Pd(II), Co(II), and Cd(II) were 194.93, 67.92, and 156.74 mg/g, respectively. The ZnO NP composite with clay removed Pb(II) ions to a maximum adsorption capacity of 14.54 mg/g using the Langmuir model. ZnO NPs had a maximum removal effectiveness of 98.4% and a maximum adsorption capacity of 47.5 mg/g at pH when they were used to remove Cu(II). At pH 4, the maximum adsorption capacity was reached, with an initial metal concentration of 8 mg/L (Mengqin et al. 2020).

10.3.5.4 Iron Oxide Nanoparticles (Fe-NPs)

Nanoscale zero-valent iron (nZVI) nanoparticles are frequently used to remove metal ions from water. There are two methods for producing nZVI: bottom-up (converting dissolved iron into nZVI via reductants) and top-down (reducing large iron particles into tiny nanoparticles using mechanical or chemical processes). Fe, Al, Ni, and Zn are among the hazardous metal complexes treated with various nZVI NPs. They are efficient at eliminating a range of metal(loid)s due to their strong reactivity to contaminants and brief half-life. Using an external magnetic field to remove impurities is known as magnetic separation, and nZVI can be utilized in this process. This nanomaterial exhibits high reactivity when exposed to different contaminants and modifications. Examples of modifications are emulsified nZVI for improved compatibility, carbon-supported nZVI for increased stability, and surface-modified nZVI for improved dispersion. The greatest reported adsorption capacity for As(V) and As(III) when ascorbic acid-coated iron oxide (Fe_3O_4) NPs were used for metal removal was 16.56 and 46.06 mg/g, respectively, based on the Langmuir isotherm. Fe_2O_3 nanoparticles in a cellulose matrix had a maximum adsorption capacity of 23.16 and 32.11 mg/g for As(V) (Langmuir isotherm) and 9.64 and 3.25 mg/g for As(III) (Freundlich isotherm). The Fe_3O_4/phenol-formaldehyde resin nanocomposite was made by hydrothermal carbonization, and it has an adsorption capacity of 216.9 mg/g for the elimination of As (Ebrahim et al. 2016).

In high pH conditions, the standard reduction potential of nZVI is lower than that of several metals, such as Pb, Cd, Ni, and Cr. This characteristic makes it easier for these metals to bind to nZVI nanoparticles. In that order, the nZVI adsorbent exhibited removal efficiencies of 4.33–5.56, 5.40–6.94, and 5.41–6.95 mg/g for Cd, Cu, and Pb. After 5 h, the adsorption capacity of nZVI was 93% when Cd (6 ppm) was removed. The maximal adsorption capacity of Cr(VI) was 244.07 mg/g for a 2-hour contact period and 221.84 mg/g for a 25-hour contact period (Debnath et al. 2014).

On the other hand, agglomeration formation or (oxy) hydroxide corrosion can make the successful application of nZVI more challenging. The efficiency of nZVI may be affected by other contaminants, such as microplastics. The metal removal potential of nZVI for Cu, Cr, Pb, and Zn dropped in the presence of microplastics. Even after being kept for an extended period, metal may still be able to be removed by ZVI NPs. In the study, nZVI NPs had almost equal potential for the removal of Pb(II) (98%) when used both fresh and after 10 months of storage (Ebrahim et al. 2016).

10.3.6 Bimetallic Nanomaterials

The two metal components that make up bimetallic nanoparticles (NPs) have special qualities such as chemical stability, reactivity, and size-dependent electrical and optical capabilities. Shape, size distribution, and composition frequently impact these attributes. Bimetallic Pd/Cu-BNPs can be created by grafting a bi-naphthyl moiety onto the Pd metallic surface, which serves as a stabilizer and can be used for effective catalytic processes (Saha et al. 2019). The preparation process strongly influences the size distributions of NPs and their suitability for recycling as catalyst materials. A number of bimetallic materials, including Ag-Cu (Hao et al. 2019), Au/Pd (Chen et al. 2019), Au-Pd@SiO_2 (Wu et al. 2019), Fe-Cu (Lozhkomoev et al. 2019), and Ag-TiO_2-Ag (Su et al. 2019), have demonstrated that a combination of two metals can exhibit distinct surface activity and dispersion of NPs as well as optical and magnetic properties. According to earlier research, several variables, including the preparation circumstances and the miscibility of the two metals, significantly impact the nanostructure of bimetallic materials. There are five kinds of bimetallic nanoparticles, each with a unique structure.

10.3.7 Composite Nanomaterials

According to Oh et al. (2017), Lozhkomoev et al. (2019), composites are solid materials made up of many phases, one of which has dimensions smaller than 100 nm, or structures with a nanoscale repetition spacing between phases (Lozhkomoev et al. 2019; Oh et al. 2017). Physical dimensions in the nanoscale size range are always used when creating composite structures. Different qualities, including flexural strength, water sorption, optical properties, wear resistance, and gloss retention, can be produced by combining multiple materials to make a composite (Zhang et al. 2019; Biswas et al. 2012; Abozaid et al. 2019). Chitosan-tripolyphosphate/TiO_2 nanocomposites can be made utilizing the Box-Behnken design, using less energy and chemicals for the reactive orange dye adsorption process (Abdulhameed et al. 2019). The materials' surface area increased from 0.156 to 2.75 m^2/g when combined to create nanocomposites. Composite materials can enhance the adsorption capacity of a material via various forms of interactions, including dipole–dipole hydrogen bonding, electrostatic interactions, n–π interactions between lone pair electrons delocalized into the π-orbital, and Yoshida H-bonding (Parker et al. 2012).

10.3.8 Zeolite and Silica-Based Nanomaterial

Because of their mesoporous structure and advantageous surface chemistry, zeolite-based NMs have found widespread application. Templates, the rate of hydrolysis, and the circumstances of the reaction can all impact the mesoporous morphology and pore size of silica. By adjusting the pH solution, template concentration, and the usage of hydrophobic chemicals, mesoporous silica nanoparticles with pore diameters ranging from 2 to 50 nm can be produced in basic or acidic environments (Kumar et al. 2018). Nanoclays, which are NPs of layered silicates with Si atoms bound tetrahedrally to octahedral $Al(OH)_3$ or $Mg(OH)_2$, are the most often utilized silica-based materials. The weak van der Waals force that typically connects the clay layer can break the intercalation polymer chain. However, because of the variations in surface energy, the majority of polymers are incompatible with the nanoclay structure (Barua et al. 2019). Numerous inorganic complexes can be used to modify nanoclay particles. According to Yılmaz Baran et al. (2019), the Schiff-base modified technique was used to create the nanoclay supported Pd complex, or Hal-Pd, an active heterogeneous catalyst for Suzuki–Miyaura coupling processes (Yılmaz Baran et al. 2019). The substance demonstrated excellent stability, maximum catalytic activity, ideal reaction time, and ease of recovery for subsequent reactions. Compared to the meta- and ortho-positions, the para-position of aryl halide substitution maximizes the catalytic activity of Hal-Pd. CNTs can also be added to other silica-based NMs to improve their chemical and physical characteristics (Saleh 2016).

10.3.9 Ceramic Nanomaterials

A kind of NP material called nano-ceramic is built of ceramics and is further categorized as heat-resistant, inorganic, and nonmetallic solids consisting of metal and nonmetal compounds with sizes less than 100 nm (Gleiter 2000). Numerous chemical and physical techniques have been investigated and documented for producing ceramic nanomedicines. These materials were discovered to have improved ferromagnetic, ferroelectric, superconductive, electro-optical, and structural qualities. Similarly, varying the doping concentration of Ti-doped $BiFeO_3$ nanoceramics can alter its structural and physical characteristics. This may result in removing oxygen vacancies and structural distortion of the materials (Tian et al. 2018). Sobierajska et al. (2018) reported on the manufacture of porous hydroxyapatite nanoceramics and their uses because of their antibacterial activity (Sobierajska et al. 2018). High-purity hydroxyapatite (HAP) and methylcellulose were co-precipitated to create the calcium hydroxyapatite porous nanoceramics (ncHAP). The sintering temperatures affect the porosity and specific surface area of ncHAP. Its antibacterial activity is influenced by the interactions between the bacteria and its surface charges. Al_2O_3-ZrO_2 nanoceramics are created by synthesizing nanoceramics at low temperatures and without pressure.

10.3.10 Semiconductor Nanomaterials

Less than 4 eV is the low bandgap energy of semiconductor nanometers. Silicon, germanium, gallium arsenide, and elements close to the so-called "metalloid staircase" on the periodic table are examples of known semiconductors. These NMs are made up of diverse chemicals from several groups, including III–V (GaAs), IV (SiO_2), and II–VI (ZnO). Due to the quantum size effect or by increasing surface area, the structural modification of these materials at the nanoscale can change their chemical and physical properties. The high porosity of the C/ZnO semiconductor demonstrated that the formation of a nanostructure is a determining factor in the materials' high electrical conductivity (Yan et al. 2019). There are two categories into which semiconductor NMs fall.

1. Pure compounds or elements present from other metals in the structure but do not undergo doping to become intrinsic semiconductors. Intrinsic semiconductors are characterized primarily by negative temperature coefficients of resistance. This indicates that the material's resistivity will drop and its conductivity will increase as the temperature rises;
2. Extrinsic semiconductors, such as type-n and type-p semiconductors, are a kind of material added to other metals by doping in their structure to boost their conductivity.

10.3.11 Polymeric Nanomaterials

Nanosized solid particles called polymeric nanoparticles (NPs) are made of either synthetic or natural polymers. As drug release controllers that detect the body, these materials are widely used in pharmaceutical and medical applications (Yang et al. 2020). NMs based on polymers comprise the following:

1. Polymeric micelles are created when amphiphilic block copolymers self-assemble in a particular solvent. Because of their special qualities—such as their nanosize, stability, biocompatibility, micellar association, and low toxicity—chitosan polymeric micelles can be employed for drug administration (Yang et al. 2020).
2. Polymeric nanoparticles (NPs) typically consist of biocompatible and biodegradable polymers with an average pore size ranging from 10 to 1000 nm. Drug delivery to particular sites is commonly used for these materials (Sur et al. 2019). The unique properties of the material created are influenced by the preparation

techniques used to create polymer nanoparticles (Gharieh et al. 2019). The preparation techniques are broadly categorized as follows: dispersion of polymers, ionic gelation of hydrophilic polymers, and polymerization of monomers.
3. Dendrimers contain 3D-shaped macromolecules and are smaller than 15 nm. These materials represent a novel class of polymeric nanomedicines (NMs) with extensive applicability in pharmaceutical and medicinal fields because of their structure, size, and multivalence (Okrugin et al. 2017).
4. Polymeric nanocomposites combine polymers and other nanofillers to provide better qualities and attributes (Fu et al. 2019).

10.3.12 Lipid-Based Nanomaterials

Because they can carry both hydrophilic and hydrophobic molecules, have minimal toxicity, may regulate medication release in human body targets, lipid-based NMs, such as liposomes, solid lipid NPs (SLNs), and nanostructure lipid carriers (NLCs), and are employed for drug delivery. SLNs offer several advantages, including site-specific targeting, cheap cost, non-toxicity, and the ability to control hydrophilic and hydrophobic molecules (Zhong and Zhang 2019). However, according to a different study, SLN materials have drawbacks such as a restricted ability to load and expel drugs since they can crystallize while being stored (Meikle et al. 2019). Compared to SLNs, NLCs are more stable and are intended for use as medication release control materials. Liposomes comprise phospholipid and cholesterol-containing molecules and range in size from 50 to 100 nm. Because of their improved bioavailability and decreased toxicity, they are suitable for drug delivery, particularly for cytotoxic medicines.

10.3.13 Metal-Organic Structures

A hybrid material made up of organic ligands and inorganic metal ions is called a metal-organic framework (MOF). Their numerous beneficial qualities are their high porosity, high surface area, ease of surface modification, and well-organized structure and configuration. Because of the strong reactivity provided by the NP dispersion on their surfaces, MOFs have been frequently exploited as support materials. Fe-BTC MOF materials provided distinct reactivity and effective immobilization of the enzyme by facilitating the enzyme's contact with the material surface (Gascón et al. 2018). MOFs can be employed as hosts and catalysts to limit metal NPs in MOF holes. These are the synthesis techniques for MOFs:

1. Nano-reactor confinement, in which the isolation of nucleation sites under constrained physical circumstances using the nanoreactor technique controls the MOFs' NP size. A particular kind of non-polar solvent that is employed to create an emulsion from nanoscale monodisperse droplets has an impact on particle size.

2. Coordination modulation, a process that creates MOFs by chemically regulating the interactions between metals and ligands; and rapid nucleation, which creates MOFs by controlling the molar ratio and concentration of reactants using processes like accelerated heating and fast precipitation (Gascón et al. 2018).

10.3.14 The Nanomaterials' Core–Shell

The material core, which makes up the majority of NPs, and the surface layer, which can branch with various tiny moieties, surfactants, metal ions, and polymeric branches, are the two main layers that make up NPs. The shell layer's, whose composition differs from that of the core. The core–shell nanostructures come in various forms, such as (a) a metal core (single sphere or collection of several tiny spheres) with a shell composed of a different metal. This shell can consist of a single layer, smaller spheres attached to a larger core sphere, or a collection of smaller core spheres.

(b) A nonmetal shell encasing a metal core.
(c) A polymer-shelled metal core.
(d) A nonmetal shell encasing a nonmetal core.
(e) A nonmetal shell encasing a polymer core.
(f) A polymer core with more than one shell or a shell composed of several polymers (Su et al. 2019). Additionally, core–shell nanoparticles (NPs) that consist of a core encased in a different shell material can be divided into the following classes:
   1. If a chemical reaction occurs between the core–shell particles, the coating type will be either physical or chemical.
   2. Sorts according to core–shell components: organic–inorganic, inorganic–organic, and inorganic–inorganic.

The purpose of the shell is to increase surface activity and preserve the chemical stability of the core. Coating-covered core particles are engineered to possess specific attributes, including optical, magnetic, and catalytic capabilities (Su et al. 2019; Gascón et al. 2018).

10.4 Preparation

10.4.1 Synthesis of Nanoparticles

Two basic methods for synthesizing nanomaterials: (1) top-down and (2) bottom-up.

Before being incorporated into the intended substance, the NPs are first extracted at the atomic level. This covers the production of NPs through colloidal dispersion and powders through the sol–gel process, which is succeeded by integration. Additional examples include deoxyribonucleic acid-scaffolding of nanoelectronics, sedimentation, reduction, green synthesis, spinning, biological synthesis, atomic layer deposition, molecular self-assembly, and vapor phase deposition of NMs (Biswas et al. 2012).

- The top-down approach
- These methods create nanoparticles by using bigger (macroscopic) structures. Attrition or millings are common top-down methods for producing nanoparticles. The manufacturing process for this method is faster and less expensive, and power is only externally controlled when structures are reduced from macroscale to nanoscale sizes. The contamination factor and control energy input are the process keys. High-speed ball milling, etching through the mask, and extreme plastic deformation equipment are commonly utilized in this technique. The top-down method has some important drawbacks, including internal stress introduction, contamination during attrition, and crystallographic damage. Nonetheless, bulk production of nanomaterials is achieved using top-down techniques (Biswas et al. 2012).
- The bottom-up approach
- The atomic level materials for the creation of nanostructure components with an additional self-assembly process are included in the bottom-up method. It costs far less to fabricate. The primary benefit of this approach is the production of monosized nanoparticles with greater purity and improved control over particle size. The process keys here are to control the nucleation and development of the particles. This approach's primary drawback is separating the freshly created particles from the previously formed particles. With this approach, the rates of nanoparticle production are very low. The fundamental building blocks of bigger, stable structures are assembled by physical forces operating at the nanoscale during self-assembly. The two main techniques used in the bottom-up method are nucleation-growth of nanoparticles from colloidal dispersion and quantum dots by epitaxial growth.

10.5 Functionalization of Nanomaterials

It is crucial to modify NMs for a variety of purposes. Because of intermolecular forces like dipole–dipole and van der Waals interactions, unmodified NPs aggregate into stable groups. NMs can be altered by adding or removing functional groups and chemical species to create particular active sites and fine-tune their surface properties. Functionalized nanomedicines (NMs) are utilized in many domains and applications because of their distinct surface interactions (Prasad et al. 2019). Covalent and non-covalent interactions can functionalize an object, such as hydrogen bonding, van der Waals, and electrostatic forces. Conditions like reaction temperature, solvents, and surfactants can be selectively changed to control specific functionalities and unique configurations. Covalent interactions include the formation of chemical bonds between atoms on the surface of NMs through chemical processes. Since functionalization is accomplished by hydrophobic interaction, π–π stacking, hydrogen bonding, and ionic bonding in addition to hydrophobic contact, the electronic configurations and characteristics of atoms on NM surfaces remain largely unchanged in non-covalent interactions (Georgakilas et al. 2012). There are a few ways to carry out the NM functionalization, including the following ones:

1. Chemical techniques include heat treatment, H_2O_2 treatment, acid treatment, alkali treatment, and modification using appropriate metal ethoxides, alkoxides, and silane coupling agents (De and Madhuri 2020).
2. Ascorbic acid, oleic acid, oleylamine, trioctylphosphine, polyvinyl alcohol, polyvinylpyrrolidone, and cetyl trimethyl ammonium bromide are examples of ligand exchange processes.
3. In situ surface modifications, which are accomplished in the NP preparation stage by the thermal breakdown of polyol, reverse micelle, and organometallic chemical approaches, help prevent aggregations; these include capping agents, surfactants, and oleic acid, which are dissolved in the reaction media. Anionic surfactants, such as Al_2O_3, TiO_2, and Fe_2O_3, disperse nanoparticles (NPs). These surfactants can also be salts (Palmqvist and Holmberg 2008).
4. Grafting synthetic polymers, encompassing the techniques of grafting through, to, and from. End-functionalized polymers are typically attached directly to NM surfaces in the grafting-to process. While a low-molecular-weight monomer is co-polymerized with any other monomer on the surfaces of NMs in the grafting-through technique, polymer chains are formed from the initiating sites of the surface in the grafting-from approach (Wang et al. 2007). Depending on the necessary uses and the physicochemical characteristics of the intended NP applications, alternative methods for modifying NPs using polymers might be developed. The synthesis process choice is crucial to produce NMs with the required characteristics. Numerous synthesis methods that fall within the following categories have been reported:
   (a) The monomers polymerization.
   (b) Using premade polymers is an option. To avoid potential problems, choosing an appropriate technique and Journal Pre-proof to get appropriate attributes is crucial.

10.6 Properties

10.6.1 Electronic Properties

Electron conduction is delocalized in bulk materials, meaning that electrons can go in any direction. The quantum effect takes center stage when the scale is lowered to the nanoscale. Since all of the dimensions in zero-dimensional nanomaterials are at the nanoscale, the electrons are restricted to three-dimensional space. As a result, there is no electron delocalization or movement freedom. Since electrons are confined in two dimensions in one-dimensional nanomaterials, electron delocalization happens along the axis of the nanotube, nanorod, or nanowire. Electron confinement causes discrete energy states to replace energy bands in conducting materials, causing them to behave as either insulators or semiconductors.

According to Shin et al. (2015), the size, surface area, chemical composition, and alteration of NPs affect their electrical characteristics (Shin et al. 2015). The addition of organic substances like ligands can enhance the various surface properties of NMs. The modification of Al_2O_3 with organic ligands, which regulate the size of the materials formed during the aggregation process, demonstrates how different types of ligand monomers utilized can also impact the structural properties of the material created (Henkel et al. 2020). Additionally, this type of alteration offers various electrical properties. As a result, adding inorganic chemicals to the polymer systems improved the polymer nanocomposites' electric characteristics. Barium titanate (BT) can be added to increase electrical qualities like electrical conductivity and the dielectric constant because of its piezoelectric capabilities and perovskite nanostructures (Hassan et al. 2019).

10.6.2 Optical Properties

Changes in the size of the nanoscale can be used to regulate the emission of visible light in nanomaterials due to quantum confinement. It has been noted that the blue shift, or shift in emission peak toward shorter wavelengths, occurs when nanomaterials shrink in size. NPs' optical characteristics, particularly those of semiconductor materials, are crucial for several uses, including photovoltaics and photocatalysis. Basic light concepts and the Beer–Lambert law can be used to establish the optical qualities. The size distribution, shape, sizes, and types of modifiers are some of the parameters that affect semiconductor nanoparticles' greater absorption of wavelengths. UV–vis spectroscopy has been used to investigate the optical characteristics of Nd-doped NiO (Saiganesh et al. 2021). Due to the interchange of electrons in the energy band and the localized electron of Nd^{3+}, the Nd-doped NiO NPs may move to a lower energy value than pure NiO. Surface alteration and metal ion doping are two examples of how nanostructure composition affects optical characteristics. The particle size of NMs affects optical characteristics, particularly the scattering phenomenon and reflectance. As the refractive index rises, the reflectance falls and rises with increasing particle size. As a result, when exposed to light, particle size can alter the patterns of scattering particles, leading to variations in spectrum reflectance (Mikhailov et al. 2023).

10.6.3 Magnetic Properties

A nanomaterial's large surface-to-volume ratio causes it to interact magnetically differently with nearby atoms, producing a magnetic property that differs from a bulk one. A ferromagnetic material has several magnetic domains, but as it gets down to the nanoscale, it only has one domain that shows super-paramagnetic phenomena. Nanoparticles have a far smaller magnetic moment than their bulk size.

The size of magnetic nanoparticles also influences the value of magnetism. The magnetization increases considerably below 20 nm grain size. Therefore, it is possible to enhance the quality of magnets made from granular magnetic materials by reducing their particle size. Applications for magnetic nanoparticles exist in the environmental and medical domains.

The particle size of NPs affects their magnetic characteristics; according to Lamouri et al. (2020), NPs with a particle size of less than 35 nm exhibit the optimum performance (Lamouri et al. 2020). The number of magnetic atoms in a single compound NP directly represents the magnetic moment value of the molecule; in multicomponent NPs, on the other hand, the number of lone pair electrons determines the magnetic value based on the valence–shell electron-pair repulsion (VSEPR) hypothesis. Particle size changes are often very slight and have little effect on the metals' lattice properties. However, in the case of metals that have surface-deposited metal oxides, mismatches between the metal's and the oxides' lattice parameters can cause the metal's lattice parameters to alter in response to changes in particle size, further exacerbating interfacial stress. As a result, if particle size changes, so will the magnetization value. Other elements like the fabrication techniques and nanostructure composition also affect the magnetic characteristics (Lakshmiprasanna et al. 2019).

10.6.4 Mechanical Properties

When nanomaterials are made, many flaws are included, affecting mechanical qualities. Because of how their atoms are arranged, some nanostructures have very different properties from their bulk structure. For example, single- or multi-walled carbon nanotubes (CNTs) have higher mechanical strength, elastic limit, and flexibility. At high temperatures, coarse-grained ceramics exhibit brittleness, while nano-phase ceramics exhibit ductility. Qi and Wang (2002) discovered that when the ratio of atom size to particle size falls below 0.01–0.1, cohesive energy also lowers. This phenomenon affects the melting point reduction (Qi and Wang 2002). Nanda and associate researchers conducted a study on the surface energy of various material scales in 2003. He stated that the surface energy of a nanoparticle is greater than that of a bulk particle and that of embedded nanoparticles (Nanda et al. 2003). Then, compared to bulk materials or microparticles, NPs exhibit distinct mechanical characteristics. Large surface area and ease of modification allow NMs to be used to boost mechanical qualities like adhesion, hardness, elastic modulus, and stress and strain. Organic compounds typically have low mechanical properties, whereas NPs from a class of inorganic compounds have mechanical qualities. Thus, adding inorganic chemicals to organic compounds is a popular way to increase their mechanical properties. By adding the metal oxide SnO_2, Bui et al. (2020) examined the mechanical characteristics of acrylic polyurethane (Bui et al. 2020). SnO_2 enhanced the mechanical properties of the polymer matrix, including adhesion, hardness, and resistance to impact and abrasion. Nevertheless, because metal oxides can lower polymer–polymer contacts, polymer-to-metal-oxide interactions, and potential agglomeration processes, adding more of them to the polymer matrix may reduce mechanical characteristics. According to other research, NP size affects mechanical characteristics because larger surface areas facilitate more interactions (Nguyen et al. 2020).

10.6.5 Thermal Properties

Thermal characteristics will change as the size gets closer to nanoscale. Higher surface energy and interatomic spacing changes cause a material's melting point drop. A material's bond strength and melting point are directly correlated. Since the surface-to-volume ratio of bulk materials is low, surface effects can be disregarded. However, in nanomaterials, the melting point is size-dependent, falling as the diameters of the particles get smaller. The explanation is that surface atoms in materials at the nanoscale have greater mobility because they are not bound in a direction perpendicular to the surface plane. Due to their enormous surface area, which allows heat exchanges to occur directly on the material's surface, NPs have better thermal characteristics than their fluid counterpart. Adding more metal oxide (SiO_2) to the polycarbonate progressively increased the materials' thermal characteristics (Nomai and Schlarb 2019). SiO_2 and other metal oxides can limit the development of polymer chains and enhance the interactions between NPs and polymers. Adding a nanofiller, with a high intrinsic thermal conductivity, influenced the thermal properties of NMs. Overall, the thermal properties of NPs depend on the large surface area, mass concentration, the ratio of energic atoms in NPs, and the fraction of NP volume dispersed (Jeon and Lee 2019).

10.6.6 The Structural Properties

The crystal structure of nanomaterials may differ from that of bulk one with different lattice parameters. Generally, gold (Au) and aluminum (Al) have face-centered cubic (FCC) crystal structures in bulk form, but in the case of nanoform, the crystal structure will be icosahedral. Indium is face-centered tetrahedral and it will be FCC if the size is less than 6.5 nm. Because of the presence of short-range core-to-core repulsion and long-range electrostatic forces, nanomaterials have less interatomic space than their bulk counterparts. In the case of Al, if the interatomic spacing decreases from 2.86 to 2.81 Å, the binding energy subsequently decreases from 3.39 to 2.77 eV. Nanostructure materials have unique characteristics that showing crystalline nature (Chupradit et al. 2022).

10.6.7 Chemical Properties

The nano-cluster's ionization energy is higher than that of the bulk materials. Because of their higher surface area to volume ratio, nanomaterials exhibit very high radical change in chemical reactivity, efficiency of chemical reactions, and pace of chemical reactions. For this reason, catalysis falls under the realm of nanomaterials. The ionization potential at nanosize is higher than that for the bulk materials (Abbasi et al. 2023).

10.7 Applications of Nanomaterials

Technology revolutions induce social changes and environmental apprehension. Peter Grutter, a physicist at McGill University, expressed his views by stating, physicist, at McGill University "I have a lot of hope that nanotechnology can get governments to put structures in place nationally and internationally that are more adaptable, so that the public could decide it wants the technology and then we could go for it." In contrast, Thomas Murray (a bioethicist and president of the Hastings Centre) states, "It is important that the applications of nanotechnology be thoughtful and appealing in their early stages." The public is looking for control of their lives. Could you, for example, use nanotechnology to empower people to practice healthy behaviors? The primary industries adopting nanotechnology are water purification, textiles, defense, energy, and cosmetics. However, nanotechnology will also be a possible tool in all other industries in a few years. Nanotechnology has applications in all conceivable areas (Epelle et al. 2022).

10.7.1 Environmental Applications

A catalyst made of silver nanoclusters can be used to screen for polluting by-products in a manufacturing plant that uses propylene oxide. Applying propylene oxide helps remove organic pollutants from groundwater and volatile organic compounds from the air while making plastics, paint, detergents, and brake fluid. Only water molecules are allowed to pass through the effective and affordable water filtration devices created by nanotechnology (Bratovcic 2019).

10.7.2 Treatment of Soil and Groundwater

The issue of oil and groundwater contamination resulting from manufacturing processes is highly intricate and concerning. Landfills, abandoned mines, underground leaks, contaminated industrial sites, and storage tanks are impacted. Heavy metals (e.g., mercury, lead, and cadmium) and organic chemicals (e.g., benzene, chlorinated solvents, and creosote) are among the pollutants found in these locations. Techniques for more specialized and affordable cleanup tools can be developed using nanotechnology. Many processes used to remove harmful pollutants are costly, time-consuming, and labor-intensive. Furthermore, it is frequently necessary to remove the contaminated region, which will harm the ecology. The application of nanotechnology can help develop technologies that can reach hard-to-reach places like aquifers and cracks and perform in-place repair, hence reducing the need for costly "pump-and-treat" procedures (Alazaiza et al. 2021).

Furthermore, by using nanoscience to change matter at the molecular level, remediation tools tailored to a specific contaminant (e.g., metal) can be created, improving the technique's sensitivity and selectivity as well as affinity and selectivity. An important concern is the quality of drinking water and the pollution that taints it. Above all, two extremely nephrotoxic metals that pose serious health hazards are arsenic and mercury. Therefore, there is an urgent need for remediation techniques that enable the quick, affordable, and efficient cleanup of contaminated water. Nanotechnology can offer fresh approaches to wastewater treatment, water desalination, and novel ways to clean and purify water from impurities (Galal et al. 2023).

10.7.3 The Impact of NPs in Water Remediation

The inability to prevent NP accumulation in the water, the difficulty of collecting these NPs after treatment, the need for superior mechanical properties to separate NPs from metals, and the possibility of new contaminants leaking into wastewater from NP complexes are the main disadvantages of using NPs to prevent metal hazards in water. Further drawbacks to using these NPs include reduced dependability, higher maintenance costs, and decreased operating efficiency. Although the technology used in the application of NPs for water cleanup is sophisticated, there are risky side effects that could harm people and the environment. Metal ions that are originally benign in bulk form can become exceedingly toxic at the nanoscale (González 2020). Other factors contributing to NPs' toxicity include their shape, crystallinity, surface charge and reactions, large surface area, solubilizing and agglomerating properties, and surface charge and reactivity (Zhong and Zhang 2019).

Nanoparticles (NPs) are small, one of the main reasons they can easily enter tissues and organs and subsequently accumulate in humans and animals. Due to deposits in their gills, crayfish exposed to CuNPs for an extended length of time may experience oxidative stress (Kermanizadeh et al. 2015). Additionally, plants exposed to high concentrations of CuNPs may experience sterility in their pollen grains (El-Abeid et al. 2024).

Even though NPs are marketed as cost-effective, some NPs, such as Au and Ag, can be too costly to be widely used in water treatment, particularly in low-income countries. To the best of our knowledge, no country has laws particularly controlling nanoparticles except those overseeing organic agriculture, where using nanotechnology is prohibited (Brar et al. 2022).

10.7.4 Mechanisms of Metal Removal by Nanoparticles

Several strategies are available for treating wastewater and eliminating metals, including chemical precipitation, oxidation-reduction, ion exchange, and adsorption. One important and beneficial condition for removing metals is the availability of a large surface area to accomplish increased metal removal in a single cycle of operation, and nanomaterials fully satisfy this criterion. They are more surface area per unit volume in adsorbents comprised of nanomaterials than in bulk substances because of their nanoscale nature. More reactive surfaces are also provided by the larger surface area in the case of chemically modified NPs. What interaction mechanisms enable the removal of metal ions from aqueous solutions is still a mystery. Four basic remediation techniques are used to remove these toxins: soil, sorption reduction, photocatalysis, and precipitation. Because of the chemical interactions between nanomaterials and metal ions, sorption is one of the most straightforward methods for removing metal ions from contaminated water. One type of nanomaterial is mesoporous silica, which traps metal ions in the adsorption mechanism using both chemical and physical interactions via its functional groups. Large adsorptive surfaces characterize the nanomaterials. Furthermore, nanoadsorbents offer the unique opportunity to be recycled through chemical processes such as desorption and several cycles of reuse (Aragaw et al. 2021).

Similar to immobilization, sorbent reduction is a technique for changing high-valent metal ions into low-valent ones. Lowering the concentration of high-valent metal ions is necessary to produce denser particles or clusters that precipitate more easily. Similarly, various nanoparticle forms are used in precipitation to convert metal ions into hydroxides, carbonates, and other insoluble precipitates. The solid residue can then be separated via filtration. The conversions of Se^{4+} into Se^{2+} and Cr^{6+} into Cr^{3+} are two typical instances of these sorption-reduction processes. While the method of photocatalytic degradation is frequently utilized to eliminate different types of organic contaminants, it has also been extensively applied to eliminate small amounts of metals. The basis of this method is based on photocatalytic processes, which are strongly influenced by the catalyst's form, its ability to absorb light, and its active sites. Several mechanisms can be involved depending on the type of metal ions and light sources. Moreover, the ion exchange mechanisms are driven by the cationic or anionic metal species that are exchanged with the ionic ligands attached to the nanomaterials (Singh et al. 2021a).

10.7.5 Techniques for Extracting Heavy Metals from Water

Over time, several methods have been developed to remove heavy metals from water or reduce their content. These methods include chemical extraction, membrane technology, ion exchange, adsorption, ultrafiltration, chemical precipitation, and electrochemical treatment. However, most are either not cost-effective, require a lot of energy to run, or are inappropriate for usage in an aqueous medium. Over time, many metallic nanoparticles have been used to immobilize heavy metals. Immobilization is lowering the contaminant's bioavailability and stopping it from penetrating the earth, plants, and soil. In nanoremediation, these are the primary nanoparticles used (Singh et al. 2021b).

10.8 Conclusion

Nanomaterials are promising for wastewater treatment due to their diverse physical and chemical characteristics. The type of nanomaterial used for wastewater treatment is chosen after considering location, availability, practicality, and economic conditions. Because they are ecologically benign, cost-effective, and have time-saving qualities, nanomaterials solve problems that arise with conventional processes. Since they have a high selectivity and concentration of sorbents, nanomaterials can effectively remove heavy metals and organics and act as disinfectants. Nanoremediation, which also lowers contaminant concentration and cleanup time, can be used to clean up more contaminated sites. Since they are regarded as immobilization carriers in biosensors and biosorbents, metal oxide-containing nanomaterials are thought to be great tools. The potential for using nanotechnology in wastewater treatment is considerable. As research into nanotechnology advances and its practical application grows, it is expected to address the world's water scarcity and pollution issues significantly.

References

Abbasi R et al (2023) Structural parameters of nanoparticles affecting their toxicity for biomedical applications: a review. J Nanopart Res 25:43

Abdel-Shafy HI, El-Khateeb M (2021) Heavy metals in citrus fruits as affected by primary treated sewage irrigation. Egypt J Chem 64(1):165–176

Abdulhameed AS, Mohammad AK-T, Jawad AH (2019) Application of response surface methodology for enhanced synthesis of chitosan tripolyphosphate/TiO2 nanocomposite and adsorption of reactive orange 16 dye. J Clean Prod 232(20):43–56

Abozaid RM et al (2019) Optical properties and fluorescence of quantum dots CdSe/ZnS-PMMA composite films with interface modifications. Opt Mater 92:405–410

Ahmed A et al (2022) Egyptian reuse standards of treated wastewater for irrigation: a comparative study of physico-chemical and agronomic controls. MEJ Mansoura Eng J 47(2):43–59

Ahsan MA et al (2020) Nanoscale nickel metal organic framework decorated over graphene oxide and carbon nanotubes for water remediation. Sci Total Environ 698:134214

Alazaiza MYD et al (2021) Recent advances of nanoremediation technologies for soil and groundwater remediation: a review. Water 13(16):2186

Alkaç İM et al (2021) Nanomaterials and their classification. In: Nanomaterials for direct alcohol fuel cells. Elsevier, pp 17–33

Aragaw TA, Bogale FM, Aragaw BA (2021) Iron-based nanoparticles in wastewater treatment: a review on synthesis methods, applications, and removal mechanisms. J Saudi Chem Soc 25(8):101280

Aversa R et al (2018) The first annotated set of scanning electron microscopy images for nanoscience. Sci Data 5(1):1–10

Bachhav PA, Shroff RM, Shirkhedkar AA (2020) Silver nanoparticles: a comprehensive review on mechanism, synthesis and biomedical applications. Asian J Pharm Res 10(3):202–212

Barrocas B et al (2016) Titanate nanotubes sensitized with silver nanoparticles: synthesis, characterization and in-situ pollutants photodegradation. Appl Surf Sci 385:18–27

Barua S et al (2019) Chapter 8—Silicon-based nanomaterials and their polymer nanocomposites. In: Nanomaterials and polymer nanocomposites. Elsevier, pp 261–305

Battaglia S et al (2020) A novel intermolecular potential to describe the interaction between the azide anion and carbon nanotubes. Diam Relat Mater 101:107533

Berradi M et al (2019) Textile finishing dyes and their impact on aquatic environs. Heliyon 5(11):e02711

Biswas A et al (2012) Advances in top–down and bottom–up surface nanofabrication: techniques, applications & future prospects. Adv Colloid Interf Sci 170:2–27

Brar KK et al (2022) Green route for recycling of low-cost waste resources for the biosynthesis of nanoparticles (NPs) and nanomaterials (NMs)-a review. Environ Res 207:112202

Bratovcic A (2019) Different applications of nanomaterials and their impact on the environment. SSRG Int J Mater Sci Eng (IJMSE) 5(1):1–7

Bui TMA et al (2020) Investigation of crosslinking, mechanical properties and weathering stability of acrylic polyurethane coating reinforced by SiO2 nanoparticles issued from rice husk ash. Mater Chem Phys 241:122445

Chai L et al (2021) A new strategy for the efficient exfoliation of graphite into graphene. New Carbon Mater 36(6):1179–1186

Chen X et al (2019) Bimetallic Au/Pd nanoparticles decorated ZnO nanowires for NO_2 detection. Sensors Actuators B Chem 289:160–168

Chu P et al (2017) On the volume resistivity of silica nanoparticle filled epoxy with different surface modifications. Compos A: Appl Sci Manuf 99:139–148

Chupradit S et al (2022) Morphological control: properties and applications of metal nanostructures. Adv Mater Sci Eng 2022

Das L et al (2020) Synthesis of hybrid hydrogel nano-polymer composite using Graphene oxide, Chitosan and PVA and its application in waste water treatment. Environ Technol Innov 18:100664

De S, Madhuri R (2020) Functionalized nanomaterials for electronics and electrical and energy industries. In: Handbook of functionalized nanomaterials for industrial applications. Elsevier, pp 269–296

Debnath B et al (2014) Nanotechnology in waste water treatment: a review. Ecol Urban Areas:563

Ebrahim SE, Sulaymon AH, Alhares HS (2016) Competitive removal of Cu2+, Cd2+, Zn2+, and Ni2+ ions onto iron oxide nanoparticles from wastewater. Desalin Water Treat 57(44):20915–20929

El-Abeid SE et al (2024) Antifungal activity of copper oxide nanoparticles derived from Zizyphus spina leaf extract against Fusarium root rot disease in tomato plants. J Nanobiotechnol 22(1):28

EL-Bady MS, Galal AAE, El-Khateeb MA (2023) Monitoring some chemical properties and human health risk assessments of Cd and Pb of Bardawil Lagoon water, north of Sinai, Egypt. Desalin Water Treat 298:35–43

El-Khateeb M (2009) Treatment of ink wastewater via heterogeneous photocatalytic oxidation. Desalin Water Treat 7(1–3):1–5

El-Khateeb M et al (2009) Use of wetlands as post-treatment of anaerobically treated effluent. Desalination 245(1–3):50–59

El-Khateeb M, Nashy E, Nayl A-AA (2021) Combining chemical coagulation process and innovative aerobic reactor for the treatment of de-hairing wastewater. Waste Biomass Valori 12:2557–2564

El-Khateeb M, Ahmed H, Ahmed N (2022) Effective granular activated carbon for greywater treatment prepared from corncobs. Egypt J Chem 65(131):255–263

Epelle EI et al (2022) Advances in the applications of nanomaterials for wastewater treatment. Environments 9(11):141

Fu S et al (2019) Some basic aspects of polymer nanocomposites: a critical review. Nano Materials Science 1:2–30

Galal AA et al (2023) Chemical properties of drinking water recovered through the application of pressure management activities in Egypt. Egypt J Chem 66(13):1945–1954

Gascón V et al (2018) Semi-crystalline Fe-BTC MOF material as an efficient support for enzyme immobilization. Catal Today 304(304):119–126

Gaur P, Banerjee S (2019) CN cross coupling: novel approach towards effective aryl secondary amines modification on nanodiamond surface. Diam Relat Mater 98:107468

Gellé A, Moores A (2019) Plasmonic nanoparticles: photocatalysts with a bright future. Curr Opin Green Sustain Chem 15:60–66

Georgakilas V et al (2012) Functionalization of graphene: covalent and non-covalent approaches, derivatives and applications. Chem Rev 112(11):6156214

Gharieh A, Khoee S, Mahdavian AR (2019) Emulsion and miniemulsion techniques in preparation of polymer nanoparticles with versatile characteristics. Adv Colloid Interf Sci 269:152–186

Gleiter H (2000) Nanostructured materials: basic concepts and microstructure. Acta Mater 48(1):1–29

González V (2020) Nanoremediation and nanobioremediation in water treatment: the search for an eco-friendly alternative. In: Handbook of nanomaterials and nanocomposites for energy and environmental applications. Springer, Cham, pp 1–27

Gu M et al (2020) The selective heavy metal ions adsorption of zinc oxide nanoparticles from dental wastewater. Chem Phys 534:110750

Hao H et al (2019) Epitaxial growth of Ag-Cu bimetallic nanoparticles via thermal evaporation deposition. Appl Surf Sci:143871

Hassan ML et al (2019) Novel cellulose nanofibers/barium titanate nanoparticles nanocomposites and their electrical properties. J Phys Org Chem 32:e3897

Hellal MS et al (2021) Technologies for wastewater treatment and reuse in Egypt: prospectives and future challenges. In: Handbook of advanced approaches towards pollution prevention and control. Elsevier, pp 275–310

Henkel C et al (2020) Mixed organic ligand shells: controlling the nanoparticle surface morphology toward tuning the optoelectronic properties. Nano Micro Small:1903729

Hur JU et al (2019) Fabrication of SnO2-decorated Fe3O4 nanoparticles with anionic surface modification. Ceram Int 45(17):21395–21400

Jeon H, Lee K (2019) Effect of gold nanoparticle morphology on thermal properties of polyimide nanocomposite films. Colloids Surf A Physicochem Eng Asp 579:123651

Kermanizadeh A et al (2015) Nanomaterial translocation–the biokinetics, tissue accumulation, toxicity and fate of materials in secondary organs–a review. Crit Rev Toxicol 45(10):837–872

Kim C, Lee H (2018) Light-assisted surface reactions on metal nanoparticles. Cat Sci Technol 8(15):3718–3727

Kotia A et al (2020) Carbon nanoparticles as sources for a cost-effective water purification method: a comprehensive review. Fluids 5(4):230

Krishchenko I et al (2020) Resonant optical phenomena in heterogeneous plasmon nanostructures of noble metals: a review. Theor Exp Chem 56(2):67–110

Kumar S, Malik MM, Purohit R (2018) Synthesis of high surface area mesoporous silica materials using soft templating approach. Mater Today Proc 5:4128–4133

Lakshmiprasanna HR et al (2019) Effect of Pr^{3+}-doping on the structural, elastic and magnetic properties of Mn–Zn ferrite nanoparticles prepared by solution combustion synthesis method. Chem Data Collect 24:100273

Lamouri R et al (2020) Size effect on the magnetic properties of CoFe2O4 nanoparticles: a Monte Carlo study. Ceram Int 46(6):8092–8096

Li Z et al (2019) Carbon-based functional nanomaterials: preparation, properties and applications. Compos Sci Technol 179:10–40

Lisha K, Anshup, Pradeep T (2009) Towards a practical solution for removing inorganic mercury from drinking water using gold nanoparticles. Gold Bull 42:144–152

Liu Z et al (2018) A green and facile approach to the efficient surface modification of alumina nanoparticles with fatty acids. Appl Surf Sci 447:664–672

Lo S-I et al (2012) Gold nanoparticle–aluminum oxide adsorbent for efficient removal of mercury species from natural waters. Environ Sci Technol 46(5):2724–2730

Lozhkomoev AS et al (2019) Fabrication of Fe-Cu composites from electroexplosive bimetallic nanoparticles by spark plasma sintering. Vacuum 170(108980)

Meikle TG, Strachan JB, Conn CE (2019) Chapter 8—Lipid nanomaterials for the delivery of antimicrobial peptides. In: Gurtler V, Ball AS, Soni S (eds) Methods in microbiology. Academic Press, pp 173–195

Mikhailov MM, Yuryev SA, Lapin AN (2023) Investigating changes in the diffuse reflectance spectra of BaSO4 powders modified with SiO2 nanoparticles exposed to proton irradiation. Nucl Instrum Methods Phys Res, Sect B 539:62–67

Nanda K et al (2003) Higher surface energy of free nanoparticles. Phys Rev Lett 91(10):106102

Nandi A et al (2017) Reduced burden of childhood diarrheal diseases through increased access to water and sanitation in India: a modeling analysis. Soc Sci Med 180:181–192

Nehra M et al (2019) Carbon-based nanomaterials for the development of sensitive nanosensor platforms. In: Advances in nanosensors for biological and environmental analysis. Elsevier, pp 1–25

Nguyen TV, Nguyen TA, Nguyen TH (2020) The synergistic effects of Sio2 nanoparticles and organic photostabilizers for enhanced weathering resistance of acrylic polyurethane coating. J Compos Sci 4(1):23

Nhlane D, Richards H, Etale A (2020) Facile and green synthesis of reduced graphene oxide for remediation of Hg(II)-contaminated water. Mater Today Proc

Nomai J, Schlarb AK (2019) Effects of nanoparticle size and concentration on optical, toughness, and thermal properties of polycarbonate. J Appl Polym Sci 136:47634

Oh Y, Lee J, Lee M (2017) Fabrication of Ag-Au bimetallic nanoparticles by laser-induced dewetting of bilayer films. Appl Surf Sci 434:1293–1299

Okrugin BM, Neelov IM, Leermakers FAM, Borisov OV (2017) Structure of asymmetrical peptide dendrimers: insights given by self-consistent field theory. Polymer 125:292–302

Palit S, Hussain CM (2018) Green sustainability, nanotechnology and advanced materials–a critical overview and a vision for the future. Green Sustain Adv Mater Appl 2:1–18

Palmqvist L, Holmberg K (2008) Dispersant adsorption and viscoelasticity of alumina suspensions measured by quartz crystal microbalance with dissipation monitoring and in situ dynamic rheology. Langmuir 24(18):998996

Parker HL et al (2012) The importance of being porous: polysaccharide-derived mesoporous materials for use in dye adsorption. RSC Adv 2:8992–8997

Pasinszki T, Krebsz M (2020) Synthesis and application of zero-valent iron nanoparticles in water treatment, environmental remediation, catalysis, and their biological effects. Nano 10(5):917

Prasad R et al (2019) Nanobiotechnology in bioformulations. Springer

Qi W, Wang M (2002) Size effect on the cohesive energy of nanoparticle. J Mater Sci Lett 21:1743–1745

Qi Y et al (2019) Controllable synthesis of transition metal ion-doped CeO2 micro/nanostructures for improving photocatalytic performance. J Alloys Compd 782:780–788

Qin W et al (2018) Near-infrared photocatalysis based on YF_3: Yb^{3+},Tm^{3+}/TiO_2 core/shell nanoparticles. Chem Commun 2010(13)

Rajput VD et al (2023) A practical evaluation on integrated role of biochar and nanomaterials in soil remediation processes. Environ Geochem Health 45(12):9435–9449

Saha R, Arunprasath D, Sekar G (2019) Surface enriched palladium on palladium-copper bimetallic nanoparticles as catalyst for polycyclic triazoles synthesis. J Catal 377:673–683

Saiganesh S et al (2021) Structural, optical and antibacterial properties of neodymium (Nd3+) doped nickel oxide (NiO) nanoparticles using Sesbania grandiflora leaf extract. Indian J Sci Technol 14(18):1442–1451

Saleh TA (2016) Nanocomposite of carbon nanotubes/silica nanoparticles and their use for adsorption of Pb(II): from surface properties to sorption mechanism. Desalin Water Treat 57(23):10730–10744

Saleh TA, Fadillah G (2019) Recent trends in the design of chemical sensors based on graphene–metal oxide nanocomposites for the analysis of toxic species and biomolecules. TrAC Trends Anal Chem 120:115660

Sarker B et al (2021) Surface and ground water pollution: causes and effects of urbanization and industrialization in South Asia. Sci Rev 7(3):32–41

Shiau B-W et al (2018) The characteristics and mechanisms of Au nanoparticles processed by functional centrifugal procedures. J Phys Chem Solids 116:161–167

Shin SW, Song IH, Um SH (2015) Role of physicochemical properties in nanoparticle toxicity. Nano 5:1351–1365

Silva AA et al (2018) Graphene sheets produced by carbon nanotubes unzipping and their performance as supercapacitor. Appl Surf Sci 446:201–208

Singh S et al (2021a) Mechanism and kinetics of adsorption and removal of heavy metals from wastewater using nanomaterials. Environ Chem Lett 19:2351–2381

Singh A, Chaudhary S, Dehiya BS (2021b) Fast removal of heavy metals from water and soil samples using magnetic Fe3O4 nanoparticles. Environ Sci Pollut Res 28:3942–3952

Sobierajska P et al (2018) Preparation and antimicrobial activity of the porous hydroxyapatite nanoceramics. J Alloys Compd 748:179–187

Sopi SAIM, Hassan MHA (2024) A review on the extraction of silica nanoparticles from Poaceae family via sol-gel. Malays J Sci Health Technol 10(1):7–18

Su H et al (2019) Luminescence of perovskite light-emitting diodes with quasi-core/shell structure enhanced by Al–TiO2–Ag bimetallic nanoparticle. Superlattice Microst 136:106323

Sudha PN et al (2018) Nanomaterials history, classification, unique properties, production and market. In: Emerging applications of nanoparticles and architecture nanostructures. Elsevier, pp 341–384

Sumi N, Chitra K (2019) Fullerene C60 nanomaterial induced oxidative imbalance in gonads of the freshwater fish, Anabas testudineus (Bloch, 1792). Aquat Toxicol 210:196–206

Sur S et al (2019) Recent developments in functionalized polymer nanoparticles for efficient drug delivery system. Nano-Struct Nano-Objects 20:100397

Tian L, Li L (2018) A review on the strengthening of nanostructured materials. Int J Curr Eng Technol 8(02)

Tian Y, Xue F, Fu Q, Zhou L, Wang C, Gou H, Zhang M (2018) Structural and physical properties of Ti-doped BiFeO3 nanoceramics. Ceram Int 44:4287–4291

Ullah A, Lim SI (2022) Plant extract-based synthesis of metallic nanomaterials, their applications, and safety concerns. Biotechnol Bioeng 119(9):2273–2304

Vidmar J et al (2018) Investigation of the behaviour of zero-valent iron nanoparticles and their interactions with Cd2+ in wastewater by single particle ICP-MS. Sci Total Environ 634:1259–1268

Wang X et al (2007) Surface initiated graft polymerization from carbon-doped TiO_2 nanoparticles under sunlight illumination. Polymer 48(20):58348

Wang K et al (2022) TiO2/Au/TiO2 plasmonic photocatalyst with enhanced photocatalytic activity and stability under visible-light irradiation. Catal Today 397:257–264

Wu J-b et al (2019) Selective oxidation of methanol to methyl formate over bimetallic Au-Pd nanoparticles supported on SiO2. J Fuel Chem Technol 47:780–790

Xie F et al (2019) Carbon-based nanomaterials–a promising electrochemical sensor toward persistent toxic substance. TrAC Trends Anal Chem 119:115624

Xue X-y et al (2017) Nanomaterials for water pollution monitoring and remediation. Environ Chem Lett 15(1):23–27

Yan Q-L et al (2016) Highly energetic compositions based on functionalized carbon nanomaterials. Nanoscale 8(9):4799–4851

Yan X-W et al (2019) Metal–organic framework derived porous 2D semiconductor C/ZnO nanocomposite with the high electrical conductivity. Mater Lett 252:325–328

Yang X et al (2020) Selective uptake of chitosan polymeric micelles by circulating monocytes for enhanced tumor targeting. Carbohydr Polym 229:115435

Yaqoob AA et al (2020) Role of nanomaterials in the treatment of wastewater: a review. Water 12(2):495

Yaqoob S, Ullah F, Mehmood S, Mahmood T, Ullah M, Khattak A, Zeb MA (2018) Effect of waste water treated with TiO2 nanoparticles on early seedling growth of Zea mays L. Journal of Water Reuse and Desalination 8(3):424–431

Yılmaz Baran N, Baran T, Menteş A (2019) Design of highly robust halloysite nanoclay supported palladium complex as a highly active heterogeneous catalyst for construction of biaryls. Appl Clay Sci 181:105225

Zeng T et al (2019) Facile fabrication of durable superhydrophobic and oleophobic surface on cellulose substrate via thiol-ene click modification. Appl Surf Sci 493:1004–1012
Zhang C, Hu Z, Deng B (2016) Silver nanoparticles in aquatic environments: physiochemical behavior and antimicrobial mechanisms. Water Res 88:403–427
Zhang Q et al (2019) A C-coated and Sb-doped SnO2 nanocomposite with high surface area and low charge transfer resistance as ultrahigh capacity lithium ion battery anode. Mater Today Energy 13:93–99
Zhong Q, Zhang L (2019) Nanoparticles fabricated from bulk solid lipids: preparation, properties, and potential food applications. Adv Colloid Interf Sci 273:102033

Types of Water Desalination and Its Impact on Environmental Pollution and Climate Changes

11

Sana Ullah, Dilawar Khan, Murad Muhammad, Nasr Ullah, Sobia Khaliq, Asim Gulzar, and Abdul Basit

Abstract

A number of desalination methods are considered to be "emerging," meaning they have not yet seen widespread application because further development is required. Increasing the availability of water sources, decreasing the necessary amount of energy, and using fewer chemicals are the main goals for improving these technologies. Compared to traditional methods of purification, using chemicals results in less environmental damage. Both the feedwater's origin and the desalination process significantly impact ecological balance. This chapter examines the effects of two novel desalination methods—membrane distillation and thermal based—and their impacts on the environment. This chapter contrasts the effects of current technologies on the environment with those of previous desalination methods. The initial evaluation of these technologies' global impacts is

S. Ullah (✉)
Department of Agricultural Leadership, Education & Communication, College of Agricultural & Environmental Sciences, University of Georgia Athens, Athens, GA, USA
e-mail: sana.ullah@uga.edu

D. Khan · S. Khaliq · A. Gulzar
Department of Entomology, PMAS-Arid Agriculture University, Rawalpindi, Pakistan

M. Muhammad
State Key Laboratory of Desert and Oasis Ecology, Key Laboratory of Ecological Safety and Sustainable Development in Arid Lands, Xinjiang Institute of Ecology and Geography, Chinese Academy of Sciences, Urumqi, China

University of Chinese Academy of Sciences, Beijing, People's Republic of China

N. Ullah
Institute of Management Sciences (IMSciences), Peshawar, Khyber Pakhtunkhwa, Pakistan

A. Basit
Department of Horticultural Science, Kyungpook National University, Daegu, South Korea

A. A. H. Abdel Latef et al. (eds.), *Bioremediation and Nanotechnology for Climate Change Mitigation*, https://doi.org/10.1007/978-981-96-3069-1_11

conducted using qualitative methods. According to the results, the natural forward osmosis (FO) approach is the most ecofriendly option because it causes the least harm to the environment. It was better for the environment to use membrane or thermal-based desalination instead of reverse osmosis (RO) which takes up 61% of the total desalination processes. Research establishes that using waste heat to produce the thermal energy needed for desalination can reduce carbon emissions by 30%, significantly enhancing the environmental sustainability of the process. At the end, we conclude by offering some recommendations for future improvements to these novel desalination systems, aiming to maximize their potential environmental benefits on both local as well as global scales.

Keywords

Desalination · Climate change impacts · Sustainable water solutions · Pollution reduction strategies

11.1 Introduction

It is estimated that water covers around 75% of the surface of the Earth, making it a particularly abundant natural resource. The percentage of water sources appropriate for human use is only 3%. It is estimated that around one-quarter of the world's population does not have access to an appropriate supply of freshwater that satisfies their requirements in terms of both quantity and quality. Furthermore, around 80 countries are currently struggling with serious difficulties related to water usage. The situation is expected to get much more dire as a result of the drought and desertification that is occurring all around the world. It is possible that in the not-too-distant future, countries that do not now face water shortages will be required to address the issue of freshwater scarcity within their borders. According to the World Watch Institute, by the year 2025, more than two-thirds of the world's population will be threatened by the problem of water scarcity. Unless they aggressively cut their water consumption and/or find alternate water sources, every nation, especially wealthy nations, will be forced to face the consequences. Among the potential solutions to the problem of water scarcity is the process of desalination.

According to Sepehr et al. (2017), desalination removes salts and dissolved compounds from saline water. The World Health Organization has set an admissible threshold of 500 parts per million (ppm) for salinity in drinking water (Esfahani et al. 2016). This method results in salt concentrations at or below this threshold. Desalination, the removal of salts from brackish water or seawater, has experienced global growth, following a faster pace in number and capacity increase (>307%) than any other source (Nriagu et al. 2016). Desalination has proven to be an effective means of water supply since its introduction in the 1950s. This in particular is a consequence of technological progress but also irresistible/enticing. Specifically, a significant increase in the most recent consideration of the desalination capacity over the global area was identified. The concern of the consumption of water is quite interesting; it has scaled up to the highest and has been estimated at 95 million cubic meters (MCM) per day since 2005, when it was only 35 million cubic meters per day. Jones et al. (2019) also identified that the overall rate was 62%, which was higher among the young population. Twenty-five percent of this provision was meant for municipal use; the other 30% was reserved for general use. Out of the 20% population proportion that was allocated to commerce and business administration, 2% was allocated to the industry sector. Shahzad et al. (2017) have suggested that desalination remains a challenge today due to the high costs associated with producing the final product. Furthermore, desalination has a number of environmental repercussions, including the creation of waste by-products and the emission of considerable amounts of greenhouse gases, both of which have the potential to have an impact on marine habitats (Ihsanullah et al. 2021; Sepehr et al. 2017; Shahzad et al. 2017).

11.2 Types of Water Desalination Techniques

Desalination can be broken down into two distinct categories, according to Xu et al. (2013): thermal and Membrane Biorecator (MBR) or membrane-based water purification. Conventional approaches of heat integration are described by Harandi et al. (2017) as applying thermal energy to seawater making it evaporate and condensing the produced water vapor as drinkable water. From the study that was conducted by Xu et al. (2013), it was realized that thermal technologies are common where water is saline and energy cost is cheap. For example, the Middle East and the Caribbean are good examples of this when it comes to studying investment facilities. This section found that among the thermal-based technologies that are used frequently, there are three, namely multi-stage flash distillation (MSF), multi-effect distillation (MED), and vapor compression distillation. Currently, there is high usage of thermal technology, while in certain areas such as the Middle East, there is up swelling use of membrane solutions. The reason stated by Eveloy et al. (2015) for this is that while the membrane-based solutions require lesser energy, their impact on the environment is much less, and the capacity flexibility resulting from the enhanced capacity is a bonus. According to the research carried out by Jiang et al. (2017) and Xu et al. (2013), the existing literature indicates that there are many membrane technologies. This paper also presents that water purification can be achieved by several methods such as electrodialysis, reverse osmosis (RO), and ultrafiltration. At the present time, the method of water desalination that is practiced most actively in different countries is reverse osmosis (RO). Secondary methods that are also in use-are; Multi-stage Flash (MSF) and Multi-effect Distillation (MED) (Nair and Kumar 2013). One method that works with pressure is reverse osmosis, which also uses increased pressure to oppose the natural flow of water through a membrane, which is only selectively permeable. Osmosis is the natural movement of water across the selectively permeable membrane from an area of low solute concentration to an area of high solute concentration. In the case of RO systems, the high-pressure pump injects excessive pressure into the system. Desalination is the process that entails the application of pressure to make seawater pass through membranes that are not fully permeable (Kämpf and Clarke 2013). This led to the fabrication of seawater desalination water that is free from salts. Also, the water collected by the pumps must undergo pretreatment to control the number of bacteria and pollutants that may be destructive to the RO membranes (Jiang et al. 2017). Since this is the case, the RO membranes will always remain free of contamination. As Garud et al. (2011) pointed out, in the pretreatment process, it is common to use other established treatment processes like chemical addition followed by coagulation, filtration, and sedimentation processes. This is followed next by the pretreatment where highly effective high-pressure pumps are used to exert high pressure after the pretreatment process is done. According to the study by Chong et al. (2015), osmotic pressure, membrane resistance, and channel flow are some of the challenges that membrane systems face when allowing for the movement of water. The membrane systems involve a pressure vessel that houses a polyamide thin-film composite semi-permeable membrane. Following the findings of Kämpf and Clarke (2013) and Garud et al. (2011), this particular stand is designed with a mini opening termed a pore through which water molecules can penetrate but not salts or other forms of contamination. After the RO process is complete, two different streams are produced. These streams move the saline solution and the water that has been purified from salt. After the desalination process, the water is taken through post-treatment in a bid to purify the water further. For this reason, it is fitting to choose the type of post-treatment according to the quality of the water to be desalinated and the intended purpose of the water. Some of the processes that are encompassed by this category include processes that involve the treatment of water, alteration of water pH, purification of water, and the remineralization of water (Garud et al. 2011; Szymoniak et al. 2022).

11.3 Desalination of Water Through Thermal Processes and Their Effects on the Environment

Thermal desalination has effects on the environment similar to those of membrane desalination systems in terms of environmental impacts (Eis). However, the release from thermal desalination facilities has a more considerable impact because the brine's temperature is higher than the surrounding temperature (Elsaid et al. 2020). There might also be the possibility that the increased effect has something to do with the increased volume of the brine stream which is up to five times larger than that of the produced stream in membrane desalination for a desalination plant of a similar capability (Van der Bruggen and Vandecasteele 2002). In this regard, it will indicate that even though the value of "S" is lower, the salinity level shown is by far quite high. Perhaps the brine stream volume will be significantly larger compared to the amount of freshwater obtainable with membrane desalination.

Still, as mentioned earlier, thermal desalination usually has a lower recovery ratio when compared to the membrane. In this regard, the inputs of water in the desalination facilities owned and operational by MSF are vast and lots of input water is needed to produce a certain amount of water. Concerning the behavior, MED plants have similar behavior to the other conventional plants; but they are not very exhaustive in terms of water consumption mostly for cooling as taken by El-Ghonemy (2012). In their general analysis, Manju and Sagar (2017) opined that the desalination system is more likely to trap and transfer aquatic species. This is what is considered in the evaluation of the system if the lens through which it is viewed is the environmental lens. In that case, the environmental impacts associated with the disposal of brine in the said thermal desalination systems are equally challenging as those observed with membrane desalination technology. Based on the effects mentioned above, flow rate, salinity, and temperature are the flow properties that give the two aforesaid substances their major differences. From Lattemann and Höpner (2008), it is postulated that the volumetric flow rate associated with the brine in the membrane desalination can range from five to six times that of the brine stream. This is an indication that the recovery ratio has fallen over this period and/or that there is a greater need for cooling. Therefore, brine resulting from using thermal-based desalination processes is normally less salty than brine resulting from seawater reverse osmosis (SWRO) that falls between 50 g/L depending on feedwater salinity of 35 g/L. According to the findings of the studies carried out by Dawoud (2012), Lattemann and Höpner (2008), and Zhou et al. (2013), it has been noted that the temperature of brine is frequently elevated by 5–15 °C in comparison to the temperature of the ocean that is around it.

No doubt, one of the most complex and pressing problems experienced by the majority of desalination units is the appearance of living organisms. This beckons the use of biocides in the process of disinfecting feedwater as confirmed by Zhou and his colleagues as stated in 2013. Al-Agha and Mortaja (2005) made some recommendations on using chlorides and hinted that chlorine is one of the most frequently used biocides in desalination centers that utilize both MSF and MED. El-Ghonemy (2012) stated that the material is added into the feedwater streams at adequate rates to ensure concentration in the range of 2 ppm all through the process. After this, it decreases to between 0.1 and 0.5 ppm due to communication between live organisms and the organic matter of the feedwater and degradation of non-life compounds. Oh and Jang (2016) explained that only fake desalination plants do not have residual chlorine concentrations in water. Based on this, chlorine was considered as the most appropriate biocide that should be used in disinfection. This decision was made because SS chlorine is very effective and relatively cheaper than other disinfectants. To a reasonable degree, the toxicity of the substance under consideration has been proved by various studies done in toxicology. In the case of desalination plants that are categorized as MSF and MED, the scale growth in the concentrated stream increases at times beyond particular levels of salt saturation (Hao et al. 2019). The formation of scale is called scale creation. The major type of scale according to Al-Hamzah and Fellows (2015) is calcium carbonate. But the scale can also contain calcium sulfate and magnesium hydrate. From the point of view of the existing techniques, the presence of scales complicates heat exchange and increases pressure losses—all this requires more pumping capacity. Various sources, including Sherif (2013), made it clear that several essential and performance factors describe the creation of a scale. Some practical parameters that affect the system include the type of ions present within the feedwater, the nature of the temperature at which the system operates, and the rate at which dissolved salts build up solid precipitate. Belkin et al. (2017) have shown that some salts' mobility rises when precipitation in solutions of higher supersaturation is increased. This is due to the episodes of rain that endured for a long period. One can interfere with the formation of scale by using chemicals that will counter the action of $CaCO_3$, applying antiscalants, or using a combination of both. Indeed, since the attack on calcium carbonate $CaCO_3$ can be done through performing a chemical interaction with acids, the scaling can be prevented due to this interaction. On the other hand, if the stoichiometric reaction of this type is desired, it is possible to bring the concentrations of the acids in the water, which is regarded as input to about 20–50 ppm. Two large issues plague the system: cost problems, which result from expensive acid to manage, and financial problems, which rise due to this; and mechanical challenges, which are experienced due to wear out in the pipeline of the plant process, which cuts the life span of the plant. These two are causing the system to be plagued with massive problems that highlighted two issues of leadership and politicization. Both of these are making it progressively more complex to employ the system. This leads to a decrease in the negative consequences that acidity brings about. From this case, it can be seen that it is good to see this new development from an environmental perspective. Thus, it is possible to apply antiscalants in small and irregular quantities, from 1 to 2 μg/L, in order to prevent scaling differently. According to Belkin et al. (2017), antiscalants are effective agents since they reduce nucleation and crystal growth, leading to crystal scale formation. Analyzing various studies, it was concluded that polyphosphates, phosphonates, and organic polymers are the primary antiscalants, as reported by Petersen et al. (2018). An approximate biodegradation time frame of 1 month is reflected through antiscalants, which are still of great importance to the ecology. Petersen et al. (2018) carried out a study and they found that none of these substances are dangerous to fish or any other invertebrate. Also, the quantities of such substances are significantly lower concerning the setting would bring negative connotations, if used. In MSF and MED desalination plants, many phase changes are experienced by the seawater to evolving foam. These phase shifts include more evaporation and condensation concerning foam formation. Another cause of foam generation is the concentration of dissolved materials in seawater; it has been discovered that such materials when concentrated form foams. El-Ghonemy (2012) states that these substances are generated through the breakdown of algae waste. The use of antifoaming agents is crucial. Its necessity depends on the life cycle of the treated organisms and the feedwater quality. According to Murphy et al. (2002), alkylated polyglycols and fatty acids are used treatments to control foam formation aiming to prevent its occurrence. Typically, these chemicals are present at a concentration of around 0.1 parts per million (ppm), leading to an 80% decrease in foam production. These substances hinder foam formation by inhibiting the interaction between steam and water. From a standpoint, fatty acids and their esters are considered nontoxic and they show no harmful effects on marine life.

Thermal desalination plants have a heavy impact on the environment because they emit greenhouse gases into the atmosphere. Every factory needs lots of energy and consumes heat directly, which are often generated by burning fossil fuels. The type of desalination technique chosen is critical concerning the amount of energy used and associated emissions of greenhouse gases (Becker et al. 2010; Raluy et al. 2005). When thermal desalination processes consume less energy, there is a corresponding reduction in greenhouse gas emissions. For instance, a 20% reduction in energy consumption can lead to a 15% decrease in carbon emissions. Furthermore, studies have shown that utilizing waste heat to generate the thermal energy required for desalination can reduce carbon emissions by up to 30%, making the process significantly more environmentally friendly (Li et al. 2021).

11.4 Environmental Effects of Membrane Water Desalination

There are a number of factors that can be attributed to the environmental effects of membrane desalination systems. These factors include the origin of the water that is being treated, the initial treatment of the water, the desalination process itself, the disposal of the concentrated brine, and the use of power. Brine discharge is the primary carrier of environmental contaminants (EIs), and it is accountable for practically all liquid waste. This is because brine discharge contains chemicals utilized in pretreatment, by-products of corrosion, brine from reverse osmosis (RO) desalination, and brine from filter backwash.

It is generally agreed upon that the brine produced by desalination procedures is the most significant waste stream in terms of its volume and quality. Within the membrane desalination framework, the brine stream's ratio to the product water stream is typically anywhere between one and two. It has been reported by Henthorne and Boysen (2015) and Qasim et al. (2019) that this phenomenon is frequently observed in desalination systems that use seawater reverse osmosis (SWRO). Salinity, temperature, acidity/basicity (pH), heavy metals, and residual chemicals are the key factors that contribute to the environmental consequences of brine on the marine environment and aquatic life (Liu et al. 2013; Mannan et al. 2019; Zhou et al. 2013). These factors are mostly responsible for brine's negative effects on the marine environment. The majority of the negative effects that brine has on the environment can be attributed to these components. In the majority of environmental impact studies, the primary focus is on analyzing the characteristics of the brine at the time of its release and the subsequent mixing of the brine with seawater in the mixing zone. This procedure is carried out to accurately evaluate the influence of the brine on the surrounding environment. During the membrane desalination process, there is a minor increase in temperature; thus, it is essential to emphasize that the influence of temperature and pH on brine is insignificant. Furthermore, as was made abundantly evident, the temperature of the brine is practically identical to the temperature of the feedwater, with the brine temperature being responsible for the greatest fluctuation of 2 °C. Both the friction that takes place in the channels of the RO module and the dissipation of heat that takes place in high-pressure pumps could be responsible for the disparity. According to Lattemann and Höpner (2008), Miller et al. (2015), and Zhou et al. (2013), the buffering capacity of the brine is taken into consideration while neutralizing it prior to release. It is necessary to gently acidify the feedwater in order to prevent the formation of scales, which are reliant on the pH of the water. Additionally, the brine is usually made chemically stable before environmental discharge. To keep the pH in the range of 6.7, we have to inject acid into the product water as part of our routine practice, and there will not be more than 170 ppm of acid for each cubic meter of PW. Sodium hydroxide (NaOH) solution is used to neutralize the acidic water to make the acidity level at 67 ppm number, which is the lowest (Khawaji et al. 2007). However, it needs to be noted that this chemical is the one that gives the desired effect. Saltiness is the term used to refer to the salinity level of the brine stream encountered in the marine environment. The brine in the SWRO desalination plant has 1.5–2 times the salinity of the saltwater that is supplied to the plant. The recovery rate of a facility may differ depending on the facility and could be anywhere from 35 to 50% . Lattemann and Höpner (2008), Tarnacki et al. (2012), as well as Zhou et al. (2013) give an idea that the salinity of the brine is between 65 and 85 g/L. The marine environment can undergo a series of physical, chemical, and biological changes due to the existence of salinity. The environment impact assessment (EIA) is an assessment done during the desalination plant construction phase to predict the environmental impact. The function of this assessment is to decrease and cut off the unconstructive effects of the ecosystem in the neighboring vicinity of these facilities (Elsaid et al. 2020). One of the seawater reverse osmosis (SWRO) desalination systems that has the potential to produce 0.33 million cubic meters (MCM) of water per day and 127 MCM of water annually was examined by Miller et al. (2015). At the site of the plant's water outlet and at distances of 0.5 and 1 km, the scientists were amazed to find out that the salt content at the bottom of the sea ranged from 41.9 to 39.2 g/L. After the test was conducted, the salinity was determined to be 39.1 g/L in the open sea from 5 km, and the two seas were compared. A line was there that was found to be parallel on the surface of the seawater.

When separating debris from seawater, the common guideline is the use of agents like ferric chloride and alum-coordination, in addition to flocculants. The sequence initiates with multimedia filters, as outlined by Shenvi et al. (2015), followed by cartridge filters. Ultrafiltration (UF) and nanofiltration (NF) are two extremely popular methods that are mostly used for the pretreatment of water before the process of reverse osmosis (RO) salt removal. This viewpoint is corroborated by Ma et al. (2007) and Voutchkov (2010). The aim is to develop feedwater quality through fouling reduction on the RO membrane. According to Khawaji et al. (2007), the most common input levels of ferric chloride are 0.2–1.4 parts per million per cubic meter. However, the introduction of polyelectrolytes at dosages of 1–5.2 parts per million per cubic meter has been found.

The production of scale that is caused by salts that have a low solubility is sometimes dealt with by using antiscalant. This concept remains applicable because calcium, barium, and strontium sulfates are a part of the groups of these salts. At the same time, calcium and magnesium carbonates are the main components. In their study, Penate and García-Rodríguez (2012) have confirmed that applying these strategies effectively minimizes problems with inactivity. The accuracy of this statement is especially high in situations where the rate of recuperation is accelerated. According to the recommendations of an antiscalant formulator, the dosage should be between 1 and 3 parts per million per cubic meter. According to Khawaji et al. (2007), the dosage is established based on the combined factors of the feedwater quality and the desired recovery rate.

The desalination process is one of the processes that consumes a lot of energy and relies mainly on fossil fuel plants. The world has already demonstrated a lot of concern about desalination facilities and the effect they could have on climate change, especially their impact on the environment. These measures are intensive enough that we need to separate the reduction in greenhouse gas emissions from the reduction in fossil energy consumption. Decreased greenhouse gas emissions and decreased fossil fuel consumption can be achieved only by using renewable energy. There is a variety of pollutants in the environment related to the desalination sector, as it consumes a great deal of energy. It contributes to atmospheric air pollution since it is the most important source of thermal activities of man-made plants in the environment. It is responsible for producing large quantities of carbon dioxide, nitrogen oxides, sulfur dioxide, and particulate matter. The exact values of energy consumption in SWRO desalination are also highly variable and dependent on each specific case, as mentioned by Kim et al. (2019). An environmental study was conducted by Mezher et al. (2011) using natural gas as a fuel for the steam cycle, the internal combustion engine, and the combined cycle. At the same time, they calculated the typical carbon dioxide issued during SWRO desalination. According to Tarnacki et al. (2012), SWRO has a 1.77 kg-CO_2eq/m^3 value for the global warming potential (GWP), expressed in CO_2. According to them, the acidification potential is about 25 g of sulfur dioxide equivalent to per cubic meter. Ameen et al. (2018) found that the carbon footprint of a SWRO desalination plant with a capacity of 0.63 million cubic meters per day ranges between 2.3 and 2.5 kilos of carbon dioxide per cubic meter. In another study by Heihsel et al. (2019), nearly 20 desalination plants combining SWRO and BWRO were assessed for their carbon footprint. With a collective capacity amounting to 1.736 million cubic meters per day, these plants produced an astonishing 1193 kilotons of carbon dioxide equivalent annually. The study by Heihsel et al. (2019) found that the standard amount of carbon dioxide released per cubic meter of water is 1.9 kg.

11.5 Environmental Impacts of Intake in the Desalination Process

Big desalination and power facilities typically utilize open water intakes. These intakes can be located either on the surface or below the water's surface. The location of these intakes might be either on the surface of the water or below the surface of the water column. Surface intakes are watercourses that are excavated from the uppermost layer of water or tranquil and protected areas such as bays or lagoons.

Surface intakes are also known as surface waterways. These canals were constructed to transport water into the pumping station, and their primary function is to guide water in that direction. According to Kress (2019), the standard process begins with basic bar screens, moves on to prescreens, and finally arrives at wire mesh screens, with efficiency improvements of up to 20% in debris removal at each stage. Both a vertical riser and an offshore intake head will be installed on the submerged open intakes. These components are typically positioned between 2 and 6 m above the water surface, which can reduce sediment intake by 15–25%, enhancing overall system efficiency. As an additional point of interest, they have a screening procedure comparable to a surface intake. The screening mechanisms that are utilized in open water systems are extremely important because they restrict the number of large marine species that enter the facility. As a result, they minimize the damage caused to the environment and protect the plant's infrastructure. There are a number of advantages that come with open water intakes, the most important of which are their uncomplicated construction, their cost-effectiveness, and their capacity to accommodate significant amounts of water.

There are two types of subsurface intakes: those located beneath sediment on coastal beaches or the seafloor offshore. Pipelines are another name for subsurface intakes. When the strata have a high permeability and porosity, subsurface intakes are the most suitable water intake technique because they guarantee hydraulic continuity in the collection zone. This strategy is especially effective when applied in broad areas with dense collection zones and strong recharge rates. One of the most significant benefits of subsurface intake is that it improves the quality of feedwater and reduces the required pretreatment. As a result, it reduces the amount of chemicals that are used and the impact that it has on the environment. According to Dehwah and Missimer (2017), one of the disadvantages of this circumstance is the ever-increasing costs that are related to building and maintenance responsibilities. Because they do not directly interact with the marine environment, subsurface intakes have lower environmental impacts (EIs) in comparison to other types of intakes. They dwell at the deepest depths of the ocean. There are two distinct forms of subsurface intakes:soil wells and infiltration galleries. Specific configurations can be employed for intake wells in small-to-medium-scale reverse osmosis (RO) desalination facilities. These configurations can be single or multiple wells, vertical, horizontal, or inclined orientations with directed or radial flow patterns. Even though small-scale seawater reverse osmosis (SWRO) plants have started using infiltration galleries as a reliable source for medium-to-large-scale SWRO desalination plants, it is a recent trend observed in recent years—with infiltration galleries actively participating in the facility's operations at present. In terms of water collection, infiltration galleries exhibit greater promise due to their shallow depth and expansive horizontal reach—a feature unique to this gallery type. The distinctive characteristic of infiltration galleries sets them apart from others. According to Dehwah and Missimer (2017), these galleries typically comprise several layers including permeable sand, gravel, screens, and geotextile fabrics. Similarly, Shahabi et al. (2015) tested the environmental and economic effects of applying a 35,000 m^3/day SWRO plant in two methods: open-intake (SWRO-PI) and beach-well intake.

Study findings indicated that there are many benefits beach well intake has over open intake: lower energy consumption, lifts that can be attached further upstream closer to the sea, and a greater amount of lift that can be achieved with it than the pipe itself. This without doubt demonstrates that the energy needed for underground intakes is substantially lower than that needed in the case of open intakes.

11.6 The Environmental Repercussions that Result from Emissions from Desalination

Using seawater to desalinate plants, these plants produce brine, which is then expelled into the ocean together with the outfall structures. This is known as brine discharge. Open systems located near the shore where they come from and submerged systems far offshore provide two different types of intakes (Kress 2019). For the outflow of brine into the ocean to achieve maximum dilution and distribution, a system operating at peak efficiency is crucial. The rate of salt discharge should be kept as low as possible—typically under 50 g/L—to minimize environmental impact. In future installations, even though overall production capacities must increase by up to 40% due to rising global water deficits, seawater can still be treated to meet potable water standards, with recovery rates reaching up to 85% in advanced desalination plants. This activity is helpful to reduce the adverse effects on the marine environment, which could be impacted by collateral damage. Both the construction of an outfall and the selection of its location require careful consideration of factors such as salinity and temperature, which can vary by up to 10–15% in certain regions. Additionally, the potential influence of diesel fuel contamination, which can increase local toxicity levels by 20–30%, must be mitigated to protect marine biodiversity. Depending on what sort of water is actually obtained from a desalination process, how these characteristics happen to look may differ. The brine produced by thermal desalination facilities, on the other hand, will have a temperature between 5 and 15 °C greater than the temperature of the environment it belongs to. According to Lattemann and Höpner (2008) and Tarnacki et al. (2012), the salinity of the water will only increase by 20–50% when compared to the number of water molecules that were initially present. Both temperature and salinity affect the buoyancy of the brine stream, which means that these two factors have opposing effects on the density of the brine stream. According to the inverse relationship between increasing warmth and decreasing density, the brine will become positively buoyant as a result. When the salt content in the brine increases, the density of the brine also increases, resulting in a decrease in the buoyancy of the brine. Therefore, it is paramount that the discharge system of every desalination plant is properly designed (Clark et al. 2018; Kress 2019). Besides, compared with closed-loop installations, open ones are usually simpler to build and have lower construction, operation, and maintenance fees. Open systems also have a lower overall cost. In addition, open outfalls can be deployed in conjunction with local power stations and wastewater treatment plants to mitigate these facilities' negative effects on the environment. This procedure entails adjusting and balancing streams, which ultimately results in achieving temperature and salinity levels that are in perfect balance with one another. Underwater outfalls demand a significant financial investment for their construction, management, and maintenance, as does the development of advanced structures (Kress 2019). This commitment is also required for the production of advanced structures. Submerged outfalls, much like submerged intakes, call for substantial financial resources to be made available.

11.7 Conclusion

The process of desalination is an extremely beneficial technique that can transform the seemingly limitless water resources found in oceans into water that can be consumed. However, it is of the utmost importance to continue implementing the water conservation measures that are now in place. This assertion is especially valid when one takes into consideration the limitations that currently exist in the manufacturing capacity of desalination facilities. Furthermore, finding sources of drinkable water that are more cost-effective than traditional desalination facilities is of the utmost need. The significance of this cannot be overstated, especially in economically disadvantaged nations, where the effects of climate change and water scarcity can have severe repercussions. Water reuse and recycling, particularly in the agricultural sector, can effectively address water needs while simultaneously improving both food and water security. This is especially true in the agricultural sector. The agricultural industry is one that bears this specific significance. When it comes to desalination operations, finding a solution to the problem of brine production is of the utmost importance. Although brine is carefully diluted before being discharged into the ocean, even minute variations from the typical salinity levels can significantly impact marine organisms and the ecosystems in which they live. Previously, there was a widespread idea that the ocean's immensity rendered it immune to severe damage caused by human activity. Now, however, this belief has been disproved. Despite this, the appearance of problems such as ocean acidification demonstrates that this perspective is inaccurate. It has been shown through empirical evidence that even low pollution levels can gradually build up and have far-reaching consequences on a global scale. Because of this, it is very necessary to exercise the utmost caution when engaged in the construction and management of enormous initiatives such as desalination plants, as these activities have the potential to result in serious consequences. In the end, desalination technology provides individuals all over the world with several substantial benefits. In addition to addressing the public health concerns that are brought about by the use of polluted surface water, the need for freshwater will be satisfied, water security will be enhanced, excessive extraction of groundwater will be decreased, and the need for freshwater will be fulfilled. It can settle disagreements between states as well as those that arise between governments regarding the distribution of water resources. As a result, it is of the utmost importance to improve the technology while simultaneously addressing the numerous environmental and health implications that are linked with desalination methods. Implementing brine discharge control measures that are more efficient, in conjunction with enhancing the efficiency of desalination plants, would improve the cost-effectiveness and sustainability of desalination as a viable alternative for meeting the freshwater requirements of the global civilization.

References

Al-Agha MR, Mortaja RS (2005) Desalination in the Gaza strip: drinking water supply and environmental impact. Desalination 173(2):157–171

Al-Hamzah AA, Fellows CM (2015) A comparative study of novel scale inhibitors with commercial scale inhibitors used in seawater desalination. Desalination 359:22–25. https://doi.org/10.1016/j.desal.2014.12.027

Ameen F, Stagner JA, Ting DSK (2018) The carbon footprint and environmental impact assessment of desalination. Int J Environ Sci 75(1):45–58. https://doi.org/10.1080/00207233.2017.1389567

Becker N, Lavee D, Katz D (2010) Desalination and alternative water-shortage mitigation options in Israel: a comparative cost analysis. J Water Resource Prot 2(12):1042

Belkin N, Rahav E, Elifantz H, Kress N, Berman-Frank I (2017) The effect of coagulants and antiscalants discharged with seawater desalination brines on coastal microbial communities: a laboratory and in situ study from the southeastern Mediterranean. Water Res 110:321–331. https://doi.org/10.1016/j.watres.2016.12.013

Chong TH, Loo SL, Fane AG, Krantz WB (2015) Energy-efficient reverse osmosis desalination: effect of retentate recycle and pump and energy recovery device efficiencies. Desalination 366:15–31. https://doi.org/10.1016/j.desal.2015.01.017

Clark GF, Knott NA, Miller BM, Kelaher BP, Coleman MA, Ushiama S, Johnston EL (2018) First large-scale ecological impact study of desalination outfall reveals trade-offs in effects of hypersalinity and hydrodynamics. Water Res 145:757–768. https://doi.org/10.1016/j.watres.2018.08.071

Dawoud MA, Al Mulla MM (2012) Environmental impacts of seawater desalination: Arabian Gulf case study. Int J Environ Sustain 1(3):22–37

Dehwah AH, Missimer TM (2017) Seabed gallery intakes: investigation of the water pretreatment effectiveness of the active layer using a long-term column experiment. Water Res 121:95–108. https://doi.org/10.1016/j.watres.2017.05.014

El-Ghonemy AMK (2012) Future sustainable water desalination technologies for the Saudi Arabia: a review. Renew Sustain Energy Rev 16(9):6566. https://www.iqpc.com/media/7596/2839.pdf

Elsaid K, Sayed ET, Abdelkareem MA, Baroutaji A, Olabi YA (2020) Environmental impact of desalination processes: mitigation and control strategies. Sci Total Environ 740:140125. https://doi.org/10.1016/j.scitotenv.2020.140125

Esfahani IJ, Rashidi J, Ifaei P, Yoo C (2016) Efficient thermal desalination technologies with renewable energy systems: a state-of-the-art review. Korean J Chem Eng 33:351–387. https://link.springer.com/content/pdf/10.1007/s11814-015-0296-3.pdf

Eveloy V, Rodgers P, Qiu L (2015) Hybrid gas turbine–organic Rankine cycle for seawater desalination by reverse osmosis in a hydrocarbon production facility. Energy Convers Manag 106:1134–1148. https://doi.org/10.1016/j.enconman.2015.10.019

Garud RM, Kore SV, Kore VS, Kulkarni GS (2011) A short review on process and applications of reverse osmosis. Univers J Environ Res Technol 1(3)

Hao J, Li L, Zhao W, Wu X, Xiao Y, Zhang H, Wang X (2019) Synthesis and application of CCQDs as a novel type of environmentally friendly scale inhibitor. ACS Appl Mater Interfaces 11(9):L9277–9282. https://pubs.acs.org/doi/pdf/10.1021/acsami.8b19015

Harandi HB, Rahnama M, Javaran EJ, Asadi A (2017) Performance optimization of a multi stage flash desalination unit with thermal vapor compression using genetic algorithm. Appl Therm Eng 123:1106–1119. https://doi.org/10.1016/j.applthermaleng.2017.05.170

Heihsel M, Lenzen M, Malik A, Geschke A (2019) The carbon footprint of desalination: an input-output analysis of seawater reverse osmosis desalination in Australia for 2005–2015. Desalination 454:71–81. https://doi.org/10.1016/j.desal.2018.12.008
Henthorne L, Boysen B (2015) State-of-the-art of reverse osmosis desalination pretreatment. Desalination 356:129–139. https://doi.org/10.1016/j.desal.2014.10.039
Ihsanullah I, Atieh MA, Sajid M, Nazal MK (2021) Desalination and environment: a critical analysis of impacts, mitigation strategies, and greener desalination technologies. Sci Total Environ 780:146585. https://doi.org/10.1016/j.scitotenv.2021.146585
Jiang S, Li Y, Ladewig BP (2017) A review of reverse osmosis membrane fouling and control strategies. Sci Total Environ 595:567–583. https://doi.org/10.1016/j.scitotenv.2017.03.235
Jones E, Qadir M, van Vliet MT, Smakhtin V, Kang SM (2019) The state of desalination and brine production: A global outlook. Sci Total Environ 657:1343–1356. https://doi.org/10.1016/j.scitotenv.2018.12.076
Kämpf J, Clarke B (2013) How robust is the environmental impact assessment process in South Australia? Behind the scenes of the Adelaide seawater desalination project. Mar Policy 38:500–506. https://doi.org/10.1016/j.marpol.2012.08.005 Get rights and content
Khawaji AD, Kutubkhanah IK, Wie JM (2007) A 13.3 MGD seawater RO desalination plant for Yanbu Industrial City. Desalination 203(1–3):176–188
Kim J, Park K, Yang DR, Hong S (2019) A comprehensive review of energy consumption of seawater reverse osmosis desalination plants. Appl Energy 254:113652. https://doi.org/10.1016/j.apenergy.2019.113652
Kress N (2019) Marine impacts of seawater desalination: science, management, and policy. Elsevier, Amsterdam
Lattemann S, Höpner T (2008) Environmental impact and impact assessment of seawater desalination. Desalination 220(1–3):1–15
Li Y, Chen X, Xu Y, Zhuo Y, Lu G (2021) Sustainable thermal-based desalination with low-cost energy resources and low-carbon footprints. Desalination 520:115371. https://doi.org/10.1016/j.desal.2021.115371
Liu TK, Sheu HY, Tseng CN (2013) Environmental impact assessment of seawater desalination plant under the framework of integrated coastal management. Desalination 326:10–18. https://doi.org/10.1016/j.desal.2013.07.003
Ma W, Zhao Y, Wang L (2007) The pretreatment with enhanced coagulation and a UF membrane for seawater desalination with reverse osmosis. Desalination 203(1–3):256–259
Manju S, Sagar N (2017) Renewable energy integrated desalination: a sustainable solution to overcome future fresh-water scarcity in India. Renew Sustain Energy Rev 73:594–609. https://doi.org/10.1016/j.rser.2017.01.164
Mannan M, Alhaj M, Mabrouk AN, Al-Ghamdi SG (2019) Examining the life-cycle environmental impacts of desalination: a case study in the State of Qatar. Desalination 452:238–246. https://doi.org/10.1016/j.desal.2018.11.017
Mezher T, Fath H, Abbas Z, Khaled A (2011) Techno-economic assessment and environmental impacts of desalination technologies. Desalination 266(1–3):263–273. https://doi.org/10.1016/j.desal.2010.08.035
Miller S, Shemer H, Semiat R (2015) Energy and environmental issues in desalination. Desalination 366:2–8. https://doi.org/10.1016/j.desal.2014.11.034
Murphy WR, Blain DA, Galiano-Roth AS, Galvin PA (2002) Benefits of synthetic lubricants in industrial applications. J Synth Lubr 18(4):301–325. https://doi.org/10.1002/jsl.3000180406
Nair M, Kumar D (2013) Water desalination and challenges: the middle east perspective: a review. Desalination and Water Treatment 51(10–12):2030–2040. https://doi.org/10.1080/19443994.2013.734483
Nriagu J, Darroudi F, Shomar B (2016) Health effects of desalinated water: role of electrolyte disturbance in cancer development. Environ Res 150:191–204. https://doi.org/10.1016/j.envres.2016.05.038. Get rights and content

Oh JH, Jang A (2016) Application of chlorine dioxide (ClO2) to reverse osmosis (RO) membrane for seawater desalination. J Taiwan Inst Chem Eng 68:281–288. https://doi.org/10.1016/j.jtice.2016.09.006
Peñate B, García-Rodríguez L (2012) Current trends and future prospects in the design of seawater reverse osmosis desalination technology. Desalination 284:1–8. https://doi.org/10.1016/j.desal.2011.09.010
Petersen KL, Paytan A, Rahav E, Levy O, Silverman J, Barzel O, ... Bar-Zeev E (2018) Impact of brine and antiscalants on reef-building corals in the Gulf of Aqaba–Potential effects from desalination plants. Water Res 144:183–191. https://doi.org/10.1016/j.watres.2018.07.009
Qasim M, Badrelzaman M, Darwish NN, Darwish NA, Hilal N (2019) Reverse osmosis desalination: a state-of-the-art review. Desalination 459:59–104. https://doi.org/10.1016/j.desal.2019.02.008. Get rights and content
Raluy RG, Serra L, Uche J (2005) Life cycle assessment of water production technologies-part 1: life cycle assessment of different commercial desalination technologies (MSF, MED, RO)(9 pp). Int J Life Cycle Assess 10:285–293. https://doi.org/10.1065/lca2004.09.179.1
Sepehr M, Fatemi SMR, Danehkar A, Mashinchian Moradi A (2017) Application of Delphi method in site selection of desalination plants. Glob J Environ Sci Manag 3(1):89–102. https://doi.org/10.22034/gjesm.2017.03.01.009
Shahabi MP, McHugh A, Ho G (2015) Environmental and economic assessment of beach well intake versus open intake for seawater reverse osmosis desalination. Desalination 357:259–266. https://doi.org/10.1016/j.desal.2014.12.003
Shahzad MW, Burhan M, Ng KC (2017) Pushing desalination recovery to the maximum limit: membrane and thermal processes integration. Desalination 416:54–64. https://doi.org/10.1016/j.desal.2017.04.024
Shenvi SS, Isloor AM, Ismail AF (2015) A review on RO membrane technology: developments and challenges. Desalination 368:10–26. https://doi.org/10.1016/j.desal.2014.12.042
Sherif ESM (2013) Electrochemical investigations on the corrosion inhibition of aluminum by 3-amino-1, 2, 4-triazole-5-thiol in naturally aerated stagnant seawater. J Ind Eng Chem 19(6):1884–1889. https://doi.org/10.1016/j.jiec.2013.02.026
Szymoniak L, Claveau-Mallet D, Haddad M, Barbeau B (2022) Application of magnesium oxide media for remineralization and removal of divalent metals in drinking water treatment: a review. Water 14(4):633. https://doi.org/10.3390/w14040633
Tarnacki K, Meneses M, Melin T, Van Medevoort J, Jansen A (2012) Environmental assessment of desalination processes: reverse osmosis and Memstill®. Desalination 296:69–80. https://doi.org/10.1016/j.desal.2012.04.009
Van der Bruggen B, Vandecasteele C (2002) Distillation vs. membrane filtration: overview of process evolutions in seawater desalination. Desalination 143(3):207–218. https://doi.org/10.1016/S0011-9164(02)00259-X
Voutchkov N (2010) Considerations for selection of seawater filtration pretreatment system. Desalination 261(3):354–364. https://doi.org/10.1016/j.desal.2010.07.002
Xu P, Cath TY, Robertson AP, Reinhard M, Leckie JO, Drewes JE (2013) Critical review of desalination concentrate management, treatment and beneficial use. Environ Eng Sci 30(8):502–514. https://doi.org/10.1089/ees.2012.0348
Zhou J, Chang VWC, Fane AG (2013) An improved life cycle impact assessment (LCIA) approach for assessing aquatic eco-toxic impact of brine disposal from seawater desalination plants. Desalination 308:233–241. https://doi.org/10.1016/j.desal.2012.07.039

Recycling Environment and Bioremediation

12

Kouadri Mohamed El Amine, Aliane Samia, Kouachi Amina, Oukid Djalel, Badani Hadjer, Bakhtaoui Zoubeyda, and Youcef Achira Djamal

Abstract

This chapter delves into the pivotal role of recycling and bioremediation in fostering environmental sustainability. It underscores the urgency of addressing waste management challenges and highlights the transformative potential of these practices in achieving a cleaner, healthier environment. Exploring the multifaceted dimensions of recycling, the chapter emphasizes its significance in resource conservation, waste reduction, and environmental cleanliness. Recyclable materials, including plastics, metals, and paper, are examined alongside recycling initiatives' challenges, innovations, and collection methods. Chemical recycling and community-based programs emerge as promising solutions, while the importance of standardized sorting systems and advanced technologies is underscored in overcoming contamination issues. Furthermore, the chapter underscores the effectiveness of bioremediation techniques, such as phytoremediation and composting, in purifying the environment from pollutants. By

K. M. El Amine (✉) · K. Amina
Department of Agronomic Sciences, Laboratory Research on Biological Systems and Geomatics (LRSBG), University Mustapha Stambouli of Mascara, Mascara, Algeria
e-mail: m.kouadri@univ-mascara.dz

A. Samia · O. Djalel · B. Hadjer
Department of Biology, Laboratory of Geo-environment, and Space Development (LGEDE), University Mustapha Stambouli of Mascara, Mascara, Algeria

B. Zoubeyda
Department of Agronomic Sciences, Laboratory of Sustainable Management of Natural Resources in Arid and Semi-Arid Regions, University Centre Salhi Ahmed, Naama, Institute of Science, Naama, Algeria

Y. A. Djamal
Department of Biotechnology, Laboratory of Aquaculture and Bioremediation "AquaBior", University Oran1 Ahmed Ben Bella, Oran, Algeria

A. A. H. Abdel Latef et al. (eds.), *Bioremediation and Nanotechnology for Climate Change Mitigation*, https://doi.org/10.1007/978-981-96-3069-1_12

integrating recycling and bioremediation, the chapter advocates for more sustainable waste management practices, which curtail the spread of contaminants and promote economic viability. Studies showcasing the successful bioremediation and recycling of diverse waste types underscore the imperative of sustainable waste management for environmental well-being. Highlighting bioremediation's cost-effectiveness and environmental friendliness, the chapter explores various techniques like ex-situ and in-situ bioremediation, demonstrating their efficiency in cleansing contaminated sites. It stresses the importance of ongoing research, education, and the integration of physicochemical and biological treatments in overcoming challenges and paving the way for a cleaner, more sustainable future through recycling and bioremediation practices.

In essence, this chapter serves as a comprehensive guide to harnessing the combination of recycling and bioremediation for environmental resilience and sustainable development. It calls for concerted efforts from stakeholders across sectors to embrace these practices and usher in a new era of waste management excellence and environmental stewardship.

Keywords

Waste management · Environmental sustainability · Resource conservation

12.1 Introduction

Today, our world is experiencing significant climate changes, posing an imminent threat to both ecosystems and human well-being as pollution escalates across various fronts. Resounding calls from health and environmental organizations echo the urgency of this crisis, underscoring its tangible and far-reaching consequences. The rapid decline in biological diversity highlights the gravity of the situation, with an estimated one million plants and animal species facing extinction due to habitat destruction and rising global temperatures (IPBES 2019). Since 1850, each decade has surpassed the previous in terms of temperature, exacerbated by the proliferation of plastic waste, particularly in our oceans, where pollution levels have increased (UNEP 2021). With an estimated 150 million metric tons of plastic choking our seas, only 9% of all plastic waste ever produced has been recycled, a figure expected to rise by 70% by 2050 (Jambeck et al. 2015).

Furthermore, the impact of pollution and climate change on human health is concerning. According to the World Health Organization (WHO), air pollution causes approximately seven million premature deaths yearly worldwide (WHO 2021). Rising temperatures and extreme weather events worsen respiratory diseases, cardiovascular problems, and heat-related illnesses, disproportionately affecting vulnerable populations. Furthermore, contaminated water sources and contaminants increase the risk of infectious diseases, creating significant public health challenges (UNEP 2020). This staggering reality reveals the inadequacy of our current waste management systems, urging the pursuit of more effective and sustainable solutions. Recycling and bioremediation are heralded as beacons of hope in combating environmental degradation. These initiatives have yielded success stories worldwide, demonstrating their ability to naturally purify polluted habitats on land and at sea (Tyagi and Kumar 2021).

This chapter delves into a comprehensive examination of environmental sustainability, focusing on the symbiotic relationship between waste management, bioremediation, and the mitigation of pollution. Through an in-depth analysis of recycling processes, bioremediation techniques, and their combined potential, we aim to elucidate practical strategies for addressing the challenges posed by climate change and pollution. By exploring these approaches' benefits, limitations, and future prospects, we seek to inspire proactive initiatives and policy interventions toward a more resilient and ecologically balanced future.

12.2 Recycling in Waste Management

Recycling is a highly efficient method of managing waste materials (Leverenz et al. 2002). This transformative process aims to create new products from recycled materials and to prevent the loss of potentially usable resources, reducing energy usage and lowering resource consumption (Bom et al. 2017; EPA 2014). Different types of material are included in this broad approach, for example, packaging waste, nuclear waste, mixed waste, and single material recycling. Recycling will play an essential role in ensuring the country's energy security, creating employment, and stimulating the growth of its recycling industry. In contrast, it is also a contributing factor to energy conservation. Various factors, including attitude, perceived behavioral control, low behavioral costs, and adherence to social standards, impact an individual's behavior toward recycling. Recycling is critical in managing waste, highlighting the circular economy and sustainability strategies for dealing with it. Increasing resource efficiency, minimizing the impact on the environment, and delivering economic benefits will play an important role, as well as fostering good practices in waste management. Recycling processes can be carried out for various types of waste, bringing a wide range of environmental benefits.

12.2.1 Historical Overview

Recycling has a rich history, dating back to ancient times when people used to reuse materials such as metal, glass, and textiles. However, the 1970s were marked by significant progress in modern recycling, coinciding with an environmental movement and a growing awareness about the finite nature of natural resources. In the early 1970s, the first recycling program curbside in the United States was introduced, and its first recycling plant opened in 1972. This is critical in integrating recycled material into solid waste management practices worldwide. Recycling has adapted to address the environmental issues caused by the growing waste generation and resource depletion. The advancement of recycling technology has strengthened the enforcement and execution of recycling policies and regulations in numerous countries. Recycling is widely recognized today as the most critical strategy to reduce waste, conserve resources, and mitigate environmental impacts (Bom et al. 2017). The continued commitment to environmental protection is reflected in the development of recycling practices. Recycling is not only an essential element of waste management, but it also adapts to current challenges. With a view to broader aspects of waste management, plastic packaging recycling and the facilitation of precise information as well as encouragement toward sorting and recycling, this focus has expanded beyond its historical roots (Lee et al. 2018; Reijonen et al. 2021; Roithner and Rechberger 2020).

12.2.2 Waste Types

Multiple categories of waste can be recycled, each requiring specific procedures and presenting unique difficulties. Plastic packaging waste, including bottles, bags, and wrappers, can be transformed into new plastic products, although the current recycling rate for plastic packaging is only 9% (Reijonen et al. 2021). Nuclear waste, a distinct classification, goes through reprocessing to transform used nuclear fuel into fresh fuel and recover unused plutonium and uranium. This process aims to optimize the energy extraction from the initial uranium source (Furnari 2021). Mixed waste, which consists of a combination of items such as paper, glass, and organic waste, undergoes sorting procedures to facilitate efficient recycling, thereby offering a systematic approach to waste management. Single-material recycling refers to the targeted recycling of certain materials, such as aluminum, steel, and glass. These materials are collected separately and then converted into new goods. Composting food waste transforms it into a valuable substance that enhances soil quality. However, at present, only a mere 3% of food waste is subjected to this beneficial process. Organic materials such as leaves and grass clippings, which comprise green and carpet waste, can be effectively recycled by composting. This process transforms them into soil enhancers that can be used for various other purposes. Ultimately, residual waste, including non-recyclable plastics, glass, and metals, must be disposed of via landfills or other waste management techniques (Lawrence 2016). To enhance recycling rates, it is crucial to focus on managing household waste, recycling plastic packaging, and providing clear guidance and assistance for waste separation and recycling (Reijonen et al. 2021).

12.2.3 Classification and Origin of Wastes

Waste classification is the first and most important step in recycling, recovering, reducing, and disposing of waste. Existing classifications are based especially on their origin, composition, and danger or other properties, depending on the local or regional legislation. Nevertheless, regardless of the conventional classifications cited above, wastes can be classified and, therefore, collected according to the treatment or management process to be undergone. So, here we should make the difference between:

12.2.3.1 Recyclable Waste
Waste is considered recyclable when it is possible to recover and reuse it or make a new/reusable object. Among these, solid waste and wastewater are predominating.

12.2.3.2 Biodegradable
Biodegradable waste includes organic materials that microorganisms break down into simpler substances, which can then be safely reintegrated into the natural environment. These materials are often treated through bioremediation processes.

12.2.3.3 Compostable
Compostable materials are made up of organic elements that can break down into natural elements in a particular environment. Since the item breaks down into natural elements, it does not harm the environment and will turn into the soil shortly.

The waste production rate generally depends on the living style, geographical conditions, population, government policies, social stability, and climatic conditions of an area (Hoornweg and Bhada-Tata 2012).

12.2.4 Benefits of Recycling

12.2.4.1 Environmental Impact
Recycling has a significant ecological impact, providing several advantages that support sustainability. First and foremost, it decreases greenhouse gas emissions by reducing the necessity for collecting and processing fresh raw materials, hence reducing the energy consumption linked to these procedures. Significantly, recycling aluminum can conserve 95% of the energy needed for fresh manufacturing, while recycling steel can conserve 65%. Moreover, recycling is essential for conserving natural resources as it enables the reuse of materials and decreases the need to extract and process new raw materials. By doing so, we guarantee the long-term viability of our planet's resources and minimize the amount of waste sent to landfills, preserving land and mitigating the release of methane—a highly potent greenhouse gas (Arora and Mishra 2020). The comprehensive reduction in energy usage during recycling procedures and enhanced food safety achieved through methods such as anaerobic digestion and aerobic composting of restaurant food waste highlight the diverse environmental advantages (Zhang et al. 2019). In addition, recycling enhances resource retrieval, enabling the reuse of valuable materials such as metals and organic substances as inputs for producing new products (Chiew et al. 2015).

12.2.4.2 Economic Advantages
Recycling provides significant economic benefits, including employment opportunities, decreased expenses for waste management, and the realization of profits. A

significant advantage is the capacity for generating employment opportunities in the recycling sector, encompassing tasks such as collecting, sorting, processing, and manufacturing. An examination of textile recycling, for example, showcased the employment prospects created by gathering, categorizing, and transforming textiles and producing fresh textile products (Leal Filho et al. 2019). Moreover, recycling helps decrease waste management expenses by reducing waste disposed of in landfills and reducing the need for new raw materials. Within the domain of rubber recycling, research has emphasized the potential for cost reduction through the decrease in waste transported to landfills and the diminished need for new raw materials (Oliveira Neto et al. 2016). Furthermore, engaging in recycling can be a lucrative endeavor for companies. A study investigating the use of reverse logistics for recycling building and demolition waste reveals that businesses can generate financial benefits by recycling solid waste and selling the resulting recycled products, such as iron, wood, paper, and plastics (Oliveira Neto and Correia 2019). Finally, promoting recycling and embracing sustainable waste management techniques may contribute to developing a more environmentally friendly future and enhance economic prosperity.

12.2.4.3 Social Benefits

Recycling offers various social advantages, such as fostering community involvement and enhancing the overall quality of life. Citizen participation in recycling projects can promote environmental responsibility and collective action, contributing to the establishment of sustainable and cohesive societies. Moreover, recycling has the potential to enhance public health by diminishing pollution and the necessity for burning, which can emit hazardous poisons into the atmosphere. Furthermore, recycling might present a favorable circumstance for establishing social enterprises and the generation of employment, specifically within the recycling industry. Recycling is fundamental in fostering robust and dynamic cultures prioritizing health and vitality (Bom et al. 2017; Leal Filho et al. 2019; Rose et al. 2019).

12.2.5 Recyclable Materials

Recyclable materials, such as paper, plastics, glass, and metals, vary in their recycling procedures and environmental impacts. Paper recycling entails collecting, sorting, and processing used paper through pulping and de-inking, yielding new paper products. This technique aids in the preservation of natural resources, the reduction of energy use, and the mitigation of greenhouse gas emissions (Demenko 1976). Plastic recycling involves collecting, sorting, cleaning, shredding, and melting plastic materials to create plastic pellets. These pellets are then used in the creation of new products, which helps to conserve resources and minimize environmental damage (Inagaki et al. 2019). Glass recycling encompasses the activities of collecting, sorting, and processing used glass containers, so promoting the preservation of resources and reducing the ecological consequences of glass waste, which may require thousands of years to break down (Bisinella et al. 2017). Metal recycling involves collecting, grouping, and processing spent metals and producing fresh metal products, contributing to resource preservation and the mitigation of environmental damage (Oliveira Neto et al. 2016).

12.2.6 Recycling Processes

Recycling involves a series of steps that vary depending on the recycled material. However, the general process involves collection, sorting, processing, and manufacturing.

12.2.6.1 Collection

The initial stage of the recycling process involves collecting materials that can be recycled. These methods include curbside collection, drop-off collection, or deposit systems. Curbside collection is a standard method of waste collection in which recyclable materials are collected from households and businesses at the curbside. The materials are usually placed in a designated recycling bin or bag and collected by a waste management company. Curbside collection is convenient for households and businesses, eliminating the need to transport materials to a recycling center. However, it can be expensive for waste management companies, as specialized equipment and personnel are required to collect and transport the materials (Liu et al. 2022). Drop-off collection involves collecting recyclable materials at designated drop-off locations, such as recycling centers or community collection points. This method is often used for materials that are difficult to collect through curbside collection, such as electronics or hazardous waste. The drop-off collection is cost-effective for waste management companies, eliminating the need for specialized equipment and personnel. However, it can be inconvenient for households and businesses, requiring them to transport materials to the drop-off location (Lamsali and Ariffin 2013; O'Dwyer et al. 2022). Deposit systems, also known as container deposit schemes (CDS), involve the payment of a deposit on certain types of containers, such as beverage containers. The deposit is reimbursed upon the container's return to a specified collecting location. Deposit systems are designed to encourage the return of containers for recycling and reduce litter. They are often used in countries with high recycling rates, such as Germany and Sweden. Deposit systems can effectively increase recycling rates, but they require a significant investment in infrastructure and administration (Berck et al. 2020; O'Dwyer et al. 2022). Innovations in collecting methods encompass the application of intelligent technology to enhance the effectiveness of curbside collection, the establishment of novel drop-off sites, and the introduction of deposit systems in previously unexplored areas. Additionally, there is a growing trend toward the use of community-based recycling programs, which involve the active participation of local communities in waste management and recycling initiatives (Liu et al. 2022; O'Dwyer et al. 2022).

12.2.6.2 Sorting, Processing, and Manufacturing

The collected materials are sorted by type and grade at the sorting facility. This involves separating different materials, such as paper, plastics, glass, and metals, into different categories. The sorting process can be done manually or using advanced technologies like optical sorters and magnetic separators. Once the materials have been sorted, they undergo processing to make them ready for recycling. This process entails the purification, fragmentation, and fusion of the ingredients to generate unprocessed substances that can be utilized to produce novel goods. The processing technique differs according to the specific material being recycled. As an illustration, paper undergoes the process of pulping and de-inking, whereas polymers are subjected to cleaning, shredding, and melting. The last stage in the recycling process involves the production of fresh goods utilizing recycled materials. Recycled materials are primary resources for producing new items, including paper, plastic containers, glass bottles, and metal products. Efficient recycling processes heavily rely on adequate sorting facilities and innovative technologies. These facilities and technologies effectively segregate and process diverse resources, guaranteeing their recycling into new products without impurities. Utilizing adequate sorting infrastructure and cutting-edge technology can enhance recycling rates, save precious resources, diminish greenhouse gas emissions, generate employment opportunities, and yield cost savings and public health advantages. Recycling procedures encompass stages that differ according to the recycled material. Efficient recycling processes rely heavily on adequate sorting facilities and innovative technology, which is essential for promoting a more sustainable and circular economy (Park et al. 2022; Roithner and Rechberger 2020).

12.2.6.3 Case Studies on How Recycled Materials Are Used in Manufacturing

A review of the socio-economic advantages of textile recycling found that recycled textiles can produce new clothing, insulation, and carpet padding. Using recycled textiles can reduce the need for new raw materials and contribute to a more sustainable and circular economy (Leal Filho et al. 2019). A study on the environmental and economic advantages of adopting reverse logistics for recycling construction and demolition waste found that recycling can generate profits for companies that recycle solid waste and sell recycled materials such as iron, wood, paper, and plastics. Additionally, recycling can reduce the environmental impact of waste management and create new opportunities for companies (Oliveira Neto and Correia 2019). An overview of the paper recycling process in Iran found that recycled paper can produce new paper products, such as cardboard and tissue paper. Using recycled paper can help conserve natural resources and reduce waste (Abdollahbeigi 2021). In general, recycled materials can be used to manufacture many items, such as garments, thermal insulation, carpet underlay, and paper products.

12.2.7 Public Awareness and Participation

12.2.7.1 Exploration of Educational Programs Promoting Recycling Awareness

Education initiatives that promote knowledge of recycling can have a pivotal impact on promoting sustainable strategies for managing waste.

Training program for primary school students: An investigation conducted in Iran demonstrated the efficacy of implementing the health promoting schools model as an intervention to enhance the recycling habits of elementary school students. The intervention encompassed didactic workshops, promotional initiatives, and interactive forums targeting parents and educators. The program enhanced the pupils' awareness, attitude, performance, and engagement in source-separated recycling (Taghdisi et al. 2022).

Sustainable integration of solar energy, behavior change, and recycling practices in educational institutions: A research study introduced a complete framework that addresses sustainable energy conservation, behavior modification, and recycling activities, specifically in educational institutions. The framework incorporates solar photovoltaic systems, fostering student engagement in their upkeep, recycling harvested water for plant irrigation, and utilizing organic waste as a natural fertilizer. The method sought to foster a cohort of environmentally conscious persons actively engaging in environmental stewardship (Altassan 2023).

Public perceptions and practices of recycling: The research in Wyoming utilized a mail-back survey to investigate the public engagement landscape and the elements influencing citizens' recycling practices and attitudes. The study revealed that over 80% of the participants in the survey expressed that their primary reason for recycling was their concern for the environment. Furthermore, these individuals strongly emphasized the significance of recycling and reported high satisfaction levels with their recycling efforts. The report proposed that Laramie should implement a proactive educational strategy, incentive policies, and a master plan in order to achieve a waste diversion rate of 40% by 2030. This may be accomplished by fostering greater public engagement in planning and enhancing community outreach initiatives (Bom et al. 2017).

12.2.7.2 Examples of Successful Community Engagement in Recycling

Manchester recycling for all: This study examined strategies to promote recycling engagement across economically disadvantaged regions of Manchester, England. The study showcased that recycling rates can be enhanced in "challenging communities" by allowing inhabitants to make informed decisions regarding alternatives. Implementing "back-alley bring sites" (BABS) proved a highly effective alternative to curbside collections, resulting in over double the typical recyclable items. The strategy incentivized non-recycling families in these localities to initiate recycling practices (Williams and Culleton 2009).

Public perceptions and practices of recycling in the City of Laramie, Wyoming: A study conducted in Laramie, Wyoming, revealed that over 80% of the participants in the survey expressed that their primary reason for recycling was their concern for the environment. Additionally, these individuals demonstrated a strong sense of the significance of recycling and reported high satisfaction levels with their recycling efforts. The study proposed that Laramie implement a proactive educational policy, incentive policies, and a Master Plan to achieve a 40% waste diversion rate by 2030. This can be accomplished by fostering greater public engagement in the planning process and enhancing community outreach programs (Bom et al. 2017).

Design of e-waste recycling programs considering participant motivations: This study in the State of Mexico aimed to ascertain the characteristics and incentives of persons who participated in an innovative e-waste recycling event. The data collected from a survey of participants indicate that their educational attainment, overall environmental concern, and involvement in pro-environmental actions are more significant than non-participants. The study found that the impact of social connections, a solid dedication to the community, and the proximity of permanent electronic waste collection sites to one's residence all substantially influenced recycling habits (Arroyo-López 2012).

These examples demonstrate the potential of community-based initiatives to address environmental challenges and promote sustainable waste management practices. By fostering local engagement, promoting autonomy, and supporting youth involvement, these initiatives can help create more sustainable communities and contribute to a more environmentally friendly future.

12.2.8 Government Policies and Regulations

Policies significantly affect recycling and waste management practices across many levels of governance, influencing the path toward more sustainable and circular economies. The impact of these policies may vary from one country to another. Several African nations need precise national rules and strict measures for adequate e-waste collection and recycling, necessitating comprehensive interventions to address this deficiency correctly (Bimir 2020). Finland's national waste management rules are crucial in guiding the efforts to recycle plastic packaging. These regulations prioritize costs and conducive conditions to promote active household participation in recycling projects (Reijonen et al. 2021). As seen in California, enacting deposit-return recycling schemes demonstrates a pragmatic strategy for improving overall recycling rates, specifically by modifying refund amounts to encourage increased participation (Berck et al. 2020). Moreover, implementing regulations that promote a circular economy, such as incorporating carbon expenses into material prices and ensuring product standards align with recyclability, is essential for progressing toward climate neutrality and encouraging the adoption of effective recycling methods (Sun et al. 2021). These examples highlight legislative frameworks' varied and crucial functions in creating sustainable waste management practices, making a vital contribution to achieving more circular and environmentally aware economies.

12.2.8.1 The Role of Legislation in Shaping Recycling Practices

Legislation is crucial in influencing recycling practices at all levels of government. At a local level, a study conducted in Manchester, England, at the municipal level showcased the efficacy of inventive strategies, such as "back-alley bring sites" (BABS), in promoting recycling engagement, especially in economically disadvantaged regions. This community-driven endeavor has shown remarkable efficacy in surpassing curbside pick-ups, leading to a significant upsurge in retrieving recyclable materials and promoting recycling habits among previously non-compliant households (Williams and Culleton 2009). On the national stage, a study conducted in Finland investigated the determinants that impact plastic packaging recycling within the country's waste management system. The findings emphasized the substantial influence of costs and facilitating conditions on home recycling behavior, indicating the necessity for policymakers to reassess the dissemination of information in order to improve plastic package recycling (Reijonen et al. 2021). Although no specific instances of international legislation have been found, it is worth noting that international agreements, such as the Basel Convention on hazardous waste movements, specifically tackle the issue of electronic waste recycling. This illustrates the broader influence of global policies in advocating for optimal recycling and waste management methods. Legislation at many levels of governance, including municipal, national, and international, has significant influence over recycling activities. It directly affects behavior and establishes a structure for sustainable waste management methods.

12.2.9 Technological Innovations

Emerging technologies such as AI and robotics are increasingly leveraged in waste management to improve efficiency, accuracy, and sustainability.

12.2.9.1 AI in Waste Management

Waste Sorting: AI-powered robots and machines can sort and separate different types of waste, including recyclables, organic waste, and non-recyclables. AI algorithms can identify and sort materials based on their characteristics, improving the accuracy of the sorting process (Bimir 2020).

Predictive Analytics: AI can analyze waste generation patterns and predict future waste generation, allowing for better planning and resource allocation in waste management (Rashidul et al. 2019).

Optimization: AI algorithms can optimize waste collection routes, leading to fuel savings, reduced emissions, and improved efficiency in waste collection and transportation (Hashemi 2015).

12.2.9.2 Robotics in Waste Management

Automated Sorting: Robotic arms and machines can be used to automate the sorting and separation of waste materials in recycling facilities, improving the speed and accuracy of the process (Lee et al. 2018).

Waste Collection: Robotic systems can be used for automated waste collection in urban environments, reducing the need for manual labor and improving the efficiency of waste collection (Oliveira Neto and Correia 2019).

Maintenance and Inspection: Robots can be used for maintenance and inspection tasks in waste management facilities, improving safety and reducing the risk of accidents for human workers.

12.2.9.3 Impacts on Efficiency and Sustainability

Emerging technologies, such as AI and robotics, have become essential in revolutionizing waste management processes, leading to substantial improvements in efficiency and sustainability. AI and robotics optimize waste management processes by automating critical functions such as sorting, collection, and transportation. This leads to faster and more precise operations, reducing costs and improving overall efficiency (Leal Filho et al. 2019; Roithner and Rechberger 2020). Furthermore, these technologies enhance the sustainability of waste management by optimizing resource retrieval and reducing waste generation. Artificial intelligence algorithms, such as those used to optimize waste collection routes, lead to fuel savings, decreased emissions, and improved waste collection and transportation efficiency. Robotics enhances the automation of waste material sorting and separation in recycling plants, increasing efficiency and accuracy while reducing the risk of contamination (Roithner and Rechberger 2020; Sun et al. 2021). Furthermore, the environmental consequences of waste management methods are alleviated by advocating for the preservation of resources and minimizing the release of greenhouse gases. Artificial intelligence algorithms can predict future waste production, allowing for improved waste management organization and distribution of resources. Robotics decreases the need for physical labor by automating waste collecting and sorting, improving human workers' safety (Arora and Mishra 2020; Zhang et al. 2019). AI and robotics are being used increasingly to improve waste management by making it more effective, precise, and sustainable. This leads to a new era of innovative and efficient waste management techniques.

12.2.10 Global Perspectives

12.2.10.1 Comparative Analysis of Recycling Practices in Different Regions

The recycling practices exhibit substantial variation among regions, as demonstrated by a comparative analysis of multiple research projects. A study conducted in Middle Eastern countries, where natural resources are abundant, emphasized the need to urgently tackle the issue of rapid waste generation and depletion of resources by focusing on the reuse and recycling of materials in architecture. The study exhibited case studies demonstrating novel methods of reutilizing rejected materials, highlighting the significance of sustainable practices in the area (Hashemi 2015). Meanwhile, in the European Union, the circular economy package is designed to lead the way in achieving recycling objectives. A study highlighted the importance of integrating quantitative and qualitative evaluations to mitigate quality degradation in recycling operations. It also introduced a novel indicator to provide more comprehensive information on the quantity and purity of recycling outputs (Roithner and Rechberger 2020). In Finland, a study examined the elements that affect plastic packaging recycling and found that costs and conducive conditions significantly impact this process. Additionally, the study emphasized the importance of reevaluating how information is provided to households to assist recycling efforts better (Reijonen et al. 2021). In Manchester, England, initiatives using "back-alley bring sites" successfully improved recycling rates in challenging communities (Williams and Culleton 2009). Conversely, Brazil demonstrated reverse logistics's economic and environmental benefits in recycling buildings and demolition debris, emphasizing increased profitability and minimized environmental harm (Oliveira Neto and Correia 2019). These diverse studies offer insights into the creative initiatives, policy implementations, and economic and environmental benefits driving recycling practices globally, emphasizing the importance of a holistic approach to achieving sustainable and effective waste management.

12.2.10.2 International Collaborations and Agreements for Global Waste Management

Global waste management is effectively tackled through the crucial involvement of international alliances and agreements. Analyzing e-waste management techniques in many African nations highlights the difficulties encountered, specifically the need for explicit national policies. The study promotes integrating interventions involving many actors and coordination with various stakeholders to decrease the amount of e-waste and improve the capacity for safe recycling. It highlights the importance of a joint endeavor (Bimir 2020). The fourth International Forum on Sustainable Future in Asia emphasizes the cooperative endeavors of governments, the corporate sector, individuals, and non-governmental organizations in fulfilling worldwide obligations concerning waste management, climate change, biodiversity, and environmental well-being. The forum aims to achieve sustainable societies by promoting research collaboration and bridging the gap between science and policy, strongly emphasizing collaborative actions (fourth NIES Inter. Forum 2019). The study on coordinated measures for high-quality recycling emphasizes the significance of global policy actions in pursuing climate neutrality. This includes integrating carbon pricing into material prices, harmonizing product standards with recyclability, enhancing consumer consciousness, and promoting investment in sorting and recycling infrastructure. The study highlights the importance of setting explicit objectives and delineating roles for successful implementation worldwide (Sun et al. 2021). These sources emphasize the importance of global cooperation and synchronized policy actions in efficiently tackling the complex difficulties of managing waste worldwide. Coordinated interventions, joint research endeavors, and unified policies are emphasized in achieving sustainable waste management practices worldwide.

12.2.11 Future Trends

The chapter explores future recycling and waste management trends, shedding light on emerging practices and approaches to tackle environmental challenges and advance sustainability. It underscores the global shift toward sustainable practices, stressing the importance of aligning with Sustainable Development Goals (SDGs) and addressing unsustainable consumption patterns (Rodiger-Vorwerk 2018). This growing momentum toward sustainability highlights the urgent need for innovative strategies to manage waste effectively.

Efforts are underway to bolster measures for adapting to and mitigating climate change, particularly in light of its impact on natural disasters and threats posed to cultural heritage sites (Bosher et al. 2020). This indicates a pivot toward resilience-building and climate adaptation strategies within waste management practices to mitigate environmental risks.

Recognizing the significance of e-waste recycling initiatives, there is a concerted effort to reduce e-waste accumulation and promote sustainability (Arroyo-López 2012). This underscores the necessity for targeted interventions and community engagement to elevate recycling rates and instill pro-environmental behaviors.

Emphasis is placed on understanding household recycling behavior to inform the design of effective recycling programs and policies (Yusop and Othman 2019). This signals a shift toward behaviorally informed waste management strategies to foster sustainable practices at the household level.

Innovative approaches, such as involving children in the design of recycling facilities, are regarded as promising avenues for instilling environmental consciousness and sustainable behaviors from an early age (Siu et al. 2020). This signifies a move toward participatory and inclusive waste management initiatives to embed sustainability values in future generations.

The journey toward climate neutrality necessitates a comprehensive circular transformation, where waste reduction, reuse, and recycling are integral components (Sun et al. 2021). This underscores the importance of coordinated policy actions and investments in recycling infrastructure to drive systemic change toward a more sustainable future.

In pursuit of more effective waste management and recycling practices, various solutions and innovative approaches are being explored to address existing challenges and drive progress toward sustainability. Notably, the development of standardized sorting systems holds promise in reducing contamination levels and improving recycling rates (Khan and Charles 2023). Additionally, advanced sorting technologies, such as optical sorters and magnetic separators, offer precise and efficient sorting capabilities (Lee et al. 2018).

Education and awareness campaigns are pivotal in promoting proper sorting practices among the general public, aiming to reduce contamination resulting from human error and lack of awareness (Federigi et al. 2020; Zhang et al. 2019). Moreover, effective sorting facilities are crucial for ensuring materials are properly separated and processed, minimizing contamination levels and environmental impact (Khan and Charles 2023).

Community-based initiatives play a vital role in sustainable waste management, encompassing education programs, infrastructure development, partnerships, and technological innovation integration (Flaspohler et al. 2003; Carey et al. 2005; Grotherr et al. 2020). These initiatives foster collaboration and community involvement, contributing to a cohesive and sustainable waste management ecosystem.

Looking ahead, emerging technologies such as chemical recycling offer promising solutions to address challenges associated with sorting plastics. Chemical recycling breaks down plastics into chemical constituents, offering a viable solution for producing new products from recycled materials (Massardier et al. 2023). Innovations in textile recycling techniques, including chemical and mechanical processes, enable the incorporation of recycled textiles into various products, supporting sustainable fashion initiatives and the circular economy (Leal Filho et al. 2019).

Furthermore, technological advancements like devulcanization are revolutionizing rubber recycling, expanding the utilization of recycled rubber in diverse applications (Oliveira Neto et al. 2016). These innovations drive progress toward a more sustainable future by reducing waste, promoting resource efficiency, and mitigating environmental impact.

In summary, these future trends underscore the evolving landscape of recycling and waste management, highlighting the importance of sustainability, innovation, and community engagement in shaping a resilient and environmentally conscious future.

12.3 Bioremediation: Nature's Cleanup Crew

The emergence of environmental pollutants poses a significant threat to both human health and natural ecosystems (Bala et al. 2022b). Telluric microorganisms play pivotal roles in promoting plant growth, recycling nutrients, controlling insects, maintaining soil fertility, and reducing pollutants. Bioremediation stands out as an ecologically sustainable and effective strategy for treating and degrading various pollutants (Radhakrishnan et al. 2023). Defined as using living organisms, primarily microorganisms, bioremediation aims to transform environmental contaminants into less toxic forms (Vidal 2001). Positioned within environmental biotechnology, bioremediation emerges as the most promising alternative for waste degradation through biological assistance. It showcases notable flexibility in harnessing the abilities of microorganisms and plants to accumulate, detoxify, degrade, or remove environmental contaminants (Abatenh et al. 2017; Hibor et al. 2017). Throughout bioremediation, harmful substances undergo degradation or detoxification, providing organisms with the necessary nutrients for their functions. Enzymes, crucial actors in each degradation stage, belong to the family of oxidoreductases, lyases, transferases, and hydrolases (Bala et al. 2022b). As bioremediators, we can utilize indigenous microorganisms from a contaminated area or microorganisms isolated elsewhere and introduced to the polluted site. To ensure efficient bioremediation, these microorganisms must possess enzymes capable of attacking pollutants and converting them into harmless forms. Environmental conditions must also favor microbial growth and activities for optimal results (Vidal 2001).

12.3.1 Bioremediation Types

Bioremediation, a versatile approach, finds application through various methods, each offering distinct effectiveness. In the next section, we delve into the details of some of the most common and efficient strategies employed in bioremediation.

12.3.1.1 Ex-Situ Bioremediation

Ex-situ bioremediation involves transporting pollutants from the contaminated site to another location for treatment. This approach requires thorough consideration of factors such as the cost of treatment, depth of pollution, type, and degree of pollutants, geographical location, and geological characteristics of the contaminated site (Azubuike et al. 2016).

Biopile

Biopiling stands out as a highly utilized technology for remediating a diverse range of petrochemical pollutants in soils and sediments. Referred to interchangeably as bioheaps, biocells, or biomounds, this method involves heaping contaminated soil or dried sediments, stimulating aerobic microorganisms biodegrading activity by creating optimal or near-optimal growth conditions within the pile (Germaine et al. 2015). Aeration, whether forced or passive, plays a crucial role, with air pumps ensuring effective air distribution but requiring power. In colder climates, integrating a heating system into a biopile extends the treatment season by increasing temperatures, accelerating microbial activities, enhancing contaminant availability, and ultimately expediting biodegradation (Scanscarter et al. 2009; Azubuike et al. 2016). The efficiency of biopiling is further enhanced by incorporating bulking agents such as straw, sawdust, or wood chips. Extreme air temperatures leading to soil dryness can slow down bioremediation, as overly dry soils are more likely to vaporize than be broken down by living organisms. Notably, the success of the biopile method heavily relies on bioavailable organic carbon (BOC). In mesophilic conditions (30 °C to 4 °C), soils contaminated by petroleum have been effectively treated with low aeration, utilizing *alpha*, *beta*, and *gamma proteobacteria* to remove total petroleum hydrocarbon (TPH) (Bala et al. 2022b). This innovative technology is also deployed in sub-Antarctic regions to treat diesel-contaminated soil, achieving an impressive 93% removal of total petroleum hydrocarbon (TPH) within a year. Biopiling efficiently treats large volumes of polluted soils and conserves space, making it a space-saving alternative to other ex-situ technologies like land farming. Additionally, it boasts advantages such as robust engineering, cost-effectiveness in maintenance and operation, and suitability for remote sites lacking power supply (Azubuike et al. 2016).

Windrows

Windrows rely on the periodic rotation of stacked polluted soils to stimulate microbial degradation of pollutants (Alori et al. 2022). Aeration, leachate management, and biotransformation of toxic soil are achieved through acclimation, biological treatment, and mineralization, enhancing the overall efficiency of bioremediation. Notably, the biopile method surpasses the windrow technique in removing hydrocarbons from soil. However, the windrow method's periodic turning makes it less suitable for treating soils contaminated with volatile pollutants. Intriguingly, the windrow technique can contribute to the emission of greenhouse gas (CH_4) due to the anaerobic conditions generated within the piled polluted soil (Bala et al. 2022b).

Land Farming

Landfarming, a straightforward method, involves excavating and spreading prepared beds with periodic tilling until pollutants are degraded. The primary objective is to stimulate indigenous biodegradative microbial populations, facilitating aerobic degradation of pollutants. However, a notable limitation is the soil's superficial treatment depth (10–20 cm). Despite this, landfarming offers several advantages, including reduced monitoring and maintenance costs, positioning it as a disposal alternative for cleanup liabilities (Vidal 2001). Widely employed to remediate crude oil-contaminated soil, landfarming effectively reduces oil concentrations through bacterially mediated biodegradation. While volatilization, abiotic processes, and fungal-mediated processes may also play a role, the method has demonstrated significant success in removing petroleum hydrocarbons on a large scale (Brown et al. 2017). Textile recycling encounters obstacles such as insufficient infrastructure, challenges separating fibers, and a poor value for recycled textiles.

Bioreactor

Bioreactor-based treatment emerges as a compelling approach, offering an optimally controlled environment for the biodegradation of hydrocarbon-polluted sites. This method eliminates rate-limiting factors such as oxygen supply, optimal pH and temperature, and specific nutrient requirements. Bioreactors, essentially tanks where living organisms engage in biological reactions, come in various operating modes including batch, fed-batch, sequencing batch, continuous, and multistage (Chikere et al. 2012). The parameters of bioreactors are tailored to meet the optimal growth needs of microbial populations. Contaminated samples whether dry matter or slurry can be introduced into a bioreactor. The bioreactor technique offers numerous advantages, providing excellent control over bioprocess parameters such as temperature, pH, agitation and aeration rates, and substrate and inoculum concentrations. This technology is versatile, finding application in treating soil or water contaminated with volatile organic compounds (VOCs) like benzene, toluene, ethylbenzene, and xylene (BTEX). Bioreactor methods have demonstrated high removal potential across a broad range of pollutants, including total petroleum and polyaromatic hydrocarbons, sulfonated amines, total nitrogen, naphthalene, nano fullerenes and nanosilver, dibromoneopentyl glycerol, carbofuran, and 2,4,6-trinitr ophenylmethylnitramine (Azubuike et al. 2016).

12.3.1.2 In-Situ Bioremediation

In-situ bioremediation encompasses various techniques aimed at treating contaminants directly at the polluted site, where they undergo degradation under natural conditions. Distinguished by their non-invasive nature, these techniques eliminate the need for excavation, demanding fewer transportation efforts and physical displacement than ex-situ bioremediation methods (Paul et al. 2021).

Bioventing

Bioventing is a technology that entails the controlled stimulation of airflow through the distribution of oxygen to the unsaturated zone, intensifying bioremediation by boosting the activities of the indigenous microbial community. This process enhances bioremediation by providing nutrients and moisture to facilitate the microbial transformation of pollutants into less harmful forms (Azubuike et al. 2016). Two pivotal criteria for the success of bioventing are the maintenance of aerobic conditions and the attainment of reasonable biodegradation rates (Paul et al. 2021).

Biosparging

The biosparging method closely resembles bioventing, involving air injection into the soil subsurface to enhance the biodegradation rate of pollutants by indigenous bacteria (Paul et al. 2021). In contrast to bioventing, air injection in biosparging occurs in the saturated zone, causing volatile organic compounds to migrate upward to the unsaturated zone, thereby improving biodegradation (Azubuike et al. 2016). The effectiveness of biosparging hinges on two critical factors: the biodegradability of pollutants and soil porosity. Under favorable conditions, bacteria produce metabolites, including metal-absorbing materials that interact with pollutants, leading to their precipitation. Introducing oxygen creates aerobic conditions conducive to the degradative action of indigenous microbes (Paul et al. 2021). Studies have reported the successful removal of diesel and kerosene from water supplies through biosparging. The removal of organic pollutants like BTEX can be achieved using various technologies, including adsorption, microbial degradation, biosparging, PRBs (permeable reactive barriers), and the utilization of modified or synthesized zeolites (Bala et al. 2022b). This effectiveness is evidenced by increased levels of dissolved oxygen, redox potential, nitrate, sulfate, and total culturable heterotrophs, coupled with a decrease in dissolved ferrous ions, sulfide, methane, and total aerobes and methanogens (Azubuike et al. 2016).

Bioslurping

Bioslurping, a relatively new subsurface remediation method, combines elements from three distinct techniques: vacuum-enhanced pumping, soil vapor extraction, and bioventing (Gidarakos and Aivaliori 2007). This method is employed for the recovery of free products, particularly light nonaqueous phase liquids (LNAPLs), enabling the remediation of capillary, unsaturated, and saturated zones. Bioslurping is also effective for remediating volatile and semi-volatile organic compounds. The design employs a "slurp" that extends into the free product layer, akin to a straw drawing liquid from a vessel, extracting liquids (free products and soil gas). The pumping mechanism induces an upward movement of LNAPLs to the surface, facilitating their separation from water and air. Once free products are removed completely, the system seamlessly transitions into a conventional bioventing system. It is important to note that excessive soil moisture limits air permeability, reducing oxygen transfer rates, and subsequently impacting microbial activities. Bioslurping is not suitable for remediating low-permeability soils (Azubuike et al. 2016). This technology contributes to cost savings by reducing groundwater volume resulting from operations, minimizing storage, treatment, and disposal costs (Bala et al. 2022b). Challenges such as establishing a vacuum on deep, highly permeable sites and dealing with fluctuating water tables can create saturated soil lenses, posing a significant difficulty for aeration (Azubuike et al. 2016).

Mycoremediation

Fungi are widespread across diverse habitats, including terrestrial, aquatic, desert, tropical rainforests, freshwater, marine water, and deep-sea sediments. Although it was initially estimated that fungal species ranged from 2.2 to 3.8 million, only 148,000 species have been identified, with approximately 1882 species added to the fungal kingdom. Distinct criteria such as biochemistry, physiology, and metabolic capacities are employed to differentiate between fungal species, making mycoremediation a promising and eco-friendly strategy for environmental cleanup. Fungi employ various mechanisms, including biosorption, precipitation, biotransformation, and sequestration, to remediate environments from pollutants. This approach proves effective for heavy metals, persistent organic pollutants, and emerging contaminants (Kumar et al. 2021). Mycoremediation, as an approach using fungi for remediating polluted sites, leverages the adaptability of fungi to hostile environments, including pH, nutrient availability, temperature, and high metal concentrations (Sen et al. 2023). Fungi secrete enzymes during their life cycle, contributing to their efficacy in water and soil bioremediation processes since the 1980s (Kamal et al. 2023). Mycoremediation has gained prominence as a biological treatment for wastewater containing pharmaceuticals and personal care products (PPCPs) (Malik et al. 2023). Absorption, a key remediation mechanism, involves the physicochemical process where pollutants concentrate on the biosurface of fungi. Specific aminopolysaccharides in the cell wall, such as chitosan or chitin, ensure absorption, with glucosamine functional groups in chitosan acting as metal absorbent sites for chelation or complexation, facilitating the reduction and recovery of heavy metals from polluted sites. *Trichoderma*, *Penicillium*, and *Aspergillus* primarily use absorption to remediate copper (Cu) and cobalt (Co). Additionally, biosorption in filamentous fungi like *Phoma* sp. plays a significant role in removing pharmaceutical pollutants. Filamentous *Mucor* has demonstrated potential as a mycoremediator for eliminating cyanobacterial toxins from aquatic ecosystems due to its high uptake of toxins into fungal cells. In environments polluted by crude oil, the effectiveness of bioremediation is often limited by the availability of microorganisms with complementary substrate specificity for degrading various hydrocarbons. Filamentous fungi, known for their ability to degrade hydrocarbon pollutants, contribute significantly to the degradation of oil, with *Aspergillus oryzae* and *Mucor irregularis* utilizing an enzymatic arsenal comprising laccase, manganese peroxidase, and lignin peroxidase (Ghosh et al. 2023).

Phytoremediation

Phytoremediation, a sustainable strategy, utilizes plants to extract and remove pollutants or reduce their soil bioavailability (Yan et al. 2020). Plants absorb ionic compounds from the soil through their root system, making this ecofriendly technology applicable to both organic and inorganic pollutants in soil, water, and air (Lone et al. 2008). Economically and environmentally favorable phytoremediation involves green plants containing, sequestering, or detoxifying pollutants from contaminated soil and water. Various mechanisms like degradation (rhizodegradation, phytodegradation), accumulation (phytoextraction, rhizofiltration), dissipation (phytovolatilization), and immobilization (hydraulic control and phytostabilization) are employed for degrading, removing, or immobilizing contaminants. The choice of mechanisms depends on the pollutants, with plants utilizing one or more to reduce their concentrations (Kafle et al. 2022). Efficiency in phytoremediation varies based on plant morphology, physiology, and anatomical characteristics affecting ion uptake mechanisms (Shehata et al. 2019). Plants involved should tolerate high concentrations of xenobiotics, accumulate or biodegrade contaminants, grow rapidly, adapt easily, produce ample biomass with high disease resistance, and withstand environmental challenges. Phytoremediation is highly regarded as an equipment-free, effective, non-invasive, economical, socially acceptable, and ecologically significant technique (Mocek-Płócinial et al. 2023). Plant selection considers root depth, the nature of contaminants and soil, and regional climate (Laghlimi et al. 2015). Three main strategies for heavy metal remediation in plants are phytoextraction, phytostabilization, and phytovolatilization. Notably, *Lactuca sativa L.* is a metallophyte species known for hyperaccumulating various heavy metals (Cioica et al. 2019). *Piriformospora indica*'s presence enhances *Medicago sativa*'s tolerance, improving photosynthetic processes in phenanthrene and cadmium-contaminated soils (Li et al. 2020).

12.4 Combining Recycling and Bioremediation

Since hazardous waste is being carelessly and unrestrainedly dumped into the ecosystem, there is increased interest in the environmental management of the toxic wastes worldwide. To prevent contamination of ground and surface water, air, and soil, wastes are treated before being released into the open environment in liquid, gaseous, and solid forms (Kumar and Bharadvaja 2019; Abdelbasir et al. 2018).

Reusing and recycling these wastes are thought to be the most effective way to implement integrated solid waste management systems within the framework of a circular economy. In this regard, the new strategy will concentrate on different types of solid waste in order to increase resource efficiency, recycling, and reuse in order to promote economic sustainability (Sayara et al. 2020).

The separation, decomposition, and stabilization of organic waste through various natural cycles form the basis of organic waste recycling. However, recycling causes the release of potentially harmful compounds into the atmosphere, including dibenzo-p-dioxins, polychlorinated ethers, polychlorinated biphenyls, and polychlorinated dibenzofurans. Bioremediation, which combines environmentally friendly technologies, is the process of treating the environment using biological methods by using biological compounds to break down, eliminate, and/or transform environmental contaminants into less harmful or non-toxic forms (Bharathi et al. 2023; Amritha and Anilkumar 2016) but the microbial technologies are easier to control and maintain since they may operate in environments with mild temperatures (Bharathi et al. 2023).

Biobased ingredients can be used in a variety of industrial cleaning processes, including metal recovery and "green" recycling (Bharathi et al. 2023; Amritha and Anilkumar 2016). Even while the organic fraction—mostly household waste—makes up a sizable portion of the waste produced, it can be recycled and utilized as a possible source of plant nutrients rather than being thrown away or mishandled (Sayara et al. 2020).

Physicochemical recycling is not as effective as bioremediation. A comparison between physicochemical and bioremediation techniques indicates a number of benefits, including the total breakdown of pollutants or their transformation into innocuous compounds, which stops the spread of contaminants (Aparicio et al. 2022).

When compared to individual chemical or biological treatments, the implementation of this integrated treatment has proved to be advantageous, very economical, and respectful, and many researchers around the world have achieved significant results in removing heavy metals from the environment (Selvi et al. 2019).

Composting is a waste recycling technique that relies on the aerobic biological decomposition of organic matter. Before the process begins, organic waste must be separated from the rest of the waste stream either through source separation or centralized mechanical separation. This method not only improves soil structure and water retention but also supplies essential nutrients for plant growth and reduces dependence on fossil fuel–based fertilizers (Wei et al. 2017; Kumari et al. 2021).

However, the production of dangerous byproducts and hard-to-treat metal precipitates has severely limited this technique. Nevertheless, combining biological treatment with treatment is preferable since it is advantageous from an economic and environmental standpoint. When compared to individual chemical or biological treatments, the implementation of this integrated treatment has proved to be advantageous, very economical, and respectful, and many researchers around the world have achieved significant results in removing heavy metals from the environment (Selvi et al. 2019).

For example, in wastewater contaminated with Cr, standalone physical-chemical or biological treatment is no longer as economical or environmentally friendly as combined chemical-biological treatment, which effectively reduces Cr concentrations in the effluent. Three combined chemical-biological methods have been proposed for Cr treatment: (a) chemical treatment at stoichiometric levels that are both economical and effective, followed by biological treatment as a polishing step; (b) biological treatment first, followed by chemical treatment as a polishing step; and (c) a staged reactor system that integrates chemical precipitation with subsequent biological treatment. In a study, *Fusarium chlamydosporum* was able to recover 99.3% of total chromium and 98.4% of Cr(VI). Despite the growing popularity of these approaches, their success will depend on the responsible and environmentally sound selection of non-toxic chemicals (Ahmed et al. 2016; Selvi et al. 2019).

So this biodegradable portion could be recycled and turned into a potential source of plant nutrients instead of being wasted through improper disposal or treatment; in line with this attitude and taking into account that the organic fraction makes up a significant portion of the trash generated, primarily household waste (Sayara et al. 2020).

In conclusion, combining techniques like physicochemical degradation and bioremediation is a very effective way to remove waste-induced toxicity and renew energy.

12.5 Challenges and Limitations

Despite their immense potential in managing environmental pollution and mitigating the impacts of climate change, recycling and bio-remediation are not free of limitations and face gaps and too many obstacles. Although these approaches offer promising and encouraging solutions, a whole series of difficulties hamper their full effectiveness and widespread adoption (Bala et al. 2022a; Jabbar et al. 2022). This section examines the complex web of technological, social, economic, and regulatory constraints that prevent the optimization of recycling and bioremediation practices.

It is, therefore, vital to understand these difficulties to develop effective strategies for overcoming them. By tackling the limitations and encouraging continuous progress, we can unlock the true potential of recycling and bioremediation in creating a cleaner, more sustainable future for our planet.

Plastic recycling encounters hurdles attributable to the lack of a uniform sorting and recycling system, resulting in reduced recycling rates and increased waste generation (Massardier et al. 2023). Similarly, textile recycling faces challenges such as insufficient infrastructure, difficulties in fiber separation, and limited market demand for recycled textiles (Leal Filho et al. 2019). The complex nature of textile fibers adds intricacy to the separation and recycling process.

Moreover, recycling rubber presents several obstacles, including complexities in separating rubber components, low market value for recycled rubber, and the need for enhanced recycling infrastructure (Oliveira Neto et al. 2016). These challenges stem from the diverse compositions of rubber materials and the limited applications for recycled rubber products.

Although recycling is an important aspect of waste management, it is essential to understand that not all materials can be recycled infinitely. Additionally, recycling can result in high energy costs, and recycled materials may be of lower quality than virgin materials.

Contamination poses a significant challenge in both recycling and bioremediation processes, profoundly impacting the quality of recycled materials and the effectiveness of bioremediation efforts. In recycling, contamination can arise at various stages, from collection to sorting and processing, due to factors such as the absence of standardized sorting systems, human error, and inadequate sorting facilities (Khan and Charles 2023; Zhang et al. 2019). Insufficient awareness about recycling practices among the general public exacerbates contamination issues, resulting in impurities in recycled materials (Federigi et al. 2020). This prevalence of recycling contamination, fueled by public confusion over acceptable items and inadequate sorting practices, highlights the urgent need for comprehensive education initiatives and enhanced collection systems.

In the realm of bioremediation, contaminants serve as substantial energy sources for aerobic heterotrophic organisms, wherein the oxidation state of carbon profoundly impacts energy yield, thereby influencing microbial degradation incentives (Boopathy 2000; Kebede et al. 2021). Moreover, the efficacy of bioremediation is significantly influenced by various microbial characteristics, including biomass, metabolic pathways, population density and diversity, enzymatic activities, biosurfactant production, and competitive interactions (Boopathy 2000; Kebede et al. 2021). These factors collectively shape the microbial community's ability to degrade contaminants and determine the overall success of bioremediation efforts.

Additionally, several factors such as chemical, physical, and biological characteristics, along with soil type, nitrogen source, carbon availability, and the type of microorganisms (Garg et al. 2012), all influence bioremediation's process and effectiveness. Moreover, environmental factors such as pH, temperature, moisture content, oxygen availability, salinity, nutrient availability, and the presence of electron acceptors intricately modulate bioremediation processes (Boopathy 2000; Kebede et al. 2021). These environmental conditions are fundamental determinants of microbial growth, metabolic activity, and enzymatic function, collectively impacting the efficiency and extent of bioremediation.

Bala et al. (2022b) consider that every organic and inorganic pollutant cannot be remediated, the ex-situ control of volatile organic pollutants may be difficult, lack of "clean" site standards, highly permeable soil is required and the environmental application of rDNA is still restricted.

According to (Sharma 2021), the effectiveness of in situ bioremediation depends on suitable conditions for microbial activity. However, several difficulties can hinder this process. For example, some contaminants may persist or transform into more toxic or mobile intermediates during bioremediation. In addition, high concentrations of heavy metals and organic compounds can inhibit the activity of indigenous microorganisms, requiring the introduction of adapted micro-organisms. In addition, injection wells used for bioremediation can clog due to excessive microbial growth if nutrients are not properly balanced. Some contaminants, known as recalcitrant compounds, cannot be biodegraded even with adapted microbes. In some cases, pollutants can be transformed into less harmful but persistent compounds.

In addition, bio-remediation, on the other hand, can be a slow process, and eliminating all contaminants may not be possible. Moreover, a limited understanding of bioremediation procedures can engender doubt or anxiety regarding the safety of introducing microorganisms into the environment.

Changes influence the long-term sustainability of recycling markets in demand for recycled materials. Moreover, when considering recycling, it is important to account for the environmental impact of transporting recyclable materials. The introduction of genetically modified micro-organisms for bio-remediation purposes needs to be closely monitored to prevent any unexpected consequences on ecosystems (Bala et al. 2022a).

The setup and maintenance of recycling infrastructure require significant investment. Moreover, fluctuating commodity prices can make recycling economically challenging. In comparison to conventional cleaning methods, bioremediation projects may not always be the most cost-effective option. Additionally, regulatory frameworks for bioremediation may need to be developed or improved to provide clear guidelines and incentives (Jabbar et al. 2022).

Addressing challenges in recycling and bioremediation, such as meeting international obligations, fostering community engagement, and boosting recycling rates, is crucial for transitioning toward more sustainable waste management practices (Arroyo-López 2012; Reijonen et al. 2021). To surmount these obstacles, stakeholders must enhance cooperation, reevaluate information dissemination strategies, and invest in recycling and bioremediation infrastructure, thereby facilitating meaningful progress in waste management endeavors.

Despite these challenges, recycling and bioremediation remain essential techniques for waste management and environmental protection. To ensure the long-term sustainability of these practices, continued research, development, and public education are necessary to overcome these limitations.

12.6 Conclusion

In conclusion, it is evident that our planet is grappling with the detrimental effects of escalating pollution and climate change. Urgent action is imperative to address these pressing environmental challenges. Sustainable solutions, particularly recycling and bioremediation, are crucial in combating pollution and restoring ecological balance. Recycling plays a crucial role in conserving resources, reducing waste, and promoting a cleaner environment, with innovative approaches like chemical recycling and community-based programs emerging to address challenges. Bioremediation, with its ability to degrade persistent contaminants and rejuvenate contaminated environments, holds significant promise in mitigating the impacts of pollution.

Despite challenges faced in both recycling and bioremediation, continued research, education, and the integration of physicochemical and biological treatments are essential in overcoming limitations and creating a more sustainable future. By fostering ongoing research, promoting education, and advocating for proactive intervention, we can pave the way for a cleaner, healthier planet. It is imperative that we prioritize the well-being of both current and future generations by adopting sustainable practices and embracing innovative solutions to protect our environment.

References

Abatenh E, Gizaw B, Tsegaye Z, Wassie M (2017) The role of microorganisms in bioremediation—a review. Open J Environ Biol 2(1):038–046

Abdelbasir SM, El-Sheltawy CT, Abdo DM (2018) Green processes for electronic waste recycling: a review. J Sustain Metallurgy 4:295–311

Abdollahbeigi M (2021) An overview of the paper recycling process in Iran. J Chem Rev 3(1):284

Ahmed E, Abdulla HM, Mohamed AH, El-Bassuony AD (2016) Remediation and recycling of chromium from tannery wastewater using combined chemical–biological treatment system. Process Saf Environ Prot 104:1–10

Altassan A (2023) Sustainable integration of solar energy, behavior change, and recycling practices in educational institutions: a holistic framework for environmental conservation and quality education. Sustain For 15(20):15157. https://doi.org/10.3390/su152015157

Alori ET, Gabasawa AI, Elenwo CE, Agbeyegbe OO (2022) Bioremediation techniques as affected by limiting factors in soil environment. Front Soil Sci 2:937186

Amritha PK, Anilkumar PP (2016) Development of landscaped landfills using organic waste for sustainable urban waste management. Procedia Environ Sci 35:368–376

Aparicio JD, Raimondo EE, Saez JM, Costa-Gutierrez SB, Alvarez A, Benimeli CS, Polti MA (2022) The current approach to soil remediation: a review of physicochemical and biological technologies, and the potential of their strategic combination. J Environ Chem Eng 10(2):107141

Arora NK, Mishra J (2020) COVID-19 and importance of environmental sustainability. Environ Sustain 3(2):117–119. https://doi.org/10.1007/s42398-020-00107-z

Arroyo-López PE (2012) Diseño de programas de reciclaje de e-waste considerando las motivaciones del participante: Un estudio exploratorio en el estado de México. Psychology 3(1):63–74. https://doi.org/10.1174/217119712799240251

Azubuike CC, Chikere CB, Okpokwasili GC (2016) Bioremediation techniques-classification based on site of application: principles, advantages, limitations and prospects. World J Microbiol Biotechnol 32(180):1–18. https://doi.org/10.1007/s11274-016-2137-x

Bala S, Garg D, Thirumalesh BV, Sharma M, Sridhar K, Inbaraj BS, Tripathi M (2022a) Recent strategies for bioremediation of emerging pollutants: a review for a green and sustainable environment. MPDI Toxics 10(484):1–23. https://doi.org/10.3390/toxics10080484

Bala S, Garg D et al (2022b) Recent strategies for bioremediation of emerging pollutants: a review for a green and sustainable environment. Toxics 10(8):484

Berck P, Stears M, Taylor RLC, Trachtman C, Villas-Boas SB (2020) Reduce, reuse, redeem: deposit-refund recycling programs in the presence of alternatives *

Bharathi SD, Dilshani A, Rishivanthi S, Khaitan P, Vamsidhar A, Jacob S (2023) Resource recycling, recovery, and xenobiotic remediation from E-wastes through biofilm technology: a review. Appl Biochem Biotechnol 195(9):5669–5692

Bimir MN (2020) Revisiting e-waste management practices in selected African countries. J Air Waste Manag Assoc 70(7):659–669. https://doi.org/10.1080/10962247.2020.1769769

Bisinella V, Götze R, Conradsen K, Damgaard A, Christensen TH, Astrup TF (2017) Importance of waste composition for life cycle assessment of waste management solutions. J Clean Prod 164:1180–1191. https://doi.org/10.1016/j.jclepro.2017.07.013

Bom UB, Belbase S, Bibriven Lila R (2017) Public perceptions and practices of recycling in the City of Laramie in Wyoming 2, p. 11. https://doi.org/10.20944/preprints201706.0037.v1

Boopathy R (2000) Factors limiting bioremediation technologies. Biores Technol 74:63–67
Bosher L, Kim D, Okubo T, Chmutina K, Jigyasu R (2020) Dealing with multiple hazards and threats on cultural heritage sites: an assessment of 80 case studies. Disaster Prevention Manag 29(1):109–128. https://doi.org/10.1108/DPM-08-2018-0245
Brown DM, Okoro S, Gils JV, Spanning RV, Bonte M, Hutchings T, Linden O, Egbuche U, Bruun KB, Smith JWN (2017) Comparison of landfarming amendments to improve bioremediation of petroleum hydrocarbons in Niger Delta soils. Sci Total Environ 596–597(2017):284–292. https://doi.org/10.1016/j.scitotenv.2017.04.072
Carey TS, Howard DL, Goldmon M, Roberson JT, Godley PA, Ammerman A (2005) Developing effective interuniversity partnerships and community-based research to address health disparities. Acad Med 80:1039
Chiew YL, Spångberg J, Baky A, Hansson PA, Jönsson H (2015) Environmental impact of recycling digested food waste as a fertilizer in agriculture—a case study. Resour Conserv Recycl 95:1–14. https://doi.org/10.1016/j.resconrec.2014.11.015
Chikere C, Chikere BO, Okpokwasili GC (2012) Bioreactor-based bioremediation of hydrocarbon-polluted Niger Delta marine sediment, Nigeria. 3 Biotech 2:53–66. https://doi.org/10.1007/s13205-011-0030-8
Cioica N, Tudora C, Iuga D, Deak G, Matei M, Negy EM, Gyorgy Z (2019) A review on phytoremediation as an ecological method for *in situ* clean up of heavy metals contaminated soils. E35 Web Conf 12:10. https://doi.org/10.1051/e3sconf/201911203024
Demenko LS (1976) Economic advantages of recycling the secondary materials resources of the synthetic fibre industry. Fibre Chem 8:89
EPA (2014) Municipal solid waste generation, recycling, and disposal in the united states: facts and figures for 2012. EPA. https://www.epa.gov/sites/default/files/2015-09/documents/2012_msw_fs.pdf
Federigi I, Verani M, Balestri E, Castelli A, De Battisti D, Maltagliati F, Menicagli V, Lardicci C, Carducci A (2020) Beach contamination by marine litter: application of DPSIR (driver, pressure, state, impact, response) analysis. In: The 3rd international electronic conference on environmental research and public health
Flaspohler P, Wandersman A, Keener D, Maxwell KN, Ace A, Andrews A, Holmes B (2003) Promoting program success and fulfilling accountability requirements in a statewide community-based initiative: challenges, Progress, and lessons learned. J Prev Interv Community 26(2):37–52. https://doi.org/10.1300/J005v26n02_04
Furnari D (2021) Metallurgical investigation of tie rod used for lifting ferro-alloy during steel making: a safety issues. J Steel Struct Constr 6:2472
Garg SK, Tripathi M et al (2012) Strategies for chromium bioremediation of tannery effluent. Rev Environ Contam Toxicol 217:75–140
Germaine KJ, Byrne J, Liu X, Keohane J, Culhane J, Lally RD, Kiwanuka S, Ryan D, Dowing DN (2015) Ecopiling: a combined phyoremediation and passive biopiling system for remediating hydrocarbon impacted soils at field scale. Front Plant Sci 5(756):1–6. https://doi.org/10.3389/fpls.2014.00756
Ghosh S, Rusyn I, Dmytruk O, Dmytruk KV, Onyeaka H, Gryzenhout M, Gafforov Y (2023) Filamentous fungi for sustainable remediation of pharmaceutical compounds, heavy metal and oil hydrocarbons. Front Bioeng Biotechnol 11:1–20. https://doi.org/10.3389/fbioe.2023.1106973
Gidarakos E, Aivaliori M (2007) Large scale and long term application of bioslurping: the case of a creek petroleum refinery site. J Hazard Mater 149(19):574–481. https://doi.org/10.1016/j.jhazmat.2007.06.110
Grotherr C, Vogel P, Martin S (2020) Multilevel Design for Smart Communities - the case of building a local online neighborhood social community. In: Hawaii international conference on system sciences. https://hdl.handle.net/10125/64023. 978-0-9981331-3-3 (CC BY-NC-ND 4.0)
Hashemi T (2015) Reused and recycled materials in architecture: selected case studies in the Middle East countries (Doctoral dissertation, Politecnico di Torino). Chicago

Hibor RM, Gavrillescu M, Tavares T, Laviers L, Olivieri G (2017) Bioremediation: an overview on current practices, advances, and new perspectives in environmental pollution treatment (editorial). Biomed Res Int 2017:6327610. https://doi.org/10.1155/2017/6327610

Hoornweg D, Bhada-Tata P (2012) What a waste: a global review of solid waste management. World Bank, Washington, DC. https://openknowledge.worldbank.org/handle/10986/17388

Inagaki M, Tsumura T, Kinumoto T, Toyoda M (2019) Graphitic carbon nitrides (g-C3N4) with comparative discussion to carbon materials. In: Carbon, vol 141. Elsevier, pp 580–607. https://doi.org/10.1016/j.carbon.2018.09.082

IPBES (2019) Global assessment report on biodiversity and ecosystem services of the intergovernmental science-policy platform on biodiversity and ecosystem services (version 1). Zenodo. IPBES. https://doi.org/10.5281/zenodo.6417333

Jabbar NM, Alardhi SM et al (2022) Challenges in the implementation of bioremediation processes in petroleum-contaminated soils: a review. Environ Nanotechnol Monit Manag 18:100694

Jambeck JR, Geyer R, Wilcox C, Siegler TR, Perryman M, Andrady A, Narayan R, Law KL (2015) Plastic waste inputs from land into the ocean. Science 347(6223):768–771. https://doi.org/10.1126/science.1260352

Kafle A, Timilsina A, Gautam A, Adhikari K, Bhattarai A, Aryal N (2022) Phytoremediation: mechanisms, plant selection and enhancement by natural and synthetic agents. Environ Adv 8:1–18. https://doi.org/10.1016/j.envadv.2022.100203

Kamal N, Parchad J, Saharan B, Kayasth M, Mudgal V, Duhan JS, Mandal BS, Sadh PK (2023) Ecosystem protection through myco-remediation of chromium and arsenic. J Xenobiot 13:159–171. https://doi.org/10.3390/jox13010013

Kebede G, Tefese T, Abda ME, Kamaraj M, Assefa F (2021) Factors influencing the bacterial bioremediation of hydrocarbon contaminants in the soil: mechanisms and impacts. Hindawi J Chem 2021:1–17. https://doi.org/10.1155/2021/9823362

Khan N, Charles KJ (2023) When water quality crises drive change: a comparative analysis of the policy processes behind major water contamination events. Expo Health 15(3):519–537. https://doi.org/10.1007/s12403-022-00505-0

Kumar L, Bharadvaja N (2019) Enzymatic bioremediation: a smart tool to fight environmental pollutants. In: Smart bioremediation technologies. Academic Press, pp 99–118

Kumar A, Yadav AN, Mondal R, Kour D, Subrahmanyam G, Shabnam AA, Khan SA, Yadav KK, Sharma GK, Cabral-Pinto M, Fagodiya RK, Gupta DK, Hota S, Maylan SK (2021) Myco-remediation: a mechanistic understanding of contaminants alleviation from natural environments and future prospect. Chemosphere 284:1–22. https://doi.org/10.1016/j.chemosphere.2021.131325

Kumari A, Aich AR, Kumari S, Mohanty S (2021) Greenways for solid waste management. In: Handbook of solid waste management: sustainability through circular economy. Singapore, Springer, pp 1–40

Laghlimi M, Baghdad B, El Hadi H, Bouabdli A (2015) Phytoremediation mechanisms of heavy metals contaminated soils: a review. Open J Ecol 5:375–388. https://doi.org/10.4236/oje.2015.58031

Lamsali H, Ariffin AS (2013) A theoretical investigation on end-of-life (EOL) product collection methods in a closed-loop logistics network. Management 2013(7):440–447. https://doi.org/10.5923/j.mm.20130307.15

Lawrence ND (2016) Commercial recycling & waste management

Leal Filho W, Ellams D, Han S, Tyler D, Boiten VJ, Paco A, Moora H, Balogun AL (2019) A review of the socio-economic advantages of textile recycling. In: Journal of cleaner production, vol 218. Elsevier, Amsterdam, pp 10–20. https://doi.org/10.1016/j.jclepro.2019.01.210

Lee S, Ha D, Kim H (2018) A definition of recycling waste disposal environment from the viewpoint of user-oriented design

Leverenz H, Tchobanoglous G, Spencer BD (2002) Recycling. In: Handbook of solid waste management, 2nd edn. Mcgraw-hill. https://doi.org/10.1036/0071356231

Li L, Zhu P, Wang X, Zhang Z (2020) Phytoremediation effect of *Medicago sativa* colonized by *Piriformospora indica* in the phenanthrene and cadmium co-contaminated soil. BMC Biotechnol 20(20):1–14. https://doi.org/10.1186/s12896-020-00613-2

Liu X, Qian S, Ma W (2022) Estimating and mitigating the congestion effect of Curbside Pick-ups and drop-offs: a causal inference approach. Transportation Sci 58:355. http://arxiv.org/abs/2206.02164

Lone MI, He Z, Stoffella PJ, Yang X (2008) Phytoremediation of heavy metal polluted soils and water: progress and perspectives. J Zhejiang Univ Sci B 9(3):210–220. https://doi.org/10.1631/jzus.B0710633

Malik S, Bora J, Nag S, Sinha S, Mondal S, Rustagi S, Hazza R, Kumar H, Rajput VD, Minkina T, Saidier NS, Almutary AG (2023) Fungal-based remediation in the treatment of anthropogenic activities and pharmaceutical-pollutant-contaminated wastewater. MDPI Water 15:1–24. https://doi.org/10.3390/w15122262

Massardier V, Belhaneche-Bensemra N, Lazaric N (2023) Editorial: alternative building blocks and new recycling routes for polymers: challenges for circular economy and triggers for innovations. Front Mater 10:1152494. https://doi.org/10.3389/fmats.2023.1152494

Mocek-Płócinial A, Mencel J, Zakrzewski W, Roszkowski S (2023) Phytoremediation as an effective remedy for removing trace elements from ecosystems. MDPI Plants 12:1–29. https://doi.org/10.3390/plants12081653

O'Dwyer C, Zaman A, Breadsell JK (2022) The uptake of container deposit schemes: a case study in Perth, Western Australia. Sustainability (Switzerland) 14(19):11863. https://doi.org/10.3390/su141911863

Oliveira Neto GC, Correia JMF (2019) Environmental and economic advantages of adopting reverse logistics for recycling construction and demolition waste: a case study of Brazilian construction and recycling companies. Waste Manag Res 37(2):176–185. https://doi.org/10.1177/0734242X18816790

Oliveira Neto G, Tucci H, Pinto L, Costa I, Rodrigues Leite R, Cardoso de Oliveira Neto G, Nieves Pujol Tucci H, Fernando Rodrigues Pinto L (2016) Economic and environmental advantages of rubber recycling. Springer, Cham, pp 818–824. https://doi.org/10.1007/978

Park E, Han C, Son SH, Lee MS, Kim YH (2022) Overview on Pyrometallurgical recycling process of spent Lithium-ion battery. Resour Recycling 31(3):27–39. https://doi.org/10.7844/kirr.2022.31.3.27

Paul O, Jasu A, Lahiri D, Nag M, Ray RR (2021) *In situ* and *ex situ* bioremediation of heavy metals the present scenario. J Environ Eng Landsc Manag 29(4):454–469. https://doi.org/10.3846/jeelm.2021.15447

Radhakrishnan A, Balaganesh P, Vasudevan M, Natarajan N, Chauhan A, Arora J, Ranjan A, Rajput VD, Sushkova S, Minkina T, Basniwal RK, Kapardar R, Srivastav R (2023) Bioremediation of hydrocarbon pollutants: recent promising sustainable approaches, scope, and challenges. Sustain For 15(7):1–18. https://doi.org/10.3390/su15075847

Rashidul I, Jobair AR, Sajjad H (2019) A comparative study on waste-to-energy (WTE) technologies for municipal solid waste (MSW) management in Chittagong, Bangladesh. In: 1st international conference on advances in science, engineering and robotics technology. ICASERT

Reijonen H, Bellman S, Murphy J, Kokkonen H (2021) Factors related to recycling plastic packaging in Finland's new waste management scheme. Waste Manag 131:88–97. https://doi.org/10.1016/j.wasman.2021.05.034

Rodiger-Vorwerk T (2018) Circular economy and waste management: trends and needs from the perspective of developing countries-Tania Rodiger-Vorwerk-German Federal Ministry for Economic Cooperation and Development. Arch Sci 2(2):1

Roithner C, Rechberger H (2020) Implementing the dimension of quality into the conventional quantitative definition of recycling rates. Waste Manag 105:586–593. https://doi.org/10.1016/j.wasman.2020.02.034

Rose D, Heller MC, Roberto CA (2019) Position of the Society for Nutrition Education and Behavior: the importance of including environmental sustainability in dietary guidance. J Nutr Educ Behav 51(1):3–15.e1. https://doi.org/10.1016/j.jneb.2018.07.006

Sayara T, Basheer-Salimia R, Hawamde F, Sánchez A (2020) Recycling of organic wastes through composting: process performance and compost application in agriculture. Agronomy 10(11):1838

Scanscarter D, Zeeb B, Koch I, Reimer K (2009) Bioremediation of diesel-contaminated soil by heated and humidified biopile system in cold climate. Cold Reg Sci Technol 55(1):167–173. https://doi.org/10.1016/j.coldregions.2008.07.004

Selvi A, Rajasekar A, Theerthagiri J, Ananthaselvam A, Sathishkumar K, Madhavan J, Rahman PK (2019) Integrated remediation processes toward heavy metal removal/recovery from various environments-a review. Front Environ Sci 7:66

Sen K, Llewellyn M, Taheri B, Turner RJ, Berglund T, Maloney K (2023) Mechanism of fungal remediation of wetland water: *Stropharia rygosoannulata* as promising fungal species for the development of biofilters to remove clinically important pathogenic and antibiotic resistant bacteria in contaminated water. Front Microbiol 14:1–10. https://doi.org/10.3389/fmicb.2023.1234586

Sharma I (2021) Bioremediation techniques for polluted environment: concept, advantages, limitations, and prospects. Trace in the environment: new approaches and recent advances. B. o. Demand

Shehata SM, Badawy RK, Aboulsoud YIE (2019) Phytoremediation of some heavy metals in contaminated soil. Bull Nat Res Cent 43(189):1–15. https://doi.org/10.1186/s42269-019-0214-7

Siu KWM, Wong YL, Lam MS (2020) Importance of involving children in designing recycling facilities: a case study of Hong Kong children collecting and recycling recyclables. Adv Intell Syst Computing 968:83–90. https://doi.org/10.1007/978-3-030-20470-9_10

Sun X, Lettow F, Neuhoff K (2021) Climate neutrality requires coordinated measures for high quality recycling. DIW Weekly Rep 11:187. https://doi.org/10.18723/diw_dwr:2021-26-1

Taghdisi MH, Estebsari F, Gholami M, Hosseini AF, Milani AS, Abolkheirian S, Rahimi Khalifeh Kandi Z (2022) A training program of source-separated recycling for primary school students: applying the health promoting schools model. Appl Environ Educ Commun 21(1):102–117. https://doi.org/10.1080/1533015X.2021.2001392

Tyagi H, Kumar N (2021) Bioremediation: principles and applications in environmental management. In: Saxena G, Kumar V, Shah MP (eds) Bioremediation for environmental sustainability. Elsevier, Amsterdam, pp 3–28. ISBN 9780128205242. https://doi.org/10.1016/B978-0-12-820524-2.00001-8

UNEP (2020) Global environment outlook—GEO-6: healthy planet, healthy people. United Nations Environment Programme. https://www.unep.org/resources/global-environment-outlook-6. Accessed 25 Mar 2024

UNEP (2021) Plastic waste pollution. United Nations Environment Programme. https://www.unep.org/interactive/beat-plastic-pollution/. Accessed 25 Mar 2024

Vidal M (2001) Bioremediation: an overview. Pure Appl Chem 73:1163–1172

Wei Y, Li J, Shi D, Liu G, Zhao Y, Shimaoka T (2017) Environmental challenges impeding the composting of biodegradable municipal solid waste: a critical review. Resour Conserv Recycl 122:51–65

WHO (2021) Ambient air pollution: health impacts. World Health Organization, Geneva. https://www.who.int/health-topics/air-pollution#tab=tab_1. Accessed 25 Nov 2024

Williams ID, Culleton A (2009) Manchester recycling for all: increasing participation in recycling by offering choice and alternatives to low recycling communities. 2009: Twelfth International Waste Management and Landfill Symposium, Sardinia, Italy 05–09 Oct 2009, Sardinia

Yan A, Wang Y, Tan SN, Yosof MML, Ghosh S, Ghen Z (2020) Phytoremediation: a promising approach for revegetation of heavy metal-polluted land. Front Plant Sci 11(359):1–15. https://doi.org/10.3389/fpls.2020.00359

Yusop YM, Othman N (2019) Linking the Malaysia's solid waste management policy instruments with household recycling behavior. Int J Acad Res Progress Educ Dev 8(4):474. https://doi.org/10.6007/ijarped/v8-i4/6567

Zhang Z, Han W, Chen X, Yang N, Lu C, Wang Y (2019) The life-cycle environmental impact of recycling of restaurant food waste in Lanzhou, China. Appl Sci 9(17):3608. https://doi.org/10.3390/app9173608

Overview of Plastic and Bioremediation: Present Investigations and Future Outlook

13

Mai M. Labib

Abstract

Plastic and bioremediation are a rapidly emerging subject of study with numerous potential applications. The demand for more environmentally friendly and sustainable solutions to the problem of plastic waste motivates research on this subject. Further research should focus on developing more effective and economical cost strategies for bioremediation and plastic recycling. Furthermore, research must focus on developing innovative methods and strategies to reduce plastic waste at its origin. Additional study is required to improve comprehension of the environmental and health consequences of plastic and bioremediation. Recent advancements in plastic recycling include the development of microbial enzymatic degradation methods that break down plastics into their basic monomers, including bacteria, fungi, and microalgae capable of degrading various plastic polymers. Advancements in biodegradable plastics are increasing, providing alternatives that can naturally dissolve without causing environmental harm. Enhanced plastic recycling techniques can yield substantial economic advantages by minimizing raw material expenses and diminishing dependence on fossil fuels for plastic manufacturing. This review includes biodegradable plastics and microorganisms that facilitate the degradation of manufactured and natural plastics. Furthermore, the application of metagenomic and bioengineering methodologies for plastic bioremediation. In addition, the application of genetic engineering, bioinformatics, and synthetic biology to improve plastic bioremediation. Ultimately, upcoming prospects and challenges encompass competition, regulation, and escalating costs, with emerging technologies to address these issues.

M. M. Labib (✉)
Department of Bioinformatics and Computer Networks, Agricultural Genetic Engineering Research Institute (AGERI), Agricultural Research Centre (ARC), Cairo, Egypt
e-mail: mailabib@ageri.sci.eg

A. A. H. Abdel Latef et al. (eds.), *Bioremediation and Nanotechnology for Climate Change Mitigation*, https://doi.org/10.1007/978-981-96-3069-1_13

Keywords

Plastic and bioremediation · Enzymatic plastic degradation · Metagenomic · Genetic engineering · Bioinformatics and · Synthetic biology bioengineering methodologies for plastic bioremediation

13.1 The Scope and Environmental Impact of Plastic Pollution

13.1.1 An Overview of Plastic Pollution

Over the last several decades, the fast increase in plastic manufacturing has led to a concerning buildup of plastic trash in the environment. The global production of plastic exponentially increased from two million tons in the 1950s to 400 million tons in 2015, with half of this growth occurring in the preceding 13 years. Consequent to this exponential rise, an astonishing 6.3 billion metric tons of plastic waste have been deposited in landfills and the environment, making a substantial contribution to plastic pollution (Mazhandu et al. 2020). The widespread distribution of plastic garbage in different ecosystems, whether terrestrial or aquatic, poses severe risks to the health of animals and, by extension, humans (Bidashimwa et al. 2023). Eriksen et al. (2023) report that the exponential growth in global plastic production and the inadequate waste disposal systems in many regions, especially developing nations, are contributing to the alarming trend of plastic leakage into the marine environment at an unsustainable rate. Current estimates place the amount of plastic waste entering the oceans each year between 4.8 and 12.7 million metric tons, and this figure is expected to continue rising in the coming decade unless significant interventions are put in place (MacLeod et al. 2021). Microplastics in the human diet come mostly from fisheries products. Contaminated seafood has the potential to assimilate a range of detrimental substances, such as microplastics, which may lead to many adverse health effects. Piskuła and Astel (2023) have identified several consequences linked to these events, including disturbances in the endocrine system, inflammation, propensity for carcinogenesis, and heightened susceptibility to cardiovascular diseases, neurological disorders, and reproductive diseases. Increased seafood consumption exacerbates the susceptibility to health issues resulting from plastic-based pollutants since they accumulate in the food chain and undergo amplification (Nakei et al. 2022).

13.1.2 The Chemical Composition of Plastics

Around the turn of the millennium, the global production of synthetic polymers reached an all-time high of 22.7 trillion kilograms. A potential new era has commenced, as referred to by Boctor et al. (2024), who name it "The Age of Plastic." Plastic, once heralded as a triumph in material science, may be directly responsible for the present global environmental catastrophe. Plastics may be defined as versatile materials composed of long-chain polymer molecules interconnected with less than 10% of total isocyanate groups. Plastics are lightweight, robust, chemically resistant, and easily manufacturable. Moreover, plastics show a low density and a commendable electrical insulating capacity. The characteristics seen in plastics may be attributed to the chemical engineering of their internal structures during the prior century. Recent research has been dedicated to exploring the utilization of biopolymers and their by-products. This interest in environmentally friendly polymers and the recognition that several beneficial plastic properties are already found in various organisms, including marine species, has sparked this activity (Tickner et al. 2021). We need to concur on what plastic is before we can discuss its chemical composition. The word "plastic" is not a chemical molecule but a popular way to describe a class of compounds known as "polymers." We may define "plastic" as any material that can be bent, cut, or physically transformed into a finished product. Both the noun "plasticos" and the verb "plassein" meaning "to form or mold" in Greek are the exact origins of the noun. The chemical industry has relied on plastic as a tool since its creation. Buildings, ships, and airplanes may be more affordably constructed using plastics than other materials due to their low density, cheap cost, and malleability. The inherent characteristics of plastics have facilitated the identification of novel applications, resulting in their extensive adoption in contemporary society. Throughout the years, they have emerged in many applications, including miniature electrical components, building materials, automotive and aerospace parts, and several others. Worldwide, the production of synthetic polymers has reached 22.7 trillion kilograms as of the turn of the millennium. The current era may be appropriately designated as "The Age of Plastic" because of the many environmental issues caused by plastic use, despite its status as a significant "success" in material development (Boctor et al. 2024).

13.1.3 Definition and Significance of Plastics

Except for materials obtained from natural sources, the quality and appearance of almost all the materials we use daily have greatly declined. Is glassware appropriate for use in homes or hospitals where illness transmission is a consideration? Also, wouldn't it be dangerous and hard to handle if ceramic components were used to make lightweight and incredibly resistant utensils? Are glass, paper, or aluminum the only acceptable materials for packaging when we require it to be robust, lightweight, stereotype-friendly, and have a short lifespan? Our suggestion recognizes the importance of plastic materials in both the present and future contexts. Although plastics are ubiquitous in many aspects of life, this does not discourage people from making judgments about them. Plastic-based products, whether formed as components or fibers, are ubiquitous in our everyday existence. The seat belt is made from a substance similar to nylon film, whereas almost 95% of domestic appliances utilize plastic components, and packaging constitutes 18% of the weight of a vehicle. Due to innovative uses that exploit the abundant physical and chemical properties of plastic materials, plastic is increasingly used as a substitute for or in conjunction with other materials, such as wood, steel, ceramics, and stone (Lase et al. 2021).

13.1.4 The Impact of Plastic Pollution on Cancer

The globe confronts two significant health challenges that are particularly difficult, as they are in opposition yet simultaneously jeopardize all global populations. Cancer rates are worsening, with a projected global increase in incidence from 14 million cases in 2012–22 million by 2030. Secondly, and less apparent to the public, are the increasing apprehensions regarding the prevalence and consequences of plastic garbage in the environment. These issues encompass the pervasive existence of microplastics in aquatic as well as atmospheric environments, the presence of toxic substances in plastics that can be transmitted to humans through ingestion or inhalation, and the unresolved degradation of plastics, which may entail the release of enduring hazardous chemicals into the ecosystem (Shi et al. 2022). Plastic pollution is associated with a heightened risk of cancer. Plastic has hazardous compounds that can seep into the environment, so they contaminate food and water supplies. These substances may induce carcinogenic alterations in cells. Microplastics, diminutive plastic particles measuring less than 5 millimeters, have been detected in many environments and within human bodies. These particles can transport detrimental substances into the body, potentially resulting in inflammation and other health complications (Ziani et al. 2023). The accumulation of microplastics may pose long-term health hazards, including cancer. Besides cancer, microplastics are linked to respiratory problems upon inhalation and may induce hormonal disturbances due to their interaction with endocrine-disrupting substances. Ingestion may also result in gastrointestinal issues, potentially causing discomfort or obstructions in the digestive tract. Moreover, microplastics have been associated with developmental problems in children, impacting growth and cognitive abilities.

In 2019, certain facets of plastic trash were examined, revealing research studies on the inadvertent toxicological and biological impacts of micro-sized plastic particles. A study assessed the capacity of human cells to absorb chitosan-coated microplastic particles (Lee et al. 2023). The utilization of human epithelial colorectal cancer cells may provide valuable insights into occupational exposure for which knowledge is lacking (Pelegrini et al. 2023). The historical trend in colorectal cancer was examined due to the potential concerns posed by microplastics to cancer patients. The alarming correlation between the global rise in colorectal cancer and the growth in plastic waste output raises the critical question of whether an undisclosed or unexamined connection exists between colorectal cancer and plastics (Bruno et al. 2024).

Nearly two decades ago, studies revealed that the majority of prevalent plastics are incapable of biodegradation when discarded into the environment. This research has resulted in the classification of plastic pollution into progressively severe categories: macroplastics, microplastics, and nanoplastics, each demonstrating distinct behaviors and presenting varying dangers. The components of plastics research necessary for a comprehensive understanding of environmental consequences encompass the identification of the polymer, the determination of additives, the examination of polymer behavior including surface alterations and interactions with contaminants, the analysis of affected organisms over time along with the mechanisms of impact transfer across trophic levels, and the formulation of an environmental risk assessment utilizing this information alongside potential transformations of the plastic (Valavanidis 2024). Consequently, it is unsurprising that, despite the abundance of articles containing the phrases "plastic" and "pollution" in their titles, only a small subset addresses the relevant keywords in conjunction with "cancer" (Dey et al. 2024).

Brynzak-Schreiber et al. 2024 discovered that brief studies demonstrated that exposure to spherical polystyrene (PS) particles did not influence cell proliferation or cell cycle distribution. Contrary to the hypothesis posited by Brill-Karniely et al. (2020), which suggested that more deformable and "soft" cell types would exhibit greater particle uptake, SW620 cells demonstrated a reduced uptake after 24 h for both 0.25 and 1 μm PS particles compared to SW480 cells (Armistead et al. 2020). Given that the HT29 cell line originates from a primary adenocarcinoma classified as Duke's Stage C and exhibits an elastic modulus comparable to that of SW620 (Armistead et al. 2020), it was postulated that particle uptake in HT29 cells would surpass that in SW480 cells, which are derived from a Duke's Stage B primary adenocarcinoma. Nonetheless, the findings of our investigation refute this notion, as after 24 h of incubation, HT29 cells exhibited the minimal uptake of 1 μm PS particles in comparison to other cell lines. Furthermore, reduced absorption rates of 0.25 μm PS particles into HT29 cells were noted in comparison to SW480 cells. HCT116 cells had the highest absorption rates for both 0.25 and 1 μm PS particles. The absence of a standardized classification of the HCT116 cell line in the literature concerning its cancer aggressiveness complicates the correlation of HCT116 cell results with those of other cell lines. The elevated uptake rates in HCT116 cells may be attributed to their classification within the mesenchymal consensus molecular subtype 4 (CMS4), potentially rendering them more aggressive due to the upregulation of genes associated with epithelial-mesenchymal transition (EMT) and signatures linked to the activation of transforming growth factor β (TGFβ) (Guinney et al. 2015). Furthermore, particles smaller than 1 μm may facilitate cell motility, potentially fostering metastasis. This discovery corresponds with recent studies indicating that micro- and nanoplastics (MNPs) may affect cellular activity and may facilitate disease progression. Results demonstrated the persistence and bioaccumulation of MNPs in colorectal cancer cell lines, fulfilling two of the three criteria in toxicology and under the REACH regulation (https://environment.ec.europa.eu/topics/chemicals/reach-regulation_en) for potentially hazardous substances.

13.2 Bioremediation: Microorganisms Catalyze the Enzymatic Decomposition of Plastic

Bioremediation is presently being investigated by scientists as a potential solution to the growing problem of environmental plastic pollution. The identification of microorganisms, such as bacteria, fungi, and microalgae, that are capable of degrading a variety of plastic polymers has been the subject of numerous recent studies, such as those conducted by Dunn and Welden (2023), Viel et al. (2023), and Nakei et al. (2022). The environmental endurance of plastic polymers may be significantly diminished if the resistant bonds are hydrolyzed by the specialized enzymes or metabolic pathways of these bacteria. Dunn and Welden (2023) and Moharir and Kumar (2019) reference studies demonstrate that certain bacteria secrete enzymes capable of cleaving the ester chemical bonds in polyethylene terephthalate, a commonly used polymer in packaging and textiles. Furthermore, numerous fungi have been demonstrated to decompose polyurethane, a material with diverse industrial applications as mentioned by Moharir and Kumar in 2019. Moreover, the microorganisms' extensive range and evident effectiveness in terrestrial and marine environments further emphasize their biological repair capability (Bhavsar et al. 2023). Identification of plastic-degrading microbes like *Ideonella sakaiensis*, *Pseudomonas putida*, and *Bacillus cereus* in diverse settings points to naturally existing microbial communities may be offering a solution to the worldwide issue of plastic pollution (Cai et al. 2023). Identifying certain bacteria that are able to break down plastic presents researchers with a hopeful future for naturally existing microbial communities to solve the global plastic pollution issue (Li et al. 2023; Cai et al. 2023). Microbial species such as *Bacillus cereus*, *Ideonella sakaiensis*, and *Pseudomonas putida* can break down different forms of plastic. A recent study indicates that microbes can break down most plastic polymers. Bacteria, fungi, and microalgae are included under this classification. Bacteria with certain metabolic pathways or enzymes may break down polyethylene terephthalate (PET), a ubiquitous plastic used in textiles and packaging, into its monomeric components. A range of bacteria, including *Bacillus cereus*, *Ideonella sakaiensis*, and *Pseudomonas putida*, can break down different forms of plast-eating bacteria with certain metabolic pathways or enzymes may degrade PET, a common plastic used in textiles and packaging, into its monomeric components. A variety of bacteria, including *Bacillus cereus, Ideonella sakaiensis,* and *Pseudomonas putida*, may degrade various types of plastics. Empirical research suggests that bacteria can break down most plastic polymers. This group includes fungal, bacterial, and microalgal species. The individual constituents of PET, a commonly used plastic in packaging and textiles, may be broken down by bacteria with specific metabolic pathways or enzyme-bacteria with certain metabolic pathways or enzymes may degrade PET, a common plastic used in textiles and packaging, into its monomeric components. A variety of bacteria, including *Bacillus cereus, Ideonella sakaiensis,* and *Pseudomonas putida*, may degrade various types of plastics. According to empirical studies, bacteria can degrade the majority of plastic polymers. This category consists of fungal, bacterial, and microalgal species. PET, a prevalent material in packaging and textiles, may be broken down by bacteria via certain metabolic pathways or enzymes. Bacteria are capable of degrading the majority of plastic polymers. Species of microalgae, fungi, and bacteria all fall under this umbrella. The primary polymer used in packaging and textiles, PET, may be degraded by bacteria via specific metabolic pathways and enzyme-hydrolyze PET, the major polymer used in packaging and textiles, may be destroyed by bacteria via particular metabolic pathways and enzymes. This category includes microalgae, fungi, and bacterial and fungal PET-active enzymes (PETases), a popular plastic used in packaging and textiles, may be degraded into monomeric components by bacteria with certain metabolic pathways or enzymes. As an outcome, the material's resistance to environmental conditions diminishes greatly (Zrimec et al. 2021; Jones et al. 2023). The broad and adaptable polyurethane resin may be susceptible to degradation by enzymes and metabolic pathways found in some fungal species. Plastic-degrading bacteria have contributed to substantial progress in biological remediation. Microorganisms may be used to reduce plastic pollution, as shown by Zrimec et al. (2). The fact that *Bacillus cereus*, *Ideonella sakaiensis*, and *Pseudomonas putida* have been shown to digest plastic in various settings suggests that microbial communities that already exist in the environment might provide a potential remedy to the global issue of plastic pollution (Cai et al. 2023). Researchers have found microbes that can degrade plastic, leading them to believe that certain naturally occurring bacterial communities might help reduce plastic pollution (Li et al. 2023; Cai et al. 2023).The findings suggest that bacteria might play a role in reducing plastic pollution (Zrimec et al. 2). Research has shown that some species of bacteria can degrade plastic in many environments; these include *Bacillus cereus*, *Ideonella sakaiensis*, and *Pseudomonas putida* (Cai et al. 2023). As a result, these microbial communities might hold the secret to ending the worldwide pollution problem. In light of the recent finding of microorganisms that can break down plastic, researchers certain populations of naturally occurring bacteria will help reduce plastic pollution (Li et al. 2023; Cai et al. 2023). Among the bacteria that could break down different kinds of plastic in unique methods are *Bacillus cereus*, *Ideonella sakaiensis*, and *Pseudomonas putida*. Recent studies suggest that microbes can break down several plastic polymers. This class covers bacteria, fungi, and microalgae. PET is ubiquitous in packaging and textiles; microbes with certain metabolic pathways or enzymes may decompose it into its monomeric components. Some examples of bacteria that break down plastic are *Ideonella sakaiensis, Pseudomonas putida*, and *Bacillus cereus*. Cai et al. (2023) have academics believing that the solution to the worldwide plastic pollution problem might be the natural microbial populations. Researchers think that colonies of naturally existing bacteria might solve the global problem of plastic pollution (Li et al. 2023; Cai et al. 2023). Microorganisms such as *Ideonella sakaiensis*, *Pseudomonas putida*, and *Bacillus cereus* may degrade many forms of plastic, including PET and PU, via specific routes. New research indicates that microbes may destroy a broad variety of plastic polymers. This category contains microalgae, fungi, and bacteria. Microorganisms with certain metabolic pathways or enzymes may depolymerize polyethylene terephthalate (PET), a commonly used plastic in packaging and textiles, into monomeric components. Consequently, the substance's environmental persistence is greatly decreased (Zrimec et al. 2021; Allemann et al. 2024). Polyurethane is a versatile and generally dispersed chemical; indeed, certain fungi likely have metabolic routes and enzymes competent to break it down. Knowing these bacteria can break down plastic marks a historic turning point in bioremediation. This finding suggests that the features of microbes might be used to solve the global plastic pollution issue (Zrimec et al. 2). Finding plastic-degrading species in different environments, including *Ideonella sakaiensis, Pseudomonas putida*, and *Bacillus cereus*, suggests that naturally occurring microbial communities might be very important in tackling the global plastic pollution problem (Cai et al. 2023). Finding several microbes that can break down plastic has given researchers hope that naturally occurring microbial populations solve the worldwide plastic pollution problem (Li et al. 2023; Cai et al. 2023). Below are several instances of bacteria capable of metabolically decomposing various forms of plastic: The microorganisms *Ideonella sakaiensis*, *Pseudomonas putida*, and *Bacillus cereus* have distinct mechanisms for breaking down various forms of plastic, including PET and PU. Empirical studies have shown that bacteria can break down various plastic polymers. Organisms belonging to this category include bacteria, fungi, and microalgae. Specific metabolic pathways or enzymes in bacteria may facilitate the depolymerization of polyethylene terephthalate (PET), a widely used plastic in packaging and textiles, into its monomeric constituents. Nets. Hence, the material's ability to withstand the environment significantly reduces (Zrimec et al. 2021; Allemann et al. 2024). Jones et al. are anticipating Jones et al. 2023 with great anticipation. The ubiquitous and adaptable polyurethane may be susceptible to degradation by enzymes and metabolic pathways found in certain fungi. Because these microbes can degrade plastic, bioremediation has taken a giant leap forward. Based on these results, it seems that bacteria might solve the worldwide problem of plastic pollution (Zrimec et al. 2021). It has been shown that certain bacteria may break the ester bonds of polyethylene terephthalate, an enzyme-releasing plastic often used in packaging and textiles (Pérez-García et al. 2023). Using enzymes that break down PET polymers into their component components, such as PETase and MHETase, might reduce the amount of this ubiquitous plastic in the environment. Bhavsar et al. (2023) came to a similar conclusion, discovering that particular fungi may degrade polyurethane (PU), a versatile polymer, via certain metabolic pathways and enzymatic activity. A significant development in bioremediation has been the identification of bacteria that can degrade plastic; this finding offers the potential to harness the innate power of bacteria to mitigate the global plastic pollution crisis (Dunn and Welden 2023). Despite much progress in the lab, bringing microbial plastic degradation to the real world is still very difficult. The effectiveness of bacteria in decomposing plastic may be affected by environmental factors such as temperature, pH, nutrient availability, and other pollutants (Dunn and Welden 2023). Therefore, studies are being conducted to learn how these microorganisms function in different environments and to find ways to make them more resilient and adaptable (Yang et al. 2023). For example, Cai et al. (2023) discovered that the content and concentration of plastic trash in a certain habitat might affect the plastic-degrading capacity of bacteria. Based on the findings of Li et al. (2023), it is hypothesized that microbial communities, which comprise bacteria like *Ideonella sakaiensis* and fungi like *Pestalotiopsis microspora*, could experience dynamic changes in reaction to variations in plastic pollution levels. These changes could eventually result in the emergence of more specialized and efficient organisms that break down plastic. Dunn and Welden (2023) and Verschoor et al. (2022) state that there is great potential for incorporating microbial plastic degradation into all-encompassing waste management systems since researchers worldwide are constantly pushing the boundaries of this field. Researchers can find better and more sustainable ways to deal with the urgent problem of plastic pollution by using microorganisms' diverse capabilities and inherent adaptability, like the fungus *Pestalotiopsis microspora* and the bacteria *Ideonella sakaiensis* (Bertocchini and Arias 2023).

13.2.1 The Global Diversity of Microbial Plastic Degradation

The world's diverse microbial community has a wealth of neglected potential to address the plastic pollution problem. Shilpa et al. (2022) highlight that plastic-degrading bacteria are present in many different environments, which imply that nature has likely developed various solutions to this serious environmental problem. Through the study of the microbiome worldwide, scientists have discovered a plethora of bacteria, fungi, and archaea that can break down various types of plastic (Viel et al. 2023) as shown in Table 13.1. This growing body of information on the taxonomic variety of microorganisms that break down plastic is essential for building efficient bioremediation schemes (Kim et al. 2022). As scientists seek to discover the fundamental processes, enhance the degradation process, and tap into the whole potential of these bacteria, this growing list of microbes that can break down plastic lays the groundwork for future study and development. A more sustainable and circular approach to plastic waste management may be achieved by using this wide variety of microbes to produce more effective and efficient ways of breaking down different plastic polymers (Tao et al. 2023). Furthermore, complex plastic structures that could be resistant to certain microbial species can be broken down by microbial consortia due to their synergistic interactions (Salinas et al. 2023). To combat the plastic pollution challenge, scientists may tap into the full potential of the global microbiome by encouraging the expansion and collaboration of these particular bacteria (Viel et al. 2023).

13.2.2 Metagenomic Approaches to Plastic Bioremediation

Research into microbial communities' metabolic capability and genetic variety has been facilitated to an unprecedented degree by developments in high-throughput DNA sequencing technology, which have shaken up the discipline of microbial ecology. Until now, metagenomic methods have been beneficial in the search for new microbes and enzymes capable of degrading plastic (Wani et al. 2023a, b;

Table 13.1 Bacteria and Fungi that involved in plastic degradation

Type of Plastic	Organism breaking down plastic	References
Bacteria		
Bacillus subtilis	PS, PE	Asmita et al. (2015)
Streptococcus, pseudomonas sp. and *bacillus* sp.	Low density PE	Vignesh et al. (2016)
Pseudomonas putida	PE	Jumaah (2017)
Staphylococcus arlettae	PE	Kathiresan (2003)
Bacillus subtilis	PE	Jumaah (2017)
Pseudomonas fluorescencs	PE	Jumaah (2017)
Bacillus amylolyticus	PE	Jumaah (2017)
Pseudomonas	PE	Kathiresan (2003)
Bacillus firmus	PE	Jumaah (2017)
Staphylococcus sp.	PE	Kathiresan (2003)
Micrococcus sp.	PE	Kathiresan (2003)
Moraxella sp.	PE	Kathiresan (2003)
Bacillus cereus NBAII B	PE	Sharma et al. (2014)
Streptococcus sp.	PE	Kathiresan (2003)
Fungi		
Aspergillus Niger	PE	Mohan and Suresh (2015)
Aspergillus sp. and *fusarium* sp.	Low-density PE	Vignesh et al. (2016)
Aspergillus glaucus	PE	Kathiresan (2003)
Aspergillus Niger	PE	Asmita et al. (2015)
Aspergillus oryzae	PE	Indumathi and Gayathri (2016)
Aspergillus oryzaeA5,1 (MG77950	Low-density PE	Muhonja et al. (2018)
Aspergillus flavus	PE	Mohan and Suresh 2015
Aspergillus foetidus	PE	Mohan and Suresh (2015)

Edwards et al. 2022). Dunn and Welden (2023) state that scientists may learn more about the complex network of microbial interactions that control the breakdown of plastic polymer by examining the genetic blueprints of whole microbial communities. This will reveal the hidden variety of plastic-degrading capabilities. Based on the distribution of identified species and their apparent effectiveness in controlled settings, there is significant potential for microbial solutions to plastic bioremediation. However, questions remain regarding the efficacy of these microbes beyond laboratory circumstances (Shilpa et al. 2022; Temporiti et al. 2022). Imperfect waste management, the resistant nature of plastic polymers, and the mass manufacturing of plastics all contribute to the environmental buildup of plastic trash (Dunn and Welden 2023). Innovative solutions to this worldwide problem have recently received more attention in light of these difficulties and the mounting evidence of the possible negative impacts of plastic waste. Investigating microbial communities for their ability to break down various plastic polymers is one such promising strategy. According to Dunn and Welden (2023), nature has already developed a vast variety of remedies to this urgent environmental problem since recent research has shown that plastic-degrading microorganisms are widely distributed throughout different ecosystems. Bacteria, fungi, and archaea are just a few of the microbes scientists have found with the genetic and metabolic capacity to break down various types of plastic as mentioned by Kaur et al. 2023. As scientists strive to discover the fundamental processes, enhance the degradation process, and tap into the whole potential of these microorganisms, this growing list of plastic-degrading bacteria serves as a basis for future study and development (Salinas et al. 2023; Zhang et al. 2022). Researchers may develop more efficient and effective methods of decomposing different types of plastic polymers by using this diverse array of microorganisms (Shilpa et al. 2022; Verschoor et al. 2022). This will lead to a more sustainable and circular approach to managing plastic waste. In addition, microbial consortia may enhance plastic-degrading capacities via synergistic interactions. This allows for the breakdown of complex plastic structures resistant to individual microbial species, including *Pestalotiopsis microspore* and *Ideonella sakaiensis*. Researchers can harness the full potential of the microbiome as a whole to combat plastic pollution by encouraging the growth and cooperation of certain specialized microbes (Viel et al. 2023; Kaur et al. 2023). In addition, microbial consortia can enhance plastic-degrading capabilities through synergistic interactions, allowing for the breakdown of complex plastic structures that may be resistant to individual microbial species (Cai et al. 2023; Viel et al. 2023; Viel et al. 2023; Kaur et al. 2023). There has been a lot of success with microbial plastic breakdown in the lab, but getting it into the real world is still a huge hurdle. Regarding temperature, pH, nutrient availability, and other contaminants, several environmental variables might affect the efficacy of bacteria that break down plastic (Cai et al. 2023). Research is now underway to better understand how these microorganisms function in various environments and to find ways to make them more resilient and adaptable (Wang et al. 2022; Albright et al. 2021). Orlando et al. (2023) highlight that developing scalable and cost-effective technologies is crucial for integrating microbial plastic degradation into complete waste management systems. Additionally, regulatory and socioeconomic concerns must be taken into account. There is great hope for the future of waste management techniques that include these cutting-edge technologies for microbial plastic breakdown, which the scientific community worldwide is still trying to pin down (Danso et al. 2019) (Zrimec et al. 2021). Researchers can create more efficient and eco-friendly ways to deal with the urgent problem of plastic pollution by using microbes' natural flexibility and varied talents (Rüthi et al. 2023). Despite these encouraging developments, major obstacles remain to overcome before the results may be used in the actual world (Dunn and Welden 2023). Table 13.2 shows the genes and enzymes that are involved in plastic degradation in each organism.

Table 13.2 Genes and enzymes are involved in plastic degradation

Microorganism	Gen Bank accession number	Gene	Enzyme	Plastic	Reference
Thermobifida alba AHK11	AB445476.2	est1 and est2	Esterase	PET	Hu et al. (2010)
Thermobifida fusca DSM4432	HQ147787.1	cut1	Cutinase	PET	Herrero-Acero et al. (2011)
Fusarium oxysporum	JH658423.1	FOQG_139	Cutinase	PET	Dimarogona et al. (2015)
Thermobifida cellulosilytica DSM4453	HQ147786.1	cut2	Cutinase	PET	Hu et al. (2010)
Ideonella kaiensisstrain 201-F	_BBYR01000074	ISF6_RS23955*	PETase from the dienelactone hydrolase family protein	PET	Yoshida et al. (2016)
Thermobifida fusca	FR727681.1	Cut-2.KW3	Cutinase	PET	Danso et al. (2018)
Saccharomonos poraviridis AHK19	AB728484.1	cut190	Cutinase from the alpha/beta hydrolase family protein	PET	Kawai et al. (2014)
Thermobifida alba DSM:431	HQ147784.1	cut1	Cutinase	PET	Danso et al. (2018)
Thermomonos poracurvata DSM 431	NC_013510.1	TCUR_RS06300	Triacylglycerol lipase	PET	Danso et al. (2018)
Polyangium brachysporum strain DSM7029	NZ_CP011371.1	AAW51_RS12360	PET12 of the dienelactone hydrolase family prote	PET	Danso et al. (2018)
Oleispira Antarctica RB-8	FO203512	lipA	Lipase PET5	PET	Danso et al. (2018)
Paenibacillus sp. *strainDK*	MK045309.1	alkB	Alkane mono oxygenase A	PE	Bardají et al. (2019)
Pseudomonas chlororaphis	AF069748.1	pueA	Polyurethanase esterase A	PU	Stern and Howard (2000)
Pseudomonas chlororaphis	EF175556	pueA and pueB	Polyurethanase A and polyurethanase B	PU	Howard et al. (2007)
Comamonas acidovorans TB-35	AB009606.1	pudA	Polyurethane ester	PU	Nomura et al. (1998)

13.2.3 Bioengineering Plastic-Degrading Microbes

Researchers have used biotechnology to improve microorganisms' capabilities and study naturally existing bacteria that degrade plastic (Tao et al. 2023). Researchers have used genetic engineering to develop or enhance the production of enzymes that break down polymers including polyethylene terephthalate and polyurethane. According to Pérez-García et al. (2023), microbe engineering aims to create plastic-degrading platforms that are more stable, efficient, and particular to certain substrates. In comparison to naturally occurring bacteria, these engineered ones may be able to decompose plastic trash more quickly, with greater efficiency, and in a broader variety of environments (Zrimec et al. 2021; Wei and Zimmermann 2017; Ali et al. 2023). Nevertheless, concerns about biosafety and regulation around the use of GMOs in large-scale waste management systems need to be further investigated (Trump et al. 2023). Although there is no denying the promise of microbial solutions for plastic bioremediation, it is still a big problem to put what we learn in the lab into practice (Shilpa et al. 2022).

13.2.4 Enzymatic Degradation of Plastic Polymers

In the fight to break down stubborn polymers made of plastic, microbial enzymes have come into the spotlight. The capacity of these specialized biocatalysts, made by various microbes, to degrade intricate plastic structures has opened up exciting new avenues for the environmentally responsible recycling and disposal of plastic. As stated by Zampolli et al. (2022), the development of an enzyme that can break down polyethylene terephthalate (PET), a material used extensively in packaging and textiles, is a prime example. Extensive studies have been conducted on PETase and its derivatives because they provide a potential solution to the ongoing problem of PET waste building up in the environment. Soong et al. (2022) and Kaur et al. (2023) note that the microbial toolbox for plastic bioremediation is being actively expanded by investigating enzymes that may target additional hazardous polymers including polyurethane and polystyrene. Enzymes like this have uses beyond decomposing polymeric polymers. Turning plastic trash into a resource instead of a liability, some enzymes selectively break polymer chains, allowing the recovery of precious monomers or other high-value chemicals according to Pérez-García et al. (2023) and Orlando et al. (2023).

13.2.5 Novel Plastic Biodegradation Pathways

The isolation of new enzymes that break down plastic has been made possible by the discovery of a vast array of microbes that can break down various polymers, including both manmade and naturally occurring plastics (Allemann et al. 2024). Many of these new methods for biodegrading plastics rely on the cooperative efforts of numerous microbial populations to degrade the polymers' intricate architecture

(Cf et al. 2021). In order to develop better bioremediation solutions, scientists need to understand better the metabolic pathways and enzymatic mechanisms that microbes use to break down plastic. Research in areas with high levels of plastic pollution or biopolymers with comparable structures has uncovered novel microbial species and enzymes that can break down plastic more effectively (Priya et al. 2021; Cai et al. 2023). More efficient and large-scale solutions will need our growing knowledge of the metabolic flexibility, taxonomic variety, and enzymatic pathways of microbes that may break down plastic (Cai et al. 2023). To drive the development of more effective and scalable solutions for the bioremediation of plastic waste, our knowledge of the taxonomic diversity, metabolic versatility, and enzymatic mechanisms of microorganisms that break down plastic must continue to advance (Viel et al. 2023; Dunn and Welden 2023; Chow et al. 2022).

13.2.6 The Potential of Engineered Microbial Strains

Researchers have made great strides in microbial plastic bioremediation by identifying and studying naturally occurring microbes that can break down plastic. However, there is great potential for future advancements in this area through the engineering of genetically modified organisms that can break down plastic even more effectively (Cai et al. 2023). Researchers may create microbial strains with improved enzyme synthesis, plastic-degrading efficiency, and modified metabolic pathways by using contemporary molecular biology and synthetic biology methods (Kaur et al. 2023). Improving the pace and amount of plastic breakdown and expanding the ability to target a larger spectrum of polymers may be achieved via the creation of these tailored microbial systems (Kaur et al. 2023). Furthermore, according to Mukherjee and Koller (2023), new possibilities for the efficient and long-term management of industrial-scale plastic waste may be unlocked by incorporating these engineered microorganisms into scalable bioremediation processes, such in-situ or ex-situ treatment systems. Finally, the worldwide plastic pollution catastrophe has tremendous potential for resolution within microbial plastic bioremediation. Microorganisms such as *Pseudomonas* and *Ideonella sakaiensis* have potential to decompose plastic (Kaur et al. 2023).

13.2.7 Plastic Degradation in Marine Environments

According to Dunn and Welden (2023), microbial-based bioremediation techniques face a distinct obstacle in maritime habitats where plastic garbage is prevalent. Finding perfect plastic-degrading microorganisms that can survive in diverse and intricate marine ecosystems is of the utmost importance. According to Dutta and Bandopadhyay (2022) and CF et al. (2021), several microorganisms may find it difficult to thrive and degrade plastic in marine habitats because of high salinity, changing temperatures, and restricted nutrition availability. To successfully bioremediate the world's oceans and coastal areas, where huge amounts of plastic pollution accumulate and endanger marine ecosystems and biodiversity, it is crucial to overcome these environmental hurdles and develop plastic-degrading microbial strains adapted to marine conditions (Cai et al. 2023; Viel et al. 2023). *Pseudomonas* and *Ideonella sakaiensis* are two examples of marine bacteria that have recently come to light as having the ability to break down different types of plastic polymers (Jesus and Alkendi 2023). According to Cai et al. (2023), these bacteria have shown that they can decompose polymers using enzymes and use plastic as a carbon and energy source. The capacity of marine microorganisms to break down plastic is crucial since there is a lot of trash floating around in the seas, and it threatens marine life and ecosystems (Allemann et al. 2024). The effects of marine plastic pollution are far-reaching; it kills marine life, messes with food webs, and, via microplastics, may affect human health. To combat this mounting problem and safeguard ocean health, it is essential to tap into the abilities of marine microbes to decompose and treat this plastic trash (Bertocchini and Arias 2023; Ziani et al. 2023).

13.3 The Application of Genetic Engineering and Artificial Biology to Enhance Bioremediation

According to Dunn and Welden (2023), genetic engineering and synthetic biology have facilitated the development of more effective microbes that break down plastic. This may enhance the probability of bioremediation as a viable approach to further ameliorate the problem of plastic pollution. As our knowledge of microbes breaks down plastic on a large-scale, pragmatic basis rises, it will be important to use our expertise to reduce the environmental harm plastic causes (Jesus and Alkendi 2023). This is fantastic for bioremediation as a weapon against plastic contamination (Shilpa et al. 2022). Genetic engineering and synthetic biology have created fascinating new opportunities for producing modified bacteria with enhanced plastic-degrading capacity. Modern tools in these fields might enable scientists to develop distinctive microbial strains that rapidly and effectively break down plastic polymers. Targeting targeted genetic changes that could give engineered microbes better enzyme production, specialized metabolic pathways, and increased environmental resilience helps them to break down plastic far more effectively than naturally occurring microbes, claims Ekanayaka et al. (2022). Cai et al. (2023) found by including these bio-engineered bacteria in plastic bioremediation systems that these approaches would be more pragmatic generally as a whole solution to the global plastic pollution problem. If they could also make garbage management more scalable, reasonably priced, and ecologically sustainable, then solving the pervasive problem of plastic waste would be further advanced (Venkatesan et al. 2022) source.

13.3.1 Enzyme Engineering Advancements and Bioprospecting

Complementing the inherent biodegradability of microorganisms, new advances in enzyme engineering have greatly enabled flexible, selective, and efficient plastic breakdown. Scientists have achieved considerable advancement by creating enzymes with better stability, higher catalytic activity, and more specificity to a greater spectrum of plastic polymers (Pérez-García et al. 2023; Atanasova et al. 2023). Integration of synthetic biology, directed evolution, and protein engineering produces synthetic enzyme variants. These modifications may provide a more precise breakdown of plastic than its original structural forms. Kaur et al. (2023) demonstrate that these altered enzymes significantly expedite and improve plastic polymers' degradation, potentially expediting the bioremediation process for plastic waste. Apart from the developments in enzyme engineering, bioprospecting has given crucial knowledge on the range of bacteria capable of breaking down plastic and their prospective applications. Research of the world's microbiome has lately revealed a wide range of novel microbial species and the enzymes they generate to degrade plastic (Jesus and Alkendi 2023). Our increasing awareness of the range of bacteria that break down plastic and their habitats will enable us to choose and use the optimal organisms or enzymes for bioremediation in many situations (Cf et al. 2021).

13.4 The Relationship Between Microbial Diversity and Plastic Degradation

Analysis of the microbiome indicates a substantial potential for natural microbial communities to degrade plastic via an intricate network of bacteria in many environments. A study by Zhai et al. (2023) examined the impact of plastic pollution levels on the distribution and abundance of microorganisms involved in plastic degradation in different environments. Microbial colonies seem to be developing new metabolic routes and enzymes to break down these refractory polymers, enabling adaptation to plastic waste. Analyzing the taxonomic diversity and functional capability of these plastic-breaking microorganisms will assist scientists in better grasping the dynamics and processes of microbial plastic bioremediation (Cai et al. 2023; Dunn and Welden 2023). Among other significant outcomes of these investigations are the discovery of many plastic-breaking bacteria from several taxa. Such variety underscores the need to implement a comprehensive strategy to harness and use the microbiota's capacity to break down plastic (Cai et al. 2023; Ekanayaka et al. 2022). Novel bioremediation methods that use the natural ability of microbial populations to adjust and respond to environmental changes, such as the introduction of plastic contaminants, show great potential (Dunn and Welden 2023). Researching the metabolic diversity and taxonomic distribution of plastic-breaking bacteria in the microbiome might enable scientists to grasp the benefits of natural processes, which will help them produce more effective bioremediation solutions. With this knowledge in hand, new biotechnological approaches may be created to capitalize on the inherent capacities of microbial populations and thus more fully and effectively address the global plastic pollution issue (Dunn and Welden 2023). If scientists can identify and define significant microorganisms that break down plastic, their metabolic paths, and the elements influencing their proliferation in environments contaminated with plastic, improved and more flexible microbial systems that can break down a wider spectrum of plastic polymers can be developed. The major goal of this broad investigation is to understand how the microbiota breaks down plastic. Long-term bioremediation plans developed from this data might potentially help reduce plastic pollution's effects on ecosystems (Allemann et al. 2024; Cai et al. 2023). Modern bioinformatics tools and procedures have now enabled researchers to better understand the metabolic pathways, enzymatic processes, gene profiles, and genomic sequencing that bacteria employ to degrade plastic. This research suggests that bioremediation strategies could be developed by identifying which plastic polymers degrade more rapidly and generate more adaptable microbial strains, as per Viel et al. (2023). The environmental implications and ecological ramifications of these microbial-driven solutions can be assessed through bioinformatics analysis, which is essential for their sustainable deployment and to mitigate the risk of damage to natural ecosystems (Cifuentes et al. 2023).

13.5 The Role of Bioinformatics and Synthetic Biology in Plastic Bioremediation

13.5.1 Bioinformatics and Plastic Bioremediation

Synthetic biology's capacity to create microbes with improved plastic-degrading skills has made it a potent tool in plastic bioremediation. Genetic engineering has allowed scientists to build bacteria with altered metabolic pathways and new genes that more efficiently break down different plastic types (Cai et al. 2023; Kaur et al. 2023). One promising option is the use of synthetic microbial consortia, which are groups of bioengineered microbes designed to degrade plastic waste in a complementary and synergistic manner. The present methodology exploits the synergistic properties of many microbial strains (Wani et al. 2023a, b; Ali et al. 2023; Priya et al. 2021) to efficiently degrade a wider variety of plastics and transform their constituents into useful chemicals or biofuels. Synthetic biology makes finding novel plastic-degrading enzymes and exploring other enzymatic paths feasible. Studying the genetic potential of communities of naturally existing bacteria and evaluating their metagenomes can help researchers identify and design novel biocatalysts that break down plastic more effectively (Li et al. 2023; Cai et al. 2023). Creating highly specialized microbes is only one of many ways synthetic biology may be used in plastic bioremediation. By using bioinformatics and computational modeling, plastic-degrading systems may be fine-tuned for optimal performance and the lowest environmental effect (Zrimec et al. 2021; Priya et al. 2021; Ali et al. 2023; Wei and Zimmermann 2017). Although microbial plastic degradation has great promise, some issues must be resolved before this technology may reach its maximum capability. Plastic bioremediation systems that are scalable, environmentally benign, and financially feasible will need constant research and development in disciplines such as synthetic biology, microbial ecology, and bioinformatics (Kaur et al. 2023).

13.5.2 Plastic Degradation: Bioinformatics and Enzymes

Our knowledge of the enzymes and methods bacteria utilize to break down plastic substances has advanced thanks in great part to bioinformatics. Analysis of metagenomic and genomic data has produced various new enzymes that break down plastic identified and described. This finding has been highly beneficial to Dunn and Welden (2023) as it has shown the taxonomic variety and metabolic pathways linked to this process. Bioinformatics has recently uncovered new enzymes that can break down plastic. These enzymes may originate from marine microbes and are catalyzed by. One may now study and control these enzymes to increase their efficiency and applicability. Wani et al. (2023a, b) advise that by combining computational analysis with experimental validation, scientists should be able to identify and use sensible bioremediation techniques. Bioinformatics, machine learning, and artificial intelligence (AI) algorithms might be utilized to develop and manufacture microbe consortiums and enzymes that degrade plastic more efficiently (Jesus and Alkendi 2023). Viel et al. (2023) explain a more sustainable and circular economy as employing microbial plastic bioremediation systems to maximize efficiency and environmental performance. Modern bioinformatics methods and approaches (Tao et al. 2023; Cai et al. 2023) will help researchers better understand the genetic makeup, metabolic pathways, and enzymatic activity of microorganisms engaged in plastic degradation. This information enables us to create bioremediation systems capable of breaking down many plastic polymers and promoting more flexible and efficient microorganisms. Moreover, bioinformatics tools might be used to evaluate the ecological consequences and environmental effects of these created solutions. This strategy will help us to guarantee its use in an eco-friendly and non-disruption of natural ecosystems way (2023). Conducting research may improve the efficacy and environmental quality of these biotechnological solutions, thus ensuring that microbial plastic bioremediation techniques help produce a more sustainable and circular economy. Advanced bioinformatics tools and technologies including data analytics, computational modeling, and genomic sequencing, could help researchers to grasp better the metabolic pathways, enzymatic mechanisms, and genetic traits of microorganisms engaged in plastic degradation (Cai et al. 2023). Equipped with this understanding, we can create more efficient and flexible bioremediation methods that break down different plastic polymers and foster microorganisms. Bioinformatics studies might be utilized to evaluate their environmental impacts and ecological repercussions to guarantee the sustainable deployment of these microbial-driven solutions with minimum disruption to natural ecosystems (Cifuentes et al. 2023; Mukherjee and Koller 2023).

13.6 Future Opportunities and Difficulties

Even with great progress in bioremediation, some issues still need to be resolved before it is used generally. Improving the scalability and efficiency of methods for decomposing plastic (Verschoor et al. 2022) presents a main difficulty. Experiments conducted in controlled laboratory settings have shown that bacteria or enzymes may degrade plastic. Still, significant advancements are required before these findings can be successfully and effectively used on a large scale. Thorough consideration of several factors is necessary to ensure the practicality and cost-effectiveness of these solutions. These include the availability of appropriate substrates, the optimization of reaction conditions, and the inclusion of bioremediation into the present waste management system (Patel et al. 2020). Examining the breakdown products and by-products generated during bioremediation operations requires more study, focusing on their long-term environmental destiny and possible effects. The goal is mostly to reduce the environmental persistence of plastics. Examining the by-products' possible buildup or transformation can help reduce the total environmental effect (Danso et al. 2019). Building comprehensive laws and regulatory systems that support and advance bioremediation technologies will help further accelerate the general implementation of these solutions (Francocci et al. 2020). Despite all these obstacles, bioremediation's promise to solve the worldwide plastic pollution problem is still quite attractive. As scientists investigate the many powers of bacteria and enzymes, it is clear that bioremediation might greatly help to solve plastic pollution. Simultaneously, technological developments help to enable more general use of these solutions (Zrimec et al. 2021; Cf et al. 2021). Targeting a larger spectrum of plastic polymers, researchers have made major progress in the creation of enzymes with improved stability, substrate selectivity, and catalytic activity. Urbanek et al. (2021) have shown that the introduction of synthetic enzyme versions that degrade polyethylene terephthalate (PETase) and methoxyethanolamine (MHETase), both of which occur naturally in fungus, has increased the effectiveness and selectivity of plastic breakdown. Bioprospecting, in conjunction with developments in enzyme engineering, has yielded significant insights into the diverse range of bacteria capable of decomposing plastic and the potential applications of these enzymes (Bhavsar et al. 2023). By examining the worldwide microbiome, researchers have discovered several previously undisclosed microbial species along with the matching enzymes that break down plastic structures. These investigations have enhanced our understanding of their ecological distribution and taxonomic diversity. Scientists have developed altered enzyme types to hasten plastic breakdown relative to naturally occurring enzymes. Synthetic biology, guided evolution, and protein engineering are some of the methodologies that fall under this area of research. The use of synthetic enzymes in the bioremediation of plastic waste might be beneficial since they dissolve polymers more effectively than existing procedures (Ali et al. 2023). Despite the encouraging findings reported in the study, using bioremediation techniques remains difficult (Wani et al. 2023a, b). Researchers are continually looking for unique solutions to increase the resistance and efficiency of bacterial breakdown of plastics. Research on this constantly changing field is being conducted by Chu et al. 2020 and Cai et al. 2023. Kaur et al. (2023) underline the need to look at microbial consortia as synergistic interactions between many species could help each other to increase their plastic-degrading abilities. Future developments in synthetic biology and genetic engineering might potentially help to maximize the possibilities of bioremediation as a solution to plastic pollution by allowing the manufacturing of modified microorganisms with amazing plastic-degrading capacity (Viel et al. 2023). New prospects shown by present research point to a bright future for bioremediation in tackling plastic pollution (Dunn and Welden 2023; Kaur et al. 2023). Cai et al. (2023), Viel et al. (2023), and Allemann et al. (2024) concur that implementing large-scale, practical applications based on our understanding of microbial plastic degradation is the most effective strategy for mitigating the environmental impacts of plastic waste. The bacteria's widespread presence and apparent efficacy in marine and terrestrial habitats further emphasize the bioremediation potential. The discovery of several organisms that can break down plastic in various environments raises the possibility that colonies of naturally occurring microbes might be the solution to the worldwide problem of plastic pollution. Recent studies indicate findings from Dunn and Welden (2023), CF et al. (2021), and Zrimec et al. (2021). Nevertheless, numerous obstacles remain to be addressed before the laboratory results can be applied on a large scale. A significant challenge is that these bacteria cannot be optimized for factors such as nutrient availability, pH levels, and temperature outside of a controlled laboratory environment (Yamamoto et al. 2022). However, new methods to improve the efficiency and resilience of bacteria that break down plastic are being investigated in this field, and the pace of research is accelerating. As part of this effort, researchers are looking at microbial consortiums, in which different species work together to break down plastic more effectively (Cai et al. 2023). One approach under research to enhance the breakdown of plastic polymers is microbial consortia. Many bacteria collaborate among these groupings to break down the material. Microbial consortia may break down a larger range of plastic types than individual species may by employing complementing metabolic pathways and specialist enzymes from diverse bacteria (Ru et al. 2020). Though in many research bioremediation methods have demonstrated encouraging results, their application is still challenging (Wani et al. 2023a, b). New methods to allow the bacterial breakdown of plastics to be more effective and resistant are continually being researched by researchers. Investigating this dynamic area closely falls to the disciplines of Chu et al. (2020) and Cai et al. (2023). Kaur et al. (2023) underline the need to look at microbial consortia as synergistic interactions between many species could help each other to increase their plastic-degrading abilities. Future developments in synthetic biology and genetic engineering might potentially help to maximize the possibilities of bioremediation as a solution to plastic pollution by allowing the manufacturing of modified microorganisms with amazing plastic-degrading capacity (Viel et al. 2023).

13.7 The Circular Economy and Plastic Waste Management

The incorporation of microbial solutions into general waste management techniques not only promotes the growth of plastic-breaking bacteria but also aids in the creation of a sustainable and regenerative economy. According to Ding and Zhu (2023), integrating biological and mechanical waste management systems allows for the recovery, repurposing, and optimum use of plastic resources. Integration of mechanical recycling with microbial degradation in complete systems has the potential to improve the efficiency of plastic waste handling, claims Yue et al. (2023). This promotes the cyclical flow of plastic goods across the economy and assists vital components to be retrieved. For instance, mixing microbial degradation with chemical or thermal recycling helps one to recover valuable elements from plastic waste. These parts could then be recycled and used in the industry. In line with this, Allemann et al. (2024) discovered that incorporating plastic-breaking microorganisms into waste sorting and separation systems might help to maximize recycling or biodegradability. Lastly, less plastic garbage would make its way into the surroundings. These all-encompassing approaches addressing the complete lifetime of plastics enable one to attain a more sustainable and circular economy. Employing the integration of mechanical and biological technologies over the plastic waste management system will make optimal recovery, reuse, and proper disposal of plastic resources possible. By lowering waste, improving material circular flow, and recovering significant plastic components, this multifarious approach at last helps build a more sustainable and environmentally aware company model (Hendrickson et al. 2024). As the study on microbial plastic breakdown develops, our knowledge of the many types of bacteria capable of breaking down plastics is also developing. Future research will be focused on enhancing our comprehension of the enzymes involved in microbial plastic degradation seeks to identify methods for increasing the efficiency of this process, ultimately providing scalable solutions to the pervasive problem of plastic pollution. Additionally, the expanding catalog of microorganisms capable of plastic degradation establishes a foundation for further investigations into the enzymatic mechanisms at activity. Employing a more sustainable and circular approach to handling plastic waste, researchers want to assist in lessening the negative consequences of this long-lasting pollution on the ecology (Diana et al. 2023). Scientists are discovering new approaches to address the global plastic pollution issue by using bacteria's metabolic diversity and innate resilience. Bioengineering, along with the enhancement of microbial strains and the examination of naturally occurring bacteria that decompose plastic, assists researchers in refining and augmenting existing methodologies for plastic waste remediation and degradation (Viel et al. 2023). Microbially driven solutions have significant promise for decreasing the environmental effect of plastic and steering us toward a more circular economy, where plastic may be recycled or disposed of ethically, as stated by Wei and Zimmermann (2017), Dunn and Welden (2023), and Priya et al. (2021).

13.7.1 Economic Prospective of Plastic Bioremediation

It is essential to address the question of whether or not extensive, economically viable use of plastic bioremediation techniques is justified. Scientists and lawmakers have to weigh the likely financial and environmental rewards of these bioremediation methods against their expenses of research, implementation, and maintenance (Shakya et al. 2020; Patel et al. 2020). Biotechnological treatments' cost determines whether plastic bioremediation is financially feasible. Factors influencing the overall cost might include genetic engineering charges, bioreactor design and construction costs, and process scalability (Dunn and Welden 2023; Chalermthai et al. 2023). The economic research should also include the expected income from recycled or reused plastic products, the possible savings from conventional trash handling, and environmental restoration costs. Through bioremediation, one may create valuable goods or by-products such as recovered polymers from plastic or high-value chemicals, therefore somewhat offsetting the systems' upfront and ongoing expenses (Shakya et al. 2020; Johnston et al. 2022). Table 13.3 for example shows the Antigua and Barbuda's (2019) plastic waste leakage rates (tons annually) per type of plastic polymer and per sector. Plastic bioremediation systems may also be more economically viable by removing environmental cleaning and remediation tasks and by reducing the requirement for landfilling, incineration, and other traditional waste disposal methods (Yue et al. 2023). Good plastic bioremediation solutions rely on carefully balancing technical advancement, regulatory oversight, and economic viability. Collaboration among academics, policymakers, and business leaders is necessary to develop and implement economically feasible and environmentally sustainable bioremediation methods to tackle the global plastic pollution problem (Anand et al. 2023). Indeed, the identification and use of bioremediation methods for treating waste pollution provide a practical solution to the global plastic crisis. Scientists are using the capacity of plants and bacteria to break down and remove dangerous chemicals from plastic (Cai et al. 2023), therefore advancing the development of sustainable and eco-friendly ways of handling plastic waste. Still, the successful adoption and use of these bioremediation techniques rely on a whole strategy that includes scientific and technological components along with economic,

Table 13.3 Antigua and Barbuda's (2019) plastic waste leakage rates (tons annually) per type of plastic polymer and per sector

Plastic Polymer	Commercial waste leakage rates (ton/year)	Household leakage rates (ton/year)	Tourism leakage rates (ton/year)
PET	24.0	73.6	26.8
PVC	77.8	73.3	3.9
HDPE	16.7	50.1	24.8
PP	7.0	25.2	8.2
LDPE	28.4	32.7	7.3
PS	4.5	19.1	4.3
Other	52.9	96.6	16.2
Total	211.4	370.6	91.5

legal, and social aspects (Dunn and Welden 2023). The growing awareness of the development of bioremediation technology follows from the recent recognition of the global plastic pollution problem. Scientists are significantly addressing this important environmental issue by using the capacity of plants and microbes to break down and eradicate toxins generated by plastic (Cai et al. 2023). Thanks to genetic and metabolic engineering techniques, plastic polymers can be broken down by microbial enzymes and metabolic paths that Jesus and Alkendi (2023) and CF et al. (2021) can reveal and maximize. Researchers have also investigated plant-based bioremediation methods like phytoremediation and rhizoremediation and found that they may effectively absorb and break down plastics (Priya et al. 2021; Sheth et al. 2019; Jesus and Alkendi 2023). Still, the wide use of these bioremediation techniques relies on strong regulatory frameworks to guarantee their effectiveness, safety, and environmental sustainability. If one wants to properly manage plastic waste streams passing through bioremediation processes using genetically modified bacteria and introducing non-native plant species, governing organizations and regulatory agencies will have to develop comprehensive rules and standards, according to Dunn and Welden (2023). To safeguard both humans and the environment, these legislative systems should address concerns regarding potential ecological impacts, ensure the regulation and containment of modified organisms, and mandate the monitoring and verification of plastic degradation rates and by-products, as per Viel et al. (2023) and Bidashimwa et al. (2023). An additional critical factor is the financial feasibility of substantial plastic bioremediation initiatives. The bioremediation systems' development, deployment, and maintenance costs should be balanced against the potential financial and environmental benefits (Dunn and Welden 2023; Anand et al. 2023). The total cost may be influenced by the scalability of the process, the expenditures associated with genetic engineering, and the design of the bioreactor. Consequently, these methods may become more financially viable when we integrate the revenue streams generated from recycling or repurposing damaged plastic products with the savings achieved through conventional garbage management and environmental restoration (Zou et al. 2023; Uekert et al. 2023). To identify long-term, ecologically acceptable bioremediation solutions to the global plastic pollution crisis, it will be essential to establish close collaboration among academics, government officials, and industry leaders (Global Plastics Outlook: Policy Scenarios to 2060 2022). Promising results of general plastic bioremediation research show that plants and microbes can break down and eradicate plastic pollution. Strong regulatory frameworks to consider their cost and for different parties to collaborate so that they may be accepted generally and sustained for a long time would enable us to make these methods successful (Sh.2 eth et al. 2019).

13.7.2 The Life Cycle Evaluation of Plastic Bioremediation

Orlando et al. (2023) assert that properly investigating the environmental consequences of microbial plastic bioremediation calls for a thorough life cycle assessment method. This comprehensive evaluation of the bioremediation process should take into account the energy and resource inputs needed for degradation as well as the environmental benefits or trade-offs associated with the end-products or by-products (Baldera-Moreno et al. 2022; García-Depraect et al. 2021). Thorough life cycle studies allow policymakers and scientists to pinpoint the best ways to include microbial-based plastic breakdown into waste management systems. Through means of microbial plastic bioremediation technologies, which this study might optimize, a more sustainable and circular economy may be reached (Villegas-Méndez et al. 2022).

13.7.3 Phytoremediation of Plastic Contaminants Surprising

As most studies on plastic bioremediation have centered on bacteria, new studies have shown that plants may help remove and break down plastic pollutants from the environment (Dunn and Welden 2023). Especially useful for tackling plastic pollution in terrestrial environments (Dunn and Welden 2023), phytoremediation methods founded on plants could enhance microbial bioremediation. An et al. (2023) and Šerá et al. (2022) claim that some plant species including sunflowers, bamboo, and switchgrass can accumulate and degrade different kinds of plastic polymers, including polyethylene and polypropylene, using their metabolic activities and the action of related microbial communities. Through their metabolic processes and the activities of related microbial communities in their rhizosphere, some plant species have shown the capacity to accumulate and break down several types of plastic polymers, including polyethylene and polypropylene (Salinas et al. 2023). Plants acting as bioremediation of plastic pollution are especially appealing when it comes to soil and terrestrial environments, where plastic waste can build up and risk ecosystem health and productivity (Jesus and Alkendi 2023). Wei and Zimmermann 2017; Dunn and Welden 2023; Zrimec et al. 2021). Phytoremediation is a supplementary technique of microbial-based bioremediation solutions that purposefully employ genetically engineered or optimized plant varieties while leveraging plants' natural abilities. Combining the advantages of microbial and plant-based systems under an integrated strategy is indicated to assist in addressing the worldwide plastic pollution problem (Wani et al. 2023a, b).

13.8 Hybrid Plastic Bioremediation Methods

Combining microbial and plant-based bioremediation technologies into hybrid systems which are already rather promising as solutions to plastic pollution may assist in making plastic degrading activities even more efficient and scalable. Combining the complementing capacities of plants and microbes might provide very remarkable synergistic solutions (Zrimec et al. 2021). For instance, a greater spectrum of plastic polymers may be broken down more totally if synthetic microbes were paired with certain types of plants. Furthermore, including these hybrid systems in scalable bioremediation systems in-situ or ex-situ treatment facilities may provide fresh chances for the efficient and long-term management of industrial-scale plastic waste, asserts Rüthi et al. (2023). One way to try to solve the global plastic pollution issue is using the development and optimization of hybrid plastic bioremediation systems. Plant-based as well as microbiologically based approaches include these. Most often used methods in stirred tanks and membrane bioreactors are enzymatic hydrolysis. One waste product of biodegradation are microbial biomass. Using organic polymers as energy and growth substrates produces waste products. Microorganisms' enzymatic energy collecting effectively breaks down natural polymers. Fascinatingly, microbes have developed also these ancient routes of breakdown for use in breaking down synthetic plastic polymers. For instance, *Ideonella sakaiensis* 201-F6 was shown to flourish just on PET. Two recently discovered enzymes catalyzed in harmless monomers by PET hydrolysis include PETases and MHETases [such as ethylene glycol (EG) and terephthalic acid (TPA) (Tamoor et al. 2021).

13.9 Scalability and Implementation Challenges

Although bioremediation techniques might assist in reducing plastic pollution, some issues still have to be handled before they can be applied and developed (Bidashimwa et al. 2023). Microbial and plant-based breakdown systems aiming to be efficient find great difficulty in the fact that many plastic polymers are innately resistant to and stable. Research is under way to find and create microorganisms and enzymes enabling of more effective breakdown of plastic. Another area of study is looking into metabolic pathways and synergistic microbial communities that can speed up the biodegradation of these long-lasting compounds (Priya et al. 2021; Nisha et al. 2020). Creating bioremediation techniques that might be used on an industrial basis and yet be fairly cheap is another vital concern. Combining existing waste management methods with contemporary monitoring and control technologies (Bakan et al. 2021; Villegas-Méndez et al. 2022) improves reactor designs, boosting bioremediation systems' efficiency and lifetime. Moreover, the effective use of bioremediation techniques for plastic pollution relies on a complete strategy incorporating legal, economical, and societal challenges. Cooperation among academics, lawmakers, industry, and the public will be necessary to overcome these challenges and allow bioremediation to be widely adopted to respond to the worldwide plastic pollution disaster (Dunn and Welden 2023; Viel et al. 2023). Scientists are making great progress in tackling the worldwide plastic pollution problem using plant and bacterial capacities to breakdown and eradicate plastic pollutants. Jesus and Alkendi (2023) and CF et al. (2021) foresee a bright future for sustainable plastic waste management if hybrid systems incorporating microbial and plant-based bioremediation techniques are developed using ways for their scalable and cost-effective implementation. These bioremediation approaches will be even more important as they define a better, more sustainable future (Cf et al. 2021; Dunn and Welden 2023; Priya et al. 2021; Sheth et al. 2019).

13.10 Plastics Bioremediation Policies

Strong legal frameworks have to be developed to guarantee plastic bioremediation methods' environmental sustainability, safety, and effectiveness (Kaur et al. 2023). The use of genetically modified organisms, the introduction of non-native plant species, and the efficient management of plastic waste streams flowing through bioremediation depend on strict laws and directions from regulatory and regulating organizations. These legislative frameworks should assist in addressing problems with likely ecological repercussions (Bidashimwa et al. 2023). They also demand surveillance and confirmation of plastic breakdown rates and by-products as well as assurance of the containment and control of modified organisms. Through the danger of invading plant or microbial species disturbing native ecosystems, one approach these control systems should solve problems regarding the possible ecological effects of imported species. According to Thomas (2023) and Poland et al. (2021), laws for their containment and control will assist in halting the uncontrollably spread of genetically modified organisms or escape. These models should also include the monitoring and validation of plastic breakdown rates and by-products to further guarantee the safe and effective remedial action of plastic waste and analyze the environmental impact of bioremediation approaches (Malik et al. 2023).

13.10.1 Social and Ethical Dilemmas

Though plastic bioremediation has made great progress in science and technology, we should not disregard the more general social and ethical challenges these solutions raise. Using genetically modified organisms or non-native plant species for bioremediation concerns many about the likely ecological disturbances and unintended repercussions (Trump et al. 2023; Gene editing in agrifood systems 2022). Researchers, authorities, and lawmakers must tightly collaborate to build thorough frameworks addressing these problems and guarantee the safe, effective, and ecologically responsible use of plastic bioremediation technology, as already mentioned in Sheth et al. (2019) and CF et al. (2021). Research, technology, control, and socioeconomic challenges all of which should be considered in a cooperative and interdisciplinary endeavor define effective plastic bioremediation solutions (Lee et al. 2022). At least, creating bioremediation techniques for garbage presents one feasible answer to the worldwide plastic dilemma. Adopting and implementing these technologies effectively rely on resolving the many scientific, legal, economic, and social issues accompanying their broad implementation (Capetillo et al. 2022).

13.11 Increasing Understanding About Ecologically Suitable Bag Usage Includes Ceramic Products

Encouraging people to use ceramic and paper bags for plastic cutlery can assist in lessening our reliance on this resource. Among biodegradable materials, their processes could demand cardboard, paper bags, and environmentally friendly containers. Better waste management systems combined with recycling and upcycling projects might assist in generally lower plastic waste, Orlando et al. (2023) observe. Combining these approaches with the advancement of bioremediation technology would provide a comprehensive and varied solution for the worldwide plastic pollution issue. Combining many approaches for waste reduction, promoting sustainable alternatives, creating effective waste management systems, and monitoring the most current findings on microbes and enzymes that break down plastic might assist one in approaching a more comprehensive plan (Yue et al. 2023). Zrimec et al. (2021) argue that by reducing the negative effects of plastic waste producing plastic recovery and essential aspects of a circular economy, this method may bring in a more sustainable future. Plastic bioremediation guides us toward a brighter future. Demand for long-term, all-encompassing solutions for the worldwide plastic pollution challenge is steadily growing. New advancements in plastic bioremediation provide encouraging methods to handle this problem (García-Depraect et al. 2021; Zhu et al. 2022). Recent research (Cf et al. 2021) implies that bacteria might be the secret to unleashing the value of a wide range of plastics, from simple polymers like polyethylene to more sophisticated ones. Finding the best microorganisms and enzymes for plastic decomposition is one primary goal. Researchers have found many likely possibilities after looking at several habitats, including those with high concentrations of plastic and those with high levels of natural polymers. Researchers have identified fresh bacterial and fungal strains capable of effectively breaking down different types of plastic polymers and looked at the metabolic pathways and enzymatic reactions supporting biodegradation. Scientists are also changing these bacteria to further break down plastic, which might lead to more effective and easier-to-scale-up bioremediation processes (Antranikian and Streit 2022; Qin et al. 2021; Cf et al. 2021). Moreover, the benefits of ideas for the circular economy include plastic bioremediation. One may make the idea of a trash-free economy true through biological processes changing plastic waste into usable chemicals, fuels, or other commodities. Researchers are still trying to make plastic bioremediation feasible by tackling its technological and logistical issues (Sheth et al. 2019). More broad use of plastic bioremediation technology will be facilitated by building legislative incentives and policy frameworks supporting its usage. Although plastic bioremediation is still in its early years of development, Dunn and Welden (2023) argue that a multidisciplinary approach, including microbiology, biochemistry, engineering, and policy-making, is required to fully use these creative ideas and negotiate towards a more sustainable, plastic-free future.

13.12 Including Plastic Bioremediation's Sustainability

There cannot be any progress toward a more sustainable global economy without including plastic bioremediation in broader sustainability projects. Building closed-loop systems that recycle and reuse old plastic (Damayanti et al. 2022) would help researchers increase the positive effects of the bioremediation methods on the economy and the environment. Combining plastic bioremediation with other circular economy techniques such as recycling, upcycling, and reusing lets one optimize plastic resources while lowering waste. By means of a comprehensive approach, we will be able to create a plastic ecosystem that can withstand time, thus reducing pollution and optimizing material recycling (Vidal et al. 2024). Moreover, recycling and reusing old plastics might motivate the development of fresh goods, supporting the expansion of a more circular economy. According to Yue et al. (2023), researchers may completely solve the global plastic problem by integrating plastic bioremediation technologies with more general sustainability objectives.

13.13 Searching New Microbial and Enzymatic Pathways

Scientists are continuously looking for fresh enzymes and microorganisms that can break down different kinds of plastic. This category includes naturally occurring microbes that break down related polymers, including cutin and lignin, and those that thrive in environments heavy in plastic trash (Cai et al. 2023). More effective and versatile bioremediation systems will result from genetically and metabolically changing these organisms to break down plastic even more (Priya et al. 2021). Plastic bioremediation will develop depending on the ongoing search for new bacteria and enzymes able to degrade a greater spectrum of plastic polymers. Plastic bioremediation might be expanded to manage a wider spectrum of plastic polymers as metabolic and genetic engineering advances, optimizing these pathways and raising the efficiency of plastic-degrading capabilities. If researchers find and design fresh microbial and enzymatic systems for plastic bioremediation, Mukherjee and Koller (2023) claim they could find better solutions to the worldwide plastic pollution problem.

13.14 Increasing and Restoring Cost-Effectiveness

Improving the cost-effectiveness and scalability of the methods will greatly help to enable plastic bioremediation on a large scale. Mostly, researchers concentrate on efforts to optimize process parameters, build more reasonably priced bioreactors, and find means to introduce bioremediation into current waste management systems. Researching creative reactor topologies such as modular designs or continuous-flow systems (Chung et al. 2021) might increase resource utilization and throughput. To improve the efficiency of plastic breakdown and save operating costs, researchers are also striving to tune process parameters like pH, temperature, and nutrient levels. Combining bioremediation with other waste management systems such as mechanical recycling or energy recovery might also assist in providing synergies and improved economic feasibility. Researchers intend to make plastic bioremediation more sensible and fairly priced for broad applications by tackling operational and technological problems (Roux and Varrone 2021).

13.15 Increasing Integrated Bioremediation Approaches

Future studies might investigate the use of many bioremediation methods including plant-based systems with their corresponding microbiomes, genetically engineered organisms with more plastic-degrading potential, and other microbial consortia. The goal would be to reach synergistic effects and consequently improve the general effectiveness of plastic cleaning (Verschoor et al. 2022; Priya et al. 2021). Malik et al. (2023) and Priya et al. (2021) propose that by combining these additional techniques, researchers may address a greater spectrum of plastic polymers and provide more thorough solutions for managing plastic pollution. Should plastic bioremediation pay attention to these developing trends and areas of study, it might continue to evolve and provide more reasonable, scalable, long-term solutions for the worldwide plastic pollution problem. Researchers can combine bioremediation with other waste management techniques, pinpoint new ways to break down plastic polymers utilizing microorganisms and enzymes and use technical advances to make bioremediation more economical (Verschoor et al. 2022). Plastic bioremediation has significant potential to move us towards a circular economy in the future and aid in reducing the environmental effect of plastic waste by means of interdisciplinary initiatives complying with the United Nations Sustainable Development Goals (Orlando et al. 2023; Zhu et al. 2022).

13.16 Plastic Research: Interdisciplinary Cooperation

Among others, microbiology, biochemistry, genetic engineering, ecology, and sustainability authorities must work in a multidisciplinary, multidimensional fashion to solve the worldwide plastic pollution catastrophe. Plastic bioremediation techniques may be better developed and used when researchers from several fields work together (Dunn and Welden 2023; Priya et al. 2021; Sheth et al. 2019; Cf et al. 2021). By combining these complementary approaches to create a more complete and efficient waste management system, Dunn and Welden (2023) advise researchers to maximize plastic recovery and re-purposing in one vital area where bioremediation and other waste management technologies including chemical recycling, thermal conversion, and circular economy approaches may cooperate. Effective implementation of plastic bioremediation technology depends on partnerships among academics, businesses, and government agencies to better align research targets with practical issues (Yue et al. 2023; Soong et al. 2022).

13.17 Changing Patterns: Plastic Bioremediation

Plastic bioremediation is one dynamic and fascinating area of study seeing rapid development. Finding and improving new kinds of microbes and enzymes able to break down a wider range of plastic polymers is a major focus (Sheth et al. 2019; Cf et al. 2021; Priya et al. 2021). In quest of bacteria that could break down challenging plastics, scientists are looking at various types of places, including landfills and areas where polymers are naturally abundant (Sheth et al. 2019). Expanding the repertoire of organisms and enzymes that break down plastic would help researchers manage a wider spectrum of plastic waste and increase the efficiency of bioremediation initiatives (Tao et al. 2023). Furthermore, according to Dunn and Welden (2023), CF et al. (2021), Priya et al. (2021), and Sheth et al. (2019), genetic and metabolic engineering can be used to maximize these paths, improve the efficiency of plastic-degrading capabilities, and extend the scope of plastic bioremediation to address a broader range of plastic polymers.

13.18 Plastic Bioremediation in Reliance of UN Sustainable Development Goals

Future applications of plastic bioremediation hold great promise to assist in addressing the worldwide plastic pollution problem. Investigating fresh microbiological and enzymatic pathways can help researchers find a better solution to the plastic waste; bioremediation is more reasonably priced and scalable and may be coupled with more general sustainability activities (Wani et al. 2023a, b). Researching the gut microbiome of many animals for new plastic-degrading microbes could uncover further bioremediation paths (Jesus and Alkendi 2023). Working on ways to increase the efficacy of microorganisms and enzymes breaking down plastic, synthetic biology and genetic engineering researchers are looking for more bioremediation paths. Research aiming at the required scale and cost-effectiveness should primarily concentrate on bioreactor design optimization, process efficiency enhancement, and investigate ways to combine bioremediation with existent waste management infrastructure (Wang et al. 2020). Developing closed-loop systems and the circular economy are two additional major sustainability projects that must be coupled with plastic bioremediation. Should plastic bioremediation align with UN Sustainable Development Goals, including Goals 12 and 14 it might have a major impact on the next projects aiming at sustainability (Orlando et al. 2023). All things considered, plastic bioremediation has great potential to solve the worldwide plastic pollution problem moving forward. Globally deployable, full solutions may be produced via consistent innovation, interdisciplinary cooperation, and scalable, reasonably priced emphasis. More effective and complete solutions to the plastic waste problem can be acquired by means of the field of plastic bioremediation's investigation of new microbial and enzymatic pathways, optimization of process efficiency, and integration with other waste management strategies (Verschoor et al. 2022). More notably, with Goals 12 and 14, coordinating these bioremediation technologies with the

United Nations Sustainable Development Goals would assist them in creating a more sustainable and regenerative economy. The future of plastic bioremediation rests on a deliberate and coordinated effort to reduce pollution and rehabilitate ecosystems damaged by plastic (Antranikian and Streit 2022).

13.19 Plastic Waste Management and Circular Economy

More than only creating plastic-breaking bacteria makes a sustainable and regenerative economy dependent on microbial solutions in general waste management techniques too. Ding and Zhu (2023) found that by combining biological and mechanical waste management techniques, plastic resources may be recovered, recycled, and subsequently used to their best potential. According to Yue et al. (2023), complete systems combining mechanical recycling with microbial degradation may increase the efficiency of plastic waste processing. This promotes the recovery of crucial components and helps permit the circular flow of plastic goods throughout the economy. For example, mixing microbial degradation with chemical or thermal recycling enables one to recover useful components from plastic trash. These components may then be recycled back into use in production. Following this, Allemann et al. (2024) found that including plastic-breaking microbes in rubbish sorting and separation systems might aid in optimizing recycling or biodegradability. From this, finally, less plastic waste would find its way into the surroundings. These all-encompassing strategies considering plastics' whole lifecycle help one to reach a more sustainable and circular economy. The integration of mechanical and biological technologies into the plastic waste management system will enable the ideal recovery, reuse, and appropriate disposal of plastic resources. This multi-pronged strategy at last helps create a more sustainable and environmentally conscious corporate model by reducing waste, enhancing material circular flow, and recovering important plastic components (Hendrickson et al. 2024). As a study on microbial plastic breakdown develops, our knowledge of the many types of bacteria capable of breaking down plastics is also developing. Future research focused on enhancing our understanding of the enzymes involved in this process seeks to optimize microbial plastic degradation, ultimately identifying scalable solutions to the pervasive problem of plastic pollution, while also expanding the catalog of microorganisms capable of plastic breakdown. A expanding array of bacteria capable of degrading plastic supports future research focused on enhancing our understanding of the enzymatic mechanisms involved, improving the efficiency of microbial plastic degradation, and ultimately developing scalable solutions to the global challenge of plastic pollution. Researchers aim to mitigate the detrimental impacts of persistent pollution on the environment via a more sustainable and circular approach to plastic waste management (Diana et al. 2023). Using bacteria's metabolic variety and natural adaptability, scientists are finding more creative ways to solve the world's plastic pollution problem. By means of bioengineering, improved microbial strains and studying and increasing naturally existing bacteria that break down plastic aid researchers to enhance and extend current strategies for plastic waste cleaning and breakdown (Viel et al. 2023.). Microbially driven solutions show great potential for reducing the environmental impact of plastic and guiding us towards a more circular economy and future whereby plastic can be recycled or disposed of responsibly, according to Wei and Zimmermann (2017), Dunn and Welden (2023), and Priya et al. (2021).

References

Albright M, Louca S, Winkler DE, Feeser KL, Haig S, Whiteson K, Emerson J, Dunbar J (2021) Solutions in microbiome engineering: prioritizing barriers to organism establishment. ISME J 16(2):331–338. https://doi.org/10.1038/s41396-021-01088-5

Ali S, Bukhari DA, Rehman A (2023) Call for biotechnological approach to degrade plastic in the era of COVID-19 pandemic. Saudi J Biol Sci 30(3):103583–103583. https://doi.org/10.1016/j.sjbs.2023.103583

Allemann MN, Tessman M, Reindel J, Scofield GB, Evans P, Pomeroy RS, Burkart MD, Mayfield SP, Simkovsky R (2024) Rapid biodegradation of microplastics generated from bio-based thermoplastic polyurethane. Sci Rep 14(1):6036. https://doi.org/10.1038/s41598-024-56492-6

An R, Liu C, Wang J, Jia P (2023) Recent advances in degradation of polymer plastics by insects inhabiting microorganisms. Polymers 15(5):1307–1307. https://doi.org/10.3390/polym15051307

Anand U, Dey S, Bontempi E, Ducoli S, Vethaak AD, Dey A, Federici S (2023) Biotechnological methods to remove microplastics: a review. Environ Chem Lett 21(3):1787–1810. https://doi.org/10.1007/s10311-022-01552-4

Antranikian G, Streit WR (2022) Microorganisms harbor keys to a circular bioeconomy making them useful tools in fighting plastic pollution and rising CO_2 levels. Extremophiles 26(1):10. https://doi.org/10.1007/s00792-022-01261-4

Armistead FJ, Gala De Pablo J, Gadêlha H, Peyman SA, Evans SD (2020) Physical biomarkers of disease progression: on-chip monitoring of changes in mechanobiology of colorectal cancer cells. Sci Rep 10(2020):1–10. https://doi.org/10.1038/s41598-020-59952-x

Asmita K, Shubham Singh T, Tejashree S (2015) Isolation of plastic degrading micro-organisms from soil samples collected at various locations in Mumbai, India. Int Res J Environment Sci 3:77–85. (PDF) Aiding the Environment: Microorganisms for the Degradation of Plastic in Soil and Water. https://www.researchgate.net/publication/343345220_Aiding_the_Environment_Microorganisms_for_the_Degradation_of_Plastic_in_Soil_and_Water. Accessed 18 Oct 2024

Atanasova N, Paunova-Krasteva T, Kambourova M, Boyadzhieva I (2023) A thermostable lipase isolated from Brevibacillus thermoruber strain 7 degrades Ɛ-Polycaprolactone. BioTech 12(1):23–23. https://doi.org/10.3390/biotech12010023

Bakan B, Bernet N, Bouchez T, Boutrou R, Choubert J, Dabert P, Duquennoi C, Ferraro V, García-Bernet D, Gillot S, Méry J, Rémond C, Steyer J, Trably É, Trémier A (2021) Circular economy applied to organic residues and wastewater: research challenges. Waste and Biomass, Valorization 13(2):1267–1276. https://doi.org/10.1007/s12649-021-01549-0

Baldera-Moreno Y, Pino V, Farrés A, Banerjee A, Gordillo F, Andler R (2022) Biotechnological aspects and mathematical modeling of the biodegradation of plastics under controlled conditions. Polymers 14(3):375–375. https://doi.org/10.3390/polym14030375

Bardají DKR, Furlan JPR, Stehling EG (2019) Isolation of a polyethylene degrading Paenibacillus sp. from a landfill in Brazil. Arch Microbiol 5:699–704. https://doi.org/10.1007/s00203-019-01637-9. (PDF) Aiding the Environment: Microorganisms for the Degradation of Plastic in Soil and Water. https://www.researchgate.net/publication/343345220_Aiding_the_Environment_Microorganisms_for_the_Degradation_of_Plastic_in_Soil_and_Water. Accessed 18 Oct 2024

Bertocchini F, Arias CF (2023) Why have we not yet solved the challenge of plastic degradation by biological means? PLoS Biol 21(3):e3001979–e3001979. https://doi.org/10.1371/journal.pbio.3001979
Bhavsar P, Bhave M, Webb HK (2023) Solving the plastic dilemma: the fungal and bacterial biodegradability of polyurethanes. World J Microbiol Biotechnol 39(5):122. https://doi.org/10.1007/s11274-023-03558-8
Bidashimwa D, Hoke T, Huynh TB, Narkpitaks N, Priyonugroho K, Ha TT, Burns AM, Weissman A (2023) Plastic pollution: how can the global health community fight the growing problem? BMJ 8(Suppl 3):e012140–e012140. https://doi.org/10.1136/bmjgh-2023-012140
Boctor J, Pandey G, Xu W, Murphy DV, Hoyle FC (2024) Nature's plastic predators: a comprehensive and bibliometric review of Plastivore insects. Polymers 16:1671
Brill-Karniely Y, Dror D, Duanis T, Assaf Y, Goldstein O, Schwob T, Millo N, Orehov T, Stern M, Jaber N, Loyfer M, Vosk-Artzi H, Benyamini D, Bielenberg T, Kaplan Y, Buganim M, Reches OB (2020) Triangular correlation (TrC) between cancer aggressiveness, cell uptake capability, and cell deformability. Sci Adv 6:eaax2861. https://doi.org/10.1126/sciadv.aax2861
Bruno A, Dovizio M, Milillo C, Aruffo E, Pesce M, Gatta M, Chiacchiaretta P, Di Carlo P, Ballerini P (2024) Orally ingested micro—and Nano-plastics: a hidden driver of inflammatory bowel disease and colorectal cancer. Cancers 16:3079. https://doi.org/10.3390/cancers16173079
Brynzak-Schreiber E, Schögl E, Bapp C, Cseh K, Kopatz V, Jakupec MA, Weber A, Lange T, Toca-Herrera JL, del Favero G, Wadsak W, Kenner L, Pichler V (2024) Microplastics role in cell migration and distribution during cancer cell division. Chemosphere 353:141463
Cai Z, Li M, Yin H, Wang X, Huang Y, Li TJ, Gong H, Yan M (2023) Biological degradation of plastics and microplastics: a recent perspective on associated mechanisms and influencing factors. Microorganisms 11(7):1661–1661. https://doi.org/10.3390/microorganisms11071661
Capetillo AA, Bauer F, Chamináde C (2022) Emerging technologies supporting the transition to a circular economy in the plastic materials value chain. Circ Econ Sustain 3(2):953–982. https://doi.org/10.1007/s43615-022-00209-2
Cf SF, Rebello S, Aneesh EM, Sindhu R, Binod P, Singh S, Pandey A (2021) Bioprospecting of gut microflora for plastic biodegradation. Bioengineered 12(1):1040–1053. https://doi.org/10.1080/21655979.2021.1902173
Chalermthai B, Charoensuppanimit P, Nootong K, Olsen BD, Assabumrungrat S (2023) Techno-economic assessment of co-production of edible bioplastic and food supplements from spirulina. Sec Rep 13(1):10190. https://doi.org/10.1038/s41598-023-37156-3
Chow J, Pérez-García P, Dierkes RF, Streit WR (2022) Microbial enzymes will offer limited solutions to the global plastic pollution crisis. Microbiol Biotechnol 16(2):195–217. https://doi.org/10.1111/1751-7915.14135
Chu N, Liang Q, Jiang Y, Zeng RJ (2020) Microbial electrochemical platform for the production of renewable fuels and chemicals. Biosensors Bioelectronics 150:111922. https://doi.org/10.1016/j.bios.2019.111922
Chung JW, Gerner G, Ovsyannikova E, Treichler A, Baier U, Libra JA, Krebs R (2021) Hydrothermal carbonization as an alternative sanitation technology: process optimization and development of low-cost reactor. Open Res Europe 1:139–139. https://doi.org/10.12688/openreseurope.14306.1
Cifuentes IEM, Degenhardt J, Neumann-Schaal M, Jehmlich N, Ngugi DK, Öztürk B (2023) Comparative biodegradation analysis of three compostable polyesters by a marine microbial community. Appl Environ Microbiol 89(12):e0106023. https://doi.org/10.1128/aem.01060-23
Damayanti D, Saputri DR, Marpaung DSS, Yusupandi F, Sanjaya A, Simbolon YM, Asmarani W, Ulfa M, Wu H (2022) Current prospects for plastic waste treatment. Polymers 14(15):3133–3133. https://doi.org/10.3390/polym14153133
Danso D, Schmeisser C, Chow J, Zimmermann W, Wei R, Leggewie C, Li X, Hazen T, Streit WR (2018) New insights into the function and global distribution of polyethyleneterephthalate (PET)-degrading bacteria and enzymes in marine and terrestrial metagenomes. Appl Environ Microbiol 8:1–13. https://doi.org/10.1128/AEM.02773-17. (PDF) Aiding the Environment: Microorganisms for the Degradation of Plastic in Soil and Water. https://www.researchgate.net/

publication/343345220_Aiding_the_Environment_Microorganisms_for_the_Degradation_of_Plastic_in_Soil_and_Water. Accessed Oct 18 2024

Danso D, Chow J, Streit WR (2019) Plastics: environmental and biotechnological perspectives on microbial degradation. Appl Environ Microbiol 85(19):e01095. https://doi.org/10.1128/aem.01095-19

Dey S, Veerendra GTN, Babu PA, Manoj AP, Nagarjuna K (2024) Degradation of plastics waste and its effects on biological ecosystems: a scientific analysis and comprehensive review. Biomed Mater Devices 2(1):70–112

Diana Z, Virdin J, Valiyaveettil S, Li H, Rittschof D (2023) Editorial: emerging challenges and solutions for plastic pollution, vol 10. Frontiers Media. https://doi.org/10.3389/fmars.2023.1162680

Dimarogona M, Nikolaivits E, Kanelli M, Christakopoulos P, Sandgren M, Topakas E (2015) Structural and functional studies of a Fusarium oxysporum cutinase with polyethyleneterephthalate modification potential. Biochim Biophys Acta Gen Subj 11:2308–2317. https://doi.org/10.1016/j.bbagen.2015.08.009. (PDF) Aiding the Environment: Microorganisms for the Degradation of Plastic in Soil and Water. https://www.researchgate.net/publication/343345220_Aiding_the_Environment_Microorganisms_for_the_Degradation_of_Plastic_in_Soil_and_Water. Accessed 18 Oct 2024

Ding Q, Zhu H (2023) The key to solving plastic packaging wastes: design for recycling and recycling technology. Polymers 15(6):1485–1485. https://doi.org/10.3390/polym15061485

Dunn RA, Welden N (2023) Management of environmental plastic pollution: a comparison of existing strategies and emerging solutions from nature. Water Air Soil Pollution 234(3):201. https://doi.org/10.1007/s11270-023-06190-2

Dutta B, Bandopadhyay R (2022) Biotechnological potentials of halophilic microorganisms and their impact on mankind. Beni-Suef Univ J Basic Appl Sci 11:75. https://doi.org/10.1186/s43088-022-00252-w

Edwards S, León-Zayas R, Ditter R, Laster H, Sheehan G, Anderson OD, Beattie T, Mellies JL (2022) Microbial consortia and mixed plastic waste: Pangenomic analysis reveals potential for degradation of multiple plastic types via previously identified PET degrading bacteria. Int J Mol Sci 23(10):5612–5612. https://doi.org/10.3390/ijms23105612

Ekanayaka AH, Tibpromma S, Dai D, Xu R, Suwannarach N, Stephenson SL, Dao C, Karunarathna SC (2022) A review of the fungi that degrade plastic. J Fungi 8(8):772–772. https://doi.org/10.3390/jof8080772

Eriksen M, Cowger W, Erdle LM, Coffin S, Villarrubia-Gómez P, Moore CJ, Carpenter EJ, Day RH, Thiel M, Wilcox C (2023) A growing plastic smog, now estimated to be over 170 trillion plastic particles afloat in the world's oceans—urgent solutions required. Plos One 18(3):e0281596–e0281596. https://doi.org/10.1371/journal.pone.0281596

Francocci F, Trincardi F, Barbanti A, Zacchini M, Sprovieri M (2020) Linking bioeconomy to redevelopment in contaminated sites: potentials and enabling factors. Front Environ Sci 8:144. https://doi.org/10.3389/fenvs.2020.00144

García-Depraect O, Bordel S, Lebrero R, Santos-Beneit F, Börner RA, Börner T, Muñoz R (2021) Inspired by nature: microbial production, degradation and valorization of biodegradable bioplastics for life-cycle-engineered products. Biotechnol Adv 53:107772–107772. https://doi.org/10.1016/j.biotechadv.2021.107772

Gene Editing and Agrifood Systems (2022) https://doi.org/10.4060/cc3579en

Global Plastics Outlook: Policy Scenarios to 2060 (2022). https://www.oecd.org/publications/global-plastics-outlook-aa1edf33-en.htm

Guinney J, Dienstmann R, Wang X, De Reyniès A, Schlicker A, Soneson C, Marisa L, Roepman P, Nyamundanda G, Angelino P, Bot BM, Morris JS, Simon IM, Gerster S, Fessler E, De Sousa F, Melo E, Missiaglia E, Ramay H, Barras D et al (2015) The consensus molecular subtypes of colorectal cancer. Nat Med 2015:1350–1356. https://doi.org/10.1038/nm.3967

Hendrickson TP, Bose B, Vora N, Huntington T, Nordahl SL, Helms BA, Scown CD (2024) Paths to circularity for plastics in the United States. One Earth 7(3):520–531. https://doi.org/10.1016/j.oneear.2024.02.005

Herrero-Acero E, Ribitsch D, Steinkellner G, Gruber K, Greimel K, Eiteljoerg I, Trotscha E, Wei R, Zimmermann W, Zinn M, Cavaco-Paulo A, Freddi G, Schwab H, Guebitz G (2011) Enzymatic surface hydrolysis of PET: effect of structural diversity on kinetic properties of cutinases from Thermobifida. Macromolecules 12:4632–4640. https://doi.org/10.1021/ma200949p. (PDF) Aiding the Environment: Microorganisms for the Degradation of Plastic in Soil and Water. https://www.researchgate.net/publication/343345220_Aiding_the_Environment_Microorganisms_for_the_Degradation_of_Plastic_in_Soil_and_Water. Accessed 18 Oct 2024

Howard GT, Mackie RI, Cann IKO, Ohene-Adjei S, Aboudehen KS, Duos BG, Childers GW (2007) Effect of insertional mutations in the pueA and pueB genes encoding two polyurethanases in Pseudomonas chlororaphis contained within a gene cluster. J Appl Microbiol 6:2074–2083. https://doi.org/10.1111/j.1365-2672.2007.03447.x. (PDF) Aiding the Environment: Microorganisms for the Degradation of Plastic in Soil and Water. https://www.researchgate.net/publication/343345220_Aiding_the_Environment_Microorganisms_for_the_Degradation_of_Plastic_in_Soil_and_Water. Accessed 18 Oct 2024

Hu X, Thumarat U, Zhang X, Tang M, Kawai F (2010) Diversity of polyester-degradingbacteria in compost and molecular analysis of a thermoactive esterase from Thermobifida alba AHK119. Appl Microbiol Biotechnol 2:771–779. https://doi.org/10.1007/S00253-010-2555-X. (PDF) Aiding the Environment: Microorganisms for the Degradation of Plastic in Soil and Water. https://www.researchgate.net/publication/343345220_Aiding_the_Environment_Microorganisms_for_the_Degradation_of_Plastic_in_Soil_and_Water. Accessed 18 Oct 2024

Indumathi A, Gayathri T (2016) Plastic Degrading ability of Aspergillus oryzae isolated from the garbage dumping sites of Thanjavur, India. Int J Curr Microbiol App Sci 3:8–13. (PDF) Aiding the Environment: Microorganisms for the Degradation of Plastic in Soil and Water. https://www.researchgate.net/publication/343345220_Aiding_the_Environment_Microorganisms_for_the_Degradation_of_Plastic_in_Soil_and_Water. Accessed 18 Oct 2024

Jesus RD, Alkendi R (2023) A minireview on the bioremediative potential of microbial enzymes as solution to emerging microplastic pollution. Front Microbiol 13:1066133. https://doi.org/10.3389/fmicb.2022.1066133

Johnston B, Adamus G, Ekere AI, Kowalczuk M, Tchuenbou-Magaia F, Radecka I (2022) Bioconversion of plastic waste based on mass full carbon backbone polymeric materials to value-added Polyhydroxyalkanoates (PHAs). Bioengineering 9(9):432–432. https://doi.org/10.3390/bioengineering9090432

Jones GR, Wang HS, Parkatzidis K, Whitfield R, Truong NP, Anastasaki A (2023) Reversed controlled polymerization (RCP): depolymerization from well-defined polymers to monomers. Am Chem Soc 145(18):9898–9915. https://doi.org/10.1021/jacs.3c00589

Jumaah OS (2017) Screening of plastic degrading bacteria from dumped soil area. IOSR J Environ Sci Toxicol Food Technol 5:93–98. (PDF) Aiding the Environment: Microorganisms for the Degradation of Plastic in Soil and Water. Available from: https://www.researchgate.net/publication/343345220_Aiding_the_Environment_Microorganisms_for_the_Degradation_of_Plastic_in_Soil_and_Water. Accessed 18 Oct 2024

Kathiresan K (2003) Polythene and plastics-degrading microbes from the mangrove soil. Rev Biol Trop 3:629–634

Kaur K, Sharma S, Shree N, Mehrotra R (2023) Recent advancements and mechanism of plastics biodegradation promoted by bacteria: a key for sustainable remediation for plastic wastes. Biosci Biotechnol Res Asia 20(1):1–12. https://doi.org/10.13005/bbra/3063

Kawai F, Oda M, Tamashiro T, Waku T, Tanaka N, Yamamoto M, Mizushima H, Miyakawa T, Tanokura M (2014) A novel Ca 2+-activated, thermo stabilized polyesterasecapable of hydrolyzing polyethyleneterephthalate from Saccharomonos poraviridis AHK190. Appl Microbiol Biotechnol 24:10053–10064. https://doi.org/10.1007/s00253-014-5860-y. (PDF) Aiding the Environment: Microorganisms for the Degradation of Plastic in Soil and Water. https://www.researchgate.net/publication/343345220_Aiding_the_Environment_Microorganisms_for_the_Degradation_of_Plastic_in_Soil_and_Water. Accessed 18 Oct 2024

Kim SH, Lee JW, Kim JS, Lee W, Park MS, Lim YW (2022) Plastic-inhabiting fungi in marine environments and PCL degradation activity. Antonie Van Leeuwenhoek 115(12):1379–1392. https://doi.org/10.1007/s10482-022-01782-0

Lase IS, Ragaert K, Dewulf J, De Meester S (2021) Multivariate input-output and material flow analysis of current and future plastic recycling rates from waste electrical and electronic equipment: the case of small household appliances. Resour Conserv Recycl 174:105772

Lee H, Shim JE, Park IH, Choo KS, Yeo M (2022) Physical and biomimetic treatment methods to reduce microplastic waste accumulation. Mol Cell Toxicol 19(1):13–25. https://doi.org/10.1007/s13273-022-00289-z

Lee Y, Cho J, Sohn J, Kim C (2023) Health effects of microplastic exposures: current issues and perspectives in South Korea. Yonsei Med J 64(5):301–308. https://doi.org/10.3349/ymj.2023.0048. PMID: 37114632; PMCID: PMC10151227

Li A, Sheng Y, Cui H, Wang M, Wu L, Song Y, Yang R, Li X, Huang H (2023) Discovery and mechanism-guided engineering of BHET hydrolases for improved PET recycling and upcycling. Nat Commun 14(1):4169. https://doi.org/10.1038/s41467-023-39929-w

MacLeod M, Arp HPH, Tekman MB, Jahnke A (2021) The global threat from plastic pollution. American association for the. Adv Sci 373(6550):61–65. https://doi.org/10.1126/science.abg5433

Malik S, Maurya A, Khare SK, Srivastava KR (2023) Computational exploration of bio-degradation patterns of various plastic types. Polymers 15(6):1540–1540. https://doi.org/10.3390/polym15061540

Mazhandu ZS, Muzenda E, Mamvura TA, Belaid M, Nhubu T (2020) Integrated and consolidated review of plastic waste management and bio-based biodegradable plastics: challenges and opportunities. Sustainability 12(20):8360–8360. https://doi.org/10.3390/su12208360

Mohan SK, Suresh B (2015) Studies on biodegradation of plastics by aspergillus sp.isolated from dye effluent enriched soil. Indo Am J Pharm Sci 12:1636–1639. (PDF) Aiding the Environment: Microorganisms for the Degradation of Plastic in Soil and Water. https://www.researchgate.net/publication/343345220_Aiding_the_Environment_Microorganisms_for_the_Degradation_of_Plastic_in_Soil_and_Water. Accessed 18 Oct 2024

Moharir RV, Kumar S (2019) Challenges associated with plastic waste disposal and allied microbial routes for its effective degradation: a comprehensive review. J Clean Produc 208:65–76. https://doi.org/10.1016/j.jclepro.2018.10.059

Muhonja CN, Makonde H, Magoma G, Imbuga M (2018) Biodegradability of polyethylene by bacteria and fungi from Dandora dumpsite Nairobi-Kenya. PLoS One 7:1–17. (PDF) Aiding the Environment: Microorganisms for the Degradation of Plastic in Soil and Water. https://www.researchgate.net/publication/343345220_Aiding_the_Environment_Microorganisms_for_the_Degradation_of_Plastic_in_Soil_and_Water. Accessed 18 Oct 2024

Mukherjee A, Koller M (2023) Microbial PolyHydroxyAlkanoate (PHA) biopolymers—intrinsically natural. Bioengineering 10(7):855–855. https://doi.org/10.3390/bioengineering10070855

Nakei MD, Misinzo G, Tindwa H, Semu E (2022) Degradation of polyethylene plastic bags and bottles using microorganisms isolated from soils of Morogoro, Tanzania. Front Microbiol 13:1077588. https://doi.org/10.3389/fmicb.2022.1077588

Nisha M, Montazer Z, Sharma P, Levin DB (2020) Microbial and enzymatic degradation of synthetic plastics. Front Microbiol 11:580709. https://doi.org/10.3389/fmicb.2020.580709

Nomura N, Shigeno-Akutsu Y, Nakajima-Kambe T, Nakahara T (1998) Cloning and sequence analysis of a po lyurethane esterase of Comamonas acidovorans TB-35. J Ferment Bioeng 4:339–345. https://doi.org/10.1016/S0922-338X(99)89001-1. (PDF) Aiding the Environment: Microorganisms for the Degradation of Plastic in Soil and Water. https://www.researchgate.net/publication/343345220_Aiding_the_Environment_Microorganisms_for_the_Degradation_of_Plastic_in_Soil_and_Water. Accessed 18 Oct 2024

Orlando M, Molla G, Castellani P, Pirillo V, Torretta V, Ferronato N (2023) Microbial enzyme biotechnology to reach plastic waste circularity: current status, problems and perspectives. Int J Mol Sci 24(4):3877–3877. https://doi.org/10.3390/ijms24043877

Patel HK, Kalaria RK, Khimani M (2020) Nanotechnology: a promising tool for bioremediation. Elsevier, Amsterdam, pp 515–547. https://doi.org/10.1016/b978-0-12-821014-7.00020-4

Pelegrini K, Pereira TCB, Maraschin TG, Teodoro LDS, Basso NRDS, De Galland GLB, Bogo MR (2023) Micro-and nanoplastic toxicity: a review on size, type, source, and test-organism implications. Sci Total Environ 878:162954

Pérez-García P, Chow J, Costanzi E, Gurschke M, Dittrich J, Dierkes RF, Molitor R, Applegate V, Feuerriegel G, Tete P, Danso D, Thies S, Schumacher J, Pfleger C, Jaeger K, Gohlke H, Smits SHJ, Schmitz RA, Streit WR (2023) An archaeal lid-containing feruloyl esterase degrades polyethylene terephthalate. Commun Chem 6(1):193. https://doi.org/10.1038/s42004-023-00998-z

Piskuła P, Astel A (2023) Microplastics in commercial fishes and by-catch from selected FAO major fishing areas of the southern Baltic Sea. Animals 13(3):458–458. https://doi.org/10.3390/ani13030458

Poland TM, Juzwik J, Rowley A, Huebner CD, Kilgo JC, Álvarez-López V, Olson DH, Pearson DE, Progar RA, Rabaglia RJ, Rothlisberger JD, Runyon JB, Sing SE (2021) Management of landscapes for established invasive species. Springer, Cham, pp 133–184. https://doi.org/10.1007/978-3-030-45367-1_7

Priya A, Dutta K, Daverey A (2021) A comprehensive biotechnological and molecular insight into plastic degradation by microbial community. J Chem Technol Biotechnol 97(2):381–390. https://doi.org/10.1002/jctb.6675

Qin Z, Mou J, Chao CY, Chopra SS, Daoud WA, Leu S, Ning Z, Tso CY, Chan CK, Tang S, Hathi Z, Haque MA, Wang X, Lin CSK (2021) Biotechnology of plastic waste degradation, recycling, and valorization: current advances and future perspectives. ChemSusChem 14(19):4103–4114. https://doi.org/10.1002/cssc.202100752

Roux M, Varrone C (2021) Assessing the economic viability of the plastic biorefinery concept and its contribution to a more circular plastic sector. Polymers 13(22):3883–3883. https://doi.org/10.3390/polym13223883

Ru J, Huo Y, Yang Y (2020) Microbial degradation and valorization of plastic wastes. Front Microbiol 11:442. https://doi.org/10.3389/fmicb.2020.00442

Rüthi J, Cerri M, Brunner I, Stierli B, Sander M, Frey B (2023) Discovery of plastic-degrading microbial strains isolated from the alpine and Arctic terrestrial plastisphere. Front Microbiol 14:1178474. https://doi.org/10.3389/fmicb.2023.1178474

Salinas J, Carpena V, Martínez-Gallardo MR, Segado M, Estrella-González MJ, Toribio AJ, Jurado M, López-González JA, Suárez-Estrella F, López M (2023) Development of plastic-degrading microbial consortia by induced selection in microcosms. Front Microbiol 14:1143769. https://doi.org/10.3389/fmicb.2023.1143769

Šerá J, Huynh F, Ly F, Vinter Š, Kadlečková M, Krátká V, Máčalová D, Koutný M, Wallis CJ (2022) Biodegradable polyesters and low molecular weight polyethylene in soil: interrelations of material properties, soil organic matter substances, and microbial community. Int J Mol Sci 23(24):15976–15976. https://doi.org/10.3390/ijms232415976

Shakya P, Nayak A, Jogi J, Rai A, Bordoloi S, Himani K, Lade A (2020) Futuristic approach to environment cleansing: bioremediation. Int J Curr Microbiol Appl Sci 9(9):418–425. https://doi.org/10.20546/ijcmas.2020.909.053

Sharma J, Gurung T, Upadhyay A, Nandy K, Agnihotri P, Mitra AK (2014) Isolation and characterization of plastic degrading bacteria from soil collected from the dumping grounds of an industrial area. Int J Adv Innov Res 3:225–232. (PDF) Aiding the Environment: Microorganisms for the Degradation of Plastic in Soil and Water. https://www.researchgate.net/publication/343345220_Aiding_the_Environment_Microorganisms_for_the_Degradation_of_Plastic_in_Soil_and_Water. Accessed 18 Oct 2024

Sheth MU, Kwartler SK, Schmaltz E, Hoskinson SM, Martz EJ, Dunphy-Daly MM, Schultz TF, Read AJ, Eward WC, Somarelli JA (2019) Bioengineering a future free of marine plastic waste. Front Marine Sci 6:624. https://doi.org/10.3389/fmars.2019.00624

Shi Z, Lin J, Wu Y, Fu S, Wan Y, Fang Y (2022) Burden of cancer and changing cancer spectrum among older adults in China: trends and projections to 2030. Cancer. Epidemiology 76:102068

Shilpa S, Basak N, Meena SS (2022) Microbial biodegradation of plastics: challenges, opportunities, and a critical perspective. Front Environ Sci Eng 16(12):161. https://doi.org/10.1007/s11783-022-1596-6

Soong YV, Sobkowicz MJ, Xie D (2022) Recent advances in biological recycling of polyethylene terephthalate (PET) plastic wastes. Bioengineering 9(3):98–98. https://doi.org/10.3390/bioengineering9030098

Stern RV, Howard GT (2000) The polyester polyurethanase gene (pueA) from pseudomonas chlororaphis encodes a lipase. FEMS Microbiol Letters 2:163–168. https://doi.org/10.1111/j.1574-6968.2000.tb09056.x. (PDF) aiding the environment: microorganisms for the degradation of plastic in soil and water. https://www.researchgate.net/publication/343345220_Aiding_the_Environment_Microorganisms_for_the_Degradation_of_Plastic_in_Soil_and_Water. Accessed 18 Oct 2024

Tamoor M, Samak NA, Jia Y, Mushtaq MU, Sher H, Bibi M, Xing J (2021) Potential use of microbial enzymes for the conversion of plastic waste into value-added products: a viable solution. Front Microbiol 12:777727. https://doi.org/10.3389/fmicb.2021.777727. PMID: 34917057; PMCID: PMC8670383

Tao X, Ouyang H, Zhou A, Wang DS, Matlock HP, Morgan JS, Ren A, Mu D, Pan C, Zhu X, Han A, Zhou J (2023) Polyethylene degradation by a Rhodococcous strain isolated from naturally weathered plastic waste enrichment. Am Chem Soc 57(37):13901–13911. https://doi.org/10.1021/acs.est.3c03778

Temporiti MEE, Nicola L, Nielsen E, Tosi S (2022) Fungal enzymes involved in plastics biodegradation. Microorganisms 10(6):1180–1180. https://doi.org/10.3390/microorganisms10061180

Thomas SM (2023) Invasive alien species: a rising global threat that needs control. Lancet Planet Health 7(11):e875–e876. https://doi.org/10.1016/s2542-5196(23)00233-4

Tickner J, Geiser K, Baima S (2021) Transitioning the chemical industry: the case for addressing the climate, toxics, and plastics crises. Environ Sci Policy Sustain Dev 63(6):4–15. tandfonline.com

Trump BD, Cummings C, Klasa K, Galaitsi S, Linkov I (2023) Governing biotechnology to provide safety and security and address ethical, legal, and social implications. Front Genet 13:1052371. https://doi.org/10.3389/fgene.2022

Uekert T, Singh A, DesVeaux JS, Ghosh T, Bhatt A, Yadav G, Afzal S, Walzberg J, Knauer KM, Nicholson S, Beckham GT, Carpenter A (2023) Technical, economic, and environmental comparison of closed-loop recycling Technologies for Common Plastics. Am Chem Soc 11(3):965–978. https://doi.org/10.1021/acssuschemeng.2c05497

Urbanek AK, Kosiorowska KE, Mirończuk AM (2021) Current knowledge on polyethylene terephthalate degradation by genetically modified microorganisms. Front Bioeng Biotechnol 9:771133. https://doi.org/10.3389/fbioe.2021.771133

Valavanidis A (2024) Plastic pollution, unsustainable environmental problem? Industry and academic research must co-operate to short out the pervasive global problem of plastic waste. Sci Rev 1:1–42

Venkatesan R, Moorthy MS, Alagumalai K, Haldhar R, Raorane CJ, Raj V, Kim S (2022) Novel approach in biodegradation of synthetic thermoplastic polymers: an overview. Polymers 14(20):4271–4271. https://doi.org/10.3390/polym14204271

Verschoor J, Kusumawardhani H, Ram AFJ, Winde JHD (2022) Toward microbial recycling and upcycling of plastics: prospects and challenges. Front Microbiol 13:821629. https://doi.org/10.3389/fmicb.2022.821629

Vidal F, Marel ERVD, Kerr RWF, McElroy C, Schroeder N, Mitchell C, Rosetto G, Chen TTD, Bailey RM, Hepburn C, Redgwell C, Williams CK (2024) Designing a circular carbon and plastics economy for a sustainable future. Nature 626(7997):45–57. https://doi.org/10.1038/s41586-023-06939-z

Viel T, Manfra L, Zupo V, Libralato G, Cocca M, Costantini M (2023) Biodegradation of plastics induced by marine organisms: future perspectives for bioremediation approaches. Polymers 15(12):2673–2673. https://doi.org/10.3390/polym15122673

Vignesh R, Deepika RC, Manigandan P, Janani R (2016) Screening of plastic degrading microbes from various dumped soil samples. Int Res J Eng Technol 4:2493–2498. (PDF) Aiding the Environment: Microorganisms for the Degradation of Plastic in Soil and Water. https://www.researchgate.net/publication/343345220_Aiding_the_Environment_Microorganisms_for_the_Degradation_of_Plastic_in_Soil_and_Water. Accessed Oct 18 2024

Villegas-Méndez MÁ, Montañez J, Contreras-Esquivel JC, Salmerón I, Koutinas A, Morales-Oyervides L (2022) Coproduction of microbial oil and carotenoids within the circular bioeconomy concept: a sequential solid-state and submerged fermentation approach. Fermentation 8(6):258–258. https://doi.org/10.3390/fermentation8060258

Wang A, Wang H, Cheng H, Liang B, Liu W, Han J, Zhang B, Wang S (2020) Electrochemistry-stimulated environmental bioremediation: development of applicable modular electrode and system scale-up. Environ Sci Ecotechnol 3:100050–100050. https://doi.org/10.1016/j.ese.2020.100050

Wang Y, Narayanan M, Shi X, Chen X, Li Z, Natarajan D, Ma Y (2022) Plant growth-promoting bacteria in metal-contaminated soil: current perspectives on remediation mechanisms. Front Microbiol 13:966226. https://doi.org/10.3389/fmicb.2022.966226

Wani AK, Akhtar N, Naqash N, Rahayu F, Djajadi D, Chopra C, Singh R, Mulla SI, Sher F, Américo-Pinheiro JHP (2023a) Discovering untapped microbial communities through metagenomics for microplastic remediation: recent advances, challenges, and way forward. Environ Sci Pollut Res 30(34):81450–81473. https://doi.org/10.1007/s11356-023-25192-5

Wani ZA, Ahmad Z, Asgher M, Bhat JA, Sharma M, Kumar A, Sharma V, Kumar A, Pant S, Лукаткин AC, Anjum NA (2023b) Phytoremediation of potentially toxic elements: role, status and concerns. Plants 12(3):429–429. https://doi.org/10.3390/plants12030429

Wei R, Zimmermann W (2017) Microbial enzymes for the recycling of recalcitrant petroleum-based plastics: how far are we? Microb Biotechnol 10(6):1308–1322. https://doi.org/10.1111/1751-7915.12710

Yamamoto M, Takaki Y, Kashima H, Tsuda M, Tanizaki A, Nakamura R, Takai K (2022) In situ electrosynthetic bacterial growth using electricity generated by a deep-sea hydrothermal vent. ISME J 17(1):12–20. https://doi.org/10.1038/s41396-022-01316-6

Yang Z, Qiao Y, Konakalla NC, Strøbech E, Harris P, Peschel G, Agler-Rosenbaum M, Weber T, Andréasson E, Ding L (2023) Streptomyces alleviate abiotic stress in plant by producing pteridic acids. Nat Commun 14(1):7398. https://doi.org/10.1038/s41467-023-43177-3

Yoshida S, Hiraga K, Takehana T, Taniguchi I, Yamaji H, Maeda Y, Toyohara K, Miyamoto K, Kimura Y, Oda K (2016) A bacterium that degrades and assimilates poly(ethylene terephthalate). Science 6278:1196–1199. https://doi.org/10.1126/science.aad6359. (PDF) Aiding the Environment: Microorganisms for the Degradation of Plastic in Soil and Water. https://www.researchgate.net/publication/343345220_Aiding_the_Environment_Microorganisms_for_the_Degradation_of_Plastic_in_Soil_and_Water. Accessed Oct 18 2024

Yue S, Wang L, Yu B, Zhang T, Zhao Z, Li Y, Zhan S (2023) From plastic waste to treasure: selective upcycling through catalytic technologies. Adv Energ Mater 13(41):202302008. https://doi.org/10.1002/aenm.202302008

Zampolli J, Orro A, Vezzini D, Gennaro PD (2022) Genome-based exploration of Rhodococcus species for plastic-degrading genetic determinants using bioinformatic analysis. Microorganisms 10(9):1846–1846. https://doi.org/10.3390/microorganisms10091846

Zhai X, Zhang X, Yu M (2023) Microbial colonization and degradation of marine microplastics in the plastisphere: a review. Front Microbiol 14:1127308. https://doi.org/10.3389/fmicb.2023.1127308

Zhang N, Ding M, Yuan Y (2022) Current advances in biodegradation of Polyolefins. Microorganisms 10(8):1537–1537. https://doi.org/10.3390/microorganisms10081537

Zhu B, Wang D, Wei N (2022) Enzyme discovery and engineering for sustainable plastic recycling. Trends Biotechnol 40(1):22–37. https://doi.org/10.1016/j.tibtech.2021.02.008

Ziani K, Ioniță-Mîndrican C, Mititelu M, Neacşu SM, Negrei C, Moroşan E, Drăgănescu D, Preda O (2023) Microplastics: a real global threat for environment and food safety: a state of the art review. Nutrients 15(3):617–617. https://doi.org/10.3390/nu15030617

Zou L, Xu R, Wang H, Wang Z, Sun Y, Li M (2023) Chemical recycling of polyolefins: a closed-loop cycle of waste to olefins. Natl Sci Rev 10(9):1. https://doi.org/10.1093/nsr/nwad207

Zrimec J, Kokina M, Jonasson S, Zorrilla F, Zelezniak A (2021) Plastic-degrading potential across the global microbiome correlates with recent pollution trends. American Soc Microbiol 12(5):21. https://doi.org/10.1128/mbio.02155-21

Fuel Shipping and Bioremediation 14

Gorthi Vaishnavi, Tanushree Maharana, N. Neetu Bai, Yuvashree Muralidaran, and Prabhakar Mishra

Abstract

Fuel shipping is essential for the global energy supply chain, facilitating the transportation of crude oil, refined petroleum products, liquefied natural gas, and other fuels. However, fuel shipping poses significant environmental risks, including oil spills that can devastate marine ecosystems, local economies, and public health. This chapter analyzes the fuel shipping industry's critical role in the global energy supply chain and the significant environmental risks posed by fuel spills, such as those from the Exxon Valdez, Deepwater Horizon, Erika, and Prestige incidents. It emphasizes the severe impacts of these spills on marine ecosystems, economies, and public health. It also highlights bioremediation as a sustainable solution for mitigating fuel spill effects, detailing the microbial degradation of hydrocarbons and factors influencing bioremediation efficacy, including temperature, pH, salinity, and nutrient availability. Various bioremediation techniques—bioaugmentation, biostimulation, and natural attenuation—are explored, along with their integration with other cleanup methods like chemical dispersants and mechanical recovery. The importance of interdisciplinary collaboration and supportive policy frameworks for effective fuel spill cleanup is discussed, with recommendations for future research directions in bioremediation.

Keywords

Oil spills · Microbial degradation · Environmental impact · Interdisciplinary collaboration · Regulatory frameworks

G. Vaishnavi · T. Maharana · N. N. Bai · Y. Muralidaran · P. Mishra (✉)
Department of Biotechnology, School of Applied Sciences, REVA University, Bengaluru, Karnataka, India

A. A. H. Abdel Latef et al. (eds.), *Bioremediation and Nanotechnology for Climate Change Mitigation*, https://doi.org/10.1007/978-981-96-3069-1_14

14.1 Introduction

14.1.1 Overview of the Fuel Shipping Industry

The fuel shipping industry is a vital component of the global energy supply chain, responsible for transporting various types of fuels, including crude oil, refined petroleum products, and liquefied natural gas (LNG), worldwide. Maritime transportation plays a significant role in this industry, with approximately 60% of the world's oil transported by sea (IEA 2020). This transportation is facilitated by a diverse range of vessels, including oil tankers, LNG carriers, and chemical tankers, each designed to transport specific types of fuels safely and efficiently. The demand for fuel shipping has steadily increased due to rising energy consumption and international trade. For instance, global oil demand reached 100.3 million barrels per day in 2019, with maritime transportation playing a vital role in meeting this demand (IEA 2020). Similarly, the demand for LNG has increased, with LNG trade reaching 354.7 million tons in 2019 (GIIGNL 2020). Fuel shipping is crucial for meeting energy demands and supporting global economic growth and development. It facilitates international trade by ensuring the movement of energy resources from regions with abundant reserves to regions with high demand. Furthermore, fuel shipping significantly diversifies energy sources and reduces dependency on specific regions for energy imports.

14.1.2 Significance of Fuel Shipping for Global Energy Supply

Fuel shipping is of paramount importance for meeting the global demand for energy and supporting economic growth and development. In addition to transporting energy resources, fuel shipping plays a crucial role in international trade, facilitating the movement of energy resources from regions with abundant reserves to regions with high demand. The significance of fuel shipping is evident from its contribution to the global energy supply. According to the International Energy Agency (IEA), maritime transportation accounts for approximately 60% of the world's oil trade, with over 40,000 oil tankers transporting oil to various destinations (IEA 2020). Similarly, LNG shipping has become increasingly important, with LNG trade reaching 354.7 million tons in 2019 (GIIGNL 2020). Fuel shipping is also essential for ensuring energy security and resilience. By diversifying energy sources and supply routes, fuel shipping helps reduce the risk of supply disruptions and price volatility. Moreover, it plays a critical role in providing access to energy resources in remote and landlocked regions where alternative transportation options are limited.

14.1.3 Environmental Risks Associated with Fuel Shipping

Fuel shipping, integral to global energy distribution, carries substantial environmental hazards, notably fuel spills. These spills, arising from loading, transit, and unloading processes, alongside accidents like collisions and equipment malfunctions, engender severe and enduring ecological repercussions. Crude oil, a primary transported fuel, proves toxic to marine ecosystems, imperiling aquatic fauna and coastal habitats. Consequently, marine biodiversity and ecosystem dynamics suffer prolonged impairments. Moreover, oil spills engender long-term environmental degradation, persisting for years and bioaccumulating within food chains, thereby endangering human health through contaminated seafood consumption (Boehm and Page 2007; National Research Council 2003). Additionally, fuel shipping contributes to air pollution, emitting greenhouse gases, particulate matter, and sulfur oxides, exacerbating climate change, and threatening both coastal communities and marine life. Mitigating these multifaceted environmental challenges necessitates concerted efforts across industry, policy, and scientific domains (Corbett et al. 2010).

14.1.4 Introduction to Bioremediation as a Sustainable Approach to Mitigate the Environmental Impacts of Fuel Spills

Bioremediation emerges as a sustainable and cost-effective solution for addressing the environmental consequences of fuel spills, utilizing microorganisms to degrade pollutants into benign by-products. This approach, particularly effective in marine and coastal environments, offers advantages over traditional cleanup methods, boasting environmental friendliness, minimal invasiveness, and on-site applicability without extensive infrastructure requirements. Bioremediation not only facilitates the reduction of spill impacts but also holds promise for complete pollutant degradation, thereby restoring ecosystems to pre-spill conditions. Numerous studies have demonstrated the efficacy of bioremediation techniques in degrading a wide range of hydrocarbons, including crude oil, gasoline, diesel, and bunker fuel (Atlas and Hazen 2011; Head et al. 2019; Joye et al. 2016). This chapter comprehensively examines fuel shipping's environmental implications, focusing on bioremediation's application for spill mitigation. Covering standard shipped fuels, associated risks, and bioremediation principles and applications, it incorporates case studies and future perspectives to highlight successes and challenges in the field.

14.2 Fuel Shipping: Sources and Environmental Impact

14.2.1 Types of Fuels Commonly Shipped

The fuel shipping industry transports various types of fuels, including crude oil, refined petroleum products, and LNG. Crude oil is the most shipped fuel, transported from extraction sites to refineries for processing into various petroleum products. These include gasoline, diesel, jet fuel, and heating oil, which are transported from refineries to distribution centers and end-users. LNG is transported from exporting regions to import terminals, where it is repacified and distributed for use in residential, commercial, and industrial applications (Etkin 2016).

14.2.2 Common Causes of Fuel Spills During Shipping

Fuel spills during shipping can occur due to various factors, including collisions, grounding, equipment failure, and human error. Collisions between vessels can result in ruptured fuel tanks and subsequent spills. Grounding occurs when a vessel runs aground, causing damage to its hull and fuel tanks. Equipment failures, such as malfunctioning valves or pumps, can lead to unintentional fuel releases. Human error, such as improper navigation or inadequate maintenance, can also contribute to fuel spills (Etkin 2016).

14.2.3 Environmental Consequences of Fuel Spills

14.2.3.1 Impact on Marine Ecosystems

Fuel spills can contaminate marine environments, posing significant risks to marine biodiversity and ecosystem health. Crude oil and petroleum products are toxic to marine life, potentially harming fish, birds, marine mammals, and other aquatic organisms. Additionally, oil spills can damage critical habitats such as coral reefs, mangroves, and marshes, disrupting ecosystem functioning and biodiversity (Boehm and Page 2007; National Research Council 2003).

14.2.3.2 Economic Consequences for Local Communities and Industries

Fuel spills can have far-reaching economic impacts, particularly on local communities and industries dependent on marine resources. The economic costs of fuel spills may include losses in fisheries and aquaculture, damage to coastal infrastructure, and declines in tourism and recreational activities (Boehm and Page 2007; National Research Council 2003).

14.2.3.3 Public Health Concerns Associated with Exposure to Spilled Fuel

Exposure to spilled fuel and associated contaminants can pose significant risks to public health. Inhalation of fuel vapors and dermal contact with oil-contaminated water can cause respiratory problems, skin irritation, and other health issues. Additionally, consuming contaminated seafood can expose individuals to harmful pollutants, with potential long-term health consequences (Boehm and Page 2007; National Research Council 2003).

14.2.4 Case Studies of Significant Fuel Spills and Their Environmental, Economic, and Social Impacts

- The Exxon Valdez oil spill in Prince William Sound, Alaska, in 1989, remains one of the most significant oil spills in history. The spill, caused by the grounding of the Exxon Valdez oil tanker, resulted in the release of approximately 11 million gallons of crude oil into the marine environment. The environmental, economic, and social impacts of the Exxon Valdez oil spill were substantial, with long-term effects on marine ecosystems, local economies, and communities (National Research Council 2003).

- The Deepwater Horizon oil spill, also known as the BP oil spill, occurred in the Gulf of Mexico in 2010 following an explosion on the Deepwater Horizon oil rig. The spill, which lasted for 87 days, released an estimated 4.9 million barrels of oil into the Gulf of Mexico, making it the largest marine oil spill in history. The environmental, economic, and social impacts of the Deepwater Horizon oil spill were profound, affecting marine ecosystems, coastal communities, and industries reliant on the Gulf of Mexico (National Commission on the BP Deepwater Horizon Oil Spill and Offshore Drilling 2011).
- The Erika oil spill occurred off the coast of Brittany, France, in December 1999, when the oil tanker Erika sank during a storm. The spill, which released approximately 20,000 tons of heavy fuel oil into the Atlantic Ocean, had significant environmental, economic, and social impacts on the affected region. The spill led to widespread pollution of coastal areas, damage to marine ecosystems, and losses in fisheries and tourism (Lanoix et al. 2002).

The Prestige oil spill occurred off the coast of Spain in November 2002, when the oil tanker Prestige suffered a structural failure and sank in rough seas. The spill, which released approximately 63,000 tons of heavy fuel oil into the Atlantic Ocean, had devastating environmental, economic, and social consequences for the affected region (Table 14.1). The spill led to widespread contamination of coastal areas, damage to marine ecosystems, and losses in fisheries, tourism, and other coastal industries (Alvarez et al. 2006).

14.3 Bioremediation Techniques for Fuel Spills

Bioremediation is a sustainable and cost-effective approach to clean fuel spills in marine and terrestrial environments. It utilizes microorganisms to degrade or transform pollutants into less harmful substances, such as carbon dioxide and water (Etkin 2016). In the context of fuel shipping, bioremediation offers a promising solution for the cleanup of oil and fuel spills, mitigating environmental damage and restoring affected ecosystems. Bioremediation principles are based on the natural ability of microorganisms to break down organic compounds, including hydrocarbons found in fuels, into simpler and less toxic compounds through enzymatic reactions. The process of bioremediation involves the following steps:

- **Introduction of microorganisms**: Indigenous or introduced microorganisms capable of degrading the target pollutants are introduced into the contaminated environment.
- **Optimization of environmental conditions**: Environmental conditions such as temperature, pH, oxygen availability, and nutrient levels are optimized to enhance microbial activity and pollutant degradation.

Table 14.1 Summary of significant fuel spills during shipping and their environmental impacts

Sl. no.	Year	Location	Volume of spill (barrels)	Environmental impact	References
1	1989	Prince William Sound, Alaska	11 million	Devastation of marine ecosystems, loss of wildlife	Spill (2003)
2	2010	Gulf of Mexico	4.9 million	Severe damage to marine life and habitats	King et al. (2015)
3	1999	Brittany, France	20,000	Contamination of coastal ecosystems, loss of marine life	Cheng et al. (2011)
4	2002	Galicia, Spain	63,000	Environmental damage to coastal ecosystems	Palenzuela et al. (2006)
5	2002	Prestige oil spill, Spain	63,000	Severe damage to marine ecosystems, economic impact	González et al. (2006)
6	1997	South Africa	39,000	Environmental impact on marine biodiversity	Moloney et al. (2013)
7	2007	Kerch Strait, Ukraine	4000	Impact on marine ecosystems, economic damage	Kudryavtsev (2018)
8	2001	Brazil	1.3 million	Impact on marine biodiversity, economic impact	Disner and Torres (2020)
9	2019	Mauritius	1000	Environmental damage to coral reefs, marine habitats	Hebbar and Dharmasiri (2022)
10	2014	Sundarbans, Bangladesh	3500	Impact on mangrove forests, wildlife, and local communities	Lewis (1983)

- **Enhancement of microbial growth**: Microbial growth and activity are stimulated by adding nutrients, oxygen, and other growth-promoting substances.
- **Monitoring and optimization**: The progress of bioremediation is monitored through regular sampling and analysis of the contaminated site. Environmental conditions are adjusted as needed to optimize microbial activity and pollutant degradation.

Bioremediation offers several advantages over traditional cleanup methods. Bioremediation is a non-invasive and environmentally friendly approach that minimizes the use of harsh chemicals and disruption to the ecosystem. Bioremediation is often more cost-effective than traditional cleanup methods, such as chemical dispersants and mechanical recovery. Bioremediation has the potential to completely degrade pollutants, resulting in the restoration of contaminated environments to pre-spill conditions. Figure 14.1 describes the bioremediation of hydrocarbons.

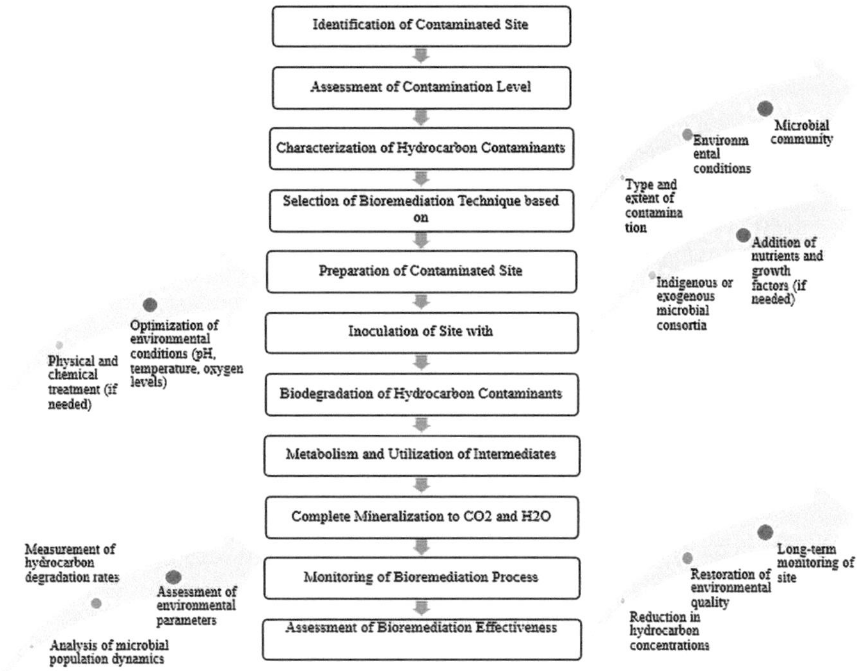

Fig. 14.1 Flowchart indicating bioremediation of hydrocarbons

14.3.1 Microbial Degradation of Hydrocarbons

14.3.1.1 Mechanisms of Hydrocarbon Degradation by Microorganisms:

Microorganisms have evolved various metabolic pathways to degrade hydrocarbons present in fuels. These metabolic pathways allow microorganisms to utilize hydrocarbons as carbon and energy sources. In the context of fuel shipping, microbial degradation of hydrocarbons is a key process in bioremediation efforts following fuel spills. Figure 14.2 describes the mechanisms of enzymatic hydrocarbon degradation.

- **Aerobic degradation**: Aerobic microorganisms utilize oxygen to break down hydrocarbons into simpler compounds through enzymatic reactions.
  The process involves the following steps:
  Hydrocarbon oxidation: Hydrocarbons are oxidized by oxygen to form fatty acids, alcohols, and other intermediate compounds.
  Intermediate metabolism: Intermediate compounds are metabolized through various enzymatic reactions, yielding carbon dioxide and water.
  Aerobic degradation is the predominant mechanism of hydrocarbon degradation in oxygen-rich environments such as surface waters and aerated soils.

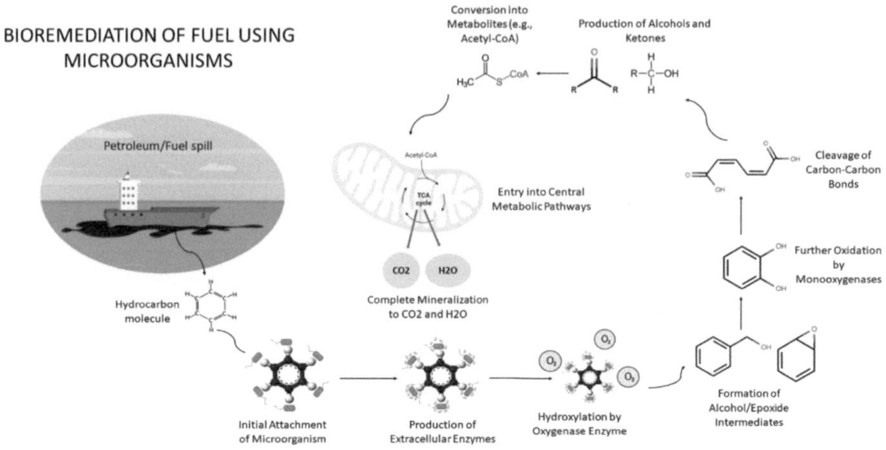

Fig. 14.2 The main mechanisms of enzymatic hydrocarbon degradation

- **Anaerobic degradation**: Anaerobic microorganisms degrade hydrocarbons without oxygen through fermentation and anaerobic respiration.
  The process involves the following steps:

Fermentation: Hydrocarbons are fermented by anaerobic microorganisms to produce organic acids, alcohols, and other fermentation products.

Anaerobic respiration: Fermentation products are further metabolized through anaerobic respiration, yielding methane, carbon dioxide, and other end-products.

Anaerobic degradation is important in environments with limited oxygen, such as deep marine sediments and anaerobic soils.

14.3.1.2 Key Microbial Species Involved in Fuel Degradation

Several microbial species have been identified as key players in the degradation of hydrocarbons present in fuels. These include bacteria and fungi.

- Pseudomonas species are well-known for their ability to degrade a wide range of hydrocarbons, including alkanes, aromatic hydrocarbons, and polycyclic aromatic hydrocarbons (PAHs) (Head et al. 2019).
- Alcanivorax species are specialized hydrocarbon-degrading bacteria commonly found in oil-contaminated environments. They are known for deleting alkanes in crude oil (Yakimov et al. 2007).
- Rhodococcus species can degrade various hydrocarbons, including alkanes, aromatic hydrocarbons, and chlorinated compounds (Whyte et al. 2002).
- Aspergillus species are common soil fungi capable of degrading a wide range of hydrocarbons, including alkanes, aromatic hydrocarbons, and PAHs (Haritash and Kaushik 2009).
- Trichoderma species are filamentous fungi known for their ability to degrade lignin and cellulose. They have also been shown to degrade various hydrocarbons present in fuels (Gupta et al. 2016).

14.3.2 Factors Affecting Bioremediation Efficacy:

Bioremediation efficacy is influenced by various environmental factors that affect microbial activity and the rate of pollutant degradation. In the context of fuel spills, several factors play a crucial role in determining the success of bioremediation efforts.

14.3.2.1 Temperature, pH, and Salinity

- Temperature significantly influences the rate of microbial metabolism and enzymatic activity (Head et al. 2019). Higher temperatures generally accelerate microbial growth and pollutant degradation, while lower temperatures may slow down or inhibit microbial activity. Optimal temperature ranges for bioremediation vary depending on the microbial species and the type of pollutant. In marine environments, temperature fluctuations due to seasonal changes can affect the efficiency of bioremediation processes (Nikolopoulou and Kalogerakis 2009).
- pH affects microbial activity and the availability of nutrients and trace elements essential for pollutant degradation. Microorganisms have optimal pH ranges for growth and activity, and extreme pH levels can inhibit microbial metabolism. For example, many hydrocarbon-degrading bacteria prefer neutral to slightly alkaline pH conditions (Nikolopoulou and Kalogerakis 2009).
- Salinity levels in marine environments can impact the composition and activity of microbial communities involved in bioremediation (Atlas 1991). High salinity levels can inhibit microbial growth and activity, limiting the efficacy of bioremediation efforts. Some halophilic microorganisms have adapted to high salinity conditions and play important roles in bioremediation processes in marine environments.

14.3.2.2 Availability of Nutrients and Oxygen

Microbial growth and activity require essential nutrients such as nitrogen, phosphorus, and trace elements (Venosa and Zhu 2003). Nutrient availability can limit microbial growth and pollutant degradation in nutrient-poor environments. The addition of nitrogen and phosphorus-containing compounds, such as nitrogen and phosphate fertilizers, can enhance microbial activity and bioremediation efficacy.

Oxygen availability is critical for aerobic hydrocarbon degradation by microorganisms. Oxygen depletion in waterlogged soils or sediments can create anaerobic conditions, limiting the efficacy of aerobic bioremediation processes. Aeration or adding oxygen-releasing compounds can enhance oxygen availability and stimulate aerobic pollutant degradation.

14.3.3 Types of Bioremediation Techniques

- Bioremediation techniques utilize natural processes to degrade pollutants and restore contaminated environments. Several bioremediation techniques have been developed to facilitate the cleanup of oil and fuel-contaminated sites in the context of fuel spills. These techniques include bioaugmentation, biostimulation, and natural attenuation.

- Bioaugmentation involves the addition of specialized microorganisms to contaminated environments to enhance the rate of pollutant degradation (Haritash and Kaushik 2009). In the case of fuel spills, specific hydrocarbon-degrading microorganisms are introduced into the contaminated site to supplement the existing microbial community and increase the rate of hydrocarbon degradation. These introduced microorganisms may include bacteria, fungi, or microbial consortia that are specifically selected for their ability to metabolize the spilled fuel components (Leahy and Colwell 1990). Bioaugmentation can significantly accelerate the cleanup process and is particularly effective in environments where indigenous microbial populations are unable to degrade the spilled fuel efficiently.
- Bio stimulation aims to enhance the activity of indigenous microorganisms by providing them with the necessary nutrients and environmental conditions to facilitate pollutant degradation (Venosa and Zhu 2003). In the context of fuel spills, biostimulation techniques involve the addition of nutrients such as nitrogen and phosphorus to stimulate the growth of hydrocarbon-degrading microorganisms. Other biostimulation strategies include the optimization of environmental factors such as temperature, pH, and oxygen levels to enhance microbial activity and pollutant degradation. Biostimulation is a cost-effective and environmentally friendly bioremediation approach and is particularly suitable for large-scale fuel spill cleanup operations (Head et al. 2019).
- Natural attenuation relies on natural processes such as microbial degradation, volatilization, and sorption to reduce the concentration of pollutants in the environment (National Research Council 1993). In the context of fuel spills, natural attenuation involves allowing indigenous microorganisms to degrade the spilled fuel over time without any human intervention. Microorganisms present in the environment naturally degrade the spilled fuel components, converting them into less harmful substances through metabolic processes. While natural attenuation is often slower than bioaugmentation and biostimulation, it is a passive and cost-effective approach to bioremediation that can be effective under certain environmental conditions.

Each bioremediation technique has advantages and limitations, and selecting the most appropriate technique depends on factors such as the nature and extent of the fuel spill, environmental conditions, and regulatory requirements (Table 14.2). Figure 14.3 describes the degradation rate of different hydrocarbons during bioremediation.

14.3.4 Advantages and Limitations of Bioremediation Compared to Other Cleanup Methods

Bioremediation, while advantageous for fuel spill cleanup, has limitations that necessitate consideration when selecting cleanup methods. Chemical dispersants rapidly break up oil, aiding its dispersion and microbial degradation, with broad application potential and cost-effectiveness. However, they contain toxic compounds harmful to marine life, disperse rather than remove oil, and efficacy can vary based on spill type and environmental conditions, with restrictions in sensitive areas. Mechanical recovery methods directly remove oil, minimizing environmental impact and avoiding toxicity concerns. Yet, they are most effective in calm, shallow waters, requiring substantial manpower and equipment, potentially disrupting ecosystems (Table 14.3). Both methods offer advantages but require careful environmental impact and efficacy assessment in specific spill scenarios (Brakstad and Daling 2015; National Research Council (NRC) 2005; Prince and Butler 2014).

14.4 Application of Bioremediation in Fuel Spill Cleanup

14.4.1 Bioremediation Strategies for Different Types of Fuel Spills

Crude oil spills pose distinctive challenges to bioremediation due to their complex composition and the diverse array of hydrocarbons they contain. Effective bioremediation strategies typically involve harnessing indigenous microorganisms capable of metabolizing various hydrocarbon compounds found in crude oil. These microorganisms, including bacteria, fungi, and archaea, have evolved specific metabolic pathways for hydrocarbon degradation (Head et al. 2019). Bioremediation approaches often employ techniques such as bioaugmentation and biostimulation to enhance the activity of indigenous hydrocarbon-degrading microorganisms. Additionally, natural attenuation processes may significantly contribute to the bioremediation of crude oil spills, particularly in coastal and marine environments (Venosa and Holder 2007).

Gasoline spills typically contain a mixture of volatile organic compounds (VOCs) such as benzene, toluene, ethylbenzene, and xylene (BTEX) and other hydrocarbons such as alkanes and aromatics. Bioremediation strategies for gasoline spills often focus on the rapid degradation of VOCs, which are highly soluble and can easily evaporate into the atmosphere. Biodegradation of gasoline components is primarily carried out by aerobic microorganisms that use these compounds as carbon and energy sources (Mrozik and Piotrowska-Seget 2010). Bioaugmentation and biostimulation techniques can be used to enhance the activity of hydrocarbon-degrading microorganisms and accelerate the biodegradation of gasoline components in contaminated soil and water.

Diesel fuel is composed of a mixture of aliphatic hydrocarbons, aromatic hydrocarbons, and other organic compounds. Bioremediation strategies for diesel spills often involve the use of indigenous microorganisms capable of degrading these hydrocarbon compounds under aerobic or anaerobic conditions. Aerobic bioremediation of diesel spills is typically carried out by bacteria that utilize aliphatic hydrocarbons and aromatic hydrocarbons as carbon and energy sources (Leahy and Colwell 1990). Anaerobic bioremediation techniques may also be employed in

Table 14.2 Case studies of successful bioremediation projects in fuel spill cleanup

Sl. no.	Location	Type of fuel spilled	Bioremediation technique used	Duration of cleanup	Environmental impact post-cleanup	References
1	Prince William Sound, Alaska	Crude oil	Natural attenuation	10+ years	Significant recovery of marine ecosystems, some long-term effects remain	Spill (2003)
2	Gulf of Mexico	Crude oil	Bioaugmentation	5 years	Improved water quality, recovery of marine life	Hazen et al. (2010)
3	Brittany, France	Crude oil	Biostimulation	2 years	Restoration of coastal ecosystems, minimal long-term effects	Cheng et al. (2011)
4	Galicia, Spain	Crude oil	Landfarming	3 years	Reduced soil contamination, restoration of affected areas	Lim et al. (2016)
5	Kerch Strait, Ukraine	Diesel	Phytoremediation	18 months	Improved water quality, restoration of coastal habitats	Nichols et al. (2014)
6	Brazil	Gasoline	In situ burning	6 months	Rapid removal of spilled fuel, minimal long-term environmental impact	Chagas-Spinelli et al. (2012)
7	Mauritius	Crude oil	Sorbent materials	1 year	Reduced contamination of coastal areas, minimal impact on marine life	Hebbar and Dharmasiri (2022)
8	Sundarbans, Bangladesh	Diesel	Biostimulation	3 years	Restoration of mangrove forests, recovery of wildlife	Nwinyi and Olawore (2017)
9	South Africa	Crude oil	Natural attenuation	5 years	Significant recovery of marine biodiversity, some residual contamination	Teske et al. (2011)
10	Brazil	Gasoline	Bioremediation using biochar	2 years	Improved soil quality, minimal impact on surrounding ecosystems	Zahed et al. (2021)

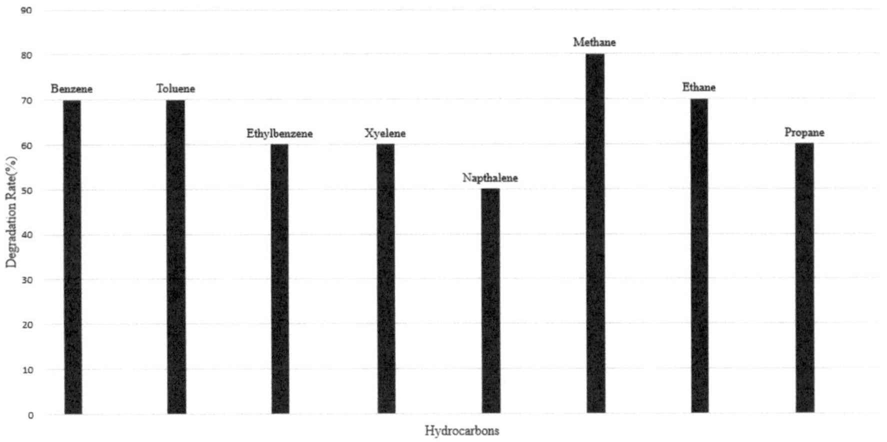

Fig. 14.3 Degradation rate of different hydrocarbons during bioremediation

environments with limited oxygen availability, such as waterlogged soils and sediments. Bioaugmentation and biostimulation techniques can be used to enhance the activity of diesel-degrading microorganisms and improve the efficiency of bioremediation processes in contaminated environments.

14.4.2 Case Studies of Successful Bioremediation Projects in Fuel Spill Cleanup

The Exxon Valdez oil spill, occurring in Prince William Sound, Alaska, in 1989, stands as a prominent event in oil spill history, prompting extensive bioremediation endeavors to mitigate environmental damage. Strategies employed included fertilizer application to stimulate indigenous hydrocarbon-degrading microorganisms and dispersant use to fragment oil into smaller droplets (Atlas and Hazen 2011). In situ bioremediation techniques further augmented natural oil attenuation in coastal and marine environments, resulting in substantial degradation of spilled oil (Bragg and Prince 2012). Similarly, the Deepwater Horizon oil spill in the Gulf of Mexico in 2010, the largest marine oil spill recorded, saw significant bioremediation efforts. The focus was on enhancing natural oil degradation by indigenous microorganisms through biostimulation techniques, providing nutrients like nitrogen and phosphorus (Passow and Ziervogel 2016). In situ bioremediation methods, including dispersant addition and microbial consortia application, expedited oil degradation in offshore and coastal areas, aiding in environmental recovery (Kimes et al. 2014). Cleanup durations post-spill vary due to factors like spill volume, environmental conditions, and remediation efficacy. The Exxon Valdez spill required several years of bioremediation efforts, witnessing gradual environmental improvements (Atlas and Hazen 2011). Likewise, post-Deepwater Horizon cleanup efforts spanned years, with ongoing environmental monitoring (Passow and Ziervogel 2016). Following

Table 14.3 Comparison of bioremediation techniques for fuel spill cleanup

Sl. no.	Technique	Principle	Advantages	Limitations	Examples	References
1	Bioaugmentation	Addition of specific microorganisms to enhance degradation	Accelerates degradation of pollutants	Requires constant monitoring and management	Deepwater Horizon oil spill	Atlas and Hazen (2011)
2	Biostimulation	Enhancement of indigenous microbial populations	Stimulates existing microbial communities	Effectiveness depends on environmental conditions	Exxon Valdez oil spill	Prince et al. (2013)
3	Natural attenuation	Relies on natural processes to degrade pollutants	Cost-effective and environmentally friendly	Slow process, may take years for complete cleanup	Erika oil spill	Sales da Silva et al. (2020)
4	Landfarming	Treatment of contaminated soil on-site	Cost-effective and minimizes soil disposal	Potential soil contamination and leaching	Prestige oil spill	Bai et al. (2024)
5	Phytoremediation	Use of plants to remove, degrade, or contain pollutants	Low cost, aesthetically pleasing, and sustainable	Limited to specific pollutants and environmental conditions	Diesel spills in marine environments	Yu et al. (2021)
6	In situ burning	Controlled burning of spilled oil	Rapid removal of oil from water surface	Air pollution and potential habitat destruction	Gulf of Mexico oil spill	Keramea et al. (2021)
7	Chemical dispersants	Break up oil into smaller droplets to enhance degradation	Rapid response and can be applied over large areas	Toxic to marine life and ecosystems	Deepwater Horizon oil spill	Clayton et al. (2020)
8	Sorbent materials	Absorbent materials used to adsorb oil	Versatile and can be used in various spill scenarios	Disposal of used sorbents can be challenging	Erika oil spill	Winquist (2014)
9	Biochar amendment	Addition of biochar to enhance biodegradation	Enhances microbial activity and reduces leaching	Long-term monitoring needed for the effectiveness	Oil spills in soil environments	Liu et al. (2015)
10	Nanoremediation	Use of nanoparticles to degrade pollutants	Highly effective in degrading contaminants	Potential environmental and health risks	Various fuel spills	Keshteli and Sheikholeslami (2019)

successful bioremediation, long-term environmental impacts are monitored. Studies post-Exxon Valdez reveal ecosystem recovery, albeit with residual oil contamination in some areas (Atlas and Hazen 2011). Similarly, post-Deepwater Horizon studies show marine and coastal ecosystem recovery, yet full environmental recovery timelines remain uncertain (Passow and Ziervogel 2016).

14.4.3 Field Trials and Laboratory Studies Evaluating the Effectiveness of Bioremediation Techniques

Field trials provide valuable insights into the effectiveness of bioremediation techniques in real-world environments. These trials typically involve the application of bioremediation agents such as microbial consortia, nutrients, and oxygen sources to contaminated sites, followed by monitoring of hydrocarbon degradation rates and environmental impacts over time. Field trials have shown that bioremediation techniques can effectively reduce the concentrations of spilled fuel components in soil, water, and sediment environments (Margesin and Schinner 2001). For example, in a field trial conducted in an alpine glacier skiing area contaminated with diesel oil, bioremediation using natural attenuation and biostimulation techniques resulted in a significant reduction in diesel oil concentrations in the soil over a period of several months (Margesin and Schinner 2001).

Laboratory studies are conducted to evaluate the efficacy of bioremediation techniques under controlled conditions. These studies often involve using microcosms or mesocosms to simulate environmental conditions found at spill sites. Laboratory experiments allow researchers to assess the effects of various factors such as temperature, pH, nutrient availability, and microbial activity on the biodegradation of spilled fuel components (Table 14.4). Laboratory studies have provided valuable insights into the mechanisms of hydrocarbon degradation by microorganisms and have helped optimize bioremediation strategies for different types of fuel spills (Haritash and Kaushik 2009). For example, in a laboratory study investigating the biodegradation of polycyclic aromatic hydrocarbons (PAHs), researchers found that certain bacterial species could degrade a wide range of PAH compounds under aerobic and anaerobic conditions (Haritash and Kaushik 2009).

14.4.4 Challenges and Limitations in the Application of Bioremediation for Fuel Spill Cleanup

One of the main challenges in the application of bioremediation for fuel spill cleanup is its effectiveness under different environmental conditions. Factors such as temperature, pH, salinity, and oxygen availability can significantly impact the activity of hydrocarbon-degrading microorganisms and the rate of biodegradation. Bioremediation techniques that are effective in one environment may not be as effective in another, making it challenging to develop universal cleanup strategies for fuel spills (Atlas and Hazen 2011). For example, in a study investigating the

Table 14.4 Techniques for monitoring and assessment of bioremediation processes

Sl. no.	Technique	Description	Application of fuel spill cleanup	References
1	Molecular techniques	PCR (polymerase chain reaction), qPCR (quantitative PCR), DGGE (denaturing gradient gel electrophoresis)	Assessment of microbial communities and their activities during bioremediation	Hazen et al. (2013)
2	Chemical analysis	GC-MS (gas chromatography-mass spectrometry), HPLC (high-performance liquid chromatography)	Quantification of hydrocarbon degradation rates and by-products	Coulon et al. (2005)
3	Biological oxygen demand	Measurement of the amount of dissolved oxygen required by aerobic organisms to break down organic material	Indicates the rate of organic matter degradation in the environment	Boonyatumanond et al. (2006)
4	Chemical oxygen demand	Measurement of the amount of oxygen required to oxidize all organic and inorganic compounds in water	Assesses the amount of organic and inorganic pollutants in water bodies	Ali et al. (2022)
5	Total petroleum hydrocarbon (TPH) analysis	Quantification of total petroleum hydrocarbons in soil or water samples	Determines the extent of hydrocarbon contamination and effectiveness of cleanup	Li and Jiang (2021)
6	Toxicity testing	Assessment of the toxicity of contaminated samples on living organisms	Evaluates the ecological risk and effectiveness of bioremediation	Arvaniti and Stasinakis (2015)
7	Microbial community analysis	DNA sequencing, metagenomics, metatranscriptomics	Identifies key microbial species and their metabolic pathways during bioremediation	Laczi et al. (2020)
8	Isotope analysis	Stable isotope probing (SIP), compound-specific isotope analysis (CSIA)	Traces the fate of pollutants and identifies biodegradation pathways	Kinnaman et al. (2007)
9	Enzyme activity assays	Measurement of enzyme activities involved in hydrocarbon degradation	Indicates microbial metabolic activity and efficiency during bioremediation	Yan et al. (2012)
10	Microcosm studies	Laboratory-scale simulations of bioremediation processes using environmental samples	Evaluates the efficacy of bioremediation techniques under controlled conditions	Haritash and Kaushik (2009)

biodegradation of diesel oil in Antarctic waters, researchers found that low temperatures and limited nutrient availability were significant factors limiting the effectiveness of bioremediation techniques in cold marine environments (Bragg and Prince 2012).

Another challenge in the application of bioremediation for fuel spill cleanup is the scale-up of bioremediation techniques for large spills. While bioremediation has been successfully used to clean up small- to moderate-sized spills, its effectiveness in large-scale spills is still a subject of debate. Scaling up bioremediation techniques to treat large volumes of spilled fuel can be logistically challenging and may require significant time and resources. In addition, the effectiveness of bioremediation may be limited in environments with high levels of contamination or low microbial activity, making it difficult to achieve complete cleanup of large spills. For example, in the case of the Deepwater Horizon oil spill, the large scale of the spill and the depth of the spill site presented significant challenges for bioremediation efforts, leading to the development of innovative cleanup techniques such as the use of dispersants and controlled burns (Prince and Butler 2014).

14.5 Monitoring and Assessment of Bioremediation Processes

14.5.1 Techniques for Monitoring Microbial Activity During Bioremediation

Molecular techniques are pivotal in monitoring microbial activity during fuel spill bioremediation. polymerase chain reaction (PCR), quantitative PCR (qPCR), denaturing gradient gel electrophoresis (DGGE), and next-generation sequencing (NGS) are commonly used for assessing microbial community diversity and abundance involved in hydrocarbon degradation (Hazen et al. 2010). PCR and qPCR enable detection and quantification of specific hydrocarbon-degrading microbial populations by targeting genes like alkane monooxygenase (alkB) and aromatic ring-hydroxylating dioxygenase (nahAc) genes (Cébron et al. 2008). DGGE and NGS offer insights into microbial community structure and dynamics, identifying key taxa in hydrocarbon degradation (Head et al. 2019).

Chemical analysis techniques such as gas chromatography-mass spectrometry (GC-MS) and high-performance liquid chromatography (HPLC) monitor fuel composition changes and metabolite formation during bioremediation. GC-MS identifies and quantifies hydrocarbons and breakdown products, assessing treatment effectiveness and degradation progress (Dombrowski et al. 2016). HPLC detects and quantifies metabolites, elucidating metabolic pathways in hydrocarbon degradation (Haritash and Kaushik 2009). For instance, GC-MS analysis in a study on gasoline-contaminated groundwater revealed decreasing BTEX concentrations over time, signifying successful bioremediation (Scow and Hicks 2005).

14.5.2 Measurement of Hydrocarbon Degradation Rates

Biological oxygen demand (BOD) gauges the dissolved oxygen consumption by microorganisms during organic matter degradation, including hydrocarbons. In bioremediation studies, BOD measures the rate of hydrocarbon degradation via oxygen consumption in contaminated water or soil samples. As hydrocarbons biodegrade, aerobic microbial respiration consumes oxygen, lowering dissolved oxygen levels. Tracking dissolved oxygen changes over time allows estimation of hydrocarbon degradation rates and treatment efficacy. For instance, in a diesel-contaminated soil study, declining BOD values over weeks indicated successful diesel oil biodegradation (Margesin and Schinner 2001).

Chemical oxygen demand (COD) quantifies the oxygen needed to oxidize organic and inorganic compounds in water. In bioremediation, COD assesses water sample organic contamination and tracks organic compound concentration changes. As hydrocarbons biodegrade, organic carbon conversion reduces COD values. For example, in gasoline-contaminated groundwater research, decreasing COD values over months indicated successful groundwater bioremediation (Scow and Hicks 2005).

Total petroleum hydrocarbon (TPH) analysis quantifies hydrocarbon concentrations in environmental samples. TPH extraction from samples using organic solvents, followed by quantification via gas chromatography-mass spectrometry (GC-MS) or high-performance liquid chromatography (HPLC), enables assessment of contamination extent and monitoring of concentration changes during bioremediation. For instance, in crude oil-contaminated soil research, TPH analysis pre- and post-bioremediation revealed significant TPH concentration reductions, indicating successful crude oil biodegradation (Atlas and Hazen 2011).

14.5.3 Assessment of Environmental Impacts Post-bioremediation

Toxicity testing is essential for assessing the environmental impacts post-bioremediation of fuel spills and ensuring that the treated site is safe for human health and the ecosystem. Various toxicity tests, including acute and chronic toxicity tests, can be conducted to evaluate the effects of residual contaminants on different organisms. These tests can include bioassays using aquatic or terrestrial organisms to determine if the bioremediation process has successfully reduced the toxicity of the contaminated site (Brar et al. 2010; Mohanty and Mukherji 2012). For instance, in a study assessing the effectiveness of bioremediation techniques for the cleanup of diesel-contaminated soil, toxicity testing using earthworms and plants was conducted to evaluate the potential adverse effects of residual contaminants. The results showed a significant reduction in toxicity levels post-bioremediation, indicating the successful remediation of the contaminated soil (Brar et al. 2010).

Ecological monitoring involves the assessment of changes in ecosystem structure and function following bioremediation activities of fuel spills. This may include monitoring changes in microbial community composition, plant growth, soil fertility, and the presence of indicator species in the treated environment (Mohanty and Mukherji 2012; Wang and Xu 2017). For example, in a study evaluating the ecological impacts of bioremediation on oil-contaminated wetlands, ecological monitoring was conducted to assess changes in plant diversity, soil microbial communities, and wildlife populations over time. The results showed that bioremediation treatments resulted in significant improvements in ecosystem health, with an increase in plant diversity and the return of native wildlife to the treated wetlands (Wang and Xu 2017).

14.5.4 Case Studies Demonstrating the Effectiveness of Monitoring and Assessment Techniques

Case Study 1 In a study conducted by Brar et al. (2010), the effectiveness of bioremediation techniques for the cleanup of diesel-contaminated soil was assessed using toxicity testing and ecological monitoring. Toxicity tests using earthworms and plants were conducted to evaluate the potential adverse effects of residual contaminants, while ecological monitoring was used to assess changes in ecosystem structure and function following bioremediation activities. The results demonstrated a significant reduction in toxicity levels and improved soil health post-bioremediation, indicating the successful remediation of the contaminated site.

Case Study 2 In another study by Wang and Xu (2017), the ecological impacts of bioremediation on oil-contaminated wetlands were evaluated using toxicity testing and ecological monitoring techniques. Toxicity tests and environmental monitoring were conducted to assess changes in plant diversity, soil microbial communities, and wildlife populations following bioremediation activities. The results showed that bioremediation treatments resulted in significant improvements in ecosystem health, with an increase in plant diversity and the return of native wildlife to the treated wetlands.

14.6 Future Perspectives and Conclusion

14.6.1 Emerging Bioremediation Technologies for Fuel Spill Cleanup

Genetically engineered microorganisms (GEMs) exhibit promising potential for fuel spill bioremediation. Designed with specific metabolic pathways, GEMs efficiently degrade hydrocarbon pollutants by targeted genetic modifications, enhancing degradation rates and environmental stress tolerance (Ron and Rosenberg 2014).

For example, genetically engineered *Pseudomonas putida* strains demonstrated enhanced degradation capabilities, effectively remediating diverse hydrocarbon pollutants (Ron and Rosenberg 2014).

Nano-bioremediation utilizes nanomaterials like nanoparticles to augment bioremediation processes by enhancing nutrient delivery, microbial activity stimulation, and contaminant bioavailability (Zhang et al. 2018). Iron-based nanoparticles, for instance, acted as electron acceptors, facilitating microbial metabolism, and promoting hydrocarbon pollutant degradation in diesel-contaminated soil (Zhang et al. 2018). Emerging technologies like phytoremediation and enzyme-assisted bioremediation offer sustainable, cost-effective solutions for fuel spill cleanup. Phytoremediation employs plants to uptake, metabolize, or sequester contaminants, while enzyme-assisted bioremediation utilizes enzymes to catalyze pollutant breakdown (Peng et al. 2017). Recombinant enzymes derived from hydrocarbon-degrading bacteria efficiently remediated oil-contaminated soil by catalyzing hydrocarbon pollutant breakdown (Peng et al. 2017).

14.6.2 Integration of Bioremediation with Other Cleanup Techniques

Integrating bioremediation with chemical dispersants and mechanical recovery techniques provides a comprehensive approach to fuel spill remediation. Chemical dispersants aid in breaking down oil slicks into smaller droplets, increasing surface area for microbial degradation (Gertler et al. 2009). Mechanical recovery methods like skimming physically remove oil from water surfaces, minimizing contamination spread (Fingas 2011). Chemical dispersants reduce oil surface tension, enhancing dispersion into water and promoting microbial access for degradation (Prince et al. 2016). In a study by Gertler et al. (2009), dispersants significantly boosted oil degradation by increasing microbial accessibility. Mechanical recovery, utilizing techniques like skimming and sorbent booms, efficiently removes oil from water surfaces, reducing contamination volume (Fingas 2011). Remote sensing methods in Fingas' study validated the effectiveness of skimming and sorbent booms in oil removal. Combining dispersants, mechanical recovery, and bioremediation enhances fuel spill cleanup effectiveness (Prince et al. 2016). Dispersants improve oil bioavailability for microbial degradation, while mechanical recovery minimizes contamination spread. Prince et al. (2016) demonstrated the enhanced efficacy of this integrated approach in crude oil spill cleanup, highlighting its synergistic effects.

14.6.3 Policy and Regulatory Frameworks Promoting the Use of Bioremediation in Fuel Spill Cleanup

International conventions and agreements, such as the International Convention on Oil Pollution Preparedness, Response and Co-operation (OPRC) by the International Maritime Organization (IMO) and guidelines from the United Nations Environment

Programme (UNEP), promote bioremediation for fuel spill cleanup. These frameworks advocate for environmentally sound methods, including bioremediation, and mandate member states to develop national contingency plans with bioremediation provisions (International Maritime Organization (IMO) 1990; United Nations Environment Programme (UNEP) 2003). National regulations and guidelines in countries like the United States, Canada, Australia, and the European Union endorse bioremediation for fuel spill cleanup. For instance, the US Environmental Protection Agency (EPA) integrates bioremediation into its National Contingency Plan (NCP) and Oil Pollution Act of 1990 (OPA), emphasizing its preference for mitigating environmental impacts (Environmental Protection Agency (EPA) 2014). Similarly, Canada, Australia, and the EU have established national regulations and guidelines promoting bioremediation for fuel spill cleanup (Environment Canada 2013; Department of the Environment and Energy 2018; European Commission 2009).

14.7 Conclusion

Despite significant advancements in bioremediation techniques for fuel spill cleanup, several areas warrant further research and innovation to improve remediation efforts' efficiency, effectiveness, and sustainability. Future research should focus on developing genetically engineered microorganisms (GEMs) with enhanced degradation capabilities for specific hydrocarbon pollutants, exploring nanomaterials and nanotechnologies for nano-bioremediation, and investigating the potential use of plants and enzymes for the remediation of fuel-contaminated environments. Additionally, there is a need for integrated cleanup approaches that combine bioremediation with other techniques such as chemical dispersants and mechanical recovery. Advanced monitoring and assessment techniques are also required to evaluate the effectiveness of bioremediation processes and assess environmental impacts post-bioremediation.

Effective fuel spill cleanup requires interdisciplinary collaboration between scientists, engineers, policymakers, and stakeholders to develop and implement innovative and sustainable remediation strategies. Interdisciplinary collaboration facilitates the exchange of knowledge, expertise, and resources, leading to more comprehensive and effective cleanup efforts. Key recommendations for promoting interdisciplinary collaboration include establishing collaborative research initiatives, promoting knowledge exchange and capacity building, engaging stakeholders in the decision-making process, and developing and implementing policy frameworks that support bioremediation and other environmentally sound cleanup techniques. By fostering interdisciplinary collaboration and promoting innovative research and development, we can advance the field of bioremediation and develop more effective and sustainable strategies for fuel spill cleanup, ultimately reducing the environmental impacts of oil spills and protecting our marine and coastal ecosystems for future generations.

Acknowledgments The authors express their profound gratitude to REVA University for providing the opportunity and platform for research.

Conflict of Interest The authors declare that there is no conflict of interest.

References

Ali H, Ahmed S, Hsini A, Kizito S, Naciri Y, Djellabi R, Ajmal Z (2022) Efficiency of a novel nitrogen-doped Fe3O4 impregnated biochar (N/Fe3O4@ BC) for arsenic (III and V) removal from aqueous solution: insight into mechanistic understanding and reusability potential. Arab J Chem 15(11):104209

Alvarez D, Velasco F, Antizar-Ladislao B (2006) Environmental impact assessment of the Prestige oil spill in the Bay of Biscay. Mar Pollut Bull 53(5-7):260–271

Arvaniti OS, Stasinakis AS (2015) Review on the occurrence, fate and removal of perfluorinated compounds during wastewater treatment. Sci Total Environ 524:81–92

Atlas RM (1991) Microbial hydrocarbon degradation—bioremediation of oil spills. J Chem Technol Biotechnol 52(2):149–156

Atlas RM, Hazen TC (2011) Oil biodegradation and bioremediation: a tale of the two worst spills in U.S. history. Environ Sci Technol 45(16):6709–6715

Bai X, Song W, Guo L, Liu R, Cao Y, Jin P, Zhang X (2024) Improvement of extraction efficiency and metabolites of pollutants from medium and low concentration organic polluted soil. Sci Energy Environ 1:5

Boehm PD, Page DS (2007) Assessing oil spill risks from marine transportation: a case study for Tampa Bay. Mar Pollut Bull 54(3):246–257

Boonyatumanond R, Wattayakorn G, Togo A, Takada H (2006) Distribution and origins of polycyclic aromatic hydrocarbons (PAHs) in riverine, estuarine, and marine sediments in Thailand. Mar Pollut Bull 52(8):942–956

Bragg JR, Prince RC (2012) Harsh environments: oil spill bioremediation in Arctic and Antarctic waters. Environ Sci Technol 46(15):7931–7940

Brakstad OG, Daling PS (2015) Biodegradation of dispersed oil in seawater is not inhibited by a commercial oil spill dispersant. Mar Pollut Bull 93(1-2):150–157

Brar SK, Verma M, Tyagi RD, Vale P, Surampalli RY (2010) Engineered strategies for the bio-degradation of xenobiotics. Crit Rev Biotechnol 30(2):75–91. https://doi.org/10.3109/07388551003665939

Cébron A, Norini MP, Beguiristain T, Leyval C (2008) Real-time PCR quantification of PAH-ring hydroxylating dioxygenase (PAH-RHDα) genes from Gram positive and Gram negative bacteria in soil and sediment samples. J Microbiol Methods 73(2):148–159

Chagas-Spinelli AC, Kato MT, de Lima ES, Gavazza S (2012) Bioremediation of a tropical clay soil contaminated with diesel oil. J Environ Manag 113:510–516

Cheng Y, Li X, Xu Q, Garcia-Pineda O, Andersen OB, Pichel WG (2011) SAR observation and model tracking of an oil spill event in coastal waters. Mar Pollut Bull 62(2):350–363

Clayton JR, Payne JR, Farlow JS, Sarwar C (2020) Oil spill dispersants: mechanisms of action and laboratory tests. CRC Press, Boca Raton

Corbett JJ, Lack DA, Winebrake JJ, Harder S, Silberman JA, Gold M (2010) Arctic shipping emissions inventories and future scenarios. Atmos Chem Phys 10(19):9689–9704

Coulon F, Pelletier E, Gourhant L, Delille D (2005) Effects of nutrient and temperature on degradation of petroleum hydrocarbons in contaminated sub-Antarctic soil. Chemosphere 58(10):1439–1448

Department of the Environment and Energy (2018) National environment protection (assessment of site contamination) measure. https://www.legislation.gov.au/Details/F2018L01432

Disner GR, Torres M (2020) The environmental impacts of 2019 oil spill on the Brazilian coast: overview. Rev Bras Gestão Ambiental Sustentabilidade 7:18
Dombrowski N, Donaho JA, Gutierrez T, Seitz KW, Teske AP, Baker BJ (2016) Reconstructing metabolic pathways of hydrocarbon-degrading bacteria from the Deepwater Horizon oil spill. Nat Microbiol 1(7):1–7
Environment Canada (2013) Guidelines for the use of bioremediation in marine oil spills. https://www.canada.ca/en/environment-climate-change/services/canadian-environmental-protection-act-registry/publications/guidelines-use-bioremediation-marine-oil-spills.html
Environmental Protection Agency (EPA) (2014) National oil and hazardous substances pollution contingency plan. https://www.epa.gov/emergency-response/national-oil-and-hazardous-substances-pollution-contingency-plan
Etkin DS (2016) Introduction to environmental management. Routledge, London
European Commission (2009) Directive 2009/147/EC of the European Parliament and of the Council of 30 November 2009 on the Conservation of Wild Birds. https://eur-lex.europa.eu/legal-content/EN/TXT/?uri=CELEX%3A32009L0147
Fingas M (2011) Review of oil spill remote sensing. Mar Pollut Bull 83(1):9–23. https://doi.org/10.1016/j.marpolbul.2014.03.042
Gertler C, Gerdts G, Timmis KN, Golyshin PN (2009) Microbial consortia in mesocosm bioremediation trial using oil sorbents, slow-release fertilizer and bioaugmentation. FEMS Microbiol Ecol 69(2):288–300. https://doi.org/10.1111/j.1574-6941.2009.00712.x
GIIGNL (2020) The LNG industry: GIIGNL annual report 2020. International Group of Liquefied Natural Gas Importers
González JJ, Viñas L, Franco MA, Fumega J, Soriano JA, Grueiro G, Albaigés J (2006) Spatial and temporal distribution of dissolved/dispersed aromatic hydrocarbons in seawater in the area affected by the Prestige oil spill. Mar Pollut Bull 53(5-7):250–259
Gupta G, Samant K, Sahu A (2016) Isolation and identification of hydrocarbon degrading bacteria from soils contaminated with crude oil spills. Indian J Exp Biol 54(1):47–52
Haritash AK, Kaushik CP (2009) Biodegradation aspects of polycyclic aromatic hydrocarbons (PAHs): a review. J Hazard Mater 169(1-3):1–15
Hazen TC, Dubinsky EA, DeSantis TZ, Andersen GL, Piceno YM, Singh N, Mason OU (2010) Deep-sea oil plume enriches indigenous oil-degrading bacteria. Science 330(6001):204–208
Hazen TC, Rocha AM, Techtmann SM (2013) Advances in monitoring environmental microbes. Curr Opin Biotechnol 24(3):526–533
Head IM, Jones DM, Röling WF (2019) Marine bioremediation: current and future perspectives. Environ Sci: Processes Impacts 21(6):1037–1055
Hebbar AA, Dharmasiri IG (2022) Management of marine oil spills: A case study of the Wakashio oil spill in Mauritius using a lens-actor-focus conceptual framework. Ocean Coast Manag 221:106103
IEA (2020) Oil 2020: analysis and forecasts to 2025. International Energy Agency, Paris
International Maritime Organization (IMO) (1990) International convention on oil pollution preparedness, response and co-operation (OPRC). https://www.imo.org/en/About/Conventions/ListOfConventions/Pages/International-Convention-on-Oil-Pollution-Preparedness,-Response-and-Co-operation-(OPRC).aspx
Joye SB, Teske AP, Kostka JE, Chanton JP (2016) Microbial dynamics following the Macondo oil well blowout across Gulf of Mexico environments. Bioscience 66(10):829–842
Keramea P, Spanoudaki K, Zodiatis G, Gikas G, Sylaios G (2021) Oil spill modeling: a critical review on current trends, perspectives, and challenges. J Marine Sci Eng 9(2):181
Keshteli AN, Sheikholeslami M (2019) Nanoparticle enhanced PCM applications for intensification of thermal performance in building: a review. J Mol Liq 274:516–533
Kimes NE, Callaghan AV, Aktas DF, Smith WL, Sunner J, Golding B, Morris PJ (2014) Metagenomic analysis and metabolite profiling of deep-sea sediments from the Gulf of Mexico following the Deepwater Horizon oil spill. Front Microbiol 5:1–16
King GM, Kostka JE, Hazen TC, Sobecky PA (2015) Microbial responses to the deepwater horizon oil spill: from coastal wetlands to the deep sea. Annu Rev Mar Sci 7:377–401

Kinnaman FS, Valentine DL, Tyler SC (2007) Carbon and hydrogen isotope fractionation associated with the aerobic microbial oxidation of methane, ethane, propane and butane. Geochim Cosmochim Acta 71(2):271–283

Kudryavtsev VN (2018) 8.11 Ocean Remote Sensing in Russia

Laczi K, Erdeiné Kis Á, Szilágyi Á, Bounedjoum N, Bodor A, Vincze GE, Perei K (2020) New frontiers of anaerobic hydrocarbon biodegradation in the multi-omics era. Front Microbiol 11:590049

Lanoix J, Bocquene G, Le Tertre A, Miramand P, Budzinski H (2002) Monitoring of Erika oil spill impacts on the benthos and marine environment. Mar Pollut Bull 44(3):205–212

Leahy JG, Colwell RR (1990) Microbial degradation of hydrocarbons in the environment. Microbiol Rev 54(3):305–315

Lewis RR (1983) Impact of oil spills on mangrove forests. In: Biology and ecology of mangroves. Springer, Dordrecht, pp 171–183

Li Y, Jiang L (2021) Comparison of the crude oil removal effects of different surfactants in electrokinetic remediation of low-permeability soil. J Environ Chem Eng 9(4):105190

Lim MW, Von Lau E, Poh PE (2016) A comprehensive guide of remediation technologies for oil contaminated soil—present works and future directions. Mar Pollut Bull 109(1):14–45

Liu WJ, Jiang H, Yu HQ (2015) Development of biochar-based functional materials: toward a sustainable platform carbon material. Chem Rev 115(22):12251–12285

Margesin R, Schinner F (2001) Bioremediation (natural attenuation and biostimulation) of diesel-oil-contaminated soil in an alpine glacier skiing area. Appl Environ Microbiol 67(7):3127–3133. https://doi.org/10.1128/AEM.67.7.3127-3133.2001

Mohanty S, Mukherji S (2012) Bioremediation of petroleum oil contaminated soil. Int J Environ Sci Technol 9(2):243–254. https://doi.org/10.1007/BF03326052

Moloney CL, Fennessy ST, Gibbons MJ, Roychoudhury A, Shillington FA, Von der Heyden BP, Watermeyer K (2013) Reviewing evidence of marine ecosystem change off South Africa. Afr J Mar Sci 35(3):427–448

Mrozik A, Piotrowska-Seget Z (2010) Bioaugmentation as a strategy for cleaning up of soils contaminated with aromatic compounds. Microbiol Res 165(5):363–375

National Commission on the BP Deepwater Horizon Oil Spill and Offshore Drilling (2011) Deep water: the Gulf oil disaster and the future of offshore drilling. National Commission on the BP Deepwater Horizon Oil Spill and Offshore Drilling

National Research Council (1993) In situ bioremediation: when does it work? National Academies Press, Washington

National Research Council (2003) Oil in the sea III: inputs, fates, and effects. National Academies Press, Washington

National Research Council (NRC) (2005) Oil spill dispersants: efficacy and effects. National Academies Press, Washington

Nichols EG, Cook RL, Landmeyer JE, Atkinson B, Malone DR, Shaw G, Woods L (2014) Phytoremediation of a petroleum-hydrocarbon contaminated shallow aquifer in Elizabeth City, North Carolina, USA. Remediat J 24(2):29–46

Nikolopoulou M, Kalogerakis N (2009) Biostimulation strategies for enhancing bioremediation of oil contaminated coastal and marine environments. Curr Opin Biotechnol 20(3):362–367

Nwinyi OC, Olawore YA (2017) Biostimulation of spent engine oil contaminated soil using Ananas comosus and Solanum tuberosum peels. Environ Technol Innov 8:373–388

Palenzuela JT, Vilas LG, Cuadrado MS (2006) Use of ASAR images to study the evolution of the Prestige oil spill off the Galician coast. Int J Remote Sens 27(10):1931–1950

Passow U, Ziervogel K (2016) Marine snow sedimented oil released during the Deepwater Horizon spill. Oceanography 29(3):118–125

Peng R-H, Xiong A-S, Xue Y, Fu X-Y, Gao F, Zhao W, Tian Y-S, Yao Q-H (2017) Microbial biodegradation of polyaromatic hydrocarbons. FEMS Microbiol Rev 40(3):774–794. https://doi.org/10.1093/femsre/fux025

Prince RC, Butler JD (2014) Chemodynamics of dispersants and dispersed oil. In: Kostka JE, Estes MA (eds) Oil spill environmental forensics case studies. Academic, New York, pp 265–301

Prince RC, McFarlin KM, Butler JD, Febbo EJ, Wang FC, Nedwed TJ (2013) The primary biodegradation of dispersed crude oil in the sea. Chemosphere 90(2):521–526

Prince RC, Coolbaugh TS, Parkerton TF, Lee H (2016) Biodegradation of dispersed oil in Arctic seawater at -1°C. Mar Pollut Bull 107(1):197–205. https://doi.org/10.1016/j.marpolbul.2016.03.036

Ron EZ, Rosenberg E (2014) Enhanced bioremediation of oil spills in the sea. Curr Opin Biotechnol 27:191–194. https://doi.org/10.1016/j.copbio.2014.02.003

Sales da Silva IG, Gomes de Almeida FC, Padilha da Rocha Silva NM, Casazza AA, Converti A, Asfora Sarubbo L (2020) Soil bioremediation: overview of technologies and trends. Energies 13(18):4664

Scow KM, Hicks KA (2005) Natural attenuation and enhanced bioremediation of organic contaminants in groundwater. Curr Opin Biotechnol 16(3):246–253. https://doi.org/10.1016/j.copbio.2005.03.004

Spill VO (2003) Long-term ecosystem response to the exxon. Science 1084282(2082):302

Teske PR, Von der Heyden S, McQuaid CD, Barker NP (2011) A review of marine phylogeography in southern Africa. S Afr J Sci 107(5):1–11

United Nations Environment Programme (UNEP) (2003) Guidelines on the use of bioremediation in the cleanup of oil spills in the marine environment. https://www.unep.org/resources/report/guidelines-use-bioremediation-cleanup-oil-spills-marine-environment

Venosa AD, Holder EL (2007) Biodegradability of dispersed crude oil at two different temperatures. Mar Pollut Bull 54(5):545–553

Venosa AD, Zhu X (2003) Biodegradation of crude oil contaminating marine shorelines and freshwater wetlands. Spill Sci Technol Bull 8(3):163–178

Wang W, Xu C (2017) Bioremediation of oil-contaminated soil: mechanisms, feasibility, and applications. Soil Biol Biochem 115:274–287. https://doi.org/10.1016/j.soilbio.2017.08.025

Whyte LG, Smits TH, Labbé D, Witholt B (2002) Characterization of a new hydrocarbon-degrading bacterium, Rhodococcus strain Q15. Appl Environ Microbiol 68(1):256–265

Winquist E (2014) The potential of ligninolytic fungi in bioremediation of contaminated soils

Yakimov MM, Timmis KN, Golyshin PN (2007) Obligate oil-degrading marine bacteria. Curr Opin Biotechnol 18(3):257–266

Yan Z, Song N, Cai H, Tay JH, Jiang H (2012) Enhanced degradation of phenanthrene and pyrene in freshwater sediments by combined employment of sediment microbial fuel cell and amorphous ferric hydroxide. J Hazard Mater 199:217–225

Yu B, Peng Y, Xu J, Qin D, Gao T, Zhu H, Dong J (2021) Phytoremediation potential of Youngia japonica (L.) DC: a newly discovered cadmium hyperaccumulator. Environ Sci Pollut Res 28:6044–6057

Zahed MA, Salehi S, Madadi R, Hejabi F (2021) Biochar as a sustainable product for remediation of petroleum contaminated soil. Curr Res Green Sustainable Chem 4:100055

Zhang Y, Yan L, Wang Y, Shen Y (2018) Advances in nano-bioremediation technology for the recovery of petroleum hydrocarbon contaminants: a review. J Environ Manag 220:293–308. https://doi.org/10.1016/j.jenvman.2018.05.055

Seas and Oceans Save by Bioremediation 15

Catia Angli Curie and Fitri Budiyanto

Abstract

Marine pollution is a global issue that coincides with the escalation in anthropogenic activity. Among the most concerning pollutants are plastic debris, organic and inorganic chemicals, and radionuclides. These pollutants are problematic due to their recalcitrant, toxicity, and long distant transport. The hazardous nature of these pollutants necessitates effective remediation strategies. Numerous solutions have been introduced to combat marine pollution, from chemical to biological agents. Bioremediation, either through human interference or natural occurrence, emerges as a viable technique to resolve the marine pollution issue cost-efficient and environmentally friendly ways. Marine bioremediation strategies, in general, have implemented biostimulation and bioaugmentation approach. Moreover, the biology agents, in removing the pollutants, may employ bioaccumulation, bioreduction, biomineralization, and biodegradation approach. This chapter discusses marine pollution, that is, plastic, organic, inorganic pollution, and radionuclides. In addition, this chapter discusses the existing bioremediation strategies desig-

Catia Angli Curie and Fitri Budiyanto contributed equally with all other contributors.

C. A. Curie
Faculty of Industrial Technology, Department of Chemical Engineering, Pertamina University, Jakarta, Indonesia
e-mail: catia.ac@universitaspertamina.ac.id

F. Budiyanto (✉)
Research Center for Vaccines and Drugs, National Research and Innovation Agency (BRIN), Bogor, Indonesia
e-mail: fitri.budiyanto@brin.go.id

A. A. H. Abdel Latef et al. (eds.), *Bioremediation and Nanotechnology for Climate Change Mitigation*, https://doi.org/10.1007/978-981-96-3069-1_15

nated to mitigate these pollutions, highlighting their mechanisms, effectiveness, and potential for optimization in real-world application. In laboratory experiment, it is feasible to optimize bioremediation outcomes, for example, achieving complete removal (100% degradation) of organic pollutant. However, implementing bioremediation in actual marine environment poses additional challenges, necessitating collaborative efforts among academia, industry, government is demanding to achieve a sustainable remediation approach.

Keywords

Marine pollution · Plastic · Organic · Inorganic · Radionuclides · Bioremediation

15.1 Introduction

Marine pollution is a growing environmental concern with its many identified contaminants such as physical, biological, chemical, and radionuclides. Exposure to these pollutants in different marine environments leads to both ecological damage and economic losses (Lestari et al. 2018, 2021). The global economic impact due to marine plastic pollution is estimated to range from \$3300 to \$33,000/ton of plastic waste annually (Xia et al. 2023). Emerging pollutants, such as plastic debris and other persistent organic pollutants, are primarily of anthropogenic origin (Alarif et al. 2023). In contrast, inorganic pollutants and radionuclides can originate from natural sources; however, human activities significantly escalate their concentration in marine environment (Budiyanto et al. 2023). To address this issue, several remediation techniques have been developed to find the best and the most suitable option to remediate the marine environment efficiently and economically (Budiyanto et al. 2021). Bioremediation is a plausible approach since it required less energy and environmental friendliness (Budiyanto et al. 2018).

Bioremediation in marine environments can be approached using in situ and ex situ strategies. In situ bioremediation engages microbial activity in their native environment without relocating the sample and vice versa for the ex situ bioremediation (Pandey et al. 2009). In situ bioremediation includes biostimulation (stimulation of the natural process) and bioaugmentation (introduction of exogenous degrading microbes) (Fig. 15.1). The biostimulation process is carried out by adding chemicals to contaminated sites such as electron donors/acceptors (glutamate, sulfate, nitrate, acetate, etc.), and nutrients (potassium, phosphorus, nitrogen, carbon, etc.) to contaminated sites (Pandey et al. 2009). However, the bioaugmentation method, or intrinsic remediation, utilizes isolates that are collected from the contamination sites. The isolates have adapted to the pollutants and developed the metabolic pathway to degrade them (Fragkou et al. 2021).

Several factors influence the bioremediation performance such as the physic-chemical properties (hydrostatic pressure, availability of oxygen and nutrients, temperature and pH) and the morphology of bottom sediment. The morphology of bottom sediment affects the distribution of pollutants and the coverage of certain

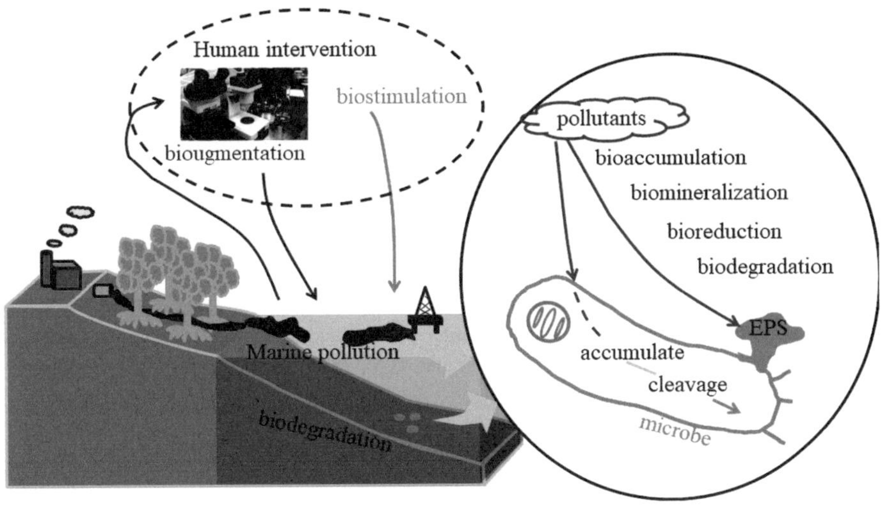

Fig. 15.1 The general schematic of marine pollution and the bioremediation strategies

pollutants, like oil, which can deplete the physicochemical properties of sediment (Geng et al. 2022). Oxygen is a crucial electron acceptor in the biodegradation process; thus, oxygen enrichment is sometimes applied using various methods, such as adding oxygen-releasing compounds, pure oxygen, and ozone injection (Fragkou et al. 2021). Moreover, the contaminants' bioavailability within their respective ecological system determines the biodegradation capability of microbes. Chemotactic microbes are sometimes introduced to improve pollutant bioavailability (Pandey et al. 2009).

Thus, this chapter is proposed to comprehensively discuss various types of marine pollution, including plastic, organic, inorganic, and radionuclides. In addition, this chapter will delve into the bioremediation strategy currently available to address these emerging pollutants in marine environment. Through this discussion, this chapter seeks to present a thorough understanding of how bioremediation can be viable, efficient, and environmentally friendly to mitigating marine pollution.

15.2 Selected Marine Pollution Types

15.2.1 Plastic Pollution Occurrence in Marine Environment

Since its discovery, plastics have become a very convenient material for use in many areas. The lightweight, durable, water-resistant, broad options, and cheap production have tempted us to massively produce and use plastics, especially for single-use purposes such as various packaging, disposable cutlery, and diapers. As a result, plastic wastes are now ubiquitous in ocean environments due to their very slow degradation rate, threatening the life of marine organisms.

Geyer et al. (2017) estimated that up to the year 2017, 8.3×10^9 metric tons of new plastics have been produced. This has brought about around 6.3×10^3 metric tons of plastic waste as of 2015. Yet, nearly 80% of the waste was only piled up in landfills or stayed in the environment, including oceans (Geyer et al. 2017). There are variations in the published amount of plastic waste entering the oceans per year estimated around 1.7 million tons of plastic waste, while a higher estimates of 10–20 million tons per year has also been published (Paul et al. 2020). More than 80% of this amount comes from the rivers (Meijer et al. 2021). Among these plastic wastes, the most abundant polymer types include: (1) polyethylene resins, such as high-density (HDPE), low-density (LDPE) and linear low-density (LLDPE); (2) polypropylene (PP); (3) polystyrene (PS); (4) polyvinyl chloride (PVC); (5) polyethylene terephthalate (PET); (6) polyurethanes (PUR); and (7) fibers from polyester, polyamide, and acrylic (Debroas et al. 2017; Geyer et al. 2017).

Plastics entering the ocean can have two main forms: large plastic particles and microplastics. Microplastics can be divided further into primary, secondary, and tertiary microplastics. Primary microplastics are those specifically produced with a size of <5 mm for particular purposes, for example, microbeads; secondary microplastics are fragments of naturally weathered plastics; and tertiary microplastics are pellets used for plastic production that enters the environment in their original state (Ugwu et al. 2021). Large plastics are often found entangling marine animals, whereas marine animals may ingest both large and microplastics. From the many studies of microplastic ingestion in marine animals, most of the microplastics ingested are in the form of fibers. This indicates that they originated from river run-offs containing remains of clothes (Ugwu et al. 2021), which is also in line with the fact that most ocean plastic wastes come from river (Meijer et al. 2021).

Exposure of polyethylene microplastics to *Pomatoschistus microps* (common goby fish) juveniles significantly lowered acetylcholinesterase (AChE) activity which could adversely affect neurofunction. This enzyme involved in the neurotransmission and aided the control of this fish's several physiological and behavioral processes (e.g., growth and reproduction) (Oliveira et al. 2013). Another study on the same type of fish also showed that simultaneous exposure of microplastics and Cr(VI) jointly decreased predatory performance and significantly reduced AChE activity (Luís et al. 2015).

Marine microplastic often carries other pollutants and altogether they can disturb endocrine function. Rochman et al. (2014) observed altered gene expression in Japanese medaka fish (*Oryzias latipes*) after exposure to an environment containing marine microplastic for 3 months (Rochman et al. 2014). It is possible that similar effects could occur in other marine animals. Another study indicated that exposure to microplastics interfered with the reproductive function and offspring performance of oysters (Sussarellu et al. 2016) as well as *Littorina* sp. (periwinkles) and *Idotea balthica* isopod (Green 2016).

In larger marine animals, the large plastic particles are often found entangling, impairing their physical functions. Plastic ingestions of various sizes have also been widely reported (Denuncio et al. 2011; de Stephanis et al. 2013; Rebolledo et al. 2013; Nelms et al. 2015; Ugwu et al. 2021). Various forms of microplastics (e.g., fibers, sheets, and fragments) are also found in the digestive tracts of dolphins (*Delphinus delphis*) and humpback whales (*Megaptera novaeangliae*) (Besseling et al. 2015; Hernandez-Gonzalez et al. 2018). Besides the negative effects of microplastics described above, ingested plastics may also block their intestinal tracts and filtering apparatus, or reduce their digestive capacity, which can cause malnutrition and eventually mortality (Nelms et al. 2015; Gola et al. 2021).

All these, and many other studies that have not been included here, show the alarming danger of plastic pollution in the seas and oceans to marine life. Since polymer fragments and/or their degradation product also bioaccumulate in their tissues (Broszeit et al. 2016), the marine food chain would cause a domino effect in microplastics bioaccumulation and emphasize their adverse effects. Microplastics and chemicals leached from the plastics degradation process (weathered plastics) can also impair our supply of fresh and clean water, thus providing another ingestion route for humans. As described above, this could negatively affect the reproductive processes and threaten wildlife populations in general and mankind.

15.2.2 Inorganic Pollutants in Marine Environment

The inorganic waste presented here will focus on ocean heavy metal pollution. Naturally, metals and heavy metals occur on earth; some are needed for the growth of plants and organisms in trace amounts. The natural distribution of heavy metals is facilitated by geochemical cycles, which also help secure nutrient provision. However, the growing numbers of metal-related industries such as mining, metal refining, battery manufacturing as well as waste power plants and incinerations (Naik and Dubey 2017) may leak out significant amounts of heavy metals into the rivers, ponds, lakes, and seas that their concentration become harmful.

Harmful heavy metals can be classified into three categories, that is, toxic metals (e.g., Hg, Pb, Cr, Cd, As, Co, Sn, Cu, Zn, Ni), precious metals (e.g., Pt, Pd, Au, Ag), and radionuclides (e.g., U, Th, Ra) (Wang and Chen 2009). Heavy metals bioaccumulate and their excretion is difficult (Joo et al. 2015). Thus, admission to the larger organisms from the heavy metal-polluted marine food chain is highly possible. The adverse effects of excessive intakes of heavy metals in humans include cancers, Alzheimer's disease, cardiovascular disease, impaired gene expressions, and mental disturbance (Joo et al. 2015; Pratush et al. 2018; Yin et al. 2019).

Among the heavy metals considered the most toxic are lead (Pb), mercury (Hg), and cadmium (Cd). Lead is naturally very rare. Yet, anthropogenic activities have spread them more to the surface. Lead pollution sources include mistreated industrial wastewater (e.g., from manufacturing batteries, ammunition, pigments, steels and alloys) and burning lead-containing gasoline (Sari and Tuzen 2008; Bilal et al. 2018). Mercury is used mostly to extract precious metals such as gold mining. Its release to the environment is mainly from gaseous Hg emissions in the form of elemental or inorganic ionic mercury, which will be deposited on land or water and in oceanic circulation (Pratush et al. 2018). As for cadmium's contamination usually comes from industrial and high risk run-offs from petroleum refining, phosphate fertilizers, inorganic pigment, stabilizers, and Cd-Ni alloy batteries manufacturing (Bilal et al. 2018; Yin et al. 2019). Apart from the above, heavy metals can also be introduced to the marine environment via accumulation on the biofilms of marine plastic wastes (Richard et al. 2019), brought by river run-offs.

15.2.3 Organic Pollutants in the Marine Environment

Carbon-based or organic substances are emerging pollution worldwide, especially in the aquatic body. Most of these pollutants are persistent to degradation, bioaccumulative, capable of long distant transport, and toxic; hence, they are categorized as persistent organic pollutants (POPs). Given that some of these compounds remain integral to industrial and daily needs, the Stockholm Convention classified these POPs into three annexes, that is, annex A (elimination), annex B (restriction), and annex C (unintentional production). Initially, the Convention listed 12 POPs, namely, dieldrin, endrin, heptachlor, hexachlorobenzene, mirex, toxaphene, polychlorinated biphenyls (PCBs), polychlorinated dibenzo-p-dioxins (PCDD), and polychlorinated dibenzofurans (PCDF) (Stockholm Convention 2019). Polychlorinated biphenyls (PCBs) are representative POPs; to date, it composes 209 congeners. Current research focuses on the biomagnification and bioaccumulation of various POPs species through the food-web (Arai and Nishi 2024).

Since numerous organic compounds released into the environment, toxicity studies on these organic compounds are ongoing. New investigated compounds are continually being registered and proposed for listing in the Convention such as chlorpyrifos, long-chain perfluorocarboxylic acids, and chlorinated paraffins (Stockholm Convention 2019). Before the 1970s, polychlorinated biphenyls (PCBs) and polybrominated diphenyl ethers (PBDEs) are widely used in various commercial and industrial products. However, their use was restricted by the Stockholm Convention in 2004 for PCBs and 2010 for PBDEs. POPs have been found to interfere with the reproduction system (Nos et al. 2024).

In marine ecosystems, POPs are mostly deposited in sediment, making mangrove ecosystem particularly vulnerable to the geochemical cycle of the compounds. In mangrove sediment in Wouri Estuary, Cameroon, two classes of POPs namely chlorinated pesticides (CLPs) and polychlorinated biphenyls (PCBs) were found at concentrations of 2.2–29.1 ng g^{-1} and 1.7–31.6 ng g^{-1}, respectively. The most abundant CLPs detected in the sediment were diphenyl trichloroethane (DDT; ≤5.0 ng g^{-1}), lindane (γ-HCH, ≤3.1 ng g^{-1}), heptachlor (≤4.3 ng g^{-1}), alachlor (≤4.9 ng g^{-1}), and endosulfan (≤18.5 ng g^{-1}) (Mbusnum et al. 2020). Due to their characteristics, POPs bioaccumulate in marine organisms, such as in endangered sea turtle *Lepidochelys olivacea*. In these turtles, POPs accumulation in the plasma was relatively higher and may alter enzymatic activity, immune cells, and biochemical and hematological in blood. Certain POPs, including γ-HCH, dichloro diphenyl dichloroethylene (DDE), diphenyl trichloroethane (DDT), αEndosulfan, Endosulfan sulfate, PCB52, and PCB118, were detected in the plasma of the sea turtle at the range of 6.1–32.5 ng mL^{-1} (Flores-Ramírez et al. 2024).

Other classes of POPs, namely brominated flame retardant (BFR) and Polyhalogenated carbazoles (PHCZs), are widely used in customer and industrial products such as electronics, furniture, textiles, indigo dyes, photoelectric material, or plastic. BFR from the class of hexabromocyclododecanes (HBCDs) and polybrominated diphenyl esters (PBDEs) has been banned by the Convention. However, perfluorooctane sulfonate (PFOS) and tetrabromobisphenol-A (TBBPA) are still in use to date. The bioaccumulation–elimination pattern of TBBPA and PFOS in aquatic organisms is affected by other substances, such as humic acid (HA). At a concentration of 1 mf L^{-1} of HA, the uptake of PFOS increased and TBBPA decreased on thick-shell mussels *Mytilus unguiculatus*. The presence of HA inhibited the accumulation of these compounds in dose-dependent manner in the mussel (Geng et al. 2024). Meanwhile, the planar structure of PHCZs resembled those polyhalogenated dibenzofurans (PCDFs) and dibenzo-p-dioxins (PCDDs). The western Pacific Oceans close to the Philippine Sea and Magellan Seamount are the sink of PHCZs. The concentration of PHCZs in the upper seawater layer is lower (0.23 ± 0.21 ng L^{-1}) compared to the bottom layer (0.65 ± 0.56 ng L^{-1}) (Hu et al. 2024).

Polycyclic aromatic hydrocarbons (PAHs) are also organic contaminants under environmental investigation. They can be generated through a multitude of natural and anthropogenic pathways. Pyrogenic PAHs are generated by the pyrolysis of organic materials under low oxygen concentration and high-temperature radiation, resulting in more stable high molecular weight PAHs with more than four aromatic rings. In contrast, petrogenic PAHs are formed through the maturation of crude oil at lower temperatures, releasing alkylated and low molecular weight PAHs with fewer than four aromatic rings (Martins et al. 2023).

Most of PAHs are pyrogenic origin; thus, the anthropogenic activity is a key factor in their spatial and temporal distribution. In the Sharam Area, Red Sea Coast, Saudi Arabia, PAHs concentrations are higher (22.09 ng kg^{-1}) due to the proximity to petrochemical industries, with two to three rings PAHs being predominant. These PAHs primarily originate from petrogenic sources in this area (Alhudhodi et al. 2022). In CanGio wetland, Vietnam, Coastal area close to ferry station exhibited higher PAHs concentration (112 ± 211 ng g^{-1}), while coastal area near oyster farms have lower concentrations (31 ± 77 ng g^{-1}). Seasonal alteration influences PAHs' spatial and temporal distribution (Thuy et al. 2021).

PAHs are transported by current in the marine environment and may be deposited in the bottom marine sediment or coastal ecosystems. The accumulation of PAHs in sediment has been documented in various regions (Fig. 15.2). For example, the total PAHs in the mangrove forest Rio de Janeiro, Brazil, ranged between 38 and 792 ng g^{-1}. In this area, alkylated PAHs, which account for 30–55% of total PAHs, predominantly originate from petroleum sources, whereas non-alkylated PAHs are mainly from the combustion source (Ceccopieri et al. 2023). PAHs moderately contaminated the Gaoqiao mangrove ecosystem, Zhanjiang, China with average value of 147.7 ± 77.7 μg kg^{-1}. Three and four-ring compounds dominated the PAHs in this area. The petrogenic sources such as incomplete combustion of fossil fuel (39.01% coal, 25.21% diesel, 10.48% gasoline-powered vehicle) and 12.72% biomass (wood

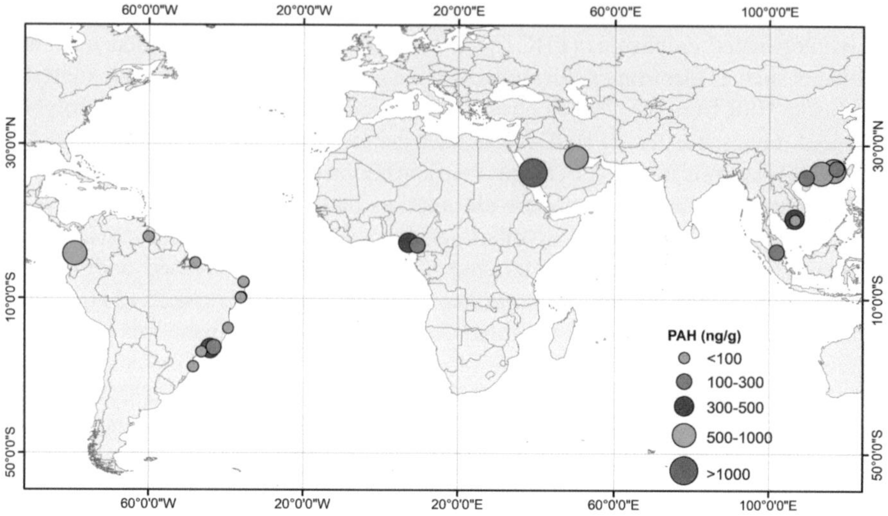

Fig. 15.2 Spatial distribution of PAHs in sediment from mangrove ecosystem in selected area. Source: Babut et al. (2019), Mbusnum et al. (2020), Thuy et al. (2021), Alhudhodi et al. (2022), Robin and Marchand (2022), Saunders et al. (2022), Armada et al. (2023), Bhatawadekar et al. (2023), Ceccopieri et al. (2023), Yan et al. (2023)

or grass) were the predominant sources of PAHs (Yan et al. 2023). Similarly, petroleum products and their combustion are also the main sources of high PAHs contamination (100–875,000 ng g^{-1}) in Goa, India. The highest concentration was detected in the Divar Island mangrove area (875,000 ng g^{-1}) and Galgibaga Beach (365,000 ng g^{-1}) (Bhatawadekar et al. 2023).

Mangroves showed bioaccumulation of PAHs in their leaves, branches, fruits, and other tissues. On Hainan Island, PAHs concentrations in the roots, branches, fruits, and leaves were recorded at 314, 335, 353, and 566 ng g^{-1}, with annual PAHs accumulation by mangrove reaching 2228 μg m^{-2}. Leaves indicated the highest PAHs accumulation due to atmospheric decomposition. The dominant PAH types in the sediment were naphthalene (73.4%), phenanthrene (3.9%), and pyrene (3.6%), while in mangrove tissue, they were phenanthrene (41.3%), fluoranthene (14.7%), and pyrene (11.4%) (Qiu et al. 2018).

PAHs undergo transformations in the environment, such as halogenation, where halogens replace hydrogen atoms in any position of the organic contaminants, resulting in a wide variety of congeners. Halogenated organic contaminants often possess high octanol–water partition coefficients (Kow) due to the properties of the halogens. This characteristic leads to significant bioaccumulation in organism tissues. In the surface sediment of the Yangtze River estuary, the concentrations of Cl-PAHs and Br-PAHs ranged from 4.50 to 18.38 ng g^{-1} and 4.80 to 61.18 ng g^{-1}, respectively. The predominant sources of Cl/Br-PAHs in the estuary include e-waste dismantling (33.6%), waste incineration (23.2%), and metal smelting (11.0%) (Zuo et al. 2024).

The accumulation of PAHs in the mangrove ecosystem alters the biotic activity in these areas. Anthropogenic pressures have altered microbial communities and assemblages at the Cayenne estuary in French Guiana. The population of Bathyarchaeota (*Candidatus* Nitrosopumilus) and *Nitrospira* genera decreased in response to organic matter enrichments. Desulfobacteraceae, Desufarculaceae, and Acanthopleuribacteraceae escalated due to PCB and dieldrin contamination, while Geobacteraceae and Chitinophagaceae escalated due to naphthalene contamination (Fiard et al. 2022). PAHs ecotoxicity influences polychaetes *Namalycastis abiuma* in the Guajará Estuari, Brazilian Amazon coast. Elevated glutathione-S-transferase (GST) activity from these polychaetes was observed at industrial sites and oil terminals during the rainy season. Lipoperoxidation of the polychaetes, however, remained consistent across seasons. Sediment PAHs concentration ranged from 17 to 386 ng g^{-1}, with higher levels recorded during the dry season. Drainage systems and the oil terminal were suspected sources of PAHs. The estimated PAHs equivalent toxicity near the oil stations was ten times that in other areas (Kawakami et al. 2023). The bioaccumulation of PAHs in marine animals from the Niger Delta was evident. The Biota-Sediment Accumulation Factors (BASFs) for various species were as follows: tilapia (0.09–0.41), periwinkles (0.42–1.73), tiger prawn (0.11–0.64), estuarine shrimp (0.03–0.46), and crabs (0.01–0.42). A BASF value above 1 indicates significant bioaccumulation. Periwinkles from the Niger Delta showed bioaccumulation of PAHs, particularly benzo(k)fluoranthene (BASF = 1.7), pyrene (1.54), and phenanthrene (1.73) (Saunders et al. 2022).

15.2.4 The Evidence of Radionuclide Pollution in the Marine Environment

Radionuclides emerge as a significant form of marine pollution, especially after the development of nuclear weapons ad nuclear power plants. Major nuclear power plant accidents, such as the Chornobyl disaster 1986 and the Fukushima Daiichi incident 2011, have led to substantial releases of radionuclides into the environment. These nuclear contaminants pose long-lasting and considerable risks to both human health and the environment. ^{137}Cs, ^{239}Pu, and ^{238}U contamination elevate the environmental degradation risk, while ^{131}I accumulates in the thyroid gland, increasing the risk of thyroid cancer (Thakur and Kumar 2024).

Radionuclides originate from both natural (primordial) and anthropogenic sources. Primordial radionuclides, such as radio-thorium, radio-uranium, and radio-potassium, are derived from the Earth's mantle or crust and have existed since before the formation of the Earth. Climate change is speculated to increase the release of these radionuclides into the marine environment, further exacerbating their presence and impact (Gwynn et al. 2024). Anthropogenic intake of radionuclides escalated following the discovery of nuclear energy and weapons in the 1950s. In the marine environment, plants and animals accumulate radionuclides, leading to increased ecological and health risks. The primordial radionuclides have a long half-life, that is, ^{40}K is 1.25 billion years, ^{232}Th is 13.5 billion years, ^{235}U is

0.7 billion years, and ^{238}U is 4.5 billion years. This primordial radionuclides experience series of decay such as ^{232}Th-^{228}Ra ($T_{1/2}$ = 5.75 years)-^{228}AC ($T_{1/2}$ = 6.15 h)-^{228}Th ($T_{1/2}$ = 1.9 years)-^{224}Ra ($T_{1/2}$ = 3.6 days)-^{220}Rn ($T_{1/2}$ = 55.6 s)-^{216}Po ($T_{1/2}$ = 0.145 s)-^{212}Pb ($T_{1/2}$ = 10.64 h)-^{212}Bi ($T_{1/2}$ = 60.55 min)-^{212}Po ($T_{1/2}$ = 294.4 ns)-^{208}Pb (stable). While the radio-uranium experiences the series of decays as ^{238}U-^{230}Th ($T_{1/2}$ = 75,400 years)-^{226}Ra ($T_{1/2}$ = 1600 years)-^{222}Rn ($T_{1/2}$ = 3.8 days)-^{218}Po ($T_{1/2}$ = 3.098 min)-^{214}Po ($T_{1/2}$ = 164.3 μs)-^{210}Po ($T_{1/2}$ = 138.4 days)-^{206}Pb (stable). The polonium isotope ^{210}Po is one of the most radiotoxic compounds that accumulate and promote tissue damage (Tan et al. 2023).

The 2011 Fukushima Daiichi nuclear power plant accident released 80% of its radioactivity into the ocean. Immediate countermeasures included the implementation of the Advanced Liquid Processing System (ALPS), which detected over 62 types of radionuclides (such as ^{3}H, ^{14}C, ^{60}Co, ^{90}Sr, ^{99}Tc, ^{106}Ru, ^{125}Sb, ^{129}I, ^{134}Cs, and ^{137}Cs) in the storage tanks. Starting in March 2023, the Japanese government began gradually releasing the water from these tanks (about 30,000 tons/year) into the ocean. This process is expected to continue for the next 30 years, during which the Japanese government will monitor the radionuclide levels in the ocean (Maderich et al. 2024).

Two radioiodines were useful for the marine quality assessment in the Japan waters, that is, the short-live ^{131}I ($t_{1/2}$ = 8.02 days) and long-lived ^{129}I ($T_{1/2}$ = 15.7 million years). ^{129}I is released from the nuclear power plant; thus, the ratio of $^{129}I/^{127}I$ was applicable to assess the marine quality. The sediment–water partition coefficient (Kd) of ^{129}I is comparable to that of stable iodine, ranging from 225 to 329 L kg^{-1}. The analysis of this radioiodine found that the concentration factor of ^{129}I in freshwater fish is higher than marine fish in the water bodies close to Fukushima. The intestinal tissue contained higher ^{129}I ($15.50 \pm 6.57 \times 10^{12}$ at kg^{-1}) compared to muscle (1.97 ± 1.18 at kg^{-1}) (Teien et al. 2023).

Radiocesium is another important radionuclide for monitoring purposes. It shares chemical characteristics with K^{+} and is highly soluble in seawater (Abbasi 2023). The atmospheric testing of nuclear weapons increased atmospheric ^{137}Cs, and its fallout was influenced by stratospheric–tropospheric exchange within the 40–70° latitude range. The highest ^{137}Cs concentration in the North Pacific Ocean before Fukushima Daiichi nuclear power plant accident was observed in 1967 (over 500 PBq) (Tsumune et al. 2023).

The monitoring of radionuclides has expanded to encompass other water bodies. In the ocean off the southeast coast of Jeju Island, Korea, no significant temporal fluctuations have been observed in the activity of ^{90}Sr ($T_{1/2}$= 28.91 year) in seawater, ranging from 0.57 to 1.0 Bq m^{-3}. These concentrations remain below the guidelines set by the International Atomic Energy Agency (IAEA). ^{90}Sr, which behaves similarly to calcium and emits high-energy β particles, tends to accumulate in mammalian bone, potentially leading to bone cancer and leukemia (Kim et al. 2023). In the Chaun Bay, East Siberian Sea, radionuclide levels indicate natural occurrences with no discernible anthropogenic contributions. Concentrations of ^{137}Cs, ^{40}K, ^{226}Ra, ^{232}Th ranged from 0.5–4.7, 535–991, 16.5–39.3, and 23.7–77.9 Bq kg^{-1}, respectively (Ulyantsev et al. 2023).

In the Adriatic Sea, Croatia, deeper marine sediment contains naturally occurring radionuclides, while anthropogenic influences affect radionuclide levels in shallower marine sediment. ^{137}Cs (2.2 ± 0.1 Bq kg^{-1}), ^{238}U (34 ± 2 Bq kg^{-1}), and ^{226}Ra (25 ± 2 Bq kg^{-1}) were significantly correlated with organic matter (Mikelić et al. 2023). In the Mediterranean Sea, seaweed concentrations of ^{137}Cs are four times higher compared to mussels. Specifically, oarweed *Laminaria digitata* contained 106.29 to 252.38 mBq kg^{-1} of ^{137}Cs, while *Mytillus galloprovincialis* contained 12.94–101.84 mBq kg^{-1} ^{137}Cs (Abbasi et al. 2023). Table 15.1 listed some records of radionuclides in the aquatic environment.

Table 15.1 The identified and recorded value of radionuclide in aquatic environment

Location	Radionuclide type	Concentration	References
Sea water (in Bq L^{-1})			
Marine area close to Fukushima Daichii Nuclear Power Plant	^{137}Cs	0.016–0.160	Abbasi (2023)
	^{134}Cs	0.076–0.024	
Jeju Island, Korea	^{90}Sr	717–909	Kim et al. (2023)
Sediment (in Bq kg^{-1})			
Chaun Bay, East Siberian Sea	^{232}Th	23.7–77.9	Ulyantsev et al. (2023)
	^{226}Ra	16.5–39.3	
	^{40}K	535–991	
	^{137}Cs	0.5–4.7	
Kaštela Bay, Croatia	^{232}Th	5.1–43	Mikelić et al. (2023)
	^{226}Ra	5.4–690	
	^{40}K	15–540	
	^{137}Cs	0.1–12	
	^{238}U	6.7–600	
Vaan Island, India	^{232}Th	23.47–32.64	Krishnan et al. (2024)
	^{40}K	124.66–207.82	
	^{238}U	17.99–29.05	
Koswari Island, India	^{232}Th	11.09–33.55	
	^{40}K	93.33–219.91	
	^{238}U	13.43–30.97	
Organism			
Turkey-Cyprus	^{137}Cs (*Laminaria digitata*)	84.2–236.05 mBq kg^{-1}	Abbasi et al. (2023)
	^{137}Cs (*Dictyota dichotoma*)	106.29–252.38 mBq kg^{-1}	
Marine area close to Fukushima Daichii Nuclear Power Plant	^{137}Cs (Marine organism)	8.52–852 Bg kg^{-1}/ Bq L^{-1}	Abbasi (2023)

15.2.5 The Exposure of Marine Pollutants to Some Selected Marine Endangered Species

15.2.5.1 Pollutant Exposure to Sea Turtle

The growing concern over plastic pollution underscores the vulnerability of sea turtles to this environmental threat. Sea turtles' feeding habits significantly contribute to their exposure to plastic, as plastics often resemble jellyfish, a common part of their diet (Yaghmour et al. 2021). An analysis of gastrointestinal contents from sea turtles in Ceará, Brazil, revealed that the solid samples were predominantly composed of plastics (76.05%), followed by fabrics (12.18%) and oily materials (Feitosa et al. 2024). Ingesting macro- and mega-plastics can have severe consequences for sea turtles, including death. Mortality is primarily due to respiratory arrest (62.5%) and pulmonary edema (12.5%) (Feitosa et al. 2024).

On shorelines, plastic debris can disrupt sea turtle nesting grounds. This issue has been observed on North Island of Qilianyu in the South China Sea, where green sea turtle *Chelonia mydas* nesting sites are particularly affected. The accumulation rate of plastic debris in these areas is approximately 11.30 ± 7.73 g m^{-2} month^{-1}. The distribution patterns of plastic debris are influenced by wind direction and speed, with higher accumulation rates occurring during tropical cyclones and the southwest monsoon season (Zhang et al. 2024).

In addition to the significant impact of large plastic debris on sea turtles, microplastics (MPs) also present a concerning environmental threat. A study utilizing fecal samples from green sea turtles *C. mydas* on Isla Blanca in the Yucatan Peninsula, Mexico, employed a non-invasive method to observe microplastics. The findings indicated an abundance of MPs ranging from 10 to 89 MP g^{-1}, predominantly comprising polyester, polypropylene, PVC, and nylon (Aranda et al. 2024). These MPs can have detrimental effects on sea turtle populations, potentially even contaminating their eggs. A separate study examined 350 undeveloped loggerhead sea turtle *Caretta caretta* eggs from the northwest coast of Florida, revealing the presence of approximately 510 microplastics. The MPs in these eggs were primarily polyethylene (11%) and polystyrene (7%) (Curl et al. 2024).

Other contaminants also pose significant threats to sea turtle health and populations. An analysis of fecal specimens from sea turtles along the southern coast of Pernambuco, Brazil, revealed that some green turtles *C. mydas*. had ingested oil. This ingestion is linked to the turtles' feeding habits, particularly their consumption of algae. The analysis found small oil particles (1–3 cm) adhered to the food content, especially algae, in the feces (Fig. 15.3) (da Silva et al. 2024). Oil contamination introduces various chemicals, including heavy metals, into marine environments. Mercury levels in the muscle and liver tissues of green turtles from Northeast Brazil ranged from 10.1 to 8569 ng g^{-1} and 81 to 3135 ng g^{-1}, respectively. The study found no significant correlation between mercury concentration and body size across all observed areas, suggesting that feeding habits are the primary factor in mercury accumulation in these turtles (Verzele et al. 2024). These metals are likely to be transferred to turtle eggs. Metal exposure has been shown to negatively affect the eggs of green sea turtles, raising concerns about embryonic development.

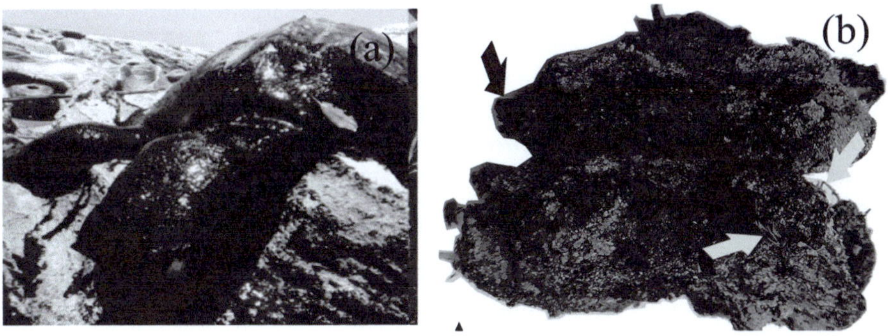

Fig. 15.3 The incident of oil spill coating a sea turtle (**a**), and the feces of sea turtle contain oil (**b**). Copied with permission from da Silva et al. (2024), Feitosa et al. (2024)

Long-term contaminant exposure may, therefore, have detrimental effects on future generations of the species (Morão et al. 2024).

15.2.5.2 Pollutant Exposure to Whale Shark (*Rhincodon typus*)

Rhincodon typus, commonly known as the whale shark, is a slow-growing, long-lived, and highly mobile species. In 2016, the International Union for Conservation of Nature (IUCN) upgraded its status from vulnerable to endangered on the Red List of Threatened Species (Reynolds et al. 2022). As filter feeders, whale sharks are particularly susceptible to pollutant exposure, making them effective sentinels for marine pollution. Their primary diet consists of phyto- and zooplankton from the lower levels of the food web, which readily absorb pollutants (Marsili et al. 2023). Skin biopsies of ten *R. typus* specimen in the Baja California Peninsula revealed significant concentrations of organochlorine pesticides (1578.1 ng g^{-1}) and polycyclic aromatic hydrocarbons (279.4 ng g^{-1}) (Villagómez-Vélez et al. 2024).

Additionally, the species' skeletal muscle accumulates heavy metals such as arsenic (As), cadmium (Cd), mercury (Hg), and lead (Pb). In an analysis of two stranded specimens in Baja California, Mexico, metals were detected in various organs, including the liver, kidney, stomach, testicles, filtering patches, heart, gills, brain, epidermis, and skeletal muscle. The liver exhibited the highest concentrations of As, Cd, Hg, and Pb with a value of 33.7, 17.50, 0.056, and 0.055 μg g^{-1}, respectively (Pancaldi et al. 2019). Meanwhile, an analysis of twenty whale shark specimens from the Gulf of Tadjoura, Djibouti, highlighted the biomagnification of heavy metals and organic pollutants within the species. The Biomagnification Factors (BMF) from plankton to whale sharks were observed to be 0.2–6.0 for molybdenum (Mo), 0.02–8.01 for nickel (Ni), and 0.03–10.6 for chromium (Cr). Furthermore, polychlorinated biphenyls (PCBs) were found in significant concentrations in the skin, reaching 624–957 ng g^{-1}. The detected PCB congeners were primarily tetra-CB (41%) and penta-CB (23%) (Boldrocchi et al. 2020).

15.2.5.3 Pollutant Exposure to Whale

Whales, as the largest marine mammals, are particularly threatened and vulnerable to pollution. An experiment using baleen plates from four whale species—North Atlantic right whale (*Eubalaena glacialis*), Northern minke whale (*Balaenoptera acutorostrata*), fin whale (*Balaenoptera physalus*), and humpback whale (*Megaptera novaeangliae*)—revealed that plastics were captured within the baleen racks, posing a risk of damage and clogging (Werth et al. 2024). The increasing plastic pollution in the ocean represents a significant and escalating threat to whale populations.

Valuable samples collected from stranded and decayed whale bodies have provided profound insights into the impact of pollution on whales. For instance, pilot whales, which are part of the human diet in the Faroe Islands, have shown significant mercury accumulation over the years. Data from 1977 to 2015 indicated that mercury (total Hg and MeHg) concentrations varied with the whales' length, age, sex, and reproductive state. Early data showed a MeHg to total Hg ratio of 50%, whereas later years recorded a ratio of 80–100%. Notably, fetuses exhibited lower MeHg levels (20–30%) compared to their mothers (Hoydal et al. 2024). In another study, stranded beaked whales (Ziphiidae) in Australia from 2007 to 2022 demonstrated significant metal accumulation. The highest mercury concentration (386 mg kg^{-1}) was found in the liver tissue of *Mesoplodon layardii*, while the highest cadmium concentration (478 mg kg^{-1}) was detected in the liver of *Ziphius cavirostris* (Palmer et al. 2024).

The exposure of persistent organic pollutants (POPs) to whales also has detrimental effects. For instance, organohalogen compounds like polybrominated diphenyl ethers (PBDEs) in St. Lawrence Estuary belugas *Delphinapterus leucas* have been linked to elevated cortisol levels in the skin. There is a positive correlation between PBDE exposure and thyroid hormone T3 levels (Jolicoeur et al. 2024). Similarly, polycyclic aromatic hydrocarbons (PAHs) pose a significant threat to killer whales *Orcinus orca*. Stranded specimens in British Columbia have shown high PAH concentrations in various tissues, including the liver and skeletal muscle. Additionally, there is efficient maternal transfer of low molecular weight PAHs, such as naphthalene, dibenzothiophene, and C3-fluorenes, to fetuses (Lee et al. 2023).

15.2.5.4 Pollutant Exposure to Dolphin

Hector's dolphin *Cephalorhynchus hectori hectori* is the only dolphin species listed as endangered by the IUCN Red List. However, studies on the effects of pollutants on this species are limited. To the best of our knowledge, the report by Stockin et al. (2010) is the most recent publication detailing the concentrations of persistent organic pollutants (POPs) in this species. This study found that maternal transfer of summed PCBs and DDTs was estimated at 4.3 and 5.7%, respectively (Stockin et al. 2010). Fortunately, numerous publications on pollutant effects in other dolphin species provide a broader understanding of the issue.

Per- and polyfluoroalkyl substances (PFAS) disrupt the cortisol levels in bottlenose dolphins *Tursiops truncates*, affecting their stress axis. This axis is crucial for short-term environmental stress responses and managing the diurnal cycle, and it is permanently programmed during embryonic development, later adapting to environmental changes (Bennett et al. 2024). An investigation of PFAS levels along the southeast coast of Australia, conducted by Foord et al. (2024) from 2006 to 2021, revealed that the Burrunan dolphin *Tursiops australis* had the highest PFAS concentration at 9750 ng g^{-1}. In contrast, the short-beaked dolphin *Delphinus delphis* and common bottlenose dolphin *T. truncates* exhibited lower PFAS levels, each less than 100 ng g^{-1} (Foord et al. 2024). Additionally, PFAS accumulation was observed in striped dolphins *Stenella coeruleoalba* from the northwest Mediterranean Sea, with liver concentrations ranging from 254 to 7010 ng g^{-1} between 1990 and 2021 (Garcia-Garin et al. 2023).

In Brazilian waters, approximately 86% of hepatic tissue samples from 50 Franciscana dolphins *Pontoporia blainvillei* showed PAH accumulation, with the highest concentration reaching 1055.6 ng g^{-1} lipid weight. Maternal transfer of PAHs was evident, as higher phenanthrene concentrations were detected in a fetus and two neonates (Santos-Neto et al. 2024). Similarly, Atlantic spotted dolphins in Southeastern Brazil exhibited accumulation of PCBs, *p*, *p′* isomers (DDD, DDE, DDT), and PBDEs, with concentrations of 97.0, 11.0, and 1.6 μg g^{-1} lipid weight, respectively. Other organic contaminant such as pesticides also recorded (Oliveira-Ferreira et al. 2024). Maternal transfer of PAHs was also documented in bottlenose dolphins *Tursiops truncates* in San Diego, California, as calf serum levels increased, while milk and maternal serum levels decreased over time. Figure 15.4 showed how Noren et al. (2024) team collect the milk and serum sample from the specimen by non-invasive methods (Noren et al. 2024).

The pollution affecting humpback dolphins in the Pearl River Estuary has declined, suspected to be due to contamination by endocrine-disrupting chemicals (EDCs) in the area. Data spanning from 2005 to 2019 revealed a correlation between EDCs and the decline in dolphin population. These chemicals disrupt sex hormones, thereby affecting birth rates and further exacerbating population decline. Among the EDCs analyzed, alkylphenols emerged as the most impactful contaminant, while conventional EDCs such as DDTs, chromium, and zinc also demonstrated significant effects (Luo et al. 2024).

Fig. 15.4 Non-invasive methods for sample collection. Copied with permission from (Noren et al. 2024)

15.3 Bioremediation Strategy

15.3.1 Bioremediation of Plastic Pollution in the Marine Environment

15.3.1.1 Bioremediation by Marine Animals

Large organisms such as bivalves, sea cucumbers, sea turtles, and fishes, as well as top predators, ingest microplastics and larger plastic particles through their regular feeding routes (Rebolledo et al. 2013; Alava 2019). In the short term, this would eliminate fractions of plastic waste from the water column. However, the ingested plastics accumulate in the organisms, threatening their lives as described above.

On land, the guts of earthworms (*Lumbricus terrestris*), waxworms, and mealworms were found to possess microbes that can degrade polyethylene (PE) and polystyrene (PS), showing their potential for bioremediation (Yang et al. 2014; Brandon et al. 2018; Huerta Lwanga et al. 2018). Unfortunately, to the best of our knowledge, similar studies based on marine animals have never been done. The many cases of plastic and microplastics ingestion by sea animals showed that the plastics remained in their body (Rebolledo et al. 2013; Besseling et al. 2015; Hernandez-Gonzalez et al. 2018). This indicates that even if there is some degree of biodegradation inside the large organism, the rate is extremely slow. Considering the whole marine food chain, such bioaccumulation will eventually reach terrestrial predators including humans.

15.3.1.2 Bioremediation by Microorganisms

In the ocean, plastic waste may be degraded through combinations of the following routes: (1) photodegradation by UV-light resulting in cracks of the surface and molecular weight reduction, (2) abiotic hydrolysis which will break the polymer chain and reduce molecular weight, and (3) biodegradation by the microorganisms that colonize the surface of the plastic waste (Viel et al. 2023). The extent of the degradation varies depending on the polymer structure (e.g., molecular constituents, crystallinity, molecular weight, hydrophobicity, additives) and conditions of the surrounding environment (salinity, pH, sunlight exposure, etc.) (Yuan et al. 2020). The first two routes aid microbial attachment, such as by releasing carbonyl residues that can be used as microbial carbon sources (Syranidou et al. 2019). This facilitates the third route which is plastic biodegradation by marine microorganisms.

Bacteria, marine fungi, and microalgae can attach and colonize the surface of plastic waste forming biofilms, then corrode them as they use the plastic/microplastic waste as their carbon source (Delacuvellerie et al. 2019). The biofilms typically contain colonies of microorganism, cell secretions, as well as organic and inorganic compounds from the surrounding water body or the leaching of the degraded plastics (Yuan et al. 2020). Such biofilm may reduce the buoyancy of the plastic wastes, allowing them to sink into ocean sediments (Harrison et al. 2014; Miao et al. 2021). The microbial biofilm may also deteriorate the plastic wastes through enzymatic activity, which causes the polymers to break down into smaller molecules (Ganesh Kumar et al. 2020).

Plastic degradation by the microbes commonly occurs via hydrolysis, initiated by the secreted enzymes (Devi et al. 2015). Microbial degradation of microplastics normally starts extracellularly because the particle size is too large to diffuse through the cell membrane. Once the size is small enough, it may enter the cell membrane and be hydrolyzed intracellularly (Yuan et al. 2020).

Plastic/microplastic wastes tend to attract and support the growth of distinct microbial structures with certain enzyme secretion and metabolic pathways, which will drive the ecological function (Syranidou et al. 2019). Factors affecting the type of the microbial community residing on the plastic wastes include polymer type, seasons, and geographical location (Harrison et al. 2014; Delacuvellerie et al. 2019). As such, biodegradation of marine plastic waste by naturally existing microbial consortia could be one of the preferable pathways for ocean bioremediation. Hopefully, This route can degrade the polymer without accumulating the toxic compounds in their tissues. Microbes such as *Klebsiella pneumonia*, *Alcanivorax borkumensis*, fungus *Parengyodontium album*, and *Roseibium aggregatum* ZY-1 were able to biodegrade plastic (Fig. 15.5). A summary of some of the existing studies on plastic biodegradation using marine microbes is given in Table 15.2.

Lab-scale studies on the biodegradation of microplastics are commonly done with pure bacterial cultures, isolated from the related environment. This gives the

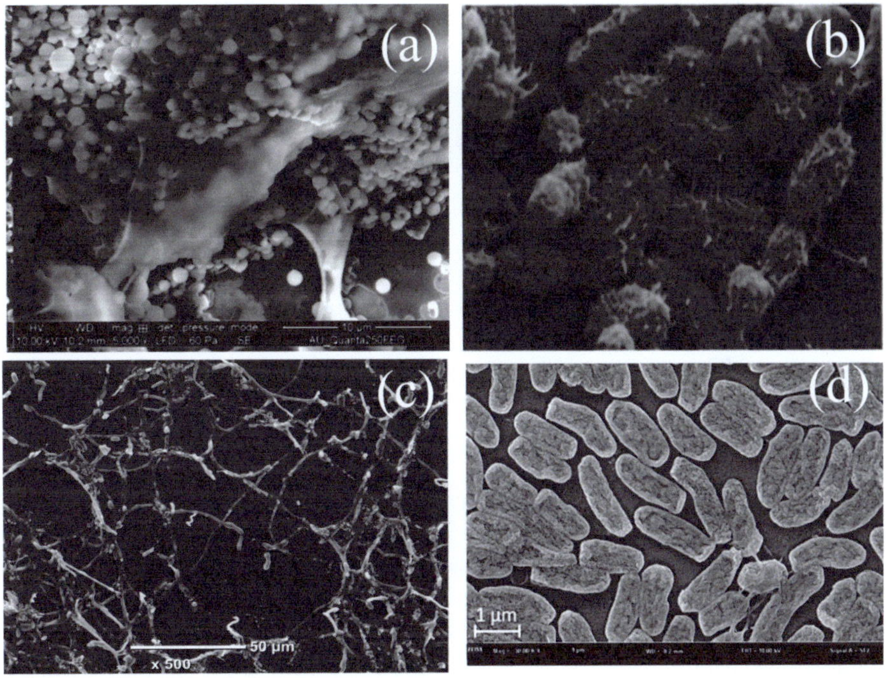

Fig. 15.5 Micrographs of (**a**) *Klebsiella pneumonia*, (**b**) *Alcanivorax borkumensis*, (**c**) fungus *Parengyodontium*, and (**d**) Roseibium aggregatum. Copied with permission from Mohanrasu et al. (2018), Delacuvellerie et al. (2019), Pan et al. (2024), Vaksmaa et al. (2024)

Table 15.2 Plastic biodegradation studies in vitro by marine

Species	Target contaminant	Results	References
Marine fungi			
Aspergillus tubingensis VRKPT1 *Aspergillus flavus* VRKPT2	HDPE	Better degrading ability of *A. flavus* VRKPT2 than *A. tubingensis* VRKPT1 6–8% HDPE weight loss in 30 days of incubation	Devi et al. (2015)
Zalerion maritimum	PE microplastics	Mass reduction of the microplastics was observed due to the use of PE particles as nutrient source for fungal growth	Paço et al. (2017)
Marine bacteria			
Brevibacillus borstelensis KY49486	HDPE	11% weight loss after 30 days	Mohanrasu et al. (2018)
Pseudomonas sp. *Arthrobacter* sp.	HDPE	The degradation ability of *Pseudomonas* sp. was higher than *Arthrobacter* sp. 12–15% HDPE weight loss in 30 days of incubation	Balasubramanian et al. (2010)
Kocuria palustris M16 *Bacillus pumilus* M27 *Bacillus subtilis* H1584	LDPE	1–1.75% LDPE weight loss after 30 days of incubation 4.6–10.3% (estimated) crystallinity decrease	Harshvardhan and Jha (2013)
Alcanivorax borkumensis	LDPE	3.5% LDPE weight loss after 80 days	Delacuvellerie et al. (2019)
Bacillus sp. *Rhodococcus* sp.	PP microplastics	4–6.4% PP weight loss after 40 days of incubation	Auta et al. (2018)
Bacillus sp.	PVC	PVC weight loss of 0.26% after 90 days of incubation Also able to degrade LDPE and HDPE	Kumari et al. (2019)
Bacillus cereus	Nylon 6 and Nylon 66	7% weight loss for nylon 66 and 2% weight loss for nylon 6	Sudhakar et al. (2007)
Marine microbial consortia			
Natural pelagic microbial consortia (the most PE-degrading consortia were dominated by *Pseudonocardia, Cellulosimicrobium,* and *Ochrobactrum*)	LDPE and HDPE	Acclimated consortia showed higher ability to degrade PE films than un-acclimated consortia Bioaugmentation improved the degrading ability	Syranidou et al. (2017)

(continued)

Table 15.2 (continued)

Species	Target contaminant	Results	References
Consortia of *Vibrio alginolyticus* and *Vibrio parahemolyticus*	PVA-LLDPE blend	20% decrease in tensile strength after 15 weeks of incubation for the blends containing 25 and 30% of PVA	Raghul et al. (2014)
Natural pelagic microbial consortia (the best consortium was dominated by *Alcanivorax* and *Ochrobactrum*)	PS	Acclimated consortia showed higher ability to degrade naturally weathered PS than un-acclimated consortia	Syranidou et al. (2019)

advantage of more straightforward and unbiased observation on the effects of growth conditions, the degradation mechanisms, and degradation progress (Janssen et al. 2002). However, degradation studies using bacterial consortiums are also increasing as they could act in symbiotic and synergistic manners, resulting in overall improved activity and tolerance for the biodegradation of plastic waste (Singh and Wahid 2015). In the following studies, biodegradation of the polymer samples was mostly confirmed from Fourier transform infrared (FTIR) analysis, polymer weight loss, and observation of plastic surface topology through scanning electron microscopy (SEM) analysis.

15.3.1.3 High-Density Polyethylene (HDPE) Biodegradation

Fungal strains, identified as *Aspergillus tubingensis* VRKPT1 and *Aspergillus flavus* VRKPT2, were proven to be able to use the HDPE as their carbon source without any pre-treatment/pro-oxidant additives. Devi et al. (2015) isolated these fungal strains from polyethylene waste dumped in the coastal area of Gulf of Mannar, India, and screened them for HDPE degrading efficiency. The degradation study was performed on commercially available HDPE films that have been cut into small strips. *A. flavus* VRKPT2 was shown to have better degrading ability than *A. tubingensis* VRKPT1. Following 30 days of incubation, the weight loss of the HDPE was 8.51 ± 0.1% and 6.02 ± 0.2%, respectively. The addition of mineral oil to the medium aided the attachment of the fungi to HDPE surface and sped up biofilm formation by promoting hydrophobic interactions. A slight increase of the biodegradation rate to a weight loss of 9.34 ± 0.2% and 6.88 ± 0.1%, respectively, was then observed (Devi et al. 2015).

Still from the waste dumped in the coastal area of the Gulf of Mannar, Balasubramanian et al. (2010) isolated *Arthrobacter* sp. and *Pseudomonas* sp. bacteria that can degrade HDPE. FTIR analysis showed alteration in the functional group(s) and/or modification of side chain(s), indicating the occurrence of the degradation. Biodegradation by *Pseudomonas* sp. was more pronounced than *Arthrobacter* sp., as shown by the HDPE weight loss of 15 and 12%, respectively, after 30 days of incubation. This was thought to be caused by the stronger cell surface hydrophobicity of *Pseudomonas* sp. (Balasubramanian et al. 2010).

HDPE was also shown to be degraded by marine bacteria, *Brevibacillus borstelensis* KY49486. This strain was isolated and screened from marine sediments, water, and oil spills in Tamil Nadu, India. The polymer sample used were pieces of HDPE films (1 × 1 cm), and the degradation was confirmed through the appearance of acid, ester groups, aldehydes, and ketones in the FTIR spectra as well as 11.4% weight loss of the HDPE sample after 30 days of the experiment (Mohanrasu et al. 2018).

15.3.1.4 Low-Density Polyethylene (LDPE) Biodegradation

Harshvardhan and Jha (2013) isolated and screened 60 marine bacteria from the pelagic zone of Arabian Sea coast, India. They found that *Kocuria palustris* M16, *Bacillus pumilus* M27, and *Bacillus subtilis* H1584 were able to degrade LDPE. Yet, the LDPE weight loss after 30 days of incubation with the isolates was rather low, that is, 1, 1.5, and 1.75% for *K. palustris* M16, *B. pumilus* M27, and *B. subtilis* H1584, respectively. Nonetheless, the degradation was also justified from the ester, keto, vinyl, and internal double bond formation, observed from the FTIR spectra. Further, they estimated the crystallinity of the LDPE sample from FTIR analysis and observed crystallinity decrease of 10.3, 8.6, and 4.6% from the aforementioned bacteria respectively (Harshvardhan and Jha 2013). Apart from that, the fungal marine *Zalerion maritimum* was also shown to have the ability to degrade PE microplastics by using them as their nutrient source (Paço et al. 2017).

Delacuvellerie et al. (2019) compared the bacterial composition of the plastic wastes obtained around the Mediterranean Sea. After enrichment to find potentially polymer-degrading species, *Alcanivorax borkumensis* was found to grow specifically on LDPE and not in the other plastics studied (PET & PS). *A. borkumensis* in fact is a hydrocarbonoclastic bacteria that can degrade hydrocarbon. In the LDPE degradation study, the bacterium decreased LDPE weight by 3.5% after 80 days of incubation (Delacuvellerie et al. 2019).

Syranidou et al. (2017) studied the degradation of naturally weathered LDPE and HDPE plastics by indigenous marine consortia using seawater as the aqueous medium to mimic the pelagic zone. Acclimating the bacterial consortia for 6 months significantly improved their plastic-degrading ability. With non-acclimated consortia, the PE weight loss was significantly lower (0.3–0.4% after 6 months) than using acclimated consortia where it could reach 19% after 6 months. Throughout the study, the bacterial diversity changed, implying that PE waste drives the striving microbial strains. The study showed that the genera Pseudonocardia, Cellulosimicrobium, and Ochrobactrum dominated the most PE-degrading consortia. *Pseudonocardia* possess *alk-B* gene that encodes alkane 1-monooxygenase, one of the key enzymes in alkane degradation, while *Cellulosimicrobium* can degrade hydrocarbon (Nkem et al. 2016; Syranidou et al. 2017). Weathered LDPE films were degraded more efficiently than weathered HDPE films, with weight loss of 15 and 5%, respectively. This is relevant to the structure of the constituent molecules (Syranidou et al. 2017).

LDPE-based polymer blend degradation using marine bacteria was also studied in vitro by Raghul et al. (2014). The polymer blend was polyvinyl alcohol

(PVA)-LLDPE films and the degrading bacteria were consortia of *Vibrio alginolyticus* and *Vibrio parahemolyticus* isolated from marine sediment. After 15 weeks of incubation, the tensile strength of the polymer film decreased by 20% for the blend containing 25 and 30% of PVA (Raghul et al. 2014).

15.3.2 Bioremediation of Inorganic Chemicals in Marine Environment

Bioremediation of heavy metals may occur through enzymatic oxidation, enzymatic reduction, complexation, siderophores, precipitation, biosorption, and/or by bioaccumulation (Pratush et al. 2018). In the first four mechanisms, heavy metals can be transformed into less toxic compounds. Meanwhile, the last three mechanisms can reduce bioavailability of heavy metals, yet the toxicity is not reduced. Safer bioremediation strategies should be the ones that rely on the first four mechanisms.

Most of the bioremediation studies for heavy metals are focused on soil contamination (Alvarez et al. 2017). Some of the microbes were applicable for aquatic body. For example, microalgae *Galdieria sulphuraria* removed up to 94% negative charge platinum complex $PtCl_6^-$ by assistance of produced EPS by the microalgae (Sun et al. 2021). Bacterium *Bacillus xiamenensis* absorbed 216.75 mg Pb/g bacterium biomass (Mohapatra et al. 2019), while, *Acidithiobacillus ferrooxidans* removed 94% Sb(V) by biomineralization process to form Fe-Sb-O complex (Chen et al. 2025). Deep sea bacterium *Pseudomonas stutzeri* 273 removed Cd by biomineralization to form CdS nanoparticle (Ma et al. 2021). The micrograph of these four aforementioned microbes is depicted in Fig. 15.6. Potential bioremediation *agents* for mercury, cadmium, and lead contaminations in the ocean are summarized in Table 15.3.

15.3.2.1 Mercury (Hg) Bioremediation

Mercury-resistant bacteria commonly have *mer* operons on their transposons, plasmids, or chromosomes. MerT and MerP cause the bacteria to uptake Hg^{2+} to the cytoplasm, and then MerA reduces the Hg^{2+} into Hg^0, which will then diffuse out of the cell. Hg^0 is less toxic and relatively inert (Zhang et al. 2012).

Pseudomonas putida strain SP1 was isolated from the coastal seawater of China. This strain is highly resistant to $HgCl_2$ and can grow well at media containing 0.28 mM of $HgCl_2$. The minimum inhibitory concentration (MIC) was found to be 0.3 mM. It also has a high tolerance to other toxic metals such as $CdCl_2$ and $PbCl_2$ with MIC of 1 mM. Incubation of *P. putida* in pure seawater containing 0.28 mM of $HgCl_2$ resulted in 89% $HgCl_2$ removal within 48 h. Inoculation of this strain in a fresh media containing 0.28 mM $HgCl_2$ for 48 h to improve its efficiency successfully removed almost 100% of the $HgCl_2$. The strain also shows low pathogenicity suggesting its safe use for bioremediation of mercury (Zhang et al. 2012).

In a study by De et al. (2008), various mercury-resistant bacteria were isolated from Indian Coast. 16S 3RNA gene analysis characterized these isolates as *Alcaligenes faecalis*, *Bacillus pumilus*, *Bacillus* sp., *Pseudomonas aeruginosa*, and

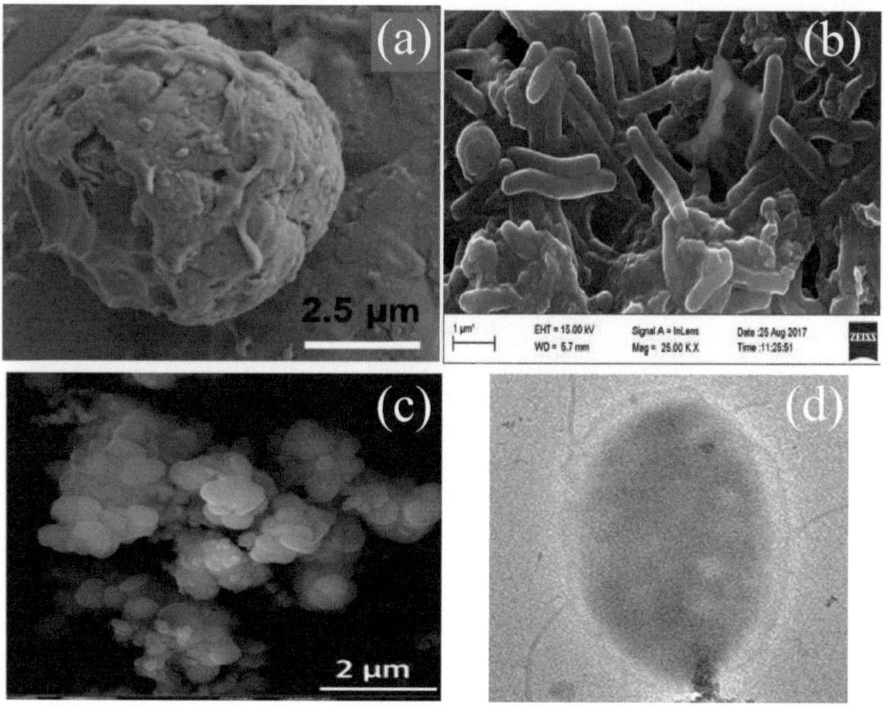

Fig. 15.6 Micrographs of (**a**) microalgae *Galdieria sulphuraria*, (**b**) bacterium *Bacillus xiamenensis*, (**c**) *Acidithiobacillus ferrooxidans*, and (**d**) deep sea bacterium *Pseudomonas stutzeri* 273. Copied with permission from Mohapatra et al. (2019), Ma et al. (2021), Sun et al. (2021), Chen et al. (2025)

Brevibacterium iodinum. All these isolates could remove Hg by volatilization, yet the degree of removal was not disclosed. *A. faecalis*, *B. pumilus*, and *P. aeruginosa* were found to have merA genes, while *Bacillus* sp. and *B.iodinum* did not (De et al. 2008).

15.3.2.2 Cadmium (Cd) Bioremediation

Biosorption, or adsorption using biomass, of heavy metals using algae has been one option to bioremediate heavy metal contaminations from rivers and seas (Yin et al. 2019). Algae are naturally abundant. Hence, biosorption using algae is inexpensive and can be performed easily in a renewable manner. Various types of algae also provide diverse functional groups and uniform distribution of binding sites on their surface (Sari and Tuzen 2008; Bilal et al. 2018). For the removal of cadmium, marine algae *Ulva* sp., *Gracillaria* sp., *Padina* sp., *Ceramium virgatum*, and *Durvillaea potatorum*, among others, have been shown to have good adsorption capacity of Cd^{2+} (Matheickal et al. 1999; Kaewsarn and Yu 2001; Sheng et al. 2004; Sari and Tuzen 2008). The reported adsorption capacity varies between studies. For example, in the study by Sheng et al. (2004), the maximum Cd^{2+} adsorption capacity

Table 15.3 Bioremediation studies using marine species to remove mercury, cadmium, and lead contaminations

Species	Target contaminant	Results	References
Marine bacteria			
Pseudomonas putida strain SP1	Hg	The strain was resistant to 280 μM $HgCl_2$ Nearly 100% removal of 280 mM $HgCl_2$ in 48 h Low virulence of the strain	Zhang et al. (2012)
Desulfovibrio desulfuricans	Cd	99.9% Cd removal from 100 ppm concentration in 7 days, in the presence of 0.05% w/v $FeSO_4$ Without $FeSO_4$, the removal was only 4.2%	Joo et al. (2015)
Alcaligenes faecalis	Cd, Hg	>70% Cd removal from 100 ppm in 72 h	De et al. (2008)
Bacillus pumilus	Hg, Pb	>88% Pb removal from 100 ppm in 96 h	
Bacillus sp.	Hg	Does not possess merA gene, but able to remove Hg through volatilization	
Pseudomonas aeruginosa	Cd, Hg, Pb	>75% Cd removal from 100 ppm in 72 h >98% Pb removal from 100 ppm in 96 h	
Brevibacterium iodinum	Hg, Pb	>87% Pb removal from 100 ppm in 96 h	
Marine algae			
Ulva lactuca	Pb^{2+}	Monolayer adsorption capacity of 34.7 mg ions/g algae	Sari and Tuzen (2008)
Padina sp.	Cd^{2+}	Adsorption capacity of 0.75 mmol/g algae (84 mg/g)	Sheng et al. (2004)
Sargassum sp.		Adsorption capacity of 0.76 mmol/g algae (85 mg/g)	
Ulva sp.		Adsorption capacity of 0.58 mmol/g algae (65 mg/g)	
Gracilaria sp.		Adsorption capacity of 0.3 mmol/g algae (33.7 mg/g)	

was 85, 84, 65, and 33.7 mg/g for *Sargassum* sp., *Padina* sp., *Ulva* sp., and *Gracillaria* sp., respectively. In another study, the maximum Cd^{2+} adsorption capacity was 0.53 mmol/g or 60 mg/g, and 90% of the adsorption capacity was filled within 35 min (Kaewsarn and Yu 2001).

Desulfovibrio desulfuricans can remove up to 99.9% Cd^{2+} from 100 ppm $Cd(NO_3)_2$ solution in an artificial sea water medium in 7 days. Yet, to reach this rate of removal, the medium should be added with 0.05% w/v $FeSO_4$. *D. desulfuricans* is a sulfate-reducing anaerobic bacterium that reduces sulfate to sulfide. The sulfide ions either convert to H_2S or bind with heavy metal ions to form metal sulfides, which then precipitate. Without $FeSO_4$, the removal was only 4.2%. At 200 ppm

Cd^{2+} concentration, the removal decreased to 45.8% (Joo et al. 2015). *A. faecalis* and *P. aeruginosa* isolated from Indian Coast was found to be able to remove Cd^{2+} with 80.8 and 82.6% removal from 100 ppm solution in 72 h. Here, the Cd removal mechanism is expected to be from the adsorption of the metal in the extracellular polymeric substance (De et al. 2008).

15.3.2.3 Lead (Pb) Bioremediation

Marine algae *Ulva lactuca* was studied to adsorb Pb^{2+} and Cd^{2+} ions. The adsorption occurs through chemisorption with monolayer adsorption capacity of 34.7 and 29.2 mg ions/g algae, for Pb^{2+} and Cd^{2+}, respectively (Sari and Tuzen 2008). Such biosorption provides a simple and economical way to remove lead and other heavy metals from the water column. However, the toxic nature of the heavy metals remains the same. Hence, consumption of the algae, whose surface contains heavy metals, by marine animals or humans would still be harmful.

The study by De et al. (2008) also explored the use of marine bacteria for lead removal. *P. aeruginosa*, *B. pumilus*, and *B. iodinum* showed more than 98, 88, and 87% Pb removal from 100 ppm Pb solution in 96 h. The removal mechanism for *B. pumilus* and *B. iodinum* seemed to be through PbS precipitation, whereas for *P. aeruginosa* the removal seemed to occur via Pb entrapment in the extracellular polymeric substances (De et al. 2008).

15.3.3 Bioremediation of Organic Chemicals in Marine Environment

15.3.3.1 The Search for Biodegrader Agents at the Laboratory Level

Bioremediation is a versatile technique capable of degrading a broad spectrum of organic pollutants, ranging from easily degraded compounds like simple alkanes to recalcitrant organic compounds such as high molecular weight PAHs. Microbes are crucial in breaking down complex and higher molecular weight molecules into smaller, simpler ones. For instance, *Hortaea* sp. has been shown to degrade four-ring PAHs like chrysene into 1-hydroxy-2-naphthoic acid and pyrene into phthalic acid (Al Farraj et al. 2019). During the biodegradation process, various gene markers are upregulated. These markers can be identified and utilized to investigate the occurrence of biodegradation and the probability of different mechanisms of action in the environment. For example, *Pelagerythrobacter* sp., has been found to contain numerous genes responsible for PAH degradation, such as RHD, C120, G120, and phthalate 4,5-dioxygenase, suggesting its ability to degrade PAHs through various mechanisms (Li et al. 2024).

Halotolerant or halophilic microbes play a significant role in the bioremediation of organic pollution in marine environments. Microbes can be classified into four categories based on their salt tolerance or requirement, namely, nonhalophile (<0.2M NaCl), halotolerant (0.2–0.5 M NaCl), moderate halophile (0.5–2.5M NaCl), and extreme halophile (2.5–5.5M NaCl) (Shivanand and Mugeraya 2011). Halophiles survive in saline conditions by employing at least two strategies:

accumulating ions such as Na^+, K^+, and Cl^- from the environment, and biosynthesizing organic osmolytes within their cytoplasm (Shivanand and Mugeraya 2011). Microbial strains isolated from contaminated sites often develop survival strategies by utilizing organic contaminants as their carbon source. Therefore, research efforts focus on isolating microbes from contaminated sites. For example, *Idiomarina piscisalsi* isolated from oil-contaminated coastal sediment in the Arabian Gulf demonstrated the capability to degrade PAHs (Nzila et al. 2018).

Microbes typically degrade simple organic pollutants more rapidly than complex ones. Marine *Pseudoalteromonas* sp. can aerobically biodegrade 90% of cyclohexyl acetic acid, a type of naphthenic acid, within 192 h. This degradation process involves activating acetyl-CoA and hydroxymethylglutarate-CoA lyase to cleave the molecule (Zan et al. 2022). While *Halomonas shengliensis* and *H. smyrnensis* required 18 days to degrade only 40–52% of pyrene (Budiyanto et al. 2018).

Specific microbes employ unique pathways to degrade PAHs. In the degradation of Benzo[a]pyrene (BaP), halotolerant *Bacillus thuringiensis* and *B. pacificus* primarily utilize hydroxylation as the main mechanism. This process involves the oxidation of the aromatic ring of BaP, resulting in the production of cis-dihydrodiol intermediates. Subsequently, these intermediates are cleaved to generate carboxylic acids, phenolics, carbonyl compounds, or simple alcohols (Muralidharan et al. 2021). Typical microbes activate specific gene markers to produce enzymes and proteins tailored for cleaving contaminants. The halophilic actinobacterium *Zhihengliuella* sp. activates epoxide hydrolase B and 9,10-monooxygenase to biodegrade phenanthrene. Within 168 hours, it degrades 87% of phenanthrene, producing intermediates such as phthalic acids (Mishra et al. 2020).

The degradation of BaP by the halophilic bacteria *Pontibacillus chungwhensis* was governed by two crucial heterogeneous genes: epoxide hydrolase (EH) and the co-expression of monooxygenase CYP102. The primary gene involved in initiating BaP degradation is CYP102, which converts BaP into 4,5-epoxide-BaP. Subsequently, EH facilitates the transformation of 4,5-epoxide-BaP into BaP-trans-4,5-dihydrodiol (Qian et al. 2024). Polyextremotolerant yeast *Rhodotorula mucilaginosa* can degrade 80% of phenanthrene and BaP within 10 days under hypersaline conditions (1M NaCl). The cleavage of the BaP ring results in the production of 2-hydroxymuconic semialdehyde pyrene through meta-oxidation facilitated by catechol 2,3-dioxygenase activity (Martínez-Ávila et al. 2021).

Innovations have been introduced to enhance bioremediation strategies employing microbes. The utilization of bacterial consortia has shown promise in improving biodegradation outcomes. Halotolerant bacteria *Vibrio alginolyticus* MMKVG1 and MMKVG2, isolated from contaminated sediment in Ramanathapuram, demonstrated the ability to degrade high concentrations of both low and high molecular weight PAHs. When used in combination, these strains achieved up to 90% PAH degradation, with maximum degradation observed at 250 mg L^{-1} PAHs by day 2 (Muralidharan et al. 2023). Another consortium composed of halotolerant bacteria, including *Stenotrophomonas maltophilia*, *Enterobacter cloacae*, and *Ochrobactrum* sp., exhibited the capability to degrade a wide range of PAHs. This consortium achieved complete degradation of phenanthrene within four days at 3% NaCl

(Arulazhagan et al. 2010). Additionally, a consortium comprising hydrocarbon-degrading and biosurfactant-producing strains successfully degraded 90% of pyrene within 10 days of incubation, while also producing biosurfactants. Strains such as *Pseudomonas putida* and *Delftia tsuruhatensis* exhibited near-complete degradation of phenanthrene (96.1%), naphthalene (99.83%), and benzoate (100%). The biosurfactants produced by this consortium were identified as heptapeptides with distinct lipid groups (Ibrar and Yang 2022).

The synergistic metabolic processes of bacterial consortia are believed to enhance PAHs biodegradation. Another may utilize metabolites produced by one type of bacteria within the consortium to initiate degradation pathways. In a study involving consortia comprising anammox culture, primarily composed of *Candidatus kuenenia*, and the PAHs degrader *Pseudomonas stutzeri*, it was found that nitrite could be removed from the culture media during phenanthrene degradation. PAH-degrading bacteria produce nitrite, which can inhibit further biodegradation processes. However, the growth of the anammox culture was inhibited by up to 91.8–97.5% by 4 mM phenanthrene (Zhang et al. 2023). The co-culture of yeast *Basidioascus persicus* and *Pseudomonas putida* improved pyrene degradation by 21%. When *B. persicus* alone was involved, it could degrade 79% of pyrene, with a half-life of 9.33 days and a decomposition rate of 0.074 day^{-1} (Kamyabi et al. 2018).

Modifications to the substrate have been found beneficial in enhancing bioremediation outcomes. For instance, the addition of a co-substrate (a mixture of PAHs) may improve biodegradation performance. *Acinetobacter johnsonii* exhibited nearly complete degradation of pyrene when naphthalene was added to the medium, as naphthalene activates key enzymes (catechol 2,3-dioxygenase and salicylate hydroxylase) in the bacterium responsible for degrading PAHs (Jiang et al. 2018). The addition of an inorganic substrate like Fe(III) enhanced phenanthrene degradation by *Kocuria oceani* from 67.37 to 83.91%. Fe(III) added to the medium stimulates riboflavin excretion, acting as a shuttle to facilitate electron transfer from phenanthrene to Fe(III). Additionally, the strain produces organic acids that delay Fe reoxidation. The H_2O_2 excreted by the strain prompts an extracellular Fenton reaction, producing OH radicals for minor phenanthrene degradation (Bai et al. 2023).

Modifying process conditions is also an alternative method to enhance biodegradation performance. Deep-sea bacteria *Vibrio* sp. LQ2 from the South China Sea produce biosurfactants from phospholipids with strong stability and high surface activity. Inoculating this strain into biochar improved biodiesel removal by up to 94.7% through biodegradation and adsorption, compared to just 54.4% removal with the free strain in seawater. The activity of genes related to biodegradation in the immobilized treatment, such as *CYP450-1* and *alkB*, was 15.2 and 3.8 times higher, respectively, compared to the free cell treatment (Zhou et al. 2021).

The physicochemical properties of the media, such as temperature, pH, and salinity, profoundly influence biodegradation performance, making optimization essential. A consortium of halophilic bacteria (57.5% *Methylophaga*, 18.2% *Marinobacter*, 15.2% *Thalassospira*) demonstrated nearly complete degradation of phenanthrene across varying salinities (1–10%). The consortium degraded 97.78%

of phenanthrene within 3 days at 1% salinity, while at 10% salinity, complete degradation occurred within 7 days. Multiple genes involved in PAH degradation were detected throughout degradation, including aldolase, dehydrogenase, and dioxygenase (Fan et al. 2022). However, the ability of a consortium of halophilic bacteria CY-1 to degrade phenanthrene was negatively impacted by salinity, with optimal degradation observed at 3% salinity (complete degradation within 5 days). Conversely, at 0.1% salinity, no degradation was observed, while at 20% salinity, complete degradation occurred after 20 days. The metabolic degradation of phenanthrene by the consortium involved deoxygenation at the C1 and C2 positions of phenanthrene in the upstream pathways, followed by various downstream pathways such as protocatechuic acid, gentisic acid, and catechol pathways. *Marinobacter* was identified as the dominant bacteria during upstream degradation, while *Halomonas* was dominant in producing intermediate products during downstream degradation (Wang et al. 2018). Various microbes capable of degrading PAHs are listed in Table 15.4. Figure 15.7 depicted *Pseudomonas aeruginosa* PFL-P1, *Bacillus subtilis* EB1, *P. furukawaii* PPS-19, and *Micrococcus* sp. strain 2A capable in degrading organic contaminants.

15.3.3.2 Microcosm Approach for Organic Pollutant Bioremediation

Microcosm investigations for bioremediation of organic pollutants often utilize natural sediment or seawater as a microbial source without isolating them in artificial environments. This approach is crucial for understanding biodegradation capabilities and mechanisms in specific areas or before upscaling. In one study, microcosms from marine sediment biodegraded 69.53% of BaP (initial conc 200 mg L^{-1}) within 45 days with a 3.09 mg L^{-1} day^{-1} removal rate. This BaP degradation was enhanced to 71.91–95.67% by the addition of co-solvents (acetone and dichloromethane). Dichloromethane initially stimulated a co-metabolic effect on BaP, while acetone alone did not lead to BaP biodegradation; instead, the biodegradation effect was initiated by the co-metabolic effect of dichloromethane. During the biodegradation process, the abundance of Marinobacter, Methylophaga, Porticoccus, and the sulfate-reducing bacteria such as Desulfococcus, Desulfosarcina, Desulfobacteraceae, and Desulfuromonas inclined (Leng et al. 2021).

In a microcosm biodegradation study using produced water, the biotransformation half-lives of PAHs ranged from 8 to over 100 days, while phenols and alkylated phenols exhibited half-lives of 5–70 days. Among the PAHs, benzo(b)fluoranthene, benzo(e)pyrene, and benzo(g,h,i)perylene had the longest biotransformation half-lives (>100 days). In contrast, other PAHs showed half-lives of less than 50 days, except for chrysene (64.6 days). The abundance of oil-degrading and heterotrophic microbes peaked throughout the experiment, increasing by over 1000 times after 14 days of incubation (Lofthus et al. 2018).

After incubating in situ burn residue from marine diesel for 6 weeks, there was a 26% decrease in *n*C16–*n*C24 alkanes in weathered burn residue and an 8% decrease in unweathered burn residue. A significant decrease (2–24%) was observed in fluorine, naphthalene, and their alkylated homologues from burning weathered diesel. Microbes identified in the unweathered burn residue included genera like

Table 15.4 Non-halophilic and halophilic microbes capable in PAHs bioremediation

Species	Target contaminant	Source
Miscellaneous microbes		
Acinetobacter johnsonii	Almost complete degradation of 400 mg L^{-1} pyrene within 157 h	Jiang et al. (2018)
Klebsiella sp.	Phenanthrene, 88.4% degradation within 7 days	Cao et al. (2023)
Pseudomonas stutzeri	46.1% of phenanthrene removal within 13 days	Zhang et al. (2023)
Mycobacterium sp. WY10	100% phenanthrene degradation within 60 h, 83% pyrene degradation within 72 h	Sun et al. (2019)
Rhodococcus wratilaviensis	100, 40, and 28% degradation of 280 μM phenanthrene, 50 μM pyrene, 40 μM benzo[a]pyrene, respectively	Subashchandrabose et al. (2019)
Kocuria oceani	83.91% phenanthrene degradation within 15 days	Bai et al. (2023)
Marinobacterium georgiense	6% pyrene and 16.4% phenanthrene degradation within 26 days	Walton and Buchan (2024)
Rhodospirillaceae EZ35	14.7% pyrene d degradation within 26 days	
Bacillus-Clostridium SE165	14.8% pyrene, 16.6% phenanthrene d degradation within 26 days	
Flavobacteraceae EZ40	16.1% pyrene degradation within 26 days	
Alteromonas macleodii EZ55	24% phenanthrene degradation within 26 days	
R. pomeroyi	23.8% pyrene degradation within 26 days	
S. stellate	24.6% pyrene degradation within 26 days	
Citreicella sp.	29.6% pyrene, 33.1% phenanthrene degradation within 26 days	
Sulfitobacter sp.	34.4% phenanthrene degradation within 26 days	
Ruegeria sp.	52.2% phenanthrene degradation within 26 days	
Halotolerant bacteria		
Cryptococcus sp.	Complete degradation of phenanthrene and anthracene after 30 days	Al Farraj et al. (2020)
Halomonas sp.	>95% degradation of phenanthrene and anthracene after 30 days	
Hypersaline bacteria		
Methylophaga, *Marinobacter*, *Thalassospira*	Phenanthrene, 97.78% degradation within 3 days, 1% salinity.	Fan et al. (2022)

(continued)

Table 15.4 (continued)

Species	Target contaminant	Source
Halomonas caseinilytica	76% of coronene was degraded within 80 days, 10% salinity	Okeyode et al. (2023)
Zhihengliuella sp.	87% phenanthrene degradation within 168 h, 5% salinity	Mishra et al. (2020)
Hortaea sp.	92% pyrene degradation within 25 days, 77% chrysene degradation within 20 days, 1% salinity	Al Farraj et al. (2019)
Pontibacillus chungwhensis	30% benzo[a]pyrene and 100% phenanthrene degradation within 6 days	Qian et al. (2024)
Bacillus thuringiensis	70.7% BaP degradation within 21 days	Muralidharan et al. (2021)
B. pacificus	76.76% BaP degradation within 21 days	
Pelagerythrobacter sp.	100% phenanthrene degradation, 9 days, 3–5% salinity	Li et al. (2024)
Halomonas shengliensis	40% pyrene degradation, 18 days, 20% salinity	Budiyanto et al. (2018)
Halomonas smyrnensis	52% pyrene degradation, 18 days, 20% salinity	

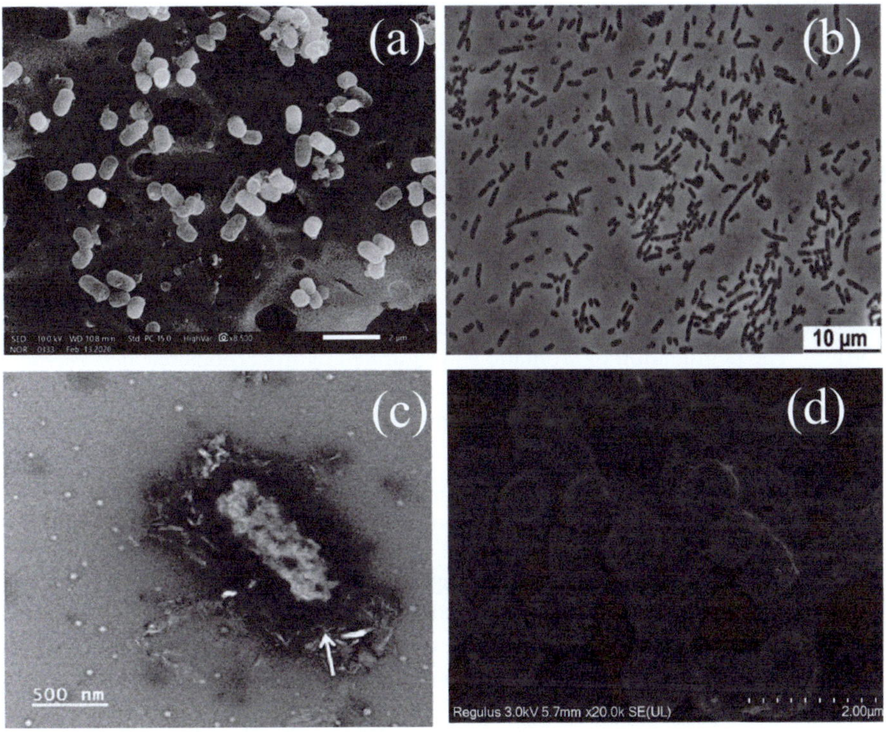

Fig. 15.7 Micrographs of (**a**) *Pseudomonas aeruginosa* PFL-P1, (**b**) *Bacillus subtilis* EB1, (**c**) *Pseudomonas furukawaii* PPS-19, and (**d**) *Micrococcus* sp. strain 2A. Copied with permission from Kumari et al. (2022), Kumar et al. (2023), Vandana and Das (2023), Huang et al. (2024)

Sulfitobacter (aromatic degrader); *Parvibaculales*, *Oleispira*, and *Oleibacter* (alkane degrader), as well as *Sphingorhabdus* and *Colwellia* (alkane and aromatic degrader). In the weathered burn residue, there was a shift in genera dominance, with *Oliespira* and *Colwellia* at day 7 to *Alcanivorax*, *Oleibacter*, *Olleya*, *Oleispira*, and *Colwellia* at day 21–47 (Pyke et al. 2023).

15.3.3.3 Mesocosm Bioremediation of Organic Pollutants

Since the late 1980s, mesocosm or in situ bioremediation techniques have garnered significant attention (Madsen 1991). However, due to the complex environmental factors involved, their effectiveness in degrading pollutants differs from laboratory-based approaches. In the supratidal zone, for instance, oil biodegradation may proceed more slowly due to higher salinity resulting from increased evaporation. This salinity variance alters the microbial composition in sediment through natural selection processes. Consequently, hydrocarbonoclastic activity is impeded by elevated salt concentrations, leading to a decrease in the degradation rate constant of oil from 75% (at 90 g L^{-1} salt) to 90% (at 160 g L^{-1} salt) within a 76-day observation period. Additionally, the abundance of halophilic Archaea, particularly Halobacteriaceae, tends to increase in areas with higher salinity and oil contamination (Khalil et al. 2021).

In situ biodegradation, akin to laboratory methods, can be enhanced through technique modifications, such as immobilizing introduced microbes in targeted areas. Following an oil spill incident in the Bohai Sea, indigenous bacteria from the sediment, including *Salfobacillus*, *Pseudomonas putida*, and *Acinetobacter calcium acetate*, were immobilized onto granular zeolite coated with poly-γ glutamic acids. After 210 days, 68.015% of PAHs and 60.99% of *n*-alkanes in the sediment were biodegraded, with three- and four-ring PAHs showing degradation rates of 84.44 and 26.62%, respectively (Wang et al. 2020). In a pilot plant setup in the intertidal region of Tianjin, China, 66.5% of heavy oils were successfully removed over 100 days (degradation rate = 0.018 day^{-1}) using an immobilized laccase-bacteria consortium. The degradation of aromatic and saturated hydrocarbons reached 78.7 and 79.2%, respectively, which was higher compared to the control (65.1 and 64.9%, respectively). The control pool was dominated by *Cyclocasticus* sp., *Alcanivorax* sp., *Marinobacter* sp., and *Idiomarina* sp., with a peak abundance of 49.6% observed at day 30, gradually decreasing thereafter. Meanwhile, the immobilized laccase-bacteria consortium was dominated by *Bacillus* sp., and *Alcanivorax* sp. (Dai et al. 2020).

In a mesocosm experiment simulating an oil spill along the Brazilian coast, 33.8% of alkanes were removed from sediment within 7 days. However, with the daily addition of low-cost NPK fertilizer, alkane removal increased significantly to 83.4% (de Souza et al. 2024). Furthermore, the biodegradation of PAHs in coastal sediment samples containing *Marinobacter* was enhanced by the addition of electron acceptors (sulfate and bicarbonate) and electron donors (such as lactate and acetate). The addition of acetate and bicarbonate resulted in 89.7% PAH degradation, while the addition of lactate and sulfate showed 87.10% degradation (Chen et al. 2024).

In an in situ mesocosm experiment conducted in Lanshan District, Rizhao, China, approximately 88% of total petroleum hydrocarbons (TPH) were biodegraded within 3 weeks. The addition of dispersant improved the biodegradation process of TPH, with key degradation microbes identified from genera such as *Glaciecola*, *Lentibacter*, *Roseobacter*, and *Marivita*. The dispersant treatment elevated and accelerated TPH biodegradation, with enhanced development of carbon and chemotaxis metabolism genes (*fade*, *fadeJ*, and *cheA*) (Zhou et al. 2023). Oil dispersant does not affect the microbe community event at low temperature (5 °C). The use of oil dispersant did not impact microbial communities even at low temperatures (5 °C). The biotransformation rates of saturated alkanes (C_{10}–C_{36}) (ranging from 1.4 to 45.2 days), naphthalene (6.1–12.5 days), and 2- to 5-ring PAHs (36.6 to >100 days) were similar with and without dispersion. The peak microbial abundance was observed at 14–21 days, predominantly composed of *Oleispira* and *Colwellia* (Brakstad et al. 2018).

In addition to oil contamination, intrinsic bacteria from marine sediment can degrade decabromodiphenyl ether (BDE-209), a member of the polybrominated diphenyl ether (PBDE) family, even at high concentrations. In an anaerobic mesocosm experiment, approximately 70% of BDE-209 (initial concentration 5 μmol) was degraded within 90 days. The presence of total organic carbon (TOC) in the sediment may accelerate the degradation of organic contaminants. During the degradation process, 35 products were identified, indicating a preference for the debromination pathway. Debromination of BDE-209 via simultaneous *ortho*- and *meta*-substitution results in nonaBDEs, octaBDEs, heptaBDEs, hexaBDEs, and pentaBDEs. The bacteria responsible for this degradation are primarily dominated by Acetobacterium from the phylum Firmicutes, with its abundance increasing to 18.36% after 90 days of the experiment. Other identified bacteria include Acinetobacter, Citrobacter, Pseudomonas, and Spaerachaeta (Zhu et al. 2019). Moreover, the biodegradation of BDE-209 by *Micobacterium Y2* upregulated genes encoding ATP-binding cassette (ABC) transporters, glutathione-S-transferase, and haloacid dehalogenase, which contributed to BDE-209 degradation. Neutralization of BDE-209 toxicity was facilitated by ribosomal proteins, oligoribonuclease, ribonuclease E, and heat shock proteins (Yu et al. 2020).

15.3.3.4 Naturally Occurring Bioremediation of Organic Pollutants

Numerous studies have highlighted the natural microbial activity in contaminated marine environments, demonstrating nature's efforts to restore balance. The Deepwater Horizon oil spill in April 2010 stands out as a notable environmental disaster, prompting extensive post-accident investigations. During the blowout, a significant amount of dispersant was injected into surface and deep waters to facilitate oil dispersion into microdroplets, ranging from 6 to 300 μm in size, observed at depths of 1000–1400 m. This dispersion aimed to accelerate the biodegradation of crude oil and hazardous compounds contained within (Driskell and Payne 2018). In the surface water oil slick during the Deepwater Horizon spill, the predominant genera identified were *Alteromonas*, *Pseudoalteromonas*, and *Cycloclasticus*, all known for their oil biodegradation capabilities. *Alteromonas* secreted exopolysaccharides (EPS), which, in conjunction with cells, bind oil to form marine oil snow (MOS), contributing to oil dispersion. *Cycloclasticus* are capable of degrading alkane (532–949 ug g^{-1} oil of C_{14} and C_{15}) and aromatic hydrocarbons (603–16,629 μg g^{-1} oil of naphthalene, phenanthrene, and their derivatives) within the oil. *Alteromonas* bio-degrades 436–1135 μg g^{-1} oil of C_{12}–C_{15} and 553–6888 μg g^{-1} oil of naphthalene and its derivatives (Gutierrez et al. 2018).

Investigating gene markers is a powerful tool for understanding natural biodegradation processes. Marker genes such as *pahE, pahAc, nagE* (from *Comamonas testosterone*), *bphE* (from *Hyphomonas oceanitis*), *nahE* (from *Pseudomonas*) are valuable for observing aerobic biodegradation activities in the environment (Liang et al. 2023a). In 2007, the Taean coast in Korea experienced heavy oil contamination. The bacterial community in this area was dominated by the phyla Bacteroidetes, Firmicutes, and Proteobacteria, which constituted over 93% of the community. The PAH biodegradation performance at this site was assessed using PAH catabolism genes such as aromatic ring hydroxylating dioxygenase and naphthalene dioxygenase. Key PAH-degrading microbes identified included *Pseudoalteromonas agarivorans*, *Rhodococcus soli*, *Cobetia marina*. These microbes demonstrated significant PAH-degrading activity, contributing to the natural bioremediation of the contaminated site (Lee et al. 2018).

Thermophilic and mesophilic bacterial consortia in the bottom sediment of Guaymas Basin effectively biodegrade C_6–C_{12}, C_{16}–C_{18} alkanes, as well as crude oil. The microbial community in this region is predominantly composed of Deltaproteobacteria, which account for 47.48% of the degrading bacteria. Functional genes involved in anaerobic hydrocarbon biodegradation, such as benzyl succinate synthase (for aromatic hydrocarbons) and alkyl succinate synthase (for alkanes), have been identified in the sediment (Liang et al. 2023b). Metabolomic studies on coastal microbial mats have identified numerous bacterial genes with potential for hydrocarbon degradation. These include sulfide-oxidizing genes in *Wengzhouxiangellaceae*; sulfur-oxidizing genes in *Halieaceae*, *Pseudomonadales*, *Xanthomonadales*, *Rhodobacteraceae*, and Phototrophic potential in *Erythrobacteraceae* (Vigneron et al. 2024).

Natural biodegradation processes also occur in extreme marine environments such as the polar regions. In the mesopelagic zone of an Arctic fjord at 615–650 m depths, the half-lives of hydrocarbon biodegradation ranged from 238.3 ± 89.3 to 238.8 ± 74.1 days. These rates are faster than those in the epipelagic zone (10–20 m), where the half-lives were 304.7 ± 119.5 and 275.5 ± 106.1 days. The microbial community in the mesopelagic zone was initially dominated by *Oleispira* spp. for the first 8–37 days, while genera *Ulvibacter* spp., *Pseudohongiella* spp., *Pseudofulvibacter* spp., *Kordia* spp., *Cycloclasticus* spp. *Colwellia* spp. and *Arcobacter* spp. dominated the community between 100 and 379 days. The bacterial density was higher in the mesopelagic zone (8.1 ± 0.1 $\log_{10}$ 16S rRNA genes/cm^2) compared to the epipelagic zone (7.3 ± 0.4 $\log_{10}$ 16S rRNA genes/cm^2) (Kampouris et al. 2023). In the pristine, sub-zero waters of Godthaab Fjord, Greenland, the half-life for *n*-alkane biodegradation was only 7 days, with the bacterial community dominated by *Oleispira antarctica*. However, in sea ice, the population of *Oleispira* was 25–100 times lower, resulting in non-observable *n*-alkane degradation. This demonstrates how microbial activity and biodegradation rates can significantly vary based on environmental conditions and microbial community composition (Vergeynst et al. 2019). The common microbes identified in the marine environment capable of degrading organic pollutants are listed in Table 15.5.

15.3.4 Bioremediation of Radionuclide in Marine Environment

Certain terrestrial and freshwater microbes and plants exhibit the ability to absorb and tolerate radionuclides, making them promising agents for phytoremediation. For instance, the aquatic plant *Fontinalis antipyretica* in Portugal has shown remarkable potential by accumulating 4979 mg kg^{-1} of uranium. Microbes play a crucial role in enhancing bioaccumulation, biomineralization, bioreduction, and biosorption processes, which can affect the mobility and toxicity of radioactive uranium (Thakur and Kumar 2024). The *Geobacteraceae* family is well-known for its metal-reducing capabilities, even in hypersaline environments. For example, adding acetate to sediments has been found to improve uranium removal significantly. In one study, the U(VI) concentration decreased from 14 to less than 2 μM within 37 days following acetate addition (Nevin et al. 2003).

Enzymatic uranium bioreduction can occur extracellularly, in the periplasm, cytoplasm, or at the outer membrane of microbes. Among the common bacteria capable of reducing U(VI) are the genera *Geobacter* and *Desulfovibrio*. Geobacter's reduction of U(VI) is primarily facilitated by cytochromes, such as the c-type cytochrome OmcZ from *Geobacter sulfurreducens* (Rogiers et al. 2022). *Geobacter uraniireducens* strain $Rf4^T$ can reduce U(VI) using various electron acceptors, including fumarate, malate, anthraquinone-2,6-disulfonate, Mn(IV), and Fe(III) (Shelobolina et al. 2008). Numerous marine actinobacteria have been found to thrive in radionuclide-contaminated sites near nuclear power plants, such as Jaitapur, India. The microbial community in these areas includes genera like *Streptomyces* (40%), *Rhodococcus* (9%), *Nocardiposis* (8%), *Nocardia* (7%), *Micromonospora* (7%), *Pseudocardia* (5%), *Saccaromonospora* (4%), *Saccharopolyspora* (4%), *Corynebacterium* (3%), *Dietzia* (2%), and *Kocuria* (1%). Notably, *Actinomadura* sp. in this site secretes 243.7 mg L^{-1} of exopolysaccharides (EPS), which contributes to the remediation of radiouranium (Sivaperumal et al. 2022).

Bacillus sp. strain TK2d removed >99% of ^{90}Sr from an initial concentration 1.0 × 10^{-3} mol L^{-1} by biomineralization to produce $SrCO_3$ within 4 days. The strain mineralized ^{90}Sr by first adsorbing it onto the cell wall surface and EPS. As a ureolytic bacterium, it hydrolyzes urea internally and releases NH_4^+ and HCO_3^- ions. These ions then react with the accumulated Sr on the cell wall, forming $SrCO_3$ crystals that eventually detach from the cell wall as they grow (Horiike et al. 2017). The marine actinobacterium *Streptomyces* sp. produces exopolysaccharides (EPS) capable of absorbing 82% of Sr^{2+} from 100 mg L^{-1} solution. The EPS was composed of carbohydrates, proteins, nucleic acids, and miscellaneous compounds (Kamala et al. 2020).

Table 15.5 Common bacteria types identified from marine environment capable of biodegrading PAHs (Lee et al. 2018; Dai et al. 2020; Khalil et al. 2021; Wang et al. 2021)

Class	Order	Family	Genus
α-Proteobacteria	*Rhizobioles*	*Methylobacteriaceae*	*Methylobacterium*
	Rhodobacterales	*Rhodobacteraceae*	*Paracoccus*
			Roseovarius
	Sphingomonadales	*Erythrobacteraceae*	*Croceicoccus*
			Erythrobacter
			Novosphingobium
			Altererythrobacter
		Sphingomonadaceae	*Sphingopyxis*
			Sphingomonas
β-Proteobacteria	*Rhodocyclales*	*Azonexaceae*	*Dechloromonas*
γ-Proteobacteria	*Aeromonadales*	*Aeromonadaceae*	*Oceanisphaera*
	Oceanospirillales	*Hahellaceae*	*Hahella*
		Halomonadaceae	*Cobetia*
			Halomonas
		Alcanivoracaceae	*Alcanivorax*
	Pseudomonadales	*Moraxellaceae*	*Psychrobacter*
			Acinetobacter
		Pseudomonadaceae	*Pseudomonas*
	Vibrionales	*Vibrionaceae*	*Vibrio*
	Thiotrichales	*Piscirickettsiaceae*	*Cycloclasticus*
	Alteromonadales	*Alteromonadaceae*	*Marinobacter*
		Idiomarinaceae	*Pseudidiomarina*
			Idiomarina
Actinobacteria	*Corynebacteriales*	*Dietziaceae*	*Dietzia*
		Nocardiaceae	*Rhodococcus*
		Mycobacteriaceae	*Mycobacterium*
	Micrococcales	*Cellulomonadaceae*	*Cellulomonas*
		Microbacteriaceae	*Herbiconiux*
			Microbacterium
			Agrococcus
		Micrococcaceae	*Arthrobacter*
			Kocuria
		Sanguibacteraceae	*Sanguibacter*
	Micromonosporales	*Micromonosporaceae*	*Micromonospora*
	Propionibacteriales	*Nocardioidaceae*	*Nocardioides*
	Streptomycetales	*Streptomycetaceae*	*Streptomyces*
Firmicutes	*Bacillales*	*Bacillaceae*	*Bacillus*
		Planococcaceae	*Planococcus*
			Planomicrobium
		Staphylococcaceae	*Staphylococcus*
			Exiguobacterium
Clostridia	*Eubacteriales*	*Eubacteriales incertae sedis*	*Fusibacter*
	Clostridiales	*Clostridiales*	*Acidaminobacter*
Chitinophagia	Chitinophagales	*Chitinophagaceae*	*Dinghuibacter*
			Chitinophaga
Flavobacteria	Flavobacteriales	*Flavobacteriaceae*	*Flavobacterium*

Marine archaea can immobilize radiocesium through biomineralization mechanisms. For instance, *Halococcus* sp. ENMS8 has been shown to remove 76.46% of Cs from an initial concentration of 100 mmol L^{-1}. This strain's exopolysaccharides (EPS) can absorb 76.4% of Cs from a 1 g L^{-1} Cs (Kannan and Sivaperumal 2023).

15.4 Conclusion and Future Strategies

Calculating the economic loss of pollution proves challenging. To illustrate the economic impact of pollution in marine environments, let's consider oil spill incidents, which are tangible and observable. The complexity of cleanup operations varies depending on factors such as spill size, distance from shorelines, season, currents, and oil type. In significant oil spill disasters, conventional cleanup procedures can cost up to 98 million euros for severe spills and 13 million euros for moderate spills (Tedesco et al. 2024). Conventional pollution remediation protocols, particularly for oil spills, typically involve containment, collection, and/or adsorption. However, these protocols often lack clear guidelines for in situ and ex situ processes, leading to complications and exacerbating costs. The cost of bioremediation varies from $5 to $300/m^3$, while physicochemical treatment can cost around $600/m^3$ (Tedesco et al. 2024). Regarding soil remediation, costs vary based on the duration of the remediation process, with short-term technology ranging from $39 to $331 per ton, medium-term from $22 to $131 per ton, and long-term from $8 to $131 per ton (Ahmed et al. 2022).

The Blue Growth strategy, as exemplified by initiatives in Europe, represents an emerging perspective in addressing marine pollution sustainably. Within this framework, bioremediation emerges as a crucial tool worthy of promotion. This assertion stems from the versatility of bioremediation across various sectors of the economy, spanning from primary to quinary. However, the practical application of bioremediation entails a complex process encompassing design, implementation, monitoring, and evaluation of efficacy. In the design phase, research and development play pivotal roles in selecting appropriate microbes and optimizing technology for microbial growth and pollutant removal. Implementation varies from laboratory-scale experiments to real-world applications, highlighting the importance of technology for scaling up and servicing processes. Subsequently, monitoring and evaluation are essential to assess the efficiency and broader impacts of bioremediation technologies. From this perspective, bioremediation offers a compelling strategy, not only providing environmentally friendly approaches to pollution management but also creating new job opportunities. Given the multidisciplinary nature of bioremediation processes, collaboration across academia and industry is imperative (Tedesco et al. 2024).

Despite the promising results observed in laboratory and ex situ experiments, bioremediation encounters several challenges when implemented in actual marine environments. While studies have demonstrated high satisfaction with the effectiveness of bioremediation across various types of pollutants, the process tends to be more time-consuming in real-world scenarios. For instance, research by Bianco et al. (2020) highlighted the cost-effectiveness of bioremediation (€228 m^{-3}) compared to alternative methods such as soil washing (€371 m^{-3}) and low-temperature thermal desorption (€1782 m^{-3}) for the removal of polycyclic aromatic hydrocarbons (PAHs) from marine sediment(Bianco et al. 2020). However, the transition from laboratory conditions to natural marine environments often results in slower degradation rates. This disparity arises because the optimized conditions for microbial growth established in vitro may not accurately reflect the environmental conditions present in situ. Moreover, in some cases, introduced microbial strains fail to thrive due to their inability to adapt to the new environmental conditions or compete effectively with native microbial communities. Consequently, ongoing research is crucial to refine and optimize bioremediation processes in marine settings, thereby ensuring more efficient and timely remediation of pollution.

References

Abbasi A (2023) Bioaccumulation and risk assessment of radiocesium in the Northwest Pacific Ocean from Fukushima Dai-ichi Nuclear Power Plant accident. Mar Pollut Bull 192:114994. https://doi.org/10.1016/j.marpolbul.2023.114994

Abbasi A, Almousa N, Zakaly HMH (2023) The distribution of radiotoxic 137Cs concentrations in seaweed and mussel species in the Mediterranean Sea. Mar Pollut Bull 197:115737. https://doi.org/10.1016/j.marpolbul.2023.115737

Ahmed IB, Nwaichi EO, Ugwoha E et al (2022) Cost reduction strategies in the remediation of petroleum hydrocarbon contaminated soil. Open Res 5:21. https://doi.org/10.12688/openresafrica.13383.1

Al Farraj DA, Hadibarata T, Yuniarto A et al (2019) Characterization of pyrene and chrysene degradation by halophilic Hortaea sp. B15. Bioprocess Biosyst Eng 42:963–969. https://doi.org/10.1007/s00449-019-02096-8

Al Farraj DA, Hadibarata T, Yuniarto A et al (2020) Exploring the potential of halotolerant bacteria for biodegradation of polycyclic aromatic hydrocarbon. Bioprocess Biosyst Eng 43:2305–2314. https://doi.org/10.1007/s00449-020-02415-4

Alarif WM, Budiyanto F, Bawakid NO et al (2023) The occurrence of microplastic in marine ecosystems in the Middle East: a review. Reg Stud Mar Sci 67:103208. https://doi.org/10.1016/j.rsma.2023.103208

Alava JJ (2019) Chapter 47: Ocean pollution and warming oceans: towards ocean solutions and natural marine bioremediation. In: Predicting future oceans: sustainability of ocean and human systems amidst global environmental change. Elsevier, Amsterdam, pp 495–518

Alhudhodi AH, Alduwais AK, Aldhafeeri ZM et al (2022) Contamination assessment of mangrove ecosystems in the Red Sea Coast by polycyclic aromatic hydrocarbons. Int J Environ Res Public Health 19:5474. https://doi.org/10.3390/ijerph19095474

Alvarez A, Saez JM, Costa JSD et al (2017) Actinobacteria: current research and perspectives for bioremediation of pesticides and heavy metals. Chemosphere 166:41–62. https://doi.org/10.1016/j.chemosphere.2016.09.070

Arai R, Nishi Y (2024) Biomass-based stochastic model for the transfer of polychlorinated biphenyls in a marine food web. Sci Total Environ 912:169270. https://doi.org/10.1016/j.scitotenv.2023.169270

Aranda DA, Sindou P, Cauich Rodriguez JV et al (2024) A non-invasive method of microplastics pollution quantification in green sea turtle Chelonia mydas of the Mexican Caribbean. Mar Pollut Bull 200:116092. https://doi.org/10.1016/j.marpolbul.2024.116092

Armada D, Martinez-Fernandez A, Celeiro M et al (2023) Assessment of the bioaccessibility of PAHs and other hazardous compounds present in recycled tire rubber employed in synthetic football fields. Sci Total Environ 857:159485. https://doi.org/10.1016/j.scitotenv.2022.159485

Arulazhagan P, Vasudevan N, Yeom IT (2010) Biodegradation of polycyclic aromatic hydrocarbon by a halotolerant bacterial consortium isolated from marine environment. Int J Environ Sci Technol 7:639–652. https://doi.org/10.1007/BF03326174

Auta HS, Emenike CU, Jayanthi B, Fauziah SH (2018) Growth kinetics and biodeterioration of polypropylene microplastics by Bacillus sp. and Rhodococcus sp. isolated from mangrove sediment. Mar Pollut Bull 127:15–21. https://doi.org/10.1016/j.marpolbul.2017.11.036

Babut M, Mourier B, Desmet M et al (2019) Where has the pollution gone? A survey of organic contaminants in Ho Chi Minh city/Saigon River (Vietnam) bed sediments. Chemosphere 217:261–269. https://doi.org/10.1016/j.chemosphere.2018.11.008

Bai B, Zhang L, Dong H, Huang Y (2023) Coupled Fe(III) reduction and phenanthrene degradation by marine-derived Kocuria oceani FXJ8.057 under aerobic condition. J Hazard Mater 459:132237. https://doi.org/10.1016/j.jhazmat.2023.132237

Balasubramanian V, Natarajan K, Hemambika B et al (2010) High-density polyethylene (HDPE)-degrading potential bacteria from marine ecosystem of Gulf of Mannar, India. Lett Appl Microbiol 51:205–211. https://doi.org/10.1111/j.1472-765X.2010.02883.x

Bennett BJ, Aung MT, Boonstra R et al (2024) Investigation of the link between per- and polyfluoroalkyl substances and stress biomarkers in bottlenose dolphins (Tursiops truncatus). Environ Sci Technol 58:9061–9070. https://doi.org/10.1021/acs.est.3c06979

Besseling E, Foekema EM, Van Franeker JA et al (2015) Microplastic in a macro filter feeder: Humpback whale Megaptera novaeangliae. Mar Pollut Bull 95:248–252. https://doi.org/10.1016/j.marpolbul.2015.04.007

Bhatawadekar VC, Damare SR, Garg A (2023) Risk assessment of polycyclic aromatic hydrocarbon infested sediments along the coast of Goa, India. Environ Sci Pollut Res 30:93114–93125. https://doi.org/10.1007/s11356-023-28838-6

Bianco F, Monteverde G, Race M et al (2020) Comparing performances, costs and energy balance of ex situ remediation processes for PAH-contaminated marine sediments. Environ Sci Pollut Res 27:19363–19374. https://doi.org/10.1007/s11356-020-08379-y

Bilal M, Rasheed T, Sosa-Hernández JE et al (2018) Biosorption: an interplay between marine algae and potentially toxic elements—a review. Mar Drugs 16:65. https://doi.org/10.3390/md16020065

Boldrocchi G, Monticelli D, Butti L et al (2020) First concurrent assessment of elemental- and organic-contaminant loads in skin biopsies of whale sharks from Djibouti. Sci Total Environ 722:137841. https://doi.org/10.1016/j.scitotenv.2020.137841

Brakstad OG, Ribicic D, Winkler A, Netzer R (2018) Biodegradation of dispersed oil in seawater is not inhibited by a commercial oil spill dispersant. Mar Pollut Bull 129:555–561. https://doi.org/10.1016/j.marpolbul.2017.10.030

Brandon AM, Gao SH, Tian R et al (2018) Biodegradation of polyethylene and plastic mixtures in mealworms (larvae of Tenebrio molitor) and effects on the gut microbiome. Environ Sci Technol 52:6526–6533. https://doi.org/10.1021/acs.est.8b02301

Broszeit S, Hattam C, Beaumont N (2016) Bioremediation of waste under ocean acidification: reviewing the role of Mytilus edulis. Mar Pollut Bull 103:5–14. https://doi.org/10.1016/j.marpolbul.2015.12.040

Budiyanto F, Thukair A, Al-Momani M et al (2018) Characterization of halophilic bacteria capable of efficiently biodegrading the high-molecular-weight polycyclic aromatic hydrocarbon pyrene. Environ Eng Sci 35:616–626. https://doi.org/10.1089/ees.2017.0244

Budiyanto F, Harmesa L, Bayu A (2021) Bioelectrochemical system for managing the metal's issues. In: Singh L, Mahapatra DM, Thakur S (eds) Bioremediation, nutrients, and other valuable products recovery: using bioelectrochemical systems. Elsevier, Radarweg, pp 41–82

Budiyanto F, Harmesa L et al (2023) Metals profile in deep-sea sediment from an active tectonic region around Simeulue Island, Aceh, Indonesia. Mar Pollut Bull 192:114983. https://doi.org/10.1016/j.marpolbul.2023.114983

Cao D, Huo Y, Zhang L et al (2023) Aerobic biodegradation of phenanthrene by a newly isolated Klebsiella sp. DS-1 from wastewater. J Water Process Eng 53:103820. https://doi.org/10.1016/j.jwpe.2023.103820

Ceccopieri M, Farias CO, Araújo M et al (2023) Aliphatic and polycyclic aromatic hydrocarbons in surface sediments as a tool for the assessment of the contamination status of mangrove forests in Rio de Janeiro (Brazil). Mar Pollut Bull 192:115049. https://doi.org/10.1016/j.marpolbul.2023.115049

Chen Q, Li Z, Chen Y et al (2024) Effects of electron acceptors and donors on anaerobic biodegradation of PAHs in marine sediments. Mar Pollut Bull 199:115925. https://doi.org/10.1016/j.marpolbul.2023.115925

Chen L, Wang Y, Liu H et al (2025) Different fates of Sb(III) and Sb(V) during the formation of jarosite mediated by Acidithiobacillus ferrooxidans. J Environ Sci 147:342–358. https://doi.org/10.1016/j.jes.2023.12.006

Curl LF, Hurst SA, Pomory CM et al (2024) Assessing microplastics contamination in unviable loggerhead sea turtle eggs. Sci Total Environ 912:169434. https://doi.org/10.1016/j.scitotenv.2023.169434

da Silva MF, Zanardi-Lamardo E, Rojas LAV et al (2024) Traces of oil in sea turtle feces. Mar Pollut Bull 200:116088. https://doi.org/10.1016/j.marpolbul.2024.116088

Dai X, Lv J, Yan G et al (2020) Bioremediation of intertidal zones polluted by heavy oil spilling using immobilized laccase-bacteria consortium. Bioresour Technol 309:123305. https://doi.org/10.1016/j.biortech.2020.123305

de Souza ES, Becker BR, Vieira-da-Motta O, de Almeida JC (2024) Oil spill bioremediation strategies in Brazilian tropical seawater—the role of polycyclic aromatic hydrocarbons degradation. Clean Soil Air Water 52:2300081. https://doi.org/10.1002/clen.202300081

de Stephanis R, Giménez J, Carpinelli E et al (2013) As main meal for sperm whales: plastics debris. Mar Pollut Bull 69:206–214. https://doi.org/10.1016/j.marpolbul.2013.01.033

De J, Ramaiah N, Vardanyan L (2008) Detoxification of toxic heavy metals by marine bacteria highly resistant to mercury. Mar Biotechnol 10:471–477. https://doi.org/10.1007/s10126-008-9083-z

Debroas D, Mone A, Halle A (2017) Plastics in the North Atlantic garbage patch: a boat-microbe for hitchhikers and plastic degraders. Sci Total Environ 599:1222–1232. https://doi.org/10.1016/j.scitotenv.2017.05.059

Delacuvellerie A, Cyriaque V, Gobert S et al (2019) The plastisphere in marine ecosystem hosts potential specific microbial degraders including Alcanivorax borkumensis as a key player for the low-density polyethylene degradation. J Hazard Mater 380:120899. https://doi.org/10.1016/j.jhazmat.2019.120899

Denuncio P, Bastida R, Dassis M et al (2011) Plastic ingestion in Franciscana dolphins, Pontoporia blainvillei (Gervais and d'Orbigny, 1844), from Argentina. Mar Pollut Bull 62:1836–1841. https://doi.org/10.1016/j.marpolbul.2011.05.003

Devi RS, Kannan VR, Nivas D et al (2015) Biodegradation of HDPE by Aspergillus spp. from marine ecosystem of Gulf of Mannar. Mar Pollut Bull 96:32–40. https://doi.org/10.1016/j.marpolbul.2015.05.050

Driskell WB, Payne JR (2018) Macondo oil in northern Gulf of Mexico waters – part 2: dispersant-accelerated PAH dissolution in the deepwater horizon plume. Mar Pollut Bull 129:412–419. https://doi.org/10.1016/j.marpolbul.2018.02.057

Fan W, Jin J, Zhang Z et al (2022) Degradation of phenanthrene by consortium 5H under hypersaline conditions. Environ Pollut 308:119730. https://doi.org/10.1016/j.envpol.2022.119730

Feitosa AF, Menezes ÍB, Duarte OS et al (2024) The impact of chronic and acute problems on sea turtles: the consequences of the oil spill and ingestion of anthropogenic debris on the tropical semi-arid coast of Ceará, Brazil. Aquat Toxicol 269:106867. https://doi.org/10.1016/j.aquatox.2024.106867

Fiard M, Cuny P, Sylvi L et al (2022) Mangrove microbiota along the urban-to-rural gradient of the Cayenne estuary (French Guiana, South America): drivers and potential bioindicators. Sci Total Environ 807:150667. https://doi.org/10.1016/j.scitotenv.2021.150667

Flores-Ramírez R, Mendoza-Rivera SP, García-Grajales J et al (2024) Persistent organic pollutants in the olive ridley turtle (Lepidochelys olivacea) during the nesting stage in the "La Escobilla" Sanctuary, Oaxaca, Mexico. Environ Sci Pollut Res 31:10911–10919. https://doi.org/10.1007/s11356-024-31833-0

Foord CS, Szabo D, Robb K et al (2024) Hepatic concentrations of per- and polyfluoroalkyl substances (PFAS) in dolphins from south-east Australia: highest reported globally. Sci Total Environ 908:168438. https://doi.org/10.1016/j.scitotenv.2023.168438

Fragkou E, Antoniou E, Daliakopoulos I et al (2021) In situ aerobic bioremediation of sediments polluted with petroleum hydrocarbons: a critical review. J Mar Sci Eng 9:1003. https://doi.org/10.3390/jmse9091003

Ganesh Kumar A, Anjana K, Hinduja M et al (2020) Review on plastic wastes in marine environment – biodegradation and biotechnological solutions. Mar Pollut Bull 150:110733. https://doi.org/10.1016/j.marpolbul.2019.110733

Garcia-Garin O, Borrell A, Colomer-Vidal P et al (2023) Biomagnification and temporal trends (1990–2021) of perfluoroalkyl substances in striped dolphins (Stenella coeruleoalba) from the NW Mediterranean sea. Environ Pollut 339:122738. https://doi.org/10.1016/j.envpol.2023.122738

Geng X, Barker CH, MacFadyen A et al (2022) Oil biodegradation in permeable marine sediments: effects of benthic pore-water advection and solute exchange. J Hazard Mater 436:129211. https://doi.org/10.1016/j.jhazmat.2022.129211

Geng Q, Zou L, Liu H et al (2024) Influence of humic acid on the bioaccumulation, elimination, and toxicity of PFOS and TBBPA co-exposure in Mytilus unguiculatus Valenciennes. Sci Total Environ 923:171358. https://doi.org/10.1016/j.scitotenv.2024.171358

Geyer R, Jambeck JR, Law KL (2017) Production, use, and fate of all plastics ever made. Sci Adv 3:e1700782. https://doi.org/10.1126/sciadv.1700782

Gola D, Kumar Tyagi P, Arya A et al (2021) The impact of microplastics on marine environment: a review. Environ Nanotechnol Monit Manage 16:100552. https://doi.org/10.1016/j.enmm.2021.100552

Green DS (2016) Effects of microplastics on European flat oysters, Ostrea edulis and their associated benthic communities. Environ Pollut 216:95–103. https://doi.org/10.1016/j.envpol.2016.05.043

Gutierrez T, Morris G, Ellis D et al (2018) Hydrocarbon-degradation and MOS-formation capabilities of the dominant bacteria enriched in sea surface oil slicks during the Deepwater Horizon oil spill. Mar Pollut Bull 135:205–215. https://doi.org/10.1016/j.marpolbul.2018.07.027

Gwynn JP, Hatje V, Casacuberta N et al (2024) The effect of climate change on sources of radionuclides to the marine environment. Commun Earth Environ 5:135. https://doi.org/10.1038/s43247-024-01241-w

Harrison JP, Schratzberger M, Sapp M, Osborn AM (2014) Rapid bacterial colonization of low-density polyethylene microplastis in coastal sediment microcosms. BMC Microbiol 14:232. https://doi.org/10.1186/s12866-014-0232-4

Harshvardhan K, Jha B (2013) Biodegradation of low-density polyethylene by marine bacteria from pelagic waters, Arabian Sea, India. Mar Pollut Bull 77:100–106. https://doi.org/10.1016/j.marpolbul.2013.10.025

Hernandez-Gonzalez A, Saavedra C, Gago J et al (2018) Microplastics in the stomach contents of common dolphin (Delphinus delphis) stranded on the Galician coasts (NW Spain, 2005–2010). Mar Pollut Bull 137:526–532. https://doi.org/10.1016/j.marpolbul.2018.10.026

Horiike T, Dotsuta Y, Nakano Y et al (2017) Removal of soluble strontium via incorporation into biogenic carbonate minerals by halophilic bacterium Bacillus sp. strain TK2d in a highly saline solution. Appl Environ Microbiol 83:e00855. https://doi.org/10.1128/AEM.00855-17

Hoydal KS, Erenbjerg SV, Dam M (2024) Mercury speciation in pilot whale from Faroe Islands, 1977–2015. Environ Pollut 356:124243. https://doi.org/10.1016/j.envpol.2024.124243

Hu S, Jiang L, Jiang L et al (2024) Spatial distribution characteristics of carbazole and polyhalogenated carbazoles in water column and sediments in the open Western Pacific Ocean. J Hazard Mater 469:133956. https://doi.org/10.1016/j.jhazmat.2024.133956

Huang J, Yang C, Zhang X et al (2024) Isolation of Micrococcus sp. strain 2A from the ocean and its biodegradation of 1-naphthol in wastewater. Int Biodeterior Biodegrad 192:105815. https://doi.org/10.1016/j.ibiod.2024.105815

Huerta Lwanga E, Thapa B, Yang X et al (2018) Decay of low-density polyethylene by bacteria extracted from earthworm's guts: a potential for soil restoration. Sci Total Environ 624:753–757. https://doi.org/10.1016/j.scitotenv.2017.12.144

Ibrar M, Yang X (2022) Reconstructing polyaromatic hydrocarbons degrading pathways in the enriched bacterial consortium and their biosurfactants characterization. J Environ Chem Eng 10:107219. https://doi.org/10.1016/j.jece.2022.107219

Janssen PH, Yates PS, Grinton BE et al (2002) Improved culturability of soil bacteria and isolation in pure culture of novel members of the divisions acidobacteria, actinobacteria, proteobacteria, and verrucomicrobia. Appl Environ Microbiol 68:2391–2396. https://doi.org/10.1128/aem.68.5.2391-2396.2002

Jiang Y, Zhang Z, Zhang X (2018) Co-biodegradation of pyrene and other PAHs by the bacterium Acinetobacter johnsonii. Ecotoxicol Environ Saf 163:465–470. https://doi.org/10.1016/j.ecoenv.2018.07.065

Jolicoeur V, Houde M, Loseto L et al (2024) Variations in thyroid hormone levels in endangered St. Lawrence Estuary belugas: potential linkage with stress and organohalogen contaminant exposure. Environ Int 186:108647. https://doi.org/10.1016/j.envint.2024.108647

Joo J, Choi J-H, Kim IH et al (2015) Effective bioremediation of Cadmium (II), nickel (II), and chromium (VI) in a marine environment by using Desulfovibrio desulfuricans. Biotechnol Bioprocess Eng 20:937–941. https://doi.org/10.1007/s12257-015-0287-6

Kaewsarn P, Yu Q (2001) Cadmium(II) removal from aqueous solutions by pre-treated biomass of marine alga Padina sp. Environ Pollut 112:209–213. https://doi.org/10.1016/S0269-7491(00)00114-7

Kamala K, Sivaperumal P, Thilagaraj R, Natarajan E (2020) Bioremediation of Sr^{2+} ion radionuclide by using marine Streptomyces sp. CuOff24 extracellular polymeric substances. J Chem Technol Biotechnol 95:893–903. https://doi.org/10.1002/jctb.6018

Kampouris ID, Gründger GF, Christensen JH et al (2023) Long-term patterns of hydrocarbon biodegradation and bacterial community composition in epipelagic and mesopelagic zones of an Arctic fjord. J Hazard Mater 446:130656. https://doi.org/10.1016/j.jhazmat.2022.130656

Kamyabi A, Nouri H, Moghimi H (2018) Characterization of pyrene degradation and metabolite identification by Basidioascus persicus and mineralization enhancement with bacterial-yeast co-culture. Ecotoxicol Environ Saf 163:471–477. https://doi.org/10.1016/j.ecoenv.2018.07.098

Kannan K, Sivaperumal P (2023) Marine halophilic archaeal isolate Halococcus sp. with emphasis on bioremediation of radionuclides through extracellular biomolecules. J Chem Technol Biotechnol. https://doi.org/10.1002/jctb.7452

Kawakami SK, Mello L, Dergan AL et al (2023) Sedimentary PAH and oxidative stress biomarkers responses on Namalycastis abiuma (Polychaeta: Nereididae) from an urbanized Amazon estuary. Reg Stud Mar Sci 57:102757. https://doi.org/10.1016/j.rsma.2022.102757

Khalil CA, Fortin N, Prince RC et al (2021) Crude oil biodegradation in upper and supratidal seashores. J Hazard Mater 416:125919. https://doi.org/10.1016/j.jhazmat.2021.125919

Kim G, Do Choi S, Lim JM, Kim H (2023) Strontium-90 levels in seawater southeast of Jeju Island during 2021–2023. Mar Pollut Bull 193:115258. https://doi.org/10.1016/j.marpolbul.2023.115258

Krishnan SJA, Nishanth P, Venkatraman A et al (2024) Investigating radiation profiles and conducting risk assessment for ecological and environmental components on Vaan and Koswari islands in the Tuticorin group of islands, India. Environ Monit Assess 196:139. https://doi.org/10.1007/s10661-024-12312-y

Kumar AG, Manisha D, Rajan NN et al (2023) Biodegradation of phenanthrene by piezotolerant Bacillus subtilis EB1 and genomic insights for bioremediation. Mar Pollut Bull 194:115151. https://doi.org/10.1016/j.marpolbul.2023.115151

Kumari A, Chaudhary DR, Jha B (2019) Destabilization of polyethylene and polyvinylchloride structure by marine bacterial strain. Environ Sci Pollut Res 26:1507–1516. https://doi.org/10.1007/s11356-018-3465-1

Kumari S, Singh K, Kushwaha P, Suresh Kumar K (2022) Functional and biochemical properties of some economically important edible seaweeds. Curr Res Nutr Food Sci 10:802–816. https://doi.org/10.12944/CRNFSJ.10.2.32

Lee DW, Lee H, Lee AH et al (2018) Microbial community composition and PAHs removal potential of indigenous bacteria in oil contaminated sediment of Taean coast, Korea. Environ Pollut 234:503–512. https://doi.org/10.1016/j.envpol.2017.11.097

Lee K, Raverty S, Cottrell P et al (2023) Polycyclic aromatic hydrocarbon (PAH) source identification and a maternal transfer case study in threatened killer whales (Orcinus orca) of British Columbia, Canada. Sci Rep 13:22580. https://doi.org/10.1038/s41598-023-45306-w

Leng Q, Mu J, Yang G (2021) Efficient anaerobic bioremediation of high-concentration benzo[a]pyrene in marine environments. Environ Pollut 284:117210. https://doi.org/10.1016/j.envpol.2021.117210

Lestari, Budiyanto F, Hindarti D (2018) Speciation of heavy metals Cu, Ni and Zn by modified BCR sequential extraction procedure in sediments from Banten Bay, Banten Province, Indonesia. IOP Conf Ser Earth Environ Sci 118:012059

Lestari L, Harmesa H, Taufiqurrahman E et al (2021) Assessment of potential variability of cadmium and copper trace metals using hindcast estimates. Environ Monit Assess 193:705. https://doi.org/10.1007/s10661-021-09501-4

Li X, Cao X, Zhang Z et al (2024) Mechanism of phenanthrene degradation by the halophilic Pelagerythrobacter sp. N7. Chemosphere 350:141175. https://doi.org/10.1016/j.chemosphere.2024.141175

Liang C, Ye Q, Huang Y et al (2023a) Distribution of the new functional marker gene (pahE) of aerobic polycyclic aromatic hydrocarbon (PAHs) degrading bacteria in different ecosystems. Sci Total Environ 865:161233. https://doi.org/10.1016/j.scitotenv.2022.161233

Liang R, Davidova IA, Teske A, Suflita JM (2023b) Evidence for the anaerobic biodegradation of higher molecular weight hydrocarbons in the Guaymas Basin. Int Biodeterior Biodegrad 181:105621. https://doi.org/10.1016/j.ibiod.2023.105621

Lofthus S, Almås IK, Evans P et al (2018) Biodegradation in seawater of PAH and alkylphenols from produced water of a North Sea platform. Chemosphere 206:465–473. https://doi.org/10.1016/j.chemosphere.2018.05.006

Luís LG, Ferreira P, Fonte E et al (2015) Does the presence of microplastics influence the acute toxicity of chromium(VI) to early juveniles of the common goby (Pomatoschistus microps)? A study with juveniles from two wild estuarine populations. Aquat Toxicol 164:163–174. https://doi.org/10.1016/j.aquatox.2015.04.018

Luo D, Guo Y, Liu Z et al (2024) Endocrine-disrupting chemical exposure induces adverse effects on the population dynamics of the Indo-Pacific humpback dolphin. Environ Sci Technol 58:9102–9112. https://doi.org/10.1021/acs.est.4c00618

Ma N, Cai R, Sun C (2021) Threonine dehydratase enhances bacterial cadmium resistance via driving cysteine desulfuration and biomineralization of cadmium sulfide nanocrystals. J Hazard Mater 417:126102. https://doi.org/10.1016/j.jhazmat.2021.126102

Maderich V, Tsumune D, Bezhenar R, de With G (2024) A critical review and update of modelling of treated water discharging from Fukushima Daiichi NPP. Mar Pollut Bull 198:115901. https://doi.org/10.1016/j.marpolbul.2023.115901

Madsen EL (1991) Determining in situ available to subsequent generations. Environ Sci Technol 25:1662–1673

Marsili L, Consales G, Romano P et al (2023) A cocktail of plankton and organochlorines for whale shark in the foraging areas of Nosy Be (Madagascar). Diversity 15:911. https://doi.org/10.3390/d15080911

Martínez-Ávila L, Peidro-Guzmán H, Pérez-Llano Y et al (2021) Tracking gene expression, metabolic profiles, and biochemical analysis in the halotolerant basidiomycetous yeast Rhodotorula mucilaginosa EXF-1630 during benzo[a]pyrene and phenanthrene biodegradation under hypersaline conditions. Environ Pollut 271:116358. https://doi.org/10.1016/j.envpol.2020.116358

Martins CC, Moreira LB, Sutilli M, de Souza Abessa DM (2023) Unraveling sources of hydrocarbons in subtropical estuaries with distinct degrees of protection in the Southwestern Atlantic Ocean, Brazil. Mar Pollut Bull 195:115499. https://doi.org/10.1016/j.marpolbul.2023.115499

Matheickal JT, Yu Q, Woodburn GM (1999) Biosorption of cadmium(II) from aqueous solutions by pre-treated biomass of marine alga DurvillAea potatorum. Water Res 33:335–342. https://doi.org/10.1016/S0043-1354(98)00237-1

Mbusnum KG, Malleret L, Deschamps P et al (2020) Persistent organic pollutants in sediments of the Wouri Estuary Mangrove, Cameroon: levels, patterns and ecotoxicological significance. Mar Pollut Bull 160:111542. https://doi.org/10.1016/j.marpolbul.2020.111542

Meijer LJJ, van Emmerik T, van der Ent R et al (2021) More than 1000 rivers account for 80% of global riverine plastic emissions into the ocean. Sci Adv 7:5803. https://doi.org/10.1126/sciadv.aaz5803

Miao L, Gao Y, Adyel TM et al (2021) Effects of biofilm colonization on the sinking of microplastics in three freshwater environments. J Hazard Mater 413:125370. https://doi.org/10.1016/j.jhazmat.2021.125370

Mikelić IL, Oreščanin V, Cukrov N et al (2023) Relationships between radionuclides, metals, and sediment properties in sediment of a bay exposed to anthropogenic pressure and mixed sediment sources (Kaštela Bay, Adriatic Sea, Croatia). Mar Pollut Bull 197:115731. https://doi.org/10.1016/j.marpolbul.2023.115731

Mishra A, Rathour R, Singh R et al (2020) Degradation and detoxification of phenanthrene by actinobacterium Zhihengliuella sp. ISTPL4. Environ Sci Pollut Res 27:27256–27267. https://doi.org/10.1007/s11356-019-05478-3

Mohanrasu K, Premnath N, Siva Prakash G et al (2018) Exploring multi potential uses of marine bacteria; an integrated approach for PHB production, PAHs and polyethylene biodegradation. J Photochem Photobiol B Biol 185:55–65. https://doi.org/10.1016/j.jphotobiol.2018.05.014

Mohapatra RK, Parhi PK, Pandey S et al (2019) Active and passive biosorption of Pb(II)using live and dead biomass of marine bacterium Bacillus xiamenensis PbRPSD202: kinetics and isotherm studies. J Environ Manag 247:121–134. https://doi.org/10.1016/j.jenvman.2019.06.073

Morão IFC, Simões T, Casado RB et al (2024) Metal accumulation in female green sea turtles (Chelonia mydas) from Eastern Atlantic affects their egg quality with potential implications for embryonic development. Sci Total Environ 931:172710. https://doi.org/10.1016/j.scitotenv.2024.172710

Muralidharan M, Kavitha R, Kumar PS et al (2021) Performance evaluation and mechanism analysis of halotolerant bacterial strains and cerium oxide nanoparticle to degrade Benzo[a]pyrene. Environ Technol Innov 24:101980. https://doi.org/10.1016/j.eti.2021.101980

Muralidharan M, Gayathri KV, Kumar PS et al (2023) Mixed polyaromatic hydrocarbon degradation by halotolerant bacterial strains from marine environment and its metabolic pathway. Environ Res 216:114464. https://doi.org/10.1016/j.envres.2022.114464

Naik MM, Dubey SK (2017) Lead- and mercury-resistant marine bacteria and their application in lead and mercury bioremediation. In: Naik MM, Dubey SK (eds) Marine pollution and microbial remediation. Springer, Singapore, pp 29–40

Nelms SE, Duncan EM, Broderick AC et al (2015) Plastic and marine turtles: a review and call for research. ICES J Mar Sci 73:165–181. https://doi.org/10.1093/icesjms/fsv165

Nevin KP, Finneran KT, Lovley DR (2003) Microorganisms associated with uranium bioremediation in a high-salinity subsurface sediment. Appl Environ Microbiol 69:3672–3675. https://doi.org/10.1128/AEM.69.6.3672-3675.2003

Nkem BM, Halimoon N, Yusoff FM et al (2016) Isolation, identification and diesel-oil biodegradation capacities of indigenous hydrocarbon-degrading strains of Cellulosimicrobium cellulans and Acinetobacter baumannii from tarball at Terengganu beach, Malaysia. Mar Pollut Bull 107:261–268. https://doi.org/10.1016/j.marpolbul.2016.03.060

Noren DP, Johnson S, Boyd D et al (2024) The dynamics of persistent organic pollutant (POP) transfer from female bottlenose dolphins (Tursiops truncatus) to their calves during lactation. Sci Total Environ 907:167888. https://doi.org/10.1016/j.scitotenv.2023.167888

Nos D, Montalvo T, Cortés-Francisco N et al (2024) Sources of persistent organic pollutants and their physiological effects on opportunistic urban gulls. J Hazard Mater 465:133129. https://doi.org/10.1016/j.jhazmat.2023.133129

Nzila A, Jung BK, Kim M-C et al (2018) Complete genome sequence of the polyaromatic hydrocarbons biodegrading bacterium Idiomarina piscisalsi strain 10PY1A isolated from oil-contaminated soil. Korean J Microbiol 54:289–292. https://doi.org/10.7845/kjm.2018.8034

Okeyode AH, Al-Thukair A, Chanbasha B et al (2023) Degradation of the highly complex polycyclic aromatic hydrocarbon coronene by the halophilic bacterial strain Halomonas caseinilytica, 10SCRN4D. Arch Environ Prot 49:78–86. https://doi.org/10.24425/aep.2023.147330

Oliveira M, Ribeiro A, Hylland K, Guilhermino L (2013) Single and combined effects of microplastics and pyrene on juveniles (0+ group) of the common goby Pomatoschistus microps (Teleostei, Gobiidae). Ecol Indic 34:641–647. https://doi.org/10.1016/j.ecolind.2013.06.019

Oliveira-Ferreira N, Manhães BMR, Santos-Neto E et al (2024) Organohalogen compounds in a hotspot for chemical pollution: assessment in free-ranging Atlantic spotted dolphins (Stenella frontalis). Sci Total Environ 926:171912. https://doi.org/10.1016/j.scitotenv.2024.171912

Paço A, Duarte K, da Costa JP et al (2017) Biodegradation of polyethylene microplastics by the marine fungus Zalerion maritimum. Sci Total Environ 586:10–15. https://doi.org/10.1016/j.scitotenv.2017.02.017

Palmer N, Reichelt-Brushett A, Hall J et al (2024) Contaminant assessment of stranded and deceased beaked whales (Ziphiidae) on the New South Wales coast of Australia. Mar Pollut Bull 204:116520. https://doi.org/10.1016/j.marpolbul.2024.116520

Pan H, Yu T, Zheng Y et al (2024) Isolation, characteristics, and poly(butylene adipate-co-terephthalate) (PBAT) degradation mechanism of a marine bacteria Roseibium aggregatum ZY-1. Mar Pollut Bull 201:116261. https://doi.org/10.1016/j.marpolbul.2024.116261

Pancaldi F, Páez-Osuna F, Soto-Jiménez MF et al (2019) Trace elements in tissues of whale sharks (Rhincodon typus) stranded in the Gulf of California, Mexico. Bull Environ Contam Toxicol 103:515–520. https://doi.org/10.1007/s00128-019-02640-y

Pandey J, Chauhan A, Jain RK (2009) Integrative approaches for assessing the ecological sustainability of in situ bioremediation. FEMS Microbiol Rev 33:324–375. https://doi.org/10.1111/j.1574-6976.2008.00133.x

Paul MB, Stock V, Cara-Carmona J et al (2020) Micro- and nanoplastics-current state of knowledge with the focus on oral uptake and toxicity. Nanoscale Adv 2:4350–4367. https://doi.org/10.1039/d0na00539h

Pratush A, Kumar A, Hu Z (2018) Adverse effect of heavy metals (As, Pb, Hg, and Cr) on health and their bioremediation strategies: a review. Int Microbiol 21:97–106. https://doi.org/10.1007/s10123-018-0012-3

Pyke R, Fortin N, Wasserscheid J et al (2023) Biodegradation potential of residue generated during the in-situ burning of oil in the marine environment. J Hazard Mater 445:130439. https://doi.org/10.1016/j.jhazmat.2022.130439

Qian Z, Yang H, Li J et al (2024) The unique biodegradation pathway of benzo[a]pyrene in moderately halophilic Pontibacillus chungwhensis HN14. Chemosphere 354:141705. https://doi.org/10.1016/j.chemosphere.2024.141705

Qiu YW, Qiu HL, Li J, Zhang G (2018) Bioaccumulation and cycling of polycyclic aromatic hydrocarbons (PAHs) in typical mangrove wetlands of Hainan Island, South China. Arch Environ Contam Toxicol 75:464–475. https://doi.org/10.1007/s00244-018-0548-4

Raghul SS, Bhat SG, Chandrasekaran M et al (2014) Biodegradation of polyvinyl alcohol-low linear density polyethylene-blended plastic film by consortium of marine benthic vibrios. Int J Environ Sci Technol 11:1827–1834. https://doi.org/10.1007/s13762-013-0335-8

Rebolledo ELB, Van Franeker JA, Jansen OE, Brasseur SM (2013) Plastic ingestion by harbour seals (Phoca vitulina) in The Netherlands. Mar Pollut Bull 67:200–202. https://doi.org/10.1016/j.marpolbul.2012.11.035

Reynolds SD, Norman BM, Franklin CE et al (2022) Regional variation in anthropogenic threats to Indian Ocean whale sharks. Glob Ecol Conserv 33:e01961. https://doi.org/10.1016/j.gecco.2021.e01961

Richard H, Carpenter EJ, Komada T et al (2019) Biofilm facilitates metal accumulation onto microplastics in estuarine waters. Sci Total Environ 683:600–608. https://doi.org/10.1016/j.scitotenv.2019.04.331

Robin SL, Marchand C (2022) Polycyclic aromatic hydrocarbons (PAHs) in mangrove ecosystems: a review. Environ Pollut 311:119959. https://doi.org/10.1016/j.envpol.2022.119959

Rochman CM, Kurobe T, Flores I, Teh SJ (2014) Early warning signs of endocrine disruption in adult fish from the ingestion of polyethylene with and without sorbed chemical pollutants from the marine environment. Sci Total Environ 493:656–661. https://doi.org/10.1016/j.scitotenv.2014.06.051

Rogiers T, Van Houdt R, Williamson A et al (2022) Molecular mechanisms underlying bacterial uranium resistance. Front Microbiol 13:822197. https://doi.org/10.3389/fmicb.2022.822197

Santos-Neto EB, Manhães BMR, Oliveira-Ferreira N et al (2024) PAHs in franciscana dolphins from the Southwestern Atlantic Ocean: concentration and maternal transfer assessments. Mar Pollut Bull 203:116455. https://doi.org/10.1016/j.marpolbul.2024.116455

Sari A, Tuzen M (2008) Biosorption of cadmium(II) from aqueous solution by red algae (Ceramium virgatum): equilibrium, kinetic and thermodynamic studies. J Hazard Mater 157:448–454. https://doi.org/10.1016/j.jhazmat.2008.01.008

Saunders D, Carrillo JC, Gundlach ER et al (2022) Analysis of polycyclic aromatic hydrocarbons (PAHs) in surface sediments and edible aquatic species in an oil-contaminated mangrove ecosystem in Bodo, Niger Delta, Nigeria: bioaccumulation and human health risk assessment. Sci Total Environ 832:154802. https://doi.org/10.1016/j.scitotenv.2022.154802

Shelobolina ES, Vrionis HA, Findlay RH, Lovley DR (2008) Geobacter uraniireducens sp. nov., isolated from subsurface sediment undergoing uranium bioremediation. Int J Syst Evol Microbiol 58:1075–1078. https://doi.org/10.1099/ijs.0.65377-0

Sheng PX, Ting Y-P, Chen JP, Hong L (2004) Sorption of lead, copper, cadmium, zinc, and nickel by marine algal biomass: characterization of biosorptive capacity and investigation of mechanisms. J Colloid Interface Sci 275:131–141. https://doi.org/10.1016/j.jcis.2004.01.036

Shivanand P, Mugeraya G (2011) Halophilic bacteria and their compatible solutes – osmoregulation and potential applications. Curr Sci 100:1516–1521

Singh L, Wahid ZA (2015) Methods for enhancing bio-hydrogen production from biological process: a review. J Ind Eng Chem 21:70–80. https://doi.org/10.1016/j.jiec.2014.05.035

Sivaperumal P, Kamala K, Subramanian K et al (2022) Baseline assessment of marine actinobacterial diversity around the nuclear power plant sites, India and its application to uranium remediation. Environ Res 212:113135. https://doi.org/10.1016/j.envres.2022.113135

Stockholm Convention (2019) All POPs listed in the Stockhom Convention. https://www.pops.int/TheConvention/

Stockin KA, Law RJ, Roe WD et al (2010) PCBs and organochlorine pesticides in Hector's (Cephalorhynchus hectori hectori) and Maui's (Cephalorhynchus hectori maui) dolphins. Mar Pollut Bull 60:834–842. https://doi.org/10.1016/j.marpolbul.2010.01.009

Subashchandrabose SR, Venkateswarlu K, Naidu R, Megharaj M (2019) Biodegradation of high-molecular weight PAHs by Rhodococcus wratislaviensis strain 9: overexpression of amidohydrolase induced by pyrene and BaP. Sci Total Environ 651:813–821. https://doi.org/10.1016/j.scitotenv.2018.09.192

Sudhakar M, Priyadarshini C, Doble M et al (2007) Marine bacteria mediated degradation of nylon 66 and 6. Int Biodeterior Biodegrad 60:144–151. https://doi.org/10.1016/j.ibiod.2007.02.002

Sun S, Wang H, Chen Y et al (2019) Salicylate and phthalate pathways contributed differently on phenanthrene and pyrene degradations in Mycobacterium sp. WY10. J Hazard Mater 364:509–518. https://doi.org/10.1016/j.jhazmat.2018.10.064

Sun Y, Shi M, Lu T et al (2021) Bio-removal of $PtCl_6^{2-}$ complex by Galdieria sulphuraria. Sci Total Environ 796:149021. https://doi.org/10.1016/j.scitotenv.2021.149021

Sussarellu R, Suquet M, Thomas Y et al (2016) Oyster reproduction is affected by exposure to polystyrene microplastics. Proc Natl Acad Sci 113:2430–2435

Syranidou E, Karkanorachaki K, Amorotti F et al (2017) Development of tailored indigenous marine consortia for the degradation of naturally weathered polyethylene films. PLoS One 12:e0183984

Syranidou E, Karkanorachaki K, Amorotti F et al (2019) Biodegradation of mixture of plastic films by tailored marine consortia. J Hazard Mater 375:33–42. https://doi.org/10.1016/j.jhazmat.2019.04.078

Tan K, Cai X, Tan K, Kwan KY (2023) A review of natural and anthropogenic radionuclide pollution in marine bivalves. Sci Total Environ 896:165030. https://doi.org/10.1016/j.scitotenv.2023.165030

Tedesco P, Balzano S, Coppola D et al (2024) Bioremediation for the recovery of oil polluted marine environment, opportunities and challenges approaching the Blue Growth. Mar Pollut Bull 200:116157. https://doi.org/10.1016/j.marpolbul.2024.116157

Teien HC, Wada T, Kashparov V et al (2023) Transfer of ^{129}I to freshwater fish species within Fukushima and Chernobyl exclusion zones. J Environ Radioact 270:107269. https://doi.org/10.1016/j.jenvrad.2023.107269

Thakur A, Kumar A (2024) Emerging paradigms into bioremediation approaches for nuclear contaminant removal: from challenge to solution. Chemosphere 352:141369. https://doi.org/10.1016/j.chemosphere.2024.141369

Thuy HTT, Loan TTC, Luu PT et al (2021) Spatial and temporal variations of PAHs in surface sediments of estuarine and coast of CanGio wetland, Vietnam. Environ Sci Pollut Res 28:11962–11975. https://doi.org/10.1007/s11356-020-08523-8

Tsumune D, Bryan FO, Lindsay K et al (2023) Simulated inventory and distribution of 137Cs released from multiple sources in the global ocean. Mar Pollut Bull 197:115663. https://doi.org/10.1016/j.marpolbul.2023.115663

Ugwu K, Herrera A, Gómez M (2021) Microplastics in marine biota: a review. Mar Pollut Bull 169:112540. https://doi.org/10.1016/j.marpolbul.2021.112540

Ulyantsev A, Ivannikov S, Bratskaya S, Charkin A (2023) Radioactivity of anthropogenic and natural radionuclides in marine sediments of the Chaun Bay, East Siberian Sea. Mar Pollut Bull 195:115582. https://doi.org/10.1016/j.marpolbul.2023.115582

Vaksmaa A, Vielfaure H, Polerecky L et al (2024) Biodegradation of polyethylene by the marine fungus Parengyodontium album. Sci Total Environ 934:172819. https://doi.org/10.1016/j.scitotenv.2024.172819

Vandana, Das S (2023) Cell surface hydrophobicity and petroleum hydrocarbon degradation by biofilm-forming marine bacterium Pseudomonas furukawaii PPS-19 under different physico-chemical stressors. J Hazard Mater 457:131795. https://doi.org/10.1016/j.jhazmat.2023.131795

Vergeynst L, Christensen JH, Kjeldsen KU et al (2019) In situ biodegradation, photooxidation and dissolution of petroleum compounds in Arctic seawater and sea ice. Water Res 148:459–468. https://doi.org/10.1016/j.watres.2018.10.066

Verzele P, Rodriguez CB, Bezerra MF, Lacerda LD (2024) Regional differences in Hg contents in the liver and muscle of the green turtle (Chelonia mydas) from three foraging grounds in Northeast Brazil. Environ Sci Pollut Res 31:3917–3925. https://doi.org/10.1007/s11356-023-31407-6

Viel T, Manfra L, Zupo V et al (2023) Biodegradation of plastics induced by marine organisms: future perspectives for bioremediation approaches. Polymers 15:2673

Vigneron A, Guyoneaud R, Goñi-Urriza M (2024) Genome-centric metatranscriptomics reveals multiple co-occurring routes for hydrocarbon degradation in chronically contaminated marine microbial mats. Environ Sci Technol 58:1551–1562. https://doi.org/10.1021/acs.est.3c08386

Villagómez-Vélez SI, Noreña-Barroso E, Galván-Magaña F et al (2024) Persistent organic pollutants in whale shark (Rhincodon typus) skin biopsies from Bahía de Los Ángeles, Mexico. Bull Environ Contam Toxicol 112:18. https://doi.org/10.1007/s00128-023-03841-2

Walton JL, Buchan A (2024) Evidence for novel polycyclic aromatic hydrocarbon degradation pathways in culturable marine isolates. Microbiol Spectr 12:e0340923. https://doi.org/10.1128/spectrum.03409-23

Wang J, Chen C (2009) Biosorbents for heavy metals removal and their future. Biotechnol Adv 27:195–226. https://doi.org/10.1016/j.biotechadv.2008.11.002

Wang C, Huang Y, Zhang Z, Wang H (2018) Salinity effect on the metabolic pathway and microbial function in phenanthrene degradation by a halophilic consortium. AMB Express 8:67. https://doi.org/10.1186/s13568-018-0594-3

Wang C, He S, Zou Y et al (2020) Quantitative evaluation of in-situ bioremediation of compound pollution of oil and heavy metal in sediments from the Bohai Sea, China. Mar Pollut Bull 150:110787. https://doi.org/10.1016/j.marpolbul.2019.110787

Wang M, Sha C, Wu J et al (2021) Bacterial community response to petroleum contamination in brackish tidal marsh sediments in the Yangtze River Estuary, China. J Environ Sci 99:160–167. https://doi.org/10.1016/j.jes.2020.06.015

Werth AJ, Kahane-Rapport SR, Potvin J et al (2024) Baleen–plastic interactions reveal high risk to all filter-feeding whales from clogging, ingestion, and entanglement. Oceans 5:48–70. https://doi.org/10.3390/oceans5010004

Xia C, Cai L, Lam SS, Sonne C (2023) Microplastics pollution: economic loss and actions needed. Eco-Environ Health 2:41–42. https://doi.org/10.1016/j.eehl.2023.04.001

Yaghmour F, Al Bousi M, Al Naqbi H et al (2021) Junk food: interspecific and intraspecific distinctions in marine debris ingestion by marine turtles. Mar Pollut Bull 173:113009. https://doi.org/10.1016/j.marpolbul.2021.113009

Yan Y, Bao K, Zhao K et al (2023) A baseline study of polycyclic aromatic hydrocarbons distribution, source and ecological risk in Zhanjiang mangrove wetlands, South China. Ecotoxicol Environ Saf 249:114437. https://doi.org/10.1016/j.ecoenv.2022.114437

Yang J, Yang Y, Wu W-M et al (2014) Evidence of polyethylene biodegradation by bacterial strains from the guts of plastic-eating waxworms. Environ Sci Technol 48:13776–13784. https://doi.org/10.1021/es504038a

Yin K, Wang Q, Lv M, Chen L (2019) Microorganism remediation strategies towards heavy metals. Chem Eng J 360:1553–1563. https://doi.org/10.1016/j.cej.2018.10.226

Yu Y, Yin H, Peng H et al (2020) Proteomic mechanism of decabromodiphenyl ether (BDE-209) biodegradation by Microbacterium Y2 and its potential in remediation of BDE-209 contaminated water-sediment system. J Hazard Mater 387:121708. https://doi.org/10.1016/j.jhazmat.2019.121708

Yuan J, Ma J, Sun Y et al (2020) Microbial degradation and other environmental aspects of microplastics/plastics. Sci Total Environ 715:136968. https://doi.org/10.1016/j.scitotenv.2020.136968

Zan S, Wang J, Wang F et al (2022) A novel degradation mechanism of naphthenic acids by marine Pseudoalteromonas sp. J Hazard Mater 424:127534. https://doi.org/10.1016/j.jhazmat.2021.127534

Zhang W, Chen L, Liu D (2012) Characterization of a marine-isolated mercury-resistant Pseudomonas putida strain SP1 and its potential application in marine mercury reduction. Appl Microbiol Biotechnol 93:1305–1314. https://doi.org/10.1007/s00253-011-3454-5

Zhang Z, Sun J, Gong X et al (2023) Novel synergistic metabolic processes for phenanthrene biodegradation by a nitrate-reducing phenanthrene-degrading culture and an anammox culture. Water Res 230:119593. https://doi.org/10.1016/j.watres.2023.119593

Zhang T, Jiang Y, Li D et al (2024) Accumulation rate and sources of plastic marine litter at nesting grounds of green turtles on the North Island of Qilianyu, Xisha Islands, South China Sea. Mar Pollut Bull 203:116485. https://doi.org/10.1016/j.marpolbul.2024.116485

Zhou H, Jiang L, Li K et al (2021) Enhanced bioremediation of diesel oil-contaminated seawater by a biochar-immobilized biosurfactant-producing bacteria Vibrio sp. LQ2 isolated from cold seep sediment. Sci Total Environ 793:148529. https://doi.org/10.1016/j.scitotenv.2021.148529

Zhou Y, Wang Y, Yang L et al (2023) Microbial degradation mechanisms of surface petroleum contaminated seawater in a typical oil trading port. Environ Pollut 324:121420. https://doi.org/10.1016/j.envpol.2023.121420

Zhu X, Zhong Y, Wang H et al (2019) New insights into the anaerobic microbial degradation of decabrominated diphenyl ether (BDE-209) in coastal marine sediments. Environ Pollut 255:113151. https://doi.org/10.1016/j.envpol.2019.113151

Zuo ZC, Zhang L, Ni J et al (2024) Occurrence of halogenated organic contaminants in surface sediments of the Yangtze River estuary and its adjacent marine area. Environ Res 251:118579. https://doi.org/10.1016/j.envres.2024.118579

Artificial Intelligence as Bioremediation Predictive Tool

16

Arwa Mo'men El-Habbaa

Abstract

Currently bioinformatics and artificial intelligence (AI) can assist in the development and application of bioremediation, as microorganisms display a wide range of contaminant degradation abilities that can efficiently and effectively bring back natural environmental conditions.

With the availability of many types of AI algorithms, it has become familiar for researchers to apply the off-shelf systems to classify and mine their database. AI offers an advanced toolbox that better facilitates problem-solving in the field. The current trends in AI are; machine learning methods which help creating more efficient, reliable, accurate neural networks and tools for determining natural and xenobiotic compounds' structures and biodegradative pathways.

This chapter aims to illustrate the relation between bioremediation and bioinformatics through machine learning models that would introduce bioremediation as a predictive tool.

Keywords

Bioinformatics · Bioremediation · Machine learning · Artificial intelligence (AI)

A. Mo'men El-Habbaa (✉)
Colors Educational Hub, Cairo, Egypt

A. A. H. Abdel Latef et al. (eds.), *Bioremediation and Nanotechnology for Climate Change Mitigation*, https://doi.org/10.1007/978-981-96-3069-1_16

16.1 Introduction

16.1.1 Bioinformatics

Bioinformatics is an interdisciplinary science that concerns many other sciences like biology, biostatistics, biochemistry, physics, computational biology, and information technology; also it focuses on both cellular and molecular levels to be applied in the modern biotechnology field. The word "bioinformatics" was initially introduced in 1970 by Dutch system-biologist Paulien Hogeweg to describe the use of information (Mehmood et al. 2014). However, bioinformatics is a new field that helps with data analysis from next-generation sequencing different platforms. It began more than half a century ago, when DNA sequencing was only a pipe dream to all of us and we never thought that we would ever reveal its sequence and more of the DNA hidden messages which helped us in many clinical and non-clinical aspects. Also, at that time desktop computers were still a theory to be used on the daily basis like nowadays. Bioinformatics also known as either a single procedure or a combination of procedures is required to analyze huge amounts of high throughput data generated from many omic-techniques. Omics requisite the employment of methods with high sensitivity and specificity (Schneider and Orchard 2011).

The term "omics" refers to a growing number of fields, including genomics (the quantitative study of protein coding genes, regulatory elements, and noncoding sequences), transcriptomics (RNA, splice variant and gene expression), proteomics (e.g., focusing on protein abundance and characterization), and metabolomics (metabolites and metabolic networks) to enhance the era of post-genomic biology and medicine, such as pharmacogenomics (the quantitative study of how genetics affects a host response to drugs) and physiomics (physiological dynamics and functions of whole organisms), and nutrigenomics (a rapidly growing discipline that well focus on identifying the genetic factors that impact the body's response to diet and studies how the bioactive constituents of food affect gene expression), and phylogenomics (analysis involving genome data and evolutionary reconstructions, especially phylogenetics) and interactomics (molecular interaction networks) (Schneider and Orchard 2011). In other fields like "viromics", it is the study of the structure of the viruses also its full genome nucleotide sequences and its functional genes as well as "non-coding" regions of the viral genome. The term viromics was used not only in relation to human viral infections but it also refers to the characterization of the virome in environmental niches (Ramamurthy et al. 2017).

16.1.2 Bioremediation

Utilizing microorganisms and their by-products is the simplest definition for bioremediation. Also, it can be described as; a technique that help to clean contaminated soil or water or any ecological system by either kill or immobilize waste materials. It is possible to use this technology in situ or ex situ. Since the contaminants must be removed and transported to other places for treatment, ex situ treatment typically entails a greater expense. In situ treatment takes place on the contamination site itself; hence, no site excavation or transportation is needed. To simplify the logistics processes necessary, so we can lower the cost of the bioremediation process, making it more practical and requiring less meddling at the site (Coelho et al. 2020).

The bioremediation process is done through many strategies: (1) mineralization, (2) transformation, or (3) modification; this detoxification procedure targets the dangerous compounds (Adams et al. 2015).

Let's have a look at the mineralization process for example.

16.1.3 Mineralization

The textile industry which is a main source of various liquid and solid wastes. It was reported that it requires almost 70–150 dm^3 of water and 40 g of reactive dyes per kg of cotton as it was estimated that more than 80,000 tons/year of dyes are been used in the textile dying processes (Méndez-Martínez et al. 2012). When these pollutants enter water streams, they provide a serious ecotoxic risk and raise the possibility of bioaccumulation, which could eventually have an impact on people through the food chain (Mohana et al. 2008). The mineralization includes a process which called anaerobic-microaerophilic process; it is considered as green technology for industrial wastewater, under anerobic conditions and mineralization in hydrophilic reactor dye molecules are cleaved through synthrophic interactions of consortium BDN. This procedure has reached a great precent of efficiency 97% of color and other pollutants from textile wastewater. Recent studies showed that dye mineralization process is dependent on enzyme regulation done by certain bacteria when there are many mutagenic intermediates during anaerobic process (Balapure et al. 2016).

16.2 Artificial Intelligence

Artificial intelligence (AI) has grown in importance as a research topic in the twenty-first century across almost all disciplines, including engineering, science, medicine, education, business, accounting, finance, marketing, economics, the stock market, and many other fields. The areas of artificial intelligence are classified into sixteen categories. These categories are reasoning, programming, artificial life, belief revision, data mining, distributed AI, expert systems, genetic algorithms, systems, knowledge representation, machine learning, natural language understanding, neural networks, theorem proving, constraint satisfaction, and theory of computation. It emerged as a technological wonder with enormous potential for environmental preservation. With AI's capacity to analyze massive volumes of data, spot trends, and make deft decisions instantly, we have an unmatched chance to completely transform how we approach sustainable growth and environmental preservation.

16.2.1 Let's Have a Look at Some of AI Applications in the Environmental Conversation

(a) Satellite imaging provides vital information for environmental monitoring. But in order to find relevant trends, the sheer amount of data collected necessitates complex processing capabilities. With the help of artificial intelligence (AI) technologies, such as convolutional neural networks (CNNs), it is now possible to detect deforestation, habitat loss, and land degradation with an unparalleled level of precision by extracting information from satellite photos (Rayhan 2023).

(b) In order to save endangered species from going extinct, monitoring is required but conventional approaches to tracking wildlife can be costly, time-consuming, and have a narrow focus. Camera trap image analysis, acoustic monitoring, and GPS tracking are AI applications in revolutionized wildlife conservation. This is done by identifying individual animals, calculating population sizes, and tracking movement patterns (Rayhan 2023).

(c) One of the most urgent issues facing the world today is climate change, which has an effect on ecosystems, human communities, and economic activity. Climate simulations driven by AI have shown their potential to enhance our comprehension of intricate climatic processes and raise the precision of long-range climate projections.

Furthermore, trends and patterns in previous climate data may be found using AI-driven algorithms, which make it easier to anticipate future situations and guide policy (Rayhan 2023).

16.3 Bioremidation as a Predictive Tool

With the great ecology concern nowadays bioinformaticians and AI have made a great progress in this field and introduced many pipelines and tools for analyzing many different types of omic-data like genomics, transcriptomics, and proteomics that may help in many clinical, industrial, and environmental issues (Tucker and Duplisea 2012), as the rate of environmental degradation currently experienced throughout the world. A large amount of it is caused by the increased usage and production of fossil fuels. Oil consumption and exploration pose a threat to the health of the people on all continents (Adams et al. 2015).

Strong incentives exist for bioremediation to be used more frequently as a successful method of reducing the risk associated with hydrocarbon-impacted soils. Anew approach is introduced to this area of research by shedding the light on the numerous biological techniques for assessing microbial performance as they were solely relying on the chemical data to assess the bioremediation. Once the researcher took the decision of applying bioremediation, a routine step should be performed which is tracking progress through measuring the decay of the target compounds or the appearance of metabolites or end products (Diplock et al. 2009). These biomarkers are set in order to forecast the onset of adverse health effects so that evaluation of these very first changes can be used to prevent long-term effects at the community and population levels. (Kumar et al. 2011)

The use of the in silico technique is crucial for predicting degradation routes and saves time and money compared to simple laboratory experiments.

Bioinformaticians developed many databases that might help in such area of interest; we have many databases like:

1. BioSurfDB is a database that links other tools to enable accurate metagenomics analysis while also providing information to support biosurfactants and biodegradation investigations.
2. The Carcinogenic Potency Database (CPDB), created by the Lawrence Berkeley National Laboratory and the University of California, Berkeley, includes the results of around 6450 long-term, chronic animal cancer analyses on 1547 different compounds (Kumari and Kumar 2021).
3. MetXBioDB is a database created for the purpose of building biotransformation guidelines, training, and validating machine learning-based metabolism prediction models and creating alternative guidelines (Kumari and Kumar 2021).
4. MetaCyc includes connected chemicals, genes, and enzymes as well as information and data about metabolic pathways that occur in primary and secondary metabolism. The database can provide relevant information for the following purposes as well: (1) anticipate an organism's metabolic pathways based on its sequenced genome; (2) facilitate metabolic engineering; (3) streamline the correlation of biochemical structure; and (4) serve as a book of metabolic knowledge (Kumari and Kumar 2021).

16.3.1 Some Recent Applications That Use AI in Prediction of Bioremediation

AI helped in creating predicting systems for the release of toxic chemicals that have been released at large scale that might not be easily detected. These systems/tools take into account a number of factors, such as the enzyme's binding site or reaction processes, the specificity of the substrate, structural changes in the combination of the substrate and product, and the gap between the substrate and product. Predicted pathways are based on both biochemical and non-biochemical procedures; the biochemically based prediction method operates according to the biotransformation rule. However, the non-biochemically procedure-based prediction approach uses statistical inference techniques to determine compound reactions (Soh and Hatzimanikatis 2010).

EAWAG-BBD/PPS has a pathway prediction system that uses the EAWAG-BBD database where the system can predict the biodegradation pathway based on available biotransformation guidelines. The University of Minnesota group led by Lawrence P. Wackett has established a new public database called RAPID, which may provide information about the thermodynamic utility of potential pathways or particular enzymes. Aukema and colleagues proposed the use of combined

EAWAG-BBD and RAPID algorithms. They used these tools to forecast the first metabolism of a variety of contaminants and chemicals, such as fragrance compounds, alkyl phthalates, and resistant medicines (Aukema et al. 2017). When it comes to man-made contaminants that lack natural catabolic routes or are not yet recognized, Biochemical Network Integrated Computational Explorer (BNICE) is another pathway prediction technique that can be utilized. The program creates every possible pathway for a specific chemical or target. In a subsequent phase, the instrument eliminated every possible route for thermodynamic utility by employing the reaction's Gibbs free energies and selected appropriate, distinct thermodynamic routes (Soh and Hatzimanikatis 2010). Soh and Hatzimanikatis suggested that the pathway created by BNICE can be evaluated further using fixed pathways analysis techniques as thermodynamic-based flux balance analysis (FBA). Grow Match enables investigation of the overall impact of these distinct pathways on the functioning of metabolic networks in host organisms. FBA can anticipate changes in phenotypes, increased yield, the occurrence of gene knockouts, synthetic biology, and the biodegradation of xenobiotics. It can be applied in a number of contexts, such as the examination of distinct and unusual metabolic pathways for metabolic research, the investigation of the evolution of metabolic pathways in a variety of organisms, the selection of targets for enzyme engineering, and the analysis of compound biodegradation pathways. PathPred is a different route prediction system. It is a fully automated, knowledge-based service that keeps repeating the prediction cycle until a certain chemical is displayed, regardless of whether the compound is supplied by the user or exists in the KEGG database. PathPred has the ability to associate genetic data with forecast outcomes. Regardless of whether the enzymes required for those steps are present or not, this server offers distinct and varied reaction phases. The E-zyme tool can be used to choose a potential EC number if the enzymes are not recognized. This number can then be used to identify the gene in the genome by comparing its sequence to that of a known gene that has the same EC (Moriya et al. 2010).

References

Adams GO et al (2015) Bioremediation, biostimulation and bioaugmentation: a review. Int J Environ Bioremed Biodegrad 3(1):28–39

Aukema KG et al (2017) In silico identification of bioremediation potential: carbamazepine and other recalcitrant personal care products. Environ Sci Technol 51(2):880–888

Balapure K et al (2016) Exploring bioremediation strategies to enhance the mineralization of textile industrial wastewater through sequential anaerobic-microaerophilic process. Int Biodeterior Biodegradation 106:97–105

Coelho E et al (2020) Bioremediation of water contaminated with uranium using Penicillium piscarium. Biotechnol Prog 36(5):e30322

Diplock E et al (2009) Predicting bioremediation of hydrocarbons: laboratory to field scale. Environ Pollut 157(6):1831–1840

Kumar PK et al (2011) Design of a smart biomarker for bioremediation: a machine learning approach. Comput Biol Med 41(6):357–360

Kumari P, Kumar Y (2021) Bioinformatics and computational tools in bioremediation and biodegradation of environmental pollutants. In: Bioremediation for environmental sustainability. Elsevier, Amsterdam, pp 421–444

Mehmood MA, Sehar U, Ahmad N (2014) Use of bioinformatics tools in different spheres of life sciences. J Data Mining Genom Proteom 5(2):1

Méndez-Martínez AJ et al (2012) Electrochemical reduction and oxidation pathways for reactive black 5 dye using nickel electrodes in divided and undivided cells. Electrochim Acta 59:140–149

Mohana S et al (2008) Response surface methodology for optimization of medium for decolorization of textile dye Direct Black 22 by a novel bacterial consortium. Bioresour Technol 99(3):562–569

Moriya Y et al (2010) PathPred: an enzyme-catalyzed metabolic pathway prediction server. Nucleic Acids Res 38(2):W138–W143

Ramamurthy M et al (2017) Application of viromics: a new approach to the understanding of viral infections in humans. Virus Dis 28:349–359

Rayhan A (2023) AI and the environment: toward sustainable development and conservation

Schneider MV, Orchard S (2011) Omics technologies, data and bioinformatics principles. In: Bioinformatics for omics data: methods and protocols. Springer, Cham, pp 3–30

Soh KC, Hatzimanikatis V (2010) DREAMS of metabolism. Trends Biotechnol 28(10):501–508

Tucker A, Duplisea D (2012) Bioinformatics tools in predictive ecology: applications to fisheries. Philos Trans R Soc B Biol Sci 367(1586):279–290

Remediation Policy and Management 17

Mansoureh Tavan

Abstract

Remediation management for contaminated sites is presently a fundamental issue worldwide, which is highly important in the fields of practice, research, and policy at the national and international levels. The environmental remediation policy provides a nationally accepted program and position, and also sets visible evidence of the country's aims and concerns. The national policy for contamination remediation should define the bigness and scale of the possible risk, as well as environmental concerns should be given high priority and financial provisions established for any remediation. Traditional strategies for contaminated site remediation mostly aim to decrease contamination levels at low cost and in a short period. Traditional approach may not be sustainable due to ignoring environmental side effects, thus there is an extensive requirement for "green" or "sustainable" approaches. Sustainable and Green Remediation (SGR) address economic, social, and environmental effects, and can conserve natural resources and protect soil, water, and air quality by decreasing emissions and other waste. Therefore, where possible and efficient, phytoremediation and bioremediation approaches may be chosen as basic remedial methods.

Keywords

Remediation · Policy · Management · Sustainable · Green remediation

M. Tavan (✉)
Faculty of Agriculture, Department of Horticultural Science, Bu Ali Sina University, Hamedan, Iran

A. A. H. Abdel Latef et al. (eds.), *Bioremediation and Nanotechnology for Climate Change Mitigation*, https://doi.org/10.1007/978-981-96-3069-1_17

17.1 Introduction

A site can be contaminated with various materials, such as organic contaminants, radioactive materials, and heavy metals. Environmental contamination such as the soil, water, and air has long been acknowledged as one of the major environmental issues, faced by varied countries worldwide. Recently, new environmental provisions were legislated to address this global concern. However, there are still a considerable number of contaminated sites that need high-quality remediation technologies (Reddy and Kumar 2018). In Europe alone, potentially more than 2.5 million contaminated sites have been estimated, of which 14% need to be remediated (Huysegoms and Cappuyns 2017). Heavy metals and mineral oil are considered the main contaminants of Europe (60%), and the management cost of these contaminations is counted at 6 billion euros annually (Van Liedekerke et al. 2014; Panagos et al. 2013). In Asia and the United States, contaminated sites are also discovered daily (Van Liedekerke et al. 2014; Panagos et al. 2013). For instance, in the United States, the Environmental Protection Agency has recognized tens of thousands of these contaminations that require remediation of which 1782 are on the National Priority List (NPL) and need urgent remediation (Reddy and Adams 2015; United States Environmental Protection Agency 2016). Groundwater and soil contamination have been the main concerns at these sites. Several remediation strategies have been expanded according to the contaminated media (sediments, groundwater, soil), type of contaminants (organic compounds and heavy metals), rate of contamination (deep, sub-surficial, surficial), and other site specifications (Sharma and Reddy 2004).

To perform a national remediation strategy, sites that have been determined as contaminated require to be prioritized. Following the primary specifications of each site, a list of contaminated sites is provided, including the size and environmental specifications of the sites, their locations, the populations potentially exposed, the types and properties of the contaminants, and other related factors. The list of contaminated sites is afterward prioritized based on the level of danger to human health and the environment. Other factors, including accessibility of funds, accessibility of scientific data, accessibility of remediation techniques, potential impacts on neighboring states, and socioeconomic effects, can also strongly affect determining remediation priorities.

The remediation of contaminated sites primarily aims to decrease and manage the dangers to humans, hence creating positive environmental alterations that are useful to society. The most suitable remediation technology in contaminated sites should be selected based on the possibility of removing contamination to the needed target level within the time and cost limits defined by the project. This strategy may resolve the problem of pollution at the site, but more environmental effects are caused by the intensive activities of resources and energy in various stages of remediation, which are often ignored. For instance, the emission of toxic substances and greenhouse gases through equipment or machinery utilized during remediation programs that negatively influence the environment are not mainly noted in the selection of remedial alternatives. Concerns about these negative effects resulted in the consideration of sustainability principles when choosing between site remediation options in a way that provides pure benefits in terms of wider environmental, economic, and social effects (SuRF 2010).

Among the major issues that the world is facing nowadays, the implication of "Sustainable and Green Remediation" (SGR) has been strongly accredited by professional institutions (e.g., Sustainable Remediation Forum (SURF) and the policymakers (e.g., U.S.EPA), American Society of Testing and Materials (ASTM), Interstate Technology and Regulatory Council (ITRC) and has promoted this growing interest in creating a framework for the performance of sustainable and green regulations in contaminated site remediation (Reddy and Kumar 2018). According to ITRC, SGR is "the site-specific utilization of technologies, processes, procedures, and products that reduce contaminant risk to receivers while making decisions that are aware of environmental effects, economic impacts, and balancing community goals." In a practical concept, this definition requires that the primary aim of SGR is to decrease the contaminant rates to targeted risk-based levels to preserve the environment and human health while foretasting and minimizing the negative environmental, social, and economic effects, inside and outside the site borders, in both present and future generations that the remediation project may cause. Indeed, unlike traditional remediation that emphasizes the contaminated site itself and site users' health, SGR introduces a more comprehensive approach by investigating life-cycle effects and encompassing broader socio-economic impacts. Such a comprehensive approach causes more effective utilization of limited resources in developing countries with more contaminated sites (Hou et al. 2014; Song et al. 2018).

17.2 General Principles for Remediation

Several key principles for the remediation of contaminated sites have been depicted in Fig. 17.1. These principles should be considered in choosing an efficient remediation solution for contaminated sites. As the first initiative in Europe, contaminated Land Rehabilitation Network for Environmental Technologies (CLARINET) brought together the merged knowledge of consultants, academics, industrial landowners, technology developers, and government experts. It presented a thematic network on interdisciplinary research, incorporating economic, societal, and technological aspects of the contaminated sites' management, in which about 16 European countries were present (Vegter et al. 2002; Cappuyns 2016). A CLARINET Working Group provided a series of general decision-making principles for selecting methods and tools for the land remediation (Bardos et al. 2002). The working group expressed that sustainable development and risk management should be considered as main aspects of contaminated site management. Several tools in the site remediation projects have been developed for selecting the most "sustainable" remediation option, considering multiple factors such as the effect on human health and agricultural productivity, the cost of remediation, and the economic profit (Scholz and Schnabel 2006).

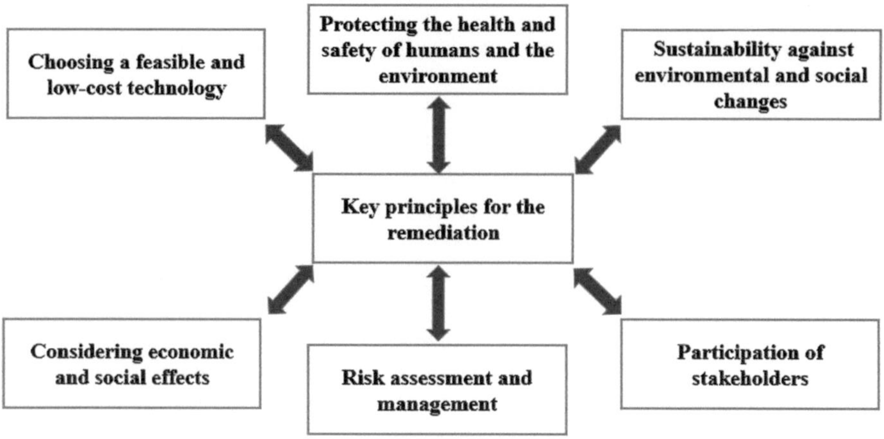

Fig. 17.1 Key principles related to the remediation

Risk management strategies are required to protect all sections of the environment (land, air, and water) and occupational and public safety and health. In risk management, the possibility of unacceptable effects on human health and the environment posed by contaminated sites is scientifically investigated, and its purpose is to support decisions about risk acceptance and actions to be efficiently taken for risk reduction (Bardos et al. 2002). A low-cost and feasible remediation technique should meet the technical and environmental standards for a specific remediation problem. Moreover, the assessment of economic and social sustainability can be facilitated by quantitative and qualitative practices in multi-criteria analysis (MCA) that indicators must be well acknowledged amongst remediation practitioners in a certain project (Hou and O'Connor 2020). Sustainable remediation provides optimal results for both present and future generations. Due to the dynamic nature of the environment, sustainable remediation strategies should be stable in response to environmental changes. Also, the regulatory and socioeconomic environment may alter. Sustainable remediation must consider these changes and provide the following characteristics: (1) compatibility to a diversity of future site development selections, (2) ability to create developing environmental and human health standards, and (3) resistance to altering geophysical conditions (Hou and Al-Tabbaa 2014).

Some remedial techniques, considering sustainability principles, can lead to more sustainable and successful management of contaminated sites. Several examples are given below (VertaseFLI Latest News 2024):

- Resource reuse and recovery: Reusing or recycling materials produced throughout remediation, such as recovered construction materials or excavated soil into site redevelopment, can maintain natural resources and decrease dependence on energy-intensive resources.
- In-situ remediation: These techniques, including chemical oxidation, soil vapor extraction, and bioremediation, clean up contaminants in the site, preventing the

requirement to excavate and transport soil, thereby decreasing energy consumption and minimizing waste production.

- Renewable energy: Using these resources, such as wind or solar power, to exploit remediation systems decreases dependence on non-renewable resources and greenhouse gas emissions.
- Green and natural substructure: Creating green and natural substructure into the remediation design, such as vegetated swales or wetlands, can improve climate resilience, stormwater management, and habitat provision while increasing the site's beauty and attractiveness to the surrounding community.

These examples show the potential of incorporating sustainability principles into the remediation of contaminated sites, contributing to more effective, responsible, and efficient management of contaminated sites.

17.3 Remediation Policy

An appointed remediation policy sets the procedures and rules to distinguish and control contamination risks nationally or internationally. Moreover, the environmental remediation policy provides a nationally accepted program and position and also sets visible evidence of the country's aims and concerns. The policy also determines lawful responsibilities for contamination, prepares financial assistance for redevelopment, significantly influences the market, and supports related research. The diversity between the policies proposes some ways that action can be more cautionary (Erdem and Nassauer 2014). Policymakers usually consider accepted national and international standards and processes to evaluate remediation needs, assess responsibilities and divide work such as sharing financial charges. Therefore, it will be necessary to set a national policy in the legal framework to establish coherent remediation approaches. Objectives and statements of policy are exhibited in Fig. 17.2.

Before updating and/or developing a national policy for the remediation of contaminated sites, people engaged in providing the policy must be conscious of any relevant law and requirement in their country, as well as any international requisite. They should consider and comprehend the following topics (Energy UN IAEA Nuclear Energy Series 2015):

1. National lawful framework and institutional structure
2. Applicable international conventions
3. Inventory of potential sites for remediation
4. Availability of resources
5. Potential transboundary issues

The existent national lawful structure and framework, as well as their appropriateness for supporting the formation of implementable policies for SGR of contaminated sites, must be considered. Also, countries must adhere to international

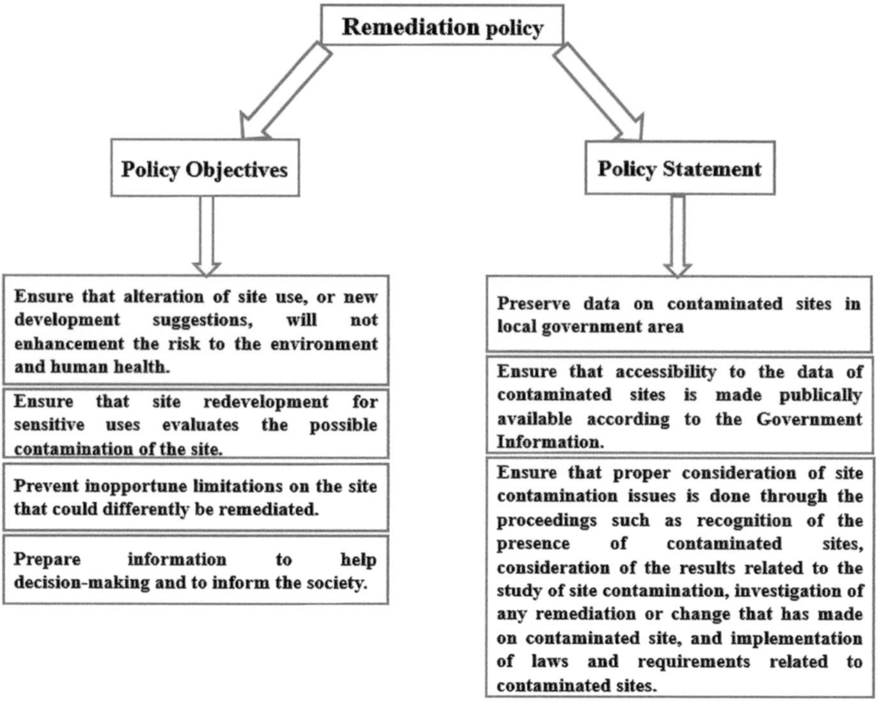

Fig. 17.2 Objectives and statement of remediation policy

conventions and their obligations due to these international instruments. In addition, the national inventory of contaminated sites that need remediation should be accessible for those interfered in the policy preparation. Also, the policymakers must consider the resources (economic, human, financial, technical, and social) present in the country to simplify the policy implementation. Transboundary issues may occur due to the migration of contamination through surface water, groundwater, or air. Therefore, it can be valuable to consider remediation solutions to deal with such challenges in the region and/or the potential of sharing existing facilities and technologies in neighboring countries.

In general, necessary considerations should be taken to deal with remediation concerns. The national policy for the remediation of contamination should define the bigness and scale of the possible risk, and environmental concerns should be given high priority, and financial provisions should be established for any remediation. It is necessary to note that the environmental contamination and the risks caused by them that need to be remediated may differ among countries, and thereby, the policy should specify these differences. Some major elements to be investigated in creating a national policy for environmental remediation consist (Energy UN IAEA Nuclear Energy Series 2015):

17.3.1 Allocation of Responsibilities

Responsibilities for contamination remediation may be allocated to the government, regulatory organizations, licensees, registrants, and other parties responsible for planning and implementing remediation. Therefore, every organization should have a specific task in the policy.

17.3.2 Provision of Resources

The national policy should establish the arrangements for determining the mechanisms for preparing funds or resources for safe management, ensuring that sufficient human resources are present for training and R&D, and supplying institutional controls and monitoring arrangements to ensure the security of the remediated sites.

17.3.3 Security and Safety Objectives

A general issue in the national policy regarding the remediation of contaminated sites is the safety objective of protecting people and the environment from the harmful impacts of hazardous contamination such as ionizing radiation, at present and in the future. Some security precautions and recommendations should be considered if necessary.

17.3.4 Public Participation and Information

The general decision-making action and the eventuating remediation solutions are of interest to a broad range of stakeholders, such as the general public. Indeed, all stakeholders will be participated in the decision-making action, regarding professional and unprofessional knowledge. The purpose is to attain a shared understanding of the status and its implications for all groups.

Policy instruments for controlling contaminated sites are divided into two main groups: (1) control and command procedures and (2) economic procedures (Christie and Teeuw 1998). Control and command procedures are utilized in many countries to begin remediation and to limit the usages to which contaminated sites may be assigned. It is especially essential when remediation is planned to match the "suitable for use" standard. A market-based procedure is an instance from an economic policy instrument. This procedure encourage market practice and retain regulatory intermediation in store for when there is no perspective of a market solution.

17.4 Remediation Strategies

Various countries may adopt different approaches to developing remediation strategies, depending on the type of contamination and the special factors of the national and site. Regulators or national institutions should carefully codify the remediation strategy to optimize resources and costs (Madear et al. 2020). Remedial objectives can be different based on the type of contamination. Organizations that are responsible for remediation usually state a remediation strategy on which the planning for remediation will be based. The strategy for remediation of site contamination shall consider at least the following:

1. List of contaminated sites and their contamination content.
2. The method of the site remediation process should be specified, and public expectations should be considered if possible.
3. The waste management process produced through remediation should be considered according to the way of classification of waste in the country.
4. Considering the funds for any type of site remediation.
5. Based on public and political perceptions, remediation measures should be prioritized and risks and available resources should be evaluated.

Therefore, before choosing a suitable strategy, it is necessary to identify the advantages and disadvantages of each viable solution. In general, remediation objectives and approaches should be pragmatic and focus on achieving a high-quality, cost-effective, timely solution.

Due to increasing global environmental problems, efforts to find more sustainable solutions in site remediation and redevelopment have enhanced remarkably. Hence, due to its social, economic and ecological importance, sustainable remediation has been highly regarded as a comprehensive approach. Sustainable remediation focuses on effective contamination management and minimizing negative social, economic, and environmental impacts of remedial actions. By incorporating sustainability principles into remediation strategies, site remediation project managers can better uphold the environment, enhance community well-being, and maintain resources.

To implement and develop a sustainable remediation strategy, different considerations and principles should be supposed, including (VertaseFLI Latest News 2024):

- Economic efficiency: Cost-effectiveness in the short and long term. This includes both direct costs of the remediation processes and indirect costs, such as maintenance, potential future debits, and persistent monitoring.
- Environmental protection: Minimizing potential damage to the natural resources, ecosystems, and biodiversity. Techniques that decrease waste production, energy consumption, and emissions are favored in sustainable remediation methods.
- Adaptive management: Sustainable remediation projects should be capable of adapting to evolving situations, including changes in site conditions, regulatory requirements, and new technological advancements. This helps ensure that the

remediation strategy remains effective and sustainable throughout the project lifecycle.

- Social responsibility: Addressing stakeholder concerns, enhancing community well-being, and meeting regulatory requirements. Investigating local job opportunities, promulgating education regarding the remediation project, and participating in free communication with stakeholders can lend to social responsibility.

By implementing these principles, those involved can establish a holistic and balanced approach to remediation that improves economic, social, and environmental outcomes.

17.5 Remediation Management

Remediation management includes the actions taken to reduce the site contamination, taking into account factors such as cost, efficiency, and ease of project implementation, as well as environmental, social, and economic consequences. Indeed, the best management techniques are the basic activities to establish environmental sustainability in the remediation of contaminated site that would not only prevent unacceptable risks to environment and human health but also support renew a site to reuse. Four basic steps can be considered for managing contaminated sites shown in the Fig. 17.3 (Ibrahim 2009).

Management strategies for the contaminated site should reflect the requirement to protect all physical and biological environment sections. During the evaluation and remediation, proper controls are needed on the site to control emissions from water, land, and air. One of the important objectives in remediation management is

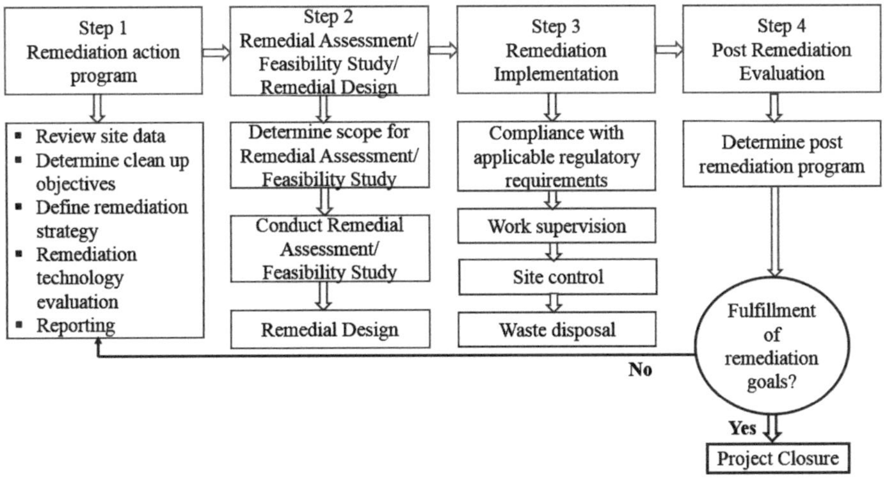

Fig. 17.3 Basic steps for the management of contaminated sites

to secure the remediated site for long-term use, where site re-use is part of the remediation strategy. A multidisciplinary procedure is necessary to the proper cleanup of contaminated sites. It is also essential to pay attention to occupational and public health and safety in adjusting every strategy to manage, assess, and remediate a contaminated site.

Best management procedures of green remediation include special processes to investigate the main elements of greener cleanups (Khan et al. 2021):

- Decrease whole energy consumption and enhance the percentage of energy from renewable resources:
- The energy requirement of the cleanup system is highly vital to assess from the point of view of green remediation. This factor emphasizes the usage of inactive energy sources to achieve all remediation aims. Moreover, energy efficient apparatuses should be applied and preserve at the highest performance to maximize efficiency.
- Decrease air contaminants and greenhouse gas emissions:
- This factor is mainly related to the air emissions created by various fuel types in any in-site or out-site cleanup practice. It emphasizes to reduce the utilization of heavy equipment needing high rates of fuel and to utilize cleaner fuels for the function of these equipment. It also considers the decrease of priority and toxic contaminants, such as particulate matter, ozone, nitrogen dioxide, carbon monoxide, lead, and sulfur dioxide with the minimization of dust output of the contaminants.
- Decrease water consumption and conserve water quality:
- Water requirement and the effects on water sources is a main constituent of green remediation through lessening the freshwater utilization and increasing water reuse throughout daily activities and cleanup operations. In addition, the adjacent water fuselage should be inhibited from effects such as nutrient loading.
- Preserve material resources and decrease waste:
- For green remediation, minimal waste production is emphasized using selected technologies. Recycling and reuse of the generated material or eliminating it from the site should be encouraged. A main concern in this regard is minimizing the extraction and disposal of natural resources. If possible, inactive sampling apparatuses should be utilized that generate minimal waste.
- Protect ecosystem and land services:
- In relation to the ecosystem and land effects, minimal invasive in-place technologies should be utilized and inactive energy technologies such as phytoremediation and bioremediation should be chosen as primary remediation where possible and efficient. This factor also requires minimizing habitat and soil destruction and reducing lighting and noise disturbance.

17.6 Conclusions

National policies and management for the remediation of contaminated site may differ in the various countries, but some elements are common among them. There is no same policy pattern in this regard, and no single pattern would be executable for the whole world. However, management and policy elements should be carefully investigated concerning the magnitude of the remediation problem. Many policies and laws related to environmental remediation are in their early stages; therefore, there is little evidence about which approach is the most possible to develop the best results. Sustainable and Green remediation (SGR) is recognized in the academic world, and unlike traditional remediation, it offers a wide scope of remediation management for contaminated sites. It should also be considered that to continue the SGR approach in managing contaminated sites, the SGR should be accepted as a new procedure of deciding for the contaminated sites remediation, where the incorporation of social, environmental, and economic elements must be seen as a key factor in decision-making.

In 2015, the United Nations developed a set of 17 goals meant to help achieve sustainable practice worldwide by 2030. These 17 goals are the sustainable development goals to create a more peaceful and prosperous world for the planet and its inhabitants. While the process of removing contamination from soils and groundwater is inherently sustainable, it only focuses on the environment and economy, while society and culture have not been the main focus during site remediation until recent years. Currently, site remediation focuses on involving all aspects of sustainability. The progress can be evaluated by evaluating a site remediation project to see which of the 17 Sustainable Development Goals it meets (Campos et al. 2022).

Additionally, the United Nations Environment Assembly (UNEA) of the United Nations Environment Programme (UNEP) is the world's highest-level decision-making body on the environment. At its third session, held in Nairobi, Kenya, from 4 to 6 December 2017, the UNEA met under the theme "Towards a pollution-free planet." As a result of the discussions among all the stakeholders present, a Ministerial Declaration and nine resolutions were adopted. In addition to the declaration, the member states adopted nine resolutions on air, water, soil, and marine pollution, as well as on the environment and health.

Other recently adopted international efforts will also contribute to tackling soil pollution more efficiently. The UN Decade on Ecosystems Restoration adopted in March 2019 refers to a wide continuum of practices that contribute to remediation of polluted sites, conservation and restoration of organic carbon in agricultural soils, and reestablishment of a diverse population of soil microorganisms that help create the natural fertility of our soils (FAO and UNEP 2021).

References

Bardos P, Lewis A, Nortcliff S, Matiotti C, Marot F, Sullivan T (2002) Review of decision support tools for contaminated land and their use in Europe. Austrian Federal Environment Agency, Vienna

Campos M, Czaplinski K, Gabriel J, Pikowski T, Verma G, Reddy KR (2022) Contribution of site remediation to sustainable development goals. In: Recent advances in sustainable environment: select proceedings of RAiSE 2022. Springer, Singapore, pp 259–272

Cappuyns V (2016) Inclusion of social indicators in decision support tools for the selection of sustainable site remediation options. J Environ Manag 184:45–56

Christie S, Teeuw R (1998) Varied policy of European Union states on contaminated land. Environ Impact Assess Rev 18:175–197

Energy UN IAEA Nuclear Energy Series (2015) Policies and strategies for environmental remediation, No. NW-G-3.1. IAEA, Vienna

Erdem M, Nassauer JI (2014) Design of brownfield landscapes under different contaminant remediation policies in Europe and the United States. Landsc J 32(2):277–292

FAO and UNEP (2021) Global assessment of soil pollution: report. Rome. ISBN: 978-92-5-134469-9, https://doi.org/10.4060/cb4894en

Hou D, Al-Tabbaa A (2014) Sustainability: a new imperative in contaminated land remediation. Environ Sci Pol 39:25–34

Hou D, O'Connor D (2020) Green and sustainable remediation: concepts, principles, and pertaining research. In: Sustainable remediation of contaminated soil and groundwater. Elsevier, Amsterdam, pp 1–17

Hou D, Al-Tabbaa A, Chen H, Mamic I (2014) Factor analysis and structural equation modelling of sustainable behaviour in contaminated land remediation. J Clean Prod 84:439–449

Huysegoms L, Cappuyns V (2017) Critical review of decision support tools for sustainability assessment of site remediation options. J Environ Manag 196:278–296

Ibrahim R (2009) Contaminated land management and control guidelines No. 3: remediation of contaminated sites. Jabatan Alam Sekitar (Department of Environment, Malaysia)

Khan MAA, Qadir Z, Asad M, Kouzani AZ, Parvez Mahmud M (2021) Environmental footprint assessment of a cleanup at hypothetical contaminated site. Appl Sci 11:4907

Madear C, Dunca EC, Soil MG (2020) Remediation strategies based on risk assessment approach. IOP Conf Ser Earth Environ Sci 1:012094

Panagos P, Van Liedekerke M, Yigini Y, Montanarella L (2013) Contaminated sites in Europe: review of the current situation based on data collected through a European network. J Environ Public Health 2013:158764

Reddy K, Adams J (2015) Sustainable remediation of contaminated sites. Momentum Press, New York

Reddy K, Kumar G (2018) Green and sustainable remediation of polluted sites: new concept, assessment tools, and challenges. ce/papers 2:83–92

Scholz RW, Schnabel U (2006) Decision making under uncertainty in case of soil remediation. J Environ Manag 80:132–147

Sharma HD, Reddy KR (2004) Geoenvironmental engineering: site remediation, waste containment, and emerging waste management technologies. Wiley, Hoboken, pp 1–961

Song Y, Hou D, Zhang J, O'Connor D, Li G, Gu Q, Li S, Liu P (2018) Environmental and socioeconomic sustainability appraisal of contaminated land remediation strategies: a case study at a mega-site in China. Sci Total Environ 610:391–401

SuRF U (2010) A framework for assessing the sustainability of soil and groundwater remediation. Contaminated land: applications in real environments (CLAIRE)

United States Environmental Protection Agency (2016) Superfund: national priorities list (NPL). https://www.epa.gov/superfund/superfundnational-priorities-list-npl. Accessed 9 February 2017

Van Liedekerke M, Prokop G, Rabl-Berger S, Kibblewhite M, Louwagie G (2014) Progress in the management of contaminated sites in Europe
Vegter J, Lowe J, Kasamas H (2002) Sustainable management of contaminated land: an overview. Umweltbundesamt GmbH, Wien
VertaseFLI Latest News (2024) Sustainable remediation strategies for contaminated sites. https://vertasefli.co.uk/sustainable-remediation-strategies-for-contaminated-sites/. Accessed 14 March 2024

Future of World Health by Bioremediation

18

T. P. Ajeesh Krishna, T. Maharajan, V. Edwin Hillary, and S. Antony Ceasar

Abstract

Urbanization, industrialization, mining, agriculture, and exploration are all major contributors to environmental pollution. Environmental pollutants affect living organisms' growth, development, physiological, biochemical, metabolic, defense, and immunity pathways. Ultimately, this leads to morbidity and mortality. Nowadays, heavy metal (HM) or metalloid contamination is a major problem in agricultural soil worldwide. It not only contaminates the soil but also hurts crop productivity, quality, and safety. HM contamination in agrarian soil may lead to health risks and hazards to humans *via* the food chain. Remediation of HM-contaminated soils is required to eliminate the associated risks and make the agricultural land resource available for crop production. Nowadays, scientists are looking for the future of world health through bioremediation strategies. Bioremediation is an emerging technology that effectively manages various environmental pollutants, especially HM-contamination. The process of bioremediation involves the utilization of biological systems for either restoring or cleaning up contaminated sites. It is a promising approach to environmental pollution management, especially eliminating HMs. This chapter presents the details of various bioremediation approaches for HMs elimination from agricultural soil. We also highlight the available innovative approaches for HM remediation technologies. Moreover, we describe the status of HM contamination in the environment and its impact on human beings. This chapter could help researchers understand the available strategies for HM bioremediation in agricultural soil for a better life.

T. P. A. Krishna · T. Maharajan · V. E. Hillary · S. A. Ceasar (✉)
Division of Plant Molecular Biology and Biotechnology, Department of Biosciences, Rajagiri College of Social Sciences, Kochi, Kerala, India

A. A. H. Abdel Latef et al. (eds.), *Bioremediation and Nanotechnology for Climate Change Mitigation*, https://doi.org/10.1007/978-981-96-3069-1_18

Keywords

Agricultural soil · Bioremediation · Environmental pollution · Heavy metals · Human health and safety

18.1 Introduction

Environmental pollution is increasing daily due to human activities, causing significant and irreversible damage to the ecosystem. Nowadays, human activities such as urbanization, industrialization, mining, agricultural activities, and exploration contribute significantly to environmental pollution (Ukaogo et al. 2020). As a result, air, water, and soil pollution pose a significant and growing threat to world health (Münzel et al. 2023). It is a global issue, and the estimated number of polluted environments has grown in recent years, prompting international efforts to remediate many of these environments. Nowadays, the widespread presence of heavy metals (HMs) in the environment has become a pressing concern worldwide, primarily due to the potential threat they pose to the safety of our food supply chain and their harmful effects on the well-being of both humans and animals (Tchounwou et al. 2012). Soil contamination with HMs, such as lead (Pb), cadmium (Cd), mercury (Hg), arsenic (As), chromium (Cr), copper (Cu), selenium (Se), nickel (Ni), silver (Ag), zinc (Zn), iron (Fe) aluminum (Al), cesium (Cs), cobalt (Co), manganese (Mn), molybdenum (Mo), strontium (Sr), and uranium (U), as a result of worldwide industrialization, has increased noticeably within the past few years. Moreover, the accumulation of toxic HMs, originating from various anthropogenic activities and natural sources (Fig. 18.1), has far-reaching consequences that necessitate immediate attention and effective measures to mitigate their adverse impacts on ecological systems and public health (Edo et al. 2024). Geological sources, mining activities, atmospheric deposition, and agricultural activities are the main sources of HM contamination, which contribute 32, 19.38, 17.57, and 31.05, respectively (Qin et al. 2022). The presence of HMs in soil poses a significant threat to human health through the consumption of contaminated agricultural productivity (Oves et al. 2012; Tauqeer et al. 2022). Furthermore, HMs can easily enter the human body via various food chains (Fig. 18.2). Many researchers have highlighted the human health issues related to HM toxicity (Reilly 2008; Jan et al. 2015; Alengebawy et al. 2021; Manwani et al. 2022). Finally, the toxicity of HMs leads to morbidity and mortality. Therefore, the researchers are trying to remove the HM contamination from agricultural soil to produce safe food.

Bioremediation is a promising approach to the removal of environmental pollution. It is a process that uses microorganisms, fungus, green plants, or their enzymes to restore a contaminated natural environment to its original state. Moreover, they help to destroy or render certain pollutants harmless through natural biological activity. It is a low-cost and low-tech approach that has been widely accepted. This chapter discusses the various bioremediation options for removing HMs from agricultural soil. We also highlight the new technologies to improve the bioremediation

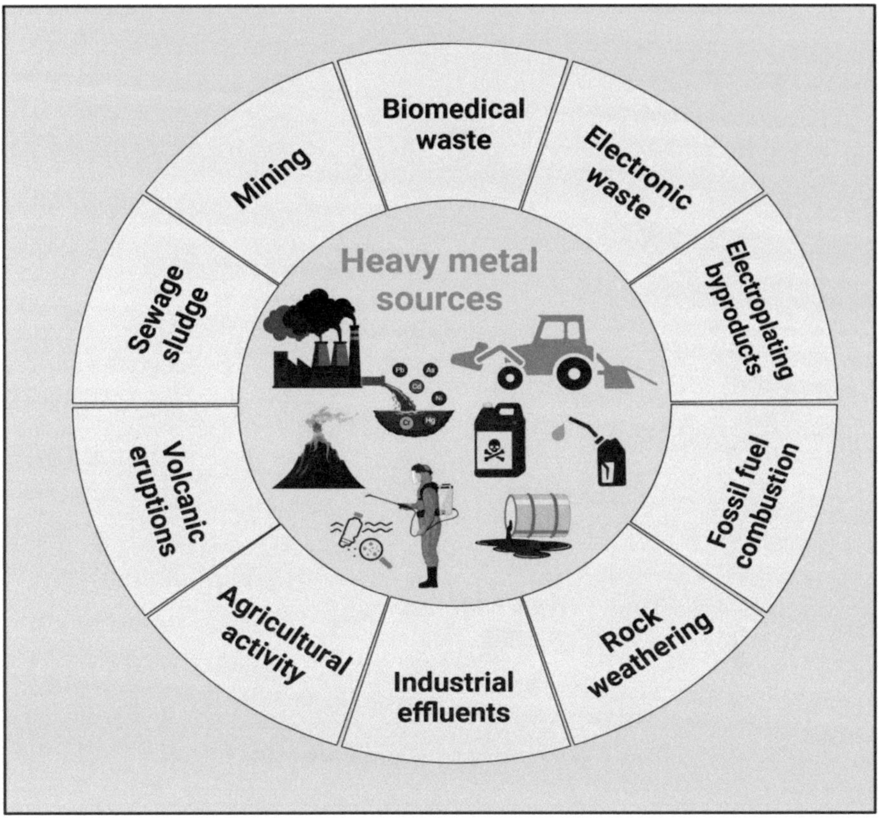

Fig. 18.1 Various anthropogenic and natural sources of heavy metals. A schematic representation of various anthropogenic and natural sources of heavy metals contaminating water, air, and soil

approaches. Furthermore, we describe the state of HM pollution in the environment and its effects on humans. We provide insight into the limitations and challenges and future directions of bioremediation of HM contamination in agricultural soil. This chapter may assist researchers in comprehending the available options for HM bioremediation in agricultural soil for a better living.

18.2 Current Challenges in Global Health

The worldwide population continues to face enormous health concerns due to worsening air, water, and soil pollution levels. Unintended exposure to various pollutants, such as HMs, air pollutants, and organic chemicals, can have various negative consequences on human bodies, resulting in the onset and progression of various diseases (Fig. 18.3). Nowadays, environmental pollution (air, water, and soil) is the leading cause of health issues and mortality worldwide (Briggs 2003). For example,

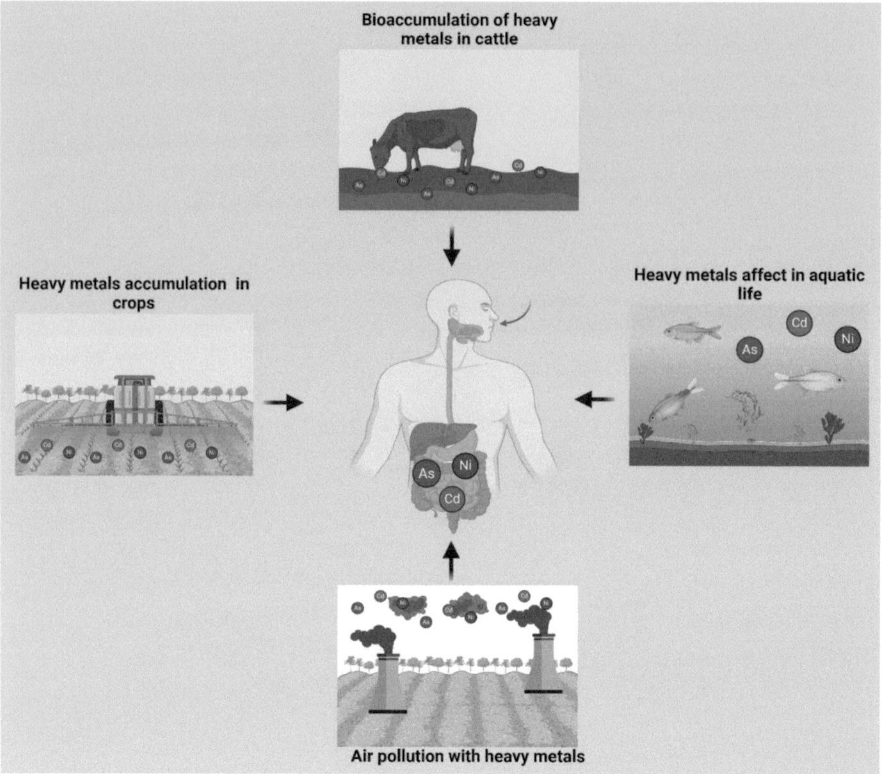

Fig. 18.2 Entry of heavy metals into the human body *via* food chains. The schematic diagram shows the heavy metal contamination in human food resources. The consumption of heavy metal-contaminated food resources may lead to significant health risks

airborne fine particulate matter increases the risk of ischemic heart disease mortality, cerebrovascular mortality, incident stroke, and incident myocardial infarction by 23, 24, 13, and 8%, respectively. Worldwide exposure to particulate matter causes over 4 million premature deaths annually (Nansai et al. 2021). HM toxicity has adverse effects on human health, which leads to hepatotoxicity, carcinogenicity, cardiovascular toxicity, neurotoxicity, and nephrotoxicity (Briffa et al. 2020). For example, Cd-induced neurotoxicity causes neurodegenerative disorders such as amyotrophic lateral sclerosis, multiple sclerosis, Alzheimer's disease, and Parkinson's disease(Branca et al. 2018). Numerous preclinical studies have shown that Cd has a negative impact on the functions of the peripheral nervous system (Miura et al. 2013) and central nervous system (Marchetti 2014). Many researchers have highlighted the human health consequences of various environmental pollutants (Madhav et al. 2020; Vandana et al. 2022; Xu et al. 2022; Shetty et al. 2023). The World Health Organization (WHO) estimates that 8.9 million people die every year from diseases induced by environmental pollution (WHO 2014, 2015). Environmental health risks are generally higher in developing countries compared

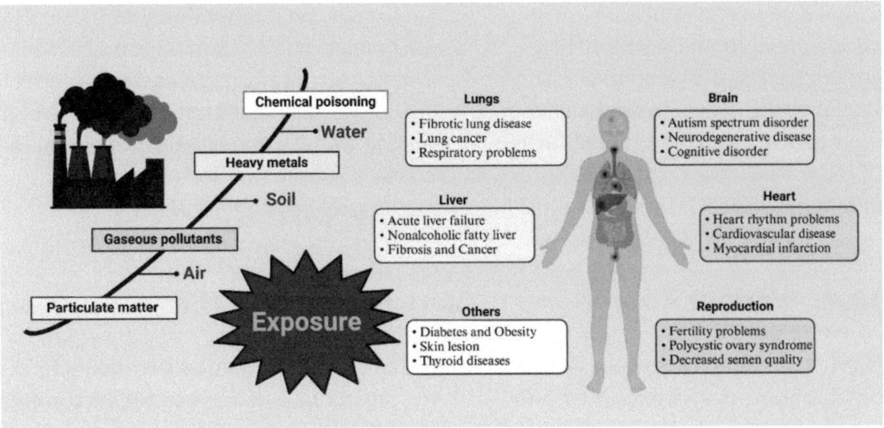

Fig. 18.3 Human health consequences related to various environmental pollutants. The effect of environmental pollutants such as chemical poisoning, heavy metals, gases, and particulate matter on the human lungs, brain, heart, and reproductive system

to low- and middle-income countries. Environmental pollution's health impacts have raised public concern. It is very urgent to reduce environmental pollution for the future of global health.

18.3 Hazardous Effects of HM

HMs find widespread use across industries, and exposure to these metals can pose significant health risks. For instance, the employment of Cr in tanning, paints, pigments, and fungicides has been associated with cancer, kidney disease, ulceration, and hair loss (Grandjean 1998). Hg present in coal, vinyl chlorides, batteries, and thermometers can trigger autoimmune disorders, depression, drowsiness, fatigue, hair loss, insomnia, memory impairment, restlessness, vision problems, tremors, anger outbursts, brain damage, as well as lung and kidney failure (Fernandes Azevedo et al. 2012). Pb, ubiquitous in plastics, paints, pipes, batteries, gasoline, and automobile exhaust, is neurotoxic and increases the risk of cardiovascular disease (Flora et al. 2012). The Cd has widely used for manufacturing fertilizers, plastics, and pigments, etc., which is carcinogenic, mutagenic, an endocrine disruptor, and can damage kidneys, lungs, weaken bones, and disrupt calcium metabolism (Genchi et al. 2020). Moreover, chemical fertilizers contain high amounts of Zn, leading to symptoms such as dizziness, fatigue, vomiting, kidney problems, and cramps (Paun et al. 2012). Se, occurring in coal and sulfur, can interfere with endocrine function, impair natural killer cells, cause liver toxicity, and gastrointestinal issues, damage the liver, kidneys, and spleen, and induce nervousness (Li et al. 2024). Ni, used in electroplating, is carcinogenic, allergenic, immunotoxic, neurotoxic, teratogenic, genotoxic, and mutagenic, and may affect fertility and cause hair loss (Schrenk et al. 2020). In some cases, trace metals like Ni, Cu, Zn, and Mo play a vital role in enzymatic and physiological activities, but higher doses of these metals also lead to many health issues in human beings. The contamination of HMs in agricultural soil is a serious issue in the current scenario due to their side effect. Because, the crops grown in HM-contaminated soil may accumulate more HMs in their edible part of the crops, further transferring into the food chain. Consumption of these food crops can cause severe health issues in humans. Therefore, there is an urgent need for a promising solution to removing HM in agricultural soil.

18.4 Various Approaches to Bioremediation of HMs

Bioremediation is the process of reclaiming degraded environments by using living organisms. It is an option that allows for the destruction of various toxins through natural biological activity and the degradation of environmental contaminants into less hazardous forms. It is also applicable to HM risks. It has proven to be a more promising and efficient approach than other alternative methods. The HM remediation *via* microbial biodegradation and phytoremediation has been widely accepted worldwide. Nowadays, nanoremediation is an emerging bioremediation approach. This section discusses the various approaches for HMs in agricultural soils. It is very helpful for scientists to understand the better bioremediation strategies for HM remediation.

18.4.1 Microbial Biodegradation

HM removal by biosorption has been widely used for the last several decades (Nachana'a Timothy 2019). Microorganisms are considered outstanding creatures for the detoxification of environmental pollutants (Jan et al. 2014). Also, it is the cheapest, simplest, and most eco-friendly cleanup method (Yadav et al. 2019; Kour et al. 2020). Microorganisms can degrade, detoxify, and even accumulate toxic chemicals as well as inorganic substances (Medfu Tarekegn et al. 2020). To ensure the efficacy of environmental pollutant detoxification, researchers worldwide are studying a broad range of microorganisms from various places and conditions (Table 18.1). Some research findings have shown that some fungal or bacterial strains can grow in HM toxicity conditions and detoxify high amounts of HMs. For example, endophytic bacteria *Bacillus* sp. can bioremediate HMs such as Cu, Cd, and Pb up to 75.78, 80.48, and 21.25%, respectively, within 24 h of incubation. Similarly, *Paenibacillus* sp. (endophytic bacterium) showed resistance to HMs such as Cu, Zn, Pb, and As up to 750, 500, 450, and 400 mg/L, respectively. Interestingly, Bhattacharya and Gupta (2013) reported that the newly isolated *Acinetobacter* sp. strain B9 showed a high level of Cr (350 mg L^{-1}) tolerance. Moreover, *Acinetobacter* sp. strain B9 was able to remove up to 67% of the Cr content in HM-contaminated soil (Bhattacharya and Gupta 2013). Similarly, the bacterial cultures such as *Sporosarcina saromensis*, *Bacillus cereus*, *Bacillus circulans*, *Bacillus subtilis*, and *Pseudomonas aeruginosa* were able to achieve high absorption efficiency of 82.5%,

Table 18.1 Microbes-mediated remediation of heavy metals. Name of the microorganisms and their category, name of the metal and observation with references are provided

Microorganism	Category	Heavy metals	Observations	References
Aspergillus terreus	Fungi	Pb	Higher accumulation capacity of Pb (59.67 mg/g)	Joshi et al. (2011)
Trichoderma viride	Fungi	Cd	Higher accumulation capacity of Cd (16.25 mg/g)	
Trichoderma longibrachiatum	Fungi	Cr	Higher accumulation capacity of Cr (0.55 mg/g)	
Aspergillus niger	Fungi	Ni	Higher accumulation capacity of Ni (0.55 mg/g)	
Pseudomonas aeruginosa	Bacteria	U	U resistance and high accumulation capacity	Choudhary and Sar (2011)
Penicillium corylophilum	Fungus	Cd, Zn, and Pb	Resistant to heavy metals such as Cd, Zn, and Pb	Mohamed and Abo-Amer (2012)
Pseudomonas aeruginosa	Bacteria	Cd, Zn, and Pb	Resistant to heavy metals such as Cd, Zn, and Pb	
Acinetobacter brisouii	Bacteria	As and Cd	High efficiencies of removal of heavy metals	Bhakta et al. (2014)
Pseudomonas abietaniphila	Bacteria	As and Cd	High efficiencies of removal of heavy metals	
Exiguobacterium aestuarii	Bacteria	As and Cd	High efficiencies of removal of heavy metals	
Planococcus rifietoensis	Bacteria	As and Cd	High efficiencies of removal of heavy metals	
Enterobacter cloacae	Bacteria	Pb	High removal rates (68.1%) of Pb	Kang et al. (2015)
Stenotrophomonas maltophilia	Bacteria	Cr	High resistance to Cr (400 mg/ml)	Raman et al. (2018)
Sporosarcina pasteurii	Bacteria	Zn, Pb, and Cd	Accumulation of higher amount of Zn, Pb, and Cd	Jalilvand et al. (2020)
Stenotrophomonas rhizophila	Bacteria	Zn, Pb, and Cd	Accumulation of higher amount of Zn, Pb, and Cd	
Variovorax boronicumulans	Bacteria	Zn, Pb, and Cd	Accumulation of higher amount of Zn, Pb, and Cd	

82%, 71.4%, 93%, and 70% of Cr, respectively (Ran et al. 2016; Nayak et al. 2018; Chaturvedi 2011; Tharannum et al. 2012). Therefore, these microorganisms could be utilized for the bioremediation process. The identification and characterization of microorganisms for HM bioremediation is a crucial process. Furthermore, it is necessary to assess the environmental impact of introducing microorganisms to the bioremediation process. Many researchers have successfully used the microorganisms to remove the HM contamination (Table 18.1). Joshi et al. (2011) have successfully utilized fungal species such as *Aspergillus awamori*,

Aspergillus flavus, *Phanerochaete chrysosporium*, and *Trichoderma viride* for the bioremediation of Pb, Cd, Cr, and Ni. In another report, the bacterium *Pseudomonas aeruginosa* and the fungi *Penicillium corylophilum* isolate showed the ability to bioremediate Cd, Zn, and Pb (Mohamed and Abo-Amer 2012). Additionally, many studies have shown that the synergistic effect of bacterial mixtures helps enhance the HM remediation process. Kang et al. (2016a, b) have reported that the synergistic effects of bacterial mixtures (*Viridibacillus arenosi* B-21, *Sporosarcina soli* B-22, *Enterobacter cloacae* KJ-46, and *Enterobacter cloacae* KJ-47) had greater efficiency (98.3% for Pb, 85.4% for Cd, and 5.6% for Cu) for the remediation of HM compared to using single strain culture treatment in soil. Similarly, the combination of three bacterial strains (*Enterococcus faecium*, *Lactiplantibacillus plantarum*, and *Limosilactobacillus fermentum*) showed a higher HM (Pb, Cd, and Ni) sorption capacity in aqueous solutions compared to their single state (Mostafidi et al. 2023). This study revealed that the synergistic effect of these three bacterial strains on the sorption of HM such as Pb, Cd, and Ni in aqueous solutions was significantly enhanced by 99.94, 99.91, and 93.75%, respectively (Mostafidi et al. 2023). Furthermore, Ashruta et al. (2014) reported efficient removal of Cr, Zn, Cd, Pb, Cu, and Co by bacterial consortia at approximately 75–85%. Microbial bioremediation is a promising approach for HM remediation. The researchers must use the microbial resources for HM bioremediation in agricultural soil to protect human health. Also, this microorganism can be used as a biofertilizer, reducing the exposure of HM to the soil as well as crops. It shows microorganisms can be used for multipurpose approaches.

18.4.2 Phytoremediation

Phytoremediation/phytoextraction is a better approach to removing HM contamination from soil (Shah and Daverey 2020). Phytoremediation is the use of plants to reduce the concentrations or toxic effects of contaminants in the environment (Etim 2012). It is a safer approach to the problem when compared to other physical and chemical methods. Moreover, it is a low-cost, efficient, and environmentally friendly solution that has gained widespread public acceptance (Babu et al. 2021). The plant species can accumulate and store HMs in the cellular organelles without any toxic effects (Kumar et al. 2017). The HM uptake and storage efficiency of plant species are classified into two types, namely (1) hyperaccumulator and (2) non-hyperaccumulator (Sytar et al. 2021). Hyperaccumulator plant species are widely used for applications in phytoremediation and phytomining. Many researchers have reported the hyperaccumulator plant species for better phytoremediation (Table 18.2). Ma et al. (2001) found that the *Pteris vittata* can accumulate more As. This study revealed that *Pteris vittata* can accumulate 1442–7526 mg kg^{-1} of As in shoots when it is growing in contaminated soil (Ma et al. 2001). Similarly, many plants are shown to have a higher accumulation of HM in contaminated soils. The Cd hyperaccumulator plant species *Thlaspi caerulescens* and *Arabidopsis halleri* can accumulate 1000 mg kg^{-1} and 157 mg kg^{-1} of Cd, respectively. Similarly, *Thlaspi*

Table 18.2 List of hyperaccumulator plant species. Details of plant name, character, and heavy metal responses with references are provided

Plant name	Type of character	Name of the metals	References
Agrostis castellana	Accumulator	As, Mn, Pb, and Zn	McCutcheon and Schnoor (2004)
Salix miyabeana	Tolerant	Ag	Guidi Nissim et al. (2014)
Brassica napus	Accumulator	Cr, Hg, Pb, Se, and Zn	Fiegl et al. (2010)
Amanita strobiliformis	Hyperaccumulator	Ag	Borovička et al. (2007)
Brassica juncea	Hyperaccumulator	Ag	Haverkamp et al. (2007)
Agrostis capillaris	Accumulator	Al, Mn, Pb, and Zn	McCutcheon and Schnoor (2004)
Sarcosphaera coronaria	Hyperaccumulator	As	Stijve et al. (1990)
Bacopa monnieri	Hyperaccumulator	Cd, and Cu	Gupta et al. (1994)
Bacopa monnieri	Accumulator	Hg and Pb	Gupta et al. (1994)
Brassica juncea	Accumulator	Cd, Cr, Pb, and U	Bennett et al. (2003)
Brassica juncea	Hyperaccumulator	Cu, Zn, Pb, and Ni	Bennett et al. (2003)
Brassica napus	Accumulator	Ag, Hg, Pb, Se, and Zn	Fiegl et al. (2010)
Vallisneria americana	Hyperaccumulator	Cd and Pb	McCutcheon and Schnoor (2004)
Hydrilla verticillata	Hyperaccumulator	Cd, Hg, and Pb	McCutcheon and Schnoor (2004)

rotundifolium (8200 mg kg^{-1}) and *Thlaspi caerulescens* (2740 mg kg^{-1}) are better Pb hyperaccumulator plant species (Baker and Brooks 1989; Baker et al. 2020). Therefore, we could use the hyperaccumulator plant species to enhance the bioremediation of HMs in contaminated soil. Similarly, microalgae can accumulate a higher amount of HMs. The microalgae can also be used to clean up the HMs from the water resources.

Understanding the processes involved in HM accumulation in the hyperaccumulator can help us better understand natural genetic variability in plant development, physiology, and adaptability to harsh environmental conditions. It is well known that the genetic components influence the HM tolerance or accumulation. Improving the genetic components can help to enhance the HM accumulation in plants. Researchers have been adopting biotechnological approaches to better understand the mechanisms of metal uptake, translocation, sequestration, and tolerance in plants for a better phytoremediation process. The metal transporters are one of the genetic components, which are helpful for metal uptake and transport. Therefore, the overexpression of key metal transporter proteins could enhance the uptake and accumulation of HMs in plants. For example, overexpression of *natural resistance-associated macrophage protein 6* (*TtNRAMP6*) genes in *Arabidopsis* improved accumulation of Cd in the roots, stems, leaves, and whole plant (Wang et al. 2019). Similarly, overexpression of *zinc-regulated, iron-regulated transporter-like protein*

3 (*HvZIP3*) in *Arabidopsis* improved the Zn accumulation in the root, shoot, and seed (Suzhen et al. 2015). Overexpression of the *heavy metal ATPase 3* (*AtHMA3*) gene in *Arabidopsis* enhanced Cd tolerance and increased its accumulation in the root and shoot (Morel et al. 2009). These findings showed that overexpression of metal transporters increased HM accumulation in plants. Many researchers have highlighted the role and utilization of metal transporters for phytoremediation (Zhang et al. 2018; Reddy et al. 2022; Yang et al. 2022; Krishna et al. 2023). Clustered regularly interspaced palindromic repeats (CRISPR) is an advanced genome-editing tool that enables the enhancement of desirable traits in plants (Basharat et al. 2018). A large number of genes responsible for HM tolerance and hyperaccumulation have already been identified in plants (Chaudhary et al. 2016; El-Sappah et al. 2023; Kim et al. 2006; Papoyan and Kochian 2004). Therefore, transcriptional regulation of desirable genes via CRISPR interference (CRISPRi) or CRISPR activation (CRISPRa) may improve the efficiency of phytoremediation. Tang et al. (2017) have demonstrated the ability of the CRISPR system to reduce Cd accumulation in rice by knocking out the *OsNRAMP5* gene. Therefore, modifications to gene expression levels could help contribute to HM tolerance and hyperaccumulation in plants (Venegas-Rioseco et al. 2021). It is a promising technique for modifying the transcription level of desirable genes without introducing any foreign genes. We hope genetic engineering and genome editing techniques can enhance HM accumulation in non-hyperaccumulator plant species. It will strengthen the phytoremediation process in the future. Scientists need to focus more on this area of research.

18.5 Limitations and Challenges of Bioremediations

Scientists are focusing on microbial remediation due to its notable benefits compared to previous techniques. This microbial remediation utilizes microorganisms' inherent ability to transform HMs into generally harmless forms (Kapahi and Sachdeva 2019; Saha et al. 2021). Microbial remediation is a straightforward, environmentally beneficial, sustainable, and very simple method to incorporate into the waste treatment process for polluted settings. Moreover, this process can be performed onsite, without the need to dig up the polluted soil. There is often no need for further treatment, and it has been proven cost-effective (Kapahi and Sachdeva 2019). Another advantage of micro-remediation is the ability to utilize indigenous microorganisms in the remediation process without requiring any intervention or alteration of the environment to promote microbial growth. However, obstacles still hinder its widespread adoption (Sun et al. 2021). The procedure is slower and requires more time than alternative therapeutic approaches. The efficacy of microbial remediation is depends on the specific microorganism's type, resistance, nature, level, and synergistic toxicity of HMs (Kang et al. 2016a). Another major drawback of microbial remediation is that it only concentrates or converts HMs into less harmful forms, without entirely removing them from the soil (Saha et al. 2021). Another major drawback of microbial remediation is the unclear regulations for determining the appropriate level or definition of a "clean" site. This confusion leads to uncertainty in the regulations regulating performance devices, making microbial remediation a complicated task (Danouche et al. 2021; Wu et al. 2021). Therefore, conducting research is necessary to develop and design microbial remediation solutions suitable for areas facing complex environmental challenges. Further elucidation of the molecular pathways involved in HM detoxification is necessary to boost the ability of microbes to accumulate HMs.

Phytoremediation is a very efficient biological method for extracting HMs from the soil. It has the benefit of being an onsite approach (RoyChowdhury et al. 2018; Raklami et al. 2021; Rajendran et al. 2022). The implementation of the vegetative cover preserves the topsoil and soil's structure, hence preserving the integrity of the soil. Furthermore, it has been recognized as a cost-effective alternative in comparison to traditional solutions because of its fewer requirements for digging, machinery, and employees (RoyChowdhury et al. 2018). This remediation technology also has several additional benefits, such as enhancing soil fertility, minimizing residue production, its applicability to various organic and inorganic pollutants, and its efficacy in addressing soil and water pollution. Besides the advantages, it has several limitations (Raklami et al. 2021). For example, this method is very slow-growing and takes months to years to remove pollution from the targeted environment fully. Another concern with this technique is the limited efficiency of cleaning the metal uptake in the roots from the soil. Recent climatic changes are also hindering phytoremediation by inhibiting plant growth and producing various biotic stresses (Rajendran et al. 2022). Therefore, to overcome these limitations, researchers can use native plants in the phytoremediation process due to their potential impact on imported species biodiversity.

18.6 Future Directions

One of the most important concerns of the twenty-first century is environmental pollution and its consequences (Table 18.3). Discovering efficient strategies to save the environment for future generations is labor-intensive. Environmental contamination, especially HM pollution, has serious consequences for soil quality, fertility, microbial biodiversity, and vegetation loss. To restore soils polluted with HMs, researchers have developed numerous physicochemical methods. Despite these strategies' frequent use and effectiveness in eliminating HMs, their delivery challenges, cost, lack of specificity, inefficiency in certain situations, and potentially major impact on soil quality are significant. Researchers have developed new biologically based technologies to remediate polluted soils. Emerging and new biotechnological solutions utilize the tolerance and capabilities of plants and microorganisms (such as bacteria, microalgae, yeast, and fungi) to detoxify and stabilize HMs. These tactics expand the possibilities for restoring polluted soils with HMs. Microbial remediation and phytoremediation are dependable, economical, effective, and environmentally feasible solutions. Microorganisms and plants have the natural capability to withstand hazardous substances and eliminate metals

Table 18.3 Major pollution types and their impact on human health. Details on the type pollution, pollutant, source of pollutant, and human health effect with references are provided

Type of pollution	Pollutant	Major sources pollutant	Health effects	References
Air	Particulate matter	Industrial processes	Respiratory and cardiovascular issues	Fiordelisi et al. (2017)
	Carbon monoxide	Combustion of fossil fuels	Respiratory illness and heart disease	Dastoorpoor et al. (2021)
	HMs	HMs-based paint and industrial processes	Neurotoxicity and heart disease	Houston (2007)
	Volatile organic compounds	Solvents and paints production	Respiratory problems and risk to cancer	Ogbodo et al. (2022)
	Sulfur dioxide	Burning of fossil fuels	Respiratory issues and heart disease	Chen et al. (2007)
Water	Pesticides	Agricultural and industrial activities	Skin and respiratory issues	Anju et al. (2010)
	HMs	Industrial, mining and agricultural activities	Nerotoxicity and increased risk of cancer	Roy et al. (2024)
	Polychlorinated biphenyls	Industrial activities	Cancer and damage to reproductive and immune systems	Carpenter (2006)
	Chlorine	Industrial activities	Allergic diseases and respiratory issues	Kanikowska et al. (2018)
Soil	HMs	Industrial, mining and agricultural activities	Neurological disorders and decreased immune system function	Jamal et al. (2013)
	Dioxins	Industrial activities	Cancer and imbalance of hormones	Matés et al. (2010)
	Polycyclic aromatic hydrocarbons	Improper disposal of petroleum and burning of fossil fuels	Cancer and metabolic disorders	Khan et al. (2021)

from the environment. The bacteria employ a range of methods, including extracellular and intracellular sequestration, the synthesis of metal chelators, precipitation, enzymatic detoxification, and volatilization. Combining the microbial remediation technique with phytoremediation technologies like phytostabilization, phytoextraction, and phytovolatilization could potentially achieve efficient reclamation. Engineered microorganisms and plants have the potential to improve bioremediation efficacy by overcoming the constraints associated with every single approach.

A single approach cannot be successful or sufficient for the operational reclamation of soils contaminated with HMs. Improving the integration of various methods, including physical, chemical, and biological techniques, is crucial for achieving highly efficient and comprehensive phytoremediation in the future. The research should concentrate on evaluating the impact of mixing various microorganisms on the effectiveness of bioremediation. This includes investigating microbial remediation's synergistic effects when combined with organic and inorganic chelating additives. In addition, it is necessary to investigate metagenomics methods and microbial metabolic studies to identify potential metal resistance and detoxification genes that may be activated in other species to enhance their unique performance. Metagenomics approaches must focus on studying the evolution of microbial communities during bioremediation. This research should aim to identify the most effective strategies to improve their survival, considering their potential battle with native microbial populations. Additionally, it is crucial to address the current inadequacy of their survivability when released into the environment during field trials. Furthermore, additional genetic research is necessary to understand the metabolic pathways and processes that microorganisms and plants employ to tolerate and detoxify HMs.

Pollution is increasing day by day due to human activities. The impact of the pollution is primarily transmitted through the food chain. Nowadays, air, water, and soil pollution have major impacts on human health, triggering many diseases and leading to high morbidity and mortality rates. Inadequate waste management is a major factor contributing to increased health concerns and pollution. Researchers estimate that pollution causes approximately 9 million deaths annually, accounting for one in six deaths globally. Therefore, pollution is a major concern in developing countries. To address these problems and promote a healthy ecosystem, it is essential to implement comprehensive solutions and promote global cooperation in the future.

18.7 Conclusion

Pollution from hazardous HMs is considered one of the world's most important environmental challenges. HM contamination in agricultural soil especially represents a threat to food safety and human health. The plants grown in HM-contaminated soil may accumulate more HMs in the edible parts of the crops, and it is possible to transfer them into the food chain. It is a major concern for global health's future. We need an eco-friendly, low-cost, and efficient approach for HM remediation in agricultural soil. Bioremediation has emerged as a promising approach to reducing the contamination of HMs in agricultural soil using various biological agents, including microorganisms (bacteria, fungi, and microalgae) and plants (phytoremediation). Microorganisms and plants have the ability to accumulate a significant amount of HMs from agricultural soils, which they can either use in their metabolic activities or convert into less toxic forms. Understanding the molecular mechanism of accumulating a higher amount of HM without any toxicity is very essential. Therefore, scientists should be more focused on this area of research. With the progress in current research, we hope that the genetic engineering and genome editing approaches will contribute to manipulating the HM accumulation efficiency at the genome level of biological agents used for bioremediation. Engineered plants or microorganisms gain the ability to carry a greater amount of HM, as well as an enzymatic attribute that enhances HM uptake. It is a promising and sustainable solution for improving the bioremediation process. Therefore, scientists should promote the use of engineered plants or microorganisms to improve HM removal in water and soil. It will contribute to achieving the future of world health through bioremediation.

Acknowledgments We sincerely thank Rajagiri College of Social Science, Kochi, Kerala for providing the research facilities and support.

Ethics Approval and Consent to ParticipateNot applicable.

Consent for Publication All authors have agreed for the publication.

Availability of Data and Materials Not applicable to this article.

Competing Interests The authors declare that they have no competing interests.

Funding This work was financially supported by Rajagiri College of Social Sciences (Autonomous), Kerala, India, under Seed Money for Faculty Minor Research.

Author Contributions All authors conceptualized and wrote the manuscript. SAC critically revised the manuscript for publication.

References

Alengebawy A, Abdelkhalek ST, Qureshi SR, Wang M-Q (2021) Heavy metals and pesticides toxicity in agricultural soil and plants: ecological risks and human health implications. Toxics 9:42

Anju A, Ravi SP, Bechan S (2010) Water pollution with special reference to pesticide contamination in India. J Water Resour Prot 2(5):1–7

Ashruta GA, Nanoty V, Bhalekar U (2014) Biosorption of heavy metals from aqueous solution using bacterial EPS. Int J Life Sci 2:373–377

Babu SM, Hossain MB, Rahman MS et al (2021) Phytoremediation of toxic metals: a sustainable green solution for clean environment. Appl Sci 11:10348

Baker AJ, Brooks R (1989) Terrestrial higher plants which hyperaccumulate metallic elements. A review of their distribution, ecology and phytochemistry. Biorecovery 1:81–126

Baker AJM, McGrath SP, Reeves RD, Smith JAC (2020) Metal hyperaccumulator plants: a review of the ecology and physiology of a biological resource for phytoremediation of metal-polluted soils. In: Phytoremediation of contaminated soil and water. CRC Press, Boca Raton, pp 85–107

Basharat Z, Novo LA, Yasmin A (2018) Genome editing weds CRISPR: what is in it for phytoremediation? Plants 7(3):51

Bennett LE, Burkhead JL, Hale KL et al (2003) Analysis of transgenic Indian mustard plants for phytoremediation of metal-contaminated mine tailings. J Environ Qual 32:432–440

Bhakta JN, Munekage Y, Ohnishi K et al (2014) Isolation and characterization of cadmium-and arsenic-absorbing bacteria for bioremediation. Water Air Soil Pollut 225:1–10

Bhattacharya A, Gupta A (2013) Evaluation of Acinetobacter sp. B9 for Cr (VI) resistance and detoxification with potential application in bioremediation of heavy-metals-rich industrial wastewater. Environ Sci Pollut Res 20:6628–6637

Borovička J, Řanda Z, Jelínek E et al (2007) Hyperaccumulation of silver by Amanita strobiliformis and related species of the section Lepidella. Mycol Res 111:1339–1344

Branca JJ, Morucci G, Pacini A (2018) Cadmium-induced neurotoxicity: still much ado. Neural Regen Res 13(11):1879–1882
Briffa J, Sinagra E, Blundell R (2020) Heavy metal pollution in the environment and their toxicological effects on humans. Heliyon 6:e04691. https://doi.org/10.1016/j.heliyon.2020.e04691
Briggs D (2003) Environmental pollution and the global burden of disease. Br Med Bull 68:1–24. https://doi.org/10.1093/bmb/ldg019
Carpenter DO (2006) Polychlorinated biphenyls (PCBs): routes of exposure and effects on human health. Rev Environ Health 21(1):1–24
Chaturvedi MK (2011) Studies on chromate removal by chromium-resistant Bacillus sp. isolated from tannery effluent. J Environ Prot 2(1):76
Chaudhary K, Jan S, Khan S (2016) Heavy metal ATPase (*HMA2*, *HMA3*, and *HMA4*) genes in hyperaccumulation mechanism of heavy metals. Plant Metal Interact 1:545–556
Chen TM, Kuschner WG, Gokhale J, Shofer S (2007) Outdoor air pollution: nitrogen dioxide, sulfur dioxide, and carbon monoxide health effects. Am J Med Sci 333(4):249–256
Choudhary S, Sar P (2011) Uranium biomineralization by a metal resistant *Pseudomonas aeruginosa* strain isolated from contaminated mine waste. J Hazard Mater 186:336–343. https://doi.org/10.1016/j.jhazmat.2010.11.004
Danouche M, El Ghachtouli N, El Arroussi H (2021) Phycoremediation mechanisms of heavy metals using living green microalgae: physicochemical and molecular approaches for enhancing selectivity and removal capacity. Heliyon 7:e07609
Dastoorpoor M, Riahi A, Yazdaninejhad H, Borsi SH, Khanjani N, Khodadadi N, Mohammadi MJ, Aghababaeian H (2021) Exposure to particulate matter and carbon monoxide and cause-specific cardiovascular-respiratory disease mortality in Ahvaz. Toxin Rev 40(4):1362–1372
Edo GI, Samuel PO, Oloni GO et al (2024) Environmental persistence, bioaccumulation, and ecotoxicology of heavy metals. Chem Ecol 40:322–349
El-Sappah AH, Abbas M, Rather SA, Wani SH, Soaud N, Noor Z, Qiulan H, Eldomiaty AS, Mir RR, Li J (2023) Genome-wide identification and expression analysis of metal tolerance protein (MTP) gene family in soybean (*Glycine max*) under heavy metal stress. Mol Biol Rep 50(4):2975–2990
Etim EE (2012) Phytoremediation and its mechanisms: a review. Int J Environ Bioenergy 2:120–136
Fernandes Azevedo B, Barros Furieri L, Peçanha FM et al (2012) Toxic effects of mercury on the cardiovascular and central nervous systems. J Biomed Biotechnol 2012:949048. https://doi.org/10.1155/2012/949048
Fiegl JL, McDonnell BP, Kostel JA, et al (2010) A resource guide: the phytoremediation of lead in urban, residential soils
Fiordelisi A, Piscitelli P, Trimarco B, Coscioni E, Iaccarino G, Sorriento D (2017) The mechanisms of air pollution and particulate matter in cardiovascular diseases. Heart Fail Rev 22:337–347
Flora G, Gupta D, Tiwari A (2012) Toxicity of lead: a review with recent updates. Interdiscip Toxicol 5:47–58. https://doi.org/10.2478/v10102-012-0009-2
Genchi G, Sinicropi MS, Lauria G et al (2020) The effects of cadmium toxicity. Int J Environ Res Public Health 17:3782. https://doi.org/10.3390/ijerph17113782
Grandjean P (1998) Health significance of metal exposures. In: Wallace RB (ed) Maxcy-Rosenau-last public health and preventive medicine. McGraw-Hill Medical, New York
Guidi Nissim W, Pitre FE, Kadri H et al (2014) Early response of willow to increasing silver concentration exposure. Int J Phytoremediation 16:660–670
Gupta M, Sinha S, Chandra P (1994) Uptake and toxicity op metals in *Scirpus lacustris* L. and bacopa monnieri l. J Environ Sci Health Part A 29:2185–2202
Haverkamp RG, Marshall AT, van Agterveld D (2007) Pick your carats: nanoparticles of gold–silver–copper alloy produced in vivo. J Nanopart Res 9:697–700
Houston MC (2007) The role of mercury and cadmium heavy metals in vascular disease, hypertension, coronary heart disease, and myocardial infarction. Altern Ther Health 13(2):128–133

Jalilvand N, Akhgar A, Alikhani HA et al (2020) Removal of heavy metals zinc, lead, and cadmium by biomineralization of urease-producing bacteria isolated from Iranian mine calcareous soils. J Soil Sci Plant Nutr 20:206–219

Jamal Q, Durani P, Khan K, Munir S, Hussain S, Munir K, Anees M (2013) Heavy metals accumulation and their toxic effects. J Bio Mol Sci 1(1):27–36

Jan AT, Azam M, Ali A, Haq QMR (2014) Prospects for exploiting bacteria for bioremediation of metal pollution. Crit Rev Environ Sci Technol 44:519–560

Jan AT, Azam M, Siddiqui K et al (2015) Heavy metals and human health: mechanistic insight into toxicity and counter defense system of antioxidants. Int J Mol Sci 16:29592–29630

Joshi PK, Swarup A, Maheshwari S et al (2011) Bioremediation of heavy metals in liquid media through fungi isolated from contaminated sources. Indian J Microbiol 51:482–487. https://doi.org/10.1007/s12088-011-0110-9

Kanikowska A, Napiórkowska-Baran K, Graczyk M, Kucharski MA (2018) Influence of chlorinated water on the development of allergic diseases-An overview. Ann Agric Environ Med 25:651–655

Kang CH, Oh SJ, Shin Y et al (2015) Bioremediation of lead by ureolytic bacteria isolated from soil at abandoned metal mines in South Korea. Ecol Eng 74:402–407

Kang CH, Kwon Y-J, So J-S (2016a) Bioremediation of heavy metals by using bacterial mixtures. Ecol Eng 89:64–69

Kang CH, Kwon YJ, So JS (2016b) Bioremediation of heavy metals by using bacterial mixtures. Ecol Eng 89:64–69

Kapahi M, Sachdeva S (2019) Bioremediation options for heavy metal pollution. J Health Pollut 9:191203

Khan A, Ahsan A, Farooq MA, Naveed M, Li H (2021) Role of polycyclic aromatic hydrocarbons as EDCs in metabolic disorders. In: Endocrine disrupting chemicals-induced metabolic disorders and treatment strategies. Springer, Cham, pp 323–341

Kim DY, Bovet L, Kushnir S, Noh EW, Martinoia E, Lee Y (2006) *AtATM3* is involved in heavy metal resistance in *Arabidopsis*. Plant Physiol 140(3):922–932

Kour D, Rana KL, Yadav AN et al (2020) Microbial biofertilizers: bioresources and eco-friendly technologies for agricultural and environmental sustainability. Biocatal Agric Biotechnol 23:101487

Krishna TPA, Maharajan T, Antony Ceasar S (2023) Significance and genetic control of membrane transporters to improve phytoremediation and biofortification processes. Mol Biol Rep 50:6147–6157

Kumar SS, Kadier A, Malyan SK et al (2017) Phytoremediation and rhizoremediation: uptake, mobilization and sequestration of heavy metals by plants. In: Plant-microbe interact agro-ecological perspect. Springer, Cham, pp 367–394

Li M, Qiu L, Ai X et al (2024) Effects of selenium and cadmium on human liver and kidney functions in exposed black shale areas. GeoHealth 8:e2024GH001040. https://doi.org/10.1029/2024GH001040

Ma LQ, Komart KM, Tu C, Zhang W, Cai Y (2001) A fern that hyperaccumulates arsenic. Nature 409:579

Madhav S, Ahamad A, Singh AK et al (2020) Water pollutants: sources and impact on the environment and human health. In: Pooja D, Kumar P, Singh P, Patil S (eds) Sensors in water pollutants monitoring: role of material. Springer, Singapore, pp 43–62

Manwani S, Vanisree CR, Jaiman V et al (2022) Heavy metal contamination in vegetables and their toxic effects on human health. Sustain Crop Prod Recent Adv 181:102651

Marchetti C (2014) Interaction of metal ions with neurotransmitter receptors and potential role in neurodiseases. Biometals 27:1097–1113

Matés JM, Segura JA, Alonso FJ, Márquez J (2010) Roles of dioxins and heavy metals in cancer and neurological diseases using ROS-mediated mechanisms. Free Radic Biol Med 49(9):1328–1341

McCutcheon SC, Schnoor JL (2004) Phytoremediation: transformation and control of contaminants. Wiley, Hoboken

Medfu Tarekegn M, Zewdu Salilih F, Ishetu AI (2020) Microbes used as a tool for bioremediation of heavy metal from the environment. Cogent Food Agric 6:1783174

Miura S, Takahashi K, Imagawa T, Uchida K, Saito S, Tominaga M, Ohta T (2013) Involvement of TRPA1 activation in acute pain induced by cadmium in mice. Mol Pain 9:1744–8069

Morel M, Crouzet J, Gravot A, et al (2009) AtHMA3, a P1B-ATPase allowing Cd/Zn/co/Pb vacuolar storage in Arabidopsis. Plant Physiol 149:894–904

Mohamed RM, Abo-Amer AE (2012) Isolation and characterization of heavy-metal resistant microbes from roadside soil and phylloplane. J Basic Microbiol 52:53–65

Mostafidi M, Sanjabi MR, Mojgani N, et al (2023) Heavy metal bioremediation potential of autochthonous lactic acid bacteria for use in edible leafy vegetables. J Food Qual 2023:8730676

Münzel T, Hahad O, Daiber A, Landrigan PJ (2023) Soil and water pollution and human health: what should cardiologists worry about? Cardiovasc Res 119:440–449. https://doi.org/10.1093/cvr/cvac082

Nachana'a Timothy ETW (2019) Environmental pollution by heavy metal: an overview. Chemistry 3:72–82

Nansai K, Tohno S, Chatani S, Kanemoto K, Kagawa S, Kondo Y, Takayanagi W, Lenzen M (2021) Consumption in the G20 nations causes particulate air pollution resulting in two million premature deaths annually. Nat Commun 12(1):6286

Nayak AK, Panda SS, Basu A, Dhal NK (2018) Enhancement of toxic Cr (VI), Fe, and other heavy metals phytoremediation by the synergistic combination of native *Bacillus cereus* strain and *Vetiveria zizanioides* L. Int J Phytoremediation 20(7):682–691

Ogbodo JO, Arazu AV, Iguh TC, Onwodi NJ, Ezike TC (2022) Volatile organic compounds: a proinflammatory activator in autoimmune diseases. Front Immunol 13:928379

Oves M, Khan MS, Zaidi A, Ahmad E (2012) Soil contamination, nutritive value, and human health risk assessment of heavy metals: an overview. Springer, Cham

Papoyan A, Kochian LV (2004) Identification of Thlaspi caerulescens genes that may be involved in heavy metal hyperaccumulation and tolerance. Characterization of a novel heavy metal transporting ATPase. Plant Physiol 136(3):3814–3823

Paun S, Tudosie M, Petris R, Macovei R (2012) The effects of Zinc on human body, including on renal failure and renal transplantation. J Med Life 5:137–140

Qin Y, Zhang F, Xue S, Ma T, Yu L (2022) Heavy metal pollution and source contributions in agricultural soils developed from karst landform in the Southwestern region of China. Toxics 10(10):568

Rajendran S, Priya TAK, Khoo KS et al (2022) A critical review on various remediation approaches for heavy metal contaminants removal from contaminated soils. Chemosphere 287:132369

Raklami A, Tahiri A, Bechtaoui N et al (2021) Restoring the plant productivity of heavy metal-contaminated soil using phosphate sludge, marble waste, and beneficial microorganisms. J Environ Sci 99:210–221

Raman NM, Asokan S, Shobana Sundari N, Ramasamy S (2018) Bioremediation of chromium (VI) by Stenotrophomonas maltophilia isolated from tannery effluent. Int J Environ Sci Technol 15:207–216

Ran ZH, Bi WA, Cai QT, Li XX, Min LI, Dong HU, Guo DB, Juan WA, Chun FA (2016) Bioremediation of hexavalent chromium pollution by *Sporosarcina saromensis* M52 isolated from offshore sediments in Xiamen, China. Biomed Environ Sci 29(2):127–136

Reddy SS, Kumar P, Dwivedi P (2022) Heavy metal transporters, phytoremediation potential, and biofortification. In: Plant metal and metalloid transporters. Springer, Cham, pp 387–405

Reilly C (2008) Metal contamination of food: its significance for food quality and human health. Wiley, Hoboken

Roy B, Bhunia B, Bandyopadhyay TK, Khan SA, Nandi NB, Nath PC (2024) Water pollution by heavy metals and their impact on human health. Handb Water Pollut 25:333–352

RoyChowdhury A, Datta R, Sarkar D (2018) Heavy metal pollution and remediation. In: Green chemistry. Elsevier, Amsterdam, pp 359–373

Saha L, Tiwari J, Bauddh K, Ma Y (2021) Recent developments in microbe–plant-based bioremediation for tackling heavy metal-polluted soils. Front Microbiol 12:731723

Schrenk D, Bignami M, Bodin L et al (2020) Update of the risk assessment of nickel in food and drinking water. EFSA 18:e06268. https://doi.org/10.2903/j.efsa.2020.6268

Shah V, Daverey A (2020) Phytoremediation: a multidisciplinary approach to clean up heavy metal contaminated soil. Environ Technol Innov 18:100774

Shetty SS, Deepthi D, Harshitha H et al (2023) Environmental pollutants and their effects on human health. Heliyon 9:e19496. https://doi.org/10.1016/j.heliyon.2023.e19496

Stijve T, Vellinga EC, Herrmann A (1990) Arsenic accumulation in some higher fungi. Persoonia Mol Phylogeny Evol Fungi 14:161–166

Sun W, Cheng K, Sun KY, Ma X (2021) Microbially mediated remediation of contaminated sediments by heavy metals: a critical review. Curr Pollut Rep 7:201–212

Suzhen L, Zhou X, Li H, Liu Y, Zhu L, Guo J, Liu X, Fan Y, Chen J, Chen R (2015) Overexpression of *ZmIRT1* and *ZmZIP3* enhances iron and zinc accumulation in transgenic *Arabidopsis*. PLoS One 10(8):e0136647

Sytar O, Ghosh S, Malinska H et al (2021) Physiological and molecular mechanisms of metal accumulation in hyperaccumulator plants. Physiol Plant 173:148–166

Tang L, Mao B, Li Y, et al (2017) Knockout of OsNramp5 using the CRISPR/Cas9 system produces low Cd-accumulating indica rice without compromising yield. Sci Rep 7:14438. https://doi.org/10.1038/s41598-017-14832-9

Tauqeer HM, Turan V, Iqbal M (2022) Production of safer vegetables from heavy metals contaminated soils: the current situation, concerns associated with human health and novel management strategies. In: Advances in bioremediation and phytoremediation for sustainable soil management: principles, monitoring and remediation. Springer, Cham, pp 301–312

Tchounwou PB, Yedjou CG, Patlolla AK, Sutton DJ (2012) Heavy metal toxicity and the environment. Experientia Suppl 101:133–164. https://doi.org/10.1007/978-3-7643-8340-4_6

Tharannum S, Krishnamurthy V, Mahmood R (2012) Characterization of chromium remediating bacterium Bacillus subtilis isolated from electroplating effluent. Int J Eng Res Appl, 2, 961-966.

Ukaogo PO, Ewuzie U, Onwuka CV (2020) Environmental pollution: causes, effects, and the remedies. Elsevier, Amsterdam, pp 419–429

Vandana, Priyadarshanee M, Mahto U, Das S (2022) Mechanism of toxicity and adverse health effects of environmental pollutants. Elsevier, Amsterdam, pp 33–53

Venegas-Rioseco J, Ginocchio R, Ortiz-Calderón C (2021) Increase in phytoextraction potential by genome editing and transformation: a review. Plants 11(1):86

Wang C, Chen X, Yao Q, et al (2019) Overexpression of TtNRAMP6 enhances the accumulation of Cd in Arabidopsis. Gene 696:225–232

WHO (2014) Burden of disease from household air pollution for 2012. WHO, Geneva

WHO (2015) Million deaths annually linked to air pollution. WHO, Geneva

Wu X, Huang P, Dong C, Deng X (2021) Nickel bioaccumulation by a marine bacterium *Brevibacterium* sp. (X6) isolated from Shenzhen Bay, China. Mar Pollut Bull 170:112656

Xu H, Jia Y, Sun Z et al (2022) Environmental pollution, a hidden culprit for health issues. Eco-Environ Health 1:31–45. https://doi.org/10.1016/j.eehl.2022.04.003

Yadav AN, Singh S, Mishra S, Gupta A (2019) Recent advancement in white biotechnology through fungi. Springer, Cham

Yang Z, Yang F, Liu J-L et al (2022) Heavy metal transporters: Functional mechanisms, regulation, and application in phytoremediation. Sci Total Environ 809:151099

Zhang J, Martinoia E, Lee Y (2018) Vacuolar transporters for cadmium and arsenic in plants and their applications in phytoremediation and crop development. Plant Cell Physiol 59:1317–1325

Nanomaterial's Remarkable Characteristics Majorly Influence

19

Bawoke Mekuye, Tigabu Bekele, Gedefaw Mebratie, Beshir Awol, Birhanu Gebeyehu, Abayeneh Girma, Yenesew Workineh, Ayenew Yibeltal, and Asnaku Tefera

Abstract

Nanomaterials are becoming a research area because they are used in many fields of application, such as agriculture, green technology, artificial intelligence, pollution reduction, nanotechnology, and controlling global warming. The properties of nanomaterials vary depending on their size. This chapter reviews the

B. Mekuye (✉) · B. Awol
Department of Physics, College of Natural and Computational Sciences, Mekdela Amba University, Tulu Awulia, Ethiopia

T. Bekele
Department of Chemistry, College of Natural and Computational Sciences, Mekdela Amba University, Tulu Awulia, Ethiopia

G. Mebratie
Department of Physics, College of Natural and Computational Sciences, Mekdela Amba University, Tulu Awulia, Ethiopia

Department of Physics, College of Science, Bahir Dar University, Bahir Dar, Ethiopia

B. Gebeyehu
Department of Plant Science, College of Agriculture and Natural Resource, Mekdela Amba University, Tulu Awulia, Ethiopia

A. Girma
Department of Biology, College of Natural and Computational Sciences, Mekdela Amba University, Tulu Awulia, Ethiopia

Y. Workineh
Department of Mathematics, College of Natural and Computational Sciences, Mekdela Amba University, Tulu Awulia, Ethiopia

A. Yibeltal
Department of Physics, College of Science, Mizan-Tepi University, Tepi, Ethiopia

A. Tefera
Department of Physics, College of Science, Bahir Dar University, Bahir Dar, Ethiopia

A. A. H. Abdel Latef et al. (eds.), *Bioremediation and Nanotechnology for Climate Change Mitigation*, https://doi.org/10.1007/978-981-96-3069-1_19

nanomaterial's remarkable characteristics, which majorly influence the remediation process to avoid environmental pollutants. One of the main causes of climate change is pollution. The planet suffers from the effects of pollution in the air, water, and soil, which renders the environment unsuitable for human habitation. These include a variety of illnesses (cancer, respiratory and skin conditions, etc.), acid rain, global warming, ozone depletion, and climate change.

Nanoparticles (NPs) have been widely used in the remediation of hazardous wastes. In this chapter, we focus on the many biological origins of NPs and their role in the remediation process to reduce the environmental impact of hazardous wastewater. This chapter discusses some specific nanoparticles' photocatalytic capabilities, including metal-doped cerium, Fe-doped cerium, Mn-doped CeO_2, and CeO_2/Ag_3PO_4 nanocomposites.

Keywords

Nanomaterials · Mine rehabilitation · Remediation · Photocatalysis · Nanotechnology

19.1 Introduction

One-billionth or 10^{-9}, is indicated by the International System of Units (ISU) prefix nano. For instance, one nanokelvin (nK) is a SI unit of thermodynamic temperature equal to 10^{-9} K; a nanoliter is a billionth of a liter or a millionth of a milliliter. A billionth of a meter, a millionth of a millimeter, or a thousandth of a micrometer (μm) are all considered nanometers. The prefix "nano" is becoming increasingly common in scientific writing since it has been used to a wider range of domains of knowledge in recent decades and is now a widely accepted descriptor for most modern research (Buzea et al. 2007; Findik 2021; Pacheco-Torgal and Jalali 2011; Pal et al. 2011). Multidisciplinary scientific domains such as nanoscience and nanotechnology study the synthesis, manipulation, and use of materials, apparatuses, and systems at the nanoscale (Much et al. 2022). When compared to coarser materials with similar chemical composition, materials with external dimensions or an internal structure measured in nanometers that exhibit extra or different characteristics and behavior are sometimes referred to as nanomaterials (Lövestam et al. 2010). Both nanoobjects and nanostructured materials are included in the general term "nanomaterial" (Lövestam et al. 2010). The primary characteristic that unites all nanoparticles is size, which is defined as the length scale ranging from around 1 to 100 nm (Standardization 2015). Whereby materials possess a dimension at least nanoscale. Consequently, materials that have at least one of the three dimensions and between 1 and 100 nm in size are referred to as nanomaterials (Abdalkreem 2018; Afolalu et al. 2019; De et al. 2008; Gaffet 2011; Lövestam et al. 2010; Standardization 2015) and its volumetric spherical surface area must exceed 60 m^2/cm^3 (Gaffet 2011; Mekuye and Abera 2023). In this sense, nanomaterials fall between bulk materials and the atomic level. The surface area-to-volume ratio of nanoscale materials is significantly higher than that of bulk materials, as seen in Fig. 19.1. One important determinant of a material's reactivity is its surface area to volume ratio. There is a greater chance of interactions and reactions when the surface area grows because more atoms or molecules are exposed to the environment. Materials can become more reactive as their surface area per volume rises; for example, higher surface area and enhanced reactivity in nanostructured materials have contributed to the development of better catalysts (Roco 2023). Utilizing living organisms, such as bacteria, to eliminate or neutralize environmental pollutants is known as bioremediation.

Nanotechnology is a major component of this process. By lowering costs, maximizing environmental impact, and enhancing the process' efficacy and efficiency, nanotechnology supports bioremediation. In order to solve environmental issues, this exciting topic integrates the best aspects of sophisticated materials science with biology. As Fig. 19.1 indicates, size becomes smaller, and the surface area to volume ratio becomes larger, from 6,000,000 mm^2 to 12,000,000 mm^2. The surface area of one cube for Fig. 19.1a is 1000 nm times 1000 nm times 6 faces, which is equal to 6,000,000 mm^2. The surface area of one cube for Fig. 19.1b is 500 nm times 500 nm times 6 faces, which is equal to 1,000,000 mm^2. There are eight cubes, and the total surface area is eight times the area ofa cube, which is equal to 12,000,000 mm^2.

Compared to their corresponding particles at bigger scales, nanoparticle matter exhibits distinct chemical, physical, and biological capabilities at the nanoscale because of its remarkable characteristics (Mekuye and Abera 2023). Different from the states of matter that are liquid, solid, gaseous, plasma and gaseous are nanoparticulate matter. Because of the nanomaterials' unique optical, magnetic, and electrical properties, quantum mechanical phenomena become significant in this dimension (Mekuye and Abera 2023). Exposure to chemicals is often thought to pose a risk to human health based on the chemical's intrinsic harmfulness, or toxicity, and the

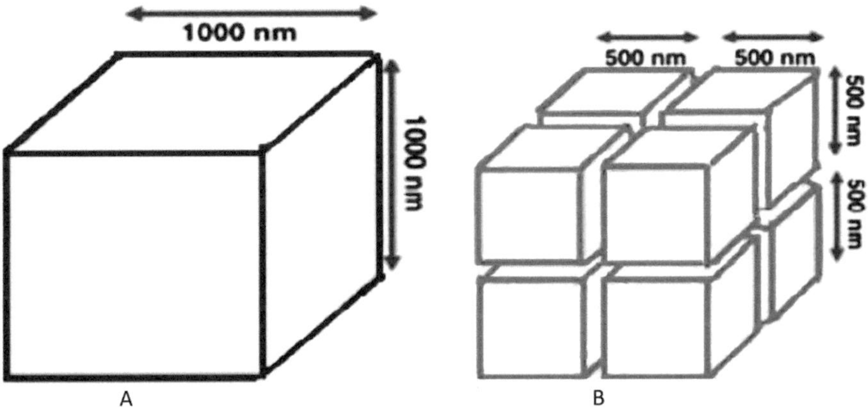

Fig. 19.1 The impact of size of nanomaterials on surface area; **a**) one cube of 1000 nm each side; **b**) eight cubes of 500 nm each side. Reprinted/adapted with permission from (Girma et al. 2024)

dose, or amount, that the chemical accumulates in a particular biological compartment, such as the lungs. The three main ways nanomaterials can enter the body are ingestion, inhalation, and skin contact (dermal contact). Over time, nanotechnology has garnered a lot of attention due to the remarkable properties of nanomaterials, such as high surface area, enhanced reactivity, surface modification, and facilitated transport, which they will employ for future industrial generation (Findik 2021; Joudeh and Linke 2022). Nanomaterials are used in many different industries, including agriculture, materials science, biomedicine, food engineering, electronics, green energy technology, telecommunication, transportation, cosmetics, and mechanical engineering, and coatings (Buzea et al. 2007; De et al. 2008; Findik 2021; Gaffet 2011).

19.2 Classification and Synthesis Methods of Nanomaterials

Based on their origin, size, pore widths, chemical composition, structural arrangement, form, and potential toxicity, nanomaterials can be divided into several classes (Fig. 19.2). Based on origin; natural and artificial; based on structural configuration; organic, carbon-based, inorganic, carbon-based, and composite materials and inorganic; based on dimensions; 3D, 2D, 1D, and 0D; based on pore diameter dimensions; based on potential toxicity; microporous, mesoporous, and mesoporous materials; persistent granular, fiber-like, and CMAR (carcinogenic mutagenic, asthmagenic, reproductive toxin) nanoparticles. Figure 19.3 illustrates the bottom-up and top-down approaches used in synthesizing nanomaterials. Examples of top-down approaches include lithography, mechanical milling or ball milling or mechanical milling, electron explosion arc discharge, sputtering, thermal breakdown, and laser ablation. Examples of bottom-up approaches include pyrolysis, spinning, sol-gel, chemical vapor deposition (CVD), and biological approaches; various plants and microorganisms (algae, fungi, and bacteria), biological templates, and various plant parts are used.

19.3 Some Selected Nanomaterials Application

Applications for nanomaterials are numerous including climate change mitigation (Banerjee et al. 2020; Teotia et al. 2023), green technology (Devatha and Thalla 2018; Thakur et al. 2022), industry environment (Sharma and Bhargava 2013), industrial and its manufacturing application (Palit and Hussain 2020; Subhan et al. 2021), medicine (Aguilar 2012; Fu 2014; Haleem et al. 2023), agriculture (Khot et al. 2012; Peters et al. 2016), and remediation of pollution (Das et al. 2015; Roy et al. 2021; Xue et al. 2017), but this study focuses on the use of nanomaterials in pollution remediation and climate change adaptation to minimize environmental risks. Pesticides, heavy metals, and chlorinated solvents are just a few of the inorganic and organic contaminants found in groundwater that are broken down and transformed by nanoscale zero-valent iron (nZVI) particles (Mishrra and Chatterjee

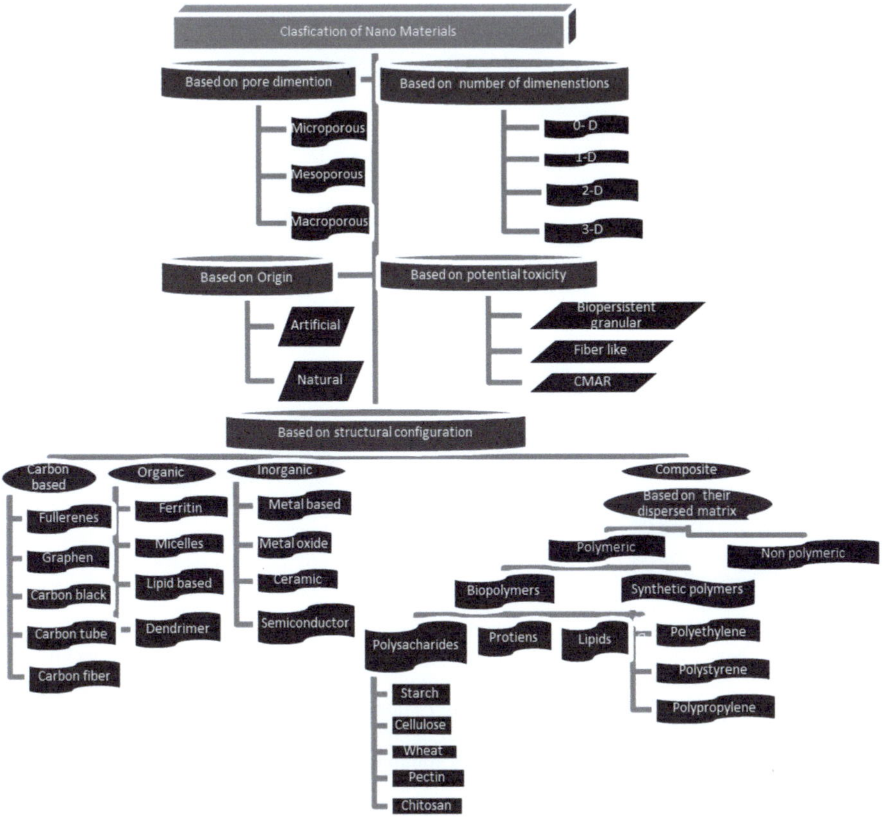

Fig. 19.2 General classification of nanomaterials based on their origin, size, pore widths, chemical content, structural configuration, shape, and possible toxicity. This figure indicates that nanomaterials may be classified in to three groups based on pore dimensions, number of dimensions, and potential toxicity; nanomaterials may be classified into two group based on origin and they may be classified in to four groups based on structural configuration

2023). Through reduction, adsorption, and catalytic processes, the nZVI particles are directly injected into the contaminated groundwater, where they disperse and interact with the target contaminants. Due to their large surface area and special adsorption abilities, carbon nanotubes are utilized to remove a wide range of contaminants from water, such as organic compounds, heavy metals, and microbiological contaminants (Upadhyayula et al. 2009). In air purification systems, titanium dioxide (TiO_2) nanoparticles are employed as photocatalysts to break down and eliminate a variety of air pollutants, including nitrogen oxides (NOx), volatile organic compounds (VOCs), and particulate matter,). The reactive oxygen species that the TiO_2 nanoparticles produce in response to UV radiation have the ability to efficiently oxidize and degrade these air contaminants. TiO_2 nanoparticles can be applied to filters, building materials, and other surfaces to provide built environments the ability to clean themselves and cleanse the air (Colangiuli et al. 2019).

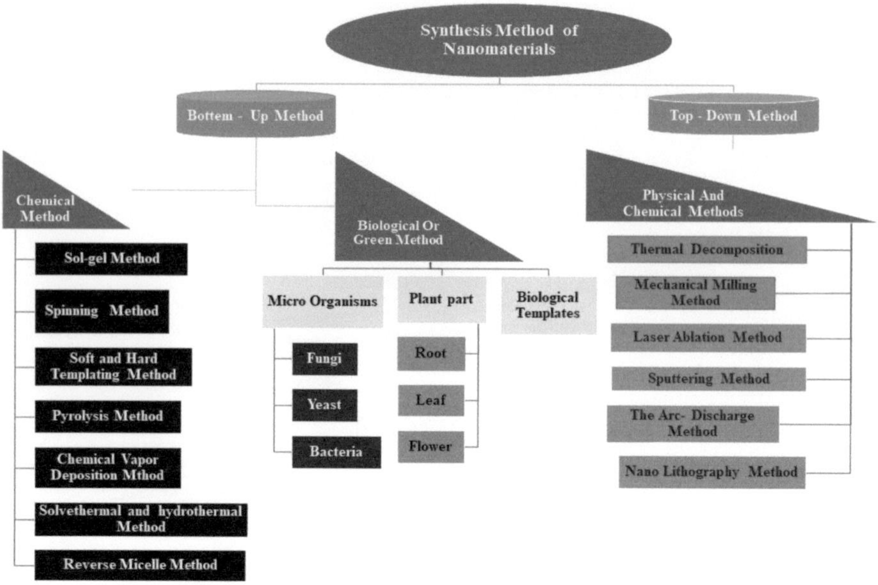

Fig. 19.3 General classification of synthesis methods of nanomaterials based on bottom-up and top-down. We have conducted a critical analysis of several nanomaterial production techniques. Both top-down and bottom-up approaches are used to create nanomaterials, as illustrated in this figure. Ball milling, lithography, sputtering, laser ablation, thermal decomposition, and electron explosion arc discharge are the best examples of the top-down approach. Bottom-up methods are classified as chemical, sol-gel, CVD, pyrolysis, spinning, pyrolysis, and biological methods; biological templates, microorganisms (fungi, bacteria, and fungi), various plant parts and biological templates are used

γ-Fe_2O_3 and magnetite (Fe_3O_4) are two examples of magnetic iron oxides nanomaterials that are used to selectively adsorb and remove heavy metal pollutants from water and soil. These nanoparticles' high surface area and magnetic qualities enable effective heavy metal extraction from contaminated media and simple separation of the metals (Khan et al. 2020; Neyaz et al. 2014). The heavy metals can be recovered and perhaps reused after the adsorption process since the nanoparticles containing the captured heavy metals are easily removed with the help of a magnetic field (Suhasini and Thiagarajan 2021). These illustrations show how successful and adaptable nanomaterials are at solving a range of environmental remediation issues, from heavy metal removal and air pollution control to groundwater and water purification. New nanomaterial-based approaches to environmentally sustainable cleanup and restoration are still being investigated through ongoing research and development.

19.3.1 Nanomaterials in Pollution Remediation

The three main categories of pollution, which are often categorized according to the environment, are air, soil, and water pollution. The technique of using microorganisms or bioremediation to remove pollutants from soil and water is known as bioremediation. The two approaches are referred to as in situ and ex situ. Methods that treat contaminated material onsite are known as in situ methods; ex situ methods involve physically removing the material to be treated elsewhere. This may be summed up as employing plants to treat environmental issues without removing contaminating material from the site and disposing of it elsewhere. Employing plants that have the ability to absorb, break down, or remove metals, solvents, pesticides, crude oil, refined fuels, and explosives, and associated polluting material, phytoremediation can be used to improve contaminated soils, water, or air. The process of humans preparing a site for usage or occupancy is known as mine remediation. This usually entails capping mine shafts, stabilizing structures, and eliminating dangerous materials. The process of returning a place to its pre-mining state is known as mine rehabilitation. The process of cleaning up pollutants from tailings dams and fixing the harm that mining operations have created is known as mine remediation. It is essential in defending the ecosystem against mining's detrimental consequences.

There are several ways to carry out mine rehabilitation, such as decontamination, revegetation, and soil stability. Economic expansion and faster industrial growth have increased energy consumption, released hazardous wastes, released poisonous gasses, and increased exhaust emissions from the automotive sector, among other environmental pollutants (Lima et al. 2016; Lottermoser 2011). Hazardous wastes include different dyes, pesticides, toxic heavy metals, and polyaromatic hydrocarbons (pyrene, chrysene, fluoranthene, naphthalene, anthracene, etc.), which pose significant environmental and health risks due to their persistence, bioaccumulation, and potential toxicity. These pollutants can contaminate water sources, soil, and air, leading to adverse effects on ecosystems and human health. Proper treatment and disposal methods, such as bioremediation, chemical neutralization, and advanced filtration techniques, are essential to mitigate their impact.

Addressing pollution and environmental contamination is a major concern. One of the many solutions that have been found and explored is the application of nanotechnology. Environmental hazardous element remediation has substantially used nanoformulations, including nanotubes, nanomembranes, nanopolymers, metal, graphene-based NPs, and silicon-based and carbon-based nanoparticles NPs (Roy et al. 2021). NPs have been proven helpful in lowering environmental pollutants with minimal consequences on ecosystems because they come from natural sources. Other living things are not harmed by the environmentally friendly green approach of NP synthesis. Because of their noticeably minimal phyto- and cytotoxic effects, the green approach of NP synthesis can be used safely in a broader range of applications, including agriculture, where they enhance plant growth and resilience against pests. This safe application paves the way for innovative approaches in bioremediation, where NPs can detoxify soil and water, leading to healthier ecosystems.

Additionally, their biocompatibility makes them suitable for medical applications, such as drug delivery and imaging, thus contributing to advancements in nanomedicine. Overall, the versatility and safety of natural NPs present an exciting opportunity for sustainable development across various fields.

In order to hasten and encourage the elimination of hazardous substances from the environment, nanomaterials have recently been combined with biological processes (Kumari and Singh 2016). Processes that involve the use of NPs along with microorganisms or plants to remove contaminants are referred to as nanobioremediation (Koul and Taak 2018). Furthermore, El-Ramady et al. categorized these kinds of procedures based on the kind of organism that was used to remove pollutants (El-Ramady et al. 2017). As a result, they gave the methods more precise names, such as microbial nanoremediation, zoo-nanoremediation, and phyto-nanoremediation. Utilizing biotechnologies in conjunction with NMs and NPs could revolutionize remediation capacities by accelerating degradation and preventing process intermediates. Nanotechnology is mainly used in bioremediation because of its small (less than 100 nm) particle size, vast surface area, and stable chemical characteristics. The present review centers on the many biological origins of nanoparticles (NPs) and their role in the remediation process to reduce the environmental impact of hazardous wastes.

19.3.2 Remediation Using Biogenic Nanoparticles to Break Down Toxic Substances

"Bioremediation" refers to remediation procedures that use biological resources, or their parts and extracts, to break down harmful substances into less harmful ones (Jayaprakash et al. 2019), as shown in Fig. 19.4. Bioremediation includes bio sorption, biological stabilization, biotransformation, biological stabilization, and bioaccumulation. These technologies use some microorganisms, including fungi and bacteria and fungi and plants as well as combinations.

The utilization of microorganisms, plants, and animals in tandem with the biofabrication of nanomaterials makes nanotechnology more environmentally friendly and sustainable. Furthermore, one possible remedy is the environmentally friendly manufacturing of nanoparticles using bacterial, fungal, and plant extract enzymes. They produce metallic nanoparticles and serve as reductive agents for the metal complex salt. Through co-precipitation or the addition of proteinaceous and bioactive components to their outer surface, these nanoparticles achieve better firmness in an aqueous environment.

Nanoformulations are excellent substitutes for eliminating environmental pollutants. They are actively employed in treating and removing hazardous contaminants and wastes in remediation operations. Metalloids, heavy metals, oils, hydrocarbons, dyes, pesticides, and hydrocarbons are some contaminants. Because heavy metals and dye pollutants limit the ability of photosynthetic organisms to take up and utilize dissolved oxygen, they pose significant risks to both terrestrial and aquatic biota. Fe-based NPs provide the qualities needed to remove heavy metals from soil

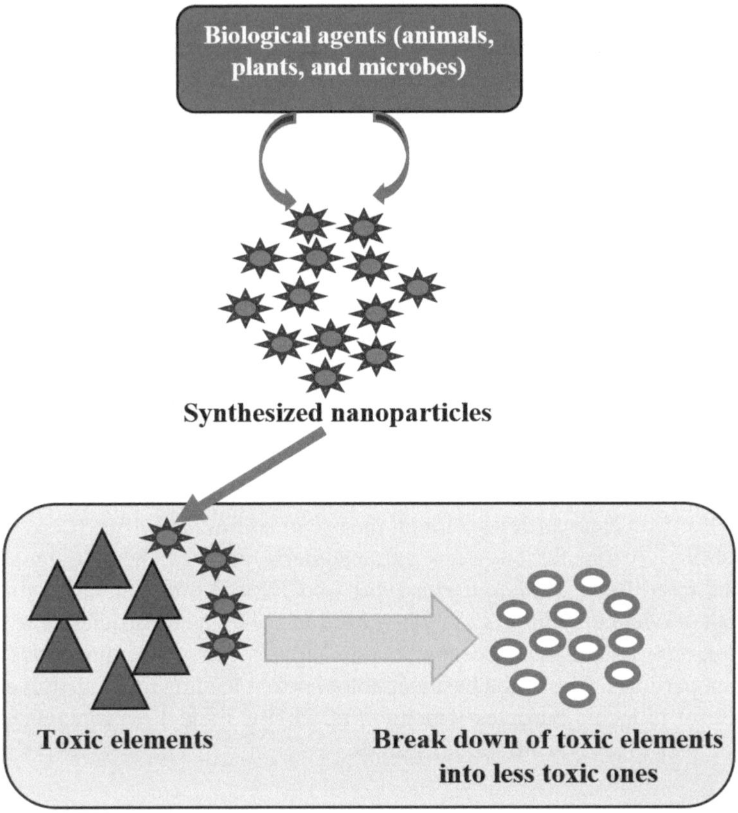

Fig. 19.4 Concept of bioremediation using green nanoparticles. Reprinted/adapted with permission from Singh et al. (2023). This figure explains the biogenic nanoparticle synthesized from biological agents like animals, plants, and microbes (bacteria, fungi, algae) and their roles in the bioremediation process (breaking down of highly toxic elements into less toxic ones)

and disinfect water (Samuel et al. 2022). Using biomass, organic compounds are mineralized into nitrogen, carbon dioxide, and water.

19.3.3 Nanoparticles of Bacteria in Remediation

Bacterial cells are good for producing nanoparticles (NPs) because they have the biomolecules needed to reduce metal ions. NPs can be synthesized in the medium, both intra- and extracellularly and have a wide range of uses. Bacillus licheniformis MTCC 9555's green-synthesized ZnO nanoflower serves as one illustration (Tripathi et al. 2014), which can degrade the pollutant dye methylene blue through photocatalysis. Enzymes that stabilize NPs are secreted by bacterial cells, preventing aggregation. Regarded as an improved oxidation process, photo-catalysis is a useful method that uses direct sun energy application to remove various organic contaminants (Wang et al. 2022). The most researched application of ZnO NPs is as photo-catalysts for treating aquatic wastewater (Liu et al. 2019). Noman et al. created copper nanoparticles using the copper-resistant Escherichia sp. SINT7. It was demonstrated that the biogenic nanoparticles could break down textile effluent and azo dye. Reactive black-5, direct blue-1, congo red, and malachite green were reduced by 83.61, 97.07, 88.42, and 90.55%, respectively, at a lower concentration of 25 mg/L. At a concentration of 100 mg/L, these reductions were reduced to 76.84, 83.90, 62.32, and 31.08%, respectively. The suspended particles as well as the chloride and phosphate ions in treated samples decreased after the industrial effluent was also treated. The effectiveness of these biogenic nanoparticles increases industries' sustainable and economical manufacturing (Noman et al. 2020). Cell-free extracts of many microbial species, including *Sphingomonas paucimobilis*, *Bacillus pumilus*, and *Bacillus paralicheniformis*, were used to create silver nanoparticles (AgNPs). Reductase and bioactive metabolic peptides, two microbial enzymes, help stabilize NPs. As a result, biosynthesized AgNPs can remove color from wastewater with up to 90% effectiveness (Lichtfouse 2005). Cheng et al. produced sulfur-iron nanoparticles without the use of additional sulfur. These nanoparticles' ability to transfer electrons extracellularly allowed them to break down Napthol Green B dye. *Pseudoalteromonas* sp. CF10-13 offers an environmentally acceptable biodegradation process when used to create nanoparticles. The generation of toxic gasses and metal complexes was suppressed by the endogenous synthesis of nanoparticles. One of the best technologies for cleaning up industrial effluents is the use of biogenic particles (Cheng et al. 2019). Table 19.1 contains citations with examples to several additional studies.

19.3.4 Fungal Nanoparticles in Remediation

Nanoparticles (NPs) have become more important due to their low cost and relatively high yield. Fungi are thought to be the most appropriate biological agents since they have a lot of mycelia and fruiting bodies (Abdul-Hadi et al. 2020) and many of the biomolecules needed for NP synthesis (Sudheer et al. 2022). Consequently, compared to other biological agents, the quantity of mycosynthesized NPs is sufficient and occurs quickly (Khandel and Shahi 2018). Many fungal species have been used to produce the NPs used in remediation through biosynthesis. Ferroferric oxide nanoparticles were synthesized from the waste substrate of *Lentigula edodes* in one study, and it was discovered that these NPs were useful in reducing contaminants, including Cr, NH4-N, Pb, Ni, and Cu (Wang et al. 2019). Numerous studies have been referenced and expanded upon. Aspergillus tubingensis (STSP 25) biofabricated iron oxide nanoparticles were isolated from the rhizosphere of Avicennia officinalis in the Sundarbans of India by Mahanty et al. With a five-cycle regeneration capacity, the produced nanoparticles were able to eliminate over 90% of the heavy metals [Pb (II), Cu (II), Ni (II), and Zn (II)] from wastewater. In endothermic processes, the metal ions were chemically adsorbed on the nanoparticle surface (Mahanty et al. 2020). Similarly, it was demonstrated that Aspergillus

Table 19.1 Numerous nanoparticles with biological origins are used in environmental cleanup

The biological resource	The organism's name	Manufactured nanoparticles	The nanoparticles' size (nm)	Source of metal ion	Applications	References
Bacteria	*Bacillus amyloliquefaciens*	Ag	20–40	$AgNO_3$	P-nitrophenol degradation by photocatalysis	Samuel et al. (2020a)
	Bacillus cereus	Ag	51	$AgNO_3$	Removal of lead and Cr	Kumari and Tripathi (2020)
	Shewanella oneidensis	Pd-Pt	13.2	Mixed salt solution of Pd(II) and Pt(IV)	Nitrophenol and azo dye degradation	Xu et al. (2018)
Fungi	*Saccharomyces cerevisiae*	Pd	32	$C_4H_6O_4Pd$	Degradation of a textile dye named direct blue 71	(Sriramulu and Sumathi 2018)
	Acaulospora mellea	Nano-zero-valent iron	69.5	–	Remediation of Zn, Pb, and Cd-contaminated soil	Cheng et al. (2021)
	Ganoderma applanatum	Au	18.7	$HAuCl_4$	Methylene blue dye reduction	Abdul-Hadi et al. (2020)
	Aspergillus tamarii	Fe_3O_4	16.5	$FeSO_4$ $7H_2O$ and $FeCl_3$ $6H_2O$	Cleaning up wastewater that contains textile dyes	El-Sharkawy et al. (2022)

(continued)

Table 19.1 (continued)

The biological resource	The organism's name	Manufactured nanoparticles	The nanoparticles' size (nm)	Source of metal ion	Applications	References
Algae	*Padina pavonica*	Fe_3O_4	23.45	$FeCl_3$	Pb removal from wastewater	El-Kassas et al. (2016)
	Sargassum acinarium	Fe_3O_4	24.5	$FeCl_3$	Elimination of lead from wastewater	El-Kassas et al. (2016)
	Chlorella sp.	ZnO	19.44	$ZnC_4H_6O_4$	Dibenzothiophene degradation	Khalafi et al. (2019)
	Chlorella pyrenoidosa *Scenedesmus obliquus*	CdSe quantum dots	6	$CdCl_2$, Cd $(NO_3)_2{\cdot}4H_2O$, Na_2SeO_3	Reusing hazardous cadmium metal	Zhang et al. (2018)
	Chlorella vulgaris	Ag	55.06	$AgNO_3$	Methylene blue dye degradation	Rajkumar et al. (2021)
	Spirulina platensis	CdS	8.4	$Cd(NO_3)_2$	Malachite green dye degradation and Cd(II) detoxification	Mandal et al. (2016)
Plants	*Amomum longiligulare*	ZnO	50	$Zn(NO_3)_2{\cdot}6H_2O$	Malachite green and methylene blue dye degradation	Liu et al. (2020)
	Anthocephalus cadamba	ZnO	167	$ZnC_4H_6O_4{\cdot}2H_2O$	Cr removal from soil	Mathin et al. (2022)
	Catharanthus roseus	Ag	58.4–97.4	$AgNO_3$	Cleaning up Cr and Cd	Verma and Bharadvaja (2022)
	Catunaregam spinosa	SnO_2	47	$SnCl_2$	Degradation of Congo red dye	Haritha et al. (2016)
	Citrus reticulata	FeO	50	$FeCl_2{\cdot}4H_2O$, $FeCl_3$	Removal of Co+2 heavy metals	Ehrampoush et al. (2015)
	Cynometra ramiflora	Fe_2O_3	58.5	$FeCl_2$, $FeCl_3$	Breakdown of Methylene Dyes	Bishnoi et al. (2018)

	Eucalyptus globulus	FeO	4.17	$Fe(NO_3)_3 \cdot 9H_2O$	Cleaning up Cr and Cd	Andrade-Zavaleta et al. (2022)
	Eucalyptus spp.	ZnO	20–40	$Zn(NO_3)_2 \cdot 6H_2O$	Congo red and malachite green dye degradation	Chauhan et al. (2020)
	Ficus benjamina	Ag	60–105	$AgNO_3$	Cd remediation	Al-Qahtani (2017)
	Jatropha curcas	TiO_2	13	$TiCl_4$	Cleanup of tannery industrial effluent and Cr	Goutam et al. (2018)
	Madhuca longifolia	CuO	30	$Cu\ (NO_3)_2 \cdot 3H_2O$	Methylene dye Degradation	Das et al. (2018)
	Ocimum tenuiflorum	Ag	32.58	$AgNO_3$	Degradation turquoise dye Degradation	Banerjee et al. (2014)
	Parthenium	Fe	100	$FeSO_4 \cdot 7H_2O$	Crystal violet dye deterioration	Rawat et al. (2021)
	Phoenix dactylifera	FeS	68	$FeSO_4 \cdot 7H_2O$	Elimination of Cr(VI) and ciprofloxacin	Bhattacharjee et al. (2021)
	Piliostigma thonningii	Ag	50–114	$AgNO_3$	Heavy metal remediation for Fe, Pb, Cu, and Mg	Shittu and Ihebunna (2017)
	Pimpinella tirupatiensis	Pd	15.4	$PdCl_2$	Degradation of Congo red dye	Narasaiah et al. (2017)
	Plumbago Zeylanica	Ag	55	$AgNO_3$	Methylene blue, phenol red dye methyl red, degradation	Roy and Bharadvaja (2019)
	Psidium guajava	Fe_2O_3-Ag	50–90	$Fe\ (NO_3)_3$ and $AgNO_3$	Cr(VI) heavy metal Remediation	Biswal et al. (2020)

(continued)

Table 19.1 (continued)

The biological resource	The organism's name	Manufactured nanoparticles	The nanoparticles' size (nm)	Source of metal ion	Applications	References
	Sapium sebiferum	Pd	5	$PdCl_2$	Removal of Pb from wastewater	Tahir et al. (2016)
	Sphagneticola trilobata	ZnO	65–80	$ZnC_4H_6O_4$	Cr heavy metal remediation	Shaik et al. (2020)
	Verbascum thapsus	Nano-zero-valent iron	40–50	$FeCl_3$	Cd remediation	Saleh et al. (2021)
	Vitex agnus-castus	SnO_2	8	$SnCl_2$	Removal of Co^{+2} heavy metals and degradation of rhodamine B	Ebrahimian et al. (2020)
	Zingiber zerumbet	ZnO	10	$ZnC_4H_6O_4{\cdot}2H_2O$	Adsorptive Pb(II) elimination	Azizi et al. (2017)

terreus S1-synthesized magnesium oxide nanoparticles (MgO NPs) that significantly removed the Cr(VI) ion, or 97.5% of wastewater. The two stages of the traditional chemical reduction process are the precipitation of Cr(III) as an insoluble hydroxide at alkaline pH and the reduction in Cr(VI) to Cr(III) by a reducing agent at acidic pH. Any salt that contains sulfur or iron can be used as the chemical reductant (Al-Qahtani 2017). According to one study, the fungus strain *Aspergillus niger* BSC-1 shows an exceptional effectiveness in removing Cr(VI) from super-magnetic iron oxide nanoparticles. This leads to microfabricated NPs selectively removing Cr(VI) ions up to >99% (Iavicoli et al. 2014). Table 19.1 contains citations with examples to several additional studies.

19.3.5 Algal Nanoparticles in Remediation

Since they actively take in metal ions from their environment and reduce them to create the corresponding NPs in both live and dried dead form, algae have been referred to as "bio-nano factories" (Pandit et al. 2022). Traditionally, appropriate capping agents or surfactants have been used to achieve a controlled growth rate and energy of producing NPs. These compounds are found in nature as residues that are hard to eliminate since they are not biodegradable fully. Because of their enormous metal-binding capability, naturally occurring biomolecules present in various algae are used for the synthesis and stabilization of NPs to get around this problem (Sharma et al. 2019a). Of the rest, Ag NPs are the most researched and widely used. $AgNO_3$ was exposed to extracts of the seaweed Enteromorpha in one investigation flexusa (Yousefzadi et al. 2014) and *Chaetomorpha linum* The extract's water-soluble constituents, including terpenoids, flavonoids, amines, and peptides, promoted the reduction in metal ions, leading to the creation of Ag NPs (Kannan et al. 2013). The green synthesis of Au NPs was carried out by the reduction in an aqueous solution of chloroauric acid using the freshwater algae *Prasiola crispa* (Sharma et al. 2014). Because they must not affect living biota, toxicity issues are crucial to take into account. Algal synthesis is a safer and more environmentally friendly method because it has little to no toxicity. They can also be cultured easily and with minimal effort, as they are regarded as "nano-reserves" (Baker et al. 2013). Since earlier techniques like redox treatment, UV degradation, activated carbon sorption, proved ineffective, many algae species have been used in remediation processes for the degradation of harmful dyes and chemicals (Sharma et al. 2019a). Seaweeds *Ulva lactuca* (Kumar et al. 2013) and *Hypnea musciformis* (Ganapathy Selvam and Sivakumar 2015), which are highly effective in breaking down methyl orange dye, were used to create Ag NPs in a green synthetic process. It was discovered that spirulina-synthesized green iron oxide nanoparticles were highly efficient at adsorbing crystal violet. NPs can be used to clean wastewater because they cause the solution to become less colored when they are treated with dye-containing water, as confirmed by analytical methods (Bhukal et al. 2022). In a different study, iron oxide nanoparticles and exopolysaccharides (EPS) derived from Chlorella vulgaris were co-precipitated. Fourier-transform infrared spectroscopy (FT-IR) study demonstrated that the functional groups of EPS successfully modified the nanoparticles. Additionally, it was found that the nanocomposite could eliminate 85% of NH_4^+ and 91% of PO_4^{3-} (Govarthanan et al. 2020). Table 19.1 has numerous additional research examples that are referenced and discussed.

19.3.6 Plant Nanoparticles in Remediation

Plant extracts have long been utilized in the synthesis of nanoparticles. Sun-dried leaves of *Cinnamomum camphora* were used to produce Ag and Au nanoparticles with diameters varying from 55 nm to 80 nm (Huang et al. 2007). Because plant extracts are non-pathogenic and may be used in a single step, they are preferable to microorganisms (Pandit et al. 2022). The majority of synthetic dyes are found in wastewater, primarily through industrial effluents. They pose a considerable risk to the ecosystem and can lead to serious health problems and natural imbalances (Wahab et al. 2019). Co-precipitation was used to biosynthesize CoO NPs through *Vitis rotundifolia*, often known as Jumbo Mascadine. This process proved successful in breaking down Acid Blue-74 (Samuel et al. 2020b).

Modern water treatment equipment and methods are more sophisticated, economical, and environmentally benign. *Salvia rosmarinus* extract-mediated TiO_2 NPs were found to efficiently decompose rhodamine B, methyl orange, and methylene blue (Silva-Osuna et al. 2022). FeO NPs were first biosynthesized from *Ruellia tuberosa* leaf extract and subsequently shown to possess antibacterial activity against a range of pathogenic bacteria, such as *Staphylococcus aureus* and *Klebsiella pneumoniae*, in addition to the ability to degrade toxic dyes (Vasantharaj et al. 2019). In one investigation, plant extracts of *Petlophorum pterocarpum* induced the formation of iron oxide (FeO, Fe_3O_4, and Fe_2O_3) particles, which were effective in removing rhodamine from wastewater (Shah et al. 2022). To effectively adsorb cadmium and chromium from wastewater, synthesized cellulose nanofiber and silver nanoparticles from *Citrus silences* (Tavker et al. 2021). According to the transmission electron microscopy images, the average diameters of cellulose nanofiber and silver nanoparticles were 47 and 32 nm, respectively. The synthesized nanocomposites exhibited remarkable adsorption capacities, with a maximum removal efficiency of 95% for cadmium and 92% for chromium, indicating their potential as effective and eco-friendly adsorbents for heavy metal removal from industrial wastewater. In a different work, Jain et al. created an iron nanomaterial from Artocarpus heterophyllus to break down the basic dye of fuchsin. Hematite, iron oxyhydroxide, and zerovalent iron were successfully synthesized, according to X-ray diffraction studies. Using a combination of iron and hydrogen peroxide, they achieved an impressive 87.5% degradation of Fuchsin basic within the first 20 min, demonstrating exceptional Fenton-like degradation (Jain et al. 2021). Table 19.1 includes further examples for several additional studies.

19.4 Mechanisms for Pollution Remediation Using Nanomaterials

As seen in Fig. 19.5, the basic concepts used in environmental pollution remediation comprise three primary approaches: physical, chemical, and biological treatments. Every strategy uses a different set of reaction mechanisms to function. Adsorption and radiation mechanisms are used in oxidation, physical treatment, and reduction mechanisms that are used in chemical treatment, and enzymatic and aerobic microbe-based disinfection are used in biological therapy (Asghar et al. 2024; Marimuthu et al. 2020).

19.4.1 Adsorption

One physical treatment method for removing and hydrating air contaminants is adsorption. Adsorption is a very useful treatment that reduces organic pollutants and has no harmful byproducts from inorganic pollutants. Adsorbents are created with unmodified, modified, and improved NMs, such as zeolites, polymers, activated carbon, and molecular sieves (MOF). It is commonly acknowledged as the primary method of eliminating heavy metals, dyes, and other contaminants. Electrostatic adsorption, ion exchange, complexation/chelation, and surface physical adsorption are four different adsorption approaches.

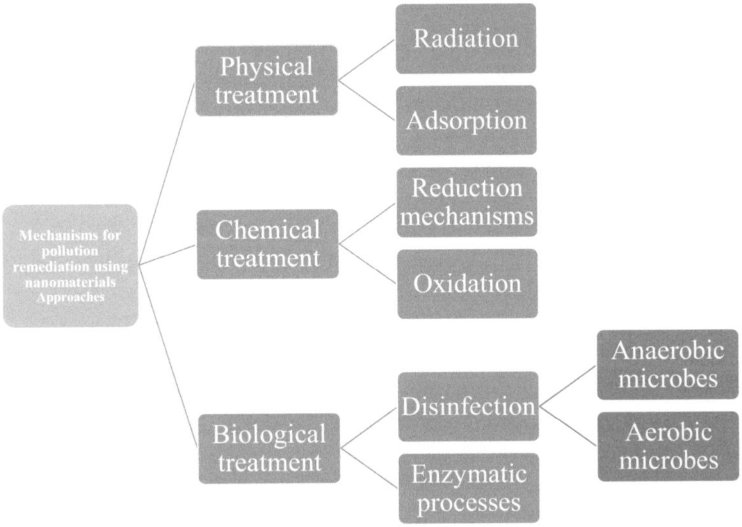

Fig. 19.5 Mechanisms for pollution remediation approaches using nanomaterials

19.4.2 Membrane Filtration

It is a primary method of separating contaminants with membranes that allow only specific materials to pass through. They function in membranes through physical capturing mechanisms such as diffusion, inertial impaction, sieving, and interception. The four basic physical capturing processes of sieving, inertial impaction, interception, and diffusion primarily perform filtration functions.

19.4.3 Disinfection

Because of their broad mechanism of action, which eliminates both organic and inorganic pollutants and renders different MOs, including bacteria, viruses, and protozoa, inactive, NMs have been employed as disinfectants. The disinfection mechanism works because NMs (such NnO^+, Ag^+, and Ti^+) stick to the lipopolysaccharide layer on the outside of the target microorganisms' cell walls. This interaction results in reactive oxygen species (ROS), which breaks down the peptidoglycan layer and causes the lipid membrane to peroxide.

19.4.4 Oxidation

NMs may oxidize both organic and inorganic substances by generating hydroxyl radicals and superoxides, two types of active oxygen species. MnO_2, Mn_2O_5, CuO, TiO_2, ZrO_2, and CeO_2 are often utilized metal oxides for oxidation. The process of dividing the reaction medium to produce oxygen-containing groups on NMs is called oxidation. Usually, an oxidizing agent and one or more inorganic acids, such as sulfuric acid (H_2SO_4) and nitric acid (HNO_3) in a refluxing state, such as H_2O_2, potassium permanganate ($KMnO_4$), sodium hypochlorite (NaOCl), are used in this process.

19.4.5 Photo-catalysis

Endocrine-disrupting chemicals, pesticides, heavy metals, nitrogen oxides (NOx), acetaldehyde, sulfur oxides (SOx), ammonia (NH_3), mercury (Hg), carbon monoxide (CO), polycyclic aromatic hydrocarbons (PAHs), and volatile organic compounds (VOCs) can all be degraded effectively and sustainably using photocatalysis (Iqbal et al. 2021; Magudieshwaran et al. 2019; Zhang et al. 2022; Zhao et al. 2019). This process accelerates the chemical reactions that transform dangerous air pollutants into safe gases, ultimately resulting in the complete mineralization of harmful contaminants. By utilizing photocatalytic materials, such as titanium dioxide (TiO_2) and other nanocomposites, this technology not only enhances the degradation efficiency but also promotes environmental sustainability by reducing the accumulation of toxic substances in ecosystems.

19.4.5.1 Photocatalytic Activities of Pristine CeO_2 Nanoparticle

Numerous non-biodegradable, oil, xenobiotic chemicals, petroleum waste, organic compounds, colors, and other substances are found in wastewater from a variety of sources, including industrial, residential, municipal, and sludge. One of the main causes of pollution is the rise in industrialization. One of the main causes of water pollution is the untreated discharge of waste materials, especially textile, chemicals, paper, pesticide industries, and fertilizer into adjacent water bodies. For decades, numerous traditional chemical, biological, and physical techniques have been employed to address this issue, including sedimentation, adsorption, coagulation, chlorination filtration, zonation, trickling filter, reverse osmosis, oxidation ditch, stabilizing ponds, and aerated lagoon (Pokharna and Shrivastava 2013). However, because they need a lot of space, expensive operating and maintenance costs, and capital investment, these are insufficient and/or inappropriate to treat wastewater streams on a wide scale (Amin et al. 2014). Second, these techniques take several days for analysis and do not break down the effluents to the level that is needed for recycling. Consequently, the development of novel hygienic purification technologies is important.

Before industrial wastewater is discharged into natural water bodies, researchers have backed the use of photocatalytic decomposition to break down and remove any harmful or nonhazardous inorganic, organic, and hard-to-biodegrade substances. The wide band gap, great stability, and nontoxicity of pristine CeO_2 are characteristics of titanium. CeO_2 has been widely chosen as a component to create complex oxides or as a dopant to enhance the performance of titania-based catalysts due to its distinct 4f electron configuration (Song et al. 2007). Pouretedal and Kadkhodaie investigated the rate of MB (20 mg/L) degradation against irradiation duration with and without CeO_2 nanoparticles (0.5 g/L) at a pH of 7 (Pouretedal and Kadkhodaie 2010). The MB molecules degraded slowly under UV-Vis irradiation without the aid of a photocatalyst. The efficiency of degradation is increased by radiation when CeO_2 nanoparticles are present. Due to CeO_2's semiconducting nature, absorption will begin at wavelengths less than 420 nm (Atkins 1978).Thus, electrons can move from the valance band to the conductance band using radiation with a wavelength of less than 420 nm.

19.4.5.2 Photocatalytic Activities of CeO_2/Ag_3PO_4 Nanoparticle

According to Song et al., the photocatalytic breakdown of a few common contaminants when exposed to visible light is used to assess the photocatalytic activity of the as-prepared samples (Song et al. 2016). It is discovered that pure CeO_2 scarcely exhibits photocatalytic activity for MB when exposed to visible light. Pure Ag_3PO_4 has good photocatalytic performance after 18 min of exposure to visible light, with a photocatalytic degradation efficiency of almost 76% of MB. In comparison to pure Ag_3PO_4, the CeO_2/Ag_3PO_4 (1 weight percent) composite degrades a comparatively high amount of MB (88%), which is enhanced by 12%. When Ag_3PO_4 and CeO_2 are combined, the photocatalytic activity increases somewhat, demonstrating that adding CeO_2 is a useful way to improve the pure Ag_3PO_4's photocatalytic performance. The heterojunction between CeO_2 and Ag_3PO_4 may be the cause of the increase in photocatalytic activity for CeO_2/Ag_3PO_4 hybrid materials.

19.4.5.3 Photocatalytic Activities of CeO_2/TiO_2 Nanoparticles

Additionally, Tju and his colleagues produced TiO_2/CeO_2 and assessed the photocatalyst's ability to degrade methyl blue (MB) dye when exposed to visible light (Tju et al. 2017).

It was suggested that the physical characteristics of the samples and the degradation of MB be correlated in order to comprehend the catalytic activity displayed by TiO_2/CeO_2 nanocomposites. In their investigation, CeO_2 loading may raise the samples' specific surface area based on BET measurement. It is evident that CeO_2/TiO_2 has a greater specific surface area than TiO_2 nanoparticles. According to reports, a rise in specific surface area may result in a larger surface active site and higher catalytic activity (Wang et al. 2011). Nevertheless, their research revealed that CeO_2/TiO_2 nanocomposites with a molar ratio of 1:0.5 have the highest catalytic activity across all catalytic tests; these findings suggested that surface area by itself is still insufficient to assess catalytic activity.

19.4.5.4 ZnO/CeO_2 Nanoparticle to Enhance Photocatalytic Performance

Additionally, Taufik and his colleagues used the sol-gel process to create ZnO/CeO_2 nanocomposites. The photocatalytic activity was evaluated using methylene blue (MB) as a model of organic contaminants when exposed to visible light (Taufik et al. 2017). Due to their high rates of electron recombination and hole formation, ZnO and CeO_2 photo catalysts still have limitations. In addition, they can only be activated by exposure to ultraviolet (UV) light (Yin et al. 2014). Taufik and his colleague addressed these issues by combining a ZnO semiconductor with another semiconductor (CeO_2), which has been shown to boost photocatalytic activity (Mageshwari et al. 2015; Shirzadi and Nezamzadeh-Ejhieh 2016).

An n-type semiconductor called cerium (IV) oxide (CeO_2) is frequently employed as a supporting photo-catalyst to boost photocatalytic activity (Contreras-García et al. 2014). By transferring electrons and holes from one semiconductor to another, the CeO_2 semiconductor paired with ZnO can suppress electron recombination and hole rate (Lamba et al. 2015). The degradation efficiency of methylene blue using ZnO/CeO_2 nanocomposite samples in ratios of 1:0.3, 1:0.5, and 1:1 when exposed to visible light. According to their results, the ZnO/CeO_2 nanocomposite with a molar ratio of 1:0.5 had the optimum degrading ability when exposed to visible light. Additionally, the degradation of methyl orange (MO) under a fluorescent lamp was used to assess the photocatalytic activity of the pure ZnO and CeO_2/ZnO catalysts (Rodwihok et al. 2020). By monitoring the shift in the MO adsorption spectra at 464 nm, the degradation of MO was quantified. CeO_2/ZnO-based photo-catalysts shown increased photocatalytic activity, degrading 94.06% of MO in 60 min. Conversely, 69.42% of MO can be broken down by photo-catalysts based on immaculate ZnO. As stated in Fig. 19.6. The findings imply that increased Ce ion doping improves MO photocatalytic degradation.

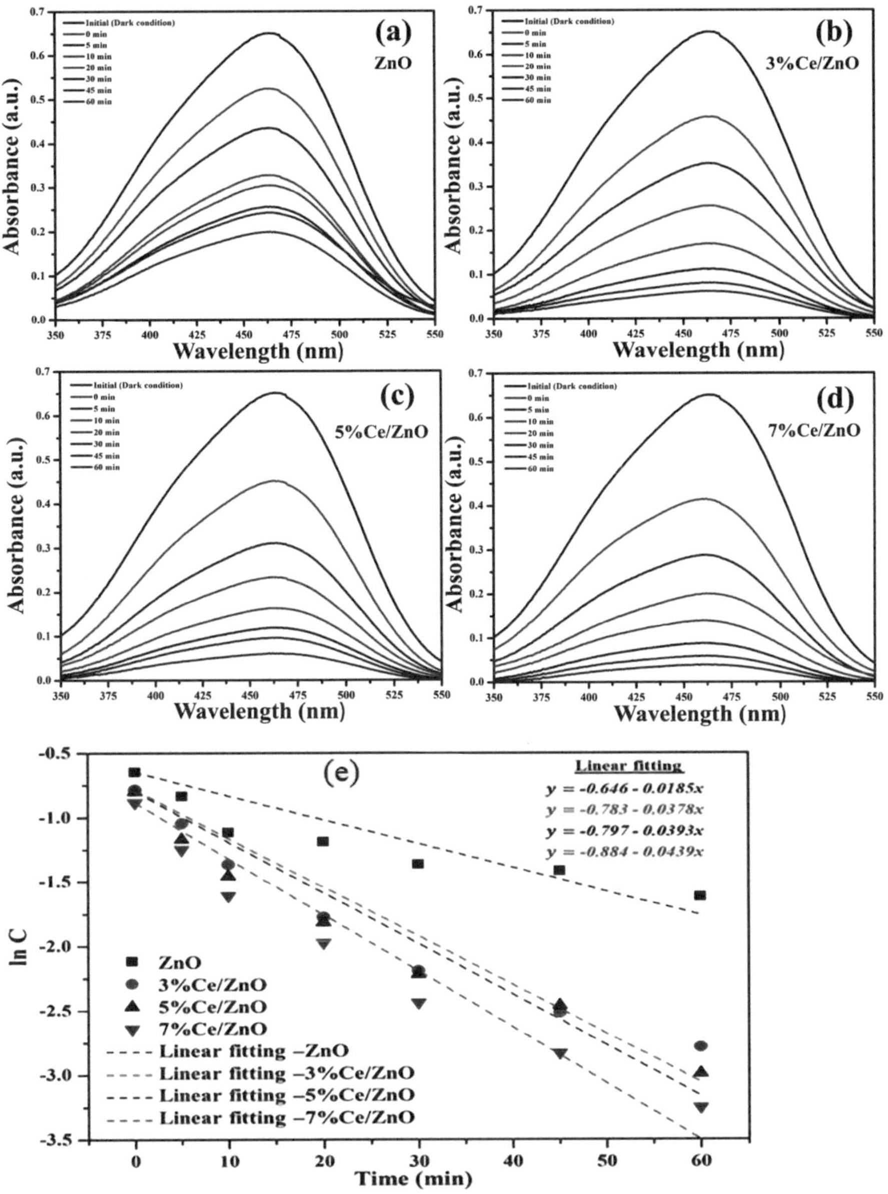

Fig. 19.6 (**a–d**) Modification of the CeO_2/ZnO catalysts' absorption spectra. (**e**) First-order kinetic adsorption curves for methyl orange photodegradation

19.4.5.5 Proposed Photocatalytic Degradation Mechanism

Many attempts are being made to identify sustainable and affordable methods for producing energy, removing pollution, or recycling in light of the growing global interest in environmental challenges like waste management, water and air

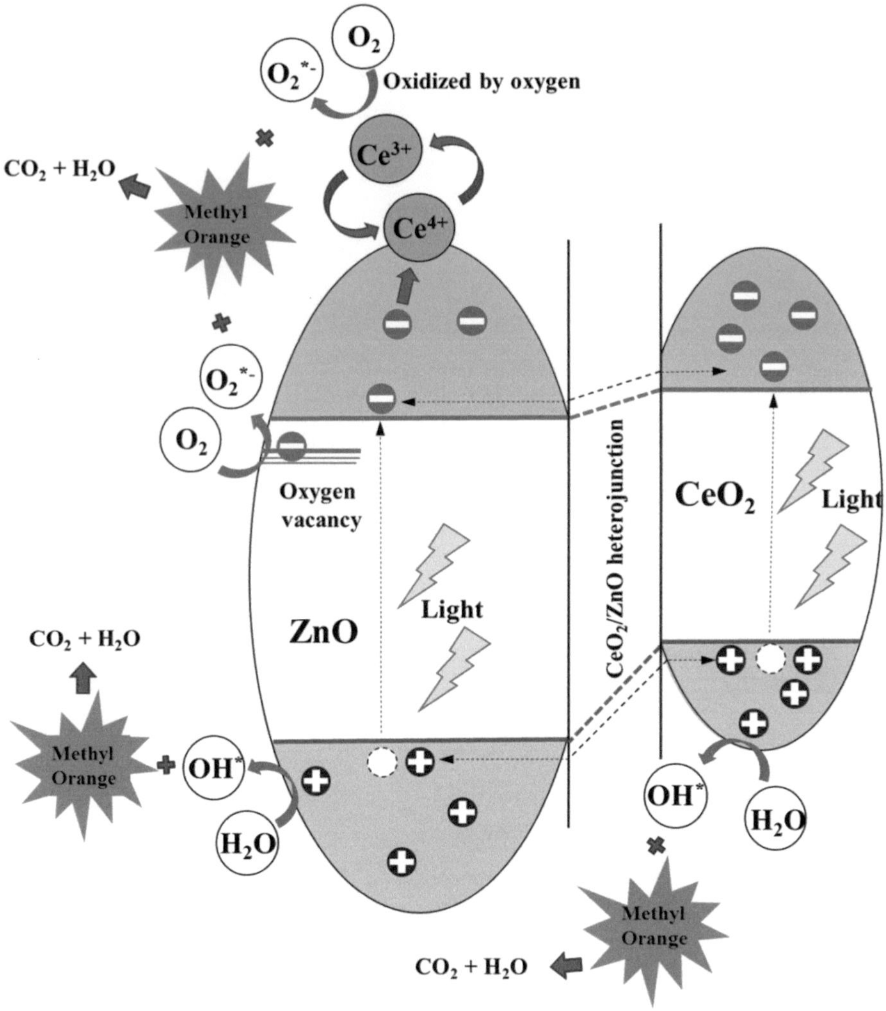

Fig. 19.7 Using CeO_2/ZnO catalysts to degrade methyl orange (MO) through electron trapping and charge transfer. Reprinted/adapted with permission from Rodwihok et al. (2020)

pollution, and global warming. In theory, all of humanity's energy needs should be satisfied by the daily amount of solar energy that the world receives. Nevertheless, solar systems are still insufficiently effective to transform and store the necessary energy. This is why solar energy is a fascinating and quickly expanding research topic that is grabbing the attention of scientists in a variety of fields, such as solar photocatalytic water treatment, hydrogen production using photocatalytic/photo-electro-catalytic water splitting, and electricity generation using photovoltaic or photo-thermal solar panels an essential step in producing energy without carbon. By forming an electron/hole pair at the photo-catalyst level, photocatalytic materials can transform an incident photon into a consumable or storable energy source.

In general, a photo-catalyst is a catalyst that, through the absorption of photons, demonstrates its catalytic characteristics when exposed to light (Dupont et al. 2020; Kabir et al. 2018).

Figure 19.7 shows a potential photocatalytic degradation mechanism of the CeO_2/ZnO catalysts. The following formulas can be used to determine the band-edge positions of the valence band (EVB) and conduction band (ECB) of metal-oxide semiconductors (Khan et al. 2017):

$$E_{\mathrm{CB}} = X - E_c - 0.5 \times E_g \tag{19.1}$$

$$E_{\mathrm{VB}} = E_g - E_{\mathrm{CB}} \tag{19.2}$$

where E_c is the free electron energy on the hydrogen electrode scale (4.5 eV), E_g is the band-gap energy (3.18 for ZnO and 3.00 eV for CeO_2), and X is the electronegativity of ZnO and CeO_2 (5.79 for ZnO and 5.56 eV for CeO_2). The computed ECB and EVB of ZnO were 2.88 eV and −0.30 eV, respectively, based on the conduction and valence band formulae. On the other hand, the estimated ECB and EVB of CeO_2 were 2.56 eV and −0.44 eV, respectively.

The band-edge potentials between CeO_2 and ZnO cause a CeO_2/ZnO heterojunction to develop, which is advantageous for photocatalytic activity since it inhibits charge recombination. Furthermore, it's possible that some Ce^{4+} ions have migrated to the ZnO's surface, which can encourage more trapped electrons through the reaction $Ce^{4+} + e^- \rightarrow Ce^{3+}$. Through the reaction $Ce^{3+} + O_2 \rightarrow Ce^{4+} + O^{2-}$, the reaction's result can react with oxygen molecules to create superoxide radicals. After that, MO is broken down by the produced radicals (superoxide and hydroxyl radicals), which results in oxidized organic compounds. In comparison to pristine ZnO, it is evident that an increase in Ce ion doping can enhance the photo-degradation reaction rate. Lastly, they draw the conclusion that Ce ion doping in CeO_2/ZnO is crucial for generating defects such oxygen vacancies and oxygen interstitials, which regulate photocatalytic activity, and for providing narrower optical energy band-gap tenability. Charge recombination in CeO_2/ZnO nanoparticles would be prevented by the carrier channel that the CeO_2/ZnO heterojunction would offer to separate photo-generated electron-hole pairs.

19.4.5.6 Activity of Metal-Doped Ceria Nanoparticles in Photo-catalysis

Further use is restricted by the Ceria nanoparticle's large band gap energy (3.2 eV). Doping with metal ions was chosen as a solution to this issue because it is thought to promote the Fenton reaction, which enhances photocatalytic activity by generating potent radicals that act as oxidizers (An et al. 2010; Khan et al. 2017). Doping with transition metal ions, such iron, lowers the band gap energy level and increases photocatalytic activity for use as a catalyst in visible regions (Chung and Park 1996). According to reports, doping semiconductors with multivalent transition-metal

(TMs) cations was thought to be an efficient way to prevent photo-generated carriers from recombining (Prabaharan et al. 2018). Through the combined effects of a reduced band gap and the introduction of intermediary bands within the prohibited gap, theoretical research revealed that those metals had the greatest potential to allow for significant optical absorption in visible or even infrared solar light (Shao 2008, 2009). The photo-activity of M-doped ceria oxide under UV and visible light has been the subject of numerous reports recently, such as Y-CeO_2 (Akbari-Fakhrabadi et al. 2015), Mn-CeO_2 (Li et al. 2015), Fe-CeO_2 (Channei et al. 2014), and Sn-CeO_2 (Kumar and Jaya 2013).

19.4.5.7 Photocatalytic Properties of Mn-Doped Ceria Nanoparticle

Since photocatalytic oxidation is triggered by UV or solar radiation, nanostructured semiconductors have a lot of promise for environmental cleanup (Feizpoor and Habibi-Yangjeh 2018). Manufacturers of leather, paint, and textiles are releasing a variety of organic contaminants into water bodies. Because of their special qualities, which include enhanced surface area and effective mass, heat, and charge transfer, nanomaterials are essential for improving energy conversion and storage applications. Semiconductor nanoparticles have shown potential in the solar cell and harvesting industries for energy applications (Feng et al. 2019). These materials can also be used to produce hydrogen, offering effective and sustainable substitutes for silicon-based solar cells. Furthermore, the optical properties of semiconductor nanoparticles can be significantly changed by metal nanoparticles (Kamalieva et al. 2018).

Among the many semiconductor materials, CeO_2 has been thoroughly studied for the removal of environmental pollutants due to its high efficiency, nontoxicity, abundance, photochemical stability, and low cost as a promising photo-catalyst. Its practical use in the photo-catalyst region is, however, limited by a few disadvantages, including as the low quantum yield in the processes, the extremely weak responsiveness to visible light, and the quick recombination rate of photo generated electron-hole pairs. In order to enhance CeO_2-based photocatalytic activity, many strategies for mitigating these limitations have been tried from a photochemistry perspective (Li et al. 2018)

Li and his colleagues used a straightforward one-step composite-hydroxide-mediated technique to create and compare pure and flower-like Mn-doped CeO_2 nanostructures (Li et al. 2018).

The light photocatalytic efficiency of CeO_2 crystalline lattices can be significantly improved by efficiently adding Mn^{2+} ions. Rhodamine blue (RhB) solution's UV/Vis spectral changes over 1% Mn-doped CeO_2 samples during the photo-degradation process clearly demonstrate that the characteristic absorption peaks corresponding to RhB rapidly decrease as the exposure time increases, indicating RhB decomposition and a noticeably lower RhB concentration.

The photo-degradation efficiencies of RhB for 1, 3, and 5% Mn-doped CeO_2 samples were around 65, 40, and 30% after 210 min of irradiation, respectively. It was clear that samples of CeO_2 doped with 1% Mn demonstrated superior photocatalytic activity for the breakdown of RhB. However, as the amount of Mn-doped

Fig. 19.8 TEM images of (**a**) CeO_2, (**b**) $Ce_{0.95}Fe_{0.05}O_2$, (**c**) $Ce_{0.90}Fe_{0.10}O_2$, (**d**) $Ce_{0.80}Fe_{0.20}O_2$ nanoparticles, (**e**) CeO_2, (**f**) $Ce_{0.95}Fe_{0.05}O_2$, and (**g**) $Ce_{0.80}Fe_{0.20}O_2$ nanoparticle SAED patterns. Channei and his colleagues demonstrated how to use a Fe-doped CeO_2 nanoparticle to decolorize methyl orange (MO) (Channei et al. 2014). Their findings demonstrated that, in comparison to undoped CeO_2, iron-modified CeO_2 films exhibited improved photocatalytic activity towards the breakdown of MO in pH 5 conditions, with 1.50 mol% Fe doping offering the fastest dye degradation. The 1.50 mol% Fe-doped CeO_2 film has the highest specific surface area, according to BET analysis, meaning that there are more active sites accessible for the MO degrading reaction (Bangash and Alam 2009)

CeO_2 increased, the photocatalytic performance of the samples declined. The findings indicate that 1% Mn dopant is the ideal Mn doping concentration in CeO_2 samples for UV light photo-catalysis. Because of the decreased average distance between trapped carriers, the Mn dopant sites might also function as effective recombination centers with higher recombination rates when the Mn doping concentration was further raised (Choi et al. 2002). The photocatalytic activity of doped CeO_2 samples may be negatively impacted by the excess Mn dopant sites, which could significantly reduce the quantity of charge carriers.

19.4.5.8 Photocatalytic Properties of Fe-Doped Ceria Nanoparticle

The particle size and surface shape of $Ce_{1-x}Fe_xO_{2-\delta}$ ($0 \leq x \leq 10$) vary significantly with different iron concentrations. As the value of x increases, the particle size tends to increase, and the surface morphology may transition from irregular shapes to more uniform structures. This variation is influenced by the synthesis conditions and the Fe doping level, impacting the overall characteristics of the nanocomposite Fig. 19.8a–f.

A wide range of particle distribution is visible in the transmission electron micrographs. According to the TEM micrograph, $Ce_{1-x}Fe_xO_{2-\delta}$, $(0 \leq x \leq 10)$ NPs have quasi-spherical forms and moderate aggregation. With Fe doping, the particle size dropped from 6.33 nm for undoped CeO_2 to 5.5 nm. The surface-to-volume ratio may rise as a result of the particle size reduction. The specific surface area (Sa) was calculated using the relation $Sa = \frac{6}{D \times \rho} \frac{cm^2}{gm}$ which was found to increase with an increase in Fe doping in CeO_2 nanoparticles. The crystal quality was also impacted by particle size, and some groups have suggested that a reduction in particle size could exacerbate defects due to oxygen vacancies and surface texture deformation (Sharma et al. 2019b).

The crystal quality was also impacted by particle size, and some groups have suggested that a reduction in particle size could exacerbate defects due to oxygen vacancies and surface texture deformation. Focusing the electron onto nanoparticles

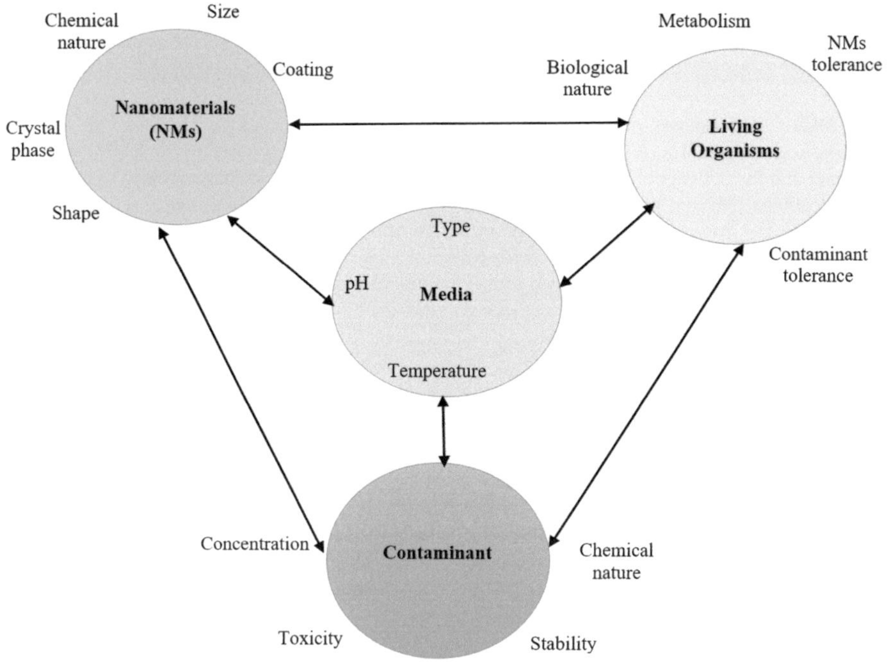

Fig. 19.9 Various factors affect how living things and nanomaterials (NMs) interact with the pollutants. Interactions are represented by two-way arrows. Reprinted/adapted with permission from Vázquez-Núñez et al. (2020). This figure demonstrates the interactions of nanomaterials, living organisms, contaminants, and media. There are different parameters that affect interactions in each group, such as nanomaterials (the size and shape of NMs, surface coating, the chemical makeup of the NMs) and contaminants (concentration, chemical nature, toxicity, and stability), the type of organism (biological nature-animals, plants, or microbes, metabolism capability, tolerance nature of the nanomaterial and contaminant), media (type, pH, temperature), and other variables that affect chemical and physical interactions between biota, NMs, and contaminants Therefore, the bioremediation of contaminants highly depends on the above-mentioned parameters

allowed for the capture of the SAED pattern. The SAED pattern, which is displayed in Fig. 19.8e–g, makes it evident that the produced nanoparticles exhibit a single-phase nature and that the results are in good agreement with XRD.

19.5 The Parameters Affect How Living Things and Nanomaterials (NMs) Interact with the Pollutants

Because bioremediation uses living species to clean up contaminated environments, a proper interaction between nanoparticles (NPs) and living things is essential. In this case, a few elements are crucial. For example, it is widely known that NP size, nanotoxicity, and nanonutrition may all affect live organisms, which can affect the bioremediation process overall.

Tan et al. claim that a number of variables influence the physical and chemical interactions between NMs, biota, and contaminants, including NM shape and size, surface coating, the chemical composition of the NMs and contaminants, the kind of organism, pH, temperature, medium, and other elements (Tan et al. 2018). These interactions are depicted in Fig. 19.9. These occurrences grow complex due to the many possible parameters impacting such interactions.

For instance, temperature and pH medium are crucial for the healthy development of biological things. These factors may, therefore, impact both the contaminant's and the NMs' stability. As an illustration, Au NPs were shown by Wang et al. to be stable in water and a buffer; however, this stability was lost at pH values of 4, 7, 8, and 10 (Wang et al. 2014).

Furthermore, Tang et al.'s research demonstrated how various synthesis techniques affected the thermal stability of Cu NPs (Tang et al. 2016). To our knowledge, no thorough research has been done in the literature on how the criteria mentioned above affect contaminant nanobioremediation. Appropriate experimental designs should be used to find out how much pH and temperature affect living organisms for the remediation of pollutants and the synergistic effect of nanomaterials (NMs) and biological agents. Controlled laboratory experiments, such as factorial designs or response surface methodology, can be implemented to systematically evaluate the interaction between pH levels, temperature variations, and the presence of NMs on the efficiency of pollutant degradation by microbial communities. Additionally, field studies may be necessary to validate laboratory findings and assess the practical implications of these factors on bioremediation processes. By employing these experimental designs, researchers can uncover optimal conditions and enhance the understanding of how environmental factors influence the effectiveness of bioremediation strategies.

A few of the anticipated outcomes of the physicochemical interactions between NMs, pollutants, and biota are shown in Fig. 19.10. Dissolution, absorption, and biotransformation are possible outcomes of NPs and biota interaction (Taylor et al. 2012). Any of the aforementioned occurrences could contribute to the pollutants' deterioration. Metabolism is also engaged in this situation. NPs have the potential to be either stimulant or poisonous to living things, which can have a biocidal or biostimulant effect. This could impact how well the organisms working in the

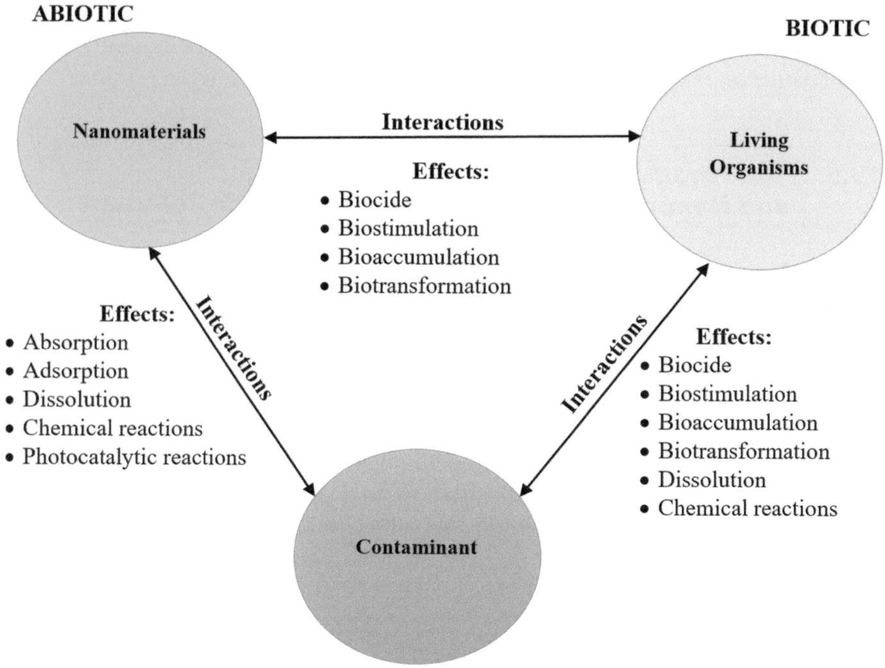

Fig. 19.10 NMs, living things, and pollutants interact physically, chemically, and biochemically during nanobioremediation processes to produce a variety of phenomena. Reprinted/adapted with permission from Vázquez-Núñez et al. (2020). This figure shows the interactions and effects between biotic and abiotic (biocide, biostimulation, bioaccumulation, and biotransformation), biotic and contaminant (biocide, biostimulation, bioaccumulation, biotransformation, dissolution and chemical reactions), and abiotic and contaminant (absorption, adsorption, dissolution, chemical reactions and photocatalytic reactions) effects observed in nanobioremediation processes. Therefore, it is crucial to deeply understand the interactions and effects of nanomaterials, living organisms, and the contaminants during nanobioremediation processes for enhancing the efficient degradation of pollutants. Soil sorption processes are critical to nanobioremediation. Soil sorption percentage varies from study to study. For example, Maida and Nalivata showed that free Al_2O_3 and Fe_2O_3 accounted for approximately 90 and 91% of the overall fluctuations in the sulphate sorption maxima, respectively (Maida and Nalivata 2016)

remediation process function. Thus, the possible synergistic impact is the benefit of using both living creatures and NPs.

Sorption involves both adsorption and absorption. In the first, the sorbent and the pollutant come into contact at the surface level. However, in the second one, the pollutant penetrates the deeper layers of the sorbent to create a solution (Vieira and Volesky 2000). There is also the possibility of another distinction. The difference between physisorption and chemisorption is that the latter just uses physical forces, while the former incorporates a chemical reaction. During sorption, the contaminants may become concentrated, sequestered, or immobilized (Vieira and Volesky 2000). To understand the nature of the NM-based adsorption processes, numerous investigations have been carried out (Sebeia et al. 2020). Therefore, mechanistic, thermodynamic, and kinetic studies are required to characterize the behavior of the

Table 19.2 Some selected nanomaterials and their application on climate change mitigation

Nanomaterial	Area of use	Application principles	Citation
Aluminum	Treatment of water	• Their catalytic and absorption qualities make them useful in water purification procedures	Alemu et al. (2014)
Palladium, platinum, nickel, manganese oxide, nano clays, iron oxides, fullerenes, iron, copper (II) oxide, cobalt, colloidal gold, cluster diamonds, cerium oxide, chromium oxide carbon black, boron oxide, calcium oxide carbon nanotubes, nano-sized composite material, nano-sized transition metal oxide	Remediation of environmental	• In environmental remediation, contaminants such as organic compounds and heavy metals are eliminated from contaminated air and water by the use of catalytic reactions and adsorption • Utilized in the electrodes of energy storage devices, including super capacitors and lithium-ion batteries • They are appropriate for these uses due to their excellent electrical conductivity and stability	Aquatar et al. (2022)
TiO_2 and Ag	The food sector	• Identifying volatile organic molecules in spite of toxicological concerns	Primožič et al. (2021)
Nano-TiO_2 Nano-Fe Nano-zero valent Fe ZnO NPs AuNPs	Abiotic stress on plants	• Enhancing the quantity of gluten and starch, as well as the growth and production of wheat • Improving growth and yield characteristics, lowering H_2O_2 and malondialdehyde (MDA) levels, and increasing chlorophyll and carotenoids • Enhancing the detrimental effects of drought stress on basil plants • Reduced Geno toxicity of Cd and Pb and an enhanced antioxidant defense system • Reduced absorption of Cd and increased absorption of melatonin	Kumari et al. (2022)

(continued)

Table 19.2 (continued)

Nanomaterial	Area of use	Application principles	Citation
Se and Cu NPs AgNPs Nanosilver Silica NPs CuNPs + potassium silicate	Biotic stress on plants	• Decreases the severity of the disease and increases both enzymatic and non-enzymatic substances • Fungicidal characteristics • The effects of antifungal • The prevention of bacterial infections • Induces pathogen tolerance, SAR, and local resistance • Pathogen tolerance is conferred by the production of both enzymatic and non-enzymatic defensive systems	Sob et al. (2020)
TiO_2, CdSe, ZnO_2, Fe_2O_3NPs	The capacity for photo catalysis	• Purification of the air	Alarifi et al. (2009)
Self-cleaning, depolluting, antimicrobial, UV protection materials	Utilizing natural materials	• Preventing pollution	Aquatar et al. (2022)
Nano-$NiFe_2O_4$	Energy efficiency can be improved with nanorefrigeration oil	• Air conditioner for residential use	Alarifi et al. (2009)
TiO_2	Avoid ice and repel water and dust thanks to the super hydrophobic/hydrophilic surface	• Self-cleaning materials and building structure protection; reduction in air pollution	Malekshahi Byranvand et al. (2013)
Nano-TiO_2 and TiO_2 NPs	The oxidation of contaminants into harmless substances is known as photocatalytic capacity	• Water purification • Photocatalytic capacity air purification	Sob et al. (2020)
TiO_2 and Ag	Foodsector	• Identifying volatile organic molecules in spite of toxicological concerns	Joudeh and Linke (2022)
ZnO, SiO_2, CuO, and Mg and Fe	Agricultural	• Fertilizer developer	Joudeh and Linke (2022)
CNTs	Environmental	• Increase the rate at which plants grow	Afolalu et al. (2019)

nanomaterial upon contact with the contaminants. Several models, such as the Freundlich and Temkin Isotherms and the Langmuir andDubinin–Radushkevich models, are explained by some writers (Matouq et al. 2015). These models depict the behavior of incorporating the biological matrix in remediation processes.

Photocatalytic reactions have the potential to decompose pollutants, contingent upon the characteristics of the NMs. The biotic systems may perform additional biotransformations on the resulting products, which would lower the pollutant concentration in the media. Furthermore, several enzymes generated by living things can break down a range of pollutants (Peixoto et al. 2011).

NPs can infiltrate contaminated zones where other entities cannot because of their size. As a result, the application fields of nanobioremediation technologies may grow (Sohail et al. 2019). When compared to alternative remedial strategies, this feature is advantageous. Other factors, meanwhile, must be taken into account, such as standardizing procedures for assessing the toxicity of nanoparticles and nanomaterials in soil and water, clarifying how they interact with biotic and abiotic components, and determining the appropriate regulatory framework in which these materials are used.

19.6 Nanotechnology Applications to Lessen the Consequences of Climate Change on the Environment

Innovations in the fields of nanotechnology and nanobiotechnology are helping the environment. Numerous sustainable solutions to numerous environmental issues have been presented by nanotechnology and nanomaterials, such as greenhouse gas emission, wastewater treatment, remediation of various pollutants, and fuel crisis, that may cause climate change, as shown in Table 19.2. As a result, nanotechnology is finding its way into environmental applications and could be the next big thing in climate change mitigation (Cerqueira et al. 2018; Change 2014; Chausali et al. 2022; Fischer and Knutti 2015).

19.6.1 Nanotechnology in Nano Bioremediation/ Environmental Remediation

One of the main causes of climate change is pollution. The world is suffering from the effects of pollution, including acid rain, ozone depletion, global warming, respiratory and skin disorders, cancer, and other conditions that render the air, water, and land unfit for human habitation.

Environmental nanotechnology (E-nano)-based products may be used for environmental remediation applications. Nanotechnology is the process of producing, modifying, and characterizing structures, instruments, and systems by controlling their shape and size to nanoscale dimensions. The following sections have covered nanotechnology-based methods for treating pollutants and harmful materials as well as controlling pollution (such as water pollution) (Fulekar et al. 2014). To address global warming and the effects of climate change, nanotechnology can convert technology into clean, green, and sustainable alternatives. Because of their special qualities, which include a greater surface area, more surface functional groups, a larger pore volume, and superior electric, mechanical, magnetic, and optical qualities, nanomaterials perform better in a variety of environmental remediation techniques. These nanomaterials with functionalized chemical groups, such as metal-organic frameworks (MOFs), nanofilms, nanocomposites, nanofibers, nanomembranes, carbon nanotubes, nanozeolites, and nanosilica, have a lot of promise for absorbing greenhouse gases (Edelstein 2001). Injecting the **Nano-Zero Valent Iron** (nZVI) particles directly into polluted soil or groundwater allows them to disperse and interact with the target contaminants in an efficient manner. nZVI-based techniques have a number of benefits over conventional groundwater restoration techniques, including quicker rates of degradation, greater versatility in terms of contaminants targeted, and reduced total expenses. The performance, delivery, and synthesis of nZVI for improved environmental cleanup applications are still being improved by ongoing research (Linley and Thomson 2021). To break down organic contaminants in water and air, photocatalytic methods employ nanoparticles such as titanium dioxide (TiO_2). These nanoparticles have the ability to convert dangerous chemicals into less damaging versions when exposed to light. Carbon nanotubes are effective at adsorbing contaminants from soil and water, making them a valuable tool for removing organic chemicals and heavy metals. To remove impurities from wastewater, magnetic nanoparticles are employed. Remediation is made easier by the fact that they are easily extracted from the water using a magnetic field once it has been treated. Nanocomposites: Certain contaminants can be targeted by combining distinct nanoparticles. Bimetallic nanoparticles, for instance, have been created to break down chlorinated substances in groundwater (Mandeep and Shukla 2020).

19.6.2 Applications of Nanotechnology in the Production of Renewable Energy

The primary energy source is conventional energy resources like fossil fuels (diesel, gasoline, and gasoline); however because of the massive emissions of greenhouse gases, these resources have harmed the ecosystem and contributed to global warming. The most effective way to meet the demand for energy has been to use non-conventional or renewable energy sources. In order to fight the energy problem and climate change, biofuels like bioethanol, biodiesel, biogas, and bio-hydrogen, as well as other unconventional energy sources including ocean, geothermal, solar, and wind energy-collectively referred to as "green energy" are essential.

The hydrolysis of lignocellulosic biomass is typically catalyzed by nanomaterials, and immobilized enzymes have been employed for increased efficiency in the manufacture of biofuel (Leo and Singh 2018). Cellulases, laccases, hemicellulases, and other enzymes have been immobilized on matrices based on metal oxide or magnetic nanoparticles. These nanomaterial-enzyme systems, called nano-catalysts, are recognized for their enhanced efficiency. Further, other renewable and non-conventional energy resources (ocean energy, solar energy, wind energy, hydrogen fuel) have become a trend, and nanotechnology is being used to improve the efficiency of these systems. To boost the efficiency of energy generation, solar cells employ thin layer solar cells, polymer solar cells, quantum dots, etc. Likewise, thermoelectric energy conversion has made use of semiconductors with nanostructures (Fulekar et al. 2014).

19.6.3 Application of Nano Technology in Sustainable and Green Architecture

International societies have been moving toward sustainable building mostly due to the energy problem and global warming. Other phrases that have been used as replacements include clean tech, green tech, and clean/green energy. To accomplish sustainability goals, it is consequently essential to comprehend the idea of smart houses and buildings.

Building materials can be made more energy-efficient by applying the concepts of nanotechnology (Mansoori 2005), composites eco-friendly coatings, nano adsorbents, polymeric structures, solar cells nano lubricants, etc. Nanostructure materials or nanoparticles have the potential to upgrade or change conventional design in sustainable or green buildings. Thus, the goal of creating green or clean energy homes may be achieved via nano smart structures or homes (Fulekar et al. 2014).

19.6.4 Nanomaterials' Application in Environmentally Friendly Technology

Green nanotechnology makes use of efficient methods for waste reduction and product recycling.

The use of each nanomaterial is being investigated for fuel cells, solar cells, metal–air batteries super capacitors, regenerative fuel cells, hydrogen energy batteries, and redox flow batteries in order to understand the reaction mechanism and improve material performance for energy storage devices (Dai et al. 2012; Li et al. 2011; Sonkar and Ganesan 2023; Tan et al. 2012).

Green nanotechnology significantly lowers trash production and introduces efficient product recycling methods. The majority of nanoproducts find use in environmental domains such as water treatment, construction materials, and renewable energy industries (Cheng and Chen 2012; Li et al. 2011; Tan et al. 2012). The creation of nanoparticles aided by chlorophyllous plant extracts is also referred to by this phrase. To enhance the reaction process and material performance of energy storage devices, nanomaterials are utilized in super capacitors, metal-air batteries, fuel cells, solar cells, regenerative fuel cells, hydrogen energy, batteries, and redox flow batteries (Li et al. 2011; Nazari et al. 2021; Stenina and Yaroslavtsev 2017).

Green technology is safe for the environment and reduces the risks associated with chemical reactions. Green nanotechnology's primary objective is to use environmentally friendly ingredients to create particles with less negative impacts. Green nanoparticle life cycle studies, which cover raw material synthesis, handling, application, and disposal, are included in environmental impact analyses. Along with creativity, sustainability efficacy, and a green and sustainable solution to a variety of problems posed in many green products guarantee a healthy living environment and industries (García-Serna et al. 2007; Gottardo et al. 2021; Nazari et al. 2021; Sarkar 2013; Sheldon 2012).

Nanotechnology plays a broad role in green energy, tackling anything from environmental clean-up to energy production, all while trying to reduce the ecological impact of human activity (Murthy et al. 2023). It plays a major role in environmental protection and sustainability by enabling innovative techniques for environmental pollution monitoring, clean-up, and mitigation. Since it provides answers for greenhouse gas management, green manufacturing, clean energy technologies, and water purification, nanotechnology has been recognized as a major player in sustainable development. For instance, oil spills can be cleaned up, contaminated soil and groundwater may be cleaned up, and air pollutants can be captured and removed using nanoparticles (Patil et al. 2016). In addition, the production of biofuels from renewable resources and improved solar energy collection and storage techniques are both possible using nanotechnology. Furthermore, goods enabled by nanotechnology, including lighter and stronger materials, can lower the amount of energy used in manufacturing and delivery (Elcock 2007).

19.7 Conclusion

Materials with external dimensions or internal structure measured in nanometers that exhibit unique or extra characteristics and behavior in comparison to coarser materials with same chemical composition are known as nanomaterials. The surface area-to-volume ratio of nanoscale materials is significantly higher than that of bulk materials. Because of its special optical, magnetic, and electrical characteristics, nanoparticle matter has extraordinary chemical, physical, and biological capabilities at the nanoscale. Ingestion, inhalation, and skin contact are the three ways that nanomaterial exposure can enter the body.

Nanotechnology has gained attention due to the remarkable properties of nanomaterials, such as enhanced reactivity, high surface area, surface modification, and facilitated transport, which they will employ for future industrial generation. Nanomaterials are categorized based on their origin, size, pore widths, chemical content, structural configuration, shape, and potential toxicity. They are used in various fields, including climate change mitigation, green technology, industry environment, medicine, agriculture, and pollution remediation. In this study, the focus is on the use of nanomaterials in pollution remediation and climate change adaptation to minimize environmental risks.

Nanomaterials can be used in bioremediation, which removes contaminants from soil and water using microorganisms. While ex situ procedures physically remove contaminated material for treatment elsewhere, in situ methods treat it onsite. Phytoremediation uses plants to improve contaminated soils, water, or air. Mine remediation involves preparing a site for human usage, while mine rehabilitation involves returning a place to its pre-mining state.

References

Abdalkreem TM (2018) Optical properties of gold and silver nanoparticles. Sudan University of Science and Technology, Khartoum

Abdul-Hadi SY, Owaid MN, Rabeea MA, Aziz AA, Jameel MS (2020) Rapid mycosynthesis and characterization of phenols-capped crystal gold nanoparticles from Ganoderma applanatum, Ganodermataceae. Biocatal Agric Biotechnol 27:101683

Afolalu S, Soetan SB, Ongbali SO, Abioye A, Oni A (2019) Morphological characterization and physio-chemical properties of nanoparticle-review. IOP Conf Ser Mater Sci Eng 640:012065

Aguilar Z (2012) Nanomaterials for medical applications. Newnes

Akbari-Fakhrabadi A, Saravanan R, Jamshidijam M, Mangalaraja RV, Gracia M (2015) Preparation of nanosized yttrium doped CeO_2 catalyst used for photocatalytic application. J Saudi Chem Soc 19:505–510

Alarifi A, Deraz N, Shaban S (2009) Structural, morphological and magnetic properties of $NiFe_2O_4$ nano-particles. J Alloys Compd 486:501–506

Alemu S, Mulugeta E, Zewge F, Chandravanshi BS (2014) Water defluoridation by aluminium oxide–manganese oxide composite material. Environ Technol 35:1893–1903

Al-Qahtani KM (2017) Cadmium removal from aqueous solution by green synthesis zero valent silver nanoparticles with Benjamina leaves extract. Egypt J Aquat Res 43:269–274

Amin MT, Alazba AA, Manzoor U (2014) A review of removal of pollutants from water/wastewater using different types of nanomaterials. Adv Mater Sci Eng 2014:825910

An H, Li J, Zhou J, Li K, Zhu B, Huang W (2010) Iron-coated TiO_2 nanotubes and their photocatalytic performance. J Mater Chem 20:603–610

Andrade-Zavaleta K, Chacon-Laiza Y, Asmat-Campos D, Raquel-Checca N (2022) Green synthesis of superparamagnetic iron oxide nanoparticles with Eucalyptus globulus extract and their application in the removal of heavy metals from agricultural soil. Molecules 27:1367

Aquatar MO, Bhatia U, Rayalu SS, Krupadam RJ (2022) Reduced graphene oxide-MnO_2 nanocomposite for CO_2 capture from flue gases at elevated temperatures. Sci Total Environ 816:151522

Asghar N, Hussain A, Nguyen DA, Ali S, Hussain I, Junejo A, Ali A (2024) Advancement in nanomaterials for environmental pollutants remediation: a systematic review on bibliometrics analysis, material types, synthesis pathways, and related mechanisms. J Nanobiotechnol 22:26

Atkins E (1978) Elements of X-ray diffraction. IOP Publishing, Bristol

Azizi S, Mahdavi Shahri M, Mohamad R (2017) Green synthesis of zinc oxide nanoparticles for enhanced adsorption of lead ions from aqueous solutions: equilibrium, kinetic and thermodynamic studies. Molecules 22:831

Baker S, Harini B, Rakshith D, Satish S (2013) Marine microbes: invisible nanofactories. J Pharm Res 6:383–388

Banerjee P, Sau S, Das P, Mukhopadhyay A (2014) Green synthesis of silver-nanocomposite for treatment of textile dye. Nanosci Technol 1:1–6

Banerjee I, Dutta S, Pohrmen CB, Verma R, Singh D (2020) Microalgae-based carbon sequestration to mitigate climate change and application of nanomaterials in algal biorefinery. Octa J Biosci 8:129–136

Bangash F, Alam S (2009) Adsorption of acid blue 1 on activated carbon produced from the wood of Ailanthus altissima. Braz J Chem Eng 26:275–285

Bhattacharjee S, Habib F, Darwish N, Shanableh A (2021) Iron sulfide nanoparticles prepared using date seed extract: green synthesis, characterization and potential application for removal of ciprofloxacin and chromium. Powder Technol 380:219–228

Bhukal S, Sharma A, Rishi D, Kumar S, Deepak B, Pal K, Mona S (2022) Spirulina based iron oxide nanoparticles for adsorptive removal of crystal violet dye. Top Catal 65:1675–1685

Bishnoi S, Kumar A, Selvaraj R (2018) Facile synthesis of magnetic iron oxide nanoparticles using inedible Cynometra ramiflora fruit extract waste and their photocatalytic degradation of methylene blue dye. Mater Res Bull 97:121–127

Biswal SK, Panigrahi GK, Sahoo SK (2020) Green synthesis of Fe_2O_3-Ag nanocomposite using Psidium guajava leaf extract: an eco-friendly and recyclable adsorbent for remediation of Cr (VI) from aqueous media. Biophys Chem 263:106392

Buzea C, Pacheco II, Robbie K (2007) Nanomaterials and nanoparticles: sources and toxicity. Biointerphases 2:17–71

Cerqueira MA, Vicente AA, Pastrana LM (2018) Nanotechnology in food packaging: opportunities and challenges. In: Nanomaterials for food packaging. Elsevier, Amsterdam, pp 1–11

Change C (2014) Synthesis report. In: Contribution of working groups I, II and III to the fifth assessment report of the Intergovernmental Panel on Climate Change. IPCC, Geneva

Channei D, Inceesungvorn B, Wetchakun N, Ukritnukun S, Nattestad A, Chen J, Phanichphant S (2014) Photocatalytic degradation of methyl orange by CeO_2 and Fe–doped CeO_2 films under visible light irradiation. Sci Rep 4:5757

Chauhan AK, Kataria N, Garg V (2020) Green fabrication of ZnO nanoparticles using Eucalyptus spp. leaves extract and their application in wastewater remediation. Chemosphere 247:125803

Chausali N, Saxena J, Prasad R (2022) Recent trends in nanotechnology applications of bio-based packaging. J Agric Food Res 7:100257

Cheng F, Chen J (2012) Metal–air batteries: from oxygen reduction electrochemistry to cathode catalysts. Chem Soc Rev 41:2172–2192

Cheng S, Li N, Jiang L, Li Y, Xu B, Zhou W (2019) Biodegradation of metal complex Naphthol Green B and formation of iron–sulfur nanoparticles by marine bacterium Pseudoalteromonas sp CF10-13. Bioresour Technol 273:49–55

Cheng P, Zhang S, Wang Q, Feng X, Zhang S, Sun Y, Wang F (2021) Contribution of nano-zero-valent iron and arbuscular mycorrhizal fungi to phytoremediation of heavy metal-contaminated soil. Nanomaterials 11:1264

Choi W, Termin A, Hoffmann MR (2002) The role of metal ion dopants in quantum-sized TiO_2: correlation between photoreactivity and charge carrier recombination dynamics. J Phys Chem 98:13669–13679

Chung K-H, Park D-C (1996) Water photolysis reaction on cerium oxide photocatalysts. Catal Today 30:157–162

Colangiuli D, Lettieri M, Masieri M, Calia A (2019) Field study in an urban environment of simultaneous self-cleaning and hydrophobic nanosized TiO_2-based coatings on stone for the protection of building surface. Sci Total Environ 650:2919–2930

Contreras-García M, García-Benjume ML, Macías-Andrés VI, Barajas-Ledesma E, Medina-Flores A, Espitia-Cabrera M (2014) Synergic effect of the TiO_2-CeO_2 nanoconjugate system on the band-gap for visible light photocatalysis. Mater Sci Eng B 183:78–85

Dai L, Chang DW, Baek JB, Lu W (2012) Carbon nanomaterials for advanced energy conversion and storage. Small 8:1130–1166

Das S, Sen B, Debnath N (2015) Recent trends in nanomaterials applications in environmental monitoring and remediation. Environ Sci Pollut Res 22:18333–18344

Das P, Ghosh S, Ghosh R, Dam S, Baskey M (2018) Madhuca longifolia plant mediated green synthesis of cupric oxide nanoparticles: a promising environmentally sustainable material for waste water treatment and efficient antibacterial agent. J Photochem Photobiol B Biol 189:66–73

De M, Ghosh PS, Rotello VM (2008) Applications of nanoparticles in biology. Adv Mater 20:4225–4241

Devatha CP, Thalla AK (2018) Green synthesis of nanomaterials. In: Synthesis of inorganic nanomaterials. Elsevier, Amsterdam, pp 169–184

Dupont E, Koppelaar R, Jeanmart H (2020) Global available solar energy under physical and energy return on investment constraints. Appl Energy 257:113968

Ebrahimian J, Mohsennia M, Khayatkashani M (2020) Photocatalytic-degradation of organic dye and removal of heavy metal ions using synthesized SnO_2 nanoparticles by Vitex agnus-castus fruit via a green route. Mater Lett 263:127255

Edelstein A (2001) Nanomaterials encyclopedia of materials: science and technology. Elsevier, Oxford

Ehrampoush MH, Miria M, Salmani MH, Mahvi AH (2015) Cadmium removal from aqueous solution by green synthesis iron oxide nanoparticles with tangerine peel extract. J Environ Health Sci Eng 13:1–7

Elcock D (2007) Potential impacts of nanotechnology on energy transmission applications and needs. Argonne National Lab (ANL), Argonne

El-Kassas HY, Aly-Eldeen MA, Gharib SM (2016) Green synthesis of iron oxide (Fe_3O_4) nanoparticles using two selected brown seaweeds: characterization and application for lead bioremediation. Acta Oceanol Sin 35:89–98

El-Ramady H, Alshaal T, Abowaly M, Abdalla N, Taha HS, Al-Saeedi AH, Shalaby T, Amer M, Fári M, Domokos-Szabolcsy É (2017) Nanoremediation for sustainable crop production. Nanosci Food Agric 5:335–363

El-Sharkawy RM, Swelim MA, Hamdy GB (2022) Aspergillus tamarii mediated green synthesis of magnetic chitosan beads for sustainable remediation of wastewater contaminants. Sci Rep 12:9742

Feizpoor S, Habibi-Yangjeh A (2018) Ternary $TiO_2/Fe_3O_4/CoWO_4$ nanocomposites: novel magnetic visible-light-driven photocatalysts with substantially enhanced activity through pn heterojunction. J Colloid Interface Sci 524:325–336

Feng H-P, Tang L, Zeng G-M, Zhou Y, Deng Y-C, Ren X, Song B, Liang C, Wei M-Y, Yu J-F (2019) Core-shell nanomaterials: applications in energy storage and conversion. Adv Colloid Interf Sci 267:26–46

Findik F (2021) Nanomaterials and their applications. Period Eng Nat Sci 9:62–75

Fischer EM, Knutti R (2015) Anthropogenic contribution to global occurrence of heavy-precipitation and high-temperature extremes. Nat Clim Chang 5:560–564

Fu PP (2014) Introduction to the special issue: nanomaterials—toxicology and medical applications. J Food Drug Anal 22:1

Fulekar M, Pathak B, Kale R (2014) Nanotechnology: perspective for environmental sustainability. In: Environment and sustainable development. Springer, Cham, pp 87–114

Gaffet E (2011) Nanomaterials: a review of the definitions, applications, health effects. How to implement secure development. arXiv preprint arXiv:1106.2206

Ganapathy Selvam G, Sivakumar K (2015) Phycosynthesis of silver nanoparticles and photocatalytic degradation of methyl orange dye using silver (Ag) nanoparticles synthesized from Hypnea musciformis (Wulfen) JV Lamouroux. Appl Nanosci 5:617–622

García-Serna J, Pérez-Barrigón L, Cocero M (2007) New trends for design towards sustainability in chemical engineering: green engineering. Chem Eng J 133:7–30

Girma A, Mebratie G, Mekuye B, Abera B, Bekele T, Alamnie G (2024) Antibacterial capabilities of metallic nanoparticles and influencing factors. Nano Select 2024:e202400049

Gottardo S, Mech A, Drbohlavová J, Małyska A, Bøwadt S, Sintes JR, Rauscher H (2021) Towards safe and sustainable innovation in nanotechnology: state-of-play for smart nanomaterials. NanoImpact 21:100297

Goutam SP, Saxena G, Singh V, Yadav AK, Bharagava RN, Thapa KB (2018) Green synthesis of TiO_2 nanoparticles using leaf extract of Jatropha curcas L. for photocatalytic degradation of tannery wastewater. Chem Eng J 336:386–396

Govarthanan M, Jeon C-H, Jeon Y-H, Kwon J-H, Bae H, Kim W (2020) Non-toxic nano approach for wastewater treatment using Chlorella vulgaris exopolysaccharides immobilized in iron-magnetic nanoparticles. Int J Biol Macromol 162:1241–1249
Haleem A, Javaid M, Singh RP, Rab S, Suman R (2023) Applications of nanotechnology in medical field: a brief review. Glob Health J 7:70–77
Haritha E, Roopan SM, Madhavi G, Elango G, Al-Dhabi NA, Arasu MV (2016) Green chemical approach towards the synthesis of SnO_2 NPs in argument with photocatalytic degradation of diazo dye and its kinetic studies. J Photochem Photobiol B Biol 162:441–447
Huang J, Li Q, Sun D, Lu Y, Su Y, Yang X, Wang H, Wang Y, Shao W, He N (2007) Biosynthesis of silver and gold nanoparticles by novel sundried Cinnamomum camphora leaf. Nanotechnology 18:105104
Iavicoli I, Leso V, Ricciardi W, Hodson LL, Hoover MD (2014) Opportunities and challenges of nanotechnology in the green economy. Environ Health 13:1–11
Iqbal A, Haq AU, Cerrón-Calle GA, Naqvi SAR, Westerhoff P, Garcia-Segura S (2021) Green synthesis of flower-shaped copper oxide and nickel oxide nanoparticles via Capparis decidua leaf extract for synergic adsorption-photocatalytic degradation of pesticides. Catalysts 11:806
Jain R, Mendiratta S, Kumar L, Srivastava A (2021) Green synthesis of iron nanoparticles using Artocarpus heterophyllus peel extract and their application as a heterogeneous Fenton-like catalyst for the degradation of Fuchsin Basic dye. Curr Res Green Sustainable Chem 4:100086
Jayaprakash K, Govarthanan M, Mythili R, Selvankumar T, Chang Y-C (2019) Bioaugmentation and biostimulation remediation technologies for heavy metal lead contaminant. In: Microbial biodegradation of xenobiotic compounds. CRC Press, Boca Raton, p 24
Joudeh N, Linke D (2022) Nanoparticle classification, physicochemical properties, characterization, and applications: a comprehensive review for biologists. J Nanobiotechnol 20:262
Kabir E, Kumar P, Kumar S, Adelodun AA, Kim K-H (2018) Solar energy: potential and future prospects. Renew Sust Energ Rev 82:894–900
Kamalieva AN, Toropov NA, Vartanyan TA, Baranov MA, Parfenov PS, Bogdanov KV, Zharova YA, Tolmachev VA (2018) Fabrication of silicon nanostructures for application in photonics. Semiconductors 52:632–635
Kannan RRR, Arumugam R, Ramya D, Manivannan K, Anantharaman P (2013) Green synthesis of silver nanoparticles using marine macroalga Chaetomorpha linum. Appl Nanosci 3:229–233
Khalafi T, Buazar F, Ghanemi K (2019) Phycosynthesis and enhanced photocatalytic activity of zinc oxide nanoparticles toward organosulfur pollutants. Sci Rep 9:6866
Khan MM, Khan W, Ahamed M, Alhazaa AN (2017) Microstructural properties and enhanced photocatalytic performance of Zn doped CeO_2 nanocrystals. Sci Rep 7:12560
Khan FSA, Mubarak NM, Khalid M, Walvekar R, Abdullah EC, Mazari SA, Nizamuddin S, Karri RR (2020) Magnetic nanoadsorbents' potential route for heavy metals removal—a review. Environ Sci Pollut Res 27:24342–24356
Khandel P, Shahi SK (2018) Mycogenic nanoparticles and their bio-prospective applications: current status and future challenges. J Nanostructure Chem 8:369–391
Khot LR, Sankaran S, Maja JM, Ehsani R, Schuster EW (2012) Applications of nanomaterials in agricultural production and crop protection: a review. Crop Prot 35:64–70
Koul B, Taak P (2018) Biotechnological strategies for effective remediation of polluted soils. Springer, Cham
Kumar KS, Jaya NV (2013) Synthesis and characterization of pure and Sn-doped CeO_2 nanoparticles. Asian J Chem 25:14267
Kumar P, Govindaraju M, Senthamilselvi S, Premkumar K (2013) Photocatalytic degradation of methyl orange dye using silver (Ag) nanoparticles synthesized from Ulva lactuca. Colloids Surf B: Biointerfaces 103:658–661
Kumari B, Singh D (2016) A review on multifaceted application of nanoparticles in the field of bioremediation of petroleum hydrocarbons. Ecol Eng 97:98–105
Kumari V, Tripathi A (2020) Remediation of heavy metals in pharmaceutical effluent with the help of Bacillus cereus-based green-synthesized silver nanoparticles supported on alumina. Appl Nanosci 10:1709–1719

Kumari A, Bhinda MS, Sharma B, Parihar M (2022) Climate change mitigation and nanotechnology: an overview. In: Sustainable agriculture reviews 53: nanoparticles: a new tool to enhance stress tolerance. Springer, Cham, pp 33–60

Lamba R, Umar A, Mehta SK, Kansal SK (2015) CeO_2ZnO hexagonal nanodisks: efficient material for the degradation of direct blue 15 dye and its simulated dye bath effluent under solar light. J Alloys Compd 620:67–73

Leo VV, Singh BP (2018) Prospectus of nanotechnology in bioethanol productions. In: Green nanotechnology for biofuel production. Springer, Cham, pp 129–139

Li X, Zhang H, Mai Z, Zhang H, Vankelecom I (2011) Ion exchange membranes for vanadium redox flow battery (VRB) applications. Energy Environ Sci 4:1147–1160

Li H, Zhou Y, Tu W, Ye J, Zou Z (2015) State-of-the-art progress in diverse heterostructured photocatalysts toward promoting photocatalytic performance. Adv Funct Mater 25:998–1013

Li P, Zhang W, Zhang X, Wang Z, Wang X, Ran S, Lv Y (2018) Synthesis, characterization, and photocatalytic properties of flower-like Mn-doped ceria. Mater Res 21:e20180167

Lichtfouse E (2005) Environmental chemistry: green chemistry and pollutants in ecosystems. Springer, Cham

Lima AT, Mitchell K, O'Connell DW, Verhoeven J, Van Cappellen P (2016) The legacy of surface mining: remediation, restoration, reclamation and rehabilitation. Environ Sci Policy 66:227–233

Linley S, Thomson NR (2021) Environmental applications of nanotechnology: nano-enabled remediation processes in water, soil and air treatment. Water Air Soil Pollut 232:59

Liu Y, Zhang Q, Xu M, Yuan H, Chen Y, Zhang J, Luo K, Zhang J, You B (2019) Novel and efficient synthesis of Ag-ZnO nanoparticles for the sunlight-induced photocatalytic degradation. Appl Surf Sci 476:632–640

Liu YC, Li J, Ahn J, Pu J, Rupa EJ, Huo Y, Yang DC (2020) Biosynthesis of zinc oxide nanoparticles by one-pot green synthesis using fruit extract of Amomum longiligulare and its activity as a photocatalyst. Optik 218:165245

Lottermoser BG (2011) Recycling, reuse and rehabilitation of mine wastes. Elements 7:405–410

Lövestam G, Rauscher H, Roebben G, Klüttgen BS, Gibson N, Putaud J-P, Stamm H (2010) Considerations on a definition of nanomaterial for regulatory purposes. Jnt Res Centre Ref Rep 80:41

Mageshwari K, Nataraj D, Pal T, Sathyamoorthy R, Park J (2015) Improved photocatalytic activity of ZnO coupled CuO nanocomposites synthesized by reflux condensation method. J Alloys Compd 625:362–370

Magudieshwaran R, Ishii J, Raja KCN, Terashima C, Venkatachalam R, Fujishima A, Pitchaimuthu S (2019) Green and chemical synthesized CeO_2 nanoparticles for photocatalytic indoor air pollutant degradation. Mater Lett 239:40–44

Mahanty S, Chatterjee S, Ghosh S, Tudu P, Gaine T, Bakshi M, Das S, Das P, Bhattacharyya S, Bandyopadhyay S (2020) Synergistic approach towards the sustainable management of heavy metals in wastewater using mycosynthesized iron oxide nanoparticles: biofabrication, adsorptive dynamics and chemometric modeling study. J Water Process Eng 37:101426

Maida J, Nalivata P (2016) Sulphate sorption and desorption characteristics of selected Malawi soils. J Soil Sci Environ Manage 7:133–142

Malekshahi Byranvand M, Nemati Kharat A, Fatholahi L, Malekshahi Beiranvand Z (2013) A review on synthesis of nano-TiO_2 via different methods. J Nanostruct 3:1–9

Mandal RP, Sekh S, Sarkar NS, Chattopadhyay D, De S (2016) Algae mediated synthesis of cadmium sulphide nanoparticles and their application in bioremediation. Mater Res Express 3:055007

Mandeep A, Shukla P (2020) Microbial nanotechnology for bioremediation of industrial wastewater. Front Microbiol 11:590631

Mansoori GA (2005) Principles of nanotechnology: molecular-based study of condensed matter in small systems. World Scientific, Singapore

Marimuthu S, Antonisamy AJ, Malayandi S, Rajendran K, Tsai P-C, Pugazhendhi A, Ponnusamy VK (2020) Silver nanoparticles in dye effluent treatment: a review on synthesis, treatment methods, mechanisms, photocatalytic degradation, toxic effects and mitigation of toxicity. J Photochem Photobiol B Biol 205:111823

Mathin SA, Raju MD, Reddy D (2022) Function of ZnO nanoparticles in the remediation of the toxic metal, seed germination, and seedling growth of the plant, synthesized by stem extracts of Anthophalus cadamba. Lett Appl NanoBiosci 12:23

Matouq M, Jildeh N, Qtaishat M, Hindiyeh M, Al Syouf MQ (2015) The adsorption kinetics and modeling for heavy metals removal from wastewater by Moringa pods. J Environ Chem Eng 3:775–784

Mekuye B, Abera B (2023) Nanomaterials: an overview of synthesis, classification, characterization, and applications. Nano Select 4:486–501

Mishrra A, Chatterjee S (2023) Nanoscale zero-valent iron (nZVI) for rejuvenation of heavily contaminated environment. In: Nanobiotechnology for bioremediation. Elsevier, Amsterdam, pp 51–74

Much RA, Winkelmann K, Hugerat M (2022) Nanochemistry for chemistry educators. Royal Society of Chemistry, London

Murthy MK, Khandayataray P, Samal D, Pattanayak R, Mohanty CS (2023) Green nanotechnology: a roadmap to long-term applications in biomedicine, agriculture, food, green buildings, coatings, and textile sectors. In: Nanomaterials: the building blocks of modern technology: synthesis, properties and applications. Springer, Cham, pp 231–261

Narasaiah P, Mandal BK, Sarada N (2017) Green synthesis of Pd NPs from Pimpinella tirupatiensis plant extract and their application in photocatalytic activity dye degradation. IOP Conf Ser Mater Sci Eng 263:022013

Nazari MA, Maleki A, Assad MEH, Rosen MA, Haghighi A, Sharabaty H, Chen L (2021) A review of nanomaterial incorporated phase change materials for solar thermal energy storage. Sol Energy 228:725–743

Neyaz N, Siddiqui WA, Nair KK (2014) Application of surface functionalized iron oxide nanomaterials as a nanosorbents in extraction of toxic heavy metals from ground water: a review. Int J Environ Sci 4:472–483

Noman M, Shahid M, Ahmed T, Niazi MBK, Hussain S, Song F, Manzoor I (2020) Use of biogenic copper nanoparticles synthesized from a native Escherichia sp. as photocatalysts for azo dye degradation and treatment of textile effluents. Environ Pollut 257:113514

Pacheco-Torgal F, Jalali S (2011) Nanotechnology: advantages and drawbacks in the field of construction and building materials. Constr Build Mater 25:582–590

Pal SL, Jana U, Manna PK, Mohanta GP, Manavalan R (2011) Nanoparticle: an overview of preparation and characterization. J Appl Pharm Sci 6:228–234

Palit S, Hussain CM (2020) Functionalization of nanomaterials for industrial applications: recent and future perspectives. In: Handbook of functionalized nanomaterials for industrial applications. Elsevier, Amsterdam, pp 3–14

Pandit C, Roy A, Ghotekar S, Khusro A, Islam MN, Emran TB, Lam SE, Khandaker MU, Bradley DA (2022) Biological agents for synthesis of nanoparticles and their applications. J King Saud Univ 34:101869

Patil SS, Shedbalkar UU, Truskewycz A, Chopade BA, Ball AS (2016) Nanoparticles for environmental clean-up: a review of potential risks and emerging solutions. Environ Technol Innov 5:10–21

Peixoto R, Vermelho A, Rosado A (2011) Petroleum-degrading enzymes: bioremediation and new prospects. Enzyme Res 2011:475193

Peters RJ, Bouwmeester H, Gottardo S, Amenta V, Arena M, Brandhoff P, Marvin HJ, Mech A, Moniz FB, Pesudo LQ (2016) Nanomaterials for products and application in agriculture, feed and food. Trends Food Sci Technol 54:155–164

Pokharna S, Shrivastava R (2013) Photocatalytic treatment of textile industry effluent using titanium oxide. Int J Recent Res Rev 2:9–17

Pouretedal H, Kadkhodaie A (2010) Synthetic CeO_2 nanoparticle catalysis of methylene blue photodegradation: kinetics and mechanism. Chin J Catal 31:1328–1334

Prabaharan DDM, Sadaiyandi K, Mahendran M, Sagadevan S (2018) Investigating the effect of Mn-doped CeO_2 nanoparticles by co-precipitation method. Appl Phys A Mater Sci Process 124:86

Primožič M, Knez Ž, Leitgeb M (2021) (Bio) nanotechnology in food science—food packaging. Nanomaterials 11:292

Rajkumar R, Ezhumalai G, Gnanadesigan M (2021) A green approach for the synthesis of silver nanoparticles by Chlorella vulgaris and its application in photocatalytic dye degradation activity. Environ Technol Innov 21:101282

Rawat S, Samreen K, Nayak AK, Singh J, Koduru JR (2021) Fabrication of iron nanoparticles using Parthenium: a combinatorial eco-innovative approach to eradicate crystal violet dye and phosphate from the aqueous environment. Environ Nanotechnol Moni Manage 15:100426

Roco MC (2023) National nanotechnology initiative at 20 years: enabling new horizons. J Nanopart Res 25:197

Rodwihok C, Wongratanaphisan D, Tam TV, Choi WM, Hur SH, Chung JS (2020) Cerium-oxide-nanoparticle-decorated zinc oxide with enhanced photocatalytic degradation of methyl orange. Appl Sci 10:1697

Roy A, Bharadvaja N (2019) Silver nanoparticle synthesis from Plumbago zeylanica and its dye degradation activity. Bioinspired Biomimet Nanobiomater 8:130–140

Roy A, Sharma A, Yadav S, Jule LT, Krishnaraj R (2021) Nanomaterials for remediation of environmental pollutants. Bioinorg Chem Appl 2021:1764647

Saleh M, Isik Z, Aktas Y, Arslan H, Yalvac M, Dizge N (2021) Green synthesis of zero valent iron nanoparticles using Verbascum thapsus and its Cr (VI) reduction activity. Bioresour Technol Rep 13:100637

Samuel MS, Jose S, Selvarajan E, Mathimani T, Pugazhendhi A (2020a) Biosynthesized silver nanoparticles using Bacillus amyloliquefaciens; application for cytotoxicity effect on A549 cell line and photocatalytic degradation of p-nitrophenol. J Photochem Photobiol B Biol 202:111642

Samuel MS, Selvarajan E, Mathimani T, Santhanam N, Phuong TN, Brindhadevi K, Pugazhendhi A (2020b) Green synthesis of cobalt-oxide nanoparticle using jumbo Muscadine (Vitis rotundifolia): characterization and photo-catalytic activity of acid Blue-74. J Photochem Photobiol B Biol 211:112011

Samuel MS, Ravikumar M, John JA, Selvarajan E, Patel H, Chander PS, Soundarya J, Vuppala S, Balaji R, Chandrasekar N (2022) A review on green synthesis of nanoparticles and their diverse biomedical and environmental applications. Catalysts 12:459

Sarkar AN (2013) Promoting eco-innovations to leverage sustainable development of eco-industry and green growth. Eur J Sustainable Dev 2:171–171

Sebeia N, Jabli M, Ghith A, Saleh TA (2020) Eco-friendly synthesis of Cynomorium coccineum extract for controlled production of copper nanoparticles for sorption of methylene blue dye. Arab J Chem 13:4263–4274

Shah Y, Maharana M, Sen S (2022) Peltophorum pterocarpum leaf extract mediated green synthesis of novel iron oxide particles for application in photocatalytic and catalytic removal of organic pollutants. Biomass Convers Biorefinery. https://doi.org/10.1007/s13399-021-02189-z

Shaik AM, David Raju M, Rama Sekhara Reddy D (2020) Green synthesis of zinc oxide nanoparticles using aqueous root extract of Sphagneticola trilobata Lin and investigate its role in toxic metal removal, sowing germination and fostering of plant growth. Inorgan Nano-metal Chem 50:569–579

Shao G (2008) Electronic structures of manganese-doped rutile TiO_2 from first principles. J Phys Chem C 112:18677–18685

Shao G (2009) Red shift in manganese-and iron-doped TiO_2: a DFT+ U analysis. J Phys Chem C 113:6800–6808

Sharma P, Bhargava M (2013) Applications and characteristics of nanomaterials in industrial environment. Res Dev 3:63–72

Sharma B, Purkayastha DD, Hazra S, Gogoi L, Bhattacharjee CR, Ghosh NN, Rout J (2014) Biosynthesis of gold nanoparticles using a freshwater green alga, Prasiola crispa. Mater Lett 116:94–97

Sharma D, Kanchi S, Bisetty K (2019a) Biogenic synthesis of nanoparticles: a review. Arab J Chem 12:3576–3600

Sharma M, Aljawfi RN, Kumari K, Chae K, Dalela S, Gautam S, Alvi P, Kumar S (2019b) Investigation of local geometrical structure, electronic state and magnetic properties of PLD grown Ni doped SnO_2 thin films. J Electron Spectrosc Relat Phenom 232:21–28

Sheldon RA (2012) Fundamentals of green chemistry: efficiency in reaction design. Chem Soc Rev 41:1437–1451

Shirzadi A, Nezamzadeh-Ejhieh A (2016) Enhanced photocatalytic activity of supported CuO–ZnO semiconductors towards the photodegradation of mefenamic acid aqueous solution as a semi real sample. J Mol Catal A Chem 411:222–229

Shittu K, Ihebunna O (2017) Purification of simulated waste water using green synthesized silver nanoparticles of Piliostigma thonningii aqueous leave extract. Adv Nat Sci Nanosci Nanotechnol 8:045003

Silva-Osuna E, Vilchis-Nestor A, Villarreal-Sanchez R, Castro-Beltran A, Luque P (2022) Study of the optical properties of TiO_2 semiconductor nanoparticles synthesized using Salvia rosmarinus and its effect on photocatalytic activity. Opt Mater 124:112039

Singh A, Tyagi P, Ranjan R, Sushkova SN, Minkina T, Burachevskaya M, Rajput VD (2023) Bioremediation of hazardous wastes using green synthesis of nanoparticles. Processes 11:141

Sob P, Alugongo A, Tengen T (2020) A conceptual model based on nanoscience and nanotechnology in design of smart home for energy efficiency. Int J Eng Res Technol 13:100–106

Sohail MI, Waris AA, Ayub MA, Usman M, Rehman MZ, Sabir M, Faiz T (2019) Environmental application of nanomaterials: a promise to sustainable future. In: Comprehensive analytical chemistry. Elsevier, Amsterdam, pp 1–54

Song S, Xu L, He Z, Chen J, Xiao X, Yan B (2007) Mechanism of the photocatalytic degradation of CI Reactive Black 5 at pH 12.0 using $SrTiO_3/CeO_2$ as the catalyst. Environ Sci Technol 41:5846–5853

Song Y, Zhao H, Chen Z, Wang W, Huang L, Xu H, Li H (2016) The CeO_2/Ag_3PO_4 photocatalyst with stability and high photocatalytic activity under visible light irradiation. Phys Status Solidi 213:2356–2363

Sonkar PK, Ganesan V (2023) Nanomaterials for sustainable energy applications. CRC Press, Boca Raton

Sriramulu M, Sumathi S (2018) Biosynthesis of palladium nanoparticles using Saccharomyces cerevisiae extract and its photocatalytic degradation behaviour. Adv Nat Sci Nanosci Nanotechnol 9:025018

Standardization IO (2015) Nanotechnologies—vocabulary—part 2: nano-objects. ISO/TS 80004-2: 2015

Stenina IA, Yaroslavtsev AB (2017) Nanomaterials for lithium-ion batteries and hydrogen energy. Pure Appl Chem 89:1185–1194

Subhan MA, Choudhury KP, Neogi N (2021) Advances with molecular nanomaterials in industrial manufacturing applications. Nanomanufacturing 1:75–97

Sudheer S, Bai RG, Muthoosamy K, Tuvikene R, Gupta VK, Manickam S (2022) Biosustainable production of nanoparticles via mycogenesis for biotechnological applications: a critical review. Environ Res 204:111963

Suhasini R, Thiagarajan V (2021) Magnetic nanomaterials for wastewater remediation. In: Nanomaterials for water treatment and remediation. CRC Press, Boca Raton, pp 279–308

Tahir K, Nazir S, Li B, Ahmad A, Nasir T, Khan AU, Shah SAA, Khan ZUH, Yasin G, Hameed MU (2016) Sapium sebiferum leaf extract mediated synthesis of palladium nanoparticles and in vitro investigation of their bacterial and photocatalytic activities. J Photochem Photobiol B Biol 164:164–173

Tan CW, Tan KH, Ong YT, Mohamed AR, Zein SHS, Tan SH (2012) Carbon nanotubes applications: solar and fuel cells, hydrogen storage, lithium batteries, supercapacitors, nanocomposites, gas, pathogens, dyes, heavy metals and pesticides. In: Environmental chemistry for a sustainable world: volume 1: nanotechnology and health risk. Springer, Cham, pp 3–46

Tan W, Peralta-Videa JR, Gardea-Torresdey JL (2018) Interaction of titanium dioxide nanoparticles with soil components and plants: current knowledge and future research needs–a critical review. Environ Sci Nano 5:257–278

Tang H, Chen X, Niu Y, Luo X, Wang Z, Chen M, Shi G (2016) Thermal stability characteristics of in situ nano-particles formed in metal melt. Mater Lett 162:261–264

Taufik A, Shabrany H, Saleh R (2017) Different heat treatment of CeO_2 nanoparticle composited with ZnO to enhance photocatalytic performance. IOP Conf Ser Mater Sci Eng 188:012038

Tavker N, Yadav VK, Yadav KK, Cabral-Pinto MM, Alam J, Shukla AK, Ali FAA, Alhoshan M (2021) Removal of cadmium and chromium by mixture of silver nanoparticles and nano-fibrillated cellulose isolated from waste peels of citrus sinensis. Polymers 13:234

Taylor A, Wilson KM, Murray P, Fernig DG, Lévy R (2012) Long-term tracking of cells using inorganic nanoparticles as contrast agents: are we there yet? Chem Soc Rev 41:2707–2717

Teotia M, Singh S, Singh BP, Akitsu T, Soni RK (2023) Combating climate change with nanoparticles. In: Nanoparticles and plant-microbe interactions. Elsevier, Amsterdam, pp 259–292

Thakur M, Sharma A, Chandel M, Pathania D (2022) Modern applications and current status of green nanotechnology in environmental industry. In: Green functionalized nanomaterials for environmental applications. Elsevier, Amsterdam, pp 259–281

Tju H, Muzakki A, Taufik A, Saleh R (2017) Photo-, sono-, and sonophotocatalytic activity of metal oxide nanocomposites TiO_2/CeO_2 for degradation of dye. AIP Conf Proc 1862:030034

Tripathi R, Bhadwal AS, Gupta RK, Singh P, Shrivastav A, Shrivastav B (2014) ZnO nanoflowers: novel biogenic synthesis and enhanced photocatalytic activity. J Photochem Photobiol B Biol 141:288–295

Upadhyayula VK, Deng S, Mitchell MC, Smith GB (2009) Application of carbon nanotube technology for removal of contaminants in drinking water: a review. Sci Total Environ 408:1–13

Vasantharaj S, Sathiyavimal S, Senthilkumar P, LewisOscar F, Pugazhendhi A (2019) Biosynthesis of iron oxide nanoparticles using leaf extract of Ruellia tuberosa: antimicrobial properties and their applications in photocatalytic degradation. J Photochem Photobiol B Biol 192:74–82

Vázquez-Núñez E, Molina-Guerrero CE, Peña-Castro JM, Fernández-Luqueño F, de la Rosa-Álvarez MG (2020) Use of nanotechnology for the bioremediation of contaminants: a review. Processes 8:826

Verma A, Bharadvaja N (2022) Plant-mediated synthesis and characterization of silver and copper oxide nanoparticles: antibacterial and heavy metal removal activity. J Clust Sci 33:1697–1712

Vieira RH, Volesky B (2000) Biosorption: a solution to pollution? Int Microbiol 3:17–24

Wahab A, Imran M, Ikram M, Naz M, Aqeel M, Rafiq A, Majeed H, Ali S (2019) Dye degradation property of cobalt and manganese doped iron oxide nanoparticles. Appl Nanosci 9:1823–1832

Wang X, Li S, Ma Y, Yu H, Yu J (2011) H_2WO_4 $H_2O/Ag/AgCl$ composite nanoplates: a plasmonic Z-scheme visible-light photocatalyst. J Phys Chem C 115:14648–14655

Wang A, Ng HP, Xu Y, Li Y, Zheng Y, Yu J, Han F, Peng F, Fu L (2014) Gold nanoparticles: synthesis, stability test, and application for the rice growth. J Nanomater 2014:3

Wang C, Tan H, Liu H, Wu B, Xu F, Xu H (2019) A nanoscale ferroferric oxide coated biochar derived from mushroom waste to rapidly remove Cr (VI) and mechanism study. Bioresour Technol Rep 7:100253

Wang H, Li X, Zhao X, Li C, Song X, Zhang P, Huo P (2022) A review on heterogeneous photocatalysis for environmental remediation: from semiconductors to modification strategies. Chin J Catal 43:178–214

Xu H, Xiao Y, Xu M, Cui H, Tan L, Feng N, Liu X, Qiu G, Dong H, Xie J (2018) Microbial synthesis of Pd–Pt alloy nanoparticles using Shewanella oneidensis MR-1 with enhanced catalytic activity for nitrophenol and azo dyes reduction. Nanotechnology 30:065607

Xue X-Y, Cheng R, Shi L, Ma Z, Zheng X (2017) Nanomaterials for water pollution monitoring and remediation. Environ Chem Lett 15:23–27

Yin H, Yu K, Song C, Huang R, Zhu Z (2014) Synthesis of Au-decorated V_2O_5@ ZnO heteronanostructures and enhanced plasmonic photocatalytic activity. ACS Appl Mater Interfaces 6:14851–14860

Yousefzadi M, Rahimi Z, Ghafori V (2014) The green synthesis, characterization and antimicrobial activities of silver nanoparticles synthesized from green alga Enteromorpha flexuosa (wulfen). J Agardh Mater Lett 137:1–4

Zhang Z, Chen J, Yang Q, Lan K, Yan Z, Chen J (2018) Eco-friendly intracellular microalgae synthesis of fluorescent CdSe QDs as a sensitive nanoprobe for determination of imatinib. Sensors Actuators B Chem 263:625–633

Zhang Y, Cheng Y, Qi H (2022) Synergistic degradation of organic pollutants on $CoFe_2O_4$/rGO nanocomposites by peroxymonosulfate activation under LED irradiation. Appl Surf Sci 579:152151

Zhao Y, Yuan B, Zheng Z, Hao R (2019) Removal of multi-pollutant from flue gas utilizing ammonium persulfate solution catalyzed by Fe/ZSM-5. J Hazard Mater 362:266–274

Benefits of Nanotechnology in Dealing with Global Warming and Climate Change

20

Sana Ullah

Abstract

With the continuous rise of global warming and its consequent effect on climate patterns, incorporating nanotechnology provides hope for yet another innovative way to battle this paradox. In this chapter, the recent research work has been compiled and analyzed to discuss different utilities of nanotechnology in dealing with climate change. Nanotechnology, which can manipulate matter at the nanoscale (10–100 nm), has significant potential to create novel solutions in numerous areas. Some of the key advances include nanomaterials for high-efficiency energy conversion, capture, and storage, which will improve solar cells or wind power. Advancements in nanotechnology could reduce carbon emissions in key sectors by up to 20%, particularly in energy-intensive industries like manufacturing and construction. Nanomaterials could split carbon dioxide (CO_2) into oxygen and carbon monoxide 60 times more efficiently than traditional methods, potentially offering a new way to reduce atmospheric CO_2. The review chapter further discusses the potential application of nanotechnology in increasing agricultural yields via nanofertilizers and nanopesticides, which is necessary for encouraging sustainable farming to help prevent climate change-related food supply threats. In addition to this, nanotechnology-based advancements in sensing and clean-up technologies have revolutionized the fields of pollution monitoring as well as water purification. These innovations tackle environmental degradation and help build resilience to challenges brought on by a changing climate. Amid these encouraging breakthroughs, the chapter also notes that more research is required to deal with any possible consequences for environmental and health issues that are an eventual consequence of nanomaterial

S. Ullah (✉)
Department of Agricultural Leadership, Education and Communication, College of Agricultural and Environmental Sciences, University of Georgia Athens, Athens, GA, USA
e-mail: sana.ullah@uga.edu

A. A. H. Abdel Latef et al. (eds.), *Bioremediation and Nanotechnology for Climate Change Mitigation*, https://doi.org/10.1007/978-981-96-3069-1_20

pollutants. The use of ethical problems supported by their deployment and the transnational policies are also debated to ensure this inclusion is a secure, sensible transformation into climate change mitigation methodologies. At the end of this chapter, it is very clear that nanotechnology embraces a huge promise to provide alternative solutions for sustainable future by acting as an advanced technology option that can challenge global warming and climate change with their ability to tackle multifaceted problems.

Keywords

Nanotechnology · Global warming · Climate change · Sustainability · Mitigation

20.1 Introduction

Global warming is the long-term increase in earth's average surface temperature based on human activities such as fossil fuel combustion and deforestation, resulting from anthropogenic sources (Sharma et al. 2023). This temperature rise has seen changes to weather patterns, leading to increased and more intense heat waves. A warming climate is melting glaciers and polar icecaps more rapidly, which also increases sea levels (Hansen et al. 2016). Intensity of rainfall threatens people living in coastal areas and low-lying lands with more flooding and salinization of sources of aquifers. The accelerated absorption of carbon dioxide (CO_2) by the oceans leads to ocean acidification, which damages marine ecosystems such as coral reefs and disrupts marine food webs (Hill and Hoogenboom 2022). Alterations in the timing and magnitude of precipitation have implications for water availability for drinking, agriculture, and industry, potentially leading to freshwater shortages in some places due to changes in regional runoff patterns fueling conflicts over these scarce resources. Changes in temperature, precipitation, and pests due to climate change lead to depressed agricultural productivity (Berhane 2018; Lemi and Hailu 2019). This can result in food scarcity and price fluctuations, particularly in vulnerable areas. Global warming and climate change are major global issues that have enormous impacts not only on human society but also on our ecosystems or the planet as a whole. Rising temperatures, virtually every natural disaster being linked to global warming, and a climate that is changing before our eyes are not distant threats but disasters happening today requiring urgent collective action at the local, national, and international level.

Nanotechnology is known as manipulating and controlling materials at the nanoscale, typically within the range of 1 to 100 nanometers, where unique quantum and surface effects emerge. On such a minute scale, materials are the subject of physical (mechanical, chemical, and optical) properties that may differ from those on larger scales. One of nanotechnology's most valuable applications is climate change and global warming (Chausali et al. 2023). Many believe that nanotechnology provides the ideal solutions to reduce global warming—by improving energy efficiency, developing renewable clean technologies, capturing and storing carbon, or supporting sustainable agriculture, including environmental monitoring and remediation.

20.2 Nanotechnology in Renewable Energy

Nanotechnology has an indispensable role in the future development of renewable energy technologies to improve their efficiency, commercial viability, and scalability. Solar cells were enhanced through nanotechnology to boost Photovoltaic (PV) technology (Wong et al. 2014). Quantum dots, nanowires, and perovskite nanoparticles are examples of nanostructured materials that can increase light absorption and help charge carrier transport within a solar cell. The use of nanomaterials enables innovation in solar technology, resulting, for example, in flexible and lightweight solar panels (Dallaev et al. 2023). This flexibility accommodates flexible form factors for use in architecture and other fields, such as the ability to coat a large area with photovoltaics or manufacture vitreous transparent solar cells that are amorphous (i.e., lacking trans crystalline regions), which is one more reason why this topic becomes novel. Nanostructured transparent conducting films (TCFs), including graphene and carbon nanotube, are used as alternative electro-conductive materials for Indium Tin Oxide (ITO) coatings (He and Tjong 2016). TCFs of this type increase light transmission and electrical conductivity, improving the efficiency of solar panels. The development of batteries with higher energy density, life cycle, and charging rates requires nanotechnology (Liu et al. 2017). Lithium-ion batteries have led to applications for nanostructured materials such as nanowires, nanotubes, and new battery types like solid state. Supercapacitors equipped with nano-engineered electrodes boast a large surface area and short charge/discharge times (Siuzdak and Bogdanowicz 2018). These features make supercapacitors attractive for dynamic applications where they require high power density and long cycling performance.

Effective nanocatalysts for hydrogen fuel cells: The catalytic activity increased at platinum nanoparticles on carbon supports (nanostructured materials), which helped decrease the cost and raise fuel cell systems' efficiency for clean energy generation (Qiao et al. 2021). It is a part of contributing the nanotechnology for development of materials now lightweight and tough.

Manufacturing wind turbine blade material: Better tensile strength to weight ratio is achieved with the help of nano-reinforced composites, which ensures an efficient and reliable wind power generation by increasing efficiency and product life cycle in case of turbine blades (Muzammil et al. 2019). On windows, nanocoatings help regulate solar heat gain and infrared radiation to keep buildings energy efficient by Withers et al. These coatings can save energy by reducing the requirement for heating and cooling in buildings.

Although nanotechnology has emerged as a potential game changer in the realm of renewable energy, the challenge lies with scaling up both high-throughput and scalable production of these nanomaterials; additional challenges include consideration for cost-effectiveness while maintaining large-scale application capability. It will be important to evaluate the environmental consequences and possible health implications of nanomaterials designed for implementation in renewable energy systems, so as they can ultimately improve sustainability. To effectively manage risks and stimulate responsible innovation, adequate regulation and norms for nanotechnology-based renewable energy technologies must be developed.

20.3 Nanotechnology in CO_2 Capture and Storage

Nanotechnology provides novel solutions for CO_2 capture and storage, a significant greenhouse gas causing global warming (Kumar et al. 2020). Metal-organic frameworks (MOFs) are a class of crystalline porous materials constructed by connecting metal ions or clusters via bridging organic linkers with the feature of large surface areas and optimal pore sizes. However, due to the advent of MOFs with pore structures and chemically functionalized surface areas well-tailored for adsorbing CO_2 molecules (Arstad et al. 2008), they are highly selective toward the removal of olefin. Their high selectivity makes them useful for separating CO_2 from gas mixtures like those in flue gases produced by power plants. Another class of porous materials is porous organic polymers (POPs), which have high surface areas with predefined or tunable pore sizes fabricated according to specific requirements (Tao et al. 2020). They have shown a high affinity for CO_2 molecules, providing promise as adsorbents for capturing CO_2 from industrial emissions and gas streams. A polymer matrix is incorporated with nanomaterials—zeolites, carbon nanotubes, or MOFs—to fabricate mixed matrix membranes (MMMs) (Yazid et al. 2022). The introduction of LNM – Layered Nanomaterials in MMMs gives several advantages for CO_2 capture that improve the gas separation properties, such as improvements in permeability and selectivity toward CO_2; these can be substantial steps to facilitate its extraction from mixed/gas streams. The nanoporous membranes loaded with nanoscale pores allow CO_2 molecules to readily penetrate whereas acting as roadblocks for the alternate gases (Wang et al. 2017). This research applies to gas separation applications, particularly in CO_2 capture from industrial emissions and natural gas processing. In the first case, nanotechnology allows us to work with very efficient catalysts for CO_2 conversion into valuable products such as fuels or chemicals (Alli et al. 2023). The carbon dioxide conversion processes will become financially attractive if the reaction rates and selectivity are increased by employing nanostructured catalysts. Some nanoparticles have been found to facilitate the transformation of CO_2 into stable carbonates or bicarbonates (Power et al. 2016). Nanominerals can be employed for capturing CO_2 in geological formations or within industrial waste streams, thereby avoiding the release of this greenhouse gas into the atmosphere.

The cost and complexity of manufacturing could also limit the application growth, as scaling nanotechnology-based CO_2 capture and storage up to industrial size is difficult. A life-cycle analysis of the environmental footprint for nanomaterials used in carbon capture and storage (CCS) technologies should be performed to evaluate whether their usage can consume enough CO_2 and if they are renewable.

Creating regulatory frameworks and standards that guarantee the safety deployment and environmental requirements is a must for nano-CCS technologies.

20.4 Nanotechnology in Agriculture and Food Security

One of the main promises is related to agriculture and food security. This area could benefit from nanotechnology providing innovative solutions for increased crop productivity, nutrient delivery systems, and problematics such as contamination that would have an environmental impact. For example, nanotechnology helps to deliver crop problems both genetically and non-genetically (Yadav et al. 2024). Using nanoencapsulation protects a low-level active compound from degradation and increases plant uptake (Pateiro et al. 2021). Nanomaterials can release nutrients, and pesticides slowly make their efficacy in longer periods with less frequent applications. Nanoformulations of fertilizers and pesticides enhance solubility, stability, and bioavailability, producing high crop yields while decreasing input usage amount (Kumar et al. 2023). Nanotechnology devices may be used to monitor the preliminary symptoms of plant diseases or stress factors such as nutrient deficiency, pests, and pathogens (Kumari et al. 2023). This allows for timely and targeted intervention. Nanostructured seed- or plant-surface coatings can protect them against pests, solar ultraviolet (UV) radiation, and harsh environmental conditions that improve the resilience of crops for better productivity (Shang et al. 2019). Nanomaterials serve to make soils more structured and facilitate increases in water retention together with the nutrients available. They can also immobilize pollutants and increase soil fertility, leading to sustainable agriculture. Nanotechnology is used in the slow and controlled release of fertilizers and soil amendments to achieve best nutrient utilization by plants and minimize environmental leaching (Elemike et al. 2019). Purified water is being used for irrigation by nanostructured membranes and filters that can remove contaminants such as heavy metals and pathogens (Joseph et al. 2023). This enhances the quality of water and helps promote sustainable agriculture. These sensors reliably track soil moisture levels and predict plant water requirements in real time to maximize irrigation efficiency and minimize wastage of precious water (Pramanik et al. 2020). Nanotechnologies in food packaging provide multiple benefits, including anti-infection and excellent barrier characteristics of food packaging material. This prolongs shelf life, slows food spoilage, and maintains the quality of foods throughout storage and transportation. The nanoscale barcodes and tags that track food products as they go through the supply chain can also improve consumers' trust, safety, and transparency in relation to their foods (Echegoyen 2015; Worku et al. 2024).

Consumer and environmental safety: Nanomaterials used in agriculture, and food safety require a careful assessment. It is also important to formulate regulatory guidelines for nanotechnology in farming, which helps govern risks and enables consideration before they are used. To ensure acceptance and to build trust with consumers, public concern on nanotechnology use in food and agriculture must be attended.

20.5 Nanotechnology in Environmental Monitoring and Remediation

Over the years, significant attention has been given to nanotechnology in environmental monitoring and remediation for innovative pollution detection techniques up to water purification process as well as soil decontamination. Nanotechnology-based sensors have high sensitivity and selectivity toward various environmental pollutants such as heavy metals, pesticides, volatile organic compounds (VOCs), and greenhouse gases (Chausali et al. 2023). This will result in the ability to watch out for environmental pollutants in real time and mitigating risks from air (gases, e.g., N_2, O_2/bio-aerosol), water (heavy metals: Pb^+, Hg^+, As^+), or soil quality, enabling clean production methods and environment-friendly end-of-pipe solutions development. The development of sensor devices has been a significant milestone in animal health given that they are compact and portable and sometimes enable continuous monitoring, particularly when being applied to remote or challenging environments using near-field communication (NFC) utilizing nanotechnology capabilities (Mamun and Yuce 2020). Nanostructured membranes have high adsorption properties to help remove contaminants in water sources such as bacteria, viruses, heavy metals, and organic pollutants (Manikandan et al. 2022). High permeability and selectivity from these water purification membranes improved the quality of drinking and agricultural waters.

Adsorbent and catalysts: These include graphene oxide, carbon nanotubes, or other types of nano-sized metal oxides. In this scenario, some types of contaminants will interact with functional groups present in nanomaterial matrix leading to removal/water purification. The best answer we can get is the end result (Wang et al. 2019). They can adsorb organic and inorganic pollutants or even catalyze the reactions that would degrade those contaminants.

Development in arid regions: Nanotechnology has the potential to tackle water scarcity issues by increasing desalination process efficiency through membrane enhancement and energy savings (Goh et al. 2013). Nano sensors are created by applying nanotechnology, which allows for developing air pollutant detection mechanisms with high levels of specificity and sensing powers (Tovar-Lopez 2023). These sensor types monitor particulate matter (PM), nitrogen oxides (NOx), sulfur dioxide (SO_2), ozone (O_3), and volatile organic compounds (VOCs) in urban and industrial environments. For uniquely designed nanotechnology sensors, we could mount them onto drones to be permanently floating above, or safe within a living room (built-in satellite) and then monitor the air quality at huge scales over massive geographical areas. To remediate contaminated soil, nanostructured materials are often employed as adsorbents for capturing and immobilizing a wide range of pollutants, including heavy metals and organic contaminants. Nanoparticles such as zero-valent iron (nZVI) and titanium dioxide (TiO_2) are used in the field of in situ remediation of contaminated sites (Wu et al. 2022). Said nanoparticles can degrade or transform contaminants through oxidation-reduction reactions and photocatalytic processes. This is commonly achieved through the design of green, eco-friendly, and sustainable nanomaterials and associated technologies such as

Graphene-Based Smart Materials (GrGSMATS) (Khan 2020). In this context, enforcing research projects that utilize renewable energy sources to produce nanomaterials or reduce energy consumption in environmental remediation applications would be imperatively significant. There is a need to carefully evaluate the environmental and health impacts of various nanomaterials utilized in environmental applications to avail safe deployment but with minimal unintended consequences. It is crucial to set up regulatory frameworks and standards for nano-based environmental solutions to minimize risks and comply with the current environmental regulations. The acceptance of nanotechnology in environmental applications hinges on public concerns for the safety and ethical issues associated with nanomaterials.

20.6 Regulatory Challenges and Ethical Consideration

However, there are some regulatory and ethical concerns about which nanotechnology can be utilized to fight against global warming. It needs a great level of responsible development and careful management. Nanomaterials can represent new associated risks, relating to their small sizes and the potential corresponding surface area increase that could unintentionally lead to adverse environmental effects or human health (Barhoum et al. 2022). Regulators require substantial frameworks to understand and mitigate these risks with a political compass. Standardized testing protocols and methodologies customized to nanomaterials must be created to evaluate if they are safe for different applications. The new properties and functionalities associated with the applications of nanotechnology are not necessarily well-served by current regulatory frameworks.

Purpose: Built regulations are required, which are sensitive to the unique properties and possible hazards of nanomaterials. An exceptional need in consistency of nomenclature and international harmonization among the various regulations for global warming mitigation should be addressed by nanotechnology solutions. Proper labeling and registration obligations will also be in place to monitor how nanomaterials are used and how they are discarded from various applications. This aids in transparency, risk communication, and informed stakeholder decision-making. Nanomaterials should be dealt with separately for disposal and end-of-life-cycle management as they are a potential source of environmental contamination. Updated regulatory guidelines are necessary to ensure that nanotechnology-based products and waste are handled, transported, and disposed of safely.

The consumer perspective on the safety and ethical aspects of nanotechnology matters for the acceptance and uptake of nano applications. Communicating in that way, transparently engaging with the public to hear these issues so they can be dealt with upfront is vital for building trust in society. Ethical considerations include balancing potential benefits and risks, equitable distribution, and minimization of harm to human health or the environment. Consequently, the need for equivalent access to nanotechnology mechanisms of global warming obstruction is indispensable (Kumari et al. 2022). There is also a question of how to ensure that affordability, accessibility, and inclusivity are maintained so that we do not exacerbate already increasing health inequities. Implementation of nanotechnology may also face financial, infrastructural, and technological constraints in developing countries. National measures are contingent on the global agreement, so ethical frameworks must encourage collaboration and empower everyone for global equitable technology transfer.

Intellectual property rights (IPR): Problems with the patenting and licensing of nanotechnology innovations can create barriers to access or dissemination. Careful balancing of the need for IPR protection while enabling technology diffusion can generate maximum societal impact. Any ethical guidelines developed to guide nano research should help ensure that the technology is used for acceptable and responsible purposes like medical treatments while ensuring it cannot be easily manipulated into military applications or surveillance technologies. Engagement among scientists, policymakers, and industrial representatives as well as civil society actors up to the public is a prerequisite for ethical guideline formulation in accordance with societal values as well as apprising sustainable development goals.

20.7 Policy Recommendations for Safe and Effective Deployment of Nanotechnology Solutions

Nanotechnology policies need to be developed to support the development and deployment of nanotech-based solutions against global warming (Pokrajac et al. 2021). Develop and modify nanotechnology application-specific regulatory frameworks addressing global warming. These frameworks should consider the novel characteristics and potential hazards of nanomaterials. Adopt a risk-based strategy to evaluate and deal with environment, health insurance policies, and basic safety implications of nanotechnology-enabled items or functions (Sargent 2011). Set protocols for risk assessment and management in the standard form. Develop and set up monitoring and surveillance networks to follow nanomaterial use and release into the environment as well as their environmental impacts. This includes setting up mobilization, reporting, and database requirements for transparency and accountability. The development and enforcement of national or even super-national safety standards for the manufacturing, handling, and disposal of nanomaterials used in global warming mitigation technologies ensure occupational safety and environmental regulations are met. Perform life-cycle assessments (LCAs) of nanotechnology-enabled products and processes, fully considering the environmental impact. Utilize LCA results for fact-based decisions and reinforcement of sustainability. Ramp up investment in nanotechnologies in the public and private sectors to solve global warming problem. Promote interdisciplinary research to tackle scientific, technical, and policy challenges. Accelerate nanotechnology inventions from research to implementation with funding programs and rewards. Promote pilot-scale demonstrations and technology validation. Create and ratify ethical principles that encourage good practice in innovation, with benefit-sharing. Engage stakeholders to address societal aspects related to nanotechnology's safety, privacy, or ethical dimensions. Promote open communication and public participation regarding the use of nanotechnologies to address global warming. This informs stakeholders, including legislators, industry, academia, and the general society, of potential benefits and risks. Engage in initiatives that harmonize international nanotechnology regulations and standards to increase global trade of novel solutions. Work in collaboration with international organizations and stakeholders to come up with universal guidelines. Facilitate technology transfer to developing countries and promote capacity building in research, development, and regulation of nanotechnology. Promote equitable access to and use of nanotechnology solutions. Develop metrics and indicators to evaluate the performance, cost-effectiveness, and broader social significance of global warming solutions that are based on nanotechnology. Ensure evaluation to measure progress and adjust policies. Employ adaptive management to account for evolving scientific information, new technological capabilities, and stakeholder input.

20.8 Conclusion

Nanotechnology is a transformative technology in our transition to halt global warming and adapt better to climate changes. Nanotechnology provides many of these benefits through its characteristics and various applications, which help make sustainable designs possible via a range of processes that uphold environmental stewardship. One, nanotechnology increases energy efficiency and supports renewable energy sources as solar power or wind. Nanostructured materials have transformed the field of solar cell technology with enhanced performance, reduced costs, and a step toward a positive clean energy shift. Moreover, higher -grade nanomaterials are used to improve the reliability and integration ability of renewable energy storage to grid. In addition, nanotechnology is essential in carbon capture and storage (CCS) technologies. Using nano-engineered materials, such as metal-organic frameworks (MOFs) and carbon nanotubes, allows for efficient CO_2 capture and sequestration from industrial processes. EN – Engineered Nanomaterials can be employed to manage greenhouse gas emissions due to its numerous benefits, and continuously work on enabling low-carbon energy systems. In addition, its applications in nanotechnology toward agricultural practices increase crop productivity and resource use efficiency through more sustainable agriculture. Taking one step back from the rounded products, members of society are generally inclined to think about vertical, box-like robots that stack inventory in giant warehouses. A Mumbai-based startup is developing nanopesticides and nanofertilizers that bring nutrients more effectively within plants while limiting environmental harm through efficient delivery systems for improved food security. Furthermore, nanosensors help detect soil health and crop condition in real time, thus supporting precision agriculture that properly manages water resources and reduces carbon emissions (from selective fertilizer application). Nanotechnology has provided new hope or energy in the area of environmental monitoring and remediation as with nanotech one can create sensors to detect pollutants, to clean water sources, etc. High-sensitivity detection of pollutants at trace levels operates at a nanomachine level using microscale chemical sensors, and removal technologies for water- and surface-bound organic contaminants are also developed based on nanostructured membranes/adsorbents. Finally, these technologies help sustainable development as they protect in particular natural habitats on land and aquatic systems that play a key role in the public health of all communities. The technology can be a double-edged sword, requiring ethical considerations and regulatory challenges to deploy nanotechnology solutions simply. The safe handling and disposal of nanomaterials are essential, as well as addressing any environmental or health impacts due to such emerging technologies. Effective regulation and appropriate ethical assessment are vital to ensure that the full benefits of nanotechnology can be realized in ways that are also used to manage risks. To sum up, the promise of nanotechnology in mitigating global warming and climate change is high. The pursuit of these next steps requires that together we continue researching, investing, and collaborating across borders. A global inclusive approach is vital for making the most out of it and ensuring a future driven by progress for future generations. Harnessing nanotechnology can help us contribute to a cleaner, sustainable world and address virtually every aspect of climate change head-on.

References

Alli YA, Oladoye PO, Ejeromedoghene O, Bankole OM, Alimi OA, Omotola EO, ... Ogunlaja AS (2023) Nanomaterials as catalysts for CO_2 transformation into value-added products: a review. Sci Total Environ 868:161547. https://doi.org/10.1016/j.scitotenv.2023.161547

Arstad B, Fjellvåg H, Kongshaug KO, Swang O, Blom R (2008) Amine functionalised metal organic frameworks (MOFs) as adsorbents for carbon dioxide. Adsorption 14:755–762. https://doi.org/10.1007/s10450-008-9137-6

Barhoum A, García-Betancourt ML, Jeevanandam J, Hussien EA, Mekkawy SA, Mostafa M, ... Bechelany M (2022) Review on natural, incidental, bioinspired, and engineered nanomaterials: history, definitions, classifications, synthesis, properties, market, toxicities, risks, and regulations. Nanomater 12(2):177. https://doi.org/10.3390/nano12020177

Berhane A (2018) Climate change and variability impacts on agricultural productivity and food security. Climate Weather Forecasting 6(240):2

Chausali N, Saxena J, Prasad R (2023) Nanotechnology as a sustainable approach for combating the environmental effects of climate change. J Agr Food Res 12:100541. https://doi.org/10.1016/j.jafr.2023.100541

Dallaev R, Pisarenko T, Papež N, Holcman V (2023) Overview of the current state of flexible solar panels and photovoltaic materials. Mater 16(17):5839. https://doi.org/10.3390/ma16175839

Echegoyen Y (2015) Nano-developments for food packaging and labeling applications. Nanotechnologies in Food and Agriculture 141–166. https://doi.org/10.1007/978-3-319-14024-7_7

Elemike EE, Uzoh IM, Onwudiwe DC, Babalola OO (2019) The role of nanotechnology in the fortification of plant nutrients and improvement of crop production. Appl Sci 9(3):499. https://doi.org/10.3390/app9030499

Goh PS, Ismail AF, Ng BC (2013) Carbon nanotubes for desalination: Performance evaluation and current hurdles. Desalination 308:2–14. https://doi.org/10.1016/j.desal.2012.07.040

Hansen J, Sato M, Hearty P, Ruedy R, Kelley M, Masson-Delmotte V, ... Lo KW (2016) Ice melt, sea level rise and superstorms: evidence from paleoclimate data, climate modeling, and modern observations that 20 C global warming could be dangerous. Atmos Chem Phys 16(6):3761–3812. https://acp.copernicus.org/articles/16/3761/2016/acp-16-3761-2016.pdf

He L, Tjong SC (2016) Nanostructured transparent conductive films: fabrication, characterization and applications. Mater Sci Eng R Rep 109:1–101. https://doi.org/10.1016/j.mser.2016.08.002

Hill TS, Hoogenboom MO (2022) The indirect effects of ocean acidification on corals and coral communities. Coral Reefs 41(6):1557–1583. https://doi.org/10.1007/s00338-022-02286-z

Joseph TM, Al-Hazmi HE, Śniatała B, Esmaeili A, Habibzadeh S (2023) Nanoparticles and nanofiltration for wastewater treatment: from polluted to fresh water. Environ Res 117114. https://doi.org/10.1016/j.envres.2023.117114

Khan SH (2020) Green nanotechnology for the environment and sustainable development. Green Materials for Wastewater Treatment 13–46. https://doi.org/10.1007/978-3-030-17724-9_2

Kumar R, Mangalapuri R, Ahmadi MH, Vo DVN, Solanki R, Kumar P (2020) The role of nanotechnology on post-combustion CO_2 absorption in process industries. Int J Low-Carbon Technol 15(3):361–367. https://doi.org/10.1093/ijlct/ctaa002

Kumar N, Samota SR, Venkatesh K, Tripathi SC (2023) Global trends in use of nano-fertilizers for crop production: advantages and constraints – a review. Soil Till Res 228:105645. https://doi.org/10.1016/j.still.2023.105645

Kumari A, Bhinda MS, Sharma B, Parihar M (2022) Climate change mitigation and nanotechnology: an overview. Sustain Agric Res 53: Nanoparticles: A New Tool to Enhance Stress Tolerance 33–60. https://doi.org/10.1007/978-3-030-86876-5_2

Kumari A, Rana V, Yadav SK, Kumar V (2023) Nanotechnology as a powerful tool in plant sciences: recent developments, challenges and perspectives. Plant Nano Biol 100046. https://doi.org/10.1016/j.plana.2023.100046

Lemi T, Hailu F (2019) Effects of climate change variability on agricultural productivity. Int J Environ Sci Nat Resour 17(1):14–20. https://juniperpublishers.com/ijesnr/pdf/IJESNR.MS.ID.555953.pdf

Liu Y, Zhou G, Liu K, Cui Y (2017) Design of complex nanomaterials for energy storage: past success and future opportunity. Acc Chem Res 50(12):2895–2905. https://pubs.acs.org/doi/epdf/10.1021/acs.accounts.7b00450?ref=article_openPDF

Mamun MAA, Yuce MR (2020) Recent progress in nanomaterial enabled chemical sensors for wearable environmental monitoring applications. Adv Funct Mater 30(51):2005703. https://doi.org/10.1002/adfm.202005703

Manikandan S, Subbaiya R, Saravanan M, Ponraj M, Selvam M, Pugazhendhi A (2022) A critical review of advanced nanotechnology and hybrid membrane based water recycling, reuse, and wastewater treatment processes. Chemosphere 289:132867. https://doi.org/10.1016/j.chemosphere.2021.132867

Muzammil WK, Rahman MM, Fazlizan A, Ismail MA, Phang HK, Elias MA (2019) Nanotechnology in renewable energy: critical reviews for wind energy. Nanotechnology: Applications in Energy, Drug and Food 49–71. https://doi.org/10.1007/978-3-319-99602-8_3

Pateiro M, Gómez B, Munekata PE, Barba FJ, Putnik P, Kovačević DB, Lorenzo JM (2021) Nanoencapsulation of promising bioactive compounds to improve their absorption, stability, functionality and the appearance of the final food products. Mol 26(6):1547. https://doi.org/10.3390/molecules26061547

Pokrajac L, Abbas A, Chrzanowski W, Dias GM, Eggleton BJ, Maguire S, ... Mitra S (2021) Nanotechnology for a sustainable future: addressing global challenges with the international network4sustainable nanotechnology. ACS Nano. https://pubs.acs.org/doi/pdf/10.1021/acsnano.1c10919

Power IM, Harrison AL, Dipple GM (2016) Accelerating mineral carbonation using carbonic anhydrase. Environ Sci Technol 50(5):2610–2618. https://pubs.acs.org/doi/epdf/10.1021/acs.est.5b04779?ref=article_openPDF

Pramanik P, Krishnan P, Maity A, Mridha N, Mukherjee A, Rai V (2020) Application of nanotechnology in agriculture. Environ Nanotechnol 4:317–348. https://doi.org/10.1007/978-3-030-26668-4_9

Qiao Z, Wang C, Zeng Y, Spendelow JS, Wu G (2021) Advanced nanocarbons for enhanced performance and durability of platinum catalysts in proton exchange membrane fuel cells. Small 17(48):2006805. https://doi.org/10.1002/smll.202006805

Sargent Jr, JF (2011) Nanotechnology and environmental, health, and safety: issues for consideration. Cornell University. https://ecommons.cornell.edu/server/api/core/bitstreams/1b7df7a3-e59e-4059-bf23-2bb8f0f9b6de/content
Shang Y, Hasan MK, Ahammed GJ, Li M, Yin H, Zhou J (2019) Applications of nanotechnology in plant growth and crop protection: a review. Mol 24(14):2558. https://doi.org/10.3390/molecules24142558
Sharma AK, Sharma M, Sharma AK, Sharma M (2023) Mapping the impact of environmental pollutants on human health and environment: a systematic review and meta-analysis. J Geochem Explor 107325. https://doi.org/10.1016/j.gexplo.2023.107325
Siuzdak K, Bogdanowicz R (2018) Nano-engineered diamond-based materials for supercapacitor electrodes: a review. Energy Technol 6(2):223–237. https://doi.org/10.1002/ente.201700345
Tao R, Ma X, Wei X, Jin Y, Qiu L, Zhang W (2020) Porous organic polymer material supported palladium nanoparticles. J Mater Chem A 8(34):17360–17391. https://pubs.rsc.org/en/content/articlehtml/2020/ta/d0ta05175f
Tovar-Lopez FJ (2023) Recent progress in micro-and nanotechnology-enabled sensors for biomedical and environmental challenges. Sensors 23(12):5406. https://doi.org/10.3390/s23125406
Wang L, Boutilier MS, Kidambi PR, Jang D, Hadjiconstantinou NG, Karnik R (2017) Fundamental transport mechanisms, fabrication and potential applications of nanoporous atomically thin membranes. Nat Nanotechnol 12(6):509–522. https://doi.org/10.1038/nnano.2017.72
Wang Y, Pan C, Chu W, Vipin AK, Sun L (2019) Environmental remediation applications of carbon nanotubes and graphene oxide: adsorption and catalysis. Nanomater 9(3):439. https://doi.org/10.3390/nano9030439
Wong KV, Perilla N, Paddon A (2014) Nanoscience and nanotechnology in solar cells. J Energy Resour Technol 136(1):014001. https://doi.org/10.1115/1.4024715
Worku MC, Feleke MG, Denekew T, Anagaw YK (2024) Radio frequency identification technology: a method of analysis of falsified pharmaceutical products: literature review. SAGE Open Med 12:20503121241272725. https://doi.org/10.1177/20503121241272725
Wu X, Wang X, Xie Y, Ren N, Ma J, Ning P (2022) Facile in-situ construction of highly dispersed nano zero-valent iron modified black TiO_2 Z-scheme recyclable heterojunction with highly efficient visible-light-driven photocatalytic activity. Appl Catal B: Environ 310:121325. https://doi.org/10.1016/j.apcatb.2022.121325
Yadav V, Pal D, Poonia AK (2024) A study on genetically engineered foods: need, benefits, risk, and current knowledge. Cell Biochem Biophys 1–16. https://doi.org/10.1007/s12013-024-01390-x
Yazid AF, Mukhtar H, Nasir R, Mohshim DF (2022) Incorporating carbon nanotubes in nanocomposite mixed-matrix membranes for gas separation: a review. Membranes 12(6):589. https://doi.org/10.3390/membranes12060589

The Solar Radiation's Absorption in the Photovoltaic Cells in Nanoparticles

21

B. Gopal Krishna, Dhriti Sundar Ghosh, Sanjay Tiwari, and M. Jagannadha Rao

Abstract

Nanostructured materials possess captivating optical features that are crucial for various solar energy usage schemes and technologies. Nanotechnology enables the creation and utilization of structures and systems with dimensions comparable to visible light's wavelength. This presents numerous opportunities to investigate the novel and frequently resonant phenomena that occur when an item's size aligns with the electromagnetic field's periodicity (as determined by the wavelength of light, λ). Several innovative renewable energy technologies are being developed in diverse designs and forms daily. This is a direct response to the growing global recognition of the need for a transition to green energy and the tight targets established for 2050. Solar direct electricity generating systems, such as photovoltaic (PV) and photovoltaic-thermal (PVT) setups, are the most popular emerging technologies in this field. This is because solar energy is widely available and limitless, and recent advancements have made solar cells more efficient and cost-effective. The study in this framework is still in progress, and nanotechnologies are being extensively studied for their potential to improve

B. Gopal Krishna (✉)
Madhya Pradesh Bhoj (Open) University, Bhopal, India

Photonics Research Laboratory, School of Studies in Electronics & Photonics, Pt. Ravishankar Shukla University, Raipur, India

D. S. Ghosh
Department of Physics, Indian Institute of Technology, Dharwad, India

S. Tiwari
Photonics Research Laboratory, School of Studies in Electronics & Photonics, Pt. Ravishankar Shukla University, Raipur, India

M. Jagannadha Rao
Department of Geology, Andhra University, Visakhapatnam, Andhra Pradesh, India

A. A. H. Abdel Latef et al. (eds.), *Bioremediation and Nanotechnology for Climate Change Mitigation*, https://doi.org/10.1007/978-981-96-3069-1_21

the performance of solar systems in different configurations and using diverse approaches. This approach is receiving more attention than any other strategy. This chapter attempts to provide a comprehensive overview of research activities involving the utilization of nanostructures, nano-enhanced materials, nanofluids, and similar technologies for solar power generation systems. These systems encompass several components, including solar cells, panel packages, and additional equipment like heat storage devices.

Keywords

Nanotechnology · Photovoltaic cells · Solar energy technologies

21.1 Introduction

The primary and enduring means of energy on Earth is solar radiation. All other known energy sources are exhaustible (fossil), limited (geothermal and gravitational), or presently not socially acceptable (e.g., nuclear, due to the concerns of proliferation). While the amount of solar power reaching the Earth is only 1.4 kW/ m^2, energy conversion efficiency becomes crucial in light-harvesting equipment. Sophisticated solar light collecting techniques can be categorized under two fundamental ideas: solar thermal and solar quantum. The former converts solar radiation into thermal energy, while the latter converts it into electrical power or chemical fuels. Examples of technologies embodying these notions are solar thermal collectors and photovoltaic systems, which harness solar energy to generate electricity or facilitate chemical reactions, such as hydrogen fuel production. The disparity in the methodologies and requirements for the light absorber can be comprehended by contrasting them with their natural counterparts. In the former case, these correspond to the atmospheric and oceanic phenomena that are influenced by solar radiation, such as wind, rain, and streams. In the latter case, they refer to the natural processes of photosynthesis and photolysis, which involve converting light energy into chemical energy and its storage in the form of sugar bonds in plants, algae, and bacteria (Willson and Mordvinov 2003; McElroy 2021). In addition to solar energy, other sources of light energy can be harnessed, although they are typically less abundant or practical. One such source is bioluminescence, which is produced by living organisms like fireflies and deep-sea creatures. Although fascinating, bioluminescence is not a practical large-scale energy source. Another source of artificial light is generated through the combustion of fuels or electrical energy in devices such as incandescent bulbs, LEDs, and lasers. These technologies convert various forms of energy into light, but their efficiency varies widely. Chemiluminescence, for example, is seen in glow sticks and is produced by chemical reactions that do not produce heat. Additionally, certain materials exhibit photoluminescence, where they absorb photons and then re-emit them as light, a principle used in fluorescent and phosphorescent lighting. While these sources are not as impactful as solar energy, they illustrate the diversity of methods by which light energy can be generated and utilized.

The Earth receives approximately 173,000 terawatts (TW) of incoming solar energy, which is roughly 10,000 times more than the world's total energy consumption of 600–700 TW per year. In terms of energy potential, solar energy constitutes a significant proportion compared to actual energy consumption, indicating that we harness only a small fraction of this vast resource. However, there are several knowledge gaps that must be addressed to fully realize the potential of solar energy.

1. Efficiency of Solar Energy Utilization: Despite the vast amount of solar energy available, the current technology for converting solar energy into usable power, such as photovoltaic cells, is not highly efficient. This creates a gap between the potential energy available and what we actually use.
2. Energy Storage: Effective storage solutions for solar energy are still under development. The knowledge gap here affects how much solar energy can be reliably used when the sun is not shining.
3. Technological Development: Advances in technology and materials science could potentially improve the efficiency of solar energy capture and conversion. Research is ongoing to bridge this gap.
4. Geographic and Economic Disparities: There is a disparity in solar energy access and utilization globally. Regions with abundant sunlight might not have the infrastructure to fully exploit this resource, and there are economic barriers to widespread adoption.

The objective of intentionally converting solar energy into heat is to maximize the absorption of all available radiation. When considering the quantum transformation of photon energy, it is important to focus on the several stages involved in the transformation process. Hence, it is necessary to carefully design the mechanisms involved in the initial absorption of photons, the creation of charge carriers, their movement, separation, and attachment. This design should aim to achieve a harmonious alignment between the spectral distribution of the incoming light, the system's ability to absorb light, and the amount of energy of the transitional and final states. The energy levels of transitions are observed to change depending on the size and shape of the active structure (Yates 2009). The issue of energy is a matter of great global significance, with both ecological and societal implications. The extraction of fossil fuels, for instance, accounts for more than 70% of the world's greenhouse gas emissions, which exacerbates climate change and causes extensive environmental degradation. Furthermore, the disparity in knowledge between wealthy and poor nations results in an imbalance in environmental stewardship. In the year 2023, the wealthiest 1% of the global population were responsible for twice the carbon emissions of the poorest 50%. This inequality not only increases pollution but also results in the waste of environmental resources, as wealthier nations often consume disproportionately, leaving poorer communities to bear the consequences of environmental degradation without the resources or knowledge to mitigate them effectively. These interconnected issues highlight the urgent need for equitable solutions in energy utilization and environmental protection.

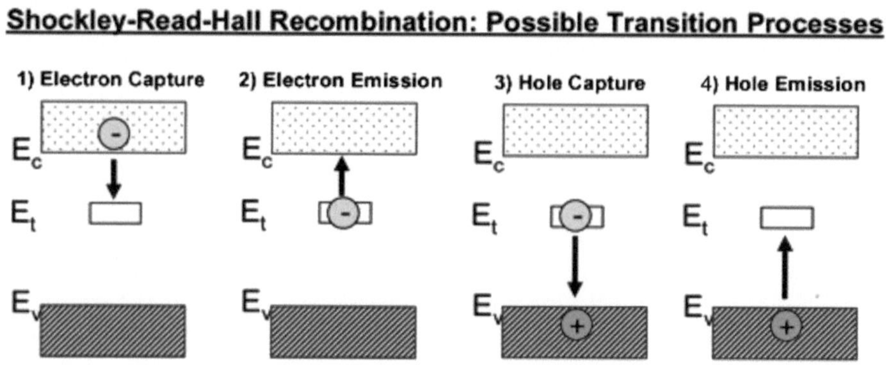

Fig. 21.1 Diagram representing the four possible electronic transition mechanisms connected to charge carrier recombination at trap sites (Yates 2009)

Figure 21.1 is a schematic diagram based on the work of Shockley, Read, and Hall, illustrating the process of trapping electrons and holes in the semiconductor. When confined, the hole or electron is destroyed at a higher rate compared to when there is no confinement, leading to shorter lifetimes for the hole. Four instances of indirect electronic transition mechanisms are depicted. Process 1 demonstrates the phenomenon of electron capture from the conduction band by a recombination center. This recombination center is initially neutral and is located within the energy gap of the semiconductor. The rate at which thermally excited electrons are captured is directly related to the density of recombination centers and the capture cross-section, which is around 10^{15} cm^2, on the order of atomic dimensions. Process 2 demonstrates the speed at which electrons are emitted from the recombination center. In a state of equilibrium, this speed will be the same as the rate at which electrons are captured. Process 3 corresponds to a hole capture process in which a confined electron combines with a hole in the valence band. The rate of this process is determined by the multiplication of the confined electron concentration and the hole concentration. Process 4 is known as hole emission, which refers to the stimulation of an electron to move from the valence band to an electron trap state, creating a hole in the valence band. When light falls on the semiconductor, the material's charge carriers increase above what is normally present in a state of thermal equilibrium. Processes 1 and 3 collectively form a recombination process that eliminates electrons and holes at the trap site, reducing the photochemical reaction rates caused by either available holes or electrons. With the advancements in nanotechnology and nanoscience, it has become feasible to manipulate the structure precisely, enabling exact adjustment of transition energy levels. By adopting this position, it becomes possible to utilize all innovative building ideas for the purpose of quantum transformation devices.

The international competition between China, the United States, and Europe for semiconductor dominance has significant implications for developing countries, particularly in terms of climate change. Semiconductor manufacturing is a highly energy-intensive process, with some estimations suggesting that the production of chips could account for approximately 3% of global electricity consumption by 2030. As the major powers compete for supremacy, they increasingly rely on rare earth minerals, which are often sourced from developing countries, leading to environmental degradation in these regions. For example, in the Congo, which supplies approximately 60% of the world's cobalt, the mining process has resulted in severe deforestation, water pollution, and soil erosion. This exploitation not only accelerates climate change but also leaves these developing countries with lasting environmental damage, while the economic benefits largely bypass the local populations. As the competition for semiconductor dominance intensifies, the environmental burden on vulnerable nations is expected to grow, exacerbating the global climate crisis.

The global focus has shifted towards sustainable energy sources due to the effects of industrialization, population growth, and increasing energy demand. Solar energy is limitless and, when harnessed appropriately, may effectively address energy problems. A photovoltaic (PV) cell can capture photons from solar radiation and transform them into electrons. Over the last 10 years, the worldwide average cost of producing electricity from PV systems has dropped by 85%. It signifies that generating electricity from solar power systems is more economically efficient than most traditional power plants in places with abundant solar energy potential (Ameur et al. 2020).

In recent years, various nations have made considerable progress in digital advancements. Estonia, commonly known as the "digital republic," has established a robust e-governance system that enables citizens to vote, pay taxes, and access public services online. Singapore has emerged as a global leader in the implementation of smart city technologies, integrating Internet of things (IoT) and artificial intelligence (AI) to optimize urban management and enhance the quality of life for its residents. In India, the Digital India initiative has transformed the country's digital landscape by bringing millions of people online through affordable Internet and mobile services, and fostering innovation in fintech and e-commerce. China's significant investment in 5G and AI technologies has bolstered its position as a technology powerhouse, with widespread adoption of digital payment systems and smart manufacturing. These countries' experiences demonstrate the transformative potential of digital expansion in driving economic growth, improving governance, and enhancing citizens' daily lives.

Figure 21.2 showcases a solar photovoltaic system. The initial two technologies consist of eight photovoltaic panels each interconnected in series. The third category is amorphous, characterized by the presence of two parallel strings. Every string consists of six panels that are connected in sequence.

Although there have been significant advancements in the techno-economic aspects, further examinations reveal that there is still ample room for improvement in terms of achieving higher efficiency and improved cost-effectiveness (Zeraatpisheh et al. 2018). Researchers have recently shown interest in utilizing nanotechnology for solar PV systems. This includes incorporating nanoparticles into PV cells, using nanofluids for photovoltaic thermal (PVT) panels, and employing nano-enhanced

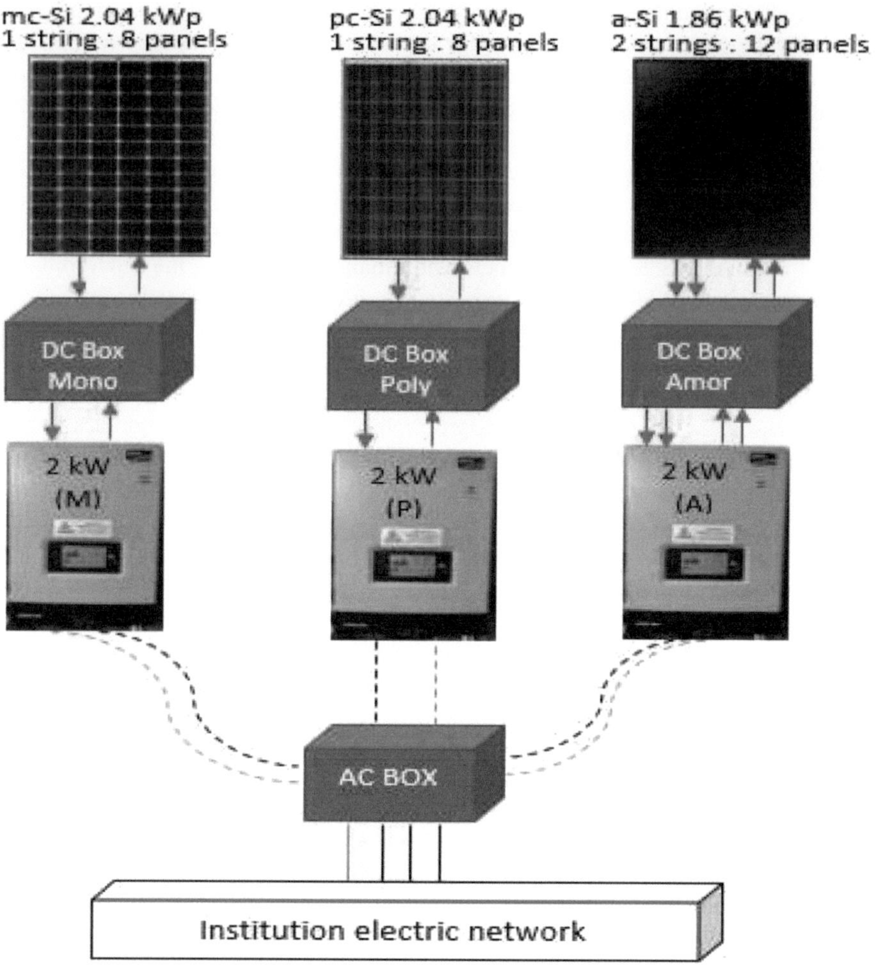

Fig. 21.2 Illustration of the photovoltaic plant configuration (Ameur et al. 2020)

phase change material, or phase change material (PCM), for photovoltaic thermal (PVT) setups.

This chapter provides a concise overview of innovative ideas for photon capture that have emerged from advancements in nanoscience and nanotechnology. Particular examples from our own research support the concepts.

21.2 Current Level of Technological Advancement

The use of well-designed optical structures has been widely acknowledged as a means to boost light absorption in Si solar cells. Traditionally, one of the initial methods to confine light within an absorbing structure is to employ substrates with textures on a scale similar to the wavelength of light. These encompass antireflection coatings that reduce reflection losses on the front surface, together with features intended to enhance light confinement within the cell. A commonly employed configuration involves the usage of pyramids or inverted pyramids that are etched into the top and sometimes the back surface of the cell. Diffractive optical designs can increase the effective thickness of cells by a factor of 4–5 (Heine and Morf 1995). These structures have broader applications because they allow for thinner absorbing layers and do not require single crystalline materials, unlike the method outlined earlier.

In dye-sensitized solar cells, the process of light trapping has been accomplished by including light scattering layers made of polycrystalline anatase. These layers have demonstrated a 10% increase in the photon-to-electricity conversion efficiency. Furthermore, the layers have demonstrated their ability to generate charge carriers actively (Ito et al. 2008).

Photonic crystals have been studied to increase the optical thickness by incorporating periodic arrays of different refractive index media into an absorbing structure, which mimics the electronic band structures found in periodic crystals. In the long term, the development of more unconventional metamaterials with optical features that cannot be found in naturally occurring materials may play a role in advancing this field.

Unlike metallic systems, nanostructured semiconductor materials exhibit enhanced light absorption due to their large and structured nature. However, due to distinct reasons, semiconductor quantum dots (QDs) are specifically intriguing for their potential use in solar devices. The cross-sections of rules are improved relative to the bulk counterpart due to momentum delocalization and relaxation of selection (Alivisatos 1996). Using impact ionization, researchers have successfully shown the formation of numerous excitons from a single photon in PbSe, CdSe, and PbS quantum dots (QDs). This breakthrough significantly increases the theoretical efficiency limit of solar cells. This method has the potential to accomplish carrier multiplication, as indicated by previous research (Shabaev et al. 2006; Paci et al. 2006).

Transport in semiconductors, especially in advanced materials like quantum dots (QDs) and nanostructures, is a critical aspect of modern solar cell technology. Semiconductor materials play a crucial role in converting light into electricity, and their transport properties determine the efficiency of this conversion. Unlike bulk materials, nanostructured semiconductors exhibit quantum confinement effects, where the movement of electrons and holes is restricted to nanoscale dimensions. This confinement leads to momentum delocalization and relaxation of selection rules, resulting in enhanced light absorption and increased interaction cross-sections. These properties make quantum dots particularly promising for photovoltaic applications, where they can potentially achieve higher efficiencies than traditional semiconductors.

Historically, semiconductor technology has driven much of the progress in electronics and energy conversion. In the mid-twentieth century, the development of the first semiconductor devices, such as transistors, revolutionized computing and communication, marking the beginning of a significant technological gap between nations that embraced semiconductor technology and those that did not. As technology advanced, the introduction of materials like silicon and gallium arsenide enabled the development of more efficient and smaller electronic devices.

The current leap in quantum dot technology highlights a new phase in this ongoing evolution. Researchers have demonstrated that by using impact ionization, multiple excitons can be generated from a single photon in QDs like PbSe, CdSe, and PbS. This advancement dramatically increases the theoretical efficiency limit of solar cells, pushing the boundaries of what is possible in energy conversion. As a result, the technological gap continues to evolve, with quantum dots paving the way for the next generation of highly efficient solar devices.

Surface and particle polaritons can be utilized to manipulate and focus light at the nanoscale scale. Plasmonic structures demonstrate significant extinction cross-sections and offer a potential method for enhancing the optical thickness of sunlight-powered systems while preserving a physically thin structure. Plasmons are electromagnetic modes that arise from the interaction between light and matter in a certain material and geometry. This connection necessitates the use of materials (medium) that possess a suitable dielectric function, which may occasionally be negative, and have particle size or surface characteristics that are smaller than the wavelength of light. The required conditions are provided by the collective oscillation of conduction electrons (plasmon-polariton) or lattice vibrations in polar crystals (phonon-polariton). Another part of the developing field of plasmonics involves the development of devices that can replace electric currents with plasmon waves. This is because plasmons have the potential to carry significantly large amounts of information that can be compressed into nanometer-sized wires. The implementation of plasmonics concepts is hindered by the limited lifespan of plasmons, which typically lasts just 10–100 fs. During this short period, plasmons either decay into regular light waves or transform electron-hole pairs and eventually vibrations (heat). The primary obstacle is to mitigate these losses in order to utilize the captured light energy for practical purposes effectively or to transmit information. Imping light on a thin film of a high refractive index semiconductor can be coupled into waveguided modes that propagate within the film through nanoparticles. The optical thickness of the semiconductor is determined by its lateral extension, whereas its physical thickness is determined by its vertical extension. Therefore, the association between the two interfering parameters is eliminated. This method has proven remarkably effective for waveguides with thicknesses in the nanoscale range (Stuart and Hall 1998; Tsai et al. 2010).

Surface and particle polaritons hold the potential to revolutionize the manipulation of light at the nanoscale, but their environmental consequences must be taken into account. Plasmonic structures, which exhibit significant extinction cross-sections, provide a promising approach to enhancing the optical thickness of sunlight-powered systems while maintaining a physically thin structure. However, the materials used in plasmonics often necessitate negative dielectric functions and specific surface characteristics, which could have unforeseen environmental effects. For instance, the extraction and processing of these materials could involve the use of harmful chemicals or energy-intensive processes, contributing to environmental degradation.

Additionally, the short lifespan of plasmons, lasting only 10–100 fs, presents a challenge. The decay of plasmons into light waves, electron-hole pairs, or heat can result in energy losses, which, if not managed efficiently, could lead to wasted energy and increased thermal pollution. As we advance in the development of plasmonic devices, which could potentially replace electric currents with plasmon waves, it is crucial to address these losses to ensure that the technology is both effective and environmentally sustainable.

The development of waveguided modes in high refractive index semiconductors, coupled with nanoparticles, represents a significant step forward. However, the environmental impact of producing and disposing of these materials must be carefully evaluated. While the potential benefits of plasmonics in terms of energy efficiency and information transmission are substantial, balancing these advancements with environmental considerations is essential for the responsible development of this technology.

Nanostructures can also be engineered to control the electromagnetic near field, restricting the energy of the field to a specific photoactive area, often drawing inspiration from the natural photosynthetic antenna system. Optical nanoantennae are tiny plasmonic particles that vibrate at frequencies that can be seen by the human eye and are close to the infrared range. Within a semiconductor, the antenna functions as the absorber of photons, with the energy being transferred to excitations of electron-hole pairs through the improved near field. Methods to achieve these structures include the use of carbon nanotubes (CNTs), zeolites, and metallic nanoparticles. Additionally, it is worth noting that optical rectennas in this scenario seek to reduce the size of the existing radio wave equivalent and convert light directly into a direct current using an integrated antenna and rectifier construction. The example demonstrates the application of plasmonic excitations in solar-powered devices by adjusting the decay to electron-hole pairs using Landau damping (Westphalen et al. 2000; Corkish et al. 2002). In these systems, the semiconductor is not necessary for generating charge carriers. Instead, the plasmonic particle aids both the absorption of light and the creation of charge carriers. A rectenna, also known as a rectifying antenna (Fig. 21.3), is designed to facilitate simpler manufacturing processes.

A device that converts electromagnetic energy traveling across space into direct current in a circuit. The device comprises multiple components, including antennas, filter circuits, and rectifying diodes or bridge rectifiers. These components are present for each antenna element or for the combined power from multiple elements.

The incorporation of solar cells with plant cells signifies a remarkable convergence of biological and technological processes whereby light energy is secured and stabilized before being transformed into chemical energy. This idea builds on mimicking photosynthesis, the natural process by which plants turn sunlight into chemical energy. By linking artificial solar cells to plant cells, scientists aim to maximize the efficiency of light absorption and stabilization. The solar cells function as a complement to the plant's chlorophyll, capturing a broader spectrum of light and transforming it into electrons. These electrons are then transferred to the

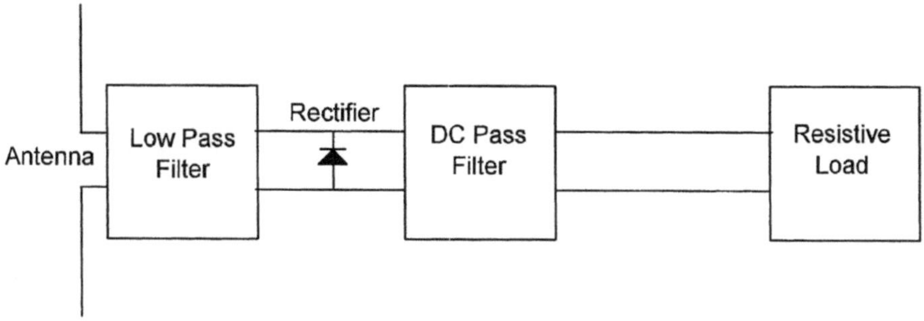

Fig. 21.3 A schematic representation of a rectenna and its associated load (Corkish et al. 2002)

plant cells, driving the biochemical reactions necessary to generate energy-rich compounds, such as ATP (adenosine triphosphate). This hybrid system could significantly improve the efficiency of bioenergy production, potentially leading to innovative ways of generating sustainable energy.

Space energy refers to the gathering and transmission of solar power from outer space. Unlike Earth-based solar energy, which is impacted by atmospheric conditions, day-night cycles, and weather, space-based solar power (SBSP) involves harnessing solar energy using satellites equipped with large solar panels. These satellites orbiting Earth continuously collect sunlight, which is then converted into electrical energy. The energy is sent back to Earth via microwave or laser beams, where it can be captured by ground-based receivers and integrated into the electrical grid.

Space energy holds the potential to provide a virtually boundless, uninterrupted supply of clean energy, potentially revolutionizing global energy systems. However, the concept faces significant technical and economic challenges, such as the development of efficient transmission systems, ensuring safety, and the high cost of launching and maintaining space-based infrastructure. If these obstacles can be overcome, space energy could become a key component of future energy solutions.

21.3 Nanotechnology

The evolution of nanotechnology Nobel Prize laureate and scientist Richard P. Feynman was the first to introduce the idea of nanotechnology. In 1974, Norio Taniguchi proposed the word "Nanotechnology" to describe the processes that take place on a nanoscale in semiconductors. Taniguchi's definition of nanotechnology is the manipulation of materials at the atomic or molecular level by processes such as separation, consolidation, and deformation (Corbett et al. 2000). Nanotechnology had significant advancement during the late 1980s and early 1990s with the invention of the scanning tunneling microscope (STM) and atomic force microscope (AFM). Prior to the 2000s, the areas of focus in research were the synthesis of nanomaterials and the attainment of precise nanostructure growth. During this time frame, a wide range of nanostructures were created, such as solid-state compounds, cylindrical and uniformly sized nanoporous membranes, well-organized two-dimensional (2D) nanoarrays, carbon nanotubes (CNTs), and carbon nanofibers (CNFs). Several technologies have been employed to create composite nanostructures. For instance, the sol-gel method, which used a two-step template process, was utilized to produce nanowires. The precise placement of the nanostructures was achieved by means of the solid-state reaction. An extensive endeavor was undertaken to actualize the commercial uses of gadgets based on nanostructures. Boroditsky et al. discovered that InGaAs and GaN are the most appropriate materials for light-emitting diodes (Boroditsky et al. 2000). Yang et al. successfully enhanced the electromagnetic interference shielding property by incorporating different carbon nanofibers (CNFs) and carbon nanotubes (CNTs) into the polystyrene matrix, achieving a level suitable for commercial use (Yang et al. 2007).

In the past 20 years, researchers have concentrated on creating three-dimensional (3D) hybrid nanostructures. Nanostructured materials have found extensive applications in several domains, including the utilization as anode materials in lithium and sodium-ion batteries. Additionally, they might be utilized to enhance the effectiveness of solar cells or gas sensors. Significantly, the key aspect is the utilization of meticulously arranged ultralong nanowire arrays, measuring up to 20 cm in length and 50 nm in diameter, for various applications such as flexible electronics, photonic devices, biochemical sensors, and solar cells (Yeon et al. 2013). WO_3 coatings have been found to enhance the effectiveness of solar water splitting in nanostructures (De Respinis et al. 2013). In 2020, Mariappan et al. utilized a 3D nickel core for the purpose of energy storage (Mariappan et al. 2020). Figure 21.4 depicts the electrochemical deposition of antimonene nanostructures on a 3D Ni foam surface, which can be utilized in energy harvesting and storage devices.

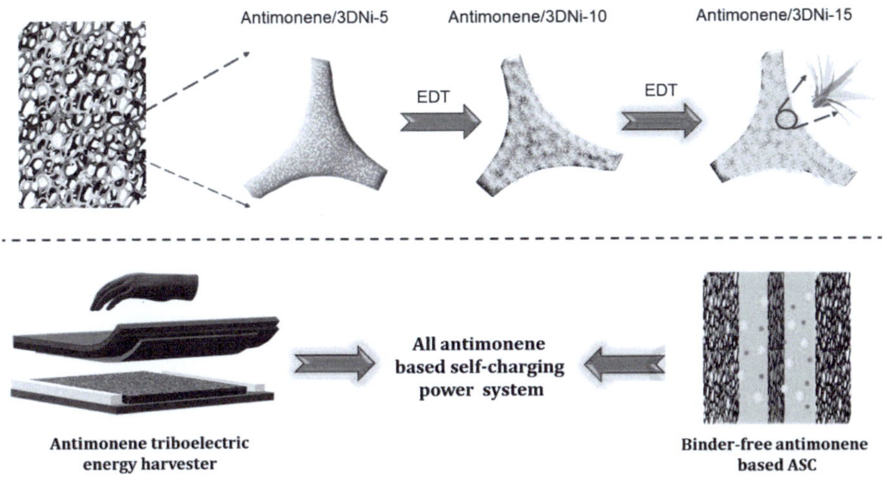

Fig. 21.4 The electrochemical deposition process is visually depicted, and its utilization in energy harvesting and electrochemical energy storage devices is illustrated (Mariappan et al. 2020)

Table 21.1 Humanity's use of various elements of energy throughout history

Time period	Element used	Application
Prehistoric times	Wood	Cooking, heating, and toolmaking
Ancient civilizations	Bronze/copper	Weapons, tools, and artifacts
Iron age	Iron	Weapons, agriculture, and infrastructure
Medieval period	Coal	Heating, blacksmithing, and early industry
18th–19th century	Steam (water)	Steam engines, industrial revolution
19th–20th century	Oil	Transportation, industrial power
Twentieth century	Uranium	Nuclear power
Twenty-first century	Lithium	Batteries for electronics, electric vehicles

The historical perspective on technology, particularly nanotechnology, unveils remarkable advancements attained by ancient civilizations. The ancient Egyptians, known for their architectural marvels and refined culture, are also credited with utilizing early forms of nanotechnology. Evidence indicates that they employed nanoparticle-based dyes to create the striking colors in their glassware, particularly in the renowned Lycurgus Cup, where the color changes depending on light exposure due to embedded nanoparticles. This sophisticated manipulation of materials showcases their advanced comprehension of chemistry and material properties, even if they didn't conceptualize it as "nanotechnology" in the modern sense. Similarly, the Indian civilization exhibited exceptional metallurgical skills in crafting weapons. The legendary Damascus steel swords, celebrated for their strength, sharpness, and distinctive patterns, were produced using techniques that involved nanostructures within the steel. The process of repeatedly heating and folding the steel imbued it with carbon nanotubes, providing these weapons with their superior qualities. This suggests that ancient Indian blacksmiths, while unaware of the term "nanotechnology," were effectively manipulating materials on a nanoscale to enhance their tools of war. In contrast, the contemporary global focus is on the lithium boom, driven by the demand for renewable energy sources and the shift toward electric vehicles (EVs). Lithium, a critical component in rechargeable batteries, is revolutionizing the electrical industry, powering everything from smartphones to EVs. As the world increasingly pursues sustainable energy solutions, lithium's role has become pivotal, representing a modern-day equivalent of the ancient reliance on advanced materials. Table 21.1 demonstrates the historical consumption of different energy sources by humanity.

21.4 Architectures of Nanomaterials

Nanostructures are systems with at least one dimension equal to or less than 100 nm. This can encompass many dimensional scales, such as 0D, indicating that the material exists on a nanometer size in all dimensions; 1D, indicating that the material exists on the nanometer level in two dimensions; 2D, indicating that the material exists on a nanometer scale in one dimension; and 3D, indicating that the material extends beyond the nanoscale in any dimension. In addition to these, another category of nanostructures known as 3D-like nanostructures, which can be created by assembling nanostructures of dimensions 0D, 1D, and 2D (Lieber 1998). There are two approaches to nanotechnology: a "bottom-up" strategy and a "top-down" technique. The "bottom-up" technique involves the construction of nanodevices using atomic or molecular components. The "top-down" technique utilizes an electron beam, intense ultraviolet, or X-ray lithography to fabricate nanodevices on silicon or alternative semiconductor components directly.

The edge and quantum confinement effects of 0D nanomaterials are further increased due to their inherent structural features, such as the surface-to-volume ratio and ultra-small diameters (Li et al. 2016). Thus far, various types of zero-dimensional nanomaterials have been suggested. Graphene quantum dots (GQDs) are tiny particles made of carbon that possess remarkable physical, chemical, and biological characteristics. One of their notable properties is photoluminescence, which enables them to emit light in various biological applications. In 2004, Xu et al. accidentally found fluorescent carbon quantum dots (CQDs) while purifying single-walled carbon nanotubes (SWCNTs) from arc-discharged soot. Due to their exceptional light emission properties, CQDs possess a range of advantageous characteristics including wide adjustability, intense luminescence, single-photon emission, biocompatibility, low toxicity, chemical stability, and ease of biomolecule modification. These properties make them suitable for applications in bioimaging, drug delivery, and other fields (Xu et al. 2004). CQDs are extensively utilized in lasing, PVs (photovoltaics), and sensing applications. Fullerene is a molecule made up of 60 carbon atoms, specifically consisting of 12 pentagons and 20 hexagons. It is also commonly referred to as C_{60} (Lu et al. 2019). Fullerene has applications in photovoltaics (PVs), biomedicine, and catalysis. Other types of 0D nanoparticles encompass noble metal nanoparticles, upconversion nanoparticles, polymer dots, and quantum dots (QDs). Figure 21.5 displays scanning electron microscopy (SEM) images of self-organized objects.

Nanostructures display remarkable diversity and functionality based on their synthesis conditions, influencing a broad range of applications from materials science to environmental technologies. One-dimensional nanofibers, synthesized in 1-propanol, exhibit high aspect ratios and are crucial in applications such as filtration, sensors, and tissue engineering. These nanofibers' elongated structure leads to a high surface area-to-volume ratio, enhancing their performance in these domains. In contrast, two-dimensional nanodisks formed in 1,4-dioxane present a flat, disc-like morphology. These nanodisks are essential in optical applications, including catalysis and electronic devices, due to their planar surfaces that can interact with light and other particles in unique ways. Their properties are often exploited in creating advanced materials with tailored functionalities. Two-dimensional windmill-like sheets, synthesized from a mixture of 1:2 2-propanol and toluene, represent another interesting nanostructure. The windmill configuration results from the interaction of solvents, creating structures with potential applications in drug delivery systems and as scaffolds in nanotechnology. Three-dimensional conical objects, formed in a 1:1 mixture of tetrahydrofuran (THF) and water, demonstrate complex structures that can be used in applications requiring three-dimensional orientation

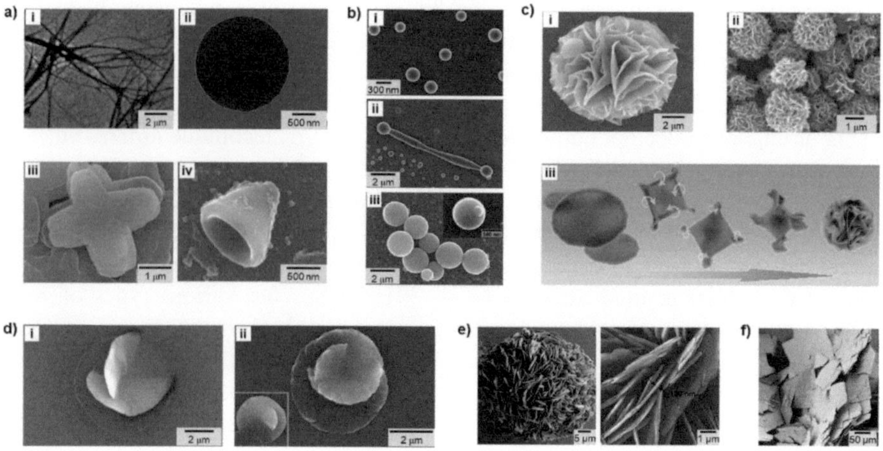

Fig. 21.5 (**a**) Scanning electron microscopy (SEM) images of self-organized objects of 5a produced in various solvents at a temperature of 20 °C: (i) One-dimensional nanofibers are formed in 1-propanol. (ii) Two-dimensional nanodisks are formed in 1,4-dioxane. (iii) Two-dimensional windmill-like sheets are formed in a mixture of 1:2 2-propanol and toluene. (iv) Three-dimensional conical objects are formed in a mixture of 1:1 THF and H_2O. (**b**) Scanning electron microscopy (SEM) images of self-assembled structures of 5a produced in a 1:1 combination of 2-propanol and toluene at a temperature of 5 °C: (i) Spherical objects acquired in a fresh state, (ii) Baton-like objects generated after 0.5 hours of ultrasonication, and (iii) Microspheres formed after one hour of ultrasonication. (**c**) Scanning electron microscope (SEM) images showing (i) flower-shaped structures formed by 5a and (ii) spherical objects formed by 5b in a solution of 1,4-dioxane. Also included is (iii) a schematic diagram illustrating the process by which the flower-shaped structures of 5a are formed. (**d**) Scanning electron microscope (SEM) pictures of (i) left-handed and (ii) right-handed spiral assemblies of compound 5a, produced in 2-(R)-butanol and 2-(S)-butanol solvents, respectively. The scanning electron microscope (SEM) images show the constructed (**e**) plate-rich gigantic particles and (**f**) sheet structures that were created by cooling their 1,4-dioxane solutions from 70 to 20 °C (Lu et al. 2019)

and high surface area, such as in sensors and catalysis. These conical shapes often exhibit unique properties due to their spatial arrangement and can influence interactions with various substances. Spherical objects, acquired in a fresh state, are universally recognized for their symmetry and uniformity. These structures are widely used in drug delivery, imaging, and as catalysts, benefiting from their predictable size and shape. Baton-like objects and flower-shaped structures, arising from spiral assemblies, represent advanced self-assembled architectures. These formations, with their intricate shapes, are significant in environmental applications, where their design can enhance performance in catalysis and pollutant removal. Overall, these diverse nanostructures illustrate how synthesis conditions dictate the morphology and functionality of nanomaterials, impacting various technological and environmental applications.

The photovoltaic and photocatalytic devices use a system that converts solar light into chemical or electrical energy through quantum processes. Solar energy is often converted into electricity through either direct or thermal conversion methods.

It is important to mention that a new process called photon-enhanced thermionic emission has been discovered. This process combines both electric and thermal conversion mechanisms and has the potential to achieve higher conversion efficiencies than traditional photovoltaic cells, possibly surpassing their theoretical limits.

The development of advanced energy conversion technologies is crucial for harnessing solar energy effectively. Traditional methods for converting solar energy into electricity include photovoltaic (PV) devices, which directly convert sunlight into electrical energy using semiconductor materials, and thermal conversion methods, which capture heat from solar radiation and use it to produce electricity via heat engines or thermoelectric materials. However, a novel process called photon-enhanced thermionic emission (PETE) represents a significant advancement in the field of energy conversion. This process combines both electric and thermal conversion mechanisms to improve overall efficiency.

PETE operates by using solar photons to excite electrons in a material. When sunlight hits the material, it excites electrons to higher energy states, generating both thermal energy and high-energy photons. Unlike traditional methods, PETE harnesses the energy from photons to boost the emission of electrons from a hot material. The material's high temperature, combined with the photon energy, allows for a more efficient conversion of solar energy. Electrons are emitted more readily due to the combined effects of thermal excitation and photon absorption.

PETE has the potential to surpass the efficiency limits of conventional photovoltaic cells. Traditional PV cells are constrained by factors like the Shockley-Queisser limit, which cap their maximum theoretical efficiency. PETE, by utilizing both thermal and photon effects, can potentially exceed these limits, offering improved performance.

Photon-enhanced thermionic emission is a promising technology that enhances the efficiency of solar energy conversion by combining the benefits of thermal and electric processes. This innovative approach has the potential to lead to breakthroughs in renewable energy technology, potentially surpassing the efficiency limits of conventional photovoltaic cells.

21.4.1 Metal Nanoparticles That Exhibit Optical Activity (Plasmonic)

As mentioned earlier, localized surface plasmon resonances (LSPRs) are the combined movements of the conducting electrons, which can lead to optically absorbed cross-sections that are several orders of magnitude larger than the geometric cross-section. Plasmons can decay in two ways: radiatively, which results in significant increases in the electromagnetic field and is utilized in techniques like surface-enhanced Raman spectroscopy, or into quasi-particles known as electron-hole pairs.

The resonance wavelength of several metals occurs throughout the range of near ultraviolet, visible, and near infrared wavelengths when considering nanostructures with diameters ranging from 20 to 200 nm. The specified spectrum range, ranging from 0.5 to 6.5 eV, encompasses the majority of the energy involved in significant chemical reactions, such as the breakdown of bonds and bond creation. This would enable the creation of conditions and the proposal of plans to improve the absorption of light in nanostructured materials, particularly solar energy.

The impact of metal nanostructures on light absorption and its repercussions for environmental sustainability, particularly in relation to solar energy, is a subject of great interest. By focusing on nanostructures with diameters ranging from 20 to 200 nm, it highlights their capacity to resonate across a broad spectrum of near ultraviolet, visible, and near infrared wavelengths. This capability spans from 0.5 to 6.5 eV, a range of utmost importance for significant chemical reactions, including bond formation and breakdown. The environmental implications of these findings are multifaceted. Firstly, enhancing light absorption in nanostructured materials can improve the efficiency of solar energy systems. The utilization of metals with optimized resonance wavelengths in enhanced solar cells can capture a greater amount of sunlight and convert it into electricity more effectively. This improvement could lead to a reduction in dependence on fossil fuels, thus decreasing greenhouse gas emissions and mitigating climate change. Additionally, better solar energy technology can contribute to more sustainable energy solutions. The increased efficiency of solar panels may lower the overall cost of renewable energy, making it more accessible and attractive to both individuals and industries. This shift towards renewable energy sources can have significant environmental benefits, including reduced air and water pollution and a smaller ecological footprint. Furthermore, advancements in nanotechnology for energy applications must be managed responsibly. While the potential for reduced environmental impact through enhanced energy efficiency is considerable, the production and disposal of nanomaterials should be carefully regulated to avoid unintended ecological consequences. Ensuring that these materials are sustainable and non-toxic is essential to fully realize their environmental benefits. In conclusion, the implications suggest a promising path towards more efficient solar energy solutions, which could play a crucial role in environmental preservation and the transition to a greener economy.

21.4.2 Geometrical Optical Resonance

Carbon materials possess distinctive optical characteristics. For instance, the relatively high refractive index and moderate dispersion are the reasons for the brilliance of diamonds. Similarly, the electron structure of amorphous carbon and graphite is responsible for their considerable absorption in the visible range (Laidani et al. 2007). The identification of novel carbon allotropes, including fullerenes, carbon nanotubes, and graphene (single layers of graphite), has prompted extensive optical investigations of nanoscale carbon structures. Nevertheless, the production and investigation of the characteristics of nanostructures in traditional carbon materials, such as glassy carbon (GC) and highly oriented pyrolytic graphite (HOPG), are particularly intriguing both on their own and for comparison.

Resonant absorption is directly connected to the sizes and heights of the nanostructures, as indicated by both the experimental and theoretical investigations. The absorption peaks undergo a redshift, transitioning from the visible range for the smallest structures to the near infrared range for the larger structures. Concurrently, the absorption peaks get more intense as the structure heights increase, whereas absorption decreases as the diameters increase. This would improve the absorption of light in nanostructured materials, for their usage in solar energy technology.

The efficiency of nanostructures in absorbing light is a topic of considerable interest, particularly in relation to solar energy technology and environmental implications. The ability of these structures to absorb light at specific wavelengths, known as resonant absorption, is influenced by factors such as size and height. As the it explains, smaller nanostructures absorb light in the visible range, while larger structures shift this absorption into the near-infrared range. Additionally, taller structures enhance the intensity of absorption, whereas larger diameters reduce it. The optimization of light absorption through nanostructured materials has the potential to revolutionize solar energy technology. By adjusting the size and shape of these nanostructures, solar panels can become more efficient at capturing a broader range of sunlight, which leads to improved energy conversion and reduced dependence on non-renewable energy sources. From an environmental standpoint, advances in nanostructure absorption could significantly decrease greenhouse gas emissions. As solar panels become more efficient, the amount of land and resources required for energy production decreases, which reduces habitat disruption and pollution associated with fossil fuels. Additionally, enhanced solar panels can facilitate the transition to a more sustainable energy system, minimizing the environmental footprint of energy production. Overall, the study of resonant absorption in nanostructures represents a vital step towards advancing solar energy technology. By improving the absorption capabilities of nanostructured materials, we can enhance the efficiency of solar panels, decrease reliance on fossil fuels, and mitigate the negative environmental impacts associated with traditional energy sources.

21.4.3 Structures That Cause Scattering

The remarkable physiochemical features of nanocrystalline TiO_2 make it an extensively utilized photocatalyst. Nevertheless, the primary limitation for its practical usage in solar energy conversion schemes is its broad bandgap of approximately 3.2 eV. The thickness of the titania layer is important for photoactivity in both pure and composite films. The activity gradually grows until it reaches a particular thickness and begins to decline. When the film thickness approaches its optimal value, which is approximately equal to the mean free path distance, it experiences increased scattering as it travels toward the surface. As a result, the photoactivity drops as the thickness exceeds the optimum level because there is greater trapping of charge carriers, even though the films have a larger space for optical absorption. In summary, this measurement effectively distinguishes the influence of thickness on charge transfer and photoactivity.

The environmental consequences of TiO_2 are influenced by its layer thickness in photocatalytic applications. Increasing the thickness to an optimal level enhances light absorption, which subsequently improves the photocatalytic performance. However, excessive thickness results in increased scattering and charge carrier trapping, which diminishes efficiency. This can lead to the inefficient use of materials and energy, raising environmental concerns related to resource utilization and waste generation. Optimizing the TiO_2 layer thickness is crucial from a sustainability perspective. Thin, well-calibrated layers minimize the need for excessive material and energy input, reducing waste and environmental footprint. Additionally, TiO_2 possesses the ability to degrade pollutants and improve air quality, offering significant environmental benefits. However, it is essential to balance these advantages with the practical limitations of TiO_2's bandgap and thickness to maximize its environmental benefits while minimizing negative impacts.

21.5 Principles of Solar Cells Solar Photovoltaic (PV) Energy Conversion

The PV energy conversion is the process of transforming sunlight into electric power. Photons, which make up light, are directed towards the surface of the semiconductor. When the frequency of the light reaches a certain value, the electrons completely leave the surface of the semiconductor. The semiconductor can be categorized into p-type and n-type. A p-n junction is created at the boundary between two distinct kinds of semiconductors. The occurrence of the photoelectric effect results in the generation of electron-hole pairs. The PV effect, encompassing the capture, creation, combination, and movement of electron holes inside the semiconductor material and contact electrodes, plays a crucial role in the physics of solar cells (Asim et al. 2012). The amount of energy absorbed at the top surface of the solar cell is as follows:

$$E = \tau_g \alpha p G W \mathrm{d}x \tag{21.1}$$

The Eq. (21.1) represents the relationship between E, τ_g, G, α, p, W, and dx. Here, τ_g refers to the glass transmissivity, G represents the solar radiation, α denotes the solar panel absorptivity, p represents the solar panel packing factor, W represents the solar cell width, and dx represents the elemental length. The effectiveness of the solar cell is dependent on its temperature. The equation for efficiency (η_T) is given by

$$\eta_T = \eta_{\mathrm{ref}}[1 - \beta_{\mathrm{ref}}\left(T - T_{\mathrm{ref}}\right) \tag{21.2}$$

Here, η_{ref} represents the efficiency at the reference temperature, β_{ref} is the efficiency temperature-coefficient, and T_{ref} is the characteristic temperature (25 °C). The resulting photovoltaic (PV) power can be calculated using the formula

$$P = G\tau_{pv}\eta_{\mathrm{ref}}A\left[1 - r_{\mathrm{ref}}\left(T - T_{\mathrm{ref}}\right)\right] \tag{21.3}$$

τ_{pv} represents the transmittance of the outer layers of the photovoltaic (PV) cell, A represents the area exposed to radiation, and r_{ref} represents the power temperature coefficient. The solar cell is a device with two terminals that generates photovoltage when exposed to sunlight during the day and acts as a diode, conducting electricity, at night in the absence of sunlight. The cells are linked sequentially and enclosed within modules to generate sufficient direct current voltages. A photovoltage power production system has several components, including cells, mechanical and electrical connections, mountings, and methods for monitoring and adjusting the electrical output. When determining the economic viability of a solar cell, three primary aspects must be considered: cost, efficiency, and working lifetime. To enhance the effectiveness of solar cells, the most efficient approach is to minimize the expenses associated with the area they occupy. The costs associated with modules include materials and systems, such as the manufacture of the cells, semiconductor materials, interconnection of the cells, packaging materials, transportation, support structures, cabling, and module mounting. Researchers typically classify solar technology into three primary generations, although some experts suggest the existence of a fourth and fifth generation based on recent innovations. The first generation of solar cells consists of crystalline Si cells, while the second generation includes thin-film solar cells like CdTe, CIGS, and AsGa. The third generation comprises new solar cell technologies such as dye-sensitized solar cells (DSSCs), perovskite solar cells (PSCs), and polymer solar cells. Crystalline silicon-based solar cells dominate the global photovoltaic market with a market share of up to 90%. This is because of their suitable bandgap, non-toxicity, abundance in materials, and advancement of technology. The efficiency of the single-junction solar cell is maximized when the energy gap is within the range of 1.35–1.5 eV (Henry 1980). Shockley and Queisser determined that the maximum theoretical efficiency of a single solar cell is 33% (Shockley and Queisser 1961). The efficiencies of commercial solar cells are lower than the theoretical levels due to surface recombination, lattice flaws, inadequate incident light absorption, and nonideal connections. Figure 21.6 includes a plot of solar-cell efficiency as a function of Eg for $C = 1000$. Both for $C = 1$ and $C = 1000$, it is assumed that the optimal cell temperature is 300 K. The greatest efficiencies are roughly 31% when C is equal to 1% and 37% when C is equal to 1000.

As the world confronts the mounting consequences of climate change, the progress of photovoltaic (PV) technology emerges as a promising solution. This scientific advancement holds the potential to transform energy production, offering a cleaner and more sustainable alternative to fossil fuels. The development of PV technology, which directly converts sunlight into electricity, has achieved remarkable progress in recent years. In 2010, the global solar photovoltaic capacity was approximately 40 gigawatts (GW); by the end of 2023, this figure had surged to over 1000 GW. This exponential growth can be attributed to the significant decrease in solar panel costs—down 90% since 2010—and the increase in efficiency, which has improved by over 50% in the last decade. The environmental benefits of PV technology are considerable. For every megawatt-hour (MWh) of electricity generated by solar panels, approximately 0.4 kg of CO_2 is emitted, compared to 1000 kg of CO_2 per MWh from coal-fired power plants. By transitioning to solar energy, we can substantially lessen greenhouse gas emissions. In 2023 alone, solar energy offset around 1.7 billion metric tons of CO_2 globally, equivalent to removing about 360

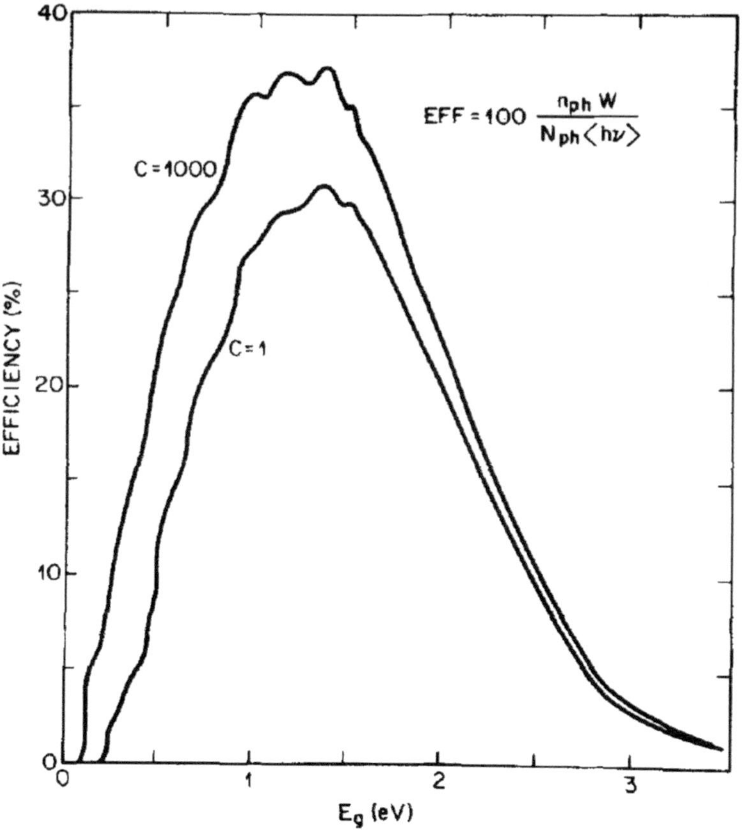

Fig. 21.6 Comparison of solar-cell efficiency to energy gap for solar concentrations of 1 sun and 1000 suns (Henry 1980)

million cars from the roads. Additionally, the deployment of solar technology contributes to job creation and economic growth. The global solar industry employed over four million people in 2023, a substantial increase from 2.8 million in 2016. As the technology continues to develop, these numbers are expected to rise, fostering further innovation and economic opportunity.

Crystalline silicon has achieved a power efficiency of 26.7% after a span of 70 years. The third-generation solar cells utilize several technologies such as perovskite solar cells (PSCs), dye-sensitized solar cells (DSSCs), organic solar cells (OSCs), and hybrid multiple-junction solar cells. PSCs, which have achieved a maximum efficiency of 25%, are becoming recognized as highly promising photovoltaic devices for the future. Due to their uncomplicated composition and affordable price, DSSC may serve as a potential substitute for silicon solar cells. Nevertheless, their effectiveness is constrained by their incapacity to utilize infrared light. The initial documented efficiency of DSSC was 7.1% (O'Regan et al. 1991). Currently, the focal points of study are enhancing these technologies' efficiency while ensuring they are ecologically benign, non-toxic, and durable. Organic solar cells (OSCs) offer several benefits, including as their uncomplicated design, suitability for large-scale applications, cost-effective solution-processing, and compatibility with flexible electronics and semi-transparent devices (Han et al. 2019). The recently revealed power conversion efficiency (PCE) has achieved a value of 15.2%. However, it is important to note that this technology may have drawbacks such as the presence of toxic halogenated solvents, sensitivity to the thickness of photovoltaic (PV) layers, and stability difficulties. An effective method to improve the efficiency of solar cells is to create two or multi-junction cells using distinct absorber layers. Theoretically, infinite junctions can achieve a maximum of 68% (Eperon et al. 2017). The most recent literature has documented an efficiency of 47.1% (using six multi-junctions), which is currently the highest reported. Nanomaterials are utilized in the production of solar panels. Solar cells based on perovskite materials perovskite solar cells have seen significant development since they were initially proposed in 2009 (Kojima et al. 2009). These cells effectively sensitize TiO_2 to enhance energy conversion efficiency by 3.8%. Subsequently, many groups have undertaken multiple endeavors to enhance efficiency. The perovskite structure is characterized by the formula ABX_3, where A represents inorganic cations such as cesium (Cs^+) and organic cations such as $CH_3NH_3^+$, $C_2H_5NH_3^+$, and B stands for metal cations such as Ge^{2+}, Pb^{2+}, and Sn^{2+}, and X stands for halide anions such as I^-, Cl^-, and Br^-. Three primary criteria that could impact the business application possibilities of PSCs are power conversion efficiency (PCE), cost, and stability. The most efficient approach for the PCE component is utilizing tandem solar cells based on PSCs, as these cells can only absorb photons with energy levels exceeding the bandgap of the semiconductor material, according to theoretical principles. The surplus energy will be discharged through thermalization. In order to surpass the Shockley-Quiesser limit, it is possible to combine several absorbers with complementary bandgaps in a multi-junction or integrated solar cell. This allows for the maximum utilization of sunlight while minimizing thermalization loss. The efficiency of the PSC is primarily determined by its architecture, which refers to the materials employed and the techniques utilized to deposit the material. The PSCs exist in two distinct forms: mesoscopic and planar architectures. The planar architectures can exist in either a regular n-i-p or inverted p-i-n layout, depending on the direction of the electric current. Efforts to enhance the power conversion efficiency (PCE) of perovskite solar cells (PSCs) have been ongoing for over a decade. The most recent research indicates that the efficiency of PSCS is 25.2% (Yoo et al. 2021). The cost of PSCs is multifaceted, encompassing factors like as energy cost, energy payback period, and raw materials choices. The low-temperature solution approach is a highly effective procedure. The instability of PSCs is attributed to the presence of moisture, exposure to UV light, and fluctuations in temperature. Humidity is regarded as a multifaceted obstacle. Multiple hypotheses have been proposed regarding the mechanisms underlying the degradation of perovskite thin films. One possibility is that the metal electrodes may undergo moisture-assisted interaction with the perovskite, resulting in a loss of stability. During the synthesis process, the temperature rises and an annealing treatment is employed to create a perovskite structure. It has been observed that perovskite breaks down at elevated temperatures. This issue could be handled by utilizing materials that provide high thermal resistance. The TiO_2 layer serves as the electron transport layer for the PSCs. However, exposure to UV light causes degradation of the layer. In order to decrease this issue, one possible solution is to employ enclosed electronics equipped with a UV filter. Another approach involves including a new and reliable compound that can expand the range of wavelengths absorbed to include the ultraviolet area. PSCs encapsulation is a highly effective method for reducing moisture and Pb leakage. Carbon nanotubes can be utilized in the construction of solar cells. Carbon nanotubes (CNTs) possess remarkable optical, electrical, and mechanical properties, making them suitable nanostructures for carrier-selective transport and for usage in collecting layers in solar cells. Additionally, CNTs are abundant in nature. The solar cell contains a transparent conducting layer, known as the top electrode, that is consistently made of indium tin oxide (ITO). The main benefit of ITO is its exceptional transparency and conductivity. Nevertheless, the material contains indium, a scarce metal. Other drawbacks of these materials include limited flexibility, increased crystal flaws, elevated cost, and instability at high temperatures. Consequently, the development of solar cells that do not include indium has been a major focus of study for future photovoltaic technologies. Because of their distinctive characteristics, carbon nanotubes (CNTs) have become appealing substitutes for indium in the field of transparent and conductive materials. Starting in 2005, scientists substituted ITO with SWCNTs in OSCs. The power conversion efficiency (PCE) of solar cells based on single-walled carbon nanotubes (SWCNTs) is superior to that of devices based on indium tin oxide (ITO). This is significant for the deposition technique since it involves the synthesis of single-walled carbon nanotubes (SWCNTs) using a solution-based procedure, which is more cost-effective compared to vacuum deposition of indium tin oxide (ITO) (Pasquier et al. 2005). As shown in Fig. 21.7a, b, bulk-heterojunction solar cells were fabricated using a 2.5 g/l solution of P3HT:PCBM 1:1.1 in chlorobenzene. An outermost coating composed of polyethylenedioxy thiophene and poly(styrene sulfonate) PEDOT:PSS, a conductive polymer blend, was applied onto SWNT-quartz substrates and ITO-glass reference substrates via spin coating. Baytron P, dissolved in a mixture of methanol and water in a ratio of 1:2, was used for this purpose.

The connections to the single-walled carbon nanotubes (SWNT) and the P3HT: PCBM layer were made using gallium-indium eutectic. The area subjected to photoactivity testing measured 0.07 cm^2. A 3 mm diameter o-ring was utilized to enclose the gallium-indium droplet and regulate the back contact region. The photovoltaic properties of the devices were assessed by utilizing glass-SWNT/PEDOT: PSS/ P3HT: PCBM 1:1/Ga structures, which were constructed using the 6 SWNT films and an ITO-glass substrate for comparison. PEDOT: PSS was utilized in the SWNT thin film devices due to its application in the reference cells as well. The devices were enhanced through the process of annealing at temperatures of up to 100 °C in the presence of air, while closely observing the changes in the characteristics of the photovoltaic device. Once the devices reached their maximum conversion efficiency, they were left to cool down to room temperature. Then, measurements of the

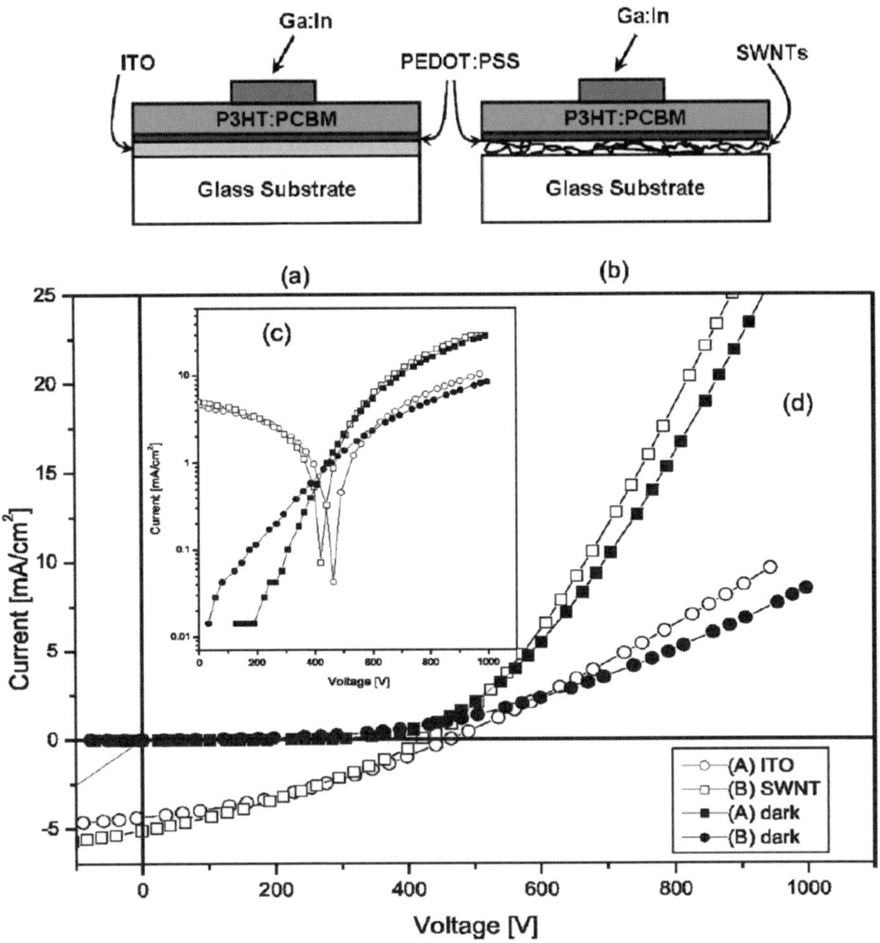

Fig. 21.7 The upper panel displays a schematic of the devices with (**a**) ITO as the anodes and (**b**) SWNT thin film. (**c**) The log I-V characteristics of the ITO and SWNT thin film devices under illumination are displayed in the inset. (**d**) The I-V curves for the reference solar cell on an ITO-glass substrate and the best solar cell employing a SWNT-glass current collector are displayed under 100 mW/cm^2 halogen white light and in the dark. (Pasquier et al. 2005)

current-voltage (I-V) were taken at different light intensities. Figure 21.7d displays the relationship between current density and voltage under dark conditions and illumination at 100 mW/cm^2 using ITO and SWNT hole collector electrodes, respectively. The logarithmic current-voltage characteristics of our devices are displayed in the inset of Fig. 21.7c.

Several variables can impact the performance of single-walled carbon nanotubes (SWCNTs) in solar cells. The cell requires optimal levels of transparency and conductivity. However, typically, the diminished conductivity of SWCNTs-electrodes is directly proportional to their transparency. An optimal approach to address this issue is to select appropriate dopants. The research on carbon nanotubes (CNTs) in solar panels primarily concentrated on perovskite solar cells (PSCs) because of their flexibility. An electron-transporting layer (ETL) is crucial for enhancing the power conversion efficiency (PCE). In n–i–p structured perovskite solar cells (PSCs), TiO_2 and SnO_2 are the most favored electron transport layers (ETLs), despite their limited electron mobility. An effective solution to address this problem is to integrate carbon-based nanoparticles. Research has indicated that the addition of a tiny quantity of carbon nanotubes (CNTs) to the titanium dioxide electron transport layer (ETL) will enhance the power conversion efficiency (PCE). Nevertheless, it is crucial to meticulously regulate the doping concentration. Introducing an excessive amount of CNTs will always lead to a decline in the device's performance. Carbon nanotubes (CNTs) can be utilized in the perovskite layer. During the deposition process, the formation of perovskite crystals often results in several grain boundaries that contain multiple defects and charge traps. The film morphology is optimized when the carbon nanotubes (CNTs) are doped with suitable functionalization.

21.6 Nanotechnologies Used to Photovoltaic Thermal (PVT) Systems

An emerging advancement in direct solar energy production involves harnessing the elevated temperature of photovoltaic (PV) cells exposed to sunlight in order to produce low-grade heat. There are two beneficial effects of this. Initially, the increase in temperature of solar cells leads to a decrease in their effectiveness and causes harm to the cells (Odeh and Behnia 2009). Consequently, the panel's temperature is decreased by removing its heat, increasing its efficiency. Furthermore, a specific quantity of thermal energy can be produced for various heating purposes, including space heating, household hot water, and as a source of heat for heat pumps. Remarkably, in terms of quantity, the heat produced by these panels, commonly known as PVT panels, exceeds their electrical energy production (Gupta and Pradhan 2021). Utilizing nanoparticles in systems can enhance the optical characteristics, expanding the range of solar energy the system can absorb (Crisostomo et al. 2017). In addition to endeavors to improve the efficiency and physical characteristics of solar cells through nanotechnology, a considerable amount of studies in this area has lately been dedicated to raising the efficiency of PVT systems using nano methods. To do this, the primary focus is on utilizing nanofluids to promote heat absorption in the collectors or employing nano-enhanced phase change materials (PCMs) to improve heat storage capacity and maintain stable temperatures in the panel. The thermal efficiency of PVT systems can be described as follows:

$$\eta_{\mathrm{PVT}} = \frac{mC_p\left(T_o - T_i\right)}{AG} \tag{21.4}$$

The Eq. (21.4) represents the thermal efficiency (η_{PVT}) of the PVT system, which is calculated using the mass flow rate (m), specific heat (C_p), collector output temperature (T_o), inlet temperature (T_i), collector area (A), and beam irradiation (G). The electrical efficiency of a PVT system is given by the equation

$$\eta_e = \frac{I_m V_m}{AG} \quad (21.5)$$

where I_m and V_m represent the current and voltage at maximum power.

21.7 Characteristics and Classification of Nanofluids

Nanofluids consist of a conventional fluid, such as water or thermal fluid, mixed with colloid nanoparticles. These nanomaterials have a size range of 1–100 nm. The introduction of nanoparticles into the fluid alters the overall characteristics (such as thermal conductivity, viscosity, and specific heat) and enhances the thermal properties, resulting in a reduction in the necessary heat transfer area over the back of the panel. Additionally, they exhibit a great capacity for absorbing solar radiation due to their small size and wide surface area. Nevertheless, nano-powders used in nanofluids may exhibit thermal instability, such as suspensions or agglomeration, as well as chemical compatibility problems with other components. Additionally, the manufacturing process can be challenging, and the cost, both in terms of materials and preparation, can be high. Nanofluids can be categorized based on three criteria: the types of nanomaterial, the composition of nanomaterial, and the type of base fluid. Nanofluids can be categorized into two types based on the nanomaterial and composition commonly employed for thermal applications: mono-nanofluids, which consist of pure metals, metal oxides, carbides, nitrides, and carbon, and hybrid nanofluids, which are composed of mixtures of nanomaterials and nanocomposites. The use of mono-nanofluids has been extensively studied in thermal processes. The nanomaterial mixture consists of a simple combination of nanoparticles (CNTs and Au, CNTs and Cu) with a base fluid. Conversely, in the case of the nanocomposite (Graphene-Multi-walled CNTs MWCNT, Al_2O_3-Cu), several nanoparticles are combined at the nanoscale (Bhanvase and Barai 2021). A typical PVT setup consists of a glass layer placed over the PV cells, a conductive plate positioned beneath the PV cells, a thermal collector serpentine located below the plate, and insulation placed beneath the thermal collector. An experimental study was conducted indoors to investigate the effects of applying a magnetic field on a photovoltaic-thermal (PVT) system. The study used Fe_2O_3-water with concentrations of 1% and 3%. Four system configurations were evaluated, and it was found that the configuration with a 3% concentration and no magnetic field showed a 24% increase in efficiency. Additionally, the configuration that alternated the magnetic field with a frequency of 50 Hz showed a 27.6% increase in efficiency compared to the configuration using water as the cooling medium (Ghadiri et al. 2015). Figure 21.8 illustrates the electrical efficiency of a ferrofluid with a concentration of 3 wt% and solar radiation intensity of 1100 W/m^2.

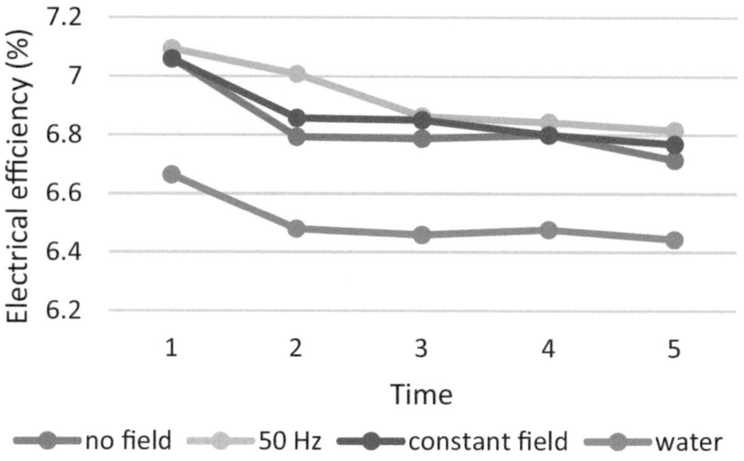

Fig. 21.8 The electrical efficiency was measured for a solution containing 3 weight percent of ferrofluid and distilled water, subjected to radiation of 600 watts per square meter and a nano ferrofluid mass flow rate of 30 L/h (Ghadiri et al. 2015)

The challenge of characterizing and classifying nanofluids lies in the significant knowledge gap regarding their environmental impacts. These fluids, containing nanoparticles, exhibit unique thermal and flow properties that enhance the efficiency of technologies, such as cooling systems and energy storage. However, the diverse nature of nanofluids, including variations in nanoparticle materials, sizes, concentrations, and fluid base types, complicates their characterization and classification. This knowledge gap has several environmental consequences. The inadequate understanding of nanofluid properties may result in inefficient use in industrial applications, leading to increased energy consumption and waste. Additionally, the long-term environmental impact of nanoparticles, such as their potential toxicity and accumulation in ecosystems, is not fully comprehended. Without precise knowledge of how these particles interact with biological systems and environmental components, there is a risk of unintended ecological consequences. To address this knowledge gap, it is crucial to conduct comprehensive studies on nanoparticle behavior, environmental interactions, and lifecycle assessments. These efforts will help mitigate potential risks and ensure that advancements in nanofluid applications contribute positively to technological progress and environmental sustainability.

A comparison was made between different nanoparticles, specifically Al_2O_3 and Cu, in different base fluids, namely water and expanded graphite (EG), with varying concentrations of 0.1/0.2/0.4 wt%, for a PVT system. The tests demonstrated that PVT had superior performance in cases when water was used as the base fluid, Cu was used as the nanoparticle, and the concentration is 0.4 wt%. An increase in nanoparticle concentration leads to a modest drop in the specific heat of the nanofluid. When using a solution of Cu-water with a concentration of 0.4 wt%, the highest achievable thermal efficiency was determined to be 76.8%, while the electrical efficiency reached a maximum of 13.7%. The utilization of Al_2O_3-water with a concentration of 0.4 wt% resulted in a peak thermal efficiency of 46.7%, indicating a 2.7% improvement in thermal efficiency compared to the usage of pure water (Rejeb et al. 2016). Al-Waeli et al. conducted an experimental study on pressure, volume, and temperature (P VT) utilizing SiC-ionized water as a nanofluid. The nanofluid was tested at various concentrations ranging from 0% to 4 wt%. During the process of preparing nanofluids, it was observed that adding 3 wt% SiC resulted in an 8.2% rise in nanofluid density, a 5.18% increase in viscosity, and a 4.3% increase in thermal conductivity. Following a rigorous six-month analysis, it was determined that the nanofluid remained stable, resulting in a mere 0.003 W/mK reduction in thermal conductivity. By utilizing a concentration of 3 wt%, the electrical efficiency of PVT was enhanced by a maximum of 24% compared to using water alone (Al-Waeli et al. 2017). PVT systems can be configured in several ways with regard to the connection between the photovoltaic (PV) cells and the thermal system. The sort of configuration can also influence the performance of a system. Hussain et al. conducted a numerical investigation on a PVT system that consisted of two heat exchangers, each using distinct fluids. The system features a parallel serpentine heat exchanger and a single-pass air heater, both of which operate autonomously. It is also equipped with a baffle sequence that effectively decreases thermal resistance. The simulations were conducted for various nanoparticles (CuO, Al_2O_3, and SiO_2) at varied concentrations, utilizing water as the solvent. The results demonstrated that the optimal concentration is 0.75%, as higher concentrations lead to aggregated nanoparticles forming. The nanofluid exhibited superior thermal stability and maximum thermal conductivity due to the presence of CuO. The overall maximum efficiency of the CuO-water (0.75 wt%) and air system was 90.3%, while the efficiency was 79.8% for the water and air system. The authors emphasized that despite the increased costs associated with pumping in both systems, the benefits of reducing the temperature of the PV cells are greater in terms of system efficiency (Imtiaz Hussain et al. 2019). Experiments were conducted in PVT systems utilizing a nanofluid consisting of a mixture of ionic liquid (IL) and water and two-dimensional MXene (Ti_3C_2) nanoparticles. The nanofluid was tested at different concentration ratios (0.05/0.1/0.2 wt%). The ideal concentration of 0.2 weight percent was determined by considering the thermal physical qualities as criteria. The thermal stability tests indicated that all samples' concentrations remained stable up to a temperature of 45 ° *C*. The findings indicated that the highest electrical efficiency reached was 14%, whereas the highest thermal efficiency reached was 81.2% (Imtiaz Hussain et al. 2019). Rubbi et al. investigated the performance of the PVT system by utilizing soybean oil as the main fluid and MXene nanoparticles at three different concentration ratios: 0.025 wt%, 0.075 wt%, and 0.125 wt%. Tests examining the thermal stability indicated that throughout the temperature range of 30 – 80 ° C, no notable alterations were seen across all concentration levels. Using soybean oil with 0.125% MXene resulted in an impressive overall thermal efficiency of 84.3%, accompanied by a surface temperature reduction of 14 °C compared to water (Rubbi et al. 2020). The performance of the MXene-soybean oil nanofluid was found to be superior to that of both MXene-palm oil and alumina-water nanofluids. An experimental investigation was conducted on various nanofluids in the PVT system, including water-based MWCNT, Al_2O_3, and TiO_2, with a concentration of 0.3 wt%. The utilization of nanofluids resulted in a significant reduction in the temperature of the PV cells. Specifically, the system employing MWCNT, Al_2O_3, and TiO_2 water-based nanofluids had temperature reductions of 48%, 37%, and 36%, respectively. The MWCNT-water, Al_2O_3-water, TiO_2-water, and pure water units attained maximum electrical efficiencies of 18%, 17.8%, 16%, and 9.5% correspondingly. The systems utilizing MWCNT, Al_2O_3, and TiO_2 water-based nanofluids had the highest overall efficiencies of around 47%, 33%, and 27%, respectively (Sangeetha et al. 2021). Sohani et al. utilized an analytical hierarchy process (AHP) to determine the most optimal nanofluid type for the PVT system. Their selection was based on parameters such as efficiency, energy gain, dependability, economic factors, and environmental considerations. The nanofluids were assessed by incorporating Al_2O_3, TiO_2, and ZnO particles into water at a concentration ratio of 0.2 wt%. The utilization of ZnO-water nanofluid is the most optimal choice when considering factors such as annual energy output, electrical efficiency, thermal efficiency, reliability, and reduction of CO_2 emissions. In terms of the payback period, ZnO-water nanofluid is inferior to the other nanofluids. The nanofluids were evaluated using the analytic hierarchy process (AHP), resulting in scores of 33.1 for ZnO-water, 26.0 for TiO_2-water, and 13.9 for Al_2O_3-water. Based on these scores, it can be concluded that ZnO-water and TiO_2-water are the most favorable options. The ZnO-water system exhibits an electrical efficiency of 14.7–47.6% and a thermal efficiency of 14.6–46%. Similarly, the TiO_2-water system shows an electrical efficiency of 14.6–46% and a thermal efficiency of 14.6–46% (Sohani et al. 2021). Abdelrazik et al. conducted a study where they used a solution of silver nanoparticles in water, known as Ag-water nanofluid, with a concentration ranging from 0.0005 to 0.05 wt%, in a PVT system. The nanofluid was analyzed using an optical filtration channel. In this scenario, the nanofluid passes over the PV cells and serves as an optical filter. When comparing the PV unit with the PVT system with optical filtering, it was found that the PV system exhibited superior electrical efficiency under low-to-medium sun intensities and climatic temperatures. Conversely, the photovoltaic system exhibits a diminished electrical efficiency when subjected to intense sun concentrations and elevated climatic temperatures. PVT with Ag-water and optical filtering is advised for applications requiring high climatic temperatures and sun concentrations, where thermal and electrical energies are needed. At a solar concentration of 5 and an environmental temperature of 45 $^\circ C$, the PVT system, which utilizes Ag-water with optical filtering, achieved thermal and electrical efficiency of 66.7% and 8.4%, respectively. Figure 21.9 displays the overall performance of the hybrid system when exposed to different weight concentrations of Ag nanoparticles. No substantial impact was detected on the PV temperature due to the varying concentration of Ag nanoparticles. Simultaneously, the substitution of the ambient temperature from 25 to 45 °C has resulted in a rise of roughly 12.5% in the PV temperature, as depicted in Fig. 21.9a. The temperature increase in the nanofluid channel was found to be occurring at a slow rate. This could be attributed to the minimal change in thermal conductivity of the nanofluid at these low concentrations of nanoparticles. When comparing the lowest and

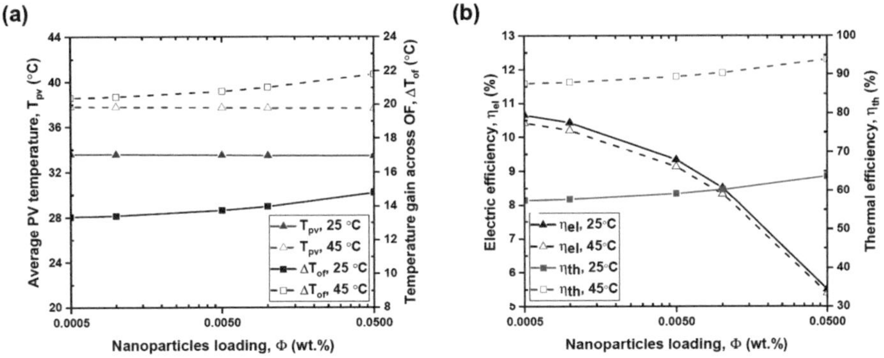

Fig. 21.9 The impact of loading nanoparticles (with a diameter of 10 nm) by weight on the thermal and electrical functionality of the hybrid system: (**a**) Influence on photovoltaic (PV) temperature and the extent of temperature rise caused by nanofluid. (**b**) Impact on thermal and electrical efficiency (Abdelrazik et al. 2021)

greatest nanoparticles loading, the heat gain has only increased by 10.6% and 7% at ambient conditions of 25 and 45 °C, respectively. Furthermore, the rise in air temperature facilitates greater heat absorption through the nanofluid channel, hence enhancing its suitability for low temperature process heating applications.

Figure 21.9b shows that there is a slight drop in electrical efficiency when the air temperature increases, although this decrease is not significant. In contrast, the rise in atmospheric temperature led to a corresponding improvement in thermal efficiency, with an average percentage increase of approximately 50.5%. This was due to the greater collection of thermal energy. In addition, the higher concentration of nanoparticles has had a detrimental impact on the electrical efficiency, as it has caused a decrease in the transmittance of the nanofluid. Approximately 48% of the electrical efficiency was reduced when comparing the lowest and highest concentrations of nanoparticles during testing at an ambient temperature of 45 °C, as shown in Fig. 21.9b (Abdelrazik et al. 2021). Table 21.2 displays a comparison of thermal and electrical efficiencies among various nanoparticles.

Copper (Cu) nanoparticles in water at a concentration of 0.4% by weight provide a high thermal efficiency of 76.8% and an electrical efficiency of 13.7%. This indicates that Cu significantly enhances both heat management and energy conversion in photovoltaic (PV) systems. In contrast, alumina (Al_2O_3) at the same concentration achieves a lower thermal efficiency of 46.7%. Although effective, it falls short of Cu's performance, and electrical efficiency data for alumina is not provided. Silicon carbide (SiC) tested at a concentration of 3% by weight shows a notable increase in electrical efficiency, reaching up to 24%. This suggests substantial improvements in the electrical conversion of energy. On the other hand, copper oxide (CuO) with a concentration of 0.75% by weight achieves the highest thermal efficiency of 90.3%, highlighting its exceptional heat absorption and transfer capabilities, though electrical efficiency is not reported. MXene (Ti_3C_2) mixed with ionic liquid and water at 0.2% by weight demonstrates high thermal efficiency (81.2%)

Table 21.2 The comparison of thermal and electrical efficiencies between different nanoparticles

Nanoparticle	Base fluid	Concentration (wt%)	Thermal efficiency	Electrical efficiency	Properties
Cu	Water	0.4	76.8%	13.7%	Highest thermal and electrical efficiency with water as base fluid
Al_2O_3	Water	0.4	46.7%	–	2.7% improvement in thermal efficiency compared to pure water
SiC	Ionized water	3	–	24%	8.2% rise in density, 5.18% increase in viscosity, 4.3% increase in thermal conductivity
CuO	Water	0.75	90.3%	–	Optimal concentration, superior thermal stability and conductivity
MXene (Ti_3C_2)	Ionic liquid and water	0.2	81.2%	14%	Stable up to 45 °C
MXene	Soybean oil	0.125	84.3%	–	Impressive thermal efficiency and surface temperature reduction
MWCNT	Water	0.3	47%	18%	Significant temperature reduction of PV cells
Al_2O_3	Water	0.3	33%	17.8%	
TiO_2	Water	0.3	27%	16%	
ZnO	Water	0.2	14.6–46%	14.7–47.6%	Most optimal based on AHP; balance of efficiency, energy gain, and environmental factors.
Ag	Water	0.0005–0.05	66.7%	8.4%	Optimal for high climatic temperatures and sun concentrations

and good electrical efficiency (14%). This shows a balanced performance in both thermal management and energy conversion. Similarly, MXene in soybean oil at 0.125% by weight provides a high thermal efficiency of 84.3%, indicating strong heat management and temperature reduction. Multi-walled carbon nanotubes (MWCNT) in water at 0.3% by weight offer a high electrical efficiency of 18% and significantly reduce PV cell temperature, enhancing overall system performance. Zinc oxide (ZnO) at 0.2% by weight presents a versatile range of efficiencies, adaptable to various needs. Lastly, silver (Ag) nanoparticles, even at low concentrations (0.0005–0.05% by weight), achieve good thermal efficiency (66.7%) but lower electrical efficiency (8.4%), making them suitable for high-temperature environments. Overall, copper and copper oxide excel in thermal performance, while MWCNT and silicon carbide lead in electrical efficiency, each providing unique advantages for optimizing PV system performance.

21.8 Utilization of Nanotechnology and Phase Change Materials (PCMs) in Solar Cells

A PCM substance undergoes a steady temperature change during the process of transitioning between different physical phases, during which it either absorbs or releases energy. The energy required for this phase transition in the material is typically much greater than the energy required to change the temperature within a specific phase. For many years, this phenomenon has been utilized as a highly effective method for storing thermal energy. The substances employed to carry out this process are referred to as phase change materials (PCMs), which can be classified into three broad categories: organic (such as paraffin and non-paraffin), inorganic (including metallics and salt hydrate), and eutectic (which encompasses combinations of organic-organic, inorganic-inorganic, and inorganic-organic substances).

The thermal system in the PV panel (PVT) is coupled to reduce the PV cells' temperature, enhancing their efficiency. Phase change materials (PCMs) are utilized in the photovoltaic thermal (PVT) system to facilitate the absorption of heat emitted by the PV cells. When the phase change material (PCM) absorbs energy, it undergoes melting at a consistent temperature. This process decreases and maintains a stable temperature for the PV cells. The length and performance of the process are contingent upon the characteristics and quantity of phase change material (PCM) and other materials utilized (Hemmat Esfe et al. 2020). Nanotechnology offers several applications for enhancing the thermophysical characteristics of phase change materials (PCMs). Nano-enhanced PCM refers to the utilization of nanoparticles in phase change materials (PCMs). Before delving into the literature on nano-enhanced phase change materials (PCMs), let's briefly review some recent studies on PVT (photovoltaic-thermal) systems equipped with PCMs and utilizing nanofluids as the cooling medium. Typically, PCM is placed behind the solar cells, between the thermal collection tubes, and on top of the insulation. The PCM in PVT can improve the electrical efficiency by 8% and the thermal efficiency by 25% under the same operating conditions (AL-Musawi et al. 2019). In their experimental study, Hassan et al. investigated the performance of a PVT system utilizing PCM (RT-35HC) with graphene/water as a nanofluid. The substance underwent testing at various concentrations (0.05/0.1/0.15%wt) and flow rates (20, 30, and 40 LPM). The concentration of 0.1% by weight exhibits optimal electrical and thermal efficiency across all measured flow rates. The highest values obtained for the overall, electrical, and thermal efficiency, as well as the reduction in PV temperature, were 61.6%, 14%, 42.2%, and 29.3 °C, respectively (for a flow rate of 40 L/min). The highest recorded efficiencies were 14.62% for total efficiency, 13.02% for electrical efficiency, and 1.68% for exergetic efficiency. These efficiencies were attained with a concentration of 0.1% by weight and a flow rate of 40 liters per minute (Wahab et al. 2020). Al-Musawi et al. investigated the utilization of SiO_2/water nanofluid with concentrations of 1% and 3% by weight, along with paraffin wax as phase change material (PCM), in PVT/PCM systems. The electrical efficiency was 14.3% and the thermal efficiency was 40%, with a weight percentage of 3%. The study also examined the impact of different melting temperatures (31, 34, and 54 ° C) on the PVT/PCM system. A rise in the melting temperature of the phase change material (PCM) results in a significant gain in thermal efficiency (by 30%) and a slight decrease in electrical efficiency (by 0.03%) (AL-Musawi et al. 2019). Naghdbishi et al. conducted a comparison between water and a mixture of 50% ethylene glycol and 50% water (EG50) as the base fluid for MWCNT nanoparticles (at concentrations of 0.1% and 0.2% by weight) in a PVT system using paraffin wax as the phase change material (PCM). The optimal performance was achieved with a concentration of 0.2% of multi-walled carbon nanotubes (MWCNT) in water. In this particular instance, the system demonstrated electrical and thermal efficiencies of 16.1% and 63.4%, respectively. The use of pure water in the PVT system with PCM exhibited superior performance compared to the utilization of MWCNT/EG50 (Naghdbishi et al. 2020). Additional experiments were conducted to investigate the performance of the PVT/PCM system using a nanofluid consisting of ZnO nanoparticles dispersed in water (0.2% by weight) and paraffin wax as the phase change material. The PV system had an average electrical output power of 92.2 W/m^2, while the PVT system had 99.6 W/m^2 and the PVT/PCM system had 104.4 W/m^2. This indicates a 13.3% increase in electrical output power (Sardarabadi et al. 2017). Figure 21.10 demonstrates that the use of a fluid-type heat recovery system (PVT) can boost the system's equivalent electrical-thermal output energy by more than 1.7 times compared to a traditional photovoltaic module. In addition, the concurrent utilization of a fluid and a PCM media amplifies the electrical-thermal outputs to a magnitude greater than twice the original value. It is important to note that the nanofluid-based collector has a minimal impact on this parameter compared to the deionized water-based collector due to the combination of electrical and thermal outputs.

Nanofluids are engineered colloidal suspensions consisting of nanoparticles in a base fluid, designed to improve thermal conductivity and heat transfer properties, making them highly effective in photothermal applications. When exposed to light, nanofluids show improved heat absorption and conversion efficiency due to the nanoparticles' high surface area and unique optical properties. Carbon nanotubes, graphene, and metal oxides like titanium dioxide and zinc oxide are commonly used in nanofluids, each contributing differently to photothermal effects. For instance, carbon nanotubes and graphene offer exceptional heat absorption and conductivity, while metal oxides like titanium dioxide and zinc oxide are valued for their stability and cost-effectiveness.

In photothermal applications, such as solar thermal energy systems and heat-based therapies, nanofluids can significantly enhance performance. The enhanced thermal conductivity of nanofluids results in better heat transfer efficiency, which is

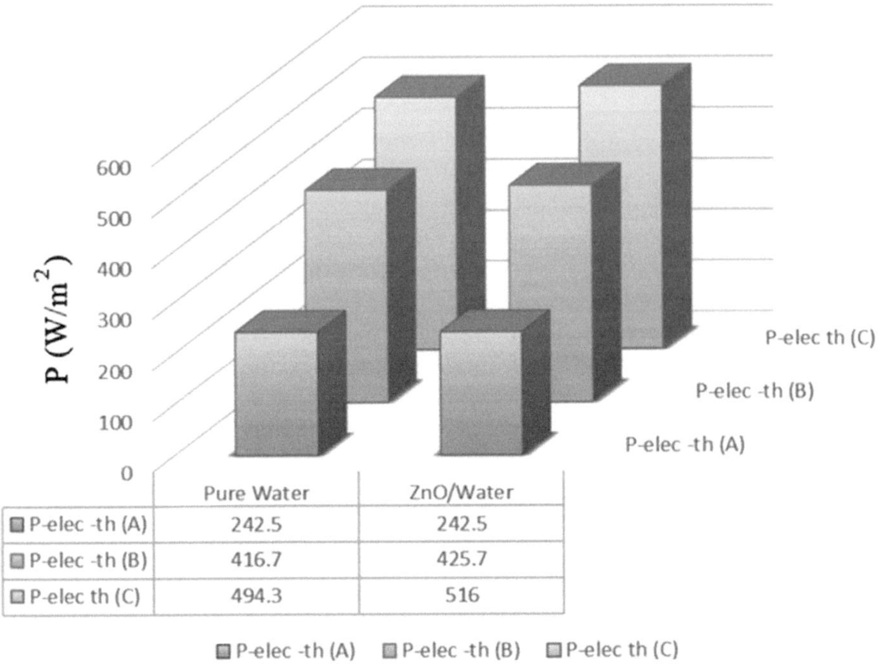

	Pure Water	ZnO/Water
P-elec -th (A)	242.5	242.5
P-elec -th (B)	416.7	425.7
P-elec th (C)	494.3	516

Fig. 21.10 The thermal-electrical power output (W/m^2) is compared for three types of collectors: (A) Conventional PV, (B) PVT nanofluid-based collector, and (C) PVT nanofluid/PCM-based collector (Sardarabadi et al. 2017)

crucial for optimizing energy systems and reducing energy consumption. This improvement can lead to more efficient solar collectors, reduced energy costs, and enhanced performance of heat-based medical treatments.

However, the benefits of nanofluids must be weighed against potential environmental concerns. The production, use, and disposal of nanomaterials can pose environmental risks. For instance, nanoparticles can be toxic to aquatic life and may accumulate in the environment, potentially leading to unforeseen ecological impacts. Moreover, the manufacturing processes for nanomaterials often involve chemicals and energy-intensive steps that contribute to environmental pollution.

To mitigate these risks, it is essential to develop sustainable practices for nanomaterial production and disposal, promote the use of environmentally benign materials, and conduct thorough risk assessments. Balancing the advantages of nanofluids in enhancing photothermal heat applications with their potential environmental impact is crucial for advancing this technology in a responsible and eco-friendly manner.

Rahmanian et al. examined a PVT/PCM system in which the nanofluid circulates across aluminum packs containing PCM. The PCM utilized was RT35-HC, the nanofluid consisted of CNT/water with a volume fraction of 0.1%, and the solar concentration was set at 5. The study examined both active cooling, which involved a nanofluid speed of 0.05 m/s, and passive cooling. The researchers concluded that employing a solitary pack for passive cooling can lower the temperature of the photovoltaic (PV) system by 18 ° *C*, while simultaneously enhancing the electrical efficiency by 12% and the thermal efficiency by 23% (Rahmanian et al. 2021). Basalike et al. studied a multilayer PVT refrigerator that utilized phase change materials (n-octadecane and caprice-palmitic acid) and a nanofluid (Al_3O_2-water 0.02%). The researchers examined the coefficient of performance (COP) and the amount of energy stored in the phase change material (PCM). It was determined that the use of paraffin produces a much higher coefficient of performance (COP) compared to the utilization of caprice-palmitic acid. The highest achieved coefficient of performance (COP) and energy storage capacities were 5 and 2.9 kilojoules (kJ), respectively (Basalike et al. 2021). The study examined the use of PCM and nanofluid as spectral filters in a concentrated PVT/PCM system. The system employed RT25 and S27 as phase change materials (PCMs) and utilized Ag/water nanofluid with a weight percentage of 0.05%. The performance analysis focused on S27 due to its favorable optical features. The nanofluid was placed on top of the PCM filter to assess its effectiveness. In this instance, the electrical efficiency was 21.5%, while the thermal efficiency was 64.9% (Yazdanifard et al. 2020). The incorporation of nano-enhanced phase change materials (PCMs) demonstrates significant potential forenhancing the efficiency of PVT systems. However, it is worth noting that most PCM materials have a low thermal conductivity, which is not ideal for efficient heat transfer. In order to boost the thermal conductivity and other characteristics of phase change materials (PCMs), numerous studies have utilized nanoparticles, resulting in the development of nano-enhanced PCMs, also known as nano-PCMs. Al-Waeli et al. conducted studies in the PVT/PCM system utilizing a nanofluid of 3% SiC in water and a nano-enhanced PCM of 0.1% SiC in paraffin. The nano-PCM achieved a thermal conductivity of 2.09 times greater than pure PCM. The studies were conducted within a flow rate range of 0.083–0.175 kg/s and an environment temperature range of $30 - 33\ ^\circ$ C. The study conducted a comparison between the suggested system and a traditional design using the same photovoltaic (PV) modules. The PVT/PCM system achieved a maximum electrical efficiency of 13.7%, while the PV system under the same conditions had an efficiency of 7.1% (Al-Waeli et al. 2020). The researchers also examined the ideal concentration ratio for the nanofluid SiC-water and nano-PCM SiC-paraffin, ranging from 0% to 4%. They found that the optimal ratio was 0.3% for the nanofluid and 0.1% for the nano-PCM. An optimized nanofluid was utilized as a coolant in a PVT/PCM system, where a nano-enhanced PCM served as a heat sink. This system was subjected to experimental testing and modelled using an artificial neural network. The findings indicated that the electrical and thermal efficiencies achieved a value of 13.3% (compared to just 8.1% for the PV) and 72%, respectively (Al-Waeli et al. 2019b). Al-Waeli et al. conducted a study on this system's technical and economic aspects, utilizing the same characteristics and estimating a lifespan of 25 years. They analyzed all the relevant elements of the system. The findings revealed an electricity expenditure of 0.112 USD per kilowatt-hour (kWh) and a payback period ranging from 4.4 to 5.3 years, signifying significant economic feasibility (Al-Waeli et al.

2019a). Fu et al. conducted an experimental study on the performance of a multilayer PVT/PCM system using water and nano-PCM (expanded graphite-paraffin). The nano-PCM was enclosed in an aluminum hoover bag and examined using three different concentrations of expanded graphite (EG), specifically 5%, 10%, and 15% by weight. The study also evaluates the impact of the PCM layer's thickness. The optimal performance was achieved by using a concentration ratio of 15% in the nano-PCM and a PCM layer thickness of 1 cm. This resulted in 18.8% and 67.1% of electrical and thermal efficiencies, respectively (Fu et al. 2021). MWCNTs were utilized in the production of nano-PCM (paraffin-based phase change material) and nanofluid (EG50-based fluid), both of which are employed in PVT/PCM systems (Sarafraz et al. 2019). A 19% enhancement in the thermal conductivity of the PCM was seen when employing a concentration of 0.3%wt. Different concentration ratios (0.1, 0.2, and 0.3%wt) were evaluated for the nanofluid. The PVT/PCM system produced a greater electrical power of 307.9 W/m^2 when the nanofluid had a MWCNT concentration of 0.2%wt. Khodadadi and Sheikholeslami examined the impact of fins in a multilayer photovoltaic thermal (PVT) system with nanofluid and nano phase change materials (PCMs). The nanofluid utilized in the experiment had a volume fraction (vf) of 0.04% and consisted of SiC particles dispersed in water. The nano-PCM employed was Al_2O_3-RT35-HC, also with a volume fraction of 0.04%. The system was tested with different numbers of fins, specifically 0, 10, 14, and 18. The electrical and thermal efficiencies were 13.2% and 70.1% for the configuration with 0 fins, 13.2% and 76.6% for the configuration with 10 fins, 13.3% and 77.1% for the configuration with 14 fins, and 13.% and 79.5% for the configuration with 18 fins, respectively (Khodadadi and Sheikholeslami 2021). Jamil et al. conducted a comparative analysis of three different nano-PCMs for photovoltaic (PV) modules. The researchers utilized the salt hydrate PT58 as the fundamental phase change material (PCM) and created the nano-PCM by including multi-walled carbon nanotubes (MWCNTs), graphene nanoplatelets (GNP), and magnesium oxide (MgO) nanoparticles at two different concentration ratios (0.25%wt and 0.5%wt). The configuration of the PVT system varies for these tests. The nano-phase change materials (PCMs) are positioned directly beneath the photovoltaic (PV) cells, without the use of a serpentine system. The nano-PCMs achieved superior performance when the concentration ratio of 0.5%wt was used. The largest temperature reduction for PV cells was 9.9 ° C for GNP-PT58, 9.8 ° C for MWCNT-PT58, and 8.4 ° C for MgO-PT58, as compared to PV panels. The electrical efficiency achieved a value of 12.1% for GNP-PT58, 12.1% for MWCNT-PT58, and 11.9% for MgO-PT58. This setup's most superior nano-PCM is the 0.5 weight percent GNP-PT58 (Jamil et al. 2021). Salem et al. investigated the application of channels for the purpose of cooling photovoltaic (PV) panels utilizing nano-phase change materials (nano-PCMs) and water. The channels can be filled with water alone, partially filled with a combination of water and nano-PCMs, or completely filled with nano-PCMs. The PCM used was calcium chloride hexahydrate, and different concentrations of the nanoparticle Al_2O_3 (0.25%, 0.5%, 0.75%, and 1% by weight) in the PCM were studied. The concentration ratio of 1% and $\lambda_{PCM} = 25\%$ yielded the highest electrical efficiency of 13.3%. The electrical efficiency obtained

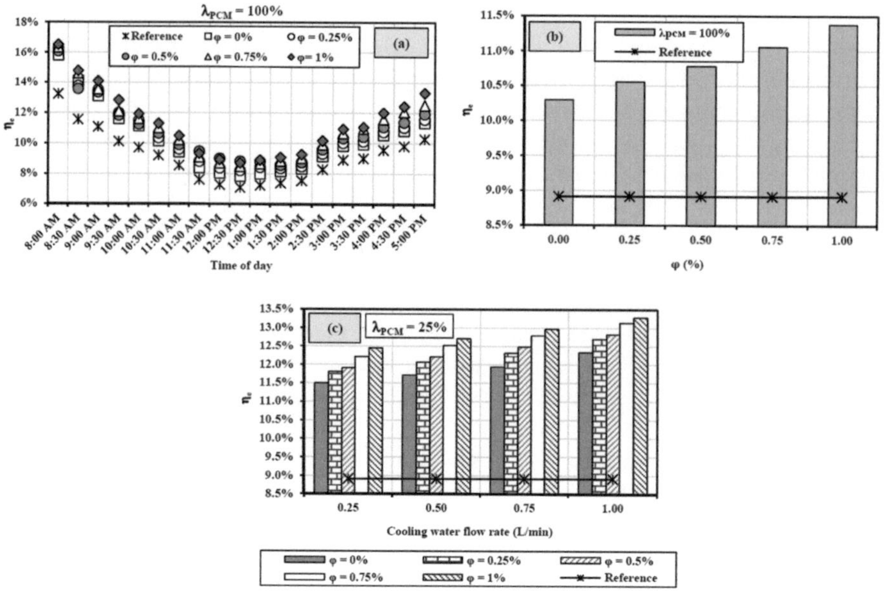

Fig. 21.11 The PV electrical efficiency is measured at various concentrations of nanoparticles, including (**a**) instantaneous values as well as (**b**) average values at $\lambda_{PCM} = 100\%$ and (**c**) $\lambda_{PCM} = 25\%$ (Salem et al. 2019)

under identical conditions (1%wt and 1 L/min water flow) was determined to be 13%, 12.7%, 12.2%, and 11.4% for $\lambda_{PCM} = 0\%$, $\lambda_{PCM} = 50\%$, $\lambda_{PCM} = 75\%$, and $\lambda_{PCM} = 100\%$, respectively. Figure 21.11 displays a sample of the panels' observed instantaneous and average electrical efficiencies at various concentrations of Al_2O_3 nanoparticles (Salem et al. 2019). Table 21.3 presents a succinct summary of the results from multiple studies on the use of nanofluids and phase change materials (PCMs) in photovoltaic thermal (PVT) systems.

Paraffin, a widely used phase change material (PCM) in photovoltaic thermal (PVT) systems, significantly influences environmental sustainability and energy efficiency. Its impact on the environment relies heavily on its production, use, and heat management. Paraffin, derived from petroleum, has a detrimental effect on the environment due to the greenhouse gas emissions and pollutants generated during its extraction and refinement processes. Moreover, paraffin-based materials are non-biodegradable, and their disposal can be problematic, causing them to persist in landfills for extended periods. Despite these environmental concerns, paraffin demonstrates considerable advantages in heat management. With its relatively high latent heat storage capacity, paraffin can absorb and release substantial amounts of heat, which stabilizes temperatures and reduces heat fluctuations, thereby enhancing the thermal performance and efficiency of PVT systems. By managing excess solar heat through its melting and solidifying phases, paraffin effectively improves the overall performance of the system. Although paraffin's heat management efficiency is a considerable advantage for energy systems, its environmental impact

Table 21.3 A concise overview of the various studies' findings regarding nanofluids and phase change materials (PCMs) in photovoltaic thermal (PVT) systems

Study	Nanofluid/PCM	Nanofluid volume fraction/ PCM concentration	Solar concentration	Cooling method	Electrical efficiency	Thermal efficiency	Key findings
Rahmanian et al. (2021)	CNT/water/ RT35-HC	CNT: 0.1%	5	Active: 0.05 m/s, passive	Active: Not provided, passive: 21.5%	Passive: 64.9%	Passive cooling reduced PV temperature by 18 °C, electrical efficiency increased by 12%, and thermal efficiency by 23%
Basalike et al. (2021)	Al_2O_3-water/n-octadecane & caprice-palmitic acid	Al_2O_3: 0.02%	Not specified	Not specified	Not specified	Not specified	Highest COP of 5 and energy storage of 2.9 kJ with paraffin
Yazdanifard et al. (2020)	Ag/water/RT25 & S27	Ag: 0.05%	Not specified	Not specified	21.5%	64.9%	Performance analysis focused on S27 due to its optical features
Al-Waeli et al. (2020)	SiC-water/ SiC-paraffin	Nanofluid: 3%, nano-PCM: 0.1%	0.083–0.175 kg/s	Not specified	13.7%	Not specified	Nano-PCM increased thermal conductivity by 2.09 times, improved PV efficiency compared to traditional designs
Al-Waeli et al. (2019b)	SiC-water/ SiC-paraffin	Nanofluid: 0.3%, Nano-PCM: 0.1%	Not specified	Not specified	13.3%	72%	Improved efficiencies with optimal nanofluid and nano-PCM ratios
Al-Waeli et al. (2019a)	Same as above	Same as above	Not specified	Not specified	Not specified	Not specified	Economic feasibility with 0.112 USD/kWh and a payback period of 4.4–5.3 years

(continued)

Table 21.3 (continued)

Study	Nanofluid/PCM	Nanofluid volume fraction/ PCM concentration	Solar concentration	Cooling method	Electrical efficiency	Thermal efficiency	Key findings
Fu et al. (2021)	Expanded graphite-paraffin	EG: 5%, 10%, 15%	Not specified	Not specified	18.8%	67.1%	Optimal performance with 15% EG and 1 cm PCM layer thickness
Sarafraz et al. (2019)	MWCNTs/ EG50-based fluid	MWCNT: 0.3%	Not specified	Not specified	307.9 W/m^2	Not specified	19% enhancement in PCM thermal conductivity with 0.3% MWCNT.
Khodadadi and Sheikholeslami (2021)	SiC-water/ Al_2O_3-RT35-HC	SiC: 0.04%	Not specified	Varying number of fins	13.2–13.3%	70.1–79.5%	Efficiency improved with increasing number of fins
Jamil et al. (2021)	MWCNTs, GNPs, MgO with PT58	0.25%wt, 0.5%wt	Not specified	Directly beneath PV cells	MWCNT: 12.1%, GNP: 12.1%, MgO: 11.9%	Not specified	GNP-PT58 had the best performance with 0.5%wt
Salem et al. (2019)	$CaCl_2 \cdot 6H_2O$/ Al_2O_3	Al_2O_3: 0.25%, 0.5%, 0.75%, 1%	Not specified	Water and/ or nano-PCMs	Max: 13.3%	Not specified	Highest efficiency with 1% Al_2O_3 and λ_PCM = 25%

cannot be disregarded. The reliance on petroleum and the challenges associated with disposal highlight the need for more sustainable alternatives. Ongoing research into biodegradable or less environmentally harmful materials is crucial for reducing the ecological footprint of phase change materials while maintaining high thermal efficiency. Embracing greener alternatives would better align with global sustainability goals, balancing energy efficiency with environmental responsibility.

21.9 Conclusion and Future Prospects

This chapter provides a comprehensive summary of the latest advancements and current research on the utilization of nanotechnology in solar cells. The presentation included a concise overview of the basics and progress of nanotechnology, an introduction to several generations of solar cells, and a discussion on potential approaches to enhance the performance of solar cells through the integration and incorporation of nanotechnology. When selecting the appropriate types of particles, it is crucial to consider stability, toxicity, and low cost, as these factors significantly impact the performance of solar cells in power conversion efficiency (PCE). Over the past five years, a significant endeavor has been to make non-silicon solar cells available for commercial use. However, several challenges remain unresolved, particularly in terms of stability, large-scale production, enhancing power conversion efficiency (PCE), and extending the lifespan of these cells. Carbon nanotubes (CNTs) have demonstrated significant enhancements in both the power conversion efficiency (PCE) of solar cells and the efficiency of phase change materials (PCMs). However, the high cost associated with CNTs poses a major obstacle. Solar cells can make use of a variety of nanoparticles in combination. The manipulation of nanoparticle morphology, namely the control of diameter and shape, will significantly impact the utilization of these particles in PVT systems. The performance of the PVT device can be enhanced through refinement of its structural design and optimizationof the chemical composition of its constituent components. Frequent investigations can be conducted to examine the performance of new configurations of PVT systems that utilize several thermal/optical layers, to establish a relationship between sun concentration ratios in PVTs. Due to the use of small light trapping structures and increased absorption cross-sections, solar systems can be made thinner than the distance charge carriers can travel while yet maintaining, or even changing, their ability to capture light effectively. In the context of nanotechnology in solar cells, environmental considerations hold significant importance. Key factors, such as the stability, toxicity, and cost of nanoparticles, greatly influence the efficiency and safety of solar cell technologies. To advance non-silicon solar cells, it is essential to address challenges while maintaining a low environmental impact. Although carbon nanotubes are effective, they are costly and pose environmental concerns. The design of solar cells using nanoparticles aims to boost power conversion efficiency (PCE) while ensuring environmental sustainability. Researchers are continuously working on innovations, such as smaller light-trapping structures and optimized nanoparticle morphologies, to improve performance with minimal ecological impact.

References

Abdelrazik AS, Saidur R, Al-Sulaiman FA (2021) Investigation of the performance of a hybrid PV/thermal system using water/silver nanofluid-based optical filter. Energy 215:119172. https://doi.org/10.1016/J.ENERGY.2020.119172

Alivisatos AP (1996) Semiconductor clusters, nanocrystals, and quantum dots. Science (1979) 271:933–937. https://doi.org/10.1126/SCIENCE.271.5251.933

AL-Musawi AIA, Taheri A, Farzanehnia A et al (2019) Numerical study of the effects of nanofluids and phase-change materials in photovoltaic thermal (PVT) systems. J Therm Anal Calorim 137:623–636. https://doi.org/10.1007/S10973-018-7972-6/METRICS

Al-Waeli AHA, Sopian K, Chaichan MT et al (2017) An experimental investigation of SiC nanofluid as a base-fluid for a photovoltaic thermal PV/T system. Energy Convers Manag 142:547–558. https://doi.org/10.1016/J.ENCONMAN.2017.03.076

Al-Waeli AHA, Kazem HA, Chaichan MT, Sopian K (2019a) Experimental investigation of using nano-PCM/nanofluid on a photovoltaic thermal system (PVT): technical and economic study. Therm Sci Eng Prog 11:213–230. https://doi.org/10.1016/J.TSEP.2019.04.002

Al-Waeli AHA, Sopian K, Yousif JH et al (2019b) Artificial neural network modeling and analysis of photovoltaic/thermal system based on the experimental study. Energy Convers Manag 186:368–379. https://doi.org/10.1016/J.ENCONMAN.2019.02.066

Al-Waeli AHA, Sopian K, Kazem HA, Chaichan MT (2020) Evaluation of the electrical performance of a photovoltaic thermal system using nano-enhanced paraffin and nanofluids. Case Stud Therm Eng 21:100678. https://doi.org/10.1016/J.CSITE.2020.100678

Ameur A, Berrada A, Loudiyi K, Aggour M (2020) Forecast modeling and performance assessment of solar PV systems. J Clean Prod 267:122167. https://doi.org/10.1016/J.JCLEPRO.2020.122167

Asim N, Sopian K, Ahmadi S et al (2012) A review on the role of materials science in solar cells. Renew Sust Energ Rev 16:5834–5847. https://doi.org/10.1016/j.rser.2012.06.004

Basalike P, Peng W, Zhang J, Lu S (2021) Numerical investigation on the performance characteristics of photovoltaic thermal refrigerator integrating phase change material and water/nanofluid as fluid medium. J Energy Storage 44:103453. https://doi.org/10.1016/J.EST.2021.103453

Bhanvase B, Barai D (2021) Introduction to nanofluids. Nanofluids Heat Mass Transf:3–42. https://doi.org/10.1016/B978-0-12-821955-3.00011-X

Boroditsky M, Gontijo I, Jackson M et al (2000) Surface recombination measurements on III–V candidate materials for nanostructure light-emitting diodes. J Appl Phys 87:3497–3504. https://doi.org/10.1063/1.372372

Corbett J, McKeown RA, Peggs GN, Whatmore R (2000) Nanotechnology: international developments and emerging products. CIRP Ann Manuf Technol 49:523–545. https://doi.org/10.1016/S0007-8506(07)63454-4

Corkish R, Green MA, Puzzer T (2002) Solar energy collection by antennas. Sol Energy 73:395–401. https://doi.org/10.1016/S0038-092X(03)00033-1

Crisostomo F, Hjerrild N, Mesgari S et al (2017) A hybrid PV/T collector using spectrally selective absorbing nanofluids. Appl Energy 193:1–14. https://doi.org/10.1016/J.APENERGY.2017.02.028

De Respinis M, De Temmerman G, Tanyeli I et al (2013) Efficient plasma route to nanostructure materials: case study on the use of m-WO3 for solar water splitting. ACS Appl Mater Interfaces 5:7621–7625. https://doi.org/10.1021/AM401936Q

Eperon GE, Hörantner MT, Snaith HJ (2017) Metal halide perovskite tandem and multiple-junction photovoltaics. Nat Rev Chem 1(12 1):1–18. https://doi.org/10.1038/s41570-017-0095

Fu Z, Liang X, Li Y et al (2021) Performance improvement of a PVT system using a multilayer structural heat exchanger with PCMs. Renew Energy 169:308–317. https://doi.org/10.1016/J.RENENE.2020.12.108

Ghadiri M, Sardarabadi M, Pasandideh-Fard M, Moghadam AJ (2015) Experimental investigation of a PVT system performance using nano ferrofluids. Energy Convers Manag 103:468–476. https://doi.org/10.1016/J.ENCONMAN.2015.06.077

Gupta SK, Pradhan S (2021) A review of recent advances and the role of nanofluid in solar photovoltaic thermal (PV/T) system. Mater Today Proc 44:782–791. https://doi.org/10.1016/J.MATPR.2020.10.708

Han YW, Jeon SJ, Lee HS et al (2019) Evaporation-free nonfullerene flexible organic solar cell modules manufactured by an all-solution process. AdEnM 9:1902065. https://doi.org/10.1002/AENM.201902065

Heine C, Morf RH (1995) Submicrometer gratings for solar energy applications. Appl Opt 34(14):2476–2482. https://doi.org/10.1364/AO.34.002476

Hemmat Esfe M, Kamyab MH, Valadkhani M (2020) Application of nanofluids and fluids in photovoltaic thermal system: an updated review. Sol Energy 199:796–818. https://doi.org/10.1016/J.SOLENER.2020.01.015

Henry HC (1980) Limiting efficiencies of ideal single and multiple energy gap terrestrial solar cells. JAP 51:4494–4500. https://doi.org/10.1063/1.328272

Imtiaz Hussain M, Kim JH, Kim JT (2019, 2019) Nanofluid-powered dual-fluid photovoltaic/thermal (PV/T) system: comparative numerical study. Energies 12, 775. https://doi.org/10.3390/EN12050775

Ito S, Murakami TN, Comte P et al (2008) Fabrication of thin film dye sensitized solar cells with solar to electric power conversion efficiency over 10%. Thin Solid Films 516:4613–4619. https://doi.org/10.1016/J.TSF.2007.05.090

Jamil F, Ali HM, Nasir MA et al (2021) Evaluation of photovoltaic panels using different nano phase change material and a concise comparison: an experimental study. Renew Energy 169:1265–1279. https://doi.org/10.1016/j.renene.2021.01.089

Khodadadi M, Sheikholeslami M (2021) Numerical simulation on the efficiency of PVT system integrated with PCM under the influence of using fins. Sol Energy Mater Sol Cells 233:111402. https://doi.org/10.1016/J.SOLMAT.2021.111402

Kojima A, Teshima K, Shirai Y, Miyasaka T (2009) Organometal halide perovskites as visible-light sensitizers for photovoltaic cells. J Am Chem Soc 131:6050–6051. https://doi.org/10.1021/JA809598R/SUPPL_FILE/JA809598R_SI_001.PDF

Laidani N, Bartali R, Gottardi G et al (2007) Optical absorption parameters of amorphous carbon films from Forouhi–Bloomer and Tauc–Lorentz models: a comparative study. J Phys Condens Matter 20:015216. https://doi.org/10.1088/0953-8984/20/01/015216

Li BL, Wang J, Zou HL et al (2016) Low-dimensional transition metal dichalcogenide nanostructures based sensors. Adv Funct Mater 26:7034–7056. https://doi.org/10.1002/ADFM.201602136

Lieber CM (1998) One-dimensional nanostructures: chemistry, physics & applications. Solid State Commun 107:607–616. https://doi.org/10.1016/S0038-1098(98)00209-9

Lu F, Neal EA, Nakanishi T (2019) Self-assembled and nonassembled alkylated-fullerene materials. Acc Chem Res 52:1834–1843. https://doi.org/10.1021/ACS.ACCOUNTS.9B00217/ASSET/IMAGES/MEDIUM/AR-2019-00217E_0008.GIF

Mariappan VK, Krishnamoorthy K, Pazhamalai P et al (2020) Antimonene dendritic nanostructures: dual-functional material for high-performance energy storage and harvesting devices. Nano Energy 77. https://doi.org/10.1016/J.NANOEN.2020.105248

McElroy MB (2021) Atmospheric history and composition. Atmos Environ:39–50. https://doi.org/10.2307/J.CTV1NXCVHQ.8

Naghdbishi A, Yazdi ME, Akbari G (2020) Experimental investigation of the effect of multi-wall carbon nanotube—water/glycol based nanofluids on a PVT system integrated with PCM-covered collector. Appl Therm Eng 178:115556. https://doi.org/10.1016/J.APPLTHERMALENG.2020.115556

O'Regan B, Graetzel M, O'Regan B, Graetzel M (1991) A low-cost, high-efficiency solar cell based on dye-sensitized colloidal TiO2 films. Nature 353:737–740. https://doi.org/10.1038/353737A0

Odeh S, Behnia M (2009) Improving photovoltaic module efficiency using water cooling. Heat Transf Eng 30:499–505. https://doi.org/10.1080/01457630802529214

Paci I, Johnson JC, Chen X et al (2006) Singlet fission for dye-sensitized solar cells: can a suitable sensitizer be found? J Am Chem Soc 128:16546–16553. https://doi.org/10.1021/JA063980H

Pasquier AD, Unalan HE, Kanwal A et al (2005) Conducting and transparent single-wall carbon nanotube electrodes for polymer-fullerene solar cells. Appl Phys Lett 87:1–3. https://doi.org/10.1063/1.2132065/936172

Rahmanian S, Rahmanian-Koushkaki H, Omidvar P, Shahsavar A (2021) Nanofluid-PCM heat sink for building integrated concentrated photovoltaic with thermal energy storage and recovery capability. Sustain Energy Technol Assess 46:101223. https://doi.org/10.1016/J.SETA.2021.101223

Rejeb O, Sardarabadi M, Ménézo C et al (2016) Numerical and model validation of uncovered nanofluid sheet and tube type photovoltaic thermal solar system. Energy Convers Manag 110:367–377. https://doi.org/10.1016/J.ENCONMAN.2015.11.063

Rubbi F, Habib K, Saidur R et al (2020) Performance optimization of a hybrid PV/T solar system using soybean oil/MXene nanofluids as a new class of heat transfer fluids. Sol Energy 208:124–138. https://doi.org/10.1016/J.SOLENER.2020.07.060

Salem MR, Elsayed MM, Abd-Elaziz AA, Elshazly KM (2019) Performance enhancement of the photovoltaic cells using Al2O3/PCM mixture and/or water cooling-techniques. Renew Energy 138:876–890. https://doi.org/10.1016/j.renene.2019.02.032

Sangeetha M, Manigandan S, Ashok B et al (2021) Experimental investigation of nanofluid based photovoltaic thermal (PV/T) system for superior electrical efficiency and hydrogen production. Fuel 286:119422. https://doi.org/10.1016/J.FUEL.2020.119422

Sarafraz MM, Safaei MR, Leon AS et al (2019) Experimental investigation on thermal performance of a PV/T-PCM (photovoltaic/thermal) system cooling with a PCM and Nanofluid. Energies 12:2572. https://doi.org/10.3390/EN12132572

Sardarabadi M, Passandideh-Fard M, Maghrebi MJ, Ghazikhani M (2017) Experimental study of using both ZnO/ water nanofluid and phase change material (PCM) in photovoltaic thermal systems. Sol Energy Mater Sol Cells 161:62–69. https://doi.org/10.1016/J.SOLMAT.2016.11.032

Shabaev A, Efros AL, Nozik AJ (2006) Multiexciton generation by a single photon in nanocrystals. Nano Lett 6:2856–2863. https://doi.org/10.1021/NL062059V

Shockley W, Queisser HJ (1961) Detailed balance limit of efficiency of p-n junction solar cells. J Appl Phys 32:510–519. https://doi.org/10.1063/1.1736034

Sohani A, Shahverdian MH, Sayyaadi H et al (2021) Selecting the best nanofluid type for a photovoltaic thermal (PV/T) system based on reliability, efficiency, energy, economic, and environmental criteria. J Taiwan Inst Chem Eng 124:351–358. https://doi.org/10.1016/J.JTICE.2021.02.027

Stuart HR, Hall DG (1998) Enhanced dipole-dipole interaction between elementary radiators near a surface. Phys Rev Lett 80:5663. https://doi.org/10.1103/PhysRevLett.80.5663

Tsai FJ, Wang JY, Kiang YW, Yang CC (2010) Absorption enhancement of an amorphous Si solar cell through surface plasmon-induced scattering with metal nanoparticles. Opt Express 18(Suppl 2). https://doi.org/10.1364/OE.18.00A207

Wahab A, Khan MAZ, Hassan A (2020) Impact of graphene nanofluid and phase change material on hybrid photovoltaic thermal system: exergy analysis. J Clean Prod 277:123370. https://doi.org/10.1016/J.JCLEPRO.2020.123370

Westphalen M, Kreibig U, Rostalski J et al (2000) Metal cluster enhanced organic solar cells. Sol Energy Mater Sol Cells 61:97–105. https://doi.org/10.1016/S0927-0248(99)00100-2

Willson RC, Mordvinov AV (2003) Secular total solar irradiance trend during solar cycles 21-23. Geophys Res Lett 30:1199. https://doi.org/10.1029/2002GL016038

Xu X, Ray R, Gu Y et al (2004) Electrophoretic analysis and purification of fluorescent single-walled carbon nanotube fragments. J Am Chem Soc 126:12736–12737. https://doi.org/10.1021/JA040082H/SUPPL_FILE/JA040082H_S.PDF

Yang Y, Gupta MC, Dudley KL (2007) Towards cost-efficient EMI shielding materials using carbon nanostructure-based nanocomposites. Nanotechnology 18:345701. https://doi.org/10.1088/0957-4484/18/34/345701

Yates JT (2009) Photochemistry on TiO2: mechanisms behind the surface chemistry. Surf Sci 603:1605–1612. https://doi.org/10.1016/J.SUSC.2008.11.052

Yazdanifard F, Ameri M, Taylor RA (2020) Numerical modeling of a concentrated photovoltaic/thermal system which utilizes a PCM and nanofluid spectral splitting. Energy Convers Manag 215:112927. https://doi.org/10.1016/J.ENCONMAN.2020.112927

Yeon J, Lee YJ, Yoo DE et al (2013) High throughput ultralong (20 cm) nanowire fabrication using a wafer-scale nanograting template. Nano Lett 13:3978–3984. https://doi.org/10.1021/NL400209N

Yoo JJ, Seo G, Chua MR et al (2021) Efficient perovskite solar cells via improved carrier management. Nature 590:587–593. https://doi.org/10.1038/s41586-021-03285-w

Zeraatpisheh M, Arababadi R, Pour MS (2018) Economic analysis for residential solar PV systems based on different demand charge tariffs. Energies 11:3271. https://doi.org/10.3390/EN11123271

Nanotechnology Reduces Global Warming

22

Sana Ullah, Dilawar Khan, Nasr Ullah, Murad Muhammad, Asim Gulzar, Abdul Basit, and Sobia Khaliq

Abstract

Nanotechnology represents a new way to address the costs of improving global warming via novel applications across industries. This chapter focuses on how nanotechnology might provide game-changing solutions to some of the biggest challenges in climate change. Before launching into the specifics of nanotechnology, it opens with a primer on what is known about controlling things at the scale of molecules and atoms. This chapter also examines some nanohyped eco-themes that seem virtually certain to slash greenhouse gases (GHGs) and boost the efficiency of energy technologies, as well as prompting a string of sustainable behaviors. Case examples illustrate the successful engineering of nanomaterials for carbon capture and storage facilities, efficient catalytic processes, and renewable energy production. A critical evaluation is also provided for the safety and

S. Ullah (✉)
Department of Agricultural Leadership, Education & Communication, College of Agricultural & Environmental Sciences, University of Georgia, Athens, GA, USA
e-mail: sana.ullah@uga.edu

D. Khan · A. Gulzar · S. Khaliq
Department of Entomology, PMAS-Arid Agriculture University, Rawalpindi, Pakistan

N. Ullah
Institute of Management Sciences (IMSciences), Peshawar, Khyber Pakhtunkhwa, Pakistan

M. Muhammad
State Key Laboratory of Desert and Oasis Ecology, Key Laboratory of Ecological Safety and Sustainable Development in Arid Lands, Xinjiang Institute of Ecology and Geography, Chinese Academy of Sciences, Urumqi, China

University of Chinese Academy of Sciences, Beijing, People's Republic of China

A. Basit
Department of Horticultural Science, Kyungpook National University, Daegu, South Korea

A. A. H. Abdel Latef et al. (eds.), *Bioremediation and Nanotechnology for Climate Change Mitigation*, https://doi.org/10.1007/978-981-96-3069-1_22

environmental implications of nanotechnology applications. By harnessing the potential of nanotechnology, this chapter underscores its pivotal role in advancing global efforts toward a more sustainable and resilient future amidst the challenges posed by climate change.

Keywords

Nanotechnology · Emissions · Climate · Mitigation · Innovation

22.1 Introduction

Global warming is a consequence of anthropogenic activities being implied as the biggest challenge of this twenty-first century with extensive future ramifications for Earth's climate system and its biotic–abiotic systems and socio-economic-ecological spheres. The evidence is incontrovertible: the Earth's average surface temperature has been rising since the Industrial Revolution, primarily as a result of increases in concentrations of greenhouse gases (GHGs) such as carbon dioxide (CO_2), methane (CH_4), and nitrous oxide (N_2O). These gases absorb and trap heat, acting as a blanket of outgoing infrared radiation, thus enhancing the greenhouse effect that heats up from lower temperatures in both the atmosphere and surface (Bothun 2019). Decades of temperature data from surface stations, ocean buoys and ships, and satellites orbiting around in space share a consistent signal: temperatures have been on an upward trend since the last century with clear regional variations and phases of rapid warming (Chin et al. 2017). At the same time, observations of disappearing glaciers, shrinking Arctic sea ice extent, and rising sea levels demonstrate measurable physical signs that this is a real global phenomenon. The effects of global warming are more important than just hotter temperatures (Singh and Singh 2012). This kind of interplay between a warming climate and the interconnected systems that help determine what changes may occur in precipitation patterns, how ecosystem dynamics will shift, or how frequently intense weather events might happen. Climate change continues to pose risks for virtually all aspects of society, including agricultural productivity and freshwater availability, biodiversity conservation, and human health (Myers and Patz 2009). To cope with these difficulties, international scientific bodies and decision- and policy-makers have increased their work to better comprehend global warming. The Intergovernmental Panel on Climate Change (IPCC) is a central provider of scientific consensus and policy advice, bringing together evidence on climate science as well as impacts and adaptations.

Nanotechnology, manipulating matter at the scale of atoms and molecules (1–100 nm), has demonstrated capabilities that are transforming environmental protection (Bhawana and Fulekar 2012). These materials have exciting properties and functionalities that are unique to their size scale, resulting in potential wide-ranging novel applications across energy generation or storage; water purification/ desalination/removal of pollution contaminants from drinking water/crop pesticides with minimal waste production sustainable agriculture strategies. The significant importance of nanotechnology in the environment is because it allows for designing new materials with tailored performance properties specific to given environmental problems. For example, nanoscale materials can greatly enhance the function of energy conversion and storage technologies like solar cells (for increasing surface area) and batteries (for better conductivity). In the field of water treatment and purification, on the other hand, nanoscale materials are capable of providing selective filtration and efficient removal processes; however, it is done (or used), promising solutions to global water scarcity as well as pollution problems (Mohmood et al. 2013). Also, nanotechnology greatly contributes to reducing air pollution, environmental, and soil pollution with valuable techniques like catalysis or by conducting environmentally friendly treatment processes known as nanoremediation (Ibrahim et al. 2016). Nanocatalysis can lead to more efficient mother, reactants and less waste, while nanoremediation has the ability for contaminant molecules of interest at environmental levels concentration level in situ transforming contaminated sites into clean ecosystems (El-Sheikh et al. 2021; Rónavári and Kónya 2021).

In agriculture, the use of nanotechnology also offers great prospects for sustainable food production and environmental friendliness. Real-time monitoring of soil health and crop conditions by nanosensors aids in reduced resource use with minimal impact on the environment (Sharma et al. 2021). On the other hand, nano-based delivery systems of fertilizers and pesticides improve their nutrient use efficiency (NUE) which translates into lower input requirements along with minimized intermediate losses to the environment (Kalia and Kaur 2019). But, in tandem with those advancements is the need to deploy nanotechnology safely and responsibly. The possible toxicity of nanomaterials and their course long-term environmental effects underline the need for comprehensive safety profiles and regulations. However, addressing these needs demands the cooperation of a range of scientists, politicians, industry actors, and the broader society to ensure that nanotechnology contributes positively to environmental sustainability. This chapter discusses nanotechnology's diverse potential in solving environmental problems that can inject a new dimension to utilization and conservation based on revolutionizing resource use, pollution alleviation, and ecosystem renovation… Through examination of current research trends, technological innovations, and emerging diverse applications, we aim to offer a unique perspective on how advancing nanotechnology could be significant in achieving more resilient and sustainable living for our planet Earth.

22.2 Nanotechnology in Renewable Energy Generation

Solar power, wind and other renewable technologies have incredible potential to reduce the de-carbonizing transition moving toward a sustainable energy future (Naudé 2011). Nanotechnology has revolutionary potential to make renewable energy technologies significantly more efficient, longer lasting and scalable (Kumar et al. 2018). Nanotechnology allows creating new, advanced materials with increased light absorption and charge carrier transport in the solar cell itself—such as quantum dots, nanowires, or perovskites (Goodnick 2018). These materials have the potential to increase the conversion efficiency and lower the cost-per-watt of solar energy. These and other nanoscale coating films enhance the flexibility of solar panels, making them adaptable for a wide range of applications, such as integration into building materials like shingles or windows. Additionally, vehicles with integrated photovoltaic body surfaces can utilize these technologies for battery charging and energy-efficient climate control systems. Roof-mounted, user-oriented photovoltaic modules improve cooling efficiency during hot seasons by reducing interior temperatures and optimizing energy use. Solar cells gain improved performance using nanostructured electrodes and transparent conductive films including materials like graphene and carbon nanotubes which provide better conductivity and transparency (Hecht et al. 2011).

Wind turbine blades are also made lighter by improving their mechanical properties using nanofibers and other nanomaterials (Merugula et al. 2010). It makes wind turbines last longer and it is good aerodynamics. Based on self-cleaning materials and anti-corrosive nanoparticles, these nanocoatings protect wind turbines' surfaces from natural aggressions, decreasing maintenance costs and increasing energy yields (Abdeen et al. 2019). Wind turbines equipped with miniaturized nanosensors constitute the closed-loop system for monitoring mechanical stress, temperature, and structural integrity in real time to optimize maintenance/day-ahead schedules, minimizing the total cost of operation while reducing major downtime costs due and large-scale component failures. It needs to assess the environmental impact and safety protocols for nanomaterials in active components associated with renewable energy technologies so that it can be deployed sustainably. At scale, there are still hurdles to deploying nanotechnology as a solution for energy consumption on an industrial level because it is not easy to integrate with today's infrastructure. Widespread adoption requires overcoming cost and regulatory hurdles.

22.3 Nanotechnology for Energy Efficiency

Nanotechnology can potentially rejuvenate energy efficiency in sectors from buildings and roadways to industrial applications (Shah et al. 2020). Nanoscale allows materials to be manipulated and gives researchers/ engineers the ability to create new solutions that will increase efficiency in energy conversion, storage, and utilization. Novel materials and structures with superior thermal and optical properties can be produced using nanotechnology (Jafarzadeh and Jafari 2021). They can be used for enhanced insulation of windows, roofs, walls, etc., to minimize heat transfer towards improved energy efficiency of buildings with minimal adverse effects in other applications like land transportation or industrial processes. Nanomaterials, such as powered nanocrystals and electrochromic films, allow sunlight to pass or block it, creating smart windows that can play with temperature. This helps control indoor temperatures and reduces the need for heating and cooling. The setup enables the creation of cost-effective nanophosphors for light-emitting diodes (LEDs) and displays technologies using nanotechnology. To boost light conversion efficiency, color rendering and longevity of lighting systems nanomaterials are being explored to reduce energy use and environmental impact. Semiconductor nanocrystals called quantum dots are used to improve the color purity and energy efficiency of displays and lighting (Huang et al. 2020). These are said to allow vibrant colors and high Pixels Per Inch (PPI) displays with less power consumption compared to traditional tech. Nanotechnology improves electrode materials in energy storage devices such as lithium-ion batteries and supercapacitors. Nanotechnology enables next-generation batteries by providing solid-state electrolytes and electrode materials (Parameswaran et al. 2023). The promise for these improvements lies in safer, cost-effective energy storage solutions with significantly less environmental impact. It, thus, becomes possible through nanotechnology to design active and selective, so-called (nano)catalysts for industrial processes such as chemical synthesis or pollution abatement. The use of nanostructured catalysts has constantly increased to enhance reaction conversion, save energy, and reduce waste production (Shen 2015). Nanofluids are suspensions of nanoparticles in heat transfer fluids that offer exceptional performance for improving the efficiency of heat exchange systems applicable to industrial processes. They increase heat transfer rates and thermal conductivity, thereby ensuring maximum energy usage during cooling or heating operations. It develops nanocomposites and nanocoatings that can help lightweight cars and planes, thereby improving fuel efficiency and lowering greenhouse gas emissions. Nanotech-supported approaches to lightweighting of components maintain their structural integrity while improving performance organs. Fuel cell efficiencies and lifetimes are increased due to nanomaterials in the catalysis of electrodes, proton exchange membranes (PEMs), and fuel storage (Li et al. 2017). This opens up clean and efficient energy conversion for transportation and stationary applications. Nanotechnology provides new classes of materials and devices, but the environmental impact and lifecycle assessment must be evaluated for sustainable deployment.

22.4 Nanotechnology for Environmental Remediation and Waste Management

By taking advantage of the special qualities of nanoparticles, nanotechnology provides novel approaches to waste management and environmental contamination, improving the sustainability, selectivity, and efficiency of remediation procedures. To rehabilitate contaminated soils, nanoparticles like carbon nanotubes (CNTs), and zero-valent iron (nZVI) are used to break down pollutants through accelerated chemical reactions and adsorption processes (Liu and Lal 2012). In order to improve soil quality and ecosystem health, organic contaminants, heavy metals, and industrial pollutants are successfully targeted using nanoremediation techniques (Rajput et al. 2022). Heavy metals and volatile organic compounds (VOCs) are among the pollutants removed from groundwater using adsorbents and filtration membranes enabled by nanotechnology. By improving sorption capacity and making it easier to remove certain pollutants, functionalized nanoparticles lower the risk of groundwater contamination. The application of nanotechnology to membrane technology improves their permeability, selectivity, and resistance to fouling in water purification (Goh et al. 2015). Nanocomposite membranes effectively remove pollutants, microbes, and pathogens from drinking water and wastewater by including nanoparticles (such as titanium dioxide and graphene oxide) (Naseem and Waseem 2022). Adsorbents such as functionalized nanoparticles and nanomaterials such as zeolites, clay minerals, and magnetic nanoparticles are used to remove pollutants from aqueous solutions, such as heavy metals, dyes, and medications. For sustainable water treatment applications, surface modification increases adsorption capacity and promotes regeneration. Through catalytic oxidation and reduction reactions, nanostructured catalysts aid in the breakdown of air pollutants (Zhu et al. 2020). Functionalized nanoparticles can reduce air pollution caused by car exhaust and industrial pollutants by improving catalytic activity, stability, and efficiency. By adding nanofibers, nanotubes, and nanoparticles to air filters, nanotechnology improves their ability to collect particulate matter (PM), allergies, and other airborne pollutants. In Heating, Ventilation, and Air Conditioning (HVAC) systems and air purifiers, nano-enabled filters increase filtration effectiveness, resistance to airflow, and durability. Nanotechnology facilitates resource recovery, recycling, and waste degradation by developing nano-enabled processes. By facilitating the selective separation and recovery of valuable elements from waste streams, nanocatalysts and nanomembranes reduce landfill waste and advance the circular economy concepts. Real-time monitoring of environmental pollutants and contaminants in waste streams is made possible by miniature nanosensors, which improve waste classification, treatment effectiveness, and regulatory compliance in solid waste management (Thakur and Kumar 2022).

22.5 Safety and Environmental Impact of Nanotechnology

Because of its capacity to work with matter at the nanoscale, nanotechnology has enormous potential to transform several industries and provide solutions to major global issues (Malik et al. 2023). Nevertheless, besides its possible advantages, there are worries about the environmental impact and safety of nanomaterials and nanotechnologies (Gottardo et al. 2021). This section delves into the intricacies of nanotechnology safety and environmental concerns, highlighting existing knowledge, obstacles, and regulatory structures. Nanomaterials' small size and high surface area-to-volume ratio allow them to have unique physical, chemical, and biological capabilities (Patel et al. 2021). These characteristics can cause them to behave differently in biological systems and the environment than their bulk equivalents (Abbas et al. 2020). Certain nanoparticles carry potential health hazards, including skin penetration, cellular damage, and inhaled toxicity (Buzea et al. 2007). It is essential to comprehend how nanoparticles interact with biological tissues and organisms to evaluate any risks and put safety precautions in place. Workers who produce, handle, or use nanomaterials may come into contact with nanoparticles (Amoabediny et al. 2009). Workplace procedures, PPE, and engineering controls that are effective are crucial for reducing occupational exposure and safeguarding worker health. When discharged into the environment, nanomaterials may change in ways that impact their bioavailability, toxicity, and mobility (Batley et al. 2013). Their environmental destiny is influenced by elements like surface changes, aggregation, and interactions with naturally occurring organic materials. When nanoparticles come into contact with both land and aquatic life, they may harm biodiversity and ecosystems (Vasyukova et al. 2021). To evaluate ecological concerns, research focuses on comprehending how nanoparticles interact with various organisms and ecosystems. Micromanaging the effects of nanotechnology on the environment and human health requires a combination of risk assessment, life-cycle analysis, and preventative measures (Furxhi et al. 2021). Effective risk management requires cooperation between scientists, industry stakeholders, regulators, and legislators. Interacting with stakeholders, the general public, academic institutions, the business community, and non-governmental organizations promotes openness, confidence, and well-informed choices about the safety and applicability of nanotechnology. Enhancing public acceptance and knowledge of nanotechnologies is achieved by transparent communication of their potential hazards and benefits (Todaro et al. 2023). The goal of outreach and education initiatives is to dispel myths and encourage the responsible application of nanotechnology.

22.6 Collaboration Opportunities Between Academia, Industry, and Government for Advancing Nanotechnology in Environmental Sustainability

Nanotechnology has a great deal of promise in solving environmental problems with creative fixes and environmentally friendly behaviors (Govindasamy et al. 2022). Applications of nanotechnology that support environmental sustainability are being advanced through corporate, government, and academic collaboration. To maximize the advantages of nanotechnology, this section examines important collaboration opportunities and tactics to promote interdisciplinary partnerships. Research collaborations among academic institutions, research centers, and industry promote knowledge sharing, creativity, and the creation of environmentally sustainable solutions enabled by nanotechnology (Pokrajac et al. 2021). These programs concentrate on solving urgent environmental problems such as water purification, pollution cleanup, and renewable energy production. Government organizations facilitate cooperative nanotechnology research by offering grants, funding possibilities, and research contracts (Tahmooresnejad and Beaudry 2019). Initiatives like consortiums and public-private partnerships (PPPs) promote industry-academia cooperation to quicken the development and commercialization of new technologies (Mete et al. 2021). The commercialization of nanotechnology-based goods and solutions, as well as technology transfer and prototype development, is made easier by cooperation between university researchers and industry partners. The industry has experience in expanding production, opening up markets, and funding cutting-edge technology. Technology transfer offices, incubators, and accelerators supported by the government help startups and Small and Medium-sized Enterprises

(SMEs) transform university research into commercially viable nanotechnology solutions (Woolley and MacGregor 2022). These programs support economic expansion, entrepreneurship, and the development of jobs in the environmental technology industry. Academic institutions, business leaders, government authorities, and non-governmental organizations (NGOs) work together to provide seminars and workshops that foster discussion, consensus-building, and the creation of legal frameworks for the responsible and safe deployment of nanotechnology in environmental applications. Governments work with academics, businesses, and other stakeholders to create standards, guidelines, and laws that address the health, safety, and environmental implications of nanomaterials and products enabled by nanotechnology (Nath et al. 2021). Unambiguous regulatory frameworks promote market uptake and guarantee adherence to environmental standards. Collaborative initiatives between academics and industry provide specialized training programs, workshops, and courses on nanotechnology applications in environmental sustainability. These courses give professionals, engineers, and researchers the know-how to tackle difficult environmental problems. Collaborations between industry and academia provide students and early-career researchers with chances for internships, co-ops, and joint research projects. The next generation of leaders in environmental science and nanotechnology is prepared by practical experience in industrial settings, which also improves technical abilities and encourages innovation. International cooperation makes it possible to share expertise, work together on research initiatives, and transfer technology in nanotechnology for environmental sustainability. Partnerships with international institutions and organizations make access to various markets, resources, and knowledge easier. Through the use of nanotechnology, cooperation between developed and developing nations advances capacity building, technological transfer, and sustainable development goals (SDGs). These collaborations address regional environmental issues and promote resilience and inclusive growth. Initiatives spearheaded by the government use funding from the business sector to further environmental sustainability through nanotechnology research and development. PPPs facilitate joint initiatives, infrastructure building, and technology application to accomplish environmental objectives and benefit society.

22.7 Conclusion

Innovating solutions across multiple sectors can address climate change through nanotechnology, which offers a potential frontier. Increasing the scalability and efficiency of solar cells, wind turbines, and energy storage devices improves renewable energy technology. Additionally, by improving lighting, vehicle economy, and building insulation, nanomaterials help save energy. Nanotechnology helps remove pollutants from soil, water, and the air in environmental remediation by using filtration membranes and nanocatalysts to increase efficiency. Moreover, nanotechnology helps with climate adaptation by creating sensors to track changes in the environment and boost agricultural resilience. To ensure the sustainable application of nanotechnology in tackling climatic concerns, however, extensive thought must be given to safety, legal frameworks, and ethical implications. Overall, nanotechnology is poised to be crucial in mitigating climate change impacts and fostering a sustainable future.

Nanotechnology is poised to revolutionize renewable energy generation by addressing key technological challenges and unlocking new opportunities for sustainable energy solutions. By leveraging nanoscale innovations, we can accelerate the global transition towards a low-carbon energy economy, mitigating climate change impacts and fostering energy independence. Nanotechnology holds significant promise for revolutionizing energy efficiency across diverse sectors, contributing to global efforts to mitigate climate change and enhance energy security. Continued research, innovation, and collaboration are essential to realize the full potential of nanotechnology in creating a sustainable energy future. Nanotechnology presents promising opportunities to address environmental challenges through innovative remediation technologies and sustainable waste management practices. Continued research, collaboration across disciplines, and stakeholder engagement are essential to maximize the benefits of nanotechnology while minimizing environmental risks, ultimately contributing to a cleaner, healthier planet. While nanotechnology offers innovative solutions to global challenges, ensuring safety and mitigating environmental risks are paramount. By integrating scientific research, robust regulatory frameworks, and stakeholder engagement, we can harness the potential of nanotechnology responsibly, contributing to sustainable development and a safer future for all. Collaborative efforts among academia, industry, and government are essential for advancing nanotechnology innovations that contribute to environmental sustainability. By leveraging complementary expertise, resources, and networks, stakeholders can accelerate the development and adoption of nanotechnology solutions, paving the way for a more resilient and environmentally friendly future.

References

Abbas Q, Yousaf B, Ali MU, Munir MAM, El-Naggar A, Rinklebe J, Naushad M (2020) Transformation pathways and fate of engineered nanoparticles (ENPs) in distinct interactive environmental compartments: a review. Environ Int 138:105646. https://doi.org/10.1016/j.envint.2020.105646

Abdeen DH, El Hachach M, Koc M, Atieh MA (2019) A review on the corrosion behaviour of nanocoatings on metallic substrates. Mater 12(2):210. https://doi.org/10.3390/ma12020210

Amoabediny GH, Naderi A, Malakootikhah J, Koohi MK, Mortazavi SA, Naderi M, Rashedi H (2009) Guidelines for safe handling, use and disposal of nanoparticles. In Journal of Physics: Conference Series. IOP Publishing. https://iopscience.iop.org/article/10.1088/1742-6596/170/1/012037/pdf

Batley GE, Kirby JK, McLaughlin MJ (2013) Fate and risks of nanomaterials in aquatic and terrestrial environments. Acc Chem Res 46(3):854–862. https://pubs.acs.org/doi/epdf/10.1021/ar2003368?ref=article_openPDF

Bhawana P, Fulekar M (2012) Nanotechnology: remediation technologies to clean up environmental pollutants. Res J Chem Sci 2(2):90–96. ISSN 2231-606X. https://isca.me/rjcs/Archives/v2/i2/15.pdf

Bothun G (2019) Greenhouse gases: properties and evolution. Affordable and Clean Energy 1–17. https://doi.org/10.1007/978-3-319-71057-0_61-1

Buzea C, Pacheco II, Robbie K (2007) Nanomaterials and nanoparticles: sources and toxicity. Biointerphases 2(4):MR17–MR71. https://doi.org/10.1116/1.2815690

Chin TM, Vazquez-Cuervo J, Armstrong EM (2017) A multi-scale high-resolution analysis of global sea surface temperature. Remote Sens Enviro 200:154–169. https://doi.org/10.1016/j.rse.2017.07.029

El-Sheikh MA, Hadibarata T, Yuniarto A, Sathishkumar P, Abdel-Salam EM, Alatar AA (2021) Role of nanocatalyst in the treatment of organochlorine compounds: a review. Chemosphere 268:128873. https://doi.org/10.1016/j.chemosphere.2020.128873

Furxhi I, Arvanitis A, Murphy F, Costa A, Blosi M (2021) Data shepherding in nanotechnology. The Initiation. Nanomater 11(6):1520. https://doi.org/10.3390/nano11061520

Goh PS, Ng BC, Lau WJ, Ismail AF (2015) Inorganic nanomaterials in polymeric ultrafiltration membranes for water treatment. Sep Purif Rev 44(3):216–249. https://doi.org/10.1080/15422119.2014.926274

Goodnick SM (2018) Nanotechnology pathways to next-generation photovoltaics. Semiconductor nanotechnology: advances in information and energy processing and storage 1–36. https://doi.org/10.1007/978-3-319-91896-9_1

Gottardo S, Mech A, Drbohlavová J, Małyska A, Bøwadt S, Sintes JR, Rauscher H (2021) Towards safe and sustainable innovation in nanotechnology: state-of-play for smart nanomaterials. Nano Impact 21:100297. https://doi.org/10.1016/j.impact.2021.100297

Govindasamy R, Gayathiri E, Sankar S, Venkidasamy B, Prakash P, Rekha K, ... Thiruvengadam M (2022) Emerging trends of nanotechnology and genetic engineering in cyanobacteria to optimize production for future applications. Life 12(12):2013. https://doi.org/10.3390/life12122013

Hecht DS, Hu L, Irvin G (2011) Emerging transparent electrodes based on thin films of carbon nanotubes, graphene, and metallic nanostructures. Adv Mater 23(13):1482–1513. https://doi.org/10.1002/adma.201003188

Huang YM, Singh KJ, Liu AC, Lin CC, Chen Z, Wang K, ... Kuo HC (2020) Advances in quantum-dot-based displays. Nanomater 10(7):1327. https://doi.org/10.3390/nano10071327

Ibrahim RK, Hayyan M, AlSaadi MA, Hayyan A, Ibrahim S (2016) Environmental application of nanotechnology: air, soil, and water. Environ Sci Pollut Res 23:13754–13788. https://doi.org/10.1007/s11356-016-6457-z

Jafarzadeh S, Jafari SM (2021) Impact of metal nanoparticles on the mechanical, barrier, optical and thermal properties of biodegradable food packaging materials. Crit Rev Food Sci Nutr 61(16):2640–2658. https://doi.org/10.1080/10408398.2020.1783200

Kalia A, Kaur H (2019) Nano-biofertilizers: harnessing dual benefits of nano-nutrient and bio-fertilizers for enhanced nutrient use efficiency and sustainable productivity. Nanoscience for Sustainable Agriculture 51–73. https://doi.org/10.1007/978-3-319-97852-9_3

Kumar A, Sadhu SD, Singh R (2018) Nanotechnology: key for sustainable future. Nanotechnol Environ Sci 759–804. https://doi.org/10.1002/9783527808854.ch24

Li Y, Yang J, Song J (2017) Structure models and nano energy system design for proton exchange membrane fuel cells in electric energy vehicles. Renew Sustain Energy Rev 67:160–172. https://doi.org/10.1016/j.rser.2016.09.030

Liu R, Lal R (2012) Nanoenhanced materials for reclamation of mine lands and other degraded soils: a review. J Nanotechnol 2012(1):461468. https://doi.org/10.1155/2012/461468

Malik S, Muhammad K, Waheed Y (2023) Nanotechnology: A revolution in modern industry. Molecules 28(2):661. https://doi.org/10.3390/molecules28020661

Merugula LA, Khanna V, Bakshi BR (2010) Comparative life cycle assessment: reinforcing wind turbine blades with carbon nanofibers. In: Proceedings of the 2010 IEEE international symposium on sustainable systems and technology (pp. 1–6). IEEE. https://ieeexplore.ieee.org/abstract/document/5507724

Mete G, Hocquet R, Sanchez F, Talebian S, Nilsson A, Choi G, … Moon J (2021) Reaching net-zero industry through public-private partnerships. Leadership Group for Industry Transition, Stockholm Environment Institution. https://www.sei.org/publications/reaching-net-zero-industry-public-private-partnerships

Mohmood I, Lopes CB, Lopes I, Ahmad I, Duarte AC, Pereira E (2013) Nanoscale materials and their use in water contaminants removal—a review. Environ Sci Pollut Res 20:1239–1260. https://doi.org/10.1007/s11356-012-1415-x

Myers SS, Patz JA (2009) Emerging threats to human health from global environmental change. Annu Rev Environ Resour 34(1):223–252. https://doi.org/10.1146/annurev.environ.033108.102650

Naseem T, Waseem M (2022) A comprehensive review on the role of some important nanocomposites for antimicrobial and wastewater applications. Int J Environ Sci Technol 19(3):2221–2246. https://doi.org/10.1007/s13762-021-03256-8

Nath H, Subhranshu SS, Manohara SR (2021) Current guidelines and regulatory challenges, insight into the legal, societal, and ethical issues of nanomaterials. In: Nanotoxicology. CRC Press, pp 445–463

Naudé W (2011) Climate change and industrial policy. Sustain 3(7):1003–1021. https://doi.org/10.3390/su3071003

Parameswaran AK, Azadmanjiri J, Palaniyandy N, Pal B, Palaniswami S, Dekanovsky L, ... Sofer Z (2023) Recent progress of nanotechnology in the research framework of all-solid-state batteries. Nano Energy 105:107994. https://doi.org/10.1016/j.nanoen.2022.107994

Patel JK, Patel A, Bhatia D (2021) Introduction to nanomaterials and nanotechnology. In Emerging technologies for nanoparticle manufacturing (pp. 3–23). Cham: Springer International Publishing. https://doi.org/10.1007/978-3-030-50703-9_1

Pokrajac L, Abbas A, Chrzanowski W, Dias GM, Eggleton BJ, Maguire S, ... Mitra S (2021) Nanotechnology for a sustainable future: addressing global challenges with the international network sustainable nanotechnology. https://doi.org/10.1021/acsnano.1c10919

Rajput VD, Minkina T, Kumari A, Shende SS, Ranjan A, Faizan M, ... Kızılkaya R (2022) A review on nanobioremediation approaches for restoration of contaminated soil. Eurasian J Soil Sci 11(1):43–60. https://doi.org/10.18393/ejss.990605

Rónavári A, Kónya Z (2021) Nanoremediation: tiny objects solving huge environmental problems. Recent Pat Nanotechnol 15(3):245–255. https://doi.org/10.2174/1872210514999201209214247

Shah KW, Huseien GF, Xiong T (2020) Functional nanomaterials and their applications toward smart and green buildings. In New Materials in Civil Engineering (pp. 395–433). Butterworth-Heinemann. https://doi.org/10.1016/B978-0-12-818961-0.00011-9

Sharma P, Pandey V, Sharma MMM, Patra A, Singh B, Mehta S, Husen A (2021) A review on biosensors and nanosensors application in agroecosystems. Nanoscale Res Lett 16:1–24. https://doi.org/10.1186/s11671-021-03593-0

Shen Y (2015) Carbothermal synthesis of metal-functionalized nanostructures for energy and environmental applications. J Mater Chem A 3(25):13114–13188. https://pubs.rsc.org/en/content/articlehtml/2021/xx/c5ta01228g

Singh BR, Singh O (2012) Study of impacts of global warming on climate change: rise in sea level and disaster frequency. Global warming—impacts and future perspective 94–118

Tahmooresnejad L, Beaudry C (2019) Collaboration or funding: lessons from a study of nanotechnology patenting in Canada and the United States. J Technol Transf 44:741–777. https://doi.org/10.1007/s10961-017-9615-7

Thakur A, Kumar A (2022) Recent advances on rapid detection and remediation of environmental pollutants utilizing nanomaterials-based (bio) sensors. Sci Total Environ 834:155219. https://doi.org/10.1016/j.scitotenv.2022.155219

Todaro NM, Gusmerotti NM, Daddi T, Frey M (2023) Do environmental attitudes affect public acceptance of key enabling technologies? Assessing the influence of environmental awareness and trust on public perceptions about nanotechnology. J Clean Prod 387:135964. https://doi.org/10.1016/j.jclepro.2023.135964

Vasyukova IA, Zakharova OV, Chaika VV, Pikula KS, Golokhvast KS, Gusev AA (2021) Toxic effect of metal-based nanomaterials on representatives of marine ecosystems: a review. Nanobiotechnol Rep 16:138–154. https://doi.org/10.1134/S2635167621020178

Woolley JL, MacGregor N (2022) The influence of incubator and accelerator participation on nanotechnology venture success. Entrep Theory Pract 46(6):1717–1755. https://doi.org/10.1177/104225872110245

Zhu S, Ho SH, Jin C, Duan X, Wang S (2020) Nanostructured manganese oxides: natural/artificial formation and their induced catalysis for wastewater remediation. Environ Sci Nano 7(2):368–396. https://pubs.rsc.org/en/content/articlehtml/2020/en/c9en01250h

Nanotechnology: A Smart Approach for Sustainable Agriculture Under Global Climate Change

23

Muhammad Tanveer Altaf, Muhammad Shaheryar, Waqas Liaqat, Amna Jamil, Momna Hayat, Hafiz Saqib Hayat, Haseeb Ahmad, and Faheem Shehzad Baloch

Abstract

Agriculture is the cornerstone of economic development for many nations and is indispensable for global food security. However, the sector faces exceptional challenges exacerbated by climate change, which disrupts ecosystems and threatens food production. It has been proposed that climate change may decrease global crop yields by an average of 8% by the middle of the century. As the world's population continues to grow, the demand for food is projected to increase significantly by 50% by 2050. To address these challenges and ensure global food security, there

M. T. Altaf (✉)
Department of Field Crops, Faculty of Agriculture, Recep Tayyip Erdoğan University, Rize/Pazar, Türkiye

M. Shaheryar · H. S. Hayat
Institute of Agronomy, Bahauddin Zakariya University, Multan, Punjab, Pakistan

W. Liaqat
Department of Field Crops, Faculty of Agriculture, Institute of Natural and Applied Sciences, Çukurova University, Adana, Türkiye

A. Jamil
Department of Horticultural Sciences, The Islamia University of Bahawalpur, Bahawalpur, Pakistan

M. Hayat
Department of Horticulture, Faculty of Agricultural Sciences and Technology, Bahauddin Zakariya University,
Multan, Punjab, Pakistan

H. Ahmad
Guangxi Key Laboratory of Agro-environment and Agro-products Safety, Key Laboratory of Crop Cultivation and Tillage, Agricultural College, Guangxi University, Nanning, China

F. S. Baloch
Department of Biotechnology, Faculty of Science, Mersin University, Mersin, Türkiye

A. A. H. Abdel Latef et al. (eds.), *Bioremediation and Nanotechnology for Climate Change Mitigation*, https://doi.org/10.1007/978-981-96-3069-1_23

is an urgent need to adopt innovative agricultural technologies. This chapter investigates the potential of nanotechnology to revolutionize agricultural practices and mitigate the adverse impacts of climate change. Nanotechnology offers a range of solutions tailored to enhance crop productivity, optimize resource utilization, and promote environmental sustainability. Nano-enabled fertilizers and pesticides hold promise in delivering nutrients effectively to plants while minimizing environmental damage. Furthermore, nanosensors provide real-time monitoring of agro-climatic conditions, enabling precise resource management. Despite its potential benefits, the integration of nanotechnology into agriculture requires careful consideration of environmental and health implications. Risks associated with nanomaterials must be thoroughly assessed to ensure responsible implementation. However, nanotechnology offers a unique chance to transform agricultural sustainability and tackle the issues presented by climate change. Through persistent research, innovation, and responsible integration, nanotechnology can lead to a resilient and sustainable agricultural future, ensuring food security for future generations.

Keywords

Nanotechnology · Agriculture · Sustainability · Climate change · Global food security

23.1 Introduction

Agriculture plays a crucial role in emerging nations' economies and global food security (Mittal et al. 2020). Climate change causes biotic and abiotic pressures in ecosystems (Liaqat et al. 2024a; Hussein and Abou-Baker 2018), altering the delicate environmental balance needed for food production and perhaps causing crop failures (Mittal et al. 2020). As the global population grows, food demand is expected to reach 70% by 2050 (Bindraban et al. 2018). Global food security and plant production require innovative and future agricultural technologies (Altaf et al. 2023; Mandal 2021). Inefficient traditional farming practices, especially fertilizer, cause environmental issues (Fellet et al. 2021). Nutrient utilization efficiency (NUE), which optimizes plant nutrient bioavailability, is used to measure agricultural productivity (Elmer and White 2016). However, typical fertilizer applications like soil surface broadcasting or via irrigation can harm the environment (Rajput et al. 2021a, b). Due to the application of chemical fertilizers, an average of 50% of nitrogen, 80% of phosphorus, and 70% of potassium are lost or become fixed in the soil. Hydroponic methods are used to grow specific crops in controlled surroundings but use a lot of water and energy, making them unsustainable (Renner et al. 2020). Hydroponic fertilizers can also pollute groundwater and cause eutrophication, endangering public health (Smith and Gilbertson 2018). Thus, alternate methods that reduce environmental effects and maintain agricultural yield are needed. Sustainable farming practices and more efficient fertilizers are needed (Pouratashi and Iravani 2012). Research and technology can help governments and farmers create sustainable agriculture and ensure food security for future generations (Mittal et al. 2020).

Innovative solutions are needed to address global agriculture issues including climate change and provide food security while protecting ecosystems, biodiversity, and the climate (Adisa et al. 2019). Best farming practices could reduce fertilizer use by 20% by 2030, sustaining 85% of global agricultural production (Faizan et al. 2021). Nanotechnology may solve these agricultural issues in this changing environment. Nano-enabled fertilizers may improve agricultural yields and agro-ecosystem health, especially in the face of climate change (Rajput et al. 2021a). These fertilizers use nanotechnology to provide nutrients to plants more efficiently, limiting losses and maximizing uptake even in drought, heat stress, or climate-induced soil degradation. Recent trends show the use of engineered nanomaterials (ENMs) in agriculture, particularly nanofertilizers and nanopesticides with enhanced delivery mechanisms. These advances enable accurate and targeted application, reducing waste and environmental effects (Singh et al. 2021a, b). Using nanotechnology, farmers can improve production, sustainability, and climate change resilience. To reduce environmental and health risks, extensive risk evaluations must accompany nanotechnology in agriculture. Nanotechnology can transform agriculture in the face of climate change through research, innovation, and responsible deployment (Fig. 23.1).

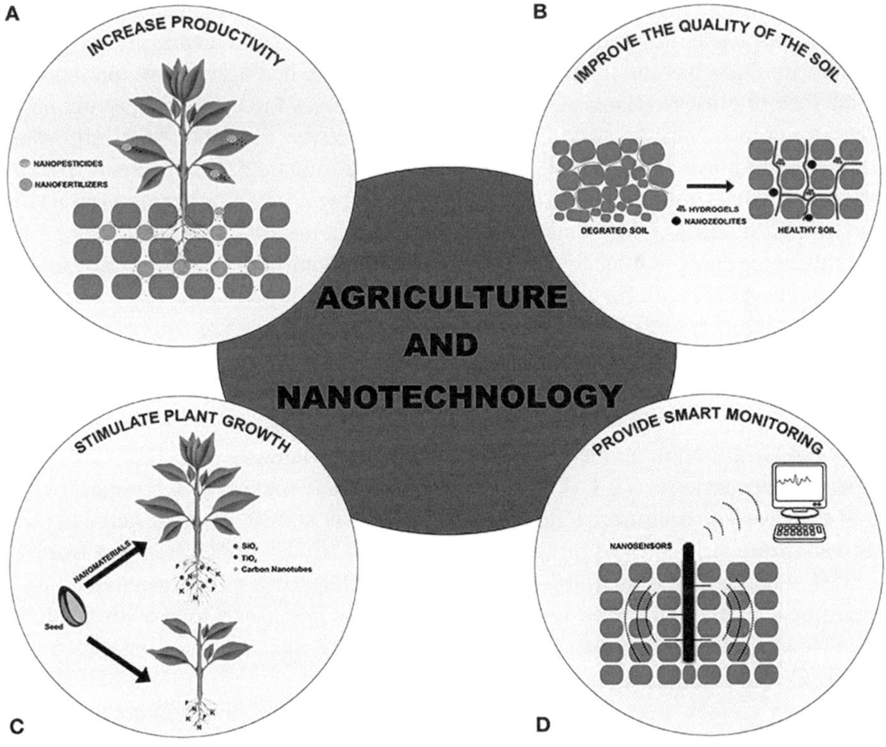

Fig. 23.1 Potential applications of nanotechnology in agriculture (Fraceto et al. 2016)

23.2 Effect of Climate Change on Agriculture Productivity

The planet faces the formidable challenge of climate change, necessitating a concerted global effort and immediate action. Long-term rainfall and temperature changes are caused by various factors, as noted by the World Meteorological Organization, Geneva (WMO, Geneva, 1992). Burning fuel has worsened the situation by changing the air (IPCC, Climate Change 2007). Since 1750, methane and carbon dioxide levels have risen significantly in the atmosphere (IPCC, Climate Change 2014). Most carbon dioxide emissions result from burning fossil fuels and industrial manufacturing processes (Sathaye et al. 2006). The atmospheric carbon dioxide concentration was 315.98 parts per million in 1959 and increased to 411.43 per million by 2019 (NOAA 2020). Sixty-five percent of air pollution is attributed to factory and car emissions, while tree-cutting accounts for 11% (IPCC, Climate Change 2014). Before 1750, minimal carbon dioxide was produced from burning fossil fuels, but industrialization significantly increased its emissions. Since 1751, approximately 1.5 trillion metric tons of CO_2 have been released (CDIAC 2020). Europe emits the highest levels of carbon dioxide, followed by Asia and North America; the United States, China, and the European Union emit the most (CDIAC 2020). The rise in carbon dioxide levels has led to improvements in plant growth and reduced heating energy requirements. However, it has also affected water resources (Tol 2013). Climate change was beneficial for most countries until 1980, but it has become detrimental for everyone since then. It is projected to worsen for both wealthy and impoverished nations in the twenty-first century (Tol 2013). Nanotechnology has the potential to significantly enhance agriculture, particularly in the face of climate change. Nanotechnology involves the use of minuscule particles to achieve various tasks. Nanofertilizers deliver nutrients precisely when needed, promoting plant growth. Nanopesticides eliminate pests in an eco-friendly manner. Nanosensors and irrigation systems aid farmers in monitoring crop health and irrigation needs. Additionally, nanomaterials can improve soil quality for crops. Overall, integrating nanotechnology in agriculture can help farmers adapt to and mitigate the effects of climate change.

Climate projections indicate that harsher conditions are on the horizon. Simulations for the mid-century period (2040–2069) in Punjab, Pakistan, suggest increased minimum and maximum temperatures during both the Kharif and Rabi seasons. For the Kharif season, a rise of 1–3.3 °C in maximum temperatures and 2–3 °C in minimum temperatures is anticipated. Similarly, the Rabi season is expected to experience a 2.1–3.5 °C increase in average maximum temperatures and 2–3 °C in average minimum temperatures. Bokhari et al. (2017) suggested that rainfall variability will be most pronounced during the Kharif season, ranging from 25 to 35%, and least during Rabi. China is also anticipating more extreme weather conditions, including higher temperatures and precipitation, along with a 0.5 °C increase in warming. Chen and Sun (2018) suggested that limiting global warming to 1.5 °C could help mitigate these extremes.

Climate change may worsen temperature and precipitation extremes. Heavy rainfall in South and East Asia may enhance river flows, whereas southern Africa and South America may have milder droughts. The upper basin of the Indus River basin may warm more than the lower basin, causing seasonal and geographical precipitation differences (Rajbhandari et al. 2015). Higher emissions may cause more warm extremes, fewer cold extremes, and enhanced precipitation extremes in the Northeast (Ning et al. 2015). Increased precipitation intensity and frequency could cause soil erosion, threatening agriculture in northeast China (Zhang et al. 2010). Over the past two decades, dry anomalies have increased cropland in developing countries by 9% (Zaveri et al. 2020). Climatic variables may reduce maize production by 3.8% and wheat production by 5.5%, threatening global food security (Lobell et al. 2011). Salinity, drought, heat, and cold stress worsen crop problems (Malhi et al. 2020). Climate change also causes water scarcity, soil fertility loss, and pest infestations (Baul and McDonald 2015). Addressing these difficulties and ensuring food security for the rising population requires mitigation and adaptation techniques. To meet global food and nutritional needs, agricultural production must rise 60% yearly from 2005/2007 to 2050, especially in poor nations (Alexandratos and Bruinsma 2012).

23.2.1 Climate Change and Climate-Smart Agriculture Technologies Economic Impact

Climate change offered some advantages initially, but its relentless warming is now harming the environment. Temperature rises exceeding 3 °C have adverse effects, with welfare losses becoming significant at 7 °C. The global social cost of carbon emissions was estimated at USD 29 per ton of Carbon (tC) in 2015, and it is projected to increase by 2% annually (Tol 2012). Implementing climate change mitigation measures in the fishery sector of the Solomon Islands could stimulate economic growth. However, agricultural markets are expected to suffer, leading to a projected reduction in global GDP by 0.26% (Costinot et al. 2016). By the 2080s, the projected climate conditions could result in a 0.2–1% annual reduction in household welfare (Ciscar et al. 2011). Additionally, every 1 °C increase in the global mean temperature is associated with a 1.2% GDP cost in market and non-market damages (Hsiang et al. 2017). If mitigation efforts focus solely on adaptation strategies, global income could decline by 23% by 2100, exacerbating income inequality (Burke et al. 2015). Furthermore, global economic growth is expected to decline by 0.28% annually (Carleton and Hsiang 2016). Table 23.1 provides a list of economic benefits associated with climate-smart agriculture methods.

23.2.2 Effect of Climate Change on Soil and Water Resources

Earth's freshwater supply constitutes only 2.5% of the total water on the planet, with a mere 0.3% available in lakes, reservoirs, and rivers. The Earth's radiation balance is influenced by atmospheric water vapor, cloud cover, and ice (Bates et al. 2008). By 2050, two-thirds of the global population are projected to face water

Table 23.1 Incremental benefits of climate-smart agriculture technologies

Location	Crop	Climate-smart technology	Enhanced efficiency	Incremental economic benefit	References
Vietnam	Rice	Site-specific nutrient management	Increased partial factor productivity of nitrogen	34 US$/ha	Pampolino et al. (2007)
Philippines				106 US$/ha	
India				168 US$/ha	
Sindh, Pakistan	Wheat	Laser land leveling	Saving of 21% irrigation water and reduced irrigation time	INR 23,250/ acre	Wagan et al. (2015)
Punjab, Pakistan	Rice-wheat cropping system	Zero tillage	Higher water productivity, saving of irrigation water, and higher fertilizer use efficiency	–	Latif et al. (2013)
Nyando basin of Kenya	Multiple crops and livestock	Stress-tolerant crop varieties	Increased household income leading to household asset accumulation and investment	Increased HH income by 83%	Ogada et al. (2020)
Semi-arid tropics of India	Groundnut	Drought-tolerant varieties	–	Increase in yield by 23%, lower variability in yield, increased share of risk benefits in total benefits	
Karnal, Haryana	Wheat	Zero tillage	Enhanced production by 1.88% and lower cultivation cost	17% reduction in variable cost	Birthal et al. (2012)
Northwestern Indo-Gangetic plains of India	Rice and wheat	Laser land leveling	Reduced irrigation time, increased yield, reduced electricity charges	Higher net income	Tripathi et al. (2013)

(continued)

Table 23.1 (continued)

Location	Crop	Climate-smart technology	Enhanced efficiency	Incremental economic benefit	References
North western India	Wheat Wheat	Zero tillage	Reduced cultivation cost, reduced GHGs emissions, and increased yield	US$ 143.5/ha/year	(Aryal et al. 2015a, b)
Upper Gangetic plains	Wheat Wheat Rice–Wheat cropping system	Site-specific nutrient management	Increased yield by 29% over farmers fertilizer practices (FFP)	US$ 97.5/ha	Aryal et al. (2015a)
Indo-Gangetic plains of India		Improved crop varieties	Increased net returns	INR 68,980/ha over FFP	Singh et al. (2015)
Tamil Nadu, India	Okra	Drip irrigation	Saving of irrigation water and electricity charges, reduced cultivation cost	Increase HH income by 16%	Mishra et al. (2017)
Punjab, India	DSR–Wheat	Direct-seeded rice	Saving of irrigation, lesser labor requirement	INR 72,711/acre	(Narayanamoorthy and Devika 2017
India	Eggplant	Drip irrigation	Reduced water, electricity and fertilizer use, and increased returns	INR 5050–INR 8100/ha over puddled transplanted rice (PTR)–Wheat	Bhullar et al. (2018)

scarcity (Gosain et al. 2006). Global warming leads to increased evaporation and altering precipitation patterns, flooding, and droughts in different regions (Open University 2016). Over 1.1 billion people worldwide lack adequate access to clean drinking water, and around 2.6 billion people do not have basic sanitation facilities (Asian Water Development Outlook 2016; Chellaney 2011). The IPCC has forecasted a significant increase in global temperatures by 2090–2099, which will impact water availability due to changes in precipitation, temperature, and evaporative demand (Brevik 2012). Changes in soil properties and processes induced by climate change pose a threat to food security (Lal 2010; Blum and Nortcliff 2013). Temperature and rainfall directly influence evapotranspiration and surface runoff, affecting water resources' spatial and temporal availability (Brevik 2013).

Climate change threatens water and food security, especially in Africa, where water is scarce. By 2020, rain-fed agricultural yields in Africa could decrease by approximately 50% due to an average temperature increase of 1–3 °C (Gomez-Zavaglia et al. 2020). Rain-fed agriculture in Sub-Saharan Africa confronts reduced rainfall and drainage, which could lower agricultural productivity (Anil 2014). While hydrological models have concentrated on climate change, studies show that land use and cover regulate hydrological responses (Warburton et al. 2011). Along with climate, land use changes and water abstractions affect stream flow dynamics (Liaqat et al. 2023a; Warburton and Schulze 2005). Climate change is expected to dramatically alter precipitation, evapotranspiration, and runoff, affecting national water supplies (Sohoulande Djebou and Singh 2016). Climate change affects terrestrial and social systems differently, with rising global temperatures driving many effects (Fig. 23.2).

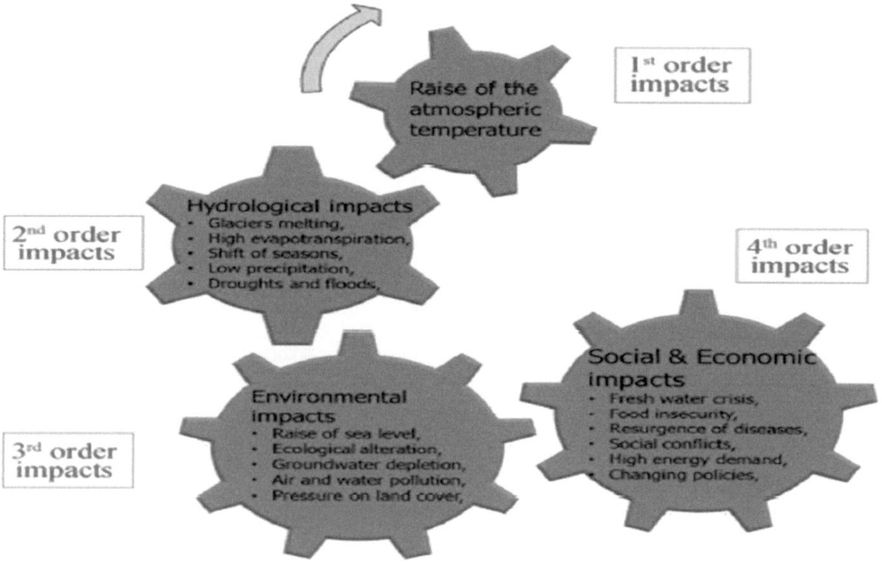

Fig. 23.2 The hierarchy of climate change impacts within the hydrological framework, as proposed by Sohoulande Djebou and Singh (2016). It emphasizes the primary impact of rising global temperatures, from which various levels of impacts are derived

23.3 Nanotechnology for Mitigating Climate Change

Human-caused climate change has accelerated during the past 200 years, causing global warming and CO_2 increases (Fischer and Knutti 2015). Weather patterns are changing due to these changes, affecting life worldwide. Global warming, the principal cause of climate change, requires immediate action. The Earth's surface temperature has risen considerably in recent decades. Temperatures have risen by 0.14 °F (0.08 °C) every decade since 1880. These rates have more than doubled in the last 40 years, reaching 0.32 °F (0.18 °C) per decade since 1981. Global mean temperature has surpassed preindustrial values by 1 °C in 2017 and is rising at 0.2 °C every decade. The second-warmest year was 2020, after 2016 (Lindsey and Dahlman 2021). The average world temperature rose by 1.11 (±0.13) °C in 2021 compared to pre-industrial levels (Naithani 2016). CO_2 and other greenhouse gases trap heat in the atmosphere, causing global warming. Natural (CO_2, methane, water vapors) and industrial (HFCs, SF6) gases deplete the ozone layer and harm living forms (IPCC, Climate Change 2014). Global warming causes glacier melting, coastal erosion, floods, droughts, pollution, and crop output losses (Lowry et al. 2019). Nanotechnology creates nanoparticles (NPs) with unique features (Chausali et al. 2022). Nanoparticles between 1 and 100 nm are used in agriculture, ecology, energy, and medicine. Nanotechnology offers sustainable alternatives to conventional tools and materials, with nanoparticles having greater surface areas per unit volume for better interaction and functioning (Chausali et al. 2021). Nanotechnology's nanofertilizers and nanopesticides boost agricultural productivity and environmental balance (Tripathi et al. 2018). Nanosensors are essential for agro-climatic monitoring and resource optimization. Nanobiosensors detect food packaging pollutants (Chausali et al. 2022). Nanocatalysts and nanolubricants reduce pollutants and energy use, enhancing environmental cleanup (Subramanian et al. 2020). Nanotechnology can reduce global warming by increasing renewable energy output and reducing fossil fuel use. Nanomaterials help fight climate change by absorbing greenhouse gasses (Yousefi 2022). Nanotechnology improves building energy efficiency, enabling green architecture (Smith 2011). Safety rules must be followed since nanomaterials may be harmful (Khan et al. 2019). Nanotechnology literature is rich, whereas climate change research is few. This chapter uses recent studies to examine nanotechnology's ability to mitigate climate change (Rezaei 2018).

23.3.1 Improving Abiotic Stress through Nanotechnology

Natural and agricultural abiotic stressors affect plants indirectly due to the physical environment. Stress reduces agricultural yield, viability, biomass, and productivity (Calanca 2017). Abiotic stress is a global environmental issue, so researchers are developing sustainable production methods for all species. Abiotic factors include drought, salinity, heavy metals, heat stress, and cold stress and cause the most damage (Khan et al. 2021). Overlapping pressures cause plants to react dynamically and complexly. If not detoxified, reactive oxygen species (ROS) created under extreme conditions can affect soil quality, fertility, and cellular activities, decreasing crop output (Khan and Khan 2017). Plant productivity has decreased due to global warming, making it harder to meet food needs (Altaf et al. 2023). To feed 9 billion people by 2050, food output must rise by 70% (Mwobobia et al. 2020). However, heavy use of pesticides and fertilizers to improve crop productivity causes environmental issues. To help plants resist abiotic challenges and increase agricultural output, environmentally friendly solutions are essential. Farming and cropping systems use agricultural and physiological strategies to reduce abiotic stress and promote plant adaptation. Nanotechnology in agriculture could boost agricultural output (Jeelani et al. 2020). NPs have distinct surface, optical, thermal, and electrical properties compared to bulk substances (Rastogi et al. 2017). NPs affect biology both positively and negatively. They can boost plant growth and development at low concentrations but alter metabolism at large dosages (Tariq et al. 2021). NPs also affect plant responses to heat, heavy metals, salinity, and drought (Iqbal et al. 2019).

23.3.2 Effect of Abiotic Stress on Plants

Abiotic pressures from climate change, global warming, human activities, and other unavoidable causes are reducing agricultural production and degrading natural resources (Ali et al. 2022; Shahzad et al. 2018). Many studies have found that heavy metals, high temperatures, droughts, and salt stressors reduce agricultural yields (Mubarik et al. 2022; Hussain et al. 2018) (Fig. 23.3). Stress causes reactions that harm plant cells. Heat, heavy metals, salt, and drought impair agricultural yield (Liaqat et al. 2023b; Godoy et al. 2021). Salinity and drought, especially damage

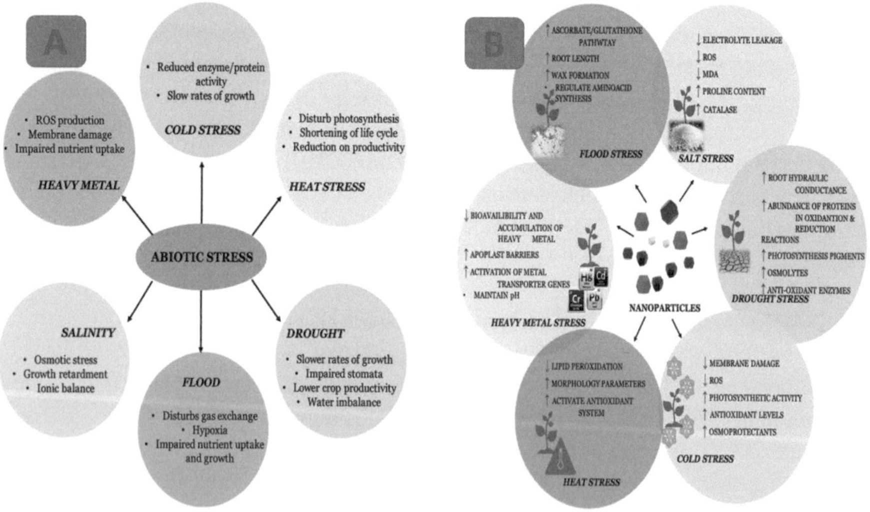

Fig. 23.3 Abiotic stress factors and their effects on plants (**a**) and nanoparticles involved in combating abiotic stress (**b**) (Al-Khayri et al. 2023)

crop physiology and biochemistry, reduce productivity (Zia et al. 2021). Abiotic stressors reduce plant development and production through oxidative and osmotic effects (Shahid et al. 2020). Salinity stress threatens crop plants' molecular, physiological, and biochemical health (Kuma et al. 2020). Excess sodium and chloride ions in plant cells during salt stress cause severe cell damage (Rajput et al. 2019a, b). Drought closes stomata, impeding photosynthesis, reducing leaf area, lowering water potential, and boosting osmolytes and reactive oxygen species (ROS), slowing plant growth (Ibrahim et al. 2019). Crop output declines depend on drought intensity and duration (Farooq et al. 2009). Water potential declines and osmosis is greatly impacted by drought and salinity stress (Hu and Schmidhalter 2005). Heavy metals, especially transition elements and groups four and five, pose serious health concerns to humans and agricultural crops (Xu et al. 2021). Heavy metal contamination costs approximately $10 billion annually (Kumar et al. 2019). Anthropogenic and industrial waste in the soil increases heavy metal buildup in crops and soil (Noman et al. 2020). Heavy metals cause free radicals and ROS, which affect protein function (Sharma et al. 2012).

23.3.3 Adaptation of Plants to Abiotic Stress

Biotechnology and omics have made comprehending plants' complex abiotic stress systems easier. Genomic, proteomic, metabolomic, and transcriptomics technologies have identified novel genes and revealed their genomic roles. Stress-specific proteins allow plants to respond transcriptionally and translationally (Cushman and Bohnert 2000). Stress responses vary by plant species and genotype, and plants may change their genes to create byproducts to reduce stress. Genomics, proteomics, metabolomics, and transcriptomics are widely employed to study stress-tolerance proteins (Vij and Tyagi 2007). Bioinformatics helps us understand plant stress responses by providing information on resistance gene expression and similarities to genes in other species. Structural and systems biology can also study how environmental stress arranges protein components. Heat stress damages reproductive organs and leaf components, causing 40% production reductions. Plants produce NO, ROS, heat shock transcription factors (HSF), and other stress-tolerant proteins and signaling components in response to heat stress (Nadeem et al. 2018). These reactions help plants adjust to heat stress and reduce crop damage.

23.3.4 Biotechnological Approaches for Abiotic Stress Tolerance

Animals have several environmental defenses that plants lack. They adapt by changing genes, proteins, and metabolism. Osmotic stress in crops can result from environmental solute concentration changes. Modern transgenic technologies and nanoparticles have enhanced nutrient absorption and identified stress response genes (Ashraf and Wu 1994). Research shows that many gene loci, some undiscovered, govern plant abiotic stress tolerance. Understanding these genes' significance in stress tolerance could lead to resilient crop cultivars. Genome studies found many stress-tolerance genes and proteins. Markers like restriction fragment length polymorphisms are employed in marker-assisted selection (MAS) to target genes. Identifying and isolating stress-related genes reveals stress tolerance biological pathways and processes. Quantitative trait locus (QTL) polygenes are highly impacted by the environment but little by qualities of interest (Broman and Speed 1999). Molecular markers help map genetic loci and quantify stress tolerance features. Their usage in agriculture presents new problems and substantially improves crops. Advanced research on nanoparticles, tissue engineering, and genetically engineered crops has led to significant results (Ran et al. 2017). Targeted nanoparticle delivery of CRISPR components seems promising. Nanosensors can improve plant disease and stress tolerance (Kwak et al. 2017).

23.4 Emerging Fields of Nanotechnology

23.4.1 The Growing Demand for Plant Resources

Due to population growth, plant-based food, shelter, and clothing are in demand. Global agriculture must increase crop production by 70% by 2050 to meet population expansion (Tyczewska et al. 2018). This has inspired scientists worldwide to develop eco-friendly crop yield systems.

23.4.2 Nanotechnology in Agriculture

Nanotechnology can reduce climate change and improve abiotic stress regulation in agriculture (Mahakham et al. 2017). Emerging nanobiotechnology uses nanoparticles (NPs) to counteract abiotic stress (Cheng et al. 2016). Researchers are investigating green nanoparticles made economically from plants, metals, or metal oxides (Iravani 2011).

23.4.3 Enhancing Crop Production with Nanoparticles

Numerous research have used nanoparticles as nanofertilizers to boost crop yield under stress (Ahmed et al. 2021). Nanoparticles improve plant development, nutrient intake, antioxidant enzyme activity, and photosynthesis by retaining nutrients better (Wang et al. 2016). Nanosilicon reduces salt stress, while iron oxide nanoparticles enhance heavy metal-contaminated soils and drought-stressed ecosystems (Adrees et al. 2019). Nanoparticles reduce abiotic stress in many crops (Song et al. 2012).

23.5 Classification of Nanoparticles

Nanoparticles are primarily classified into three categories based on their composition: organic, carbon-based, and inorganic (Ealia and Saravanakumar 2017).

23.5.1 Organic Nanoparticles

Polymers, lipids, proteins, carbohydrates, and other organic components make up organic nanoparticles (Pan and Zhong 2016). Dimer, liposome, micelle, and ferritin protein complexes are examples. These nanoparticles may be hollow, biodegradable, and non-toxic. They are thermally and electromagnetically sensitive (Ealia and Saravanakumar 2017).

23.5.2 Carbon-Based Nanoparticles

Carbon nanoparticles consist solely of carbon. Some examples include fullerenes, carbon black nanoparticles, and quantum dots. These nanoparticles find applications in drug delivery, energy storage, and bioimaging, primarily due to their high strength and electron affinity (Khan et al. 2019). Nanodiamonds and carbon nano-onions, known for their low toxicity and biocompatibility, are employed in tissue engineering and medication administration (Ahlawat et al. 2021).

23.5.3 Inorganic Nanoparticles

Inorganic nanoparticles are non-carbon or non-living particles. Nanoparticles can be metal, semiconductor, or ceramic (Khan et al. 2019). Metal nanoparticles (MNPs) are metal-only, mixed metal, or several metals (Toshima and Yonezawa 1998). Some metal nanoparticles are magnetic or optically distinctive. They're crucial for manufacturing nanodevices for numerous applications (Mody et al. 2010). Mixtures of metals and non-metals make semiconductor nanoparticles. They differ from semiconductors because bandgap changes can significantly alter their characteristics (Khan et al. 2019). These nanoparticles are beneficial in electronics and light-based chemical reactions (Gupta and Tripathi 2012). Nanoparticles of ceramics are made of oxides, carbides, phosphates, and other metals and non-metals. They can be solid, porous, or hollow and are formed by heating and cooling (Khan et al. 2019). Medical professionals utilize them since they're stable and can carry a lot. However, they're also employed in electronics, chemicals, and dye cleaning (Thomas et al. 2015).

23.6 Mechanism of Nanoparticles in Enhancing Abiotic Stress Tolerance

NPs enhance plant growth and development in many environmental situations as nano fertilizers, insecticides, herbicides, and more, while the exact mechanisms are still being explored (Siddiqui et al. 2015). NPs may reduce stress-induced damage through biochemical mechanisms like detoxification and enzymatic antioxidants (Table 23.2) (Sarraf et al. 2022). NP reactivity depends on form, size, content, surface characteristics, stability, chemical properties, purity, production method, and dosage (Burman and Kumar 2018). NP efficiency is also affected by environmental circumstances like oxidation and configuration changes (Levard et al. 2012). Studies suggest NPs protect crops from abiotic stresses (Khan et al. 2017). NPs enter plasmodesmata and go through apoplast and symplast. Plants with NPs had higher biomass, chlorophyll, photosynthetic activities, antioxidant machinery, osmolyte synthesis, and glucose levels (Singh et al. 2021a, b). NPs in plant cells can also increase protein and nitrogen levels and influence gene expression under biotic and abiotic stress (Rajput et al. 2021a, b). NPs may mimic the antioxidant defense system as nanoenzymes to reduce ROS generation in stressed settings (Sharifi et al. 2020). NP supplementation improves plant glutathione, proline, phytochelatin, and antioxidant enzyme activity (Rajput et al. 2021a, b). NPs increase the antioxidant defense system, reducing oxidative stress under salt, temperature, drought, and UV stress (Mahato et al. 2021). Enhanced characteristics from NP supplementation help plants withstand environmental obstacles. NPs enter the plant rhizosphere through lateral root synapses and reach the xylem cortex and pericycle (Dietz and Herth 2011). Metabolic processes help NPs interact with plant cells, regulating ion transport and protein levels via –SH and –COOH groups (Das and Das 2019). NPs can enter plants via root cell membrane transporters (Kurepa et al. 2009). They travel through the cuticle from roots to shoots, stems, leaves, and grains (Sharif et al. 2013). NPs combine with organelles in the plant cell cytoplasm, starting metabolic pathways needed for development and production (Zhang and Monteiro-Riviere 2009).

23.6.1 Role of SNPs in Abiotic Stress Tolerance in Plants

Plants' abiotic stress tolerance is improved with silver nanoparticles (SNPs). Extreme environmental circumstances can stunt plant growth and productivity, affecting the ecosystem. Plants have developed systemic signaling pathways to combat abiotic stressors such as ROS, hydrodynamic waves, and phytohormones (Choudhury et al. 2018). These signaling pathways let plants adapt quickly to stresses (Kollist et al. 2018). Systemic acquired resistance (SAR) activates when plant tissues sense stress signals. SAR gives tissues stress resistance (Fichman and Mittler 2020). SNPs reduce ROS, improving plant abiotic stress tolerance (Namjoyan et al. 2020).

Table 23.2 Effects of nanoparticles on plants under various stress conditions

Stress	Plant	Nanoparticles	Morphological changes under the influence of nanoparticles	Reference
Drought	*Brassica napus* L.	Fe	Increased biomass production and leaf growth	Palmqvist et al. (2017)
	Triticum aestivum L.	Se	Improved shoot and root length, leaf area, and leaf number	Tawfik et al. (2021)
	Rice	ZnO	Increased plant height, fresh weight, and dry weight	Mazhar et al. (2022a)
			Physiological response of plants under the influence of nanoparticles	
	Linum usitatissimum L.	Ti	Reduced chlorophyll damage, electrolyte leakage, lipid peroxidation and H_2O_2 accumulation	Aghdam et al. (2015)
	Brassica napus L.	Fe	Reduced MDA production	Palmqvist et al. (2017)
	Oryza sativa L.	ZnO	Decreased lipid peroxidation	Mazhar et al. (2022a)
	Linum usitatissimum L	Fe3O4	Reduced MDA and H2O2 accumulation	Mazhar et al. (2022b)
			Biochemical changes under the influence of nanoparticles	
	Oryza sativa L.	ZnO	Enhanced proline content	Mazhar et al. (2022a)
	Linum usitatissimum L	Ti	Improved protein and seed oil production	Aghdam et al. (2015)

(continued)

Table 23.2 (continued)

Stress	Plant	Nanoparticles	Morphological changes under the influence of nanoparticles	Reference
Salt			**Morphological changes under the influence of nanoparticles**	
	Lycopersicon esculentum Mill.	Si	Retained fruit quality and size	Pinedo-Guerrero et al. (2020)
	Moringa oleifera Lam	Fe3O4	Improved plant growth, number of branches, leaf area, and biomass	Tawfik et al. (2021)
	Dracocephalum moldavica L.	Ti	Increased plant height	Gohari et al. (2020)
			Physiological response of plants under the influence of nanoparticles	
	Dracocephalum moldavica L.	Ti	Enhanced nutrient uptake	Gohari et al. (2020)
	Medicago sativa L	K2SO4	Reduced electrolyte leakage	El-Sharkawy et al. (2017)
	Abelmoschus esculentus (L.) Moench	Zn	Enhanced photosynthetic pigments	Alabdallah and Alzahrani (2020)
			Biochemical changes under the influence of nanoparticles	
	Moringa oleifera Lam	Fe3O4	Increased crude protein, fiber, and minerals	Tawfik et al. (2021)
	Medicago sativa L.	K2SO	Increased proline content	El-Sharkawy et al. (2017)
	Abelmoschus esculentus (L.) Moench	Zn	Reduced proline content	Alabdallah and Alzahrani (2020)
Cadmium			**Morphological changes under the influence of nanoparticles**	
	Triticum aestivum L	Fe	Improved photosynthesis and yield	Adrees et al. (2019)
Heavy Metal			**Physiological response of plants under the influence of nanoparticles**	
	Coriandrum sativum L.	Si	Reduced MDA	Fatemi et al. (2020)

23.6.1.1 Tolerance to Heavy Metals

Nanoparticles can reduce heavy metals that damage plants. Studies show that NPs reduce the adverse effects of heavy metals such as chromium, cadmium, mercury, lead, and aluminum (Tripathi et al. 2015). NPs can reduce cadmium levels by 64.9% in rice during flowering (Chen et al. 2018). NPs also improve seed germination and plant growth, reducing aluminum toxicity in Cicer arietinum (Chandra et al. 2020). NPs reduce metal buildup, oxidative stress, and plant defense mechanisms to reduce heavy metal stress (Sousa et al. 2019). They reduce ROS by activating roots and leaf antioxidant defenses (Ali et al. 2019). NPs also boost glutathione (GSH), which detoxifies heavy metals and produces phytochelatin, lowering toxicity (Wang et al. 2015). Under cadmium stress, silicon and nitric oxide boost plant antioxidant defenses (Singh et al. 2020). NPs also increase plant root phenol and organic acid synthesis, reducing heavy metal toxicity (Sousa et al. 2019).

Cadmium (Cd) accumulates and inhibits photosynthesis, ion control, and nutrient uptake, reducing plant development (Rizwan et al. 2019). According to research, FeO NPs minimize Cd toxicity in wheat plants. Silicon nanoparticles reduce heavy metal-induced phytotoxicity in rice, pea, and wheat (Gao et al. 2018). Lead and cadmium poisoning causes cellular oxidative stress, which FeO NPs reduce to boost wheat development (Konate et al. 2017). Recent research has shown that FeO NPs can stimulate plant development and reduce heavy metal and water stress.

23.6.1.2 Tolerance to Salt Stress

Salt stress threatens crop yield, especially as many main crops are glycophytes (Munns and Tester 2008). Increased salt content and stress duration restrict plant growth, reducing photosynthesis, respiration, and biomass (Parida and Das 2005). Nanoparticles (NPs) may improve plant salt stress tolerance (Kalteh et al. 2018). In salt-stressed cucumbers, 200 mg/kg 10 nm NPs increased seed germination and vitality (Alsaeedi et al. 2018). NPs also boost plant antioxidant enzymes, reducing salt stress (Naguib and Abdalla 2019). Under salt stress, NPs increase plant cortex crystal wax deposition, preserving leaf water and chlorophyll levels and preventing salt-induced damage (Avestan et al. 2019). Changing ion concentrations regulates stomatal conductance and stabilizes transpiration under salt stress. Plant cells adapt their osmotic pressure to maintain stability under salt stress, and NPs help by boosting proline and free amino acid levels to resist salt penetration (Sabaghnia and Janmohammadi 2015). NPs can reduce ionic toxicity by decreasing sodium (Na^+) absorption and boosting potassium (K^+) absorption, affecting the K^+/Na^+ ratio in the cytoplasm (Almutairi 2016). NPs activate plant defense enzymes under salt stress, reducing ROS buildup (Naguib and Abdalla 2019). Si, FeO, and multi-walled carbon nanotubes have been demonstrated to promote plant growth, chlorophyll content, and ROS-detoxifying enzyme activity (Ye et al. 2020). NPs synthesized from plant extracts also improve seed germination, biochemical properties, and salt-stressed plant growth (Mustafa et al. 2021).

23.6.1.3 Tolerance to Drought Stress

Drought causes ecological water shortages due to warmth, light intensity, and insufficient rainfall (Liaqat et al. 2024b; Kacholi and Sahu 2018). Stress causes many changes in plant development features and metabolic reactions, affecting photosynthetic activities (Seleiman et al. 2021). Researchers are investigating phytogenic nanoparticles to mitigate drought stress on plants. Wheat, cherries, strawberries, and barley are more drought-resistant with NPs (Hellala et al. 2020). NPs increase tissue water absorption and retention, enzymatic antioxidant activity, and cell osmotic pressure, helping plants resist drought stress (Luyckx et al. 2017). They increase leaf photosynthetic pigments, proline, and carbohydrates to reduce oxidative damage of drought-induced malondialdehyde (MDA) (Zahedi et al. 2020). NPs reduce Malondialdehyde (MDA) and increase enzymatic antioxidant activity to reduce oxidative stress and delay leaf senescence, boosting photosynthesis and stomatal conductance (Namjoyan et al. 2020). NPs also boost plant growth and prevent drought stress-induced yield loss (Behboudi et al. 2018). Silicon-based NPs can stimulate shoot growth, increase chlorophyll, and reduce plant membrane lipid peroxidation to reduce drought stress (Pei et al. 2010). In drought, zinc oxide (ZnO) nanoparticles promote soybean seed germination and growth (Sedghi et al. 2013). Fe nanoparticles increase biomass production in wheat during flowering and grain filling, while titanium nanoparticles improve agronomic characteristics and yield under drought stress (Jaberzadeh et al. 2013). Silver nanoparticles reduce lentil drought stress (Hojjat 2016). Si NPs affect morpho-physiological features to improve drought resistance and recovery (Ghorbanpour et al. 2020). Under drought stress, chitosan nanoparticles increase wheat water content, photosynthetic rate, and antioxidant enzyme activity (Behboudi et al. 2019). Foliar Fe nanoparticles alleviate drought stress in safflower cultivars, while CeO nanoparticles boost soybean growth and photosynthesis (Davar-Zareii et al. 2014). Sun et al. (2017) suggest using Si NPs to give abscisic acid to plants to improve drought resilience.

23.6.1.4 Tolerance to Heat Stress

Research on trace elements, phytohormones, osmoprotectants, and signal molecules for heat stress is growing (Bhatia et al. 2021). Nanoemulsions or nanoparticles carry these compounds to targeted areas, improving their efficacy and bioavailability (Dhiman et al. 2021). Low silver nanoparticle (AgNP) concentrations protected wheat plants from heat stress, improving morphological growth. High-temperature silicon (Si) application boosts antioxidant enzymes like superoxide dismutase (SOD), peroxidase (APX), and glutathione peroxidases (GPX) while decreasing catalase (CAT) activity (Soundararajan et al. 2013). Proline accumulation, antioxidant enzyme activity, gas exchange, and photosynthetic efficiency are enhanced with foliar nanosilicon dioxide (SiO_2NPs) application (Kim et al. 2017). Titanium dioxide nanoparticles (TiO_2NPs) boost biomass output by promoting photosynthesis under mild heat stress and providing nutrients (Raliya et al. 2015). Selenium nanoparticles (SeNPs) in the soil or foliage increase antioxidant enzyme activity, gas exchange, and yield in heat-stressed crops (Djanaguiraman et al. 2018). Under high light and heat, CeO_2NPs scavenge ROS and boost photosynthesis (Heckert et al. 2008). NPs, like ROS-scavenging metalloenzymes, require the Ce^{3+}/Ce^{4+} redox pair on their surfaces (Dowding et al. 2013). Heat stress increases ROS and oxidative stress, causing plant cell death and metabolic dysfunction (Li et al. 2021). Nanotechnology may reduce agricultural heat stress (Rana et al. 2021). El-Saadony et al. (2021) found that selenium nanoparticles improve tomato and wheat development and minimize heat stress.

23.7 Nanofertilizers Versus Commercial Fertilizers

Nanoscale transporters and chemicals can precisely administer macromolecules and release agrochemicals, lowering chemical use without affecting crop production. Commercial fertilizers have greater particle sizes and lesser solubility, but nanofertilizers provide more benefits to plants (Fig. 23.4). Chemical fertilizers can pollute the environment and disturb soil life by accumulating heavy metals. Sustainable agriculture relies on nanoagrochemicals for water quality and fertilizer efficiency. Bioaccumulation and extended nanoparticle exposure raise concerns regarding their effects on edible plants and food chains (Rajput et al. 2019a, b). These nanoparticles can be absorbed and deposited in edible tissues, affecting plant functions and compositions. Nanotechnology offers creative agricultural solutions, but safety, health, ethics, and environmental effect are issues. Nanoparticle health risks are debated (Staroń et al. 2020). Despite the benefits, nanoparticle disposal is a concern due to the annual production of several hundred tons (Rajput et al. 2019a, b). Thus, recognizing the pros and cons of nanoparticles in agriculture is crucial. Scientists and professionals must handle disposal and risk concerns while appropriately using nanotechnology in agriculture.

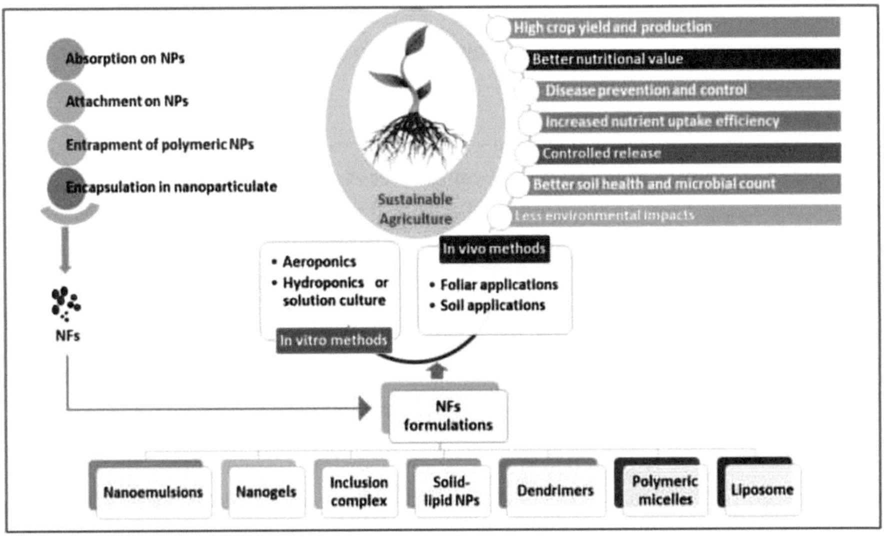

Fig. 23.4 An overview of nanofertilizer application in agriculture. *NPs* nanoparticles, *NFs* nanofertilizers (Verma et al. 2022)

23.8 Nanotechnology for Sustainable Agriculture

Nanotechnology drives agricultural output by regulating nutrients (Gruère 2012; Mukhopadhyay 2014). It promotes sustainable agriculture by monitoring water quality and pesticides (Prasad 2014). Nanomaterials have many features that make health and environmental risk assessment difficult (Prasad 2014). Chemical composition, shape, and surface structure affect toxicity (Ion et al. 2010). Nanomaterials with the same composition but various sizes or forms can be harmful. Controlling nutrients in agriculture with nanotechnology boosts productivity (Gruère 2012; Mukhopadhyay 2014). It promotes sustainable agriculture by monitoring water quality and pesticides (Prasad 2014). Nanomaterials have many features that make health and environmental risk assessment difficult (Prasad 2014). Chemical composition, shape, and surface structure affect toxicity (Ion et al. 2010). Nanomaterials with the same composition but various sizes or forms can be harmful. Nanotechnology research is crucial for agricultural sustainability. It is used in nanotubes, biosensors, and controlled delivery systems (Ion et al. 2010; Sabir et al. 2014). Nanotechnology manages resources, delivers medications to plants, and maintains soil fertility. It aids biomass and agricultural waste usage, food processing, packaging, and risk evaluation (Floros et al. 2010). Agriculture uses nanosensors to quickly detect soil and water contamination (Ion et al. 2010). Biosensors, electrochemical sensors, and optical sensors detect trace heavy metals (Ion et al. 2010). Nanomaterials degrade trash and pollutants and boost microbial efficiency. Bioremediation removes heavy metals from soil and water by utilizing beneficial bacteria, plants, fungi, and mushrooms (Dixit et al. 2015). Agricultural bioremediation removes hazardous substances to restore soil sustainably (Ion et al. 2010; Dixit et al. 2015). Nanotechnology research is crucial for agricultural sustainability. It is used in nanotubes, biosensors, and controlled delivery systems (Ion et al. 2010; Sabir et al. 2014). Nanotechnology manages resources, delivers medications to plants, and maintains soil fertility. It aids biomass and agricultural waste usage, food processing, packaging, and risk evaluation (Floros et al. 2010). Agriculture uses nanosensors to quickly detect soil and water contamination (Ion et al. 2010). Biosensors, electrochemical sensors, and optical sensors detect trace heavy metals (Ion et al. 2010). Nanomaterials degrade trash and pollutants and boost microbial efficiency. Bioremediation removes heavy metals from soil and water by utilizing beneficial microorganisms, plants, fungi, and mushrooms (Dixit et al. 2015). Environmentally eliminating harmful components from soil through agricultural bioremediation promotes soil sustainability (Ion et al. 2010; Dixit et al. 2015).

23.8.1 Nanofertilizers

Nanofertilizers have become more available in the previous decade, although large chemical companies have not yet adopted them for agricultural application. Nanoparticles such nano zinc, silica, iron, titanium dioxide, and others can be added to nanofertilizers to control release and improve quality. Al_2O_3, TiO_2, CeO_2, FeO, and ZnONPs have been studied for absorption, biological destiny, and toxicity (Zhang et al. 2016). Zinc deficiency limits agricultural output, especially in alkaline soils (Sadeghzadeh 2013). Direct proton bombardment or 18O enrichment during synthesis created 18F to label metal oxide nanoparticles (Llop et al. 2014). The nanoparticles' size, aggregation, and zeta potential in proteins and cell media have been examined. Ion beam, transmission electron, Raman, and confocal laser scanning microscopy track nanoparticle uptake and intracellular destiny (Marzbani et al. 2015). A sustainable bio-based economy using eco-efficient bio-processes and renewable bio-resources will construct twenty-first-century technology (Prasad 2014; Marzbani et al. 2015). Advances in ecology, biology, biodiversity, material science, biotechnology, and engineering can increase biomass productivity and usage of organic wastes. Faced with climate change and other challenges, smart agriculture is becoming a key approach for short- and long-term growth (Helar and Chavan 2015). It helps nations maintain vital agricultural activities (Kandasamy and Prema 2015). Research aims include expanding nanoscale resources and understanding their properties. Nanoscale inorganic materials are excellent for agricultural analysis due to the effects of quantum size and enhanced transmission (Kandasamy and Prema 2015). Gold nanoparticles (AuNPs) are valuable in biosensing due to their intrinsic characteristics, low toxicity, biocompatibility, and unique optical qualities. They are faster, more sensitive, and more flexible than biological testing for selected analytics.

23.8.2 Nanopesticides

Nanomaterials in agriculture for plant protection and food production have yet to be completely investigated, but they have shown promise. Nanomaterials could help reduce insect pests and infections in agricultural areas (Khota et al. 2012). New nanoencapsulated pesticide formulations feature gradual release, improved solubility, selectivity, permeability, and stability (Bhattacharyya et al. 2016). These formulations prolong active component efficacy against pests by preventing early degradation. They also reduce pesticide use and human exposure, making them environmentally benign for crop protection. Thus, non-toxic and effective pesticide delivery technologies are essential for increasing global food production while reducing environmental impacts (Grillo et al. 2016). Nanoencapsulation, like microencapsulation, improves chemical delivery to biological processes. Some chemical companies are selling nanoscale pesticides as "microencapsulated pesticides." Nanoscale pesticides include Syngenta's Karate ZEON and Subdue MAXX, Ospray's Chyella, and BASF's Penncap-M (Gouin 2004). Australian products including Primo MAXX, Banner MAXX, and Subdue MAXX are nanoemulsions, emphasizing the narrow line between the two (Gouin 2004). This method often uses organic nanoparticles with active agrochemicals or other compounds.

23.9 Future of Nanotechnology in Agriculture

Nanotechnology in agriculture could transform farming, crop management, and food production. A glimpse into the future:

23.9.1 Precision Farming

Nanosensors and nanodevices will help farmers more precisely monitor soil, crop, and environmental conditions. Real-time data will enhance irrigation, fertilization, and insect management, increasing crop yields and resource efficiency.

23.9.2 Nanofertilizers

Nanotechnology will create fertilizers with improved nutrition delivery. These nanofertilizers will improve plant nutrient uptake, reduce runoff and leaching, and lessen environmental impact.

23.9.3 Nanopesticides

Nano-based pesticides will deliver pest control agents precisely, reducing the need for broad-spectrum poisons. This reduces environmental pollution and pesticide resistance.

23.9.4 Smart Delivery Systems

Nanoencapsulation will manage fertilizer, pesticide, and growth regulator release. This boosts efficiency, reduces waste, and improves sustainability. Nanomaterials will help develop new crop protection measures against diseases, pests, and environmental stressors. Nano-enabled plant vaccinations, antimicrobial coatings, and smart pest control nanomaterials will make farming more resilient and sustainable.

23.9.5 Environmental Remediation

Nanotechnology will help remove soil, water, and air toxins. Catalytic nanoparticles decompose contaminants, while nanofiltration membranes purify water.

23.9.6 Nanobiosensors

Nanoscale sensors will quickly detect diseases, poisons, and contaminants in agricultural goods and ecosystems. Improving food safety, quality control, and early illness detection will protect public health. Nanotechnology will help turn agricultural waste into biofuels, bioplastics, and bioproducts. Nanocatalysts and nanoreactors will boost biomass conversion efficiency, creating a circular bioeconomy.

23.9.7 Climate Resilience

Nanotechnology will generate drought-resistant crops, sequester soil carbon, and improve resource use efficiency for climate-smart agriculture. Nanomaterial-based sensors will reveal climate stresses, enabling adaptive farming.

23.10 Limitation of Nanotechnology

Nanotechnology has great potential, but its constraints must be overcome for responsible and widespread usage. Nanomaterials are expensive to produce and implement, restricting their use in resource-constrained contexts. Nanomaterials' health and environmental effects raise worries regarding their long-term safety and sustainability. Regulatory issues further complicate market acceptance, as nanoparticles' unique features and hazards may not be effectively addressed. Responsible innovation requires addressing ethical and societal issues like privacy, security, and equity. Public perception is also important since disinformation and perceived hazards can damage nanotechnology-based products and technology credibility. To maximize nanotechnology's promise while protecting society, interdisciplinary collaboration, thorough risk assessment, clear communication, and proactive regulation are needed.

23.11 Conclusion

The effects of climate change on agriculture, weather patterns, severe temperatures, water scarcity, and soil degradation are serious and require novel responses. Farmers need quick tools to respond to changing conditions. Nanotechnology-enabled fertilizers, insecticides, sensors, and materials may solve these problems. Nano-based interventions could boost crop output, resource efficiency, and environmental effects. Nanotechnology's environmental and health risks must be assessed, yet opportunities exist. To benefit farmers worldwide, these technologies must be accessible fairly. Researchers, politicians, and stakeholders must work together to use nanotechnology in agriculture to protect ecosystems and promote sustainable food production. In the face of climate change, sustainable agriculture can be achieved by appropriately using nanotechnology and incorporating it into climate-smart agriculture methods.

23.12 Future Prospect

The integration of nanotechnology into agriculture offers significant potential to address the pressing challenges posed by global climate change. Future prospects include the development of nanomaterials to enhance crop resilience to abiotic stresses such as drought and extreme temperatures, thus boosting yields. Precision agriculture can benefit from nanosensors that provide real-time data on soil health, crop growth, and pest presence, enabling optimal resource use. Nano-based pesticides and fungicides promise targeted, environmentally friendly pest and disease management, while nanomaterials can restore soil health and improve nutrient release. Climate-smart fertilizers with slow-release properties can reduce nutrient leaching and greenhouse gas emissions. Additionally, nanotechnology can enhance water use efficiency through improved irrigation techniques and water purification, addressing water scarcity. Innovations in nano-based packaging can extend the shelf life of produce, reducing post-harvest losses. Biofortification using nanotechnology can also enhance the nutritional value of crops, addressing global micronutrient deficiencies and contributing to better health outcomes.

References

Adisa IO, Pullagurala VLR, Peralta-Videa JR, Dimkpa CO, Elmer WH, Gardea-Torresdey JL, White JC (2019) Recent advances in nano-enabled fertilizers and pesticides: A critical review of mechanisms of action. Environ Sci Nano 6:2002–2030

Adrees M, Khan ZS, Ali S, Hafeez M, Khalid S, Rehman MZU, Hussain A, Hussain K, Chatha SAS, Rizwan M (2019) Simultaneous mitigation of cadmium and drought stress in wheat by soil application of iron nanoparticles. Chemosphere 238:124681

Aghdam MTB, Mohammadi H, Ghorbanpour M (2015) Effects of nanoparticulate anatase titanium dioxide on physiological and biochemical performance of Linum usitatissimum (Linaceae) under well-watered and drought stress conditions. Braz J Bot 39:139–146

Ahlawat J, Asil SM, Barroso GG, Nurunnabi M, Narayan M (2021) Application of carbon nano onions in the biomedical field: recent advances and challenges. Biomaterials Science. Advance online publication. https://doi.org/10.1039/D0BM01476A

Ahmed T, Noman M, Manzoor N, Shahid M, Abdullah M, Ali L et al (2021) Nanoparticle-based amelioration of drought stress and cadmium toxicity in rice via triggering the stress responsive genetic mechanisms and nutrient acquisition. Ecotoxicol Environ Saf 209:111829. https://doi.org/10.1016/j.ecoenv.2020.111829

Alabdallah NM, Alzahrani HS (2020) The potential mitigation effect of ZnO nanoparticles on [Abelmoschus esculentus L. Moench] metabolism under salt stress conditions. Saudi J Biol Sci 27:3132–3137

Alexandratos N, Bruinsma J (2012) World agriculture: towards 2030/2050; ESA working paper no. 12–03; FAO: Rome, Italy

Ali S, Rizwan M, Hussain A, Rehman MZ, Ali B, Yousaf B et al (2019) Silicon nanoparticles enhanced the growth and reduced the cadmium accumulation in grains of wheat (Triticum aestivum L.). Plant Physiol Biochem 140:1–8. https://doi.org/10.1016/j.plaphy.2019.04.041

Ali A, Altaf MT, Nadeem MA, Karaköy T, Shah AN, Azeem H et al (2022) Recent advancement in OMICS approaches to enhance abiotic stress tolerance in legumes. Front Plant Sci 13:952759

Al-Khayri JM, Rashmi R, Surya Ulhas R, Sudheer WN, Banadka A, Nagella P, Aldaej MI, Rezk AA, Shehata WF, Almaghasla MI (2023) The role of nanoparticles in response of plants to abiotic stress at physiological, biochemical, and molecular levels. Plan Theory 12(2):292

Almutairi Z (2016) Effect of nano-silicon application on the expression of salt tolerance genes in germinating tomato (Solanum lycopersicum L.) seedlings under salt stress. Plant Omics 9:106–114

Alsaeedi A, El-Ramady H, Alshaal T, El-Garawani M, Elhawat N, Al-Otaibie A (2018) Exogenous silica nanoparticles improve germination and growth of cucumber by maintaining K+/Na+ ratio under elevated Na+ stress. Plant Physiol Biochem 125:164–171. https://doi.org/10.1016/j.plaphy.2018.02.006

Altaf MT, Liaqat W, Nadeem MA, Baloch FS (2023) Recent trends and applications of omics-based knowledge to end global food hunger. In: Sustainable agriculture in the era of the OMICs revolution. Springer International Publishing, Cham, pp 381–397

Anil KM (2014) Climate Change and challenges of water and food security. Int J Sustain Built Environ 3:153–165

Aryal JP, Mehrotra MB, Jat ML, Sidhu HS (2015a) Impacts of laser land leveling in rice-wheat systems of the north-western indo-gangetic plains of India. Food Sec 7:725

Aryal JP, Sapkota TB, Jat ML, Bishnoi JK (2015b) On-farm economic and environmental impact of zero-tillage wheat: a case of north-West India. Exp Agric 51:1–16

Ashraf M, Wu L (1994) Breeding for salinity tolerance in plants. Crit Rev Plant Sci 13(1):17–42

Asian Water Development Outlook (AWDO) (2016) Strengthening water security in Asia and the Pacific. Available online: https://www.adb.org/publications/asian-water-development-outlook-2016

Avestan S, Ghasemnezhad M, Esfahani M, Byrt CS (2019) Application of nano-silicon dioxide improves salt stress tolerance in strawberry plants. Agronomy 9(5):246. https://doi.org/10.3390/agronomy9050246

Bates BC, Kundzewicz ZW, Wu S, Palutikof JP (eds) (2008) Climate Change and water. Technical paper of the intergovernmental panel on Climate Change. IPCC Secretariat, Geneva

Baul TK, McDonald M (2015) Integration of indigenous knowledge in addressing climate change. Indian J Tradit Knowl 1:20–27

Behboudi F, Sarvestani TZ, Kassaee MZ, Sanavi S, Sorooshzadeh A (2018) Improving growth and yield of wheat under drought stress via application of SiO2 nanoparticles. J Agric Sci Technol 20:1479–1492. http://jast.modares.ac.ir/article-23-19814-en.html

Behboudi F, Tahmasebi-Sarvestani Z, Kassaee MZ, Modarres-Sanavy SAM, Sorooshzadeh A, Mokhtassi-Bidgoli A (2019) Evaluation of chitosan nanoparticles effects with two application methods on wheat under drought stress. J Plant Nutr 42:1439–1451. https://doi.org/10.1080/01904167.2019.1617308

Bhatia R, Gulati D, Sethi G (2021) Biofilms and nanoparticles: applications in agriculture. Folia Microbiol:1–12. https://doi.org/10.1007/s12223-021-00851-7

Bhattacharyya A, Duraisamy P, Govindarajan M, Buhroo AA, Prasad R (2016) Nano-biofungicides: emerging trend in insect pest control. In: Prasad R (ed) Advances and applications through fungal Nanobiotechnology. Springer International Publishing, Cham, pp 307–319

Bhullar MS, Singh S, Kumar S, Gill G (2018) Agronomic and economic impacts of direct seeded rice in Punjab. Agric Res J 55:236–242

Bindraban PS, Dimkpa CO, Angle S, Rabbinge R (2018) Unlocking the multiple public good services from balanced fertilizers. Food Sec 10:273–285

Birthal PS, Nigam AN, Narayanan AV, Kareem KA (2012) Potential economic benefits from adoption of improved drought-tolerant groundnut in India. Agric Econ Res Rev 25:1–14

Blum WEH, Nortcliff S (2013) Soils and food security. In: Brevik EC, Burgess LC (eds) Soils and human health. CRC Press, Boca Raton, FL, pp 299–321

Bokhari SAA, Rasul G, Ruane AC, Hoogenboom G, Ahmad A (2017) The past and future changes in climate of the rice-wheat cropping zone in Punjab, Pakistan. Pak J Meteorol 13:9–23

Brevik EC (2012) Soils and climate change: gas fluxes and soil processes. Soil Horizons 53(4):12. https://doi.org/10.2136/sh12-04-0012

Brevik EC (2013) Soils and human health: an overview. In: Brevik EC, Burgess LC (eds) Soils and human health. CRC Press, Boca Raton, FL, pp 29–56

Broman KW, Speed TP (1999) A review of methods for identifying QTLs in experimental crosses. In: Seiller-Moiseiwitsch F (ed) Statistics in molecular biology and genetics. IMS lecture notes monograph series, 33, pp 114–142

Burke M, Hsiang SM, Miguel E (2015) Global non-linear effect of temperature on economic production. Nature 527:235–239

Burman U, Kumar P (2018) Plant response to engineered nanoparticles. In: Tripathi DK, Ahmad P, Sharma S, Chauhan DK, Dubey NK (eds) Nanomaterials in plants, algae, and microorganisms, vol 1. Elsevier Academic Press

Calanca PP (2017) Effects of abiotic stress in crop production. In: Ahmed M, Stockle CO (eds) Quantification of climate variability, adaptation and mitigation for agricultural sustainability. Springer, pp 165–180

Carleton TA, Hsiang SM (2016) Social and economic impacts of climate. Science 353:9837

CDIAC (2020) Carbon dioxide information analysis center. Available online: www.cdiac.ess-dive.lbl.gov (Accessed 13 November 2020)

Chandra J, Chauhan R, Korram J, Satnami ML, Keshavkant S (2020) Silica nanoparticle minimizes aluminium imposed injuries by impeding cytotoxic agents and over expressing protective genes in Cicer arietinum. Sci Hortic 260:108885

Chausali N, Saxena J, Prasad R (2021) Nanobiochar and biochar based nanocomposites: advances and applications. J Agric Food Res 5:100191

Chausali N, Saxena J, Prasad R (2022) Recent trends in nanotechnology applications of bio-based packaging. J Agric Food Res 7:100257

Chellaney B (2011) Water: Asia's new battleground. Georgetown University Press, Washington, DC

Chen H, Sun J (2018) Projected changes in climate extremes in China in a 1.5C warmer world. Int J Climatol 38:3607–3617

Chen R, Zhang C, Zhao Y, Huang Y, Liu Z (2018) Foliar application with nano-silicon reduced cadmium accumulation in grains by inhibiting cadmium translocation in rice plants. Environ Sci Pollut Res 25(3):2361–2368. https://doi.org/10.1007/s11356-017-0681-z

Cheng HN, Klasson KT, Asakura T (2016) Nanotechnology in agriculture. In: Cheng HN et al (eds) Nanotechnology: delivering on the promise, 2. ACS, pp 233–242

Choudhury FK, Devireddy AR, Azad RK, Shulaev V, Mittler R (2018) Local and systemic metabolic responses during light-induced rapid systemic signaling. Plant Physiol 178(4):1461–1472. https://doi.org/10.1104/pp.18.01031

Ciscar JC, Iglesias A, Feyen L, Szabo L, Regemorter DV, Amelung B, Nicholls R, Watkiss P, Christensen OB, Dankers R et al (2011) Physical and economic consequences of climate change in Europe. Proc Natl Acad Sci USA 108:2678–2683

Costinot A, Donaldson D, Smith C (2016) Evolving comparative advantage and the impact of climate change in agricultural markets: evidence from 1.7 million fields around the world. J Pol Econ 124:20–25

Cushman JC, Bohnert HJ (2000) Genomic approaches to plant stress tolerance. Curr Opin Plant Biol 3(2):117–124. https://doi.org/10.1016/s1369-5266(99)00052-7

Das A, Das B (2019) Nanotechnology: a potential tool to mitigate abiotic stress in crop plants. In: De Oliveira A (ed) Abiotic and Biotic Stress in Plants. IntechOpen. https://doi.org/10.5772/intechopen.83562

Davar-Zareii F, Roozbahani A, Hosnamidi A (2014) Evaluation the effect of water stress and foliar application of Fe nanoparticles on yield, yield components and oil percentage of safflower (Carthamus tinctorious L.). Int J Adv Biol Biomed Res 2:1150–1159

Dhiman S, Yadav A, Debnath N, Das S (2021) Application of core/shell nanoparticles in smart farming: A paradigm shift for making the agriculture sector more sustainable. J Agric Food Chem 69(11):3267–3283. https://doi.org/10.1021/acs.jafc.0c05403

Dietz KJ, Herth S (2011) Plant nanotoxicology. Trends Plant Sci 16(10):582–589. https://doi.org/10.1016/j.tplants.2011.08.003

Dixit R, Wasiullah MD, Pandiyan K, Singh UB, Sahu A et al (2015) Bioremediation of heavy metals from soil and aquatic environment: an overview of principles and criteria of fundamental processes. Sustain For 7:2189–2212

Djanaguiraman M, Belliraj N, Bossmann SH, Prasad PVV (2018) High temperature stress alleviation by selenium nanoparticle treatment in grain sorghum. ACS Omega 3:2479–2491. https://doi.org/10.1021/acsomega.7b01934

Dowding JM, Seal S, Self WT (2013) Cerium oxide nanoparticles accelerate the decay of peroxynitrite (ONOO). Drug Deliv Transl Res 3:375–379. https://doi.org/10.1007/s13346-013-0136-0

Ealia SAM, Saravanakumar MP (2017) A review on the classification, characterization, synthesis of nanoparticles and their application. In: IOP conference series: materials science and engineering, vol 320. IOP Publishing, p 192019

Elmer WH, White JC (2016) The use of metallic oxide nanoparticles to enhance growth of tomatoes and eggplants in disease infested soil or soilless medium. Environ Sci Nano 3:1072–1079

El-Saadony MT, Saad AM, Najjar AA et al (2021) The use of biological selenium nanoparticles to suppress Triticum aestivum L. crown and root rot diseases induced by Fusarium species and improve yield under drought and heat stress. Saudi J Biol Sci 28:4461–4471. https://doi.org/10.1016/j.sjbs.2021.04.043

El-Sharkawy MS, El-Beshsbeshy TR, Mahmoud EK, Abdelkader NI, Al-Shal RM, Missaoui AM (2017) Response of alfalfa under salt stress to the application of potassium sulfate nanoparticles. Am J Plant Sci 8:1751–1773

Faizan M, Rajput V, Al-Khuraif A, Arshad M, Minkina T, Sushkova S, Yu F (2021) Effect of foliar fertigation of chitosan nanoparticles on cadmium accumulation and toxicity in Solanum lycopersicum. Biology 10:666

Farooq M, Wahid A, Kobayashi N, Fujita D, Basra SMA (2009) Plant drought stress: effects, mechanisms and management. Sustainability in Agriculture, Springer, pp. 153–188. https://doi.org/10.1007/978-90-481-2666-8_12

Fatemi H, Pour BE, Rizwan M (2020) Foliar application of silicon nanoparticles affected the growth, vitamin C, flavonoid, and antioxidant enzyme activities of coriander (Coriandrum sativum L.) plants grown in lead (Pb)-spiked soil. Environ Sci Pollut Res 28:1417–1425

Fellet G, Pilotto L, Marchiol L, Braidot E (2021) Tools for Nano-enabled agriculture: fertilizers based on calcium phosphate, silicon, and chitosan nanostructures. Agronomy 11:1239

Fichman Y, Mittler R (2020) Rapid systemic signaling during abiotic and biotic stresses: is the ROS wave master of all trades? Plant J 102(4):887–896. https://doi.org/10.1111/tpj.14685

Fischer EM, Knutti R (2015) Anthropogenic contribution to global occurrence of heavy-precipitation and high-temperature extremes. Nat Clim Chang 5:560–564

Floros JD, Newsome R, Fisher W, Barbosa-Cánovas GV, Chen H, Dunne CP et al (2010) Feeding the world today and tomorrow: the importance of food science and technology. Compr Rev Food Sci Food Saf 9:572–599

Fraceto LF, Grillo R, de Medeiros GA, Scognamiglio V, Rea G, Bartolucci C (2016) Nanotechnology in agriculture: which innovation potential does it have? Front EnvironSci 4:186737

Gao M, Zhou J, Liu H, Zhang W, Hu Y, Liang J, Yu H (2018) Foliar spraying with silicon and selenium reduces cadmium uptake and mitigates cadmium toxicity in rice. Sci Total Environ 631–632:1100–1108. https://doi.org/10.1016/j.scitotenv.2018.03.047

Ghorbanpour M, Mohammadi H, Kariman K (2020) Nanosilicon-based recovery of barley (Hordeum vulgare) plants subjected to drought stress. Environ Sci Nano 7:443–461. https://doi.org/10.1039/C9EN00973F

Godoy F, Olivos-Hernández K, Stange C, Handford M (2021) Abiotic stress in crop species: improving tolerance by applying plant metabolites. Plan Theory 10(2):186. https://doi.org/10.3390/plants10020186

Gohari G, Mohammadi A, Akbari A, Panahirad S, Dadpour MR, Fotopoulos V, Kimura S (2020) Titanium dioxide nanoparticles (TiO2 NPs) promote growth and ameliorate salinity stress effects on essential oil profile and biochemical attributes of Dracocephalum moldavica. Sci Rep 10:912

Gomez-Zavaglia A, Mejuto JC, Simal-Gandara J (2020) Mitigation of emerging implications of climate change on food production systems. Food Res Int 134:109256
Gosain AK, Rao S, Basuray D (2006) Climate Change impact assessment on hydrology of Indian River basins. Curr Sci 90(3):346–353
Gouin S (2004) Microencapsulation: industrial appraisal of existing technologies and trends. Trends Food Sci Technol 15:330–347
Grillo R, Abhilash PC, Fraceto LF (2016) Nanotechnology applied to bio-encapsulation of pesticides. J Nanosci Nanotechnol 16:1231–1234
Gruère GP (2012) Implications of nanotechnology growth in food and agriculture in OECD countries. Food Policy 37:191–198
Gupta SM, Tripathi M (2012) An overview of commonly used semiconductor nanoparticles in photocatalysis. High Energy Chem 46(1):1–9. https://doi.org/10.1134/S0018143912010134
Heckert EG, Karakoti AS, Seal S, Self WT (2008) The role of cerium redox state in the SOD mimetic activity of nanoceria. Biomaterials 29:2705–2709. https://doi.org/10.1016/j.biomaterials.2008.03.014
Helar G, Chavan A (2015) Synthesis, characterization and stability of gold nanoparticles using the fungus Fusarium oxysporum and its impact on seed. Int J Recent Sci Res 6:3181–3318
Hellala F, Amerb AK, El-Sayeda S, El-Azab K (2020) Mitigation the negative effect of water stress on barley by nano silica application. Plant Arch 20:3224–3231
Hojjat SS (2016) The effect of silver nanoparticle on lentil seed germination under drought stress. Int J Farming Allied Sci 5(3):208–212
Hsiang S, Kopp R, Jina A, Rising J, Delgado M, Mohan S, Rasmussen DJ, Muir-Wood R, Wilson P, Oppenheimer M et al (2017) Estimating economic damage from climate change in the United States. Science 356:1362–1369
Hu Y, Schmidhalter U (2005) Drought and salinity: A comparison of their effects on mineral nutrition of plants. J Plant Nutr Soil Sci 168:541–549. https://doi.org/10.1002/jpln.200420516
Hussain M, Farooq S, Hasan W et al (2018) Drought stress in sunflower: physiological effects and its management through breeding and agronomic alternatives. Agric Water Manag 201:152–166. https://doi.org/10.1016/j.agwat.2018.01.028
Hussein MM, Abou-Baker NH (2018) The contribution of nano-zinc to alleviate salinity stress on cotton plants. R Soc Open Sci 5:171809
Ibrahim W, Zhu YM, Chen Y, Qiu CW, Zhu S, Wu F (2019) Genotypic differences in leaf secondary metabolism, plant hormones and yield under alone and combined stress of drought and salinity in cotton genotypes. Physiol Plant 165:343–355. https://doi.org/10.1111/ppl.12862
Ion AC, Ion I, Culetu A (2010) Carbon-based nanomaterials: environmental applications. Univ Politehn Bucharest 38:129–132
IPCC, Climate Change 2007 Impacts, adaptation and vulnerability (2007) in: working group II contribution to the fourth assessment Report of the intergovernmental panel on climate change. Cambridge University Press, Cambridge, UK
IPCC, Climate Change 2014 Synthesis Report (2014) In: Pachauri RK, Meyer LA (eds), Contribution of working groups I, II and III to the fifth assessment Report of the intergovernmental panel on Climate Change. IPCC, Geneva, Switzerland
Iqbal M, Raja NI, Mashwani ZUR, Hussain M, Ejaz M, Yasmeen F (2019) Effect of silver nanoparticles on growth of wheat under heat stress. Iran J Sci Technol Trans A Sci 43(2):387–395. https://doi.org/10.1007/s40995-017-0417-4
Iravani S (2011) Green synthesis of metal nanoparticles using plants. Green Chem 13:2638–2650
Jaberzadeh A, Moaveni P, Moghadam HRT, Zahedi H (2013) Influence of bulk and nanoparticles titanium foliar application on some agronomic traits, seed gluten and starch contents of wheat subjected to water deficit stress. Notulae Botanicae Horti Agrobotanici Cluj-Napoca 41:201–207. https://doi.org/10.15835/nbha4119093
Jeelani PG, Mulay P, Venkat R, Ramalingam C (2020) Multifaceted application of silica nanoparticles: A review. SILICON 12(7):1337–1354. https://doi.org/10.1007/s12633-019-00229-y
Kacholi DS, Sahu M (2018) Levels and health risk assessment of heavy metals in soil, water, and vegetables of Dar es Salaam, Tanzania. J Chem. https://doi.org/10.1155/2018/1402674

Kalteh M, Alipour ZT, Ashraf S, Aliabadi MM, Nosratabadi AF (2018) Effect of silica nanoparticles on basil (Ocimum basilicum) under salinity stress. J Chem Health Risks 4:3. https://doi.org/10.22034/jchr.2018.544075

Kandasamy S, Prema RS (2015) Methods of synthesis of nano particles and its applications. J Chem Pharm Res 7:278–285

Khan MIR, Khan NA (2017) Reactive oxygen species and antioxidant systems in plants: role and regulation under abiotic stress. Springer

Khan MN, Mobin M, Abbas ZK, AlMutairi KA, Siddiqui ZH (2017) Role of nanomaterials in plants under challenging environments. Plant Physiol Biochem 110:194–209. https://doi.org/10.1016/j.plaphy.2016.05.038

Khan I, Saeed K, Khan I (2019) Nanoparticles: properties, applications and toxicities. Arab J Chem 12(7):908–931. https://doi.org/10.1016/j.arabjc.2017.05.011

Khan MIR, Palakolanu SR, Chopra P, Rajurkar AB, Gupta R, Iqbal N, Maheshwari C (2021) Improving drought tolerance in rice: ensuring food security through multi-dimensional approaches. Physiol Plant 172:645–668. https://doi.org/10.1111/ppl.13223

Khota LR, Sankarana S, Majaa JM, Ehsania R, Schuster EW (2012) Applications of nanomaterials in agricultural production and crop protection: a review. Crop Prot 35:64–70

Kim YH, Khan AL, Waqas M, Lee IJ (2017) Silicon regulates antioxidant activities of crop plants under abiotic-induced oxidative stress: A review. Front Plant Sci 8:510–516. https://doi.org/10.3389/fpls.2017.00510

Kollist H, Zandalinas SI, Sengupta S, Nuhkat M, Kangasjärvi J, Fichman Y (2018) Rapid responses to abiotic stress: priming the landscape for the signal transduction network. Trends Plant Sci 24(1):25–37. https://doi.org/10.1016/j.tplants.2018.10.003

Konate A, He X, Zhang Z, Ma Y, Zhang P, Alugongo G, Li Y (2017) Magnetic (Fe3O4) nanoparticles reduce heavy metals uptake and mitigate their toxicity in wheat seedling. Sustain For 9(5):790. https://doi.org/10.3390/su9050790

Kuma A, Singh S, Gaurav AK, Srivastava S, Verma JP (2020) Plant growth-promoting bacteria: biological tools for the mitigation of salinity stress in plants. Front Microbiol 11:1216. https://doi.org/10.3389/fmicb.2020.01216

Kumar S, Prasad S, Yadav KK, Shrivastava M, Gupta N, Nagar S et al (2019) Hazardous heavy metals contamination of vegetables and food chain: role of sustainable remediation approaches—A review. Environ Res 179:108792. https://doi.org/10.1016/j.envres.2019.108792

Kurepa J, Paunesku T, Vogt S, Arora H, Rabatic BM, Lu J, Smalle JA (2009) Uptake and distribution of ultrasmall anatase TiO2 alizarin red S nanoconjugates in Arabidopsis thaliana. Nano Lett 10(6):2296–2302. https://doi.org/10.1021/nl903518f

Kwak SY, Wong MH, Lew TTS et al (2017) Nanosensor technology applied to living plant systems. Annu Rev Anal Chem 10(1):113–140. https://doi.org/10.1146/annurev-anchem-061516-045310

Lal R (2010) Managing soils and ecosystems for mitigating anthropogenic carbon emissions and advancing global food security. Bioscience 60:708–721

Latif A, Shakir AS, Rashid MU (2013) Appraisal of economic impact of zero tillage, laser land levelling and bed-furrow intervention in Punjab, Pakistan. Pak J Eng Appl Sci 13:65–81

Levard C, Hotze EM, Lowry GV, Brown GE Jr (2012) Environmental transformations of silver nanoparticles: impact on stability and toxicity. Environ Sci Technol 46(13):6900–6914. https://doi.org/10.1021/es2037405

Li D, Wang M, Zhang T, Chen X, Li C, Liu Y et al (2021) Glycinebetaine mitigated the photoinhibition of photosystem II at high temperature in transgenic tomato plants. Photosynth Res 147:301–315. https://doi.org/10.1007/s11120-020-00810-2

Liaqat W, Altaf MT, Barutçular C, Zayed EM, Hussain T (2023a) Drought and sorghum: a bibliometric analysis using VOS viewer. J Biomol Struct Dyn 42:1–13

Liaqat W, Yasmin S, Barutçular C, Altaf MT (2023b) Agronomic and molecular approaches: a solution to the water and environmental crisis. A solution to the crises of soil, water, and climate in plant production, 200

Liaqat W, Altaf MT, Barutçular C, Mohamed HI, Ali Z, Khan MO (2024a) Drought stress in sorghum: physiological tools, breeding technology, omics approaches and genomic-assisted breeding-A review. J Soil Sci Plant Nutr 24:1–27

Liaqat W, Altaf MT, Barutçular C, Nawaz H, Ullah I, Basit A, Mohamed HI (2024b) Ultraviolet-B radiation in relation to agriculture in the context of climate change: a review. Cereal Res Commun 52(1):1–24

Lindsey R, Dahlman L (2021) Climate change: global temperature. https://www.climate.gov/news-features/understanding-climate/climate-change-global-temperature. Accessed 15 Mar 2021

Llop J, Estrela-Lopis I, Ziolo RF, González A, Fleddermann J, Dorn M et al (2014) Uptake, biological fate, and toxicity of metal oxide nanoparticles. Part Part Syst Charact 31:24–35

Lobell DB, Schlenker W, Costa-Roberts J (2011) Climate trends and global crop production since 1980. Science 333:616–620

Lowry GV, Avellan A, Gilbertson LM (2019) Opportunities and challenges for nanotechnology in the agri-tech revolution. Nat Nanotechnol 14:517–522

Luyckx M, Hausman JF, Lutts S, Guerriero G (2017) Silicon and plants: current knowledge and technological perspectives. Front Plant Sci 8:411. https://doi.org/10.3389/fpls.2017.00411

Mahakham W, Sarmah AK, Maensiri S et al (2017) Nanopriming technology for enhancing germination and starch metabolism of aged rice seeds using phytosynthesized silver nanoparticles. Sci Rep 7:8263. https://doi.org/10.1038/s41598-017-08669-5

Mahato DK, Mishra AK, Kumar P (2021) Nanoencapsulation for agri-food applications and associated health and environmental concerns. Front Nutr 8:663229. https://doi.org/10.3389/fnut.2021.663229

Malhi GS, Kaur M, Kaushik P, Alyemeni MN, Alsahli A, Ahmad P (2020) Arbuscular mycorrhiza in combating abiotic stresses in vegetables: an eco-friendly approach. Saudi J Biol Sci 28:1465

Mandal D (2021) Nanofertilizer and its application in horticulture. J Appl Hortic 23:70–77

Marzbani P, Afrouzi YM, Omidvar A (2015) The effect of nano-zinc oxide on particleboard decay resistance. Maderas Cienc Tecnol 17:63–68

Mazhar MW, Ishtiaq M, Hussain I, Parveen A, Bhatti KH, Azeem M, Thind S, Ajaib M, Maqbool M, Sardar T et al (2022a) Seed nano-priming with zinc oxide nanoparticles in rice mitigates drought and enhances agronomic profile. PLoS One 17:e0264967

Mazhar MW, Ishtiaq M, Maqbool M, Akram R, Shahid A, Shokralla S, Al-Ghobari H, Alataway A, Dewidar AZ, El-Sabrout AM et al (2022b) Seed priming with iron oxide nanoparticles raises biomass production and agronomic profile of water-stressed flax plants. Agronomy 12:982

Mishra AK, Khanal AR, Pede V (2017) Economic and resource conservation perspectives of direct seeded rice planting methods: evidence from India. In Proceedings of the Agricultural and Applied Economics Association's 2017 AAEA Annual Meeting, Chicago, IL, USA, 30 July–2 August 2016

Mittal D, Kaur G, Singh P, Yadav K, Ali SA (2020) Nanoparticle-based sustainable agriculture and food science: recent advances and future outlook. Front Nanotechnol 2:10

Mody VV, Siwale R, Singh A, Mody HR (2010) Introduction to metallic nanoparticles. J Pharm Bioallied Sci 2(4):282. https://doi.org/10.4103/0975-7406.72127

Mubarik MK, Hussain K, Abbas G, Altaf MT, Baloch FS, Ahmad S (2022) Productivity of sorghum (sorghum bicolar L.) at diverse irrigation regimes and sowing dates in semi-arid and arid environment. Turk J Agric For 46(1):1–18

Mukhopadhyay SS (2014) Nanotechnology in agriculture: prospects and constraints. Nanotechnol Sci Appl 7:63–71

Munns R, Tester M (2008) Mechanisms of salinity tolerance. Annu Rev Plant Biol 59:651–681. https://doi.org/10.1146/annurev.arplant.59.032607.092911

Mustafa N, Raja NI, Ilyas N, Ikram M, Mashwani ZUR, Ehsan M (2021) Foliar applications of plant-based titanium dioxide nanoparticles to improve agronomic and physiological attributes of wheat (Triticum aestivum L.) plants under salinity stress. Green Process Synth 10:246–257. https://doi.org/10.1515/gps-2021-0025

Mwobobia EG, Sichangi AW, Thiong'o KB (2020) Characterization of wheat production using earth-based observations: a case study of Meru County, Kenya. Model Earth Syst Environ 6(1):13–25. https://doi.org/10.1007/s40808-019-00699-4

Nadeem MA, Nawaz MA, Shahid MQ, Dŏgan Y, Comertpay G, Yıldız M, Baloch FS (2018) DNA molecular markers in plant breeding: current status and recent advancements in genomic selection and genome editing. Biotechnol Biotechnol Equip 32(2):261–285. https://doi.org/10.1080/13102818.2017.1400401

Naguib DM, Abdalla H (2019) Metabolic status during germination of nano silica primed Zea mays seeds under salinity stress. J Crop Sci Biotechnol 22:415–423. https://doi.org/10.1007/s12892-019-0168-0

Naithani S (2016) Plants and global climate change: a need for sustainable agriculture. Curr Plant Biol 6:1

Namjoyan S, Sorooshzadeh A, Rajabi A, Aghaalikhani M (2020) Nano-silicon protects sugar beet plants against water deficit stress by improving the antioxidant systems and compatible solutes. Acta Physiol Plant 42(10):157

Narayanamoorthy A, Devika N (2017) Economic and resource impacts of drip method of irrigation on okra cultivation: an analysis of field survey data. J Land Rural Stud 6:15–33

Ning L, Riddle EE, Bradley RS (2015) Projected changes in climate extremes over the northeastern United States. J Clim 28:3289–3310

NOAA (2020) Earth System Research Laboratory (NOAA). Available online: www.esrl.noaa.gov. Accessed 15 Dec 2020

Noman M, Shahid M, Ahmed T, Tahir M, Naqqash T, Muhammad S et al (2020) Green copper nanoparticles from a native Klebsiella pneumoniae strain alleviated oxidative stress impairment of wheat plants by reducing the chromium bioavailability and increasing the growth. Ecotoxicol Environ Saf 192:110303. https://doi.org/10.1016/S10653-019-00317-3

Ogada MJ, Rao EJO, Radeny M, Recha JW, Solomon D (2020) Climate-smart agriculture, household income and asset accumulation among smallholder farmers in Nyando basin of Kenya. World Dev Perspect 18:100203

Open University (2016) Impacts of climate change in Ethiopia. Open Learn Works, March 24, 2016

Palmqvist NM, Seisenbaeva GA, Svedlindh P, Kessler VG (2017) Maghemite nanoparticles acts as nanozymes, improving growth and abiotic stress tolerance in Brassica napus. Nanoscale Res Lett 12:631

Pampolino MF, Manguiat IJ, Ramanathan S, Gines HC, Tan PS, Chi TTN, Rajendran R, Buresh RJ (2007) Environmental impact and economic benefits of site-specific nutrient management (SSNM) in irrigated rice systems. Agric Syst 93:1–24

Pan K, Zhong Q (2016) Organic nanoparticles in foods: fabrication, characterization, and utilization. Annu Rev Food Sci Technol 7:245–266. https://doi.org/10.1146/annurev-food-041715-033215

Parida AK, Das AB (2005) Salt tolerance and salinity effects on plants: a review. Ecotoxicol Environ Saf 60:324–349. https://doi.org/10.1016/j.ecoenv.2004.06.010

Pei S, Zhao J, Du J et al (2010) Direct reduction of graphene oxide films into highly conductive and flexible graphene films by hydrohalic acids. Carbon 48(15):4466–4474. https://doi.org/10.1016/j.carbon.2010.08.006

Pinedo-Guerrero Z, Cadenas-Pliego G, Ortega-Ortiz H, González-Morales S, Benavides-Mendoza A, Valdés-Reyna J, Juárez-Maldonado A (2020) Form of silica improves yield, fruit quality and antioxidant defense system of tomato plants under salt stress. Agriculture 10:367

Pouratashi M, Iravani H (2012) Farmers' knowledge of integrated pest management and learning style preferences: implications for information delivery. Int J Pest Manag 58:347–353

Prasad R (2014) Synthesis of silver nanoparticles in photosynthetic plants. J Nanopart 2014:963961

Rajbhandari R, Shrestha AB, Kulkarni A, Patwardhan SK, Bajracharya SR (2015) Projected changes in climate over the Indus river basin using a high resolution regional climate model (PRECIS). Clim Dyn 44:339–357

Rajput V, Minkina T, Ahmed B, Sushkova S, Singh R, Soldatov M et al (2019a) Interaction of copper-based nanoparticles to soil, terrestrial, and aquatic systems: critical review of the state of the science and future perspectives. Rev Environ Contam Toxicol 252:51–96. https://doi.org/10.1007/398_2019_34

Rajput V, Minkina T, Sushkova S, Behal A, Maksimov A, Blicharska E et al (2019b) ZnO and CuO nanoparticles: A threat to soil organisms, plants, and human health. Environ Geochem Health 42:147–158. https://doi.org/10.1007/S10653-019-00317-3

Rajput VD, Minkina T, Kumari A, Singh VK, Harish H, Verma KK, Mandzhieva S, Sushkova S, Srivastava S, Keswani C (2021a) Coping with the challenges of abiotic stress in plants: new dimensions in the field application of nanoparticles. Plan Theory 10(6):1221. https://doi.org/10.3390/plants10061221

Rajput VD, Singh A, Minkina T, Rawat S, Mandzhieva S, Sushkova S, Shuvaeva V, Nazarenko O, Rajput P, Komariah et al (2021b) Nano-enabled products: challenges and opportunities for sustainable agriculture. Plan Theory 10:2727

Raliya R, Biswas P, Tarafdar JC (2015) TiO2 nanoparticle biosynthesis and its physiological effect on mung bean (Vigna radiate L.). Biotechnol Rep 5:22–26. https://doi.org/10.1016/j.btre.2014.10.009

Ran Y, Liang Z, Gao C (2017) Current and future editing reagent delivery systems for plant genome editing. Sci China Life Sci 60(5):490–505. https://doi.org/10.1007/s11427-017-9022-1

Rana RA, Siddiqui MN, Skalicky M, Brestic M, Hossain A, Kayesh E et al (2021) Prospects of nanotechnology in improving the productivity and quality of horticultural crops. Horticulturae 7:1–19. https://doi.org/10.3390/horticulturae7100332

Rastogi A, Zivcak M, Sytar O, Kalaji HM, He X, Mbarki S, Brestic M (2017) Impact of metal and metal oxide nanoparticles on plant: A critical review. Front Chem 5:78. https://doi.org/10.3389/fchem.2017.00078

Renner A, Cadillo-Benalcazar JJ, Benini L, Giampietro M (2020) Environmental pressure of the European agricultural system: anticipating the biophysical consequences of internalization. Ecosyst Serv 46:101195

Rezaei D (2018) Nanotechnology, a step toward sustainable architecture. J Eng Appl Sci 13(16):6692–6699

Rizwan M, Ali S, ur Rehman MZ, Adrees M, Arshad M, Qayyum MF et al (2019) Alleviation of cadmium accumulation in maize (Zea mays L.) by foliar spray of zinc oxide nanoparticles and biochar to contaminated soil. Environ Pollut 248:358–367. https://doi.org/10.1016/j.envpol.2019.02.031

Sabaghnia N, Janmohammadi M (2015) Effect of nano-silicon particles application on salinity tolerance in early growth of some lentil genotypes. Annals of Maria Curie-Sklodowska University, Section C, Biology 69:39. https://doi.org/10.17951/c.2014.69.2.39

Sabir S, Arshad M, Chaudhari SK (2014) Zinc oxide nanoparticles for revolutionizing agriculture: synthesis and applications. Sci World J 2014:8

Sadeghzadeh B (2013) A review of zinc nutrition and plant breeding. J Soil Sci Plant Nutr 13:905–927

Sarraf M, Vishwakarma K, Kumar V, Arif N, Das S, Johnson R et al (2022) Metal/metalloid-based nanomaterials for plant abiotic stress tolerance: an overview of the mechanisms. Plan Theory 11(3):316. https://doi.org/10.3390/plants11030316

Sathaye J, Shukla PR, Ravindranath NH (2006) Climate change, sustainable development and India: global and national concerns. Curr Sci 90:314–325

Sedghi M, Hadi M, Toluie SG (2013) Effect of nano zinc oxide on the germination of soybean seeds under drought stress. Annals of the West University of Timisoara—Series of Biology 2:73–78

Seleiman MF, Al-Suhaibani N, Ali N, Akmal M, Alotaibi M, Refay Y et al (2021) Drought stress impacts on plants and different approaches to alleviate its adverse effects. Plan Theory 10(2):259. https://doi.org/10.3390/plants10020259

Shahid MA, Sarkhosh A, Khan N, Balal RM, Ali S, Rossi L et al (2020) Insights into the physiological and biochemical impacts of salt stress on plant growth and development. Agronomy 10(7):938. https://doi.org/10.3390/AGRONOMY10070938

Shahzad B, Tanveer M, Che Z, Rehman A, Cheema SA, Sharma A et al (2018) Role of 24-epibrassinolide (EBL) in mediating heavy metal and pesticide-induced oxidative stress in plants: A review. Ecotoxicol Environ Saf 147:935–944. https://doi.org/10.1016/j.ecoenv.2017.09.066

Sharif F, Westerhoff P, Herckes P (2013) Sorption of trace organics and engineered nanomaterials on to wetland plant material. Environ Sci: Processes Impacts 15(2):267–274. https://doi.org/10.1039/C2EM30613A

Sharifi M, Faryabi K, Talaei AJ, Shekha MS, Ale-Ebrahim M, Salihi A, Nanakali NMQ, Aziz FM, Rasti B, Hasan A et al (2020) Antioxidant properties of gold nanozyme: A review. J Mol Liq 297:112004. https://doi.org/10.1016/j.molliq.2019.112004

Sharma P, Jha AB, Dubey RS, Pessarakli M (2012) Reactive oxygen species, oxidative damage, and antioxidative defense mechanism in plants under stressful conditions. J Bot 2012:1–26. https://doi.org/10.1155/2012/217037

Siddiqui MH, Al-Whaibi MH, Firoz M, Al-Khaishany MY (2015) Role of nanoparticles in plants. In: Siddiqui MH, Al-Whaibi MH, Mohammad F (eds) Nanotechnology and plant science, 1st edn. Springer, pp 19–35

Singh VK, Shukla AK, Singh MP, Majumdar K, Mishra RP, Rani M, Singh SK (2015) Effect of site-specific nutrient management on yield, profit and apparent nutrient balance under predominant cropping systems of upper Gangetic Plains. Indian J Agric Sci 85:335–343

Singh S, Prasad SM, Sharma S, Dubey NK (2020) Silicon and nitric oxide-mediated mechanisms of cadmium toxicity alleviation in wheat seedlings. Physiol Plant 174:e13065. https://doi.org/10.1111/ppl.13065. Advance online publication

Singh H, Sharma A, Bhardwaj SK, Arya SK, Bhardwaj N, Khatri M (2021a) Recent advances in the applications of nano-agrochemicals for sustainable agricultural development. Environ Sci Process Impacts 23:213–239

Singh D, Sillu D, Kumar A, Agnihotri S (2021b) Dual nanozyme characteristics of iron oxide nanoparticles alleviate salinity stress and promote the growth of an agroforestry tree, Eucalyptus tereticornis Sm. Environ Sci Nano 8:1308–1325. https://doi.org/10.1039/D1EN00040C

Smith GB (2011) Nanostructured thin films IV—green nanotechnology. In SPIE Proceedings [SPIE SPIE NanoScience + Engineering – San Diego, California], vol. 8104, 2011, 810402. https://doi.org/10.1117/12.893114

Smith AM, Gilbertson LM (2018) Rational ligand design to improve agrochemical delivery efficiency and advance agriculture sustainability. ACS Sustain Chem Eng 6:13599–13610

Sohoulande Djebou DC, Singh VP (2016) Impact of climate change on the hydrologic cycle and implications for society. Env Soc Psychol 1(1):36–49. https://doi.org/10.18063/ESP.2016.01.002

Song G, Gao Y, Wu H, Hou W, Zhang C, Ma H (2012) Physiological effect of anatase TiO2 nanoparticles on Lemna minor. Environ Toxicol Chem 31:2147–2152. https://doi.org/10.1002/etc.1933

Soundararajan P, Sivanesan I, Jo EH, Jeong BR (2013) Silicon promotes shoot proliferation and shoot growth of Salvia splendens under salt stress in vitro. Hortic Environ Biotechnol 54:311–318. https://doi.org/10.1007/s13580-013-0118-7

Sousa A, Saleh AM, Habeeb TH, Hassan YM, Zrieq R, Wadaan MAM, AbdElgawad H (2019) Silicon dioxide nanoparticles ameliorate the phytotoxic hazards of aluminum in maize grown on acidic soil. Sci Total Environ 693:133636. https://doi.org/10.1016/j.scitotenv.2019.133636

Staroń A, Długosz O, Pulit-Prociak J, Banach M (2020) Analysis of the exposure of organisms to the action of nanomaterials. Materials 13:349. https://doi.org/10.3390/ma13020349

Subramanian KS, Karthika V, Praghadeesh M, Lakshmanan A (2020) Nanotechnology for mitigation of global warming impacts. In: Venkatramanan V, Shah S, Prasad R (eds) Global climate change: resilient and smart agriculture. Springer, Singapore, pp 315–336

Sun D, Hussain HI, Yi Z, Rookes JE, Kong L, Cahill DM (2017) Delivery of abscisic acid to plants using glutathione responsive mesoporous silica nanoparticles. J Nanosci Nanotechnol 18:1615–1625. https://doi.org/10.1166/jnn.2018.14262

Tariq M, Choudhary S, Singh H, Siddiqui MA, Kumar H, Amir A, Kapoor N (2021) Role of nanoparticles in abiotic stress. In: Ahmad F, Sultan M (eds) Technology in agriculture. InTechOpen

Tawfik MM, Mohamed MH, Sadak MS, Thalooth AT (2021) Iron oxide nanoparticles effect on growth, physiological traits and nutritional contents of Moringa oleifera grown in saline environment. Bull Natl Res Cent 45:177

Thomas S, Mishra PK, Talegaonkar S (2015) Ceramic nanoparticles: fabrication methods and applications in drug delivery. Curr Pharm Des 21(42):6165–6188. https://doi.org/10.2174/1381612821666151027153246

Tol RSJ (2012) On the uncertainty about the total economic impact of climate change. Environ Res Econ 53:97–116

Tol RSJ (2013) The economic impact of climate change in the 20th and 21st centuries. Clim Chang 117:795–808

Toshima N, Yonezawa T (1998) Bimetallic nanoparticles—novel materials for chemical and physical applications. New J Chem 22(11):1179–1201. https://doi.org/10.1039/A805753B

Tripathi RS, Raju R, Thimmappa K (2013) Impacts of zero tillage on economics of wheat production in Haryana. Agric Econ Res Rev 26:101–108

Tripathi DK, Singh VP, Prasad SM et al (2015) Silicon nanoparticles (Si np) alleviate chromium (VI) phytotoxicity in Pisum sativum (L.) seedlings. Plant Physiol Biochem 96:189–198. https://doi.org/10.1016/j.plaphy.2015.07.026

Tripathi M, Kumar S, Kumar A, Tripathi P, Kumar S (2018) Agro-nanotechnology: A future Technology for Sustainable Agriculture. Int J Curr Microbiol App Sci 7:196–200

Tyczewska A, Wozniak E, Gracz J, Kuczynski J, Twardowski T (2018) Towards food security: current state and future prospects of agrobiotechnology. Trends Biotechnol 36:1219–1229. https://doi.org/10.1016/j.tibtech.2018.07.008

Verma KK, Song XP, Joshi A, Tian DD, Rajput VD, Singh M, Arora J, Minkina T, Li YR (2022) Recent trends in nano-fertilizers for sustainable agriculture under climate change for global food security. Nano 12(1):173

Vij S, Tyagi AK (2007) Emerging trends in the functional genomics of the abiotic stress response in crop plants. Plant Biotechnol J 5(3):361–380. https://doi.org/10.1111/j.1467-7652.2007.00239.x

Wagan SA, Memon QUA, Wagan TA, Memon IH, Wagan A (2015) Economic analysis of laser land leveling technology water use efficiency and crop productivity of wheat crop in Sindh, Pakistan. J Environ Earth Sci 5:21–25

Wang S, Wang F, Gao S (2015) Foliar application with nano-silicon alleviates Cd toxicity in rice seedlings. Environ Sci Pollut Res 22(4):2837–2845. https://doi.org/10.1007/s11356-014-3525-0

Wang P, Lombi E, Zhao FJ, Kopittke PM (2016) Nanotechnology: A new opportunity in plant sciences. Trends Plant Sci 21(8):699–712. https://doi.org/10.1016/j.tplants.2016.04.005

Warburton ML, Schulze RE (2005) Detection of climate change: a review of literature on changes in temperature, rainfall and stream flow, on detection methods and data problems, pp 257–274

Warburton ML, Schulze RE, Jewitt GPW (2011) Hydrological responses to joint land use change and climate change in three diverse south African catchments. Submitted to global and planetary change

World Meteorological Organization (1992) International meteorological vocabulary, 2nd edn. Geneva, Switzerland, WMO

Xu DM, Fu RB, Liu HQ, Guo XP (2021) Current knowledge from heavy metal pollution in Chinese smelter contaminated soils, health risk implications and associated remediation progress in recent decades: A critical review. J Clean Prod 286:124989. https://doi.org/10.1016/j.jclepro.2020.124989

Ye Y, Cota-Ruiz K, Hernández-Viezcas JA, Valdés C, Medina-Velo IA, Turley RS et al (2020) Manganese nanoparticles control salinity-modulated molecular responses in Capsicum annuum L. through priming: A sustainable approach for agriculture. ACS Sustain Chem Eng 8(4):1427–1436. https://doi.org/10.1021/acssuschemeng.9b05615

Yousefi R (2022) The environmental crisis and nanotechnology. Micro Nanosyst 14(3):188–190

Zahedi SM, Moharrami F, Sarikhani S, Padervand M (2020) Selenium and silica nanostructure-based recovery of strawberry plants subjected to drought stress. Sci Rep 10:17672. https://doi.org/10.1038/s41598-020-74273-9

Zaveri E, Russ J, Damania R (2020) Rainfall anomalies are a significant driver of cropland expansion. Proc Natl Acad Sci USA 117:10225–10233

Zhang LW, Monteiro-Riviere NA (2009) Mechanisms of quantum dot nanoparticle cellular uptake. Toxicol Sci 110(1):138–155. https://doi.org/10.1093/toxsci/kfp087

Zhang YG, Nearing MA, Zhang XC, Xie Y, Wei H (2010) Projected rainfall erosivity changes under climate change from multimodel and multiscenario projections in Northeast China. J Hydrol 384:97–106

Zhang Q, Han L, Jing H, Blom DA, Lin Y, Xin HL et al (2016) Facet control of gold nanorods. ACS Nano 10:2960–2974

Zia R, Nawaz MS, Siddique MJ, Hakim S, Imran A et al (2021) Plant survival under drought stress: implications, adaptive responses, and integrated rhizosphere management strategy for stress mitigation. Microbiol Res 242:126626

Strategy, Global Warming, and Nanotechnologies

24

Murad Muhammad, Nimrah Ameen, Muqadas Batool, Sana Ullah, Abdul Basit, Wen Jing Ding, Li Li, and Wen-Jun Li

Abstract

This chapter begins with an overview of the importance of addressing global warming, setting the stage for a detailed examination of its causes, effects, and current strategies for mitigation. It emphasizes the critical need for sustainable solutions and proactive measures to combat the environmental impact. A significant portion of the chapter is dedicated to the role of nanotechnology in ecological applications, climate change mitigation, and sustainable resource management. It delves into nanotechnology's innovative solutions, particularly in renewable energy solutions, carbon capture and storage (CCS) using nanomaterials, and sustainable agriculture. The transformative potential of nanotechnology in addressing environmental challenges is highlighted, underscoring its role in shaping sustainable strategies for climate change mitigation.

M. Muhammad (✉)
Xinjiang Key Laboratory of Biodiversity Conservation and Application in Arid Lands, Xinjiang Institute of Ecology and Geography, Chinese Academy of Sciences, Urumqi, People's Republic of China

University of Chinese Academy of Sciences, Beijing, People's Republic of China

State Key Laboratory of Desert and Oasis Ecology, Key Laboratory of Ecological Safety and Sustainable Development in Arid Lands, Xinjiang Institute of Ecology and Geography, Chinese Academy of Sciences, Urumqi, People's Republic of China
e-mail: Muradbotany1@uop.edu.pk

N. Ameen
Department of Botany, University of Agriculture, Faisalabad, Pakistan

M. Batool
Department of Botany (Govt. Graduate College Layyah), Bahauddin Zakariya University, Multan, Pakistan

A. A. H. Abdel Latef et al. (eds.), *Bioremediation and Nanotechnology for Climate Change Mitigation*, https://doi.org/10.1007/978-981-96-3069-1_24

Furthermore, the chapter explores policy and governance strategies, technological innovations, and international collaborations as strategic responses to global warming. It emphasizes the importance of ethical considerations and societal implications in developing and implementing climate change strategies. The role of innovation in climate change mitigation is also examined, offering insights into emerging trends and future perspectives. In conclusion, the chapter summarizes key findings and issues a compelling call to action for sustainable strategies. It advocates for proactive measures, collaborative efforts, and innovative approaches to combat climate change, emphasizing the imperative of shaping a resilient and environmentally conscious future. The chapter serves as a valuable resource for understanding the intersection of strategy, global warming, and nanotechnology in addressing one of the most pressing challenges of our time.

S. Ullah
Department of Agricultural Leadership, Education & Communication, College of Agricultural & Environmental Sciences, University of Georgia, Athens, GA, USA

A. Basit
Department of Horticultural Science, Kyungpook National University, Daegu, South Korea

W. J. Ding
Xinjiang Key Laboratory of Biodiversity Conservation and Application in Arid Lands, Xinjiang Institute of Ecology and Geography, Chinese Academy of Sciences, Urumqi, People's Republic of China

University of Chinese Academy of Sciences, Beijing, People's Republic of China

Department of Biological Sciences, Shihezi University, Shihezi, People's Republic of China

L. Li
Xinjiang Key Laboratory of Biodiversity Conservation and Application in Arid Lands, Xinjiang Institute of Ecology and Geography, Chinese Academy of Sciences, Urumqi, People's Republic of China

State Key Laboratory of Desert and Oasis Ecology, Key Laboratory of Ecological Safety and Sustainable Development in Arid Lands, Xinjiang Institute of Ecology and Geography, Chinese Academy of Sciences, Urumqi, People's Republic of China

W.-J. Li
Xinjiang Key Laboratory of Biodiversity Conservation and Application in Arid Lands, Xinjiang Institute of Ecology and Geography, Chinese Academy of Sciences, Urumqi, People's Republic of China

State Key Laboratory of Desert and Oasis Ecology, Key Laboratory of Ecological Safety and Sustainable Development in Arid Lands, Xinjiang Institute of Ecology and Geography, Chinese Academy of Sciences, Urumqi, People's Republic of China

State Key Laboratory of Biocontrol, Guangdong Provincial Key Laboratory of Plant Resources and Southern Marine Science and Engineering, Guangdong Laboratory (Zhuhai), School of Life Sciences, Sun Yat-Sen University, Guangzhou, People's Republic of China
e-mail: liwenjun3@mail.sysu.edu.cn

Keywords

Climate change · Nanotechnology · Global warming · Environmental challenges

24.1 Introduction

Climate change is a complex issue with various observable consequences, such as unprecedented weather patterns, rising sea levels, extreme weather events, and increasing global temperatures (Chakraborty et al. 2014). However, other critics argue that the usage of fossil fuels has minimal influence on the root cause of global warming. Republicans prioritize more investments and advancements in fossil fuel technologies, while Democrats prioritize the implementation of legislation to combat climate change. This argument is particularly prominent in the United States. Consumer-funded levies are the primary funding source for renewable energy; however, the UK government also supports renewable energy through initiatives such as the Renewable Heat Incentive and Energy Innovation (Connor et al. 2015).

Climate change elicits varying sentiments across different countries, with a significant proportion of individuals in Greece and South Korea perceiving it as a substantial menace (Majeed et al. 2024). The level of acceptance of climate change differs among nations such as the United States, Canada, South Africa, Australia, and the United Kingdom (Sarkodie and Strezov 2019). Factors such as the abstract nature of the idea, the absence of definitive evidence linking carbon dioxide (CO_2) and greenhouse gases (GHGs), the long-standing difficulties in reducing CO_2 and GHGs caused by human activities, the influence of powerful coal, oil, and gas industries, resistance to change and its potential disruptions, deeply held beliefs, and political affiliations contribute to the denial of climate change (Aresta and Dibenedetto 2021). Fossil fuel corporations have provided substantial financial support to conservative think tanks and charities that promote anti-climate change perspectives and actively spread uncertainty through sponsored publications and advertisements. Disinformation is also propagated via social media, as specific individuals erroneously believe that if anything is published in print or other forms of media, it is automatically accurate. The debate surrounding climate change is complex and multifaceted. The Earth is experiencing an increase in temperature due to the release of greenhouse gases caused by human activities, notably the Industrial Revolution. CO_2 is the primary greenhouse gas emitted, and it can capture and retain infrared energy (Yoro and Daramola 2020). Global warming is responsible for climate change, increased heat waves, flooding, salinity, and freezing stresses. The supply of high-quality arable land is diminishing due to the rise in the global population and the expansion of residential and commercial land utilization. The increasing population and escalating demand for freshwater are concurrently diminishing the water quality utilized for agricultural irrigation (Islam and Karim 2019).

The number of manufactured and environmental contaminants is increasing, encompassing diesel and combustion particles, heavy metals, microplastics, pesticides, herbicides, antibiotics, persistent organic pollutants, and tropospheric ozone

(Contreras Llin and Díaz-Cruz 2023). These pollutants can potentially increase levels of ultraviolet (UV) radiation, disturb the ozone layer in the stratosphere, and change the pH and salinity of soil. These environmental variables not only increase the vulnerability of plants to diseases and pests but also diminish the overall health of forest ecosystems and the pollination process facilitated by insects. With the increase in average global temperatures, there is an expectation of more occurrences of abiotic stress, which could threaten food security and global production (Calanca 2017).

Nanotechnology has great potential to solve global warming and climate change. This chapter examines nanotechnologies' strategic role in climate mitigation and adaptation. The goal is to explore how nanoscale technologies can be used to provide creative global warming solutions. Nanotechnology is being studied for renewable energy generation, carbon capture and storage (CCS), water purification, and sustainable material creation. The chapter will also examine nanotechnology research and development (R&D) trends related to climate change, identifying emerging technologies and assessing their viability, scalability, and environmental sustainability. Also necessary, the chapter will explore the strategic, ethical, regulatory, and societal implications of nanotechnology for climate change. This requires addressing safety, environmental effects, and equal access to sophisticated technologies. Nanotechnology offers precise tools for addressing pressing global challenges like climate change while maintaining biosafety standards. Its applications promise cost-effective solutions, harnessing economic value while mitigating environmental threats, making it a strategic asset in our fight against the climate crisis. This chapter addresses these multifaceted goals to thoroughly explain how nanotechnology's future might be strategically used to tackle global warming and climate change.

24.2 Understanding Global Warming

Climate change dramatically influences the occurrence of extreme weather phenomena such as hurricanes, floods, droughts, heatwaves, cold snaps, windstorms, and tornadoes. However, compound events have occurred in recent decades. These include dry-hot conditions, temperature and relative humidity events, and flooding caused by significant rainfall and storm surges. These events have significantly impacted both the social fabric and natural environments. Statistical dependence among many variables is a crucial characteristic of compound events. Research suggests that failing to consider the connection between contributing variables can lead to inaccurate estimations of the probability of compound events (Zscheischler et al. 2018). Global warming can influence the complex interaction of multiple factors contributing to compound catastrophes. Ongoing research examines the temporal evolution of the relationships between storm surge, soil moisture, and vapor pressure deficit (VPD). Modifications in interdependence can impact the occurrence rate of compound occurrences, thus influencing their level of risk concerning global warming. Hence, it is crucial to integrate the factors contributing to dependency, the correlation between the probabilities of compound occurrences, and the impact of modifying dependency on risk evaluations (Zscheischler et al. 2018).

24.2.1 Physical Processes Leading to Dependence

Multiple physical processes influence compound events. These include everyday external stimuli, feedback from the system, and conditional dependency. Examples of typical external influences include a region's vulnerability to climate change, circulation patterns, and natural cycles like the El Nino-Southern Oscillation (ENSO) (Zebiak et al. 2015). Coastal areas are at risk of simultaneous floods due to these variables' effects on heavy rainfall and high tides. During warm seasons, precipitation and temperature may negatively correlate due to significant teleconnection patterns like ENSO, which are strongly associated with simultaneous changes in both variables. Compound occurrences can also be caused by system feedback, such as when dry and hot circumstances happen simultaneously. This occurrence becomes more apparent when there is a change from a wet to a dry climate. Droughts are more likely to occur in these regions because soil moisture levels are low, reducing evapotranspiration and evaporative cooling. Conditional dependencies allow one variable to affect another's likelihood of compound events (Zhu et al. 2021).

24.2.2 Compound Flooding Events

Compound flooding in estuaries and deltas results from several meteorological, hydrologic, and oceanic factors such as wind speed, precipitation, river flow, storm surge, tide, waves, and sea level. The underestimation of joint exceedance probability for river discharge and storm surge is caused by the interdependence of both components, which consequently affects the combined likelihood of exceedance. Accurately measuring the key variables' interrelationship is crucial to successfully apply drainage and prevention techniques. The interconnectedness of the contributing elements is influenced by estuary characteristics, catchment circumstances, and large-scale climate systems (Van Niekerk et al. 2020). When assessing the possibility of compound flooding, it is crucial to meticulously account for the time gap between rainfall and storm surge. Flooding, also called compound flooding, can happen when intense precipitation is succeeded by saturated soil.

There are several scientific grounds for disastrous floods in Pakistan, Yemen, the UAE, Oman, Germany, and Iran in 2023 (Lahn and Shapland 2022). River overflow from high monsoon rains in Pakistan, exacerbated by climate change, destroyed infrastructure and agricultural land. In Yemen, cyclones and tropical storms fueled by climate change-driven sea surface temperatures caused widespread home destruction and community displacement in 2024. In 2024, urbanization's urban heat island effect and climate change's impact on extreme weather events caused urban flooding and transport disruption in the UAE (Almulhim et al. 2022). Tropical cyclones caused excessive rainfall in Oman in 2024 due to higher waters and atmospheric moisture levels, damaging coastal areas and infrastructure. In 2024, climate change increased the atmosphere's moisture-holding capacity, producing heavy rains and severe flooding in Germany, destroying homes and infrastructure and killing people. In 2024, severe rainfall, deforestation, urbanization, and climate change caused flash floods in Iran, damaging agricultural land and infrastructure and displacing residents (Kumar 2023). Discussing these floods' scientific causes and effects may dispel misunderstandings and improve understanding of climate change and extreme weather events across geographical and socioeconomic borders.

24.3 Causes and Effects of Global Warming

Hot weather may be fatal for anyone without a good way to regulate their body temperature, so we evolved with sweat glands, the capacity to walk upright on two legs, and naked skin. A common misconception is that severe weather is the only cause of heat- and cold-related deaths; in reality, preexisting conditions like heart disease and lung disease are more often to blame (Engler 2019). To understand mortality rates, one must have precise definitions of cold spells and heat waves. Climate change's direct and indirect impacts on health include harm to animals and primary producers, which in turn causes malnutrition and other health issues.

Environmental pollution is a known contributor to global warming, which has been associated with an increase in harmful health effects. Among these are the following: the proliferation of vector-borne infections, problems with breathing and allergies, occurrences of mental illness, and acts of group aggression (Upadhyay 2021). There are inconsistencies in the statistics about the relationship between climate change and mortality, and one primary reason for this is that heat waves are not universally defined. Heatwaves are described as prolonged periods of extremely high temperatures; however, the precise meaning of this term varies from one area, geography, climate, and population to another. Extreme heat and humidity have occurred over twice as often in the last 40 years, with some coastal subtropical regions reaching a maximum wet-bulb temperature of 35 °C (Coffel et al. 2017). All nations, from the wealthiest to the poorest, are not immune to the devastating effects of heatwaves on death rates. People in areas with moderate temperatures, neither too hot nor too cold, may be more vulnerable to heat waves than those with extremely hot or freezing temperatures. But there is a limit to how much heat people can tolerate before it becomes a significant health concern, with the risk of death increasing dramatically at very high temperatures.

A cold spell is defined as an extended period of very low temperatures or fast temperature drops, which forces people to take drastic precautions to protect their homes, businesses, and communities from the effects of the weather. According to official meteorological standards, a cold spell is defined as a reduction in temperature of at least 8 °C in 24 h, 10 °C in 48 h, or 12 °C in 72 h, with a daily minimum temperature of less than 4 °C (Liu et al. 2021). The phrase is vague and might not have broad national and worldwide applicability owing to temperature variations and different types of climates. Temperatures between the 2.5th and 25th percentiles are the most common causes of cold spells, seriously affecting public health (Sun et al. 2022). Cardiovascular morbidity due to cold is more common in older people, young adults, and middle-aged people. Newborns, children, pregnant women, the elderly, those with chronic illnesses, outdoor laborers, the economically disadvantaged, and city dwellers are at increased risk for heat-related illnesses.

24.3.1 Impact on the Environment and Ecosystems

Since the Industrial Revolution, global greenhouse gas emissions have significantly increased due to human activities in energy production, manufacturing, and car emissions (Malik et al. 2016). The alterations in climate have resulted in extreme weather events such as droughts, floods, and deserts. These occurrences have played a role in climate change and threaten ecosystems, water resources, biodiversity, and resilience. The Earth's average surface temperature has exhibited a consistent upward trend since the 1950s, and this pattern is projected to persist from 1980 to 2099 (Cai et al. 2021). Projections indicate that global sea levels are anticipated to increase by 0.18 to 0.59 meters between 1980 and 1999 (Simpson et al. 2014). As a result, the response of mangroves to this sea-level rise is likely to become more pronounced. In addition, atmospheric CO_2 concentrations have increased, reaching unprecedented levels in the new millennium.

Mangroves and human lives and property are significantly endangered by typhoons, windstorms, floods, heatwaves, cold spells, forest fires, and other forms of extreme weather events (Wang and Gu 2021). The reduction in the size and vitality of mangroves will increase risks to human safety and the safeguarding of coastal areas from potential hazards. Mangrove swamp plants are essential components of ecosystems that serve a vital role in mitigating coastal erosion (Kathiresan 2021). They achieve this by effectively filtering nutrients from the ground and as a protective barrier against waves, acting as a buffer between the ocean and the land. Poor environmental management can lead to pollution from various sources, such as households, farms, or urban areas. In addition, mangroves play a crucial role in mitigating the adverse impacts of climate change globally (Muhammad et al. 2023). Mangroves are very productive in biomass generation, are dependable carbon sinks, and play a crucial role in CO_2 production within coastal ecosystems. Most studies investigating the relationship between mangroves and climate change have focused on indoor simulations and remote sensing techniques (Gu et al. 2022). The global decline in plant diversity has raised concerns about its impact on ecosystem function and the provision of products and services to humans. Despite the focus of most biodiversity experiments on plant carbon sequestration (CS), there is a notable absence of understanding regarding the impact of plant variety on soil carbon storage. Soil organic carbon (SOC) is a crucial natural resource that plays a vital role in controlling the pace of climate change, guaranteeing a consistent food supply, and sustaining a robust ecosystem. Minor changes in SOC can significantly affect the global carbon cycle, as soils store three times more organic carbon than the combination of air and plants (Gerke 2022).

The impact of plant species diversity on SOC might vary depending on the study, with potential outcomes ranging from upbeat and neutral to adverse effects (Xiong et al. 2016). Plant diversity can either increase or decrease the gains and losses of soil organic carbon (SOC) in mineral soils. The balance between the carbon influx from plant litter and the carbon efflux from the microbial breakdown of plant inputs and preexisting soil organic matter determines the fluctuations in soil organic carbon. Plant species combinations can affect SOC production and accumulation by altering plant litter's decomposition and transformation, both above and below the ground.

Species-rich ecosystems tend to promote the carbon sequestration of live plants due to their diverse resource utilization and plant variety. Enhancing biomass output in species-rich communities can boost the input of plant litterfall and root materials into soils, resulting in a direct increase in SOC through the accumulation of resistant plant components. Due to the presence of a more significant amount of short-lived, superior plant debris easily decomposed by soil microbes, an increase in biomass production might indirectly result in an elevation of SOC levels through the accumulation of dead microbial biomass (Gross and Harrison 2019).

Soil microbial biomass carbon (SMBC) significantly impacts the management of SOC loss and sequestration (Mahajan et al. 2018). Although constituting a small portion of SOC, it plays a crucial role in enhancing the availability of carbon and nutrients to soil microorganisms in species-rich ecosystems. Unlike monocultures, which offer low-quality substrate inputs that are difficult for microbes to break down and support a smaller soil microbial biomass carbon (SMBC), plant communities with a high diversity of species can improve nitrogen (N) utilization efficiency by producing biomass with higher carbon-to-nitrogen (C:N) ratios (Prommer et al. 2020).

The influence of various species' roles on primary productivity, soil resident microorganisms, and litter decomposition leads to an increase in the impact of plant diversity on SOC and SMBC as the species mixtures become more diverse (Chen et al. 2020). The effect on ecosystem function is mainly determined by the diversity of functional categories rather than the amount of species. The influence of plant mixtures on soil microorganisms and the SOC they generate through chemically and physically varied plant debris may be intensified by distinct functional traits.

The increase in interspecific complementarity leads to a multiplication effect on biomass output in microorganisms and plants, as influenced by plant diversity (Scherer-Lorenzen et al. 2003). The accumulation of microbial nanomass and the addition of plant waste inputs can contribute to the temporal rise in SOC. Differences in vegetation physiology, structure, and lifespan result in varying links between SOC and soil microbial biomass carbon (SMBC) diversity and productivity. These connections, in turn, cause SOC and SMBC responses to plant variety to vary depending on the surrounding environment. In a study by Jenkins in 2002, differences in vegetation physiology, structure, and lifespan influenced varying links between soil organic carbon (SOC) and soil microbial biomass carbon (SMBC) diversity and productivity. These connections result in diverse responses of SOC and SMBC to plant variety, with soil organic carbon levels ranging from 1.5 to 5% of the soil's total mass and soil microbial biomass carbon typically accounting for about 1–5% of the total carbon pool in soil (Jenkins 2002).

24.4 Current Strategies for Mitigating Global Warming

Weather Morph: Climate Change Weather File Generator, CC World Weather Gen, and Weather Shift are just a few examples of weather generator software used in recent studies to generate climate data for the future (Jentsch et al. 2013). Energy consumption simulations often use these applications to examine historical weather files. The climatological data in these files spans over a decade and includes hourly readings for various climate factors. Then, to create future weather data for programs that mimic building energy consumption, these files are subjected to several calculations using a morphing technique. A study examining climatic predictions for several locations found that the north will benefit from milder winters in the year 2050. However, cooling energy will be in high demand across Europe, particularly in southern countries where summers can get hot and air conditioning is not enough (Horta et al. 2019). Research from Italy and Greece predicts that by 2050, cooling requirements for buildings in Mediterranean climes will have increased by 15–30% (Georgopoulou et al. 2024). Buildings in the Spanish cities of Valladolid and Valencia have had their impact on climate change evaluated. A decrease in air leakage, an increase in thermal insulation, a change in the window-to-wall ratio, the installation of sunshades, and the optimization of the Heating, Ventilation, and Air Conditioning (HVAC) system's performance could all help reduce the projected increase in cooling energy demand (Ascione 2017).

These principles will be more precise in the future thanks to technological advancements. All four models (the first two, third, and fourth) were subject to the mitigating strategies. The mitigation measures included installing paneling systems or Sustainable Agriculture and Technology Education (SATE)-type façades, regulating the air quality entering the house with CO_2 sensors, and changing the thermal insulation of the building envelope. The thermal insulation was improved with a U-value of 0.2 W/m^{2}K for the opaque envelope and 0.5 W/m^{2}K for the transparent one (Dhalla and Eng 2015). Compared to the transparent envelope, whose thermal insulation was reduced to 1.5 W/m^{2}K, the opaque envelope's insulation was decreased to 0.4 W/m^{2}K (Kienzl 1999).

24.5 Nanotechnology and Environmental Applications

The issue of global warming is currently a challenging challenge being addressed by scientists and governments worldwide. It seems probable that the Earth's average surface temperature will increase by 1.5–2 °C over the next 40–50 years (Karmalkar and Bradley 2017). If this tendency continues, Earth will become uninhabitable within the next century. Nanotechnology has various applications in mitigating the effects of climate change. Extensive research has been conducted on nanostructured materials due to their ability to capture and store greenhouse gases. The materials include metal-organic frameworks (MOFs), nanoporous carbonaceous materials, nano silica, nano zeolites, functionalized nanomaterials, and nanocomposites (Kaneti et al. 2017). Nanomaterials (NMs) can enhance material interactions, facilitate storing and transporting renewable energy fuels, absorb greenhouse gases, and possess a high surface area per unit volume. Nanocomposites, lightweight transportation materials, reduce the need for conventional fossil fuels, mitigating climate change's pace. Using nanocatalysts has two advantages: decreased fuel consumption and reduced emissions of greenhouse gases. These catalysts can retain oxygen and promote the thorough combustion of fuels. Nanomaterial-based lubricants and coatings significantly reduce engine friction and wear, resulting in a 2% drop in fuel consumption and carbon dioxide emissions (Pownraj and Valan Arasu 2021).

24.5.1 Sustainable Water Supply: Provide Clean Water for the Planet

The United States and other regions globally face the difficulty of providing safe drinking water for human consumption and purified water for diverse sectors (Jackson et al. 2001). The water demand is rising as the population grows. Still, water supplies are being strained by escalating contamination, salinization, groundwater depletion, and snowpack loss caused by climate change. Aquifer salinization is a burgeoning problem in the Gulf Coast, southern Atlantic, and Pacific coasts due to the escalating release of contaminants and nutrients from surface runoff. According to the United Nations Environment Program, there will be a decrease in freshwater availability in numerous locations, including Sub-Saharan Africa, the Middle East and North Africa, and South Asia by 2020. Nanotechnology can offer practical, economical, and environmentally sustainable solutions for providing drinkable water for human use and clean water for agricultural and industrial needs (Dasgupta et al. 2017).

24.5.2 Food Security and Sustainability: Feed the Planet

The United States spent $1165 billion on food consumption in 2008, according to the United States Department of Agriculture (USDA) in 2009 (Lee and Kilmer 2010). Food-borne illness is another expense associated with food, with an estimated cost of $152 billion annually in the United States (Scharff 2015). The globe will encounter significant difficulties in satisfying the worldwide need for sufficient food in the next four decades since the global population is projected to reach around 9 billion by 2050 (Gouel and Guimbard 2019). Effective resolutions to this predicament will necessitate a profound overhaul of agriculture, involving cultivating a greater quantity of food while simultaneously reducing the ecological footprint of the agriculture and food sectors. Additionally, it will be imperative to address the consequences of global climate change and guarantee the safety and security of the food supply. Progress in nanotechnology can significantly enhance the technologies employed for cultivating, processing, preserving, and transporting food (Nile et al. 2020).

24.6 Introduction to Nanotechnology

Nanotechnology refers to the capacity to construct materials, electronics, and systems with utmost precision at the atomic level (Nasrollahzadeh et al. 2019). Nanotechnology is the capability to manipulate matter at the atomic and molecular levels to construct complex structures with novel molecular arrangements. The National Science and Technology Council provides this definition. The objective is to harness these characteristics by attaining command over structures and devices at the atomic, molecular, and supramolecular scales and acquiring the knowledge and skills necessary to produce and utilize these devices. Nanoscience/nanotechnology, defined by the United States National Science Foundation, refers to the study of materials and systems with specific vital qualities.

1. Dimension: The size must be between 1 and 100 nanometers (nm).
2. Process: Formulated using approaches that exhibit fundamental mastery of structures' physical and chemical characteristics at the molecular level.
3. Building block property: They can be merged to create more extensive formations. Nanoscience is inherent to microbiological sciences because many bioparticles, such as enzymes and viruses, have sizes that fall within the nanometer range (Chehroudi 2006). The potential and fundamental nature of nanoscale science and technology lies in the proven observation that materials at the nanoscale exhibit distinct properties (such as chemical, electrical, magnetic, mechanical, and optical) that differ significantly from those of bulk materials. Certain qualities of these entities exhibit characteristics between those of individual atoms and molecules that make up these entities and the attributes of larger-scale materials. Studies have shown that nanoparticles have superior performance characteristics to bulk materials in identical applications. Nanotechnology has several current and anticipated applications in various fields, such as biology, health, pharmaceuticals, electronics, energy, and environmental industries. These applications primarily include bottom-up approaches like self-assembly. The use of nanotechnology in these areas is quickly growing. Global issues include over 2 billion people without safe drinking water World Health Organization/United Nations International Children's Emergency Fund(WHO/UNICEF), 690 million with chronic hunger Food and Agriculture Organization (FAO), and 2 billion with hidden hunger, lacking essential vitamins and minerals (Hendriks et al. 2023). In 2020, 155 million people were acutely food insecurity World Food Programme (FAO/WFP) (Dlamini 2020). Many places lack science, education, health, and disease management resources and infrastructure. Despite these obstacles, studies show that nanoparticles outperform bulk materials in identical applications (Baig et al. 2021). Nanotechnology has potential in biology, health, drugs, electronics, energy, and the environment. To meet global needs, bottom-up applications like self-assembly are growing rapidly.

24.7 Role of Nanotechnology in Climate Change Mitigation

Innovations in nanotechnology and nanobiotechnology are improving environmental conditions. Several ecological problems, such as wastewater treatment, fuel shortages, greenhouse gas emissions, and the removal of various substances contributing to global warming, have been proposed as viable solutions by nanotechnology (Khan et al. 2021). Nanotechnology may emerge as the preeminent tool for combating climate change, and its use in environmental contexts is on the rise. Table 24.1 presents a concise summary of how nanotechnology contributes to climate change mitigation. The table includes information about the application, description, nanotechnology contribution, and associated benefits.

24.7.1 Nanotechnology in Environmental Remediation/ Nano Bioremediation

Pollution is a significant contributor to the phenomenon of climate change. Climate change, acid rain, ozone depletion, and a slew of diseases (such as cancer, respiratory and skin ailments, and others) are only a few of the world's repercussions of air, water, and soil pollution (Gates 2023). Environmental nanotechnology (E-nano) products could be utilized in ecological cleanup initiatives because nanotechnology involves developing, altering, and describing structures, tools, and systems at the nanoscale. The following sections discuss nanotechnology methods to clean up polluted areas and remove dangerous compounds.

Table 24.1 Overview of nanotechnology's role in climate change mitigation

Application	Description	Nanotechnology contribution	Benefits	References
Renewable energy	Enhanced solar cells and batteries	Nanomaterials improve efficiency and reduce costs	Increased adoption of renewable energy	Khalid et al. (2021)
Energy efficiency	Lightweight materials and intelligent coatings	Nano-engineered materials improve energy efficiency	Reduced energy consumption and emissions	Karali et al. (2014)
Carbon capture and storage (CCS)	Efficient CO_2 capture and conversion	Nanoporous materials and catalysts enable cost-effective CCS	Decreased carbon emissions from industries	Muhammad et al. (2024)
Water purification and desalination	Advanced filtration and desalination techniques	Nanomembranes increase efficiency and reduce energy consumption	Improved access to clean water resources	Gude (2017)

24.7.2 Bioremediation of Heavy Metals and Other Pollutants from Wastewater

For all forms of life on Earth, water is a necessary component. However, getting safe drinking water has become more difficult in the current climate. Seawater, polluted freshwater, brackish water, stormwater, and wastewater are some of the unconventional resources that have been utilized to address the need for clean water as a result of the scarcity of these resources (Karimidastenaei et al. 2022). However, purifying this water to a usable state is no easy task, because traditional approaches to water and wastewater treatment have their limitations. Thus, researchers are currently concentrating on developing nanotechnology approaches that are more effective, cutting-edge, and multipurpose in purifying water (Tlili and Alkanhal 2019).

Water treatment procedures made possible by nanotechnology include adsorption, membrane separation, photocatalysis, disinfection, monitoring, and sensing, all of which use a variety of nanomaterials to destroy these contaminants (Majeed et al. 2024).

24.7.3 Greenhouse Gas Sequestration (GHGS)

As they absorb and emit radiation in the thermal infrared region, greenhouse gases raise the average global temperature, which in turn causes global warming. Industrial greenhouse gases such as perfluorocarbons (PFCs), hydrofluorocarbons (HFCs), sulfur hexafluoride (SF6), and atmospheric greenhouse gases like CO_2, nitrous oxide (NOx), ozone, methane, and volatile organic compounds (VOCs) are known to contribute to climate change (Tohka 2005). Pollutants from greenhouses have long-lasting adverse effects on ecosystems, human health, and the ozone layer. Below are some of the tactics used by GHGS to lessen the impact of global warming.

- Using non-conventional energy resources to minimize fossil fuel usage.
- Carbon management and greenhouse gas sequestration.
- Increase the efficiency of existing technologies to reduce greenhouse gas emissions.

Nanotechnology possesses the capacity to fundamentally transform technology through the development of ecologically sound and enduring remedies to address climate change and its ramifications (Aithal and Aithal 2022). Nanomaterials surpass conventional materials in several environmental tests due to their remarkable electrical, mechanical, magnetic, and optical properties, as well as their larger pore volume, greater surface area, and higher number of surface functional groups. Promising potential lies in the utilization of functionalized nanomaterials, including films, composites, fibers, MOFs, membranes, carbon nanotubes (CNTs), zeolites, nano silica, and others, for the collection of greenhouse gases (Shen et al. 2017).

24.7.4 Nanotechnological Applications in Renewable Energy Generation

The primary energy sources are fossil fuels, including diesel, gasoline, and petrol. Nevertheless, environmental degradation and global warming have been caused by their significant production of greenhouse gases. The best option for meeting energy needs is renewable or alternative energy sources. Green energy refers to the energy that comes from renewable sources, such as the sun, the wind, the ocean, geothermal heat, biofuels (such as bioethanol, biodiesel, biogas, and biohydrogen), and other unconventional sources (Demirbas and Demirbas 2016). When it comes to combating climate change and the energy crisis, these resources are invaluable.

Catalyzing the hydrolysis of lignocellulosic biomass is a typical application of nanomaterials, whereas immobilized enzymes are used to improve the efficiency of biofuel synthesis (Singhvi and Kim 2020). Matrixes composed of magnetic nanoparticles or metal oxide have been connected to enzymes, including cellulases, laccases, and hemicellulases. The enhanced effectiveness of nanocatalysts, complexes of nanomaterials and enzymes, is well known.

24.8 Policy and Governance Strategies

The term "governance" describes the change from a top-down legislative model to a bottom-up system model that promotes self-regulation and helps achieve larger objectives (Pagallo et al. 2019). A change from restrictive to facilitating forms of regulation is seen in updating policy frameworks. Any parties, regulations, standards, procedures, and tools that gather, analyze, and disseminate pertinent risk data are all part of risk governance. This is of utmost importance when stakeholder cooperation is necessary to mitigate risk. Institutional frameworks and sociopolitical culture are contextual elements that risk governance considers (Renn 2017). Assessing and managing the consequences of nanotechnology—which promises more minor, lighter, faster, and more resource-efficient devices—require strong governance and risk governance. People have a negative impression of businesses and governments due to a lack of trust in both sectors and worries about potential disruptions. Misuse or ineffective implementation of nanotechnology might result from a lack of formal and informal instruction (Miller et al. 2004).

24.9 Technological Innovations and Research

Currently, the measurements and evaluations of nanomaterials in the workplace and environment are inadequate; whereas some concerns and exposures are being addressed, others are being neglected (Pietroiusti et al. 2018). Specialized metrology techniques designed explicitly for nanoscale measurements are necessary to transport nanoparticles effectively using various methods. This is remarkably accurate in the realms of the environment and medicine. The presence of nanoparticles requires a reassessment of existing approaches for evaluating risks and implementing regulatory restrictions. Nanotechnology governance processes require specialized knowledge in sustainability and Environmental Health and Safety (EHS) (Sargent 2008). The risk management of nanotechnology is facilitated by a fragmented structure of national and international regulatory frameworks characterized by both gaps and overlaps. Organizations advocating for the public's welfare seek unbiased entities that will be more transparent in sharing information and test results. Using second- to fourth-generation nanotechnology requires a proactive and corrective approach with adaptive management for systems facing difficulties. The absence of detailed scenarios for generations two through four of nanoproducts hinders our ability to effectively address their possible long-term effects on human development. A significant deficiency in the current governance of nanotechnology risks is the lack of effective coordination among many players and stakeholders, including research institutions, industry, consumers, government regulators, civil society, and international entities (Sarma 2011).

Regulatory ambiguity hampers innovation in the industrial sector, especially for smaller enterprises. The insurance risk transfer procedure is hindered by a gap caused by the challenge of accurately assessing the risk profile of nanotechnology firms. Cognitive flaws such as overconfidence, false consensus, and status quo bias can impact risk management. The use of nanotechnology in developing new weapons is a sensitive matter due to the secrecy surrounding the technology, the wide range of potential applications, and the unforeseen consequences that may arise (Nasu and Faunce 2009). The current nanotechnology treaties are insufficient to adequately address the broader issues of humanity, particularly those about the well-being of future generations, including the third and fourth generations.

24.9.1 International Collaborations and Agreements

International collaboration between scientists has gained increasing prominence in recent years. Researchers' ability to travel internationally, the prevalence of Internet communication, shared data and research tools, formal international collaborations, transnational social capital links, and funding availability have all contributed to increased international scientific cooperation. Emerging nations like China, India, and Brazil, as well as more developed nations like the United States, engage in extensive transnational scientific collaboration (Chan and Costa 2005). For developed and developing economies alike, the US–China scientific partnership is fascinating. The United States is still at the top of most scientific domains, but other countries, especially those in Asia and Europe, have improved in research and development. Between the mid-1990s and the mid-2000s, China's research and development (R&D) spending as a proportion of Gross Domestic Product (GDP) and the number of researchers more than doubled, greatly amplifying the country's scientific influence in the last decade (Heitor and Bravo 2010). The percentage of Chinese-published papers included in the 2007 Web of Science citation indices, encompassing scientific and social science publications, was 7.5%. With this, China surpassed Japan's 7.0% and took second place (Ambekar 2022). There is growing consensus that China's rise in scientific fields is significant and that the US–China relationship is paramount worldwide. As a result of its rapid rise to prominence, China is now one of the few nations capable of performing cutting-edge research and development in the field of nanotechnology. From China's perspective, multiple outcomes can be achieved through international scientific partnerships with the United States. Working together, leading US and Chinese researchers in emerging fields like nanotechnology can improve China's scientific capacity by gaining access to state-of-the-art information and methods (Tang 2011). Emphasizing global cooperation, however, may hinder China's ability to invest in nanotechnology R&D for commercial use.

24.10 The Intersection of Nanotechnology and Climate Change

Many believe nanotechnology could solve many pressing global problems, including pollution, overpopulation, climate change, resource loss, social inequality, economic inequality, and urbanization (TRENDS G 2017). Though sustainability issues are complex, optimism and certainty usually win out. The new discipline of sustainability science has laid the theoretical and methodological groundwork necessary to tackle the long-standing, complicated sustainability problems caused by systemic flaws. Because sustainability issues are complex, finding solutions calls for a holistic strategy. One of several reasons for this omission is the inclination to mix non-biophysical hazards like diseases, violent conflicts, or economic exploitation with resource-related problems like water contamination or energy availability. Also, sometimes, people advocate nanotechnological solutions as technical answers without thinking about other options, possible downsides, or actual advancement. An issue needs to be urgent, have lasting implications, be location-specific, manifest at various spatial dimensions, and be debatable to be considered a sustainability concern (Batabyal and Nijkamp 2009). Relying on a hammer to remove a single nail will not cut it when trying to solve complex sustainability concerns. Rather, sustainability issues are seen by researchers as intricate webs of interdependent cause-and-effect linkages that offer opportunities for intervention and various solutions. Enhancement of sustainability is the primary goal of this research on nanotechnologies. These technologies have a wide range of potential applications, such as improving the efficiency of solar panels, cleaning up the environment, and purifying water and air. Remember that "green" applications account for just 10% of total nanotechnology patents (Malik et al. 2023). With its recent designation as the least sustainable metropolis internationally, Phoenix provides an excellent case study for intervention research on urban sustainability concerns. Thanks to everyone's hard work and dedication to a sustainable future, scholars, city planners, and the general public have established a sustainability-focused draft general plan and other relevant projects (Withycombe Keeler et al. 2018).

China's technological advancements significantly address climate change, showcasing a commitment to sustainable development. As the world's largest emitter of greenhouse gases, China's efforts are pivotal in the global fight against climate change. The country has invested heavily in renewable energy technologies, becoming the largest producer of solar panels and wind turbines and leading the way in electric vehicle adoption (Lee 2020). Additionally, China's ambitious reforestation projects and innovative carbon capture and storage approaches demonstrate its dedication to reducing carbon emissions. These initiatives are complemented by China's significant investments in research and development of green technologies, which aim to mitigate climate impacts domestically and contribute to global sustainability efforts. Through these multifaceted strategies, China is positioned as a crucial player in the international endeavor to combat climate change and promote environmental justice (Wang et al. 2021).

24.10.1 Nanotechnology in Renewable Energy Solutions

Solar, wind, biomass, hydrogen, and geothermal energies are all types of renewable energy sources that may produce power, heat, and light without causing harm to the environment. Fossil fuels, such as crude oil, coal, and natural gas, are non-renewable resources that contribute to pollution. According to projections, the global energy demand is anticipated to exceed 30 terawatts (TW) by 2050 and 46 TW by 2100 (Khalili et al. 2019). Nanotechnology is currently a popular subject due to its remarkable capabilities in engineering and technology. Nanomaterials, with diameters ranging from 1 to 100 nm, enable the creation of smaller information processing components in devices such as cell phones and laptop computers (Elmustafa and Sohal 2017). The shrinkage increases efficiency, more electrical storage capacity, and reduced emissions. Nanotechnology can also improve renewable energy sources. For instance, using rotor blades composed of lightweight nanomaterials can enhance the efficiency of wind generation. Nano-based precision farming can optimize biomass energy production. Nanocoatings can be used to safeguard tidal energy equipment against corrosion. Enhancing the fatigue resistance of drilling machines in geothermal energy applications is possible. Solar energy is utilized in various applications such as solar cells, solar power plants, solar collectors, and saltwater desalination, making it one of the most efficient renewable energy sources. The high surface-to-volume ratio of nanoparticles enables them to efficiently absorb a more significant amount of sunlight and convert it into usable energy. A cost-effective method of converting solar energy into usable electricity is through photovoltaic solar cells, which convert light energy into electrical current by harnessing the movement of electrons (Smestad 1998).

24.10.2 Nanomaterials for Carbon Capture and Storage

The massive CO_2 emission from various sources, including industry and power plants, is a significant factor in the world's climate and the Earth's life cycle. Along with an increase of 0.8 °C in the average surface temperature of the Earth, the concentration of CO_2 has increased from 280 to 400 parts per million (ppm) (Davis 2017). According to the Intergovernmental Panel on Climate Change (IPCC), greenhouse gas (GHG) emissions must be reduced by 50–80% by 2050 to avoid a catastrophic collapse (Davis 2016). Several methods were suggested during the Conference of the Parties (COP) 21 meeting to limit CO_2 concentrations. These included energy efficiency upgrades, using renewable and low-carbon fuels, geoengineering strategies such as afforestation, and developing carbon capture and storage (CCS) technology. Nearly 40% of the world's carbon dioxide emissions come from power plants, and experts predict that number will climb to 60% by the century's end (Dyson 2005). Twenty percent of total CO_2 emissions come from the transportation industry, with the construction and agricultural sectors each contributing about 17%. Between 1970 and 2010, the concentration of CO_2 from fossil fuel combustion increased by 10%, and that from forestry and other land uses declined by 17–11%. Because of their potential to quickly remove CO_2 from the air and thereby reduce the effects of climate change, cutting-edge technologies like CCS have attracted much attention. Reducing CO_2 emissions from different energy-intensive sectors is one way that CCS technology could help keep global warming below 2 °C. Technologies for capturing and storing carbon dioxide after combustion include coal gasification, oxyfuel combustion, and CCS before combustion. However, these technologies will need further improvements to overcome limitations like high costs, insufficient resources, lack of incentives, and outdated infrastructure (Lybbert and Sumner 2012).

24.10.3 Nanotechnology in Sustainable Agriculture and Resource Management

Worldwide, water shortage is a significant problem in agriculture, calling for more environmentally friendly irrigation practices and better water handling (Pereira et al. 2002). Water softening and wastewater treatment using nanofilters is a promising application of nanotechnology that can help with this problem. These filters are also very good at controlling the amount of desalinated water farmers use for irrigation. It is advised that irrigation water should not contain any particles larger than 50 μm, harmful salts, or heavy metals and should also have a low salinity. Reduced irrigation and fertilizer needs by 25% and significantly increased crop yields have resulted from using solar-powered nanofilters to manage desalinated water in arid and hot climates (Maftouh et al. 2023). As activators to different enzymes, micronutrients are vital to plant metabolism. In soils high in pH and calcium, chitosan nanoparticles inhibit the secretion of certain growth hormones, whereas iron oxide nanoparticles stimulate plant development. Improved plant uptake of micronutrients is possible using nano-formulated zinc, iron, and molybdenum. Using nanoemulsions in developing "smart seeds" allows for controlling when and how the seeds germinate. Novel methods for editing plant genes with nanoparticles, capsules, and nanofibers are also available thanks to nanotechnology. These materials control the passage of genetic materials by acting as carriers, holding plant genes and chemicals. Using starch nanoparticles in plant cell DNA binding and transport and nanofibers in crop engineering and medication delivery are two of their many applications (Verma et al. 2020). When used to detect pollen grain contamination from genetically modified crops, nanobiosensors can release alarms, protecting crop fields from potential harm. By combining nanotechnology and biotechnology, scientists have created three-dimensional molecular architectures that can improve crucial crops by linking and classifying important organic molecules (Muhammad et al. 2023).

Chemical fertilizers have greatly improved agricultural yields, especially for cereals, throughout the last half-century. Nevertheless, the inefficient application of these fertilizers leads to fertilizer loss, increasing production costs and polluting the environment (Savci 2012). To guarantee the sustainable use of nutrients, scientists are turning to nanotechnology to create new approaches. Nutrients or nanomaterials that carry them or add them to other substances are known as nanofertilizers. Encapsulating nutrients into nanomaterials is another way they might be created. Nanofertilizers can make farming more sustainable by increasing crop yield and quality with less wasteful use of nutrients and lower production costs. Compared to traditional fertilizers, nanofertilizers demonstrated an average improvement in the efficacy of 18–29% (Iqbal et al. 2019). Evidence shows phosphorus nanofertilizers can boost soybean development and seed production. They enhance plant metabolism and nutrient absorption through molecular transporters or nanostructured cuticle holes. This comprehensive table (Table 24.2) covers various aspects and encompasses different aspects of nanotechnology's role in sustainable agriculture and resource management. It provides detailed descriptions, outlines contributions and benefits, highlights benefits, addresses challenges, presents examples, and offers insights into current research efforts and prospects.

As we advance toward more sustainable agricultural practices, alternatives to traditional chemical fertilizers and pesticides are becoming crucial. Organic fertilizers such as compost, manure, and green manure enrich the soil with nutrients and enhance its structure, while biofertilizers use beneficial microorganisms to promote nutrient availability (Singh et al. 2020). Biopesticides, including microbial and biochemical pesticides derived from natural substances, offer targeted pest control with minimal environmental impact. Nanotechnology provides innovative solutions like nanofertilizers and nanopesticides, which improve nutrient delivery and pest management efficiency. Integrated Pest Management (IPM) combines biological control, cultural practices, and mechanical methods to manage pests sustainably (El Wakeil et al. 2017). These alternatives support plant health and yield and protect ecosystems, paving the way for a more sustainable and resilient agricultural future.

Table 24.2 Nanotechnology applications in sustainable agriculture and resource management

Application	Description	Nanotechnology contribution	Benefits	References
Precision agriculture	Targeted delivery of fertilizers and pesticides based on real-time data	Nanosensors and nanomaterials enable precision farming practices	Reduced resource use, minimized environmental impact, and increased crop yields	Bongiovanni and Lowenberg-DeBoer (2004)
Nanopesticides and nanofertilizers	Enhanced efficiency and reduced environmental impact of crop inputs	Nanoencapsulation improves the delivery and controlled release of agrochemicals	Lower chemical usage, decreased runoff, and improved crop health	Khan and Rizvi (2017)
Soil remediation	Removal of pollutants and enhancement of soil health	Nanomaterials adsorb contaminants and improve nutrient retention in soils	Remediation of polluted soils, increased soil fertility, and sustainable land use	Tripathi et al. (2017)
Water management	Efficient use and purification of water resources	Nanomembranes and nanosorbents enable effective water treatment and desalination	Conservation of water resources, increased access to clean water, and reduced pollution	Majeed et al. (2024)

24.11 Future Perspectives and Challenges

Nanotechnology has been used to generate various agricultural products, such as nanobiosensors, nanofertilizers, and nanopesticides (Rai et al. 2015). The adverse effects of nanoparticles on the growth and development of plants. Crop plants have alterations in their physical structure and biological functions due to higher levels of nanoparticles. An excessive concentration of nanoparticles in the root zone inhibits seed germination, hampers root growth, alters water and nutrient absorption, retards leaf development, and diminishes biomass production. Specifically, the toxicity of nanomaterials induces oxidative bursts, resulting in a disarray of chloroplasts, reduced photosynthesis, membrane rupture, cellular damage, and altered gene expression (Manishaa Sri 2023). The investigation of the enduring impacts of nanomaterials on ecosystems and agriculture is a subject of great fascination. To develop efficient, adaptable, stable, cost-effective, and environmentally friendly nanoparticles, it is necessary to conduct collaborative research among institutes that investigate different applications of nanomaterials.

Additionally, this would aid in addressing the deficiencies in the function, trajectory, behavior, and assessment of NMs' ecotoxicity (Lead et al. 2018). Although specific studies indicate that applying nanomaterials (NMs) to crop plants can enhance their growth and productivity, the impact may vary depending on the species involved. Therefore, thorough investigation into the screening and optimization of nanomaterials for different plant species is essential for their commercial application. The stability and properties of nanomaterials can be precisely adjusted to manipulate their behavior and enhance their efficiency (Elsabahy and Wooley 2012). To achieve this objective, it would be highly advantageous to create novel and enhanced synthesis methods that enable precise manipulation of the composition of the final product. This would significantly improve their productivity. To develop thorough remediation plans, examining the function of nanomaterials (NMs) in bioremediation is crucial. The information on the practical implementation of nano-assisted agriculture is few, and the majority of the research on this subject is derived from controlled experiments. Additional information at the operational level is necessary to implement nano-based strategies on a large scale (Westerhoff et al. 2016).

24.12 Ethical and Societal Implications of Nanotechnology

We can alter the chemical makeup of any material, synthetic or otherwise, by operating on the nanoscale. This opens up the possibility of investigating the possibility of restoring any material. The increases in design power bring up primary social and ethical considerations. We must educate everyone involved in nanotechnology thoroughly if we want it to advance in a way that is ecological, ethical, and commercially successful (Sandler 2009). In this training, students should learn about the potential benefits, limitations, and risks of nanotechnology, both now and in the future. All fields will be impacted by nanotechnology, just like by earlier advances. For example, nanotechnology will likely integrate automated diagnostics into the medical area of the healthcare industry. Fewer patients will require a physical examination, faster diagnoses, less room for human error, and more access to healthcare (Schiff et al. 2005). Nanomedicines can potentially increase the average human life expectancy, increasing the number of older adults needing medical treatment and driving up healthcare expenses in the long run. Four social objectives should be prioritized by those involved in nanotechnology: (1) gaining a thorough understanding of the forces and issues affecting individuals and societies on a local and global scale; (2) guiding societies on a local and global scale toward responsible technology utilization; (3) increasing societal awareness of technological risks and failures; and (4) encouraging informed and ethical decision-making and leadership to tackle challenges in a technological society (Floridi et al. 2018). While nanotechnology has been incredibly beneficial, it has also introduced several risks and challenges to human and environmental health. To evaluate the potential dangers of nanotechnology, the relevant regulatory bodies must establish and implement appropriate norms and processes. *Engineers of Creation* author Eric Drexler has identified four problems with nanotechnology's progress, effects, and consequences for society (Micheletto 2018): (1) The Difficulty of Technological Forecasting (knowing the bare minimum of future possibilities); (2) The Problem with Technological Forecasting: Being an Expert on the Make-Up of Matter; (3) The Difficulty with Belief and Clarity (deepened knowledge of the technical capacities); and (4) Creating Prompt and Effective Public Policies (policymaking based on understanding).

24.13 The Role of Innovation in Climate Change Mitigation

The Integrated Soil Management Protocol (ISMP) strategy encompasses various practices such as incorporating manure, mulch, compost, cover crops, nutrient management, minimal fertilizer application, zero-to-conservation tillage, and appropriate supplementary irrigation (Adenle et al. 2015). These practices are essential for achieving sustainable agriculture. To establish sustainable and productive agricultural systems, ISMP interventions require the combined use of mineral and organic fertilizers, along with their strategic management. The International Soil Management Practice (ISMP) paradigm asserts that no single factor can provide sustainable soil management (Anekwe et al. 2023). The role of ISMP in reducing agricultural carbon emissions is evident. Farmers in North and South America, Asia, and Latin America have adopted zero tillage practices. Limited access to resources, lack of knowledge, and labor limitations have caused significant opposition to adopting African conservation agriculture practices, such as zero tillage. The implementation of the Integrated Soil Management Protocol (ISMP) can lead to an increase in SOC, which in turn can enhance carbon stock through the process of carbon sequestration (CS) and help remove atmospheric CO_2 (Nair et al. 2015). Sustainable agriculture in impoverished nations depends on SOC management. By maintaining a high SOC level, the ISMP, incorporating composting technologies, shows significant potential in reducing land cultivation-related CS and GHG emissions. To enhance carbon sequestration and the functioning of agroecosystems, adaptation strategies should prioritize advancing technologies that improve the sustainable management of soil and water resources under the ISMP (Adenle et al. 2015). When addressing soil restoration from biological and physical deterioration, selecting innovative solutions to enhance soil structure and promote water conservation is essential. Adopting innovative drought-tolerant crops, modifying agricultural patterns, and implementing integrated water management measures such as drip irrigation are critical for effectively responding to climate change. Modern biotechnology can substantially enhance adaptation and mitigation strategies in response to climate change. Modern biotechnology, encompassing genetics and bioinformatics, can potentially reduce the demand for chemical fertilizers by expanding the range of nitrogen-fixing crops (Tran and Nguyen 2009). Another crucial element in combating poverty and guaranteeing food security in developing countries is the development of innovative biotech crops that resist living organisms and non-living factors such as climate change. Tissue culture, a form of traditional biotechnology, has been essential in addressing insect-related challenges and generating drought-resistant crops such as millet, sunflower, and sorghum in numerous developing countries (Krupa et al. 2017).

24.14 Conclusion

Rising atmospheric CO_2 levels and average worldwide temperatures are mostly the product of human activities, which have dominated the process of climate change. This has already harmed every living thing on Earth and could have far-reaching consequences in the future. CO_2 and other greenhouse gases trap heat in the atmosphere, devastatingly impacting ecosystems and the health of all living things. The field of nanotechnology holds excellent potential for revolutionizing various industries, including agriculture, food production, medicine, the environment, energy, catalysis, and material science, by concentrating on creating particles at the nanoscale. Because of their large surface area to volume ratio, nanomaterials are better able to interact with other substances, help transfer clean energy, and absorb gasses that contribute to climate change.

Enhancing productivity while reestablishing ecological balance, nanotechnology in agriculture offers cutting-edge farming tools like precision farming technologies, nanopesticides, and nanofertilizers. Improving food crop yield and managing environmental conditions rely on agro-climatic parameters, which can be detected using nanosensors. One novel use of nanotechnology is nanobiosensors in food packaging. They detect diseases, contaminants, and toxins linked to them.

Nanocomposites, nanocatalysts, nanocoating, and nanolubricants are a few nanostructure materials that show great promise as environmentally friendly alternatives to more conventional materials. By decomposing dyes and other colorless pollutants in wastewater streams, they help clean up the environment and reduce emissions of greenhouse gases. Although nanomaterials are more stable and resistant to degradation than tin and mercury nanoparticles, the possible harm they may cause to the environment and human health is a significant concern. Consequently, safety regulations must be strictly followed. Damage to ecosystems, agricultural output, and other sectors has directly resulted from climate change's radical alteration of the climatic system. Reducing these consequences and promoting long-term growth are two areas where nanotechnology shows excellent promise.

To combat the terrifying consequences of climate change on people worldwide, environmentally friendly technology must be implemented. Sustainable alternatives to traditional procedures are provided by nanotechnology, which includes nanolubricants, nanocatalysts, and nanosensors. Many industries use nanomaterials, including those dealing with bioenergy, wastewater treatment, environmental remediation, catalysis, and greenhouse gas sequestration. To effectively combat the effects of climate change, worldwide regulations must be put in place, and sustainable practices and products must be developed. Nanotechnology has very bright future potential.

References

Adenle AA, Azadi H, Arbiol J (2015) Global assessment of technological innovation for climate change adaptation and mitigation in developing world. J Environ Manag 161:261–275

Aithal PS, Aithal S (2022) Opportunities and challenges for green and eco-friendly nanotechnology in twenty-first century. Sustainable nanotechnology: strategies, products, and applications 31–50

Almulhim AI, Bibri SE, Sharifi A, Ahmad S, Almatar KM (2022) Emerging trends and knowledge structures of urbanization and environmental sustainability: a regional perspective. Sustain For 14:13195

Ambekar S (2022) Linking China's hegemonic intentions through its institutions in Southeast Asia

Anekwe IMS, Zhou H, Mkhize MM, Akpasi SO (2023) Effects of climate change on agricultural production in South Africa. Int J Clim Change: Impacts & Responses 15

Aresta M, Dibenedetto A (2021) The carbon dioxide revolution. Carbon Dioxide Revolut:31–43

Ascione F (2017) Energy conservation and renewable technologies for buildings to face the impact of the climate change and minimize the use of cooling. Sol Energy 154:34–100

Baig N, Kammakakam I, Falath W (2021) Nanomaterials: a review of synthesis methods, properties, recent progress, and challenges. Mater Adv 2:1821–1871

Batabyal AA, Nijkamp P (2009) Sustainable development and regional growth. Handbook of regional growth and development theories. Edward Elgar Publishing

Bongiovanni R, Lowenberg-DeBoer J (2004) Precision agriculture and sustainability. Precis Agric 5:359–387

Cai W, Santoso A, Collins M, Dewitte B, Karamperidou C, Kug J-S, Lengaigne M, McPhaden MJ, Stuecker MF, Taschetto AS (2021) Changing El Niño–southern oscillation in a warming climate. Nat Rev Earth Environ 2:628–644

Calanca PP (2017) Effects of abiotic stress in crop production. Quantification of climate variability, adaptation and mitigation for agricultural sustainability 165–180

Chakraborty S, Pattanayak A, Mandal S, Das M, Roychowdhury R (2014) An overview of climate change: causes, trends and implications. Crop improvement in the era of climate change IK. International Publishing House, New Delhi, pp 1–29

Chan L, Costa S (2005) Participation in the global knowledge commons: challenges and opportunities for research dissemination in developing countries. New Libr World 106:141–163

Chehroudi B (2006) Supercritical fluids: nanotechnology and select emerging applications. Combust Sci Technol 178:555–621

Chen X, Chen HY, Chen C, Ma Z, Searle EB, Yu Z, Huang Z (2020) Effects of plant diversity on soil carbon in diverse ecosystems: a global meta-analysis. Biol Rev 95:167–183

Coffel ED, Horton RM, De Sherbinin A (2017) Temperature and humidity based projections of a rapid rise in global heat stress exposure during the 21st century. Environ Res Lett 13:014001

Connor PM, Xie L, Lowes R, Britton J, Richardson T (2015) The development of renewable heating policy in the United Kingdom. Renew Energy 75:733–744

Contreras Llin A, Díaz-Cruz MS (2023) Microplastics retention in managed aquifer recharge systems using wastewater effluent

Dasgupta N, Ranjan S, Ramalingam C (2017) Applications of nanotechnology in agriculture and water quality management. Environ Chem Lett 15:591–605

Davis J (2016) Climate change and the scope for global greenhouse gas reductions. CABI Rev:1–11

Davis WJ (2017) The relationship between atmospheric carbon dioxide concentration and global temperature for the last 425 million years. Climate 5:76

Demirbas A, Demirbas A (2016) Future energy sources. Waste Energy for Life Cycle Assessment 33–70

Dhalla F, Eng P (2015) Investigating thermal bridging in window systems insulated with monolithic silica aerogel

Dlamini CS (2020) Contribution of forest ecosystem services toward food security and nutrition. Zero hunger. Springer, pp 179–196

Dyson T (2005) On development, demography and climate change: the end of the world as we know it? Popul Environ 27:117–149
El Wakeil N, Alkahtani S, Gaafar N (2017) Is nanotechnology a promising field for insect pest control in IPM programs? New pesticides and soil sensors. Elsevier, pp 273–309
Elmustafa SAA, Sohal HS (2017) Nanotechnology in communication engineering: issues, applications, and future possibilities. World Scientific News:134–148
Elsabahy M, Wooley KL (2012) Design of polymeric nanoparticles for biomedical delivery applications. Chem Soc Rev 41:2545–2561
Engler D (2019) Dealing with high temperatures: a survey assessing the degree of heat risk awareness and the adaptation measures applied in hospitals and retirement homes in Berlin. Hochschule für angewandte Wissenschaften Hamburg
Floridi L, Cowls J, Beltrametti M, Chatila R, Chazerand P, Dignum V, Luetge C, Madelin R, Pagallo U, Rossi F (2018) AI4People—an ethical framework for a good AI society: opportunities, risks, principles, and recommendations. Mind Mach 28:689–707
Gates A (2023) Polluted earth: the science of the earth's environment. John Wiley & Sons
Georgopoulou E, Mirasgedis S, Sarafidis Y, Giannakopoulos C, Varotsos KV, Gakis N (2024) Climate change impacts on the energy system of a climate-vulnerable Mediterranean Country (Greece). Atmos 15:286
Gerke J (2022) The central role of soil organic matter in soil fertility and carbon storage. Soil Syst 6:33
Gouel C, Guimbard H (2019) Nutrition transition and the structure of global food demand. Am J Agric Econ 101:383–403
Gross CD, Harrison RB (2019) The case for digging deeper: soil organic carbon storage, dynamics, and controls in our changing world. Soil Syst 3:28
Gu X, Zhao H, Peng C, Guo X, Lin Q, Yang Q, Chen L (2022) The mangrove blue carbon sink potential: evidence from three net primary production assessment methods. For Ecol Manag 504:119848
Gude VG (2017) Desalination and water reuse to address global water scarcity. Rev Environ Sci Biotechnol 16:591–609
Heitor M, Bravo M (2010) Portugal at the crossroads of change, facing the shock of the new: people, knowledge and ideas fostering the social fabric to facilitate the concentration of knowledge integrated communities. Technol Forecast Soc Chang 77:218–247
Hendriks S, Soussana J-F, Cole M, Kambugu A, Zilberman D (2023) Ensuring access to safe and nutritious food for all through the transformation of food systems. Science and innovations for food systems transformation 31
Horta A, Gouveia JP, Schmidt L, Sousa JC, Palma P, Simões S (2019) Energy poverty in Portugal: combining vulnerability mapping with household interviews. Energ Buildings 203:109423
Iqbal M, Umar S, Mahmooduzzafar (2019) Nano-fertilization to enhance nutrient use efficiency and productivity of crop plants. Nanomaterials and plant potential 473–505
Islam SMF, Karim Z (2019) World's demand for food and water: the consequences of climate change. Desalination-challenges and opportunities 1–27
Jackson RB, Carpenter SR, Dahm CN, McKnight DM, Naiman RJ, Postel SL, Running SW (2001) Water in a changing world. Ecol Appl 11:1027–1045
Jenkins AB (2002) Organic carbon and fertility of forest soils on the Allegheny Plateau of West Virginia. West Virginia University
Jentsch MF, James PA, Bourikas L, Bahaj AS (2013) Transforming existing weather data for worldwide locations to enable energy and building performance simulation under future climates. Renew Energy 55:514–524
Kaneti YV, Tang J, Salunkhe RR, Jiang X, Yu A, Wu KCW, Yamauchi Y (2017) Nanoarchitectured design of porous materials and nanocomposites from metal-organic frameworks. Adv Mater 29:1604898
Karali N, Xu T, Sathaye J (2014) Reducing energy consumption and CO2 emissions by energy efficiency measures and international trading: a bottom-up modeling for the US iron and steel sector. Appl Energy 120:133–146

Karimidastenaei Z, Avellán T, Sadegh M, Kløve B, Haghighi AT (2022) Unconventional water resources: global opportunities and challenges. Sci Total Environ 827:154429

Karmalkar AV, Bradley RS (2017) Consequences of global warming of 1.5 C and 2 C for regional temperature and precipitation changes in the contiguous United States. PLoS One 12:e0168697

Kathiresan K (2021) Mangroves: types and importance. Mangroves: ecology, biodiversity and management 1–31

Khalid B, Urbański M, Kowalska-Sudyka M, Wysłocka E, Piontek B (2021) Evaluating consumers' adoption of renewable energy. Energies 14:7138

Khalili S, Rantanen E, Bogdanov D, Breyer C (2019) Global transportation demand development with impacts on the energy demand and greenhouse gas emissions in a climate-constrained world. Energies 12:3870

Khan MR, Rizvi TF (2017) Application of nanofertilizer and nanopesticides for improvements in crop production and protection. Nanoscience and plant–soil systems 405–427

Khan S, Naushad M, Al-Gheethi A, Iqbal J (2021) Engineered nanoparticles for removal of pollutants from wastewater: current status and future prospects of nanotechnology for remediation strategies. J Environ Chem Eng 9:106160

Kienzl N (1999) Advanced building skins: translucent thermal storage elements. Massachusetts Institute of Technology

Krupa K, Dalawai N, Shashidhar H, Harinikumar K, Manojkumar H, Bharani S, Turaidar V (2017) Mechanisms of drought tolerance in sorghum: a review. Int J Pure Appl Biosci 5:221–237

Kumar N (2023) Urban flood mitigation strategies using geo spatial tools: a practical approach for cities and towns of developing world. Taylor & Francis

Lahn G, Shapland G (2022) Cascading climate risks and options for resilience and adaptation in the Middle East and North Africa. Chatham House, pp 463–472

Lead JR, Batley GE, Alvarez PJ, Croteau MN, Handy RD, McLaughlin MJ, Judy JD, Schirmer K (2018) Nanomaterials in the environment: behavior, fate, bioavailability, and effects—an updated review. Environ Toxicol Chem 37:2029–2063

Lee JW (2020) Green finance and sustainable development goals: the case of China. J Asian Finance Econ Bus 7:577–586

Lee J-Y, Kilmer RL (2010) At-home convenience food consumption and BMI

Liu X, He Y, Tang C, Wei Q, Xu Z, Yi W, Pan R, Gao J, Duan J, Su H (2021) Association between cold spells and childhood asthma in Hefei, an analysis based on different definitions and characteristics. Environ Res 195:110738

Lybbert TJ, Sumner DA (2012) Agricultural technologies for climate change in developing countries: policy options for innovation and technology diffusion. Food Policy 37:114–123

Maftouh A, El Fatni O, Bouzekri S, Rajabi F, Sillanpää M, Butt MH (2023) Economic feasibility of solar-powered reverse osmosis water desalination: a comparative systemic review. Environ Sci Pollut Res 30:2341–2354

Mahajan N, Jat L, Tiwari R, Yadav A (2018) Soil organic carbon fractions, soil microbial biomass carbon, and enzyme activities impacted by crop rotational diversity and conservation tillage in north west IGP: a review. Int J Curr Microbiol App Sci 7:3573–3600

Majeed M, Noor W, Rasool A, Saleem M, Bashir I, Fatima R, Gulshan AB, Sinha D, Nath S, Hassan S (2024) Nanotechnology for soil remediation and land restoration. Harnessing NanoOmics and Nanozymes for sustainable agriculture. IGI Global, pp 60–77

Malik A, Lan J, Lenzen M (2016) Trends in global greenhouse gas emissions from 1990 to 2010. Environ Sci Technol 50:4722–4730

Malik S, Muhammad K, Waheed Y (2023) Nanotechnology: a revolution in modern industry. Molecules 28:661

Manishaa Sri M (2023) Impact of titanium dioxide nanoparticles on the nutritional composition of haematococcus pluvialis. UTAR

Micheletto R (2018) Nanotechnology and future technological evolutions. Future courses of human societies. Routledge, pp 68–86

Miller JC, Serrato R, Represas-Cardenas JM, Kundahl G (2004) The handbook of nanotechnology: business, policy, and intellectual property law. John Wiley & Sons

Muhammad M, Waheed A, Wahab A, Majeed M, Nazim M, Liu Y-H, Li L, Li W-j (2023) Soil salinity and drought tolerance: an evaluation of plant growth, productivity, microbial diversity, and amelioration strategies. Plant Stress 100319

Muhammad M, Batool S, Hivare V, Li W-J, Waheed A, Sinha D (2024) Bioremediation techniques—classification, principles, advantages, limitations, and prospects. In: Microbiome-assisted bioremediation. Elsevier, pp 1–23

Nair R, Mehta C, Sharma S (2015) Carbon sequestration in soils-a review. Agric Rev 36:81–99

Nasrollahzadeh M, Sajadi SM, Sajjadi M, Issaabadi Z (2019) An introduction to nanotechnology. Interface science and technology, vol 28. Elsevier, pp 1–27

Nasu H, Faunce T (2009) Nanotechnology and the international law of weaponry: towards international regulation of nano-weapons. J Law Inf Sci 20:23–60

Nile SH, Baskar V, Selvaraj D, Nile A, Xiao J, Kai G (2020) Nanotechnologies in food science: applications, recent trends, and future perspectives. Nanomicro Lett 12:1–34

Pagallo U, Casanovas P, Madelin R (2019) The middle-out approach: assessing models of legal governance in data protection, artificial intelligence, and the Web of Data. Theory Pract Legis 7:1–25

Pereira LS, Oweis T, Zairi A (2002) Irrigation management under water scarcity. Agric Water Manag 57:175–206

Pietroiusti A, Stockmann-Juvala H, Lucaroni F, Savolainen K (2018) Nanomaterial exposure, toxicity, and impact on human health. Wiley Interdiscip Rev Nanomed Nanobiotechnol 10:e1513

Pownraj C, Valan Arasu A (2021) Effect of dispersing single and hybrid nanoparticles on tribological, thermo-physical, and stability characteristics of lubricants: a review. J Therm Anal Calorim 143:1773–1809

Prommer J, Walker TW, Wanek W, Braun J, Zezula D, Hu Y, Hofhansl F, Richter A (2020) Increased microbial growth, biomass, and turnover drive soil organic carbon accumulation at higher plant diversity. Glob Chang Biol 26:669–681

Rai M, Ribeiro C, Mattoso L, Duran N (2015) Nanotechnologies in food and agriculture. Springer

Renn O (2017) Risk governance: coping with uncertainty in a complex world. Routledge

Sandler R (2009) Nanotechnology: the social and ethical issues

Sargent JF (2008) Nanotechnology and environmental health and safety: issues for consideration. DIANE Publishing

Sarkodie SA, Strezov V (2019) Economic, social and governance adaptation readiness for mitigation of climate change vulnerability: evidence from 192 countries. Sci Total Environ 656:150–164

Sarma SD (2011) How resilient is India to nanotechnology risks? Examining current developments, capacities and an approach for effective risk governance and regulation. Eur J Law Technol 2

Savci S (2012) An agricultural pollutant: chemical fertilizer. Int J Environ Sci Dev 3:73

Scharff RL (2015) State estimates for the annual cost of foodborne illness. J Food Prot 78(6):1064–1071

Scherer-Lorenzen M, Palmborg C, Prinz A, Schulze E-D (2003) The role of plant diversity and composition for nitrate leaching in grasslands. Ecology 84:1539–1552

Schiff GD, Kim S, Abrams R, Cosby K, Lambert B, Elstein AS, ... McNutt RA (2005) Diagnosing diagnosis errors: lessons from a multi-institutional collaborative project. Advances in patient safety: from research to implementation (volume 2: concepts and methodology)

Shen X, Du H, Mullins RH, Kommalapati RR (2017) Polyethylenimine applications in carbon dioxide capture and separation: from theoretical study to experimental work. Energ Technol 5:822–833

Simpson MJ, Breili K, Kierulf HP (2014) Estimates of twenty-first century sea-level changes for Norway. Clim Dyn 42:1405–1424

Singh TB, Ali A, Prasad M, Yadav A, Shrivastav P, Goyal D, Dantu PK (2020) Role of organic fertilizers in improving soil fertility. Contaminants in agriculture: sources, impacts and management 61–77

Singhvi M, Kim BS (2020) Current developments in lignocellulosic biomass conversion into biofuels using nanobiotechology approach. Energies 13:5300

Smestad GP (1998) Education and solar conversion:: demonstrating electron transfer. Sol Energy Mater Sol Cells 55:157–178

Sun Q, Sun Z, Chen C, Yan M, Zhong Y, Huang Z, He L, Li T (2022) Health risks and economic losses from cold spells in China. Sci Total Environ 821:153478

Tang L (2011) The US-China scientific collaboration, knowledge moderation, and China's rise in nanotechnology. Georgia Institute of Technology

Tlili I, Alkanhal TA (2019) Nanotechnology for water purification: electrospun nanofibrous membrane in water and wastewater treatment. J Water Reuse Desalin 9:232–248

Tohka A (2005) The GAINS model for greenhouse gases-version 1.0: HFC, PFC and SF6

Tran LSP, Nguyen HT (2009) Future biotechnology of legumes. Nitrogen fixation in crop production 52:265–307

TRENDS G (2017) Challenges and opportunities in the implementation of the Sustainable Development Goals. United Nations Development Programme& United Nations Research Institute for Social Development

Tripathi V, Edrisi SA, Chen B, Gupta VK, Vilu R, Gathergood N, Abhilash P (2017) Biotechnological advances for restoring degraded land for sustainable development. Trends Biotechnol 35:847–859

Upadhyay RK (2021) Climate induced virus generated communicable diseases: management issues and failures. J Atmos Sci Res 4:27–50

Van Niekerk L, Adams J, James N, Lamberth S, MacKay C, Turpie J, Rajkaran A, Weerts SP, Whitfield A (2020) An estuary ecosystem classification that encompasses biogeography and a high diversity of types in support of protection and management. Afr J Aquat Sci 45:199–216

Verma ML, Dhanya BS, Rani V, Thakur M, Jeslin J, Kushwaha R (2020) Carbohydrate and protein based biopolymeric nanoparticles: current status and biotechnological applications. Int J Biol Macromol 154:390–412

Wang Y-S, Gu J-D (2021) Ecological responses, adaptation and mechanisms of mangrove wetland ecosystem to global climate change and anthropogenic activities. Int Biodeterior Biodegradation 162:105248

Wang Y, Guo C-h, Chen X-j, Jia L-q, Guo X-n, Chen R-s, Zhang M-s, Chen Z-y, Wang H-d (2021) Carbon peak and carbon neutrality in China: goals, implementation path and prospects. China Geol 4:720–746

Westerhoff P, Alvarez P, Li Q, Gardea-Torresdey J, Zimmerman J (2016) Overcoming implementation barriers for nanotechnology in drinking water treatment. Environ Sci Nano 3:1241–1253

Withycombe Keeler L, Beaudoin FD, Lerner AM, John B, Beecroft R, Tamm K, Wiek A, Lang DJ (2018) Transferring sustainability solutions across contexts through city–university partnerships. Sustain For 10:2966

Xiong D, Shi P, Zhang X, Zou CB (2016) Effects of grazing exclusion on carbon sequestration and plant diversity in grasslands of China—a meta-analysis. Ecol Eng 94:647–655

Yoro KO, Daramola MO (2020) CO2 emission sources, greenhouse gases, and the global warming effect. In: Advances in carbon capture. Elsevier, pp 3–28

Zebiak SE, Orlove B, Muñoz ÁG, Vaughan C, Hansen J, Troy T, Thomson MC, Lustig A, Garvin S (2015) Investigating El Niño-southern oscillation and society relationships. Wiley Interdiscip Rev Clim Chang 6:17–34

Zhu Y, Liu Y, Wang W, Singh VP, Ren L (2021) A global perspective on the probability of propagation of drought: from meteorological to soil moisture. J Hydrol 603:126907

Zscheischler J, Westra S, Van Den Hurk BJ, Seneviratne SI, Ward PJ, Pitman A, AghaKouchak A, Bresch DN, Leonard M, Wahl T (2018) Future climate risk from compound events. Nat Clim Chang 8:469–477

Future of Nanotechnology and Climate Change

25

Murad Muhammad, Aqsa Arooj, Sana Ullah, Abdul Wahab, Muhammad Nazim, and Abdul Basit

Abstract

As climate change worsens, finding innovative solutions to mitigate its effects is crucial. Nanotechnology, which allows for manipulating matter at the nanoscale, offers great potential for addressing the challenges posed by climate change. This chapter examines the potential of nanotechnology in combating climate change through a comprehensive review of current research and emerging trends. Nanotechnology provides new approaches for renewable energy generation and storage. Nanomaterials such as quantum dots, carbon nanotubes, and nanostructured electrodes have enhanced properties that can significantly improve the efficiency and performance of solar cells, fuel cells, and batteries.

Additionally, nanotechnology-enabled energy storage devices are crucial for overcoming the intermittency issues associated with renewable energy sources,

M. Muhammad (✉)
Xinjiang Key Laboratory of Biodiversity Conservation and Application in Arid Lands, Xinjiang Institute of Ecology and Geography, Chinese Academy of Sciences, Urumqi, People's Republic of China

University of Chinese Academy of Sciences, Beijing, People's Republic of China

State Key Laboratory of Desert and Oasis Ecology, Key Laboratory of Ecological Safety and Sustainable Development in Arid Lands, Xinjiang Institute of Ecology and Geography, Chinese Academy of Sciences, Urumqi, People's Republic of China
e-mail: Muradbotany1@uop.edu.pk

A. Arooj
University of Chinese Academy of Sciences, Beijing, People's Republic of China

Key Laboratory of Urban Pollutant Conversion, Institute of Urban Environment, Chinese Academy of Sciences, Xiamen City, Fujian, People's Republic of China

S. Ullah
Department of Agricultural Leadership, Education & Communication, College of Agricultural & Environmental Sciences, University of Georgia, Athens, GA, USA

A. A. H. Abdel Latef et al. (eds.), *Bioremediation and Nanotechnology for Climate Change Mitigation*, https://doi.org/10.1007/978-981-96-3069-1_25

making their widespread adoption easier. Nanotechnology plays an essential role in enhancing energy efficiency across various sectors. Nano-enabled materials and coatings can improve the insulation properties of buildings, reduce energy consumption in transportation through lightweight nanocomposites, and improve the efficiency of industrial processes through catalytic nanomaterials. These advancements help lower greenhouse gas emissions and promote sustainable practices.

Furthermore, nanotechnology offers innovative solutions for carbon capture, utilization, and sequestration (CCUS). Nanomaterial-based adsorbents and membranes demonstrate high selectivity and capacity for capturing carbon dioxide from industrial flue gases and ambient air. Additionally, nanocatalysts facilitate the conversion of captured CO_2 into valuable products such as fuels, chemicals, and building materials, closing the carbon loop and mitigating emissions.

With a compound annual growth rate (CAGR) of 17.6%, the global market for nanotechnology is expected to develop at a rate of USD 91.18 billion by 2024, demonstrating its significant contribution to environmental preservation. For example, nanotechnology can increase solar cell efficiency by up to 20–30% and the effectiveness of carbon capture and storage (CCS) systems, making them more commercially viable. In conclusion, nanotechnology holds immense promise for addressing the challenges of climate change by revolutionizing energy systems, enhancing resource efficiency, and fostering sustainable practices across various sectors. Nevertheless, concerted efforts are required to realize this potential while mitigating associated risks, paving the way for a more resilient and sustainable future.

Keywords

Nanotechnology · Climate change · Renewable energy · Carbon capture · Sustainability

A. Wahab
University of Chinese Academy of Sciences, Beijing, People's Republic of China

Shanghai Centre for Plant Stress Biology, CAS Centre for Excellence in Molecular Plant Sciences, Chinese Academy of Sciences, Shanghai, People's Republic of China

M. Nazim
Xinjiang Key Laboratory of Biodiversity Conservation and Application in Arid Lands, Xinjiang Institute of Ecology and Geography, Chinese Academy of Sciences, Urumqi, People's Republic of China

University of Chinese Academy of Sciences, Beijing, People's Republic of China

A. Basit
Department of Horticulture Science, Kyungpook National University, Daegu, South Korea

25.1 Introduction

Nanotechnology is primarily concerned with the manufacturing of nanoparticles (NPs), which improve the particles' biological efficacy and physical and chemical properties (Cerqueira et al. 2018; Chausali et al. 2022; Singh et al. 2017). The production of nanoparticles occurs via the implementation of nanotechnology. These particles, denoted as "nanomaterials" (NMs) and "nanoparticles" (NPs), possess a minimum one-dimensional size spanning from one to one hundred nanometers. These expressions are mutually exclusive and may be applied indiscriminately. Nanotechnology empowers the implementation of nanostructures by utilizing nanoscale devices to tackle various challenges in various sectors. These concerns may be confronted across a range of academic fields. Nanotechnology encompasses various crucial domains, such as medicine, materials science, agriculture, the environment, energy, food production, and catalysis (Chausali et al. 2022). Additional disciplines are likewise affected by nanotechnology.

Moreover, it has generated prospects for developing innovative technologies that are ecologically sustainable and have the potential to replace conventional implements across diverse sectors. Nanomaterials are characterized by a significantly larger surface area in proportion to their volume than their macroscopic counterparts. This is because nanoparticles are comprised of a more minute dimension.

Furthermore, they benefit from enhanced interaction with other substances, the absorption of greenhouse gases (GHGs), and the transfer of sustainable energy (Subramanian et al. 2020). Furthermore, the medical and health sector has substantially progressed due to the implementation of equipment and products utilizing nanotechnology to diagnose and treat illnesses (Shankar et al. 2022). This development signifies progress within the field of medicine. Nanotechnology has been implemented in the agricultural sector to develop sophisticated farming implements, such as nanopesticides, nanofertilizers (NFs), and precision farming technologies. Implementing these instruments has contributed to increased agricultural output and the restoration of ecological balance (Altabbaa et al. 2023). Addressing climate change is of the utmost importance. Climate change presents a highly dire peril to the Earth and its inhabitants. We must implement measures to alleviate its impacts and guarantee a sustainable trajectory for future generations.

The primary cause of the Earth's climate change preponderance is human activities, which have accelerated considerably in the last two centuries. Changes in the occurrence and severity of meteorological phenomena, increased atmospheric carbon dioxide (CO_2) concentrations, and the mean worldwide temperature have been documented in scientific research as indicators of global warming (GW) (Change 2014; Fischer and Knutti 2015). Our planet is grappling with the detrimental effects of climate change; furthermore, it may face a catastrophic event that will affect all living things. Anthropogenic global warming is the primary factor influencing the current climate crisis, necessitating an immediate and grave response. Over the past several decades, the average surface temperature of the Earth has increased significantly and at a rapid rate due to the phenomenon of global warming. Earth's surface temperature increase has been 0.14° (0.08°) per decade since 1880. Since 1981, the annual growth rate has been 0.32 °F (0.18 °C), significantly higher than the growth rate observed in the preceding four decades.

Additionally, scholarly investigations have unveiled that the average global temperature is escalating at a projected rate of 0.2 °C per decade. Consequently, as of 2017, it has surpassed pre-industrial temperatures by an estimated 1 °C. Based on temperature data provided by the National Oceanic and Atmospheric Administration (NOAA), 2020 was the second-highest year in recorded history; in terms of total warmth, only 2016 surpassed it. According to the World Meteorological Organization (WMO), an organization of the United Nations, the mean worldwide temperature in 2021 was 1.11 °C (±0.13 °C) higher than the highest temperatures recorded during the pre-industrial era (1850–1900) (Stuart et al. 2022). The global temperature rose by 1 °C in 2021, surpassing its pre-industrial mean for the seventh year in a row, encompassing the period from 2015 to 2021.

Carbon dioxide and additional greenhouse gases are the primary contributors to global warming (Kweku et al. 2018). A diverse range of substances comprises greenhouse gases, including water vapors, CO_2, nitrous oxide (NOx), methane (CH_4), and ozone. Furthermore, perfluorocarbons (PFCs), sulfur hexafluoride (SF6), and hydrofluorocarbons (HFCs) are classified as greenhouse gases and are produced predominantly in industrial settings (Bozorg 2022; Salawitch et al. 2017). Due to their ability to deplete the ozone layer, greenhouse gases substantially threaten the environment and the health of various organisms. Further repercussions of climate change include the gradual erosion of coastlines and glaciers, land inundation caused by excessive water levels, extended periods of drought, environmental contamination, the emergence of infectious diseases, and reduced agricultural productivity and expansion (Lowry et al. 2019; Naithani 2016; Velásquez et al. 2018).

Collaborative research and substantial investments are driving the global adoption and advancement of nanotechnology. In an effort to improve sustainability and lessen the effects of climate change, the United Nations Environment Programme (UNEP) incorporates nanotechnology into a number of its environmental activities. The production of cutting-edge nanomaterials for applications in renewable energy, pollution prevention, and water purification is one of these initiatives (Shapira and Youtie 2015). Through its Horizon 2020 initiative, which finances significant nanotechnology research focused on environmental sustainability, Europe has taken the lead at the regional level. This initiative supports projects that develop nanotechnologies for waste management and cleaner production processes. The National Nanotechnology Initiative (NNI) is the driving force behind nanotechnology initiatives in the United States. A record $2.16 billion has been asked for NNI in the President's 2024 Budget, with a focus on research and development (R&D) in nanoscale science, engineering, and technology. Since its establishment in 2001, NNI has committed more than $43 billion to the promotion of nanotechnology developments, tackling national objectives such as environmental health and safety (Roco 2023). With significant government spending on R&D, China is another pioneer in the field of nanotechnology. Numerous initiatives to use nanotechnology for sustainable development and environmental protection are supported by the Chinese Academy of Sciences (Gao et al. 2016).

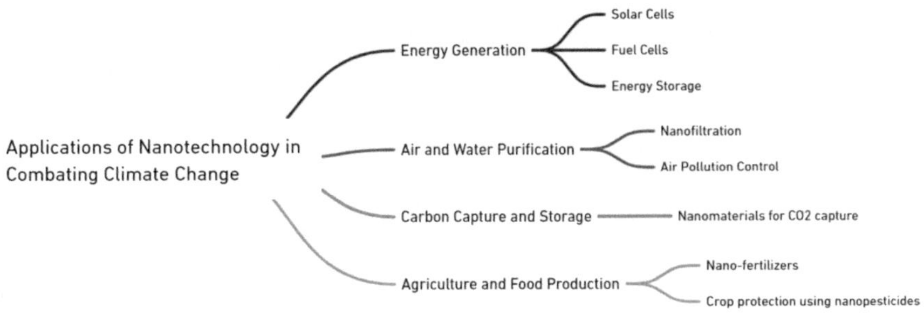

Fig. 25.1 Various applications of nanotechnology in combating climate change

Nanotechnology has great potential to solve climate change problem. Nanotechnology applications that mitigate and adapt to climate change are examined in this chapter. The main goal is to use nanoscale technology to produce creative solutions in renewable energy generation, carbon capture and storage (CCS), water purification, and sustainable material manufacture. The chapter will also examine nanotechnology research and development trends related to climate change, identifying emerging technologies and assessing their viability, scalability, and environmental sustainability. Also noteworthy is that the chapter will explore the ethical, regulatory, and societal consequences of nanotechnology for climate change. This requires addressing safety, environmental effects, and equal access to sophisticated technologies. This chapter strives to explain how nanotechnology can be used to fight climate change by addressing these numerous goals (Fig. 25.1).

25.2 Clean Energy Solutions

25.2.1 Enhancing Solar Energy Efficiency

Solar power is a highly dependable choice for sustainable energy. Various practical applications can effectively utilize solar energy, such as solar power plants, solar cells, seawater desalination, solar collectors, and more (Ahmed et al. 2022). The sun's beams hold the key to this question. The amount of solar energy that reaches the Earth's surface every hour is sufficient to supply power to the entire world for a year (Lewis 2007). Hence, solar power rapidly supplants fossil fuels as the predominant energy source worldwide. There is a correlation between the quantity of solar energy that Earth receives and the seasons. During summertime, the Earth gets the highest amount of solar energy, which is 25 units. The inherent deficiency in the absorption properties of conventional fluids results in reduced efficiency, posing a significant barrier to their practical application. Nanotechnology offers a straightforward and efficient answer to this issue (Mourad et al. 2022). To maximize solar energy collection and efficiency, it is advisable to increase the exposure of conducting surfaces to sunlight by utilizing nanoparticles with a higher surface-area-to-volume ratio. Utilizing substances such as lead-selenide is an additional approach through which nanotechnology may enhance the effectiveness of solar cells (Liu et al. 2015; Shah et al. 2022). The interaction of light with these compounds leads to the augmentation of the electron population and, consequently, the generation of electricity (Mitra and Maiti 2021).

Furthermore, the cost-effectiveness of solar technology plays a crucial role in determining its total efficiency. Solar power costs are equivalent to fossil fuels when transformed into electricity. Using semiconductor materials exhibiting the photovoltaic effect enables the conversion of solar radiation into energy through the photovoltaic process (Sampaio and González 2017). Photovoltaics refers to surfaces that directly convert sunlight into electrical energy. These usually consist of a conducting oxide layer and a catalytic platinum layer (Abdin et al. 2013). A photovoltaic solar cell is an electronic device that converts sunlight into electric current by utilizing the movement of electrons triggered by light photons (Smets et al. 2016). Solar power is highly environmentally friendly. A decrease of 1% in worldwide electricity demand achieved through a distributed solar system would lead to an annual reduction of approximately 40 million tons of carbon dioxide (Shahsavari and Akbari 2018).

25.2.2 Improving Energy Storage Devices

Electricity is quickly becoming the most common kind of power, and this change is very noticeable in energy consumption (Zou et al. 2016). Portable consumer electronics, medical equipment, electric autos, electric grids, the growing Internet of Things, and wearable technologies are driving the need for reversible energy storage and discharge capabilities as a crucial technological advancement (Agarwal et al. 2023). Solar panels, wind power generators, heat sources, triboelectric and piezoelectric generators, moving machinery, and other similar technologies can only be used to their full potential with significant advancements and expansions in energy storage technology (Tian et al. 2020).

Nanomaterials offer significantly better electrical conductivity and ionic transport than traditional materials used in batteries and supercapacitors. They also fill all the interparticle gaps, which speeds up ion diffusion and increases specific capacities (Panda et al. 2020). Electrodes made of nanomaterials have properties that make them suitable for robust and efficient energy storage since they can withstand high amounts of electrical current. Nanomaterials can achieve high power and energy densities simultaneously because of their small size, high surface-area-to-volume ratio, and limited diffusion pathways (Liu et al. 2019a). In addition, advanced production techniques like printing, spray coating, and roll-to-roll assembly can easily incorporate nanoparticles. Wearable, flexible, and foldable energy storage technologies are now within reach (Tong and Tong 2022).

Making batteries and supercapacitors using conventional methods is critical for streamlining vast applications. Compared to the current process, we can obtain better speed, potency, and extended longevity using nanoparticles (Yang et al. 2015). One way to improve the energy density of batteries is to use nanostructured silicon

(Kovalenko et al. 2011) instead of graphite. Electrodes with various sizes, shapes, and physical qualities can be made using nanoscale materials. Typical batteries, for instance, are often thought of as separate things. Nevertheless, the integration of structural components with electrodes is not hindered by any constraints, enabling their dispersion. The study by He et al. (2018) improved the performance of novel carbon materials for batteries and supercapacitors. Their carbon compounds had incredible specific capacitance of 150 F/g and energy density of 200 Wh/kg. A comprehensive analysis of nanostructured carbon materials by Pomerantseva et al. (2019) highlighted their energy storage potential. The review emphasized performance measures like 2500 m^2/g surface area and 180 F/g specific capacitance for carbon nanostructures. Carbon materials are essential to energy storage technology efficiency and effectiveness, highlighting the need for ongoing study and innovation.

Electrodes display properties like strength, durability, and flexibility while also being able to adjust to different geometries (He et al. 2018; Pomerantseva et al. 2019).

25.3 Carbon Capture and Storage

25.3.1 Nanomaterials for CO_2 Capture

Carbon capture and storage (CCS) is now recognized as a prominent solution in pursuing environmental sustainability and the fight against climate change and global warming. Due to the significant reliance on non-renewable resources for energy production, a complete shift to renewable resources is now unattainable. Implementing CCS technology is paramount, given the circumstances. Nanoparticles are anticipated to exhibit exceptional efficacy in CO_2 capture due to their extraordinary and unparalleled properties (Balasubramanian and Chowdhury 2015).

Nanomaterials are much sought after as separation supports for gas purification and capture because of their diverse physicochemical properties (Szczęśniak et al. 2017). An advantage of this is the capacity to be functionalized with various chemicals, such as surfactants, to increase their attraction to specific target molecules (Wang et al. 2021). Furthermore, their inherent characteristics contribute to their exceptional stability and endurance, while their compact size enhances their surface responsiveness. This characteristic renders them highly accessible and appropriate for diverse uses. An area of research that shows great potential is the utilization of NPs for gas cleanup. Hence, a primary objective of contemporary nanotechnology research is to comprehend this phenomenon in gas–solid interactions (Mirzaei et al. 2016). Nanomaterials, including metal-organic frameworks (MOFs), porous organic polymers, and advanced nanoporous materials, have garnered significant interest due to their chemically changeable architectures and extensive surface areas. This is due to their potential applications in adsorption storage (Liu et al. 2022). There has been a recent increase in interest in these materials following experiments that showed their ability to absorb CH_4 and CO_2. Multiple studies have demonstrated that they possess enhanced CO_2 sorption and, notably, CH_4 adsorption capacities compared to other sorbents. MOFs achieved a peak adsorption capacity of 68 units.

Currently, adsorbents made from porous materials and treated with amines or metals demonstrate the most potential in their ability to absorb greenhouse gases (Varghese and Karanikolos 2020). To attain carbon dioxide emission neutrality and tackle significant power production emission issues, it is necessary to initially decrease the cost of capturing and expand the technology to the required magnitude (Diego et al. 2017). Nanotechnology is crucial for the effective implementation of CCS (Bajpai et al. 2022), as it can potentially improve environmental conditions and economic opportunities in the future. The sequestered carbon can persist within the Earth's crust for an extensive duration, perhaps spanning trillions of years. However, soil acidification is possible in the event of a carbon leakage, and emissions are released into the atmosphere (Anwar et al. 2018; Moore et al. 2022). Hence, to attain significant environmental sustainability and address the challenges posed by global warming, it is imperative to engage in worldwide and collaborative endeavors to research, enhance, and apply novel techniques for CCS (Anwar et al. 2018).

25.3.2 Storage Solutions for Greenhouse Gases

Increased concentrations of greenhouse gases (GHGs), including carbon dioxide (CO_2), methane (CH_4), nitrous oxide (N_2O), hydrofluorocarbons (HFCs), perfluorocarbons (PFCs), and sulfur hexafluorocarbon (F6C), contribute to global warming (GW) (Markeb and Mohamed 2017). At the moment, carbon dioxide (CO_2) and methane (CH_4) comprise the majority of greenhouse gases emitted, each at a proportion of 18%. Extensive research has been conducted on CO_2 due to its substantial emission levels (Yoro and Daramola 2020). The combustion of fossil fuels for energy and electricity production is predominantly responsible for the global emission of greenhouse gases, including carbon dioxide (CO_2) and methane (CH_4), which constitute the principal source of air pollution and the greenhouse effect. As a result, an abundance of issues have surfaced, such as climate change and global warming, which are poised to significantly disrupt the standard of living for all humans (Welzer 2015). Numerous strategies have been proposed to further mitigate the excessive emission of greenhouse gases, including using carbon dioxide (CO_2) and methane (CH_4). This technological advancement can reduce atmospheric CO_2 levels while producing sustainable energy in syngas (Liu et al. 2019b). Hence, employing CO_2 and CH_4 as fuel sources is feasible to mitigate the energy challenges arising from population expansion (Saravanan et al. 2021). Four primary domains comprise applying the CO_2 and CH_4 system: plasma technology, increased catalyst systems, photocatalytic reduction, and electrochemical reduction (Sher Shah et al. 2020).

Nanoparticles are expected to exhibit exceptional and unmatched characteristics, rendering them highly efficient at capturing CO_2 and CH_4 (Vasileff et al. 2017). Presently, the most auspicious outcomes concerning the absorption of greenhouse gases (GHGs) are attained via adsorbents comprising porous supports fortified with metals or amine (Kamran and Park 2021). The Fe_3O_4-graphene and MOF-117 nanoparticles demonstrated the most effective capacity for adsorbing CO_2 due to their remarkable thermal stability and substantial porosity. Although Isoreticular Metal-Organic Framework-6 (IRMOF-)6, ordered mesoporous carbon, MOF-177, and MOF-5 have exhibited enhanced capabilities in adsorbing CH_4, their practical efficacy is considerably diminished in comparison to that of graphene or nanoparticles based on graphene (Zeleňák and Saldan 2021). One notable challenge pertains to the mitigation of costs linked to the deployment and enlargement of the technology, which is essential for effectively tackling the issue of emissions from power production and attaining neutrality in carbon dioxide and methane emissions (Bajpai et al. 2022). Further investigation is necessary to ascertain the practical, long-lasting efficacy of the nanoparticles as absorbents. Moreover, future research must prioritize exhaustive investigations into the precise mechanisms through which nanoparticles assimilate greenhouse gases. Developing a better understanding of these mechanisms will help nanoparticles capture and store greenhouse gasses more effectively, which is critical for creating scalable solutions to fight climate change (Alonso et al. 2017).

25.4 Water Purification and Desalination

25.4.1 Nanotechnology in Water Treatment

The prevalence of organic and inorganic pollutants is a significant threat to ecosystems worldwide. There has been a dramatic rise in the frequency and duration of pollution in the last several years (Kurwadkar et al. 2020). Several methods have been employed to raise the standard of naturally occurring water so people can safely consume it. Thanks to integrating cutting-edge nanotechnology into conventional process engineering, water and wastewater treatment have come a long way (Mauter et al. 2018). By facilitating water purification, reuse, and recycling, nanotechnology may increase water resources' quality, availability, and sustainability (Manikandan et al. 2022).

Although sedimentation, charcoal filtration, and sand filtration are prevalent water treatment techniques, they are impractical for large-scale applications (Zhao et al. 2019). Precise characteristics such as cost-effectiveness, photocatalytic capability, adsorption capability, and rapid adsorption and desorption kinetics have been identified as critical attributes in alternative materials (Kefeni et al. 2017). Recently, an intriguing area of study has implemented nanoparticles to eliminate contaminants. Nanomaterials have a significantly higher surface-area-to-volume ratio than conventional micrometer-sized adsorbents. This results in enhanced photocatalytic activity, a greater adsorption capacity, more effective substance removal, and quicker removal kinetics. Nanotechnology studies the precise manipulation of matter at the molecular and atomic scales. Its primary objective is utilizing scientific concepts and methods to effect substantial environmental protection progress. Technological advancements in nanotechnology have enabled scientists to fabricate nanoparticles with the desired properties (Keçili et al. 2019). Nanoparticles are utilized extensively in various manufacturing sectors due to their exceptional characteristics, including electrical, catalytic, magnetic, optical, and antibacterial actions, among others. The synthesis of biogenic nanoparticles has recently experienced a surge in prominence, resulting in the production of environmentally friendly and cost-effective nanoparticles (Elayakumar et al. 2019).

A wide variety of therapeutic uses are made possible by nanoparticles and related technologies and devices, depending on their unique properties (Guerra et al. 2018). It has many different uses, including treating heavy metal pollution, decontaminating wastewater, cleaning up solid waste, and removing hydrocarbons (HC) and radioactive materials (Rajendran et al. 2022). The worldwide effort to treat water has significantly been impacted by nanotechnology, especially those aimed at reducing water scarcity and enhancing water quality. The use of nanotechnology in South Africa's water treatment initiatives is a well-known example. Water-filtering devices based on nanotechnology have been introduced in South Africa to guarantee safe irrigation and enhance the quality of drinking water. These devices efficiently purge water from pollutants, heavy metals, and pathogens using nanoparticles. Due to the effectiveness of these initiatives, South Africa is now a pioneer in the use of nanotechnology for sustainable water management, serving as an example for other nations with comparable water-related issues.

Adding nanoparticles to aerobic digestion also makes a considerable difference (Hassanein et al. 2021). Table 25.1 summarizes various nanotechnology-enabled water treatment processes in detail. The cleaning procedure makes use of a wide variety of nanoparticles. When cleaning up polluted environments, three nanomaterials are most often used: organic, inorganic, and polymer-based (Kalita and Baruah 2020) (Fig. 25.2).

25.4.2 Desalination Technologies

The desalination techniques commonly employed in commercial applications, such as osmosis (both forward and reverse), rely on employing artificial membranes. However, these membranes have significant limitations, including reduced water flow, susceptibility to biofouling, short lifespan, diminished ability to repel water, and limited effectiveness in removing salt (Saleem et al. 2020). Thus, nanotechnology has been integrated with conventional techniques to overcome these constraints. Several nanocomposites have been used with traditional methods to enhance the results (Malik et al. 2022). Nanocomposites exhibit a range of advantageous features, including exceptional hydrophilicity, high photocatalytic and photodegradative activities, as well as excellent self-cleaning and antifouling properties (Nehra et al. 2022). Utilizing nanocomposites in conjunction with desalination procedures enhances the water flux and the removal of salt ions in membranes.

Furthermore, these membranes exhibit significantly higher stability than conventional membranes (El Batouti et al. 2021). Moreover, these nanocomposite membranes exhibit exceptional cost-effectiveness, operational flexibility, and a

Table 25.1 Different nanotechnology-mediated water treatment processes

Process	Description	References
Nanofiltration	Nanofiltration utilizes nanoscale membranes with delicate pores to separate contaminants from water based on size exclusion, effectively removing particles, bacteria, viruses, and dissolved ions	Fahimirad et al. (2021)
Nanoscale adsorbents	Nanoparticles, such as activated carbon nanoparticles or metal oxide nanoparticles, are used to adsorb contaminants from water through surface interactions, effectively removing organic compounds, heavy metals, and other pollutants	Soni et al. (2020)
Photocatalysis	Photocatalytic processes involve using semiconductor nanoparticles, such as titanium dioxide (TiO_2) or zinc oxide (ZnO), which, when exposed to light, generate reactive oxygen species that degrade organic pollutants and kill bacteria in water	Puri and Gupta (2023)
Nanomembrane filtration	Nanomembrane filtration employs thin membranes with nanoscale pores to selectively remove contaminants from water, including bacteria, viruses, heavy metals, and organic compounds while allowing clean water to pass through	Chadha et al. (2022)
Nanoscale zero-valent iron (nZVI)	Nanoscale zero-valent iron particles are used to remediate contaminated water by promoting reducing and degrading pollutants through redox reactions, effectively removing heavy metals and organic compounds	Alazaiza et al. (2022)
Nanobubble technology	Nanobubbles, nanoscale gas bubbles, enhance water treatment processes by improving gas transfer efficiency, increasing oxygen levels, and facilitating the removal of contaminants through flotation or oxidation processes	Wu et al. (2021)
Nanosensors	Nanotechnology-based sensors are used for real-time monitoring of water quality parameters, such as pH, conductivity, dissolved oxygen, and specific contaminants, enabling early detection and prompt water pollution remediation	Verma et al. (2024)
Nanocomposite membranes	Nanocomposite membranes incorporate nanoparticles into membrane matrices to enhance filtration efficiency, selectivity, and durability, providing improved removal of contaminants such as bacteria, viruses, and organic pollutants from water	Nasir et al. (2022)

Source: Ajith et al. (2021)

remarkable capacity for salt elimination. These qualities make them particularly suitable for integration with desalination procedures (Goh et al. 2016).

Desalination typically uses two types of membranes: integrated nanomembranes, or nanocomposites, and free-standing nanomembranes (Rabiee et al. 2023). Numerous nanomaterials, such as nanoparticles, nanosheets, and nanofibers, have enhanced conventional membranes in desalination procedures (Daer et al. 2015). However, traditional desalination membranes have been restricted because of their poor water recovery and requirement for high-pressure operation, especially when creating large volumes of liquid waste (Yusuf et al. 2020).

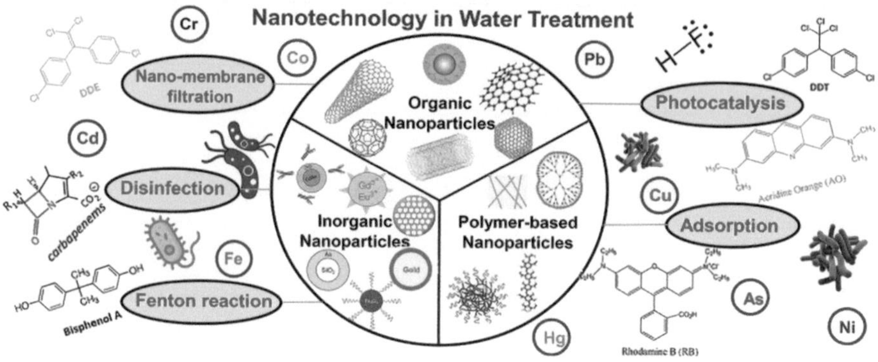

Fig. 25.2 Use of nanotechnology for wastewater treatment. (Source: Ajith et al. 2021)

Since nanotechnology uses nanoparticles to improve desalination results, it is currently the most sought-after field of study for desalination (Thakur et al. 2021). Desalination nanocomposite membranes have been made using nanomaterials such as bimetallic and bifunctional nanoparticles, Mxene-based nanocomposites, carbon nanotubes, graphene-based nanocomposites, and Halloysite nanotubes (Saleh and Hassan 2023). Graphene-, carbon nanotube-, and zeolite-based nanomaterials are continuously being developed to advance desalination technologies. These nanoparticles significantly reduce the time and cost required for the desalination process Incorporating nanotechnology into traditional desalination techniques, waste purification, and wastewater treatment could benefit from improved performance. Costs would go down, and efficiency would increase (Bhoj et al. 2021; India 2018).

Currently, the globe provides about 300 million people with daily desalination of over 95 million cubic meters of water. Desalinated water can cost anywhere from \$0.50 to \$3.00 per cubic meter (1000 L); however, this can vary significantly based on the technique employed and the local conditions (Ghaffour et al. 2013; Jones et al. 2019). For instance, desalinating saltwater typically costs between \$1.00 and \$2.00 per cubic meter, whereas desalinating brackish water can cost less, usually between \$0.50 and \$1.00 per cubic meter. Energy prices, the size of the plant, and the quality of the source water all have an impact on these expenses (Elimelech and Phillip 2011; Voutchkov 2018) (Fig. 25.3).

25.5 Smart Agriculture

25.5.1 Nanosensors for Precision Agriculture

To maximize profits, farmers consistently look for ways to reduce the cost of agricultural inputs. This goal is achieved when farmers maximize crop productivity by applying pesticides, herbicides, and fertilizers (Panhwar et al. 2019). Because of the overuse of agrochemicals, there is now a significant tradeoff between soil and

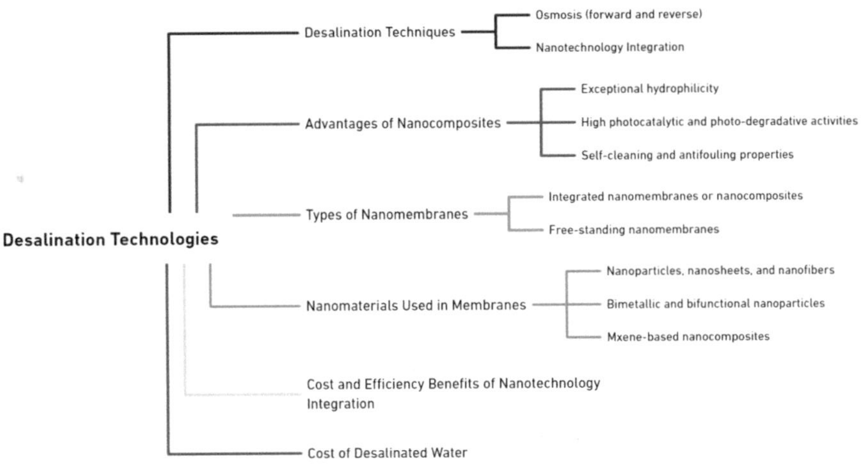

Fig. 25.3 Several varieties of desalination practices

groundwater health and increased crop output (Saha and Bauddh 2020). Due to population expansion, the world's farming lands have increased unparalleled in recent decades (Meyer and Turner 1992). The use of agrochemicals rises in tandem with the expansion of farms, promoting soil, water, and air contamination. As pollution levels continue to rise, experts are under increasing pressure to find solutions that will help farmers and the world (Wang et al. 2022). There is growing pressure on farmers to decrease their use of agrochemicals by embracing alternative agricultural practices, as this issue is becoming recognized globally (Muller et al. 2017).

Precision agriculture aims to reduce the use of agrochemicals and increase economic returns through crop-specific, site-specific treatments (Yang et al. 2016). Precision farming aims to increase harvest yields with fewer chemical inputs like fertilizers, insecticides, and herbicides (Shafi et al. 2019). Precision agriculture that uses nanotechnology measures environmental and crop-based characteristics using computers, Global Positioning System (GPS), and other remote sensing devices (Kaushal and Wani 2017). The component of nanotechnology known as nanomaterials (NM) has distinct properties that set them apart from the materials from which they originate. These materials' surface areas, cation exchangeability, and ion absorption capacities are usually much higher than their bulk equivalents (Chen et al. 2019; Kiaee et al. 2022). Precision farming uses efficient monitoring tools and methods to reduce the application of fertilizers, pesticides, and herbicides. Optimized nutrient utilization and disease resistance can be achieved through the targeted controlled release of agrochemicals, according to this approach (Duhan et al. 2017). Nanofertilizers (NFs), nanoherbicides, nanopesticides, nanoscale transporters, and nanosensors are all examples of such goods (Bratovcic et al. 2023). Precision nanotechnology-based agriculture allows farmers to decrease pesticide use without sacrificing crop yields, water and soil health, or environmental friendliness (Pramanik et al. 2020).

Nanotechnology is revolutionizing precision agriculture by enhancing agricultural yields, decreasing waste, and minimizing environmental impacts. The nanoscale material modification allows for distinct features and advantages over traditional farming practices (Acharya and Pal 2020). Providing more precise data on the locations and amounts of water and fertilizer plants requiring nanosensors to monitor soil and plant health in real-time could help farmers gain a more substantial grip over their crops (El-Chaghaby and Rashad 2023). One way to lessen the adverse effects of pesticides and fertilizers on the environment is to use nanoparticles (NPs) as delivery vehicles. Plant fibers and seeds are examples of agricultural materials that can improve their qualities by nanotechnology, increasing their resistance to pests and weathering (Shang et al. 2019). However, further study is required to determine the hazards and their effects on water, soil, and human health. Using nanotechnology in precision agriculture can transform crop growth through enhanced efficiency, sustainability, reduced waste, and environmental implications (Yadav et al. 2023b).

The distribution of these cutting-edge nanotechnologies to developing nations needs to be a top priority for international organizations in order to improve the sustainability and productivity of their agriculture. By encouraging higher resource efficiency and lowering reliance on hazardous agrochemicals, precision agriculture using nanosensors and nanomaterials can help lessen the effects of climate change (Zain et al. 2023). Since it gives developing countries the means to adapt to and lessen the negative consequences of climate change, this technology transfer is essential to attaining both environmental sustainability and global food security (Lybbert and Sumner 2012).

25.5.2 Resource Optimization in Farming

Nanotechnology can potentially transform the agro-food industry, among other exciting applications. This world is home to around seven billion people, and many developing countries lack the means to meet even the most fundamental nutritional needs (Ashraf et al. 2021). There are numerous ways in which nanotechnology could transform the production of sustainable food. For instance, it can develop more sophisticated sensors to track physical, chemical, and biological processes and qualities, enhance membranes and sorbents for decentralized water treatment and resource retrieval, and control infections more skillfully, resulting in safer food and reduced waste (He et al. 2023).

Nanotechnology in agriculture has the potential to significantly enhance crop productivity (Chhipa 2019) by lowering agricultural inputs and improving food quality. Nanotechnology offers several advantages to the agro-food industry, such as the creation of novel diagnostic and management tools, the implementation of eco-friendly insecticides and biopesticides, and the controlled release of micronutrients and fertilizers over time (Manjunatha and Devika 2023). Nanobiotechnology can enhance the nutritional quality of many crops by uncovering previously undiscovered biological characteristics. Furthermore, implementing these recommendations will prove beneficial in tackling environmental and food security issues (Mishra et al. 2017). However, subjecting these products to thorough toxicity testing before their utilization is crucial. The laws and regulations regarding this domain are also little documented, and much of the research in this sector is limited to laboratories (Hofmann et al. 2020). Moreover, experts in the industry are reluctant to make significant financial commitments in this domain due to the substantial upfront costs of production. Therefore, these worldwide issues require coordinated efforts from all perspectives (Han et al. 2021).

The following categories best describe the primary contributions of nanotechnology to the agricultural sector: This includes the following tasks: dispersing nanocide-pesticide; improving green pesticides and biopesticides by employing nanomaterials (NMs); controlling and releasing micronutrients, fertilizers, and biofertilizers with the help of NMs; transporting genetic materials for crop development with the help of NMs; and using nanobiosensors to quickly and accurately identify plant pathogens and pesticides, among other uses (Acharya and Pal 2020). Developing sensors with the ability to measure physical, chemical, or biological properties and processes, monitor chemicals, identify diseases or toxins, and monitor physical attributes is one way that nano-enabled technology contributes to improving agri-food systems' sustainability. Microbe-controlling technology can be used to reduce food waste and improve food safety. When dispersing water treatment and recovering resources, membranes and sorbents are used. Delivery products for agrochemicals are used to determine the exact time and location of application. Materials for health monitoring and improvement are used to ensure the welfare of farm animals (Rodrigues et al. 2017).

In order to protect human health and the environment, biosafety rules for agricultural nanocomposites must be put into place. To evaluate the possible dangers connected to nanomaterials, such as their toxicity, persistence in the environment, and effect on non-target organisms, biosafety protocols need to be in place (Kah et al. 2019). Regulatory frameworks should require comprehensive risk assessments, and protocols for the safe use, application, and disposal of nanocomposites should be established. This is crucial to avert unforeseen negative consequences and encourage the ethical application of nanotechnology in farming (Okeke et al. 2022).

25.6 Air Pollution Control

25.6.1 Nanoparticles in Catalytic Converters

The challenge of developing robust environmental and safety regulations is one that every developed country faces. This reason has driven the necessity of creating a virtual model of a catalytic converter before its actual implementation (Pai and Prabhu 2018). In automobile exhaust systems, catalytic converters are frequently used to reduce unwanted emissions from internal combustion engines (Manojkumar et al. 2021). To reduce the pollutants released into engines' exhaust, catalytic converters are used in various engine-powered equipment, such as generator sets, forklifts, mining equipment, cars, buses, and trains (Hamade 2018).

The catalytic converter is the primary method of controlling pollution in modern cars and chemical reactors. This product is designed to meet the increasingly stringent emission standards required by law (Wijeyakulasuriya et al. 2022). It is anticipated that the catalytic converter will achieve conversion efficiencies of at least 90% for the primary pollutants found in exhaust streams, such as NOx, hydrocarbons, and carbon monoxide (Sharma et al. 2015). Many design and operating elements affect the system's overall efficiency, which includes the system itself, the exhaust line, and the catalytic converter (Hariram et al. 2019). Temperature, flow rate, and exhaust composition are among the volatile conditions in the exhaust, making the design optimization approach especially difficult (Odenbrand et al. 1999).

A catalytic converter provides the perfect setting for enabling a chemical reaction. This technique produces less reactive gases in place of potentially harmful gaseous byproducts of internal combustion engines. Usually, the vehicle releases these gases into the surrounding atmosphere (Hamade 2018). Within the monolith, a chemical reaction is triggered by the rising temperature. All hazardous gases should ideally be subjected to a conversion procedure that produces less dangerous byproducts such as carbon dioxide, nitrogen gas, water, and other comparable substances (Othman Aljamal 2017). However, conventional catalytic converters might find it challenging to efficiently reduce such a large amount of undesired contamination if the engine exhaust contains a high concentration of undesirable pollutants, such as high levels of nitrous oxides (NOx), carbon oxides (COx), and unburned hydrocarbons (HC) (Chaudhary and Thakur 2019). In such cases, adding a catalyst that increases the catalytic converter's contact surface area may be helpful and lengthen its response time (Hamade 2018).

25.6.2 Emission Reduction Technologies

There is a rising awareness of environmental issues as urban civilization develops further. The consequent harmful and poisonous environmental pollutants are particularly problematic. To protect human health and the environment, worldwide efforts are afoot to lessen pollution. Many believe that nanotechnology will play a significant role in helping to keep the planet habitable. Over the past two decades, scientists have studied several nanomaterials in depth to see whether they can solve or at least ameliorate some of the most pressing environmental problems, such as air pollution (Mazari et al. 2021). Nanomaterials have several applications in pollution detection, environmental remediation, and energy generation as shown in Table 25.2. These include carbon-based nanomaterials, metallic nanoparticles, transition metal dichalcogenides, and metal-organic frameworks (Yadav et al. 2023a). Graphene oxide, fullerene, carbon nanotubes, and virgin graphene are all examples of carbon-based nanomaterials. These unique materials have huge pore volumes, excellent water stability, and high specific surface areas (Fatima and Mushtaq 2023). Sensors, membrane separation, catalytic degradation, energy conversion and storage, and

Table 25.2 Nanoparticle application in air pollution remediation

Application	Description	References
Nanoparticle-based filters	Nanoparticles, such as titanium dioxide or carbon nanotubes, are used in air filters to capture particulate matter (PM), volatile organic compounds (VOCs), and heavy metals	Saleem et al. (2022)
Photocatalytic degradation	Nanoparticles with photocatalytic properties, like titanium dioxide (TiO_2) or zinc oxide (ZnO), catalyze the breakdown of pollutants into harmless substances when exposed to light, such as sunlight or ultraviolet (UV) light	Ani et al. (2018)
Nanoparticle-coated surfaces	Surfaces coated with nanoparticles can capture and deactivate pollutants through adsorption and chemical reactions; examples include coating building exteriors or road surfaces with nanoparticles to reduce pollution levels in urban areas	Momba et al. (2017)
Nanoemulsions for air purification	Nanoemulsions containing nanoparticles dispersed in water are sprayed into the air to capture and neutralize pollutants, providing a cost-effective and efficient method for air purification	Saritha et al. (2022)
Nanoparticle-enhanced catalytic converters	Incorporating nanoparticles, such as cerium oxide (CeO_2) or palladium (Pd), into catalytic converters improves their efficiency in converting harmful gases like nitrogen oxides (NOx), carbon monoxide (CO), and hydrocarbons into less toxic substances	Hamill et al. (2022)
Nanoparticle-based sensors	Nanoparticles are used in sensors to detect and monitor air pollutants in real-time, providing valuable data for pollution control and management efforts	Koedrith et al. (2015)

adsorption-based pollution treatment are just a few environmental applications that have heavily used these materials (Sahoo 2017). The unusual physical chemistry properties, tiny size, excellent biocompatibility, and abundance of inexpensive carbon dots (CDs) precursors have recently attracted a lot of interest (Tejwan et al. 2021). The use of CDs has varied throughout several fields, and the number of articles discussing them has increased yearly. The following features of CDs make them very promising in the environmental domain: Firstly, their unique optical properties include, but are not limited to, strong fluorescence, remarkable stability when exposed to light, highly controllable photoluminescence, and excellent optoelectronic capabilities (Cui et al. 2021). Secondly, CDs have a wide range of beneficial physicochemical properties due to the variety of raw materials and synthetic methods used to make them. These characteristics include a high concentration of functional groups on the surface, a tiny size, an enormous surface area, strong charge transport capabilities, and the ability to donate and absorb electrons very well (Saputra et al. 2024). Thirdly, CDs are eco-friendly, inexpensive, and compatible with biological systems. Thus, the distinct properties of CDs have made them useful in several applications, including water treatment membrane fabrication, antibacterial agent treatment against hazardous microbes, and contamination monitoring and removal (Mehta et al. 2019).

To mitigate climate change, reducing emissions from industrial sources is essential, and nanotechnology offers creative alternatives. For instance, the use of nanofilters in cement and fertilizer plants significantly reduces CO_2, NOx, and SOx emissions. These filters reduce the environmental impact of these businesses by capturing and converting pollutants using cutting-edge materials, including metal-organic frameworks and titanium dioxide (Gopinath et al. 2021; Mamaghani et al. 2020). According to Pérez-Fortes et al. (2016), putting such solutions into practice not only helps cut greenhouse gas emissions but also creates a sustainable precedent for other industries.

25.7 Climate Monitoring and Sensing

25.7.1 Improved Sensors for Climate Studies

Nanoscale technologies provide more effective detection, treatment, and remediation of environmental toxins. Nanoscale devices composed of distinct nanomaterials can identify and monitor unanticipated ecological issues. Continuous and accurate surveillance of air pollution levels is an essential strategic measure in mitigating pollution in the environment. NP-based sensors enable rapid detection of air contaminants (Chaudhary et al. 2022). Achieving significant progress in this domain, researchers have successfully generated intelligent dust, a collection of exceptionally lightweight computational nanosensors capable of remaining airborne for extended periods. These nanosensors distinguish themselves from competitors due to their compact size, exceptional sensitivity, and minimal energy consumption.

Additionally, they offer a highly cost-efficient solution (Pandit et al. 2016). Carbon-based nanosensors detect pollutants without odor (Nehra et al. 2019). Graphene, a carbon-based substance, is becoming increasingly important in developing nanosensors because of its remarkable optical and electrochemical properties (Fahmy et al. 2022).

Instead of utilizing traditional environmental monitors, you can employ contemporary detectors known as solid-state gas sensors. These sensors are inexpensive, small, and portable. Moreover, they serve as a convenient and efficient means of monitoring environmental conditions since they can be easily put in any location to collect data and then disseminated to the public through a wireless network system (Yi et al. 2015). This portable solution enables the swift distribution of pollution levels across numerous sites simultaneously by utilizing Bluetooth communication technologies and GPS (David 2020). The device consists of gas sensors made of solid-state materials, which are connected to a personal digital assistant (PDA). Internet Geographic Information System (GIS) enables the Process Control Diagram (PCD) to access up-to-date air quality report data, enhancing public awareness and engagement (Franchi et al. 2023). To predict future events and determine the best approach for researching air pollution, researchers have used data collected at each monitoring site, geostatistical interpolation, and local deterministic approaches (Das et al. 2015).

25.7.2 Enhancing Climate Data Collection

Climate change is becoming an increasingly hot topic because it has such a profound impact on people worldwide. To analyze and model climate change, researchers need access to massive amounts of historical and real-time data (Urban et al. 2016). Immediate and dynamic monitoring of Earth's land, atmosphere, and oceans is possible using several Earth observation platforms. These systems provide massive, accurate, real-time data on the Earth's surface. The methods used to gather data and understand the Earth system have been significantly altered by using multiple sources of observational data from the planet. It also has an essential role in easing the way for new scientific discoveries (Pei et al. 2021). When using observation data from many platforms and sensors, data inversion and information extraction issues can be avoided (Zheng et al. 2022) compared to using data from a single sensor. In addition to overcoming the shortcomings of irregular spatiotemporal observations, the consistent and long-lasting spatial data from Earth observations from multiple sources is an essential basis for studying global change (Jia et al. 2020).

All three layers of our planet's complex system—the atmosphere, the oceans, and the land—are studied concerning climate change. Consequently, a wide range of scientific evidence is required to uncover the nuanced elements of climate change. Researchers have gathered much data through synchronous satellite aerial ground observation trials, including data from various platforms, bands, and scales (Ding and Jin 2024). Using this data, researchers have explored new ideas, technology, and approaches to understanding climate change. They have developed assimilation models combining several spatial data sources to efficiently and accurately identify important climate change aspects. As a bonus, they have created simulation platforms tailored for investigations of regional climate change research (Zhang et al. 2016). Essential geographic information data, data gained from on-site observations, and data produced by the Earth system and data assimilation models are additional examples of scientific data that can be acquired. This Earth observation data, when combined with auxiliary data, demonstrates key characteristics often associated with big data, such as volume, variety, velocity, and veracity (Guo et al. 2015).

25.8 Material Efficiency and Resource Conservation

25.8.1 Nanoengineered Materials

The search for renewable power sources has become exceedingly crucial in light of the increasing worldwide energy and environmental issues. Combining nanoparticles and enzyme immobilization techniques makes it possible to achieve economically feasible and technologically advanced sustainable energy solutions (Kim et al. 2018). Nanomaterials surpass standard materials in various settings because they can be customized according to our individual needs. Nanoparticles possess unique physicochemical features due to their remarkably high surface-area-to-volume ratio (Ndukwu et al. 2020).

Nanoengineered biomaterials have the potential to help the bioenergy industry considerably. These materials possess unique features and functionalities at the nanoscale, which can significantly enhance bioenergy processes' efficiency, scalability, and sustainability (Khoo et al. 2020). Microorganisms' cells are stimulated for biofuel production with catalysts such as nanoparticles, nanotubes, and nanosheets (Bhardwaj et al. 2021). Investigating catalysts that improve essential processes, such as breaking down biomass, catalyzing enzymatic reactions, or enhancing electrochemical activities, is vital for the synthesis of biofuels, production of biogas, and creation of microbial fuel cells (Assad et al. 2022). Functionalizing high surface area nanoparticles can potentially boost their catalytic activity. Enhancing selectivity and catalytic activity can be achieved by functionalizing with specific ligands or groups. The matrix aimed to achieve enhanced catalytic performance by pursuing objectives such as developing a more stable catalyst, facilitating efficient mass transfer between reactants and products, and improving overall catalytic functioning (Zhang et al. 2022).

By applying careful nanoscale design and engineering, bioengineered materials can maximize bioenergy systems' performance, such as biofuel and bioelectricity generation. They can prolong the life of bioenergy systems, increase the range of feedstocks that can be used, and improve the effectiveness of energy conversion processes (Jamuna et al. 2023). Further evidence of this method's sustainability comes from its capacity to reduce greenhouse gas emissions, control waste generation, and promote a circular economy by turning waste biomass into a helpful resource (Cantzler et al. 2020). Energy security and resilience are increased by constructing decentralized energy production systems made possible using nanoengineered biomaterials. According to recent studies, nanoparticles can improve the activation of immobilized enzymes during hydrolysis and esterification processes, as well as provide enhanced protection against denaturation (Pandya et al. 2022). The combined impacts of immobilized enzymes, enhanced immobilization methods, and chemically enhanced catalytic capabilities are some advantages of this approach (Satyanarayana and Reddy 2018).

25.8.2 Sustainable Product Development

Applications of active nanoparticles have been implemented in several fields, including resource conservation, air pollution control, and climate change mitigation. These nanoparticles might be applied to other industries, such as agriculture (Ashraf et al. 2021). At the moment, opinions regarding assessing the sustainability of intelligent nanomaterials and products made from them in the long run are divergent. Life cycle assessment (LCA) studies must be conducted during the early stages of their development to thoroughly understand the advantages and disadvantages of innovative nanomaterials. However, there is currently a lack of assessment of the complexity and dynamism of these materials. These materials concern environmental and human health due to their flexibility and susceptibility to environmental stimuli (Chaud et al. 2021). Experts and government agencies agree that there are currently uncertainties and a lack of clear rules when evaluating and controlling the hazards associated with intelligent nanomaterials. Due to a lack of understanding and classification, current risk assessment and regulatory frameworks are probably ill-equipped to identify and address specific sustainability and safety concerns (Jahnel 2015).

The potential of nanotechnology to support sustainable economic growth is a topic of much discussion (OECD 2013). The essence of green nanotechnology, green nanoscience, or green nanomaterials is the application of green chemistry principles to the design, production, use, and disposal of nanomaterials and nano-enabled items (Rani et al. 2022). The creation and application of safer nanomaterials, such as nitrocellulose, as an alternative to synthetic nanopolymers, as well as the adoption of synthesis methods that employ safer and renewable starting materials, reagents, and solvents, are recent advancements in this field. Functional issues with materials and smart nanoparticles are frequent. Furthermore, all problems and difficulties with nanoscale materials are addressed by intelligent nanomaterials (Thangudu 2020).

To the fullest extent possible, society and all industrial sectors will not be able to ultimately reap the benefits of intelligent nanomaterials unless developers, safety assessors, and regulators have open and continuous dialogue and information exchange. This will better address several social and ethical issues, such as those about sustainability and safety. One method to accomplish this is to create a "trusted environment" where innovators and regulators feel at ease exchanging information (Gottardo et al. 2021). Distributing modern technologies like intelligent nanomaterials to impoverished nations is complex. Fewer than 50 countries use this technology. Infrastructure and technology limitations, financial and resource constraints, legal and policy impediments, and policymaker and public ignorance are the key barriers. International collaboration and knowledge-sharing platforms, targeted training and capacity-building programs, global organization funding, and public–private partnerships can address these concerns. Creating a "trusted environment" where inventors, safety assessors, and regulators may freely share knowledge is essential to addressing social and ethical challenges and using clever nanomaterials responsibly and sustainably. Poor countries can use these technologies to solve water, energy, agriculture, healthcare, and environmental problems by overcoming these constraints.

25.9 Conclusion

25.9.1 Summary of Key Points

Global populations are experiencing adverse effects due to climate change and other environmental issues. It is imperative to address climate change without delay. To mitigate the impacts of climate change, it is essential to replace outdated practices that contribute to pollution and global warming with sustainable technologies. Ultimately, nanotechnology has demonstrated its practicality as a technology capable of supplanting conventional alternatives and providing sustainable solutions. Advancements in nanobiotechnology and nanotechnology have positive effects on the environment. Nanotechnology offers several sustainable solutions for numerous environmental challenges, such as wastewater treatment, greenhouse gas emissions, the fuel crisis, and the cleanup of different pollutants, all of which can contribute to climate change. Hence, nanotechnology is increasingly being utilized in environmental applications and has the potential to serve as a solution for mitigating the effects of global warming. Metal-organic frameworks (MOFs), carbonaceous materials, nano zeolites, nano silica, functionalized nanomaterials, nanopackaging, nanocomposites, nanosensors, nanocoatings, nanolubricants, nanometals, nanocatalysts, and similar substances are all efficient and helpful tools for environmental applications. Due to their distinctive characteristics, these nanoparticles can be utilized in several industries, such as wastewater treatment, bioenergy, environmental remediation, catalysis, sustainable materials, and greenhouse gas storage. It is now time to enact climate change policy at both regional and global levels.

Furthermore, nanotechnology possesses immense promise in developing environmentally sustainable practices, products, and procedures to mitigate the effects of climate change. Overall, nanotechnology is poised. Enacting regional and global climate change legislation and actively promoting nanotechnology and its sustainable solutions to poor countries are essential to closing this gap. International collaboration, capacity-building, and targeted finance are needed to deliver these transformative technologies in regions that need them most. Making nanotechnology accessible and empowering poor countries to use sustainable solutions can help mitigate climate change and create a more equitable and environmentally resilient future for all.

25.9.2 Future Prospects and Challenges

The application of nanotechnology has already yielded significant advantages for society, with several untapped opportunities remaining for developing and utilizing novel technologies. However, potential risks come with the benefits of nanotechnology research and development. Frequently, revolutions yield unforeseen consequences and can even engender counteractions, such as in the case of genetically modified organisms in agriculture. Our understanding of the impact of nanotechnology on the environment and human health is still incomplete. To ensure the safety and sustainability of nanotechnology and to establish it as a credible catalyst for a sustainable planet and future, it is necessary to assess and mitigate these hazards.

One of the most significant research challenges to the growth of nanotechnologies is the lack of quantification about the consequences of the circular economy on nanomaterials, industry, and economies. Standardized methods will be required to measure and develop best-practice standards in the circular nanomaterials economy and its effects to address this issue. Decision support tools will be essential for the initial design and optimization of products and production processes that use artificial intelligence (AI)-based green chemistry to eliminate waste and reduce exposure to dangerous nanomaterials. Further research is required to investigate the risks associated with nanotechnology and nanomaterials, as well as their impact on the environment, health, and safety. Systematic review approaches, evidence integration techniques utilizing machine learning, and creative approach methodologies are essential for analyzing the toxicity and exposure profiles of materials, goods, and processes. To identify barriers to progress and explore alternative policy approaches, it is necessary to assess the existing legislation and policies about nanotechnology and nanomaterials.

25.10 Conclusion

Finally, one of our most pressing problems may be better addressed thanks to the coming together of nanotechnology and climate change. When applied to energy systems, nanotechnology's adaptable solutions have the potential to usher in a new era of sustainability and resource efficiency.

An enormous amount of clean energy might be unlocked by using nanomaterials for renewable energy. A faster transition to a low-carbon economy is possible with the help of nanomaterials' unique characteristics, which include enhanced conductivity and catalytic activity. Nanotechnology is essential for carbon capture, utilization, and sequestration to economically reduce greenhouse gas emissions. We can close the carbon loop and help mitigate climate change by converting carbon dioxide into valuable goods using adsorbents, membranes, and catalysts based on sophisticated nanomaterials. Improvements in crop yields, efficiency in the use of resources, and reduced ecological footprint might result from advances in agriculture and food production powered by nanotechnology. We may build resilience to climate-related stresses and promote sustainable food systems by employing precision agriculture techniques, real-time soil condition monitoring, and nanomaterials for targeted pesticide application. It will be necessary to conduct safety evaluations, think about ethics, and establish regulatory frameworks to reduce risks and guarantee responsible development and deployment of nanotechnology to realize its potential in combating climate change. Scientists, politicians, business leaders, and the general public must work together if nanotechnology is to play a role in the fight against climate change in the future. A more robust, fair, and wealthy future is within our reach if we responsibly and sustainably use nanotechnology.

References

Abdin Z, Alim MA, Saidur R, Islam MR, Rashmi W, Mekhilef S, Wadi A (2013) Solar energy harvesting with the application of nanotechnology. Renew Sust Energ Rev 26:837–852

Acharya A, Pal PK (2020) Agriculture nanotechnology: translating research outcome to field applications by influencing environmental sustainability. NanoImpact 19:100232

Agarwal R, Singh S, Shalan AE (2023) Sustainable energy storage devices and device design for sensors and actuators applications. Sustainable energy storage in the scope of circular economy: advanced materials and device design, 225–290

Ahmed F, Aziz MSA, Palaniandy P, Shaik F (2022) A review on application of renewable energy for desalination technologies with emphasis on concentrated solar power. Sustain Energy Technol Assess 53:102772

Ajith M, Aswathi M, Priyadarshini E, Rajamani P (2021) Recent innovations of nanotechnology in water treatment: a comprehensive review. Bioresour Technol 342:126000

Alazaiza MY, Albahnasawi A, Copty NK, Bashir MJ, Nassani DE, Al Maskari T et al (2022) Nanoscale zero-valent iron application for the treatment of soil, wastewater and groundwater contaminated with heavy metals: a review. Desalin Water Treat 253(2021):194–210

Alonso A, Moral-Vico J, Markeb AA, Busquets-Fité M, Komilis D, Puntes V et al (2017) Critical review of existing nanomaterial adsorbents to capture carbon dioxide and methane. Sci Total Environ 595:51–62

Altabbaa S, Mann NA, Chauhan N, Utkarsh K, Thakur N, Mahmoud GA-E (2023) Era connecting nanotechnology with agricultural sustainability: issues and challenges. Nanotechnol Environ Eng 8(2):481–498

Ani I, Akpan U, Olutoye M, Hameed B (2018) Photocatalytic degradation of pollutants in petroleum refinery wastewater by TiO2-and ZnO-based photocatalysts: recent development. J Clean Prod 205:930–954

Anwar M, Fayyaz A, Sohail N, Khokhar M, Baqar M, Khan W et al (2018) CO2 capture and storage: a way forward for sustainable environment. J Environ Manag 226:131–144

Ashraf SA, Siddiqui AJ, Abd Elmoneim OE, Khan MI, Patel M, Alreshidi M et al (2021) Innovations in nanoscience for the sustainable development of food and agriculture with implications on health and environment. Sci Total Environ 768:144990

Assad H, Kaya S, Kumar PS, Vo D-VN, Sharma A, Kumar A (2022) Insights into the role of nanotechnology on the performance of biofuel cells and the production of viable biofuels: a review. Fuel 323:124277

Bajpai S, Shreyash N, Singh S, Memon AR, Sonker M, Tiwary SK, Biswas S (2022) Opportunities, challenges and the way ahead for carbon capture, utilization and sequestration (CCUS) by the hydrocarbon industry: towards a sustainable future. Energy Rep 8:15595–15616

Balasubramanian R, Chowdhury S (2015) Recent advances and progress in the development of graphene-based adsorbents for CO 2 capture. J Mater Chem A 3(44):21968–21989

Bhardwaj R, Sharma T, Nguyen DD, Cheng CK, Lam SS, Xia C, Nadda AK (2021) Integrated catalytic insights into methanol production: sustainable framework for CO2 conversion. J Environ Manag 289:112468

Bhoj Y, Pandey G, Bhoj A, Tharmavaram M, Rawtani D (2021) Recent advancements in practices related to desalination by means of nanotechnology. Chem Phys Impact 2:100025

Bozorg O (2022) Climate change drivers. Climate change in sustainable water resources management, 59

Bratovcic A, Hikal WM, Mehdizadeh M, Al Ahl HA, Omidi A, Adetunji CO et al (2023) Application of nanotechnology in agroecosystems: nanoparticles for improving agricultural production. Rev Agric Sci 11:291–309

Cantzler J, Creutzig F, Ayargarnchanakul E, Javaid A, Wong L, Haas W (2020) Saving resources and the climate? A systematic review of the circular economy and its mitigation potential. Environ Res Lett 15(12):123001

Cerqueira MA, Vicente AA, Pastrana LM (2018) Nanotechnology in food packaging: opportunities and challenges. Nanomaterials for food packaging, 1–11

Chadha U, Selvaraj SK, Thanu SV, Cholapadath V, Abraham AM, Manoharan M, Paramsivam V (2022) A review of the function of using carbon nanomaterials in membrane filtration for contaminant removal from wastewater. Mater Res Express 9(1):012003

Change IC (2014) Synthesis report. Contribution of working groups I. II and III to the fifth assessment report of the intergovernmental panel on climate change, 151(10.1017)

Chaud M, Souto EB, Zielinska A, Severino P, Batain F, Oliveira-Junior J, Alves T (2021) Nanopesticides in agriculture: benefits and challenge in agricultural productivity, toxicological risks to human health and environment. Toxics 9(6):131

Chaudhary R, Thakur M (2019) Exhaust gases pollution control using of bimetallic nanoparticles and bimetallic nanocomposites in catalytic converter: a review. Int Res J Pharm Med Sci 2(4):53–58

Chaudhary V, Ashraf N, Khalid M, Walvekar R, Yang Y, Kaushik A, Mishra YK (2022) Emergence of MXene–polymer hybrid nanocomposites as high-performance next-generation chemiresistors for efficient air quality monitoring. Adv Funct Mater 32(33):2112913

Chausali N, Saxena J, Prasad R (2022) Recent trends in nanotechnology applications of bio-based packaging. J Agric Food Res 7:100257

Chen Y, Zhang D, Wang D, Lu L, Wang X, Guo G (2019) A carbon-supported BiSn nanoparticles based novel sensor for sensitive electrochemical determination of Cd (II) ions. Talanta 202:27–33

Chhipa H (2019) Applications of nanotechnology in agriculture. Methods Microbiol 46:115–142. Elsevier

Cui L, Ren X, Sun M, Liu H, Xia L (2021) Carbon dots: synthesis, properties and applications. Nano 11(12):3419

Daer S, Kharraz J, Giwa A, Hasan SW (2015) Recent applications of nanomaterials in water desalination: a critical review and future opportunities. Desalination 367:37–48

Das S, Sen B, Debnath N (2015) Recent trends in nanomaterials applications in environmental monitoring and remediation. Environ Sci Pollut Res 22:18333–18344

David A (2020) Air pollution control monitoring & delivery rate escalated by efficient use of markov process in manet networks: to measure quality of service parameters. Test Engineering & Management, The Mattingley Publishing Co., Inc. ISSN, 0193–4120

Diego ME, Akram M, Bellas JM, Finney KN, Pourkashanian M (2017) Making gas-CCS a commercial reality: the challenges of scaling up. Greenhouse Gases Sci Technol 7(5):778–801

Ding H, Jin S (2024) A review of 3S technology and its applications in meteorology. 3S Technology Applications in Meteorology, 1–32

Duhan JS, Kumar R, Kumar N, Kaur P, Nehra K, Duhan S (2017) Nanotechnology: the new perspective in precision agriculture. Biotechnol Rep 15:11–23

El Batouti M, Al-Harby NF, Elewa MM (2021) A review on promising membrane technology approaches for heavy metal removal from water and wastewater to solve water crisis. Water 13(22):3241

Elayakumar K, Dinesh A, Manikandan A, Palanivelu M, Kavitha G, Prakash S et al (2019) Structural, morphological, enhanced magnetic properties and antibacterial bio-medical activity of rare earth element (REE) cerium (Ce3+) doped CoFe2O4 nanoparticles. J Magn Magn Mater 476:157–165

El-Chaghaby GA, Rashad S (2023) Nanosensors in agriculture: applications, prospects, and challenges. Handbook of nanosensors: materials and technological applications, 1–29

Elimelech M, Phillip WA (2011) The future of seawater desalination: energy, technology, and the environment. Science 333(6043):712–717

Fahimirad S, Fahimirad Z, Sillanpää M (2021) Efficient removal of water bacteria and viruses using electrospun nanofibers. Sci Total Environ 751:141673

Fahmy HM, Abu Serea ES, Salah-Eldin RE, Al-Hafiry SA, Ali MK, Shalan AE, Lanceros-Méndez S (2022) Recent progress in graphene-and related carbon-nanomaterial-based electrochemical biosensors for early disease detection. ACS Biomater Sci Eng 8(3):964–1000

Fatima T, Mushtaq A (2023) Efficacy and challenges of carbon-based nanomaterials in water treatment: a review. Int J Chem Biochem Sci 23:232–248

Fischer EM, Knutti R (2015) Anthropogenic contribution to global occurrence of heavy-precipitation and high-temperature extremes. Nat Clim Chang 5(6):560–564

Franchi F, Graziosi F, Di Fina E, Galassi A (2023) A survey of cloud-enabled GIS solutions toward edge computing: challenges and perspectives. IEEE Open Journal of the Communications Society

Gao Y, Jin B, Shen W, Sinko PJ, Xie X, Zhang H, Jia L (2016) China and the United States—global partners, competitors and collaborators in nanotechnology development. Nanomedicine 12(1):13–19

Ghaffour N, Missimer TM, Amy GL (2013) Technical review and evaluation of the economics of water desalination: current and future challenges for better water supply sustainability. Desalination 309:197–207

Goh P, Matsuura T, Ismail A, Hilal N (2016) Recent trends in membranes and membrane processes for desalination. Desalination 391:43–60

Gopinath KP, Rajagopal M, Krishnan A, Sreerama SK (2021) A review on recent trends in nanomaterials and nanocomposites for environmental applications. Curr Anal Chem 17(2):202–243

Gottardo S, Mech A, Drbohlavová J, Małyska A, Bøwadt S, Sintes JR, Rauscher H (2021) Towards safe and sustainable innovation in nanotechnology: state-of-play for smart nanomaterials. NanoImpact 21:100297

Guerra FD, Attia MF, Whitehead DC, Alexis F (2018) Nanotechnology for environmental remediation: materials and applications. Molecules 23(7):1760

Guo H-D, Zhang L, Zhu L-W (2015) Earth observation big data for climate change research. Adv Clim Chang Res 6(2):108–117

Hai FI, Nguyen LN, Nghiem LD, Liao B-Q, Koyuncu I, Price WE (2016) Trace organic contaminants removal by combined processes for wastewater reuse. Advanced treatment technologies for urban wastewater reuse, 39–77

Hamade TA (2018) Novel electrically stimulated catalytic converter prototype for replacement of conventional auto exhaust emission converters. J Shanghai Jiaotong Univ (Sci) 23:85–96

Hamill O, Artioli N, Goguet A, Collis N, Millington P, Collier J, Mantarosie L (2022) Mechanism understanding for NOx storage, release and reduction on Pt doped Ceria based catalysts. Book of Abstract-NAM 2022

Han J-W, Zuo M, Zhu W-Y, Zuo J-H, Lü E-L, Yang X-T (2021) A comprehensive review of cold chain logistics for fresh agricultural products: current status, challenges, and future trends. Trends Food Sci Technol 109:536–551

Hariram V, Seralathan S, Arunraja P, Arshad AS, Jaganathan R (2019) Numerical analysis on cone length and its effect on performance of an automotive catalytic converter. Int J Vehicle Struct Syst 11:2

Hassanein A, Keller E, Lansing S (2021) Effect of metal nanoparticles in anaerobic digestion production and plant uptake from effluent fertilizer. Bioresour Technol 321:124455

He Y, Matthews B, Wang J, Song L, Wang X, Wu G (2018) Innovation and challenges in materials design for flexible rechargeable batteries: from 1D to 3D. J Mater Chem A 6(3):735–753

He J, Xia S, Li W, Deng J, Lin Q, Zhang L (2023) Resource recovery and valorization of food wastewater for sustainable development: an overview of current approaches. J Environ Manag 347:119118

Hofmann T, Lowry GV, Ghoshal S, Tufenkji N, Brambilla D, Dutcher JR et al (2020) Technology readiness and overcoming barriers to sustainably implement nanotechnology-enabled plant agriculture. Nat Food 1(7):416–425

India D (2018) Nanotechnology based separation systems for sustainable water resources. Nanotechnology for sustainable water resources, 525

Jahnel J (2015) Addressing the challenges to the risk assessment of nanomaterials. Nano:485–521. Elsevier

Jamuna G, Yasodha S, Thamarai P, Vickram A, Swaminaathan P, Saravanan A, Yaashikaa P (2023) Design strategies, utilization and applications of nano-engineered biomaterials for the enhancement of bioenergy: a sustainable approach. Biochem Eng J 200:109104

Jia G, Zhang L, Zhu L, Xu R, Liang D, Xu X, Bao T (2020) Digital earth for climate change research. Manual of digital earth, 473–494

Jones E, Qadir M, van Vliet MT, Smakhtin V, Kang S-M (2019) The state of desalination and brine production: a global outlook. Sci Total Environ 657:1343–1356

Kah M, Tufenkji N, White JC (2019) Nano-enabled strategies to enhance crop nutrition and protection. Nat Nanotechnol 14(6):532–540

Kalita E, Baruah J (2020) Environmental remediation. In: Colloidal metal oxide nanoparticles. Elsevier, pp 525–576

Kamran U, Park S-J (2021) Chemically modified carbonaceous adsorbents for enhanced CO2 capture: a review. J Clean Prod 290:125776

Kaushal M, Wani SP (2017) Nanosensors: frontiers in precision agriculture. Nanotechnology: an agricultural paradigm, 279–291

Keçili R, Büyüktiryaki S, Hussain CM (2019) Advancement in bioanalytical science through nanotechnology: past, present and future. TrAC Trends Anal Chem 110:259–276

Kefeni KK, Mamba BB, Msagati TA (2017) Application of spinel ferrite nanoparticles in water and wastewater treatment: a review. Sep Purif Technol 188:399–422

Khoo KS, Chia WY, Tang DYY, Show PL, Chew KW, Chen W-H (2020) Nanomaterials utilization in biomass for biofuel and bioenergy production. Energies 13(4):892

Kiaee G, Dimitrakakis N, Sharifzadeh S, Kim HJ, Avery RK, Moghaddam KM et al (2022) Laponite-based nanomaterials for drug delivery. Adv Healthc Mater 11(7):2102054

Kim KH, Lee OK, Lee EY (2018) Nano-immobilized biocatalysts for biodiesel production from renewable and sustainable resources. Catalysts 8(2):68

Koedrith P, Thasiphu T, Weon J-I, Boonprasert R, Tuitemwong K, Tuitemwong P (2015) Recent trends in rapid environmental monitoring of pathogens and toxicants: potential of nanoparticle-based biosensor and applications. Sci World J 2015

Kovalenko I, Zdyrko B, Magasinski A, Hertzberg B, Milicev Z, Burtovyy R et al (2011) A major constituent of brown algae for use in high-capacity Li-ion batteries. Science 334(6052):75–79

Kurwadkar S, Kanel SR, Nakarmi A (2020) Groundwater pollution: occurrence, detection, and remediation of organic and inorganic pollutants. Water Environ Res 92(10):1659–1668

Kweku DW, Bismark O, Maxwell A, Desmond KA, Danso KB, Oti-Mensah EA et al (2018) Greenhouse effect: greenhouse gases and their impact on global warming. J Sci Res Rep 17(6):1–9

Lewis NS (2007) Toward cost-effective solar energy use. Science 315(5813):798–801

Liu S, Tang Z-R, Sun Y, Colmenares JC, Xu Y-J (2015) One-dimension-based spatially ordered architectures for solar energy conversion. Chem Soc Rev 44(15):5053–5075

Liu T, Zhang L, Cheng B, Yu J (2019a) Hollow carbon spheres and their hybrid nanomaterials in electrochemical energy storage. Adv Energy Mater 9(17):1803900

Liu Y, Tang H, Muhammad A, Huang G (2019b) Emission mechanism and reduction countermeasures of agricultural greenhouse gases—a review. Greenhouse Gases Sci Technol 9(2):160–174

Liu X, Yang H, Diao Y, He Q, Lu C, Singh A et al (2022) Recent advances in the electrochemical applications of Ni-based metal organic frameworks (Ni-MOFs) and their derivatives. Chemosphere 307:135729

Lowry GV, Avellan A, Gilbertson LM (2019) Opportunities and challenges for nanotechnology in the agri-tech revolution. Nat Nanotechnol 14(6):517–522

Lybbert TJ, Sumner DA (2012) Agricultural technologies for climate change in developing countries: policy options for innovation and technology diffusion. Food Policy 37(1):114–123

Malik S, Dhasmana A, Preetam S, Mishra YK, Chaudhary V, Bera SP et al (2022) Exploring microbial-based green nanobiotechnology for wastewater remediation: a sustainable strategy. Nano 12(23):4187

Mamaghani AH, Haghighat F, Lee C-S (2020) Role of titanium dioxide (TiO2) structural design/morphology in photocatalytic air purification. Appl Catal B Environ 269:118735

Manikandan S, Subbaiya R, Saravanan M, Ponraj M, Selvam M, Pugazhendhi A (2022) A critical review of advanced nanotechnology and hybrid membrane based water recycling, reuse, and wastewater treatment processes. Chemosphere 289:132867

Manjunatha A, Devika C (2023) Agriculture technologies: towards a sustainable future. Agricultural Sector in India, 267–275

Manojkumar R, Haranethra S, Muralidharan M, Ramaprabhu A (2021) IC engine emission reduction using catalytic converter by replacing the noble catalyst and using copper oxide as the catalyst. Mater Today Proc 45:769–773

Markeb AA, Mohamed A (2017) Environmental applications of engineered nanomaterials: synthesis and characterization

Mauter MS, Zucker I, Perreault F, Werber JR, Kim J-H, Elimelech M (2018) The role of nanotechnology in tackling global water challenges. Nat Sustain 1(4):166–175

Mazari SA, Ali E, Abro R, Khan FSA, Ahmed I, Ahmed M et al (2021) Nanomaterials: applications, waste-handling, environmental toxicities, and future challenges—a review. J Environ Chem Eng 9(2):105028

Mehta A, Mishra A, Basu S, Shetti NP, Reddy KR, Saleh TA, Aminabhavi TM (2019) Band gap tuning and surface modification of carbon dots for sustainable environmental remediation and photocatalytic hydrogen production–a review. J Environ Manag 250:109486

Meyer WB, Turner BL (1992) Human population growth and global land-use/cover change. Annu Rev Ecol Syst 23(1):39–61

Mirzaei A, Hashemi B, Janghorban K (2016) α-Fe 2 O 3 based nanomaterials as gas sensors. J Mater Sci Mater Electron 27:3109–3144

Mishra S, Keswani C, Abhilash P, Fraceto LF, Singh HB (2017) Integrated approach of agri-nanotechnology: challenges and future trends. Front Plant Sci 8:254477

Mitra S, Maiti DK (2021) Nanotechnology for green energy and sustainable future. In: Nano tools and devices for enhanced renewable energy. Elsevier, pp 521–533

Momba MN, Baloyi L, Mpenyana-Monyatsi L, Kamika I (2017) Nanotechnology-based filters for cost-effective drinking water purification in developing countries. In: Water purification. Elsevier, pp 169–208

Moore D, Heilweck M, Petros P (2022) Planetary bioengineering on earth to return and maintain the atmospheric carbon dioxide to pre-industrial levels: assessing potential mechanisms. Front Astron Space Sci 9:797146

Mourad A, Aissa A, Said Z, Younis O, Iqbal M, Alazzam A (2022) Recent advances on the applications of phase change materials for solar collectors, practical limitations, and challenges: a critical review. J Energy Storage 49:104186

Muller A, Schader C, El-Hage Scialabba N, Brüggemann J, Isensee A, Erb K-H et al (2017) Strategies for feeding the world more sustainably with organic agriculture. Nat Commun 8(1):1–13

Naithani S (2016) Plants and global climate change: a need for sustainable agriculture. Curr Plant Biol 6:1–1

Nasir AM, Adam MR, Kamal SNEAM, Jaafar J, Othman MHD, Ismail AF et al (2022) A review of the potential of conventional and advanced membrane technology in the removal of pathogens from wastewater. Sep Purif Technol 286:120454

Ndukwu M, Ikechukwu-Edeh C, Nwakuba N, Okosa I, Horsefall I, Orji F (2020) Nanomaterials application in greenhouse structures, crop processing machinery, packaging materials and agro-biomass conversion. Mater Sci Energy Technol 3:690–699

Nehra M, Dilbaghi N, Hassan AA, Kumar S (2019) Carbon-based nanomaterials for the development of sensitive nanosensor platforms. In: Advances in nanosensors for biological and environmental analysis. Elsevier, pp 1–25

Nehra S, Hasan SZ, Sharma R, Kumar D (2022) Advanced carbon-based nanomaterials for treatment of water. In: Applications of advanced nanomaterials in water treatment. CRC Press, pp 25–44

Odenbrand CI, Blanco J, Avila P, Knapp C (1999) Lean NOx reduction in real diesel exhaust with copper and platinum titania based monolithic catalysts. Appl Catal B Environ 23(1):37–44

OECD (2013) Nanotechnology for green innovation. OECD Publishing
Okeke ES, Ezeorba TPC, Mao G, Chen Y, Feng W, Wu X (2022) Nano-enabled agrochemicals/materials: potential human health impact, risk assessment, management strategies and future prospects. Environ Pollut 295:118722
Othman Aljamal M (2017) Reduction of the car toxic emissions by using double catalytic converter
Pai D, Prabhu MK (2018) Recent advances in substrate materials and thermal analysis of catalytic converters. Mater Today Proc 5(11):24221–24230
Panda PK, Grigoriev A, Mishra YK, Ahuja R (2020) Progress in supercapacitors: roles of two dimensional nanotubular materials. Nanoscale Adv 2(1):70–108
Pandit S, Dasgupta D, Dewan N, Prince A (2016) Nanotechnology based biosensors and its application. Pharm Innov 5(6, Part A):18
Pandya HN, Parikh SP, Shah M (2022) Comprehensive review on application of various nanoparticles for the production of biodiesel. Energy Sources Part A Recov Util Environ Eff 44(1):1945–1958
Panhwar QA, Ali A, Naher UA, Memon MY (2019) Fertilizer management strategies for enhancing nutrient use efficiency and sustainable wheat production. In: Organic farming. Elsevier, pp 17–39
Pei T, Xu J, Liu Y, Huang X, Zhang L, Dong W et al (2021) GIScience and remote sensing in natural resource and environmental research: status quo and future perspectives. Geogr Sustain 2(3):207–215
Pérez-Fortes M, Schöneberger JC, Boulamanti A, Tzimas E (2016) Methanol synthesis using captured CO2 as raw material: techno-economic and environmental assessment. Appl Energy 161:718–732
Pomerantseva E, Bonaccorso F, Feng X, Cui Y, Gogotsi Y (2019) Energy storage: the future enabled by nanomaterials. Science 366(6468):eaan8285
Pramanik P, Krishnan P, Maity A, Mridha N, Mukherjee A, Rai V (2020) Application of nanotechnology in agriculture. Environ Nanotechnol 4:317–348
Puri N, Gupta A (2023) Water remediation using titanium and zinc oxide nanomaterials through disinfection and photo catalysis process: a review. Environ Res 115786
Rabiee N, Sharma R, Foorginezhad S, Jouyandeh M, Asadnia M, Rabiee M et al (2023) Green (Nano) membranes: a review. Environ Res 231:116133
Rajendran S, Priya T, Khoo KS, Hoang TK, Ng H-S, Munawaroh HSH et al (2022) A critical review on various remediation approaches for heavy metal contaminants removal from contaminated soils. Chemosphere 287:132369
Rani M, Yadav J, Shanker U (2022) Environmental, legal, health, and safety issues of green nanomaterials. In: Green functionalized nanomaterials for environmental applications. Elsevier, pp 567–594
Roco MC (2023) National Nanotechnology Initiative at 20 years: enabling new horizons. J Nanopart Res 25(10):197
Rodrigues SM, Demokritou P, Dokoozlian N, Hendren CO, Karn B, Mauter MS et al (2017) Nanotechnology for sustainable food production: promising opportunities and scientific challenges. Environ Sci Nano 4(4):767–781
Saha L, Bauddh K (2020) Sustainable agricultural approaches for enhanced crop productivity, better soil health, and improved ecosystem services. Ecological and Practical Applications for Sustainable Agriculture, 1–23
Sahoo TR (2017) Polymer nanocomposites for environmental applications. Properties and applications of polymer nanocomposites: clay and carbon based polymer nanocomposites, 77–106
Salawitch RJ, Canty TP, Hope AP, Tribett WR, Bennett BF (2017) Paris climate agreement: Beacon of hope. Springer Nature
Saleem H, Trabzon L, Kilic A, Zaidi SJ (2020) Recent advances in nanofibrous membranes: production and applications in water treatment and desalination. Desalination 478:114178
Saleem H, Zaidi SJ, Ismail AF, Goh PS (2022) Advances of nanomaterials for air pollution remediation and their impacts on the environment. Chemosphere 287:132083

Saleh HM, Hassan AI (2023) Synthesis and characterization of nanomaterials for application in cost-effective electrochemical devices. Sustain For 15(14):10891

Sampaio PGV, González MOA (2017) Photovoltaic solar energy: conceptual framework. Renew Sust Energ Rev 74:590–601

Saputra AMA, Piliang AFR, Goei R, Ramadhan HR, Gea S (2024) Synthesis, properties, and utilization of carbon quantum dots as photocatalysts on degradation of organic dyes: a mini review. Catal Commun 187:106914

Saravanan A, Vo D-VN, Jeevanantham S, Bhuvaneswari V, Narayanan VA, Yaashikaa P et al (2021) A comprehensive review on different approaches for CO2 utilization and conversion pathways. Chem Eng Sci 236:116515

Saritha GNG, Anju T, Kumar A (2022) Nanotechnology-big impact: how nanotechnology is changing the future of agriculture? J Agric Food Res 10:100457

Satyanarayana T, Reddy SS (2018) A review on chemical and physical synthesis methods of nanomaterials. Int J Res Appl Sci Eng Technol 6(1):2885–2889

Shafi U, Mumtaz R, García-Nieto J, Hassan SA, Zaidi SAR, Iqbal N (2019) Precision agriculture techniques and practices: from considerations to applications. Sensors 19(17):3796

Shah AA, Khan MA, Raj V, Gupta A (2022) Intervention of nanotechnology as a tool for enhanced renewable energy application in the field of solar power harnessing. Environmental security and sustainable development; Discovery Publishing House Pvt. Ltd.: New Delhi, India

Shahsavari A, Akbari M (2018) Potential of solar energy in developing countries for reducing energy-related emissions. Renew Sust Energ Rev 90:275–291

Shang Y, Hasan MK, Ahammed GJ, Li M, Yin H, Zhou J (2019) Applications of nanotechnology in plant growth and crop protection: a review. Molecules 24(14):2558

Shankar P, Jagtap J, Sharma G, Sharma GP, Singh J, Parashar M et al (2022) A revolutionary breakthrough of bionanomaterials in tissue engineering and regenerative medicine. In: Bionanotechnology: emerging applications of bionanomaterials. Elsevier, pp 399–441

Shapira P, Youtie J (2015) The economic contributions of nanotechnology to green and sustainable growth. Green processes for nanotechnology: from inorganic to bioinspired nanomaterials, 409–434

Sharma S, Goyal P, Maheshwari S, Chandra A (2015) A technical review of automobile catalytic converter: current status and perspectives. Strat Technol Complex Environ Issues Sustain Appr 1:171–179

Sher Shah MSA, Oh C, Park H, Hwang YJ, Ma M, Park JH (2020) Catalytic oxidation of methane to oxygenated products: recent advancements and prospects for electrocatalytic and photocatalytic conversion at low temperatures. Adv Sci 7(23):2001946

Singh T, Shukla S, Kumar P, Wahla V, Bajpai VK, Rather IA (2017) Application of nanotechnology in food science: perception and overview. Front Microbiol 8:268461

Smets A, Jäger K, Isabella O, Van Swaaij R, Zeman M (2016) Solar energy: the physics and engineering of photovoltaic conversion, technologies and systems. Bloomsbury Publishing

Soni R, Pal AK, Tripathi P, Lal JA, Kesari K, Tripathi V (2020) An overview of nanoscale materials on the removal of wastewater contaminants. Appl Water Sci 10(8):189

Stuart L, Lüterbacher J, Paterson L, Deviller R, Castonguay S (2022) United In Science 2022: a multi-organization high-level compilation of the most recent science related to climate change, impacts and responses

Subramanian K, Karthika V, Praghadeesh M, Lakshmanan A (2020) Nanotechnology for mitigation of global warming impacts. Global climate change: resilient and smart agriculture, 315–336

Szczęśniak B, Choma J, Jaroniec M (2017) Gas adsorption properties of graphene-based materials. Adv Colloid Interf Sci 243:46–59

Tejwan N, Saini AK, Sharma A, Singh TA, Kumar N, Das J (2021) Metal-doped and hybrid carbon dots: a comprehensive review on their synthesis and biomedical applications. J Control Release 330:132–150

Thakur AK, Sathyamurthy R, Velraj R, Lynch I, Saidur R, Pandey A et al (2021) Sea-water desalination using a desalting unit integrated with a parabolic trough collector and activated carbon pellets as energy storage medium. Desalination 516:115217

Thangudu S (2020) Next generation nanomaterials: smart nanomaterials, significance, and biomedical applications. Applications of nanomaterials in human health, 287–312

Tian J, Chen X, Wang ZL (2020) Environmental energy harvesting based on triboelectric nanogenerators. Nanotechnology 31(24):242001

Tong C, Tong C (2022) Printed flexible electrochemical energy storage devices. Advanced materials for printed flexible electronics, 433–521

Urban MC, Bocedi G, Hendry AP, Mihoub J-B, Pe'er G, Singer A et al (2016) Improving the forecast for biodiversity under climate change. Science 353(6304):aad8466

Varghese AM, Karanikolos GN (2020) CO2 capture adsorbents functionalized by amine–bearing polymers: a review. Int J Greenhouse Gas Control 96:103005

Vasileff A, Zheng Y, Qiao SZ (2017) Carbon solving carbon's problems: recent progress of nanostructured carbon-based catalysts for the electrochemical reduction of CO2. Adv Energy Mater 7(21):1700759

Velásquez AC, Castroverde CDM, He SY (2018) Plant–pathogen warfare under changing climate conditions. Curr Biol 28(10):R619–R634

Verma G, Mondal K, Islam M, Gupta A (2024) Recent advances in advanced micro and nanomanufacturing for wastewater purification. ACS Applied Engineering Materials

Voutchkov N (2018) Energy use for membrane seawater desalination–current status and trends. Desalination 431:2–14

Wang X, He T, Hu J, Liu M (2021) The progress of nanomaterials for carbon dioxide capture via the adsorption process. Environ Sci Nano 8(4):890–912

Wang G, Shi R, Mi L, Hu J (2022) Agricultural eco-efficiency: challenges and progress. Sustain For 14(3):1051

Welzer H (2015) Climate wars: what people will be killed for in the 21st century. John Wiley & Sons

Wijeyakulasuriya S, Kim J, Probst D, Srivastava K, Yang P, Scarcelli R, Senecal PK (2022) Enabling powertrain technologies for Euro 7/VII vehicles with computational fluid dynamics. Transp Eng 9:100127

Wu J, Zhang K, Cen C, Wu X, Mao R, Zheng Y (2021) Role of bulk nanobubbles in removing organic pollutants in wastewater treatment. AMB Express 11:1–13

Yadav A, Kumar H, Sharma R, Kumari R (2023a) Synthesis, processing, and applications of 2D (nano) materials: a sustainable approach. Surfaces and Interfaces 39:102925

Yadav A, Yadav K, Ahmad R, Abd-Elsalam KA (2023b) Emerging frontiers in nanotechnology for precision agriculture: advancements, hurdles and prospects. Agrochemicals 2(2):220–256

Yang Z, Ren J, Zhang Z, Chen X, Guan G, Qiu L et al (2015) Recent advancement of nanostructured carbon for energy applications. Chem Rev 115(11):5159–5223

Yang C, Sui R, Lee WS (2016) Precision agriculture in large-scale mechanized farming. Precision Agriculture Technology for crop farming, 177–204

Yi WY, Lo KM, Mak T, Leung KS, Leung Y, Meng ML (2015) A survey of wireless sensor network based air pollution monitoring systems. Sensors 15(12):31392–31427

Yoro KO, Daramola MO (2020) CO2 emission sources, greenhouse gases, and the global warming effect. In: Advances in carbon capture. Elsevier, pp 3–28

Yusuf A, Sodiq A, Giwa A, Eke J, Pikuda O, De Luca G et al (2020) A review of emerging trends in membrane science and technology for sustainable water treatment. J Clean Prod 266:121867

Zain M, Ma H, Chaudhary S, Nuruzaman M, Azeem I, Mehmood F et al (2023) Nanotechnology in precision agriculture: advancing towards sustainable crop production. Plant Physiol Biochem 108244

Zeleňák V, Saldan I (2021) Factors affecting hydrogen adsorption in metal–organic frameworks: a short review. Nano 11(7):1638

Zhang K, Kimball JS, Running SW (2016) A review of remote sensing based actual evapotranspiration estimation. Wiley Interdiscip Rev Water 3(6):834–853

Zhang Y, Yang J, Ge R, Zhang J, Cairney JM, Li Y et al (2022) The effect of coordination environment on the activity and selectivity of single-atom catalysts. Coord Chem Rev 461:214493

Zhao W, Chen I-W, Huang F (2019) Toward large-scale water treatment using nanomaterials. Nano Today 27:11–27

Zheng J, Song X, Yang G, Du X, Mei X, Yang X (2022) Remote sensing monitoring of rice and wheat canopy nitrogen: a review. Remote Sens 14(22):5712

Zou C, Zhao Q, Zhang G, Xiong B (2016) Energy revolution: from a fossil energy era to a new energy era. Nat Gas Ind B 3(1):1–11

MIX
Papier aus verantwortungsvollen Quellen
Paper from responsible sources
FSC® C105338

If you have any concerns about our products,
you can contact us on
ProductSafety@springernature.com

In case Publisher is established outside the EU,
the EU authorized representative is:
Springer Nature Customer Service Center GmbH
Europaplatz 3, 69115 Heidelberg, Germany

Printed by Libri Plureos GmbH
in Hamburg, Germany